Basic

G·E·N·E·T·I·C·S

Basic

G·E·N·E·T·I·C·S

A Contemporary Perspective

Robert F. Weaver

UNIVERSITY OF KANSAS

·

Philip W. Hedrick

THE PENNSYLVANIA STATE UNIVERSITY

WCB **Wm. C. Brown Publishers**

Book Team

Editor *Kevin Kane*
Developmental Editor *Margaret J. Manders*
Production Editor *Kennie Harris*
Designer *David C. Lansdon*
Art Editor *Jess Schaal*
Photo Editor *Carol M. Smith*
Permissions Editor *Vicki Krug*
Visuals Processor *Joseph P. O'Connell*

 Wm. C. Brown Publishers

President *G. Franklin Lewis*
Vice President, Publisher *George Wm. Bergquist*
Vice President, Publisher *Thomas E. Doran*
Vice President, Operations and Production *Beverly Kolz*
National Sales Manager *Virginia S. Moffat*
Advertising Manager *Ann M. Knepper*
Marketing Manager *Craig S. Marty*
Editor in Chief *Edward G. Jaffe*
Managing Editor, Production *Colleen A. Yonda*
Production Editorial Manager *Julie A. Kennedy*
Production Editorial Manager *Ann Fuerste*
Publishing Services Manager *Karen J. Slaght*
Manager of Visuals and Design *Faye M. Schilling*

Interior design by Terri Webb Ellerbach

Chapter 12 opening quotation: From "LOOK THROUGH MY
WINDOW" Words and Music by JOHN PHILLIPS. © Copyright
1966 by MCA MUSIC PUBLISHING, A Division of MCA Inc.,
New York, NY 10019. USED BY PERMISSION. ALL RIGHTS
RESERVED.

Chapter 14 opening quotation: From VERSES FROM 1929 ON by
Ogden Nash. Copyright 1940 by Ogden Nash. First appeared in the
Saturday Evening Post. By permission of Little, Brown and Company.
Reprinted by permission of Curtis Brown Ltd. Copyright © 1940 by
Ogden Nash.

The credits section for this book begins on page 507, and is considered
an extension of the copyright page.

■

To Ginny and Stephanie;

And to Austin, Nancy, and R. A.

■

Brief

T·A·B·L·E

of

C·O·N·T·E·N·T·S

P·A·R·T

An Introduction

1

A Brief History   *2*

P·A·R·T

2

Mendelian Genetics

2

Fundamentals of Mendelian Genetics   *14*

3

Extensions and Applications of Mendelian Genetics   *38*

4

Chromosomes and Heredity   *70*

5

Genetic Linkage   *104*

P·A·R·T

3

Molecular Genetics

6

Chemistry of the Gene   *130*

7

An Introduction to Gene Function   *150*

8

Replication and Recombination of Genes   *170*

9

Transcription and Its Control in Prokaryotes   *192*

10

Eukaryotic Gene Structure and Expression   *216*

11

Translation   *250*

12

Gene Mutation, Transposable Elements, and Cancer   *272*

13

Genetics of Bacteria and Phages   *312*

14

Developmental Genetics   *340*

15

Gene Cloning and Manipulation   *368*

P·A·R·T

4

Population Genetics

16

An Introduction to Population Genetics   *412*

17

Extensions and Applications of Population Genetics   *438*

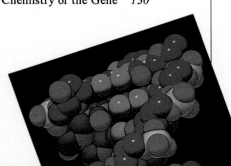

List of Boxes   *xii*
Preface   *xiii*
Acknowledgments   *xv*

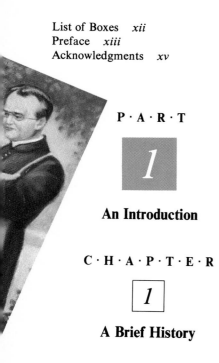

P·A·R·T

1

An Introduction

C·H·A·P·T·E·R

1

A Brief History

Learning Objectives   *3*
The Three Branches of Genetics   *4*
Transmission Genetics   *5*
  Mendel's Laws of Inheritance   *5*
  The Chromosome Theory of
    Inheritance   *5*
  Genetic Recombination and
    Mapping   *7*
  Physical Evidence for
    Recombination   *7*
Molecular Genetics   *8*
  The Discovery of DNA   *8*
  What Genes Are Made Of   *8*
  The Relationship Between Genes
    and Proteins   *8*
  What Genes Do   *9*
Why Study Genetics?   *11*
Summary   *11*

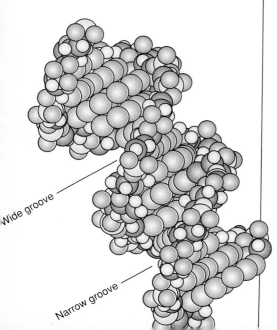

Wide groove

Narrow groove

P·A·R·T

2

Mendelian Genetics

C·H·A·P·T·E·R

2

Fundamentals of Mendelian Genetics

Learning Objectives   *15*
Principle of Segregation   *16*
  Mendel's Experimental Results   *19*
  Mendel's Explanation   *21*
Principle of Independent Assortment
  23
Probability   *25*
  Sum Rule   *25*
  Product Rule   *25*
  Conditional Probability   *26*
  Binomial Probability   *27*
  Forked-Line Approach   *29*
Goodness of Fit   *31*
Mendelian Inheritance in Humans   *33*
  Recessive Traits   *34*
  Dominant Traits   *35*
Summary   *36*
Problems and Questions   *36*

C·H·A·P·T·E·R

3

Extensions and Applications of Mendelian Genetics

Learning Objectives   *39*
Dominance   *40*
  Complete Dominance   *40*
  Incomplete Dominance   *40*
  Codominance   *40*
Lethals   *41*
Pleiotropy   *42*
Penetrance and Expressivity   *43*
Multiple Alleles   *45*
Sex-linked Genes   *48*
Multiple Genes and Epistasis   *51*

Expanded

T·A·B·L·E
of
C·O·N·T·E·N·T·S

Genotype-Phenotype Relationships   *54*
Quantitative Traits   *56*
   Mean and Variance   *58*
   Model of Quantitative Traits   *59*
Estimating Genetic Variance and
   Heritability   *61*
   Race and Intelligence   *64*
Genetic Counseling   *64*
Paternity Exclusion   *65*
Summary   *68*
Problems and Questions   *68*

Chromosomes and Heredity

Learning Objectives   *71*
Sexual Reproduction   *72*
Chromosomal Morphology   *73*
Mitosis   *76*
   Prophase   *77*
   Metaphase   *78*
   Anaphase   *78*
   Telophase   *78*
Meiosis   *78*
   Meiosis I   *79*
   Meiosis II   *80*
   Spermatogenesis and Oogenesis
      81
Genes and Chromosomes   *82*
Chromosomal Changes   *85*
   Changes in Chromosomal Structure
      85
   Changes in Chromosomal Number
      93
Chromosomes in Different Species   *98*
Chromosomes and Sex Determination
   99
Summary   *102*
Problems and Questions   *102*

Genetic Linkage

Learning Objectives   *105*
Linkage   *106*
   A Mechanistic Explanation   *106*
   Rate of Recombination   *109*
   The Linkage Map   *109*
   Three-point Crosses   *111*
   Interference   *112*
Linkage in Humans   *114*
   Somatic Cell Hybridization   *116*
Four-stranded Crossing Over   *117*
   Tetrad Analysis   *118*
Unequal Crossing Over   *122*
Map Distance and Physical Distance
   125
Summary   *125*
Problems and Questions   *125*

Molecular Genetics

Chemistry of the Gene

Learning Objectives   *131*
The Nature of Genetic Material   *132*
   Transformation in Bacteria   *132*
   The Chemical Nature of
      Polynucleotides   *137*
DNA Structure   *139*
   Experimental Background   *139*
   The Double Helix   *140*
Genes Made of RNA   *142*

Physical Chemistry of Nucleic Acids
   143
   A Variety of DNA Structures   *143*
   Separating the Two Strands of a
      DNA Double Helix   *144*
   Rejoining the Separated DNA
      Strands   *145*
   Hybridization of Two Different
      Polynucleotide Chains   *145*
   DNAs of Various Sizes and Shapes
      145
Summary   *149*
Problems and Questions   *149*

An Introduction to Gene Function

Learning Objectives   *151*
Replication   *152*
Storing Information   *155*
   Protein Structure   *155*
   Protein Function   *156*
Gene Expression   *158*
   Discovery of Messenger RNA   *159*
   Transcription   *161*
   Translation   *162*
   Coding and Anticoding Strands of
      DNA   *164*
Mutations   *165*
   Sickle-cell Disease   *165*
Colinearity of Gene and Protein   *167*
Summary   *169*
Problems and Questions   *169*

Replication and Recombination of Genes

Learning Objectives   *171*
Mechanism of DNA Replication   *172*
   Bidirectional Replication   *172*
   Unidirectional Replication   *175*
   Semidiscontinuous Replication
      175

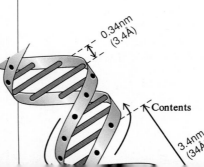

0.34nm
(3.4Å)

Contents

3.4nm
(34Å)

Priming of DNA Synthesis   *177*
"Stitching Together" Nascent DNA
Fragments   *178*
Enzymology of DNA Replication   *178*
Strand Separation   *178*
Initiation of Replication   *181*
Elongation   *182*
Termination   *185*
Fidelity of Replication   *186*
The Mechanism of Recombination   *187*
Forms of Recombination   *187*
A Model for Generalized
Recombination   *189*
Induction of DNA Strand Breaks   *190*
Summary   *190*
Problems and Questions   *191*

C·H·A·P·T·E·R

9

Transcription and Its Control in Prokaryotes

Learning Objectives   *193*
The Mechanism of Transcription   *194*
Structure of RNA Polymerase   *194*
Binding RNA Polymerase to
Promoters   *196*
Common Base Sequences in *E. coli*
Promoters   *197*
Termination of Transcription   *198*
Operons   *200*
The *lac* Operon   *200*
Temporal Control of Transcription   *206*
Modification of the Host RNA
Polymerase   *206*
The RNA Polymerase Encoded in
Phage T7   *208*
Control of Transcription During
Sporulation   *208*
Infection of *E. coli* by Phage λ   *210*
Specific DNA-Protein Interactions   *214*
Summary   *214*
Problems and Questions   *215*

C·H·A·P·T·E·R

10

Eukaryotic Gene Structure and Expression

Learning Objectives   *217*
Chromatin Structure   *218*
Nucleosomes: The First Order of
Chromatin Folding   *219*
Further Folding of the Nucleosome
String   *220*
Nonhistone Proteins   *222*
RNA Polymerases and Their Roles   *222*
Promoters   *223*
Promoters Recognized by
Polymerase II   *223*
Promoters Recognized by
Polymerase I   *226*
Promoters Recognized by
Polymerase III   *226*
Regulation of Transcription   *227*
Enhancers   *227*
Transcription Factors   *228*
Chromatin Structure and Gene
Expression   *231*
The Relationship Between Gene
Structure and Expression   *232*
Interrupted Genes   *232*
Other RNA Processing Events   *237*
Types of Gene Expression Control   *242*
Transcriptional Control   *242*
Posttranscriptional Control   *244*
Translational Control   *244*
Posttranslational Control   *245*
Mechanisms of Transcription
Control   *246*
Summary   *248*
Problems and Questions   *249*

C·H·A·P·T·E·R

11

Translation

Learning Objectives   *251*
Transfer RNA   *252*
Secondary Structure of tRNA   *252*
Three-dimensional Structure of
tRNA   *252*
Amino Acid Binding to tRNA   *254*

The Genetic Code   *255*
Nonoverlapping Codons   *255*
No Gaps in the Code   *255*
The Triplet Code   *256*
Breaking the Code   *257*
Unusual Base Pairs Between Codon
and Anticodon   *258*
The (Almost) Universal Code   *259*
Ribosomes   *260*
Self-assembly of Ribosomes   *260*
Functions of Ribosomal Proteins   *262*
Mechanism of Translation   *263*
Initiation   *263*
Elongation: Adding Amino Acids to
the Growing Chain   *265*
Termination: Release of the
Finished Polypeptide   *267*
Summary   *270*
Problems and Questions   *270*

C·H·A·P·T·E·R

12

Gene Mutation, Transposable Elements, and Cancer

Learning Objectives   *273*
Types of Gene Mutations   *274*
Somatic Versus Germ-Line
Mutations   *274*
Morphological Mutations   *275*
Nutritional Mutations   *275*
Lethal Mutations   *276*
Conditional Mutations   *276*
Effects of Mutations on the Genetic
Material   *277*
Spontaneous Mutations   *277*
Chemical Mutagenesis   *283*
Radiation-Induced Mutations   *284*
Silent Mutations   *285*
Reversion   *286*
DNA Repair   *286*
Directly Undoing DNA Damage   *287*
Excision Repair   *288*
Mismatch Repair   *290*
Coping with DNA Damage without
Repairing It   *290*
Severe Consequences of Defects in
DNA Repair Mechanisms   *292*

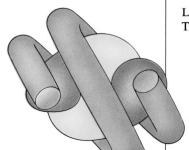

Detecting Mutagens   293
Bacterial Transposons   294
 Insertion Sequences: The Simplest
  Transposons   294
 More Complex Transposons   296
 Mechanisms of Transposition   296
 Transposons as Mutagenic Agents
  298
Eukaryotic Transposable Elements
 300
 The First Examples of Transposable
  Elements   300
 Yeast and *Drosophila* Transposons
  302
 Retroviruses   302
 Transposons and the Mutability of
  Chromosomes   304
Gene Mutations and Cancer   305
 Carcinogens as Mutagens   305
 Oncogenes   307
 The c-Ha-*ras* Gene Product: A G
  Protein-like Substance   308
Summary   310
Problems and Questions   310

C·H·A·P·T·E·R

13

Genetics of Bacteria and Phages

Learning Objectives   313
Bacterial Conjugation   314
 Bacterial Genetic Notation   314
 Bacterial Exchange of DNA   314
 Hfr Strains   317
 Mapping by Interrupted
  Conjugation   318
 Completion of Conjugation   322
Genetic Mapping with *E. Coli*   322
 F′ Plasmids   323
 The *Cis-Trans* Complementation
  Test   324
 Transduction   326

Genetics of T-even Phages   330
 Structure and Growth of T-even
  Phages   331
 Crossing Mutant T-even Phages
  334
 Fine Structure Mapping   335
 Applying the *Cis-Trans* Test to the
  rII Region   337
Summary   337
Problems and Questions   338

C·H·A·P·T·E·R

14

Developmental Genetics

Learning Objectives   341
The Genetic Basis of Cell
   Differentiation   342
 Mosaic Development Versus
  Regulative Development   343
 Determinants   346
 Inducers   346
Control of Gene Expression   347
The Nature of Determination   349
 Control of the 5S rRNA Genes in
  Xenopus laevis   350
 The Role of DNA Methylation
  351
 Determination in the Fruit Fly
  Drosophila   352
 Vertebrate Homeo Boxes   359
 Determination in the Roundworm
  360
Gene Rearrangement During
   Development   361
 Rearrangement of Immunoglobulin
  Genes   361
 Rearrangement of an Oncogene
  364
 Rearrangement of Trypanosome
  Coat Genes   364
 Sequential Activation of Members
  of a Gene Family   364
Summary   365
Problems and Questions   366

C·H·A·P·T·E·R

15

Gene Cloning and Manipulation

Learning Objectives   369
Gene Cloning   370
 The Role of Restriction
  Endonucleases   370
 Vectors   372
 Identifying a Specific Clone with a
  Specific Probe   378
 cDNA Cloning   379
Methods of Expressing Cloned Genes
 380
 Expression Vectors   381
Manipulating Cloned Genes   385
 Protein Engineering with Cloned
  Genes   385
 Using Cloned Genes as Probes
  387
 Determining the Base Sequence of a
  Gene   392
 DNA Sequencing: The Ultimate in
  Genetic Mapping   394
Practical Applications of Gene Cloning
 395
 Proteins: The New Pharmaceutical
  Frontier   395
 Improved Vaccines   396
 Intervening in Human Genetic
  Disease   396
 Using Cloned Genes in Agriculture
  397
 Protein Products from Transgenic
  Organisms   397
 Genetically Engineered Crops   398
Mapping Genetic Defects in Humans
 401
 Huntington's Disease   401
 Restriction Fragment Length
  Polymorphisms   401
Summary   408
Problems and Questions   408

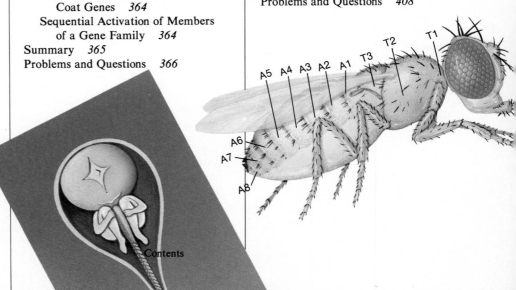

Contents

P·A·R·T

Population Genetics

C·H·A·P·T·E·R

An Introduction to Population Genetics

Learning Objectives   413
Types of Genetic Variation   414
Measuring Genetic Variation   416
The Hardy-Weinberg Principle   419
  The Chi-square Test   421
  More Than Two Alleles   422
  X-linked Genes   422
Inbreeding   423
  Self-fertilization   423
  The Inbreeding Coefficient   424
  Calculating the Inbreeding
    Coefficient   425

Mutation   426
Genetic Drift   427
Gene Flow   429
Natural Selection   431
  Selection Against a Homozygote
    431
  Heterozygous Advantage   433
Summary   436
Problems and Questions   436

C·H·A·P·T·E·R

Extensions and Applications of Population Genetics

Learning Objectives   439
Frequency of Human Genetic Disease
    440
  Mutation-Selection Balance   440
  Other Factors Affecting Disease
    Frequency   442
Molecular Evolution   444
  Neutrality   444
  Molecular Clock   445
  Phylogenetic Trees   446
  Use of mtDNA in Studying Genetic
    Relationships   447
  Multigene Families   448

Conservation Genetics   449
  Crop Species   449
  Inbreeding Depression   450
  Loss of Genetic Variation   451
Animal and Plant Breeding   452
  Artificial Selection   452
  Inbred Lines   454
  Heterosis   456
Pesticide Resistance   457
  Factors Important in Pesticide
    Resistance   459
  Examples of Pesticide Resistance
    459
  Control of Pesticide Resistance
    460
Summary   461
Problems and Questions   461

Appendix A
  Problems with Solutions   463
Appendix B
  Answers to End-of-Chapter
    Problems and Questions   471
Selected Readings   489
Glossary   493
Credits   507
Index   509

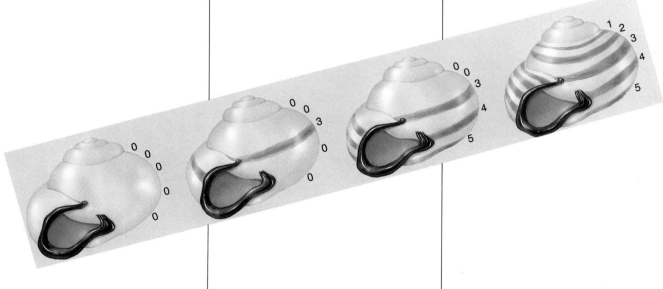

L·I·S·T of B·O·X·E·S

Chapter 2

2.1   The Father of Genetics   *16*

Chapter 3

3.1   Grandpaternity Exclusion   *67*

Chapter 5

5.1   Multigene Families   *123*

Chapter 8

8.1   Biochemists Like to Grind Things Up   *181*

Chapter 10

10.1   HeLa Cells   *229*

Chapter 11

11.1   How the Amber Mutation Got Its Name   *268*

Chapter 12

12.1   Transposition Impacts on Public Health   *298*

Chapter 13

13.1   Inadvertent Sharing of a Phage   *333*

Chapter 15

15.1   Uses of DNA Fingerprinting   *390*
15.2   Ethical Issues   *402*
15.3   Problems in Genetic Screening   *406*

Chapter 16

16.1   Industrial Melanism   *434*

Chapter 17

17.1   Lysenko—Soviet Anti-Geneticist   *458*

T his textbook—like our previous book, *Genetics*—is intended for use in a first college course in genetics. Also like *Genetics,* it follows the same organizational approach. Mendelian or transmission genetics is covered first, followed by molecular genetics. However, with *Basic Genetics* we provide a more balanced approach to the course. In *Genetics,* transmission genetics was covered in a somewhat synopsized way, allowing us to devote more space to the increasingly explosive field of molecular genetics. In *Basic Genetics,* transmission genetics has been given a fuller, more detailed treatment, while some of the more difficult and detailed molecular topics have been eliminated.

Basic Genetics also features a greater emphasis on problem solving—in both the narrative and in the problems that appear at the end of each chapter where a broader range of problems designed to test a wide variety of skills have been included. Likewise, sample problems with fully worked-out solutions are included in the book's appendix.

As with *Genetics*—in fact, even more so this time—we have tried to keep our writing concise and easy to understand. We have attempted to capture the most important basic concepts, using language that is economical, and understanding fully that each instructor will shed additional light on those topics he or she feels are in need of further explication.

THE CHAPTERS

Like all instructors, we have our prejudices. But in spite of them (or perhaps in recognition of them), we have tried to make this book easy to use no matter what order of topics an instructor prefers. Accordingly, we have organized our book in the following way:

Chapter 1 is a historical introduction to genetics. It provides a few essentials of both major branches of the discipline. After reading this chapter, students should have some appreciation for the molecular character of genes as they study the subsequent chapters on transmission genetics. If, on the other hand, an instructor chooses to start with molecular genetics, chapter 1 introduces students to the basics of heredity, providing a context for the molecules.

Chapters 2–5 constitute a unit on transmission genetics. In chapter 2, we use Mendel's experiments to introduce the principles of segregation and independent assortment. At the same time, since it is impossible to discuss these topics without dealing with probabilities, we provide a brief introduction to statistics. In chapter 3, we show that simple Mendelian analysis does not always work, and introduce concepts such as penetrance, expressivity, and epistasis that illustrate this point. Chapter 4 deals with chromosomes. The chromosome concept leads naturally to the idea of linkage and recombination and how to construct linkage maps with eukaryotes. Each chapter in this first part of the book includes a discussion of the application of genetics to human problems. For example, in chapter 2, students learn to apply the concepts of segregation and independent assortment to pedigree analysis of human genetic diseases. In chapter 4, they see how human chromosomal abnormalities can have severe consequences.

The central part of the book is a ten-chapter sequence emphasizing molecular genetics. Despite our efforts to balance the coverage of classical and molecular topics, we still believe that the direction of modern genetics research demands that the heart of a modern genetics text be molecular.

Chapters 6–12 present the fundamentals of molecular genetics. Chapters 6 and 7 provide an overview: Chapter 6 explains the structure and properties of DNA; chapter 7 introduces the three main functions of DNA in its role as the genetic material: (1) providing for faithful replication of genes; (2) containing the information for making the gene products, RNA and protein; and (3) accumulating mutations, which make evolution possible. Chapters 8–12 expand on these themes. Chapter 8 treats DNA replication in detail; chapters 9, 10, and 11 deal with gene expression, the second function of a gene. In particular, chapters 9 and 10 cover transcription in prokaryotes and eukaryotes, respectively, and chapter 11 spells out the translation process. Finally, chapter 12 deals with the third attribute of a gene—its capacity for mutation, with special sections focusing on transposable elements in mutagenesis and oncogenes in cancer.

Chapters 13–15 further illustrate specific aspects of the functions of genes. Chapter 13 is a treatment of the transmission genetics of bacteria and phages. It might reasonably be placed in the first part of the book, but since so much is known about the molecular basis of prokaryotic genetics, it seems to fit better in the molecular genetics section. Chapter 14 explores the ways that genes

govern the development of organisms. The intricate and fascinating development of the fruit fly is the major paradigm for this chapter. Chapter 15 presents the most celebrated practical application of molecular genetics. It covers gene cloning and manipulation, and shows how one branch of genetics has evolved into a kind of engineering. Our experience has been that students have fun with this topic, but since they need to master the basics first, this chapter comes last in this section. On the other hand, manipulation of cloned genes has become such a common and powerful tool of genetic analysis that some instructors will want to introduce at least parts of chapter 15 right after chapter 6, making the references to experiments using cloned genes in subsequent chapters less mysterious.

The book concludes with a two-chapter sequence on population genetics. Chapter 16 deals with introductory population genetics, including selection, genetic drift, inbreeding, mutation, and gene flow. Chapter 17 integrates these factors to discuss human diseases, molecular evolution, conservation genetics, and pesticide resistance.

The most important underlying theme of this book is its experimental approach. Genetics, like any science, is not a collection of facts, but a way of posing and answering questions about nature. Wherever possible, we have included the experimental rationale and results that have led to our present understanding of genetics. Furthermore, many of the end-of-chapter questions require students to examine and interpret hypothetical experimental results, or to design experiments to test hypotheses. We know that genetics, especially molecular genetics, can be bewildering to college students. However, we have found that if given a clear understanding of the way geneticists practice their science, students can master the subject more easily. That is the goal of this book.

Robert F. Weaver
Philip W. Hedrick

ACKNOWLEDGMENTS

We would like to thank the following reviewers of our manuscript for their many helpful suggestions:

William S. Barnes
Clarion University of Pennsylvania

J. W. Bennett
Tulane University

Edward Berger
Dartmouth College

Anna W. Berkovitz
Purdue University

Elliott S. Goldstein
Arizona State University

Jeffrey C. Hall
Brandeis University

Duane L. Johnson
Colorado State University

Paul F. Lurquin
Washington State University

Lester J. Newman
Portland State University

Robert R. Schalles
Kansas State University

Thomas P. Snyder
Michigan Technological University

Archie Waterbury
California Polytechnic State University

Oscar H. Will, III
Augustana College

P·A·R·T

1

An Introduction

C·H·A·P·T·E·R

1

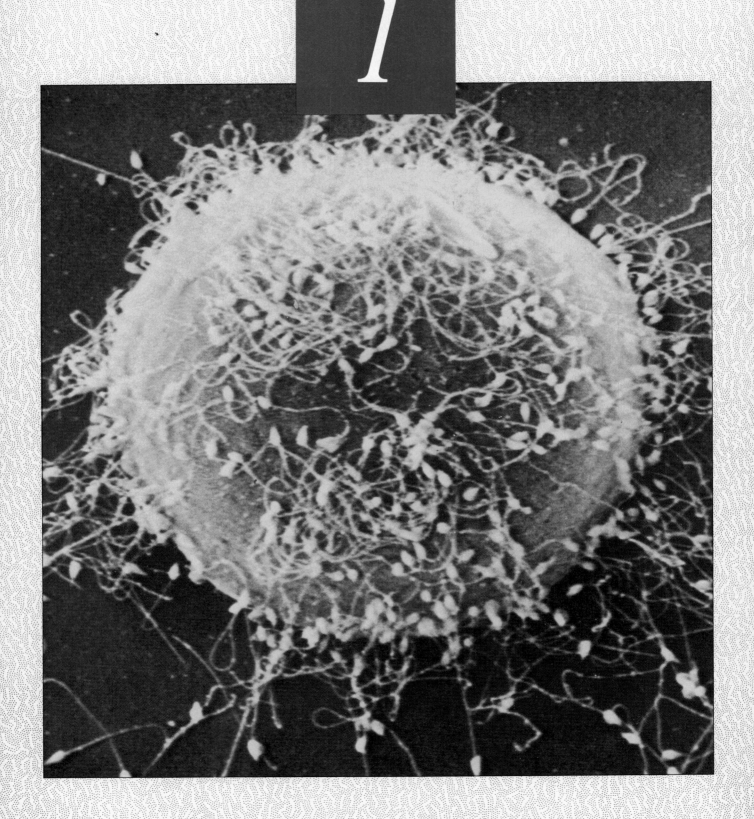

A Brief History

*T*hat is the essence of science: ask an
impertinent question and you are on the way to a
pertinent answer.

■

Jacob Bronowski
British mathematician

Learning Objectives

This chapter will introduce:

1. The three branches of genetics.
2. The fundamentals of transmission genetics.
3. The fundamentals of molecular genetics.

False-color scanning electron micrograph of the egg of a sea
urchin surrounded by sperm. The micrograph illustrates the vast
size differential between the female germ cell, the egg, and the
male germ cells, the sperm.
© *Dr. Gerald Schatten/Science Photo Library/Photo Researchers, Inc.*

beginning genetics student once walked up to his teacher after class and asked, "Which gene were you talking about today, the Mendelian gene, the DNA gene, or the population gene?" This question gets to the heart of the three main branches of modern genetics: (1) Mendelian genetics, the study of the transmission of traits from one generation to the next; (2) molecular genetics, the study of the chemical structure of genes and how they operate at the molecular level; and (3) population genetics, the study of the variation of genes between and within populations.

THE THREE BRANCHES OF GENETICS

The first branch of genetics relies mainly on the same kind of experimental approach used by Gregor Mendel in the middle of the nineteenth century. Organisms with differing traits, or *phenotypes,* are mated, and the transmission of these traits to the next generation is observed. These progeny organisms can then be mated with others of the same or different phenotypes (or even with themselves, as with Mendel's peas), and again the transmission of traits can be observed. Because Mendel pioneered this approach to genetics, we frequently call it **Mendelian** or **classical genetics.** Since it deals with the transmission of genes between generations, we also call it **transmission genetics.**

The second branch, **molecular genetics,** approaches the subject from its fundamental base: molecules. Instead of phenotypic characters, molecular geneticists examine the genes themselves. They are concerned with the molecules that compose genes, the molecules that control genes, and the molecules that are the products of genes. The molecular geneticist's job has become much easier (and more fun) since we learned how to clone genes in the 1970s. With these new techniques, we can produce a gene in large quantities in pure form and study its structure and function.

The third branch, **population genetics,** examines the extent of genetic variation within and among populations. This variation was traditionally studied on the phenotypic level, but population geneticists now focus increasingly on molecular variation in a population. Much of population genetics study is aimed at understanding the evolutionary cause of the observed genetic variation.

Population genetics techniques are also useful in describing the genetic differences between species and in learning about the process of species formation. Remember that a **species** is a reproductively isolated group, such as *Zea mays* (corn) or *Drosophila melanogaster* (fruit fly). A species name has two parts. The first part, which is capitalized (e.g., *Drosophila*), is the **genus,** and may encompass several species. The second part, which starts with a lowercase letter, designates a particular species within the genus.

Because of the power of molecular genetics, it would be natural to assume that all of modern genetics is becoming molecular. And indeed, a great deal of genetics research is now done at the molecular level. However, the other branches of genetics are in no danger of disappearing. We should bear in mind that although molecular techniques can uncover new genes, they frequently cannot reveal a gene's function. A more fruitful approach is this: A researcher first discovers an abnormal organism with interesting characteristics. The abnormality in this organism is caused by a mutated, or altered gene, so the *mutant* is the "smoke" that leads us to a new gene. Once the gene has been identified this way, we can begin to use molecular techniques to examine its structure and function in detail and perhaps to understand the evolutionary significance of variation in the gene. Actually, it is misleading to put too much emphasis on the divisions within the field of genetics. Most geneticists now approach their subject in several ways, using methods from more than one of the three branches.

The student's innocent question also points to a problem faced by teachers of genetics (and writers of genetics textbooks): how to help students see that there is no Mendelian gene, DNA gene, or population gene; rather, that they are all the same gene studied in different ways. This problem is really one of organization. Do we start by talking about the molecular aspects of the gene, so that later, when we talk about how genes are transmitted from generation to generation, students can appreciate fully what this means? Do we talk about transmission genetics first and only later explain about molecular genetics? Or do we try to talk about both at the same time?

In this book, we have decided to use the second approach: transmission genetics first, then molecular genetics, and finally, population genetics. This feels right to us for two reasons. First, transmission genetics is in some ways easiest to understand. It deals with the inheritance of characteristics that we can frequently see, and seeing is not only believing—seeing is usually understanding. Furthermore, inheritance is a concept that college students already appreciate to some degree.

A second reason for treating transmission genetics first is that it developed first historically. We like a good story, and the history of genetics is a very good one. By adopting a historical approach, we get a chance to retell that story.

No matter which part of genetics we cover first, we always wish our students had prior exposure to the other, because it would make their understanding so much richer. This is a problem that is impossible to resolve completely, but we propose to compromise by providing in this introductory chapter a brief overview of transmission and molecular genetics, so that at least the most fundamental concepts of these branches of genetics will be familiar as we look at each in detail in later chapters. In keeping with the basic plan of this book, we will present this overview in historical fashion.

Chapter 1

TRANSMISSION GENETICS

In 1865, Gregor Mendel (figure 1.1) published his findings on the inheritance of seven different traits in the garden pea. Prior to this time, inheritance was considered to occur through a blending of each trait of the parents in the offspring, but Mendel concluded that inheritance was particulate. That is, each parent contributes particles, or genetic units, to the offspring. We now call these particles *genes*. Furthermore, by carefully counting the number of progeny plants having a given phenotype, Mendel was able to make some important generalizations. (The word phenotype, by the way, comes from the same Greek root as "phenomenon," meaning "appearance"; thus, a tall pea plant exhibits the tall phenotype, or appearance.)

MENDEL'S LAWS OF INHERITANCE

Mendel saw that a gene can exist in several different forms called *alleles*. For example, the pea can have either yellow or green seeds. One allele of the gene for seed color gives rise to yellow seeds, the other to green. Moreover, one allele can be *dominant* over the other, *recessive* allele. In this case, yellow is dominant. Mendel showed this when he mated a green-seeded pea with a yellow-seeded pea. All of the progeny in the first filial generation (F_1) had yellow seeds. However, when these F_1 yellow peas were allowed to self-fertilize, some green-seeded peas reappeared. The ratio of yellow to green seeds in the second filial generation (F_2) was very close to 3:1.

Mendel concluded that the green allele must have been preserved in the F_1 generation, even though it did not affect the seed color of those peas. His explanation was that each parent plant carried two copies of the gene; that is, the parents were *diploid,* at least for the characteristics he was studying. According to this concept, *homozygotes* have two copies of the same allele, either two yellow alleles or two green alleles. *Heterozygotes* have one copy of each allele. The two parents in the first mating above were homozygotes; the resulting F_1 peas were all heterozygotes. Further, Mendel reasoned that sex cells only contain one copy of the gene; that is, they are *haploid*. Homozygotes can therefore produce sex cells, or gametes, that have only one allele, but heterozygotes can produce gametes having either allele.

This explains what happened in the matings of yellow with green peas. The yellow parent contributed a yellow gamete; the green parent, a green gamete. Therefore, all the F_1 peas got one yellow and one green allele. They had not lost the green allele at all, but since yellow is dominant, all the peas were yellow. However, when these heterozygous peas were self-fertilized, they produced yellow and green gametes in equal numbers, and this allowed the green phenotype to reappear.

FIGURE 1.1
Gregor Mendel.
Courtesy of the Department of Library Services, American Museum of Natural History, negative 219467.

G enes can exist in several different forms, or alleles. One allele can be dominant over another, thus heterozygotes having two different alleles of one gene will generally exhibit the characteristic dictated by the dominant allele. The recessive allele is not lost; it can still exert its influence when paired with another recessive allele in a homozygote.

THE CHROMOSOME THEORY OF INHERITANCE

Other scientists either did not know about or uniformly ignored the implications of Mendel's work until 1900 when three botanists, who had arrived at similar conclusions independently, rediscovered it. After that time, most geneticists accepted the particulate nature of genes, and the field of genetics began to blossom. One factor that made it easier for geneticists to accept Mendel's ideas was a beginning understanding of the nature of chromosomes, which had occurred in the latter half of the nineteenth century. Mendel had predicted that gametes would contain only one allele of each gene instead of two. If chromosomes

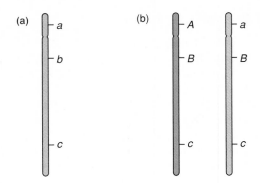

FIGURE 1.3
(*a*) A schematic diagram of a chromosome, indicating the positions of three genes—*a, b,* and *c.* (*b*) A schematic diagram of a diploid pair of chromosomes, indicating the positions of three genes—*A* (or *a*), *B*, and *c*—on each.

FIGURE 1.2
Thomas Hunt Morgan.
Courtesy of the National Library of Medicine.

carry the genes, their numbers should also be reduced by half in the gametes—and they are. Chromosomes therefore appeared to be the discrete physical entities that carried the genes.

This notion that chromosomes carry genes is the **chromosome theory of inheritance.** It was a crucial new step in genetic thinking. No longer were genes disembodied factors; now they were observable objects in the cell nucleus. Some geneticists, particularly Thomas Hunt Morgan (figure 1.2), remained skeptical of this idea. Ironically, however, it was Morgan himself who in 1910 provided the first definitive evidence for the chromosome theory.

Morgan worked with the fruit fly (*Drosophila melanogaster*), which was in many respects a much more convenient organism than the garden pea for genetic studies because of its small size, short generation time, and large number of offspring. When he mated red-eyed flies (dominant) with white-eyed flies (recessive), most but not all of the F_1 progeny were red-eyed. Furthermore, when Morgan mated the red-eyed males of the F_1 generation with their red-eyed sisters, they produced about ¼ white-eyed males, but no white-eyed females. In other words, the eye color phenotype was sex linked. It was transmitted along with sex in these experiments. How could this be?

In hindsight, the answer leaps out at us. Sex and eye color are transmitted together because the genes governing these characteristics are located on the same chromosome—the single X chromosome. (Most chromosomes, called **autosomes,** occur in pairs in a given individual, but the X chromosome is an example of a **sex chromosome** of which the female fly has two copies and the male has one.) However, Morgan was reluctant to draw this conclusion until he observed the same sex linkage with two more phenotypes: miniature wing and yellow body, also in 1910. That was enough to convince him of the validity of the chromosome theory of inheritance.

Before we leave this topic, let us reiterate two crucial points. First, every gene has its place on a chromosome. Figure 1.3*a* depicts a hypothetical chromosome and the positions of three of its genes, called *a, b,* and *c.* Second, diploid organisms such as human beings contain two copies of all chromosomes (except sex chromosomes). That means that they have two copies of most genes, and that these copies can be the same alleles (*homozygous*) or different alleles (*heterozygous*). For example, figure 1.3*b* shows a diploid pair of chromosomes with different alleles at one locus (*Aa*) and the same alleles at the other two loci (*BB* and *cc*). The **genotype,** or allelic constitution, of this organism with respect to these three genes is *AaBBcc.* Since this organism has two different alleles (*A* and *a*) in its two chromosomes at the *a* locus, it is heterozygous at that locus (Greek: *hetero* = different). Since it has the same, dominant *B* allele in both chromosomes at the *b* locus, it is homozygous dominant at that locus (Greek: *homo* = same). And since it has the same, recessive *c* allele in both chromosomes at the *c* locus, it is homozygous recessive there. Finally, because the *A* allele is dominant over the *a* allele, the phenotype of this organism would be ABc.

Chapter 1

FIGURE 1.4
Recombination in *Drosophila*. The two X chromosomes of the female are shown schematically. One of them (red) carries two normal genes: normal wing (m^+) and red eye (w^+). The other (blue) carries two mutant genes: *miniature* (*m*) and *white* (*w*). During egg formation, a recombination, or crossover, indicated by the X, occurs between these two genes on the two chromosomes. The result is two recombinant chromosomes with mixtures of the two parental genes. One is m^+w, the other is mw^+.

GENETIC RECOMBINATION AND MAPPING

It is easy to understand that genes on separate chromosomes behave independently in genetic experiments, and that genes on the same chromosome—like the genes for miniature wing (*miniature*) and white eye (*white*)—behave as if they are linked. However, genes on the same chromosome do not always show perfect genetic linkage. In fact, Morgan discovered this when he examined the behavior of the sex-linked genes he had found. For example, although *white* and *miniature* are both on the X chromosome, they only remain linked in offspring 65.5% of the time. The other offspring have a new combination of alleles not seen in the parents and are therefore called *recombinants*.

How are recombinants produced? The answer was already apparent by 1910, because microscopic examination of chromosomes during meiosis (gamete formation) had shown crossovers between *homologous* chromosomes (chromosomes carrying the same genes or alleles of the same genes). These crossovers resulted in the exchange of genes between the two homologous chromosomes. In the example above, during formation of eggs in the female, an X chromosome bearing the *white* and *miniature* alleles experienced a crossover with a chromosome bearing the red-eye and normal-wing alleles (figure 1.4). Because the crossover occurred between these two genes, it brought together the *white* and normal alleles on one chromosome and the red and *miniature* alleles on the other. Because it produced a new combination of alleles, we call this process **recombination.**

Morgan assumed that genes are arranged in a linear fashion on chromosomes, like beads on a string. This, together with his awareness of recombination, led him to propose that the farther apart two genes are on a chromosome, the more likely they are to recombine. This makes sense because there is simply more room between widely spaced genes for a crossover to occur. A. H. Sturtevant extended this hypothesis to predict that there should be a

FIGURE 1.5
Barbara McClintock.
The Bettmann Archive.

mathematical relationship between the distance separating two genes on a chromosome and the frequency of recombination between these two genes. Sturtevant collected data on recombination in the fruit fly that supported this hypothesis. This established the rationale for **genetic mapping** techniques still in use today. By the 1930s, other investigators found that the same rules applied to other **eukaryotes** (nucleus-containing organisms), including the mold *Neurospora,* the garden pea, maize (corn), and even human beings. These rules also apply to **prokaryotes,** organisms that have no nuclei.

PHYSICAL EVIDENCE FOR RECOMBINATION

Barbara McClintock (figure 1.5) finally provided a direct physical demonstration of recombination in 1931. By examining maize chromosomes microscopically, she could detect recombinations between two easily identifiable features of a particular chromosome (a knob at one end and a long extension at the other). Furthermore, whenever this physical recombination occurred, she could also detect recombination genetically. Thus, there is a direct relationship between a region of a chromosome and a gene.

FIGURE 1.6
Friedrich Miescher.
Courtesy of the National Library of Medicine.

FIGURE 1.7
Oswald Avery.
The Bettmann Archive.

The chromosome theory of inheritance holds that genes are arranged in linear fashion on chromosomes. The reason that certain traits tend to be inherited together is that the genes governing these traits are on the same chromosome. However, recombination between two chromosomes during meiosis can scramble the parental alleles to give nonparental combinations. The farther apart two genes are on a chromosome, the more likely such recombination between them will be.

•

MOLECULAR GENETICS

Studies such as those just discussed tell us important things about the transmission of genes and even about how to map genes on chromosomes, but they do not tell us what genes are made of or how they work. This has been the province of molecular genetics, which also happens to have its roots in Mendel's era.

THE DISCOVERY OF DNA

In 1869, Friedrich Miescher (figure 1.6) discovered in the cell nucleus a mixture of compounds that he called nuclein. The major component of nuclein is **deoxyribonucleic acid (DNA)**. By the end of the nineteenth century, chemists had learned the general structure of DNA and of a related compound, **ribonucleic acid (RNA)**. Both are long polymers—chains of small compounds called nucleotides. Each nucleotide is composed of a sugar, a phosphate group, and a base. The chain is formed by linking the sugars to one another through their phosphate groups.

WHAT GENES ARE MADE OF

By the time the chromosome theory of inheritance was generally accepted, geneticists agreed that the chromosome must be composed of a polymer of some kind. This would accord with its role as a string of genes. But which polymer comprised the genes? There were essentially three choices: DNA, RNA, and protein. Protein was the other major component of Miescher's nuclein; its chain is composed of links called amino acids.

Oswald Avery (figure 1.7) and his colleagues demonstrated in 1944 that DNA is the right choice. These investigators built on an experiment performed earlier by Frederick Griffith in which he transferred a genetic trait from one strain of bacteria to another. The trait was virulence, the ability to cause a lethal infection, and it could be transferred simply by mixing dead virulent cells with live avirulent cells. It was very likely that the substance that caused the transformation from avirulence to virulence in the recipient cells was the gene for virulence, because the recipient cells passed on this trait to their progeny.

What remained was to learn the chemical nature of the transforming agent in the dead virulent cells. Avery and his coworkers did this by applying a number of chemical and biochemical tests to the transforming agent and showing that it behaved just as DNA should, not as RNA or protein should.

THE RELATIONSHIP BETWEEN GENES AND PROTEINS

The other major question in molecular genetics is: How do genes work? To lay the groundwork for the answer to this question, we have to backtrack again, this time to 1902. That was the year Archibald Garrod noticed that the human disease alcaptonuria seemed to behave as a Mendelian recessive trait. It was likely, therefore, that the disease was caused by a defective, or mutated gene. Moreover, the main symptom of the disease was the accumulation of a black pigment in the patient's urine, which Garrod believed derived from the abnormal buildup of an intermediate in a biochemical pathway.

By this time, biochemists had shown that all living things carry out countless chemical reactions, and that these reactions are accelerated or catalyzed by proteins called *enzymes.* Many of these reactions take place in sequence, so that one product becomes the starting material, or substrate, for the next reaction. Such sequences of reactions are called *pathways,* and the products/substrates within a pathway are called *intermediates.* Garrod postulated that an intermediate accumulated to abnormally high levels in alcaptonuria because the enzyme that would normally convert this intermediate to the next was defective. Putting this together with the finding that alcaptonuria behaved genetically as a Mendelian recessive trait, Garrod suggested that a defective gene gives rise to a defective enzyme. To put it another way: A gene is responsible for the production of an enzyme.

(a)   (b)

FIGURE 1.8
(a) George Beadle; (b) E. L. Tatum.
Photos © AP/Wide World Photos.

Garrod's conclusion was based in part on conjecture; he did not really know that a defective enzyme was involved in alcaptonuria. It was left for George Beadle and E. L. Tatum (figure 1.8) to prove the relationship between genes and enzymes. They did this using the mold *Neurospora* as their experimental system. *Neurospora* has an enormous advantage over the human being as an experimental system because scientists are not limited to the mutations that nature provides, but can use *mutagens* to introduce mutations into genes and then observe the effects of these mutations on biochemical pathways. Beadle and Tatum found many instances where they could pin the defect down to a single step in a biochemical pathway, and therefore to a single enzyme. They did this by adding the intermediate that would normally be made by the defective enzyme and showing that it restored normal growth. By circumventing the blockade, they discovered where it was. In these same cases, their genetic experiments showed that a single recessive Mendelian gene was involved. Therefore, a defective gene gives a defective (or absent) enzyme. In other words, a gene seemed to be responsible for making one enzyme. This was the one gene-one enzyme hypothesis, which was actually not quite right because an enzyme can be composed of more than one protein chain, but a gene actually has the information for making only one protein chain.

WHAT GENES DO

We see that genes are made of DNA, and that each gene carries the information for making one protein chain. Let us now return to the question at hand: How do genes work? This is really more than one question because genes do more than one thing. First, they replicate faithfully; second, they direct the production of proteins; third, they accumulate mutations and so allow evolution. Let us look briefly at each of these activities.

FIGURE 1.9
James Watson (left) and Francis Crick.
The Bettmann Archive.

How Genes Replicate

First of all, how does DNA replicate faithfully? To answer that question, we need to know the exact structure of the DNA molecule as it is found in the chromosome. James Watson and Francis Crick (figure 1.9) provided the answer in 1953 by building models based on chemical and physical data that had been gathered in other laboratories, primarily X-ray diffraction data collected by Rosalind Franklin and Maurice Wilkins (figure 1.10).

Watson and Crick proposed that DNA is a **double helix**—two DNA strands wound around one another. More important, the bases of each strand are on the inside of the helix, and a base on one strand pairs with one on the other in a very specific way. There are only four different bases in DNA, adenine, guanine, cytosine, and thymine, which we abbreviate A, G, C, and T. Wherever we find an A in one strand, we always find a T in the other; wherever we find a G in one strand, we always find a C in the other. In a sense, then, the two strands are complementary. If we know the base sequence of one, we automatically know the sequence of the other. This complementarity is what allows DNA to replicate faithfully. The two strands come apart, and enzymes build new partners for them using the old strands as templates and following the Watson-Crick base-pairing rules (A with T, G with C).

(a)   (b)

FIGURE 1.10
(a) Rosalind Franklin; (b) Maurice Wilkins.
Photo (a) from Cold Spring Harbor. Photo (b) courtesy of Professor Wilkins, Biophysics Dept., King's College, London.

How Genes Direct the Production of Proteins

The instructions for making protein are "written" in a **genetic code** that gives the sequence of bases in a DNA strand. Two steps are required to make a protein. First, an enzyme makes a copy of one of the DNA strands; this copy is not DNA but its close cousin RNA. In the second step, this RNA (messenger RNA) carries the instructions to the cell's protein factories, called *ribosomes*. The ribosomes "read" the genetic code and put together a protein according to its instructions.

What is the nature of this genetic code? Marshall Nirenberg and Har Gobind Khorana (figure 1.11), working independently, cracked the code in the early 1960s. They found that three bases constitute a code word, called a *codon,* that stands for one amino acid. Out of the sixty-four possible three-base codons, sixty-one stand for amino acids; the other three are stop signals. The ribosomes scan a messenger RNA three bases at a time and bring in the corresponding amino acids to link to the growing protein chain. When they reach a stop signal, they release the completed protein.

How Genes Accumulate Mutations

Genes change in a number of ways. The simplest is a change of one base to another. For example, if a certain codon in a gene is GAG (for the amino acid called glutamic acid), a change to GTG converts it to a valine codon. The protein that results from this mutated gene will have a valine where it ought to have a glutamic acid. This may be one change out of hundreds of amino acids, but it can have profound effects. In fact, this specific change has occurred in the gene for one of the human blood proteins and is responsible for the genetic disorder we call sickle-cell disease. Genes can suffer more profound changes, such

FIGURE 1.11
Har Gobind Khorana (left) and Marshall Nirenberg.
The Bettmann Archive.

as deletions or insertions of large pieces of DNA. The more drastic the change, the more likely that the gene or genes involved will be totally inactivated.

In recent years, geneticists have learned to isolate genes, place them in new organisms, and reproduce them by a set of techniques collectively known as **gene cloning.** Cloned genes not only give molecular geneticists plenty of raw materials for their studies, but they can also be transplanted to plants and animals, and even to humans. Thus, they may provide powerful tools for agriculture and for intervening in human genetic diseases.

> M ost genes are made of DNA arranged in a double helix. This structure explains how genes perform their three main functions: replication, carrying information, and collecting mutations. The complementary nature of the two DNA strands in a gene allows them to replicate faithfully by separating and serving as templates for the assembly of two new complementary strands. The sequence of nucleotides in a typical gene is a genetic code that carries the information for making a protein. Changes in the sequence of bases constitutes a mutation, which usually changes the sequence of amino acids in the gene's protein product.

T ABLE 1.1	*Genetics Time Line*	
1865	Charles Darwin	Published *On the Origin of Species.*
1865	Gregor Mendel	Advanced the principles of segregation and independent assortment.
1869	Friedrich Miescher	Discovered DNA.
1900	Hugo de Vries, Carl Correns, Erich von Tschermak	Rediscovered Mendel's principles.
1902	Archibald Garrod	Noted the first genetic disease.
1902	Walter Sutton, Theodor Boveri	Proposed the chromosome theory.
1908	G. H. Hardy, Wilhelm Weinberg	Formulated the Hardy-Weinberg principle.
1910, 1916	Thomas Morgan, Calvin Bridges	Demonstrated that genes are on chromosomes.
1913	A. H. Sturtevant	Constructed a genetic map.
1927	H. J. Muller	Induced mutation by X rays.
1931	Barbara McClintock	Obtained physical evidence for recombination.
1941	George Beadle, E. L. Tatum	Proposed the one gene-one enzyme hypothesis.
1944	Oswald Avery, Colin McLeod, Maclyn McCarty	Identified DNA as the material genes are made of.
1953	James Watson, Francis Crick, Rosalind Franklin, Maurice Wilkins	Determined the structure of DNA.
1958	Matthew Meselson, Franklin Stahl	Demonstrated the semiconservative replication of DNA.
1961	Sidney Brenner, François Jacob, Matthew Meselson	Discovered messenger RNA.
1966	Marshall Nirenberg, Har Gobind Khorana	Completed the genetic code.
1972	Paul Berg	Made the first recombinant DNA in vitro.
1973	Herb Boyer, Stanley Cohen	First used a plasmid to clone DNA.

WHY STUDY GENETICS?

This, then, is a brief introduction to the field of genetics. The major milestones in this story are listed in table 1.1. The following chapters will expand on these themes and fill in the large gaps that we have necessarily left.

The study upon which you are embarking is a fascinating one. It is as ancient as our prehistoric ancestors' selection of more fruitful grains to grow and more desirable animals to breed. It is as modern as tomorrow's gene cloning experiment. And it could hardly be more important for promoting an understanding of our future world, since some of the most profound changes in medicine and agriculture over the next few decades are likely to flow from the manipulation of genes. Indeed, the very power of such manipulation is certain to provoke debate about benefits versus risks and even about whether it is appropriate for us to use such awesome tools at all. It is imperative that the public be educated about genetics in order to make intelligent decisions about these issues. In fact, F. H. Westheimer, professor emeritus of chemistry at Harvard University, commented about molecular biology (in this context, molecular genetics) as follows: "The greatest intellectual revolution of the last forty years may have taken place in biology. Can anyone be considered educated today who does not understand a little about molecular biology?"

Summary

T he field of genetics consists of three main branches: transmission genetics, the study of the transmission of traits from one generation to the next; molecular genetics, the study of the structure and expression of genes at the molecular level; and population genetics, the study of the variation of genes between and within populations. These branches are not exclusive of one another; rather, they reinforce one another.

Genes can exist in several different forms called alleles. A recessive allele can be masked by a dominant one in a heterozygote, but it does not disappear. It can express itself again in a homozygote bearing two recessive alleles.

Genes exist in a linear array on chromosomes. Therefore, traits governed by genes that lie on the same chromosome tend to be inherited together. However, crossovers between homologous chromosomes occur during meiosis, so that gametes bearing nonparental combinations of alleles can be produced. The farther apart two genes lie on a chromosome, the more likely it is that such recombination will take place between them.

Most genes are made of double-stranded DNA arranged in a double helix. One strand is the complement of the other, which means that faithful gene replication requires only that the two strands separate and acquire complementary partners. The linear sequence of bases in a gene carries the information for making a protein. Any change (mutation) in this sequence is likely to cause a corresponding change in the protein product. ∎

P·A·R·T

2

Mendelian Genetics

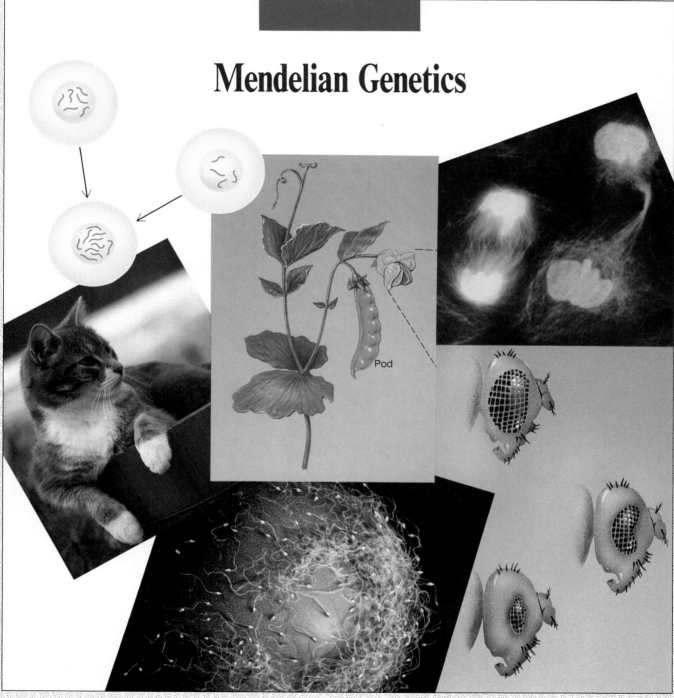

Fundamentals of Mendelian Genetics

G regor Mendel's short treatise "Experiments on Plant Hybrids" is one of the triumphs of the human mind. . . . It can give pleasure and provide insight to each new reader and strengthen the exhilaration of being in the company of a great mind at every subsequent study.

■

Curt Stern and Eva Sherwood
American geneticists

Learning Objectives

In this chapter you will learn:

1. The principle of segregation that governs the transmission of different alleles at a gene from generation to generation.

2. The principle of independent assortment that determines the transmission of genetic variants at different genes from generation to generation.

3. How to describe Mendel's principles using probability models.

4. How to apply the chi-square test to genetic data.

5. The usefulness of genetic models to identify inheritance patterns in human pedigrees.

or centuries, people realized that individual characteristics were passed on from parent to offspring. However, the mechanism of inheritance was not generally understood until this century. The prevailing idea in the mid-nineteenth century was that the characteristics from the parents somehow blended at conception to form the characteristics of the offspring. This theory of **blending inheritance** did account for the fact that offspring possessed the characteristics of both parents. However, the idea of **particulate inheritance,** as proposed by Gregor Mendel in 1866, was able to explain many other observations. Mendel's theory suggested that a specific genetic unit was passed on from generation to generation; this unit we now call a **gene.**

PRINCIPLE OF SEGREGATION

It was not an accident that Mendel was able to explain the phenomenon of inheritance, a subject that had puzzled many previous researchers. His success was based on at least three factors. First, he realized that to understand the basis of inheritance, he had to thoroughly document the results of his experiments and quantify his findings. This inclination apparently arose from his earlier interest in physics, which was even then a quantitative science. Second, Mendel's chosen experimental organism, the garden pea, was well suited for genetic investigation for several reasons: Peas have a number of contrasting characteristics in different individuals—for example, round seeds versus wrinkled seeds; garden pea flowers are large

B·O·X

2.1

The Father of Genetics

Gregor Mendel, born in 1822, grew up on his father's farm in a province of Austria. Because his family was poor, he entered a monastery in order to continue his education. During this period, he studied physics and mathematics, subjects that were sound training for his later experiments with inheritance. He was sent by his order to the University of Vienna to take examinations so that he could become an accredited teacher. However, he failed the exams and returned to the monastery where he was an "unaccredited" teacher for many years.

Beginning on his father's farm, Mendel had a lifelong interest in plants and animals, and he generally kept flowers, bees, and mice. He was fascinated by animal breeding, but because his superiors did not think it appropriate for a cleric to breed animals, he concentrated instead on raising garden peas and other plants in the monastery garden (figure B2.1). One can visit this tranquil garden in Brno, Czechoslovakia, yet today. The garden has a planting of flowers, instead of garden peas, to illustrate Mendel's principles.

Mendel was able to determine the basic principles of inheritance from his plant breeding experiments. The results of his work were reported in 1865 to the Brünn Society for the Study of Natural Science and published in German in an 1866 paper entitled "Experiments on Plant Hybrids." The paper was distributed fairly widely through libraries, but Mendel's contemporaries did not appreciate his findings, probably in part because of the

FIGURE B2.1
Gregor Mendel, early discoverer of the principles of genetics, in his garden.
The Bettmann Archive.

(making them easy to manipulate); pea plants produce large progeny numbers; and many different varieties of peas were available commercially in Mendel's time. Third, Mendel realized that he could best understand the basis of inheritance by concentrating initially on a single trait at a time, rather than on a number of traits as many of his contemporaries had tried to do.

Mendel's choice of the garden pea, *Pisum sativum,* was also fortunate in that varieties of the pea plant are capable of being developed into **true-breeding lines** in which parental plants reproduce offspring all with the same characteristics as themselves. Mendel ensured that his lines were true-breeding by growing them for two years in the monastery garden before beginning his experiments.

mathematical explanations he gave for his results. In addition, most other researchers of the time studied many characteristics simultaneously, which led to such complicated results that the basic principles of inheritance were not discernible. In fact, later on, Mendel himself apparently had some questions about the generality of his principles. When he tried to duplicate his experiments in another plant called hawkweed, the results were completely different from those in garden peas. We now know that these differences occurred because hawkweed requires a pollen grain for the egg to develop but does not include the genetic content of the pollen in the offspring.

Mendel became abbot of the monastery in 1868 and published nothing further on heredity after his 1866 masterpiece. Until his death in 1884, he spent much of his time engaged in administrative conflicts with the local government and also developed a fondness for secular pleasures, including food and cigars.

Unfortunately, Mendel died before his work was appreciated by the rest of the scientific community. In 1900, three botanists, Carl Correns of Germany, Hugo de Vries of the Netherlands, and Erich von Tschermak of Austria, rediscovered his work after each had independently reached similar conclusions. This marked the beginning of modern genetic research. Mendel's experimentation is now recognized as a classic example of carefully planned and executed scientific research and his paper as an excellent illustration of scientific genius. ∎

Garden pea varieties are true-breeding because they reproduce primarily by **self-fertilization,** meaning that both the male and female gametes come from the same individual. Self-fertilization occurs when the anthers (the "male" part of the flower containing the pollen) transfer pollen to the stigma (the "female" part) of the same flower (see figure 2.1).

In order to study inheritance, Mendel crossed, or hybridized, different true-breeding lines (see figure 2.2). He did this by transferring pollen from the anthers of one plant to the stigma of another plant. The second plant had been emasculated; that is, its anthers had been removed prior to the development of pollen so that it could not self-fertilize.

When Mendel crossed lines that differed in one characteristic, such as round seeds versus wrinkled seeds, he found that the progeny resembled one of the parents. The parental type found in the progeny is called **dominant,** while the alternative, missing characteristic is called **recessive.** (The terms dominant and recessive generally refer to characteristics, but in some cases they also refer to different forms of a gene.)

The characteristics (or traits) that Mendel used to differentiate his pea plants are portrayed in figure 2.3. For each trait, the dominant form is on the left. The physical appearance of an individual plant or animal is called its **phenotype.** For example, an individual plant may have the round seed phenotype and the violet flower phenotype. When Mendel crossed two plants that were alike except that one had round seeds and the other had wrinkled seeds, all the progeny had the dominant round seed phenotype. Crosses between these progeny are called **monohybrid crosses,** because the offspring are hybrid—they are the product of two different varieties and they have different parental contributions for a single trait. Mendel obtained the same results whether the female parent had round seeds and the male parent wrinkled seeds, or vice versa. In other words, **reciprocal matings** (matings between a female parent of the first type crossed with a male of the second type, or between a female of the second type crossed with a male of the first type) yielded equivalent progeny types. The progeny generation from such a cross is usually referred to as the F_1, or **first filial generation.**

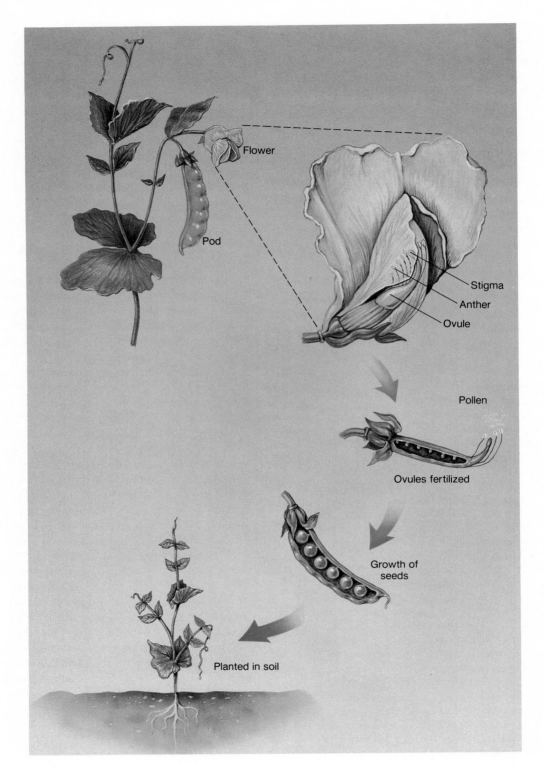

FIGURE 2.1

The anatomy and life cycle of the pea plant. Notice that the pollen and ovule (egg) are from the same flower; that is, self-fertilization is occurring.

From Linda R. Maxson and Charles H. Daughtery, Genetics: A Human Perspective, *2d ed. Copyright © 1989 Wm. C. Brown Publishers, Dubuque, Iowa. All Rights Reserved. Reprinted by permission.*

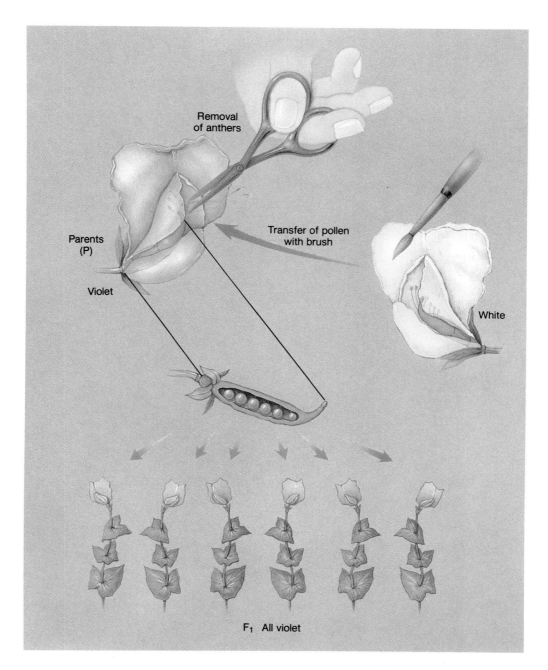

FIGURE 2.2
Artificial cross of a white-flowered (male parent) with a violet-flowered (female parent) pea plant. In this case, all F_1 progeny are like the dominant parental type, having violet flowers.

MENDEL'S EXPERIMENTAL RESULTS

What were the experimental results that first allowed Mendel to understand the principles of heredity, concepts that continued to elude his contemporaries and other researchers until 1900? After obtaining F_1 progeny, he allowed those individuals to self-fertilize and then observed the phenotypic numbers of the **second filial generation,** or F_2. In this generation, he observed both of the original parental phenotypes but found approximately three times as many of the dominant parental type as the recessive parental type, or 75% dominant and 25% recessive. These findings allowed Mendel to reject the theory of blending

FIGURE 2.3
The seven characteristics in garden peas that Mendel studied. The dominant form is on the left in each case.

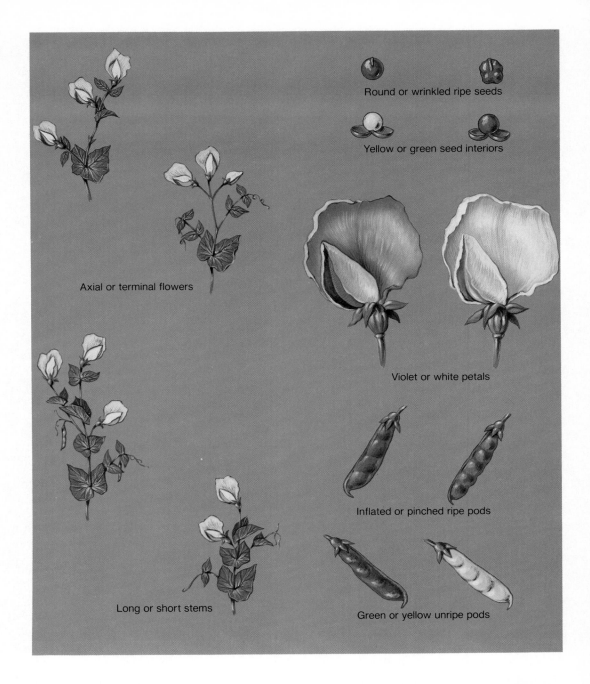

Round or wrinkled ripe seeds

Yellow or green seed interiors

Axial or terminal flowers

Violet or white petals

Inflated or pinched ripe pods

Long or short stems

Green or yellow unripe pods

inheritance because (1) the F_1 was not intermediate in phenotype and (2) the recessive phenotype had not been lost, since it reappeared in the F_2 progeny.

A summary of Mendel's results from the seven different monohybrid crosses is given in table 2.1. For all of these crosses, the ratio of dominant to recessive individuals in the F_2 was very close to 3:1. For example, in the F_2 of the cross between plants with round versus wrinkled seeds, there were 5,474 round seeds and 1,850 wrinkled seeds, giving an F_2 ratio of 2.96 round to 1 wrinkled—extremely close to 3:1.

In addition, Mendel self-fertilized the individual F_2 plants and observed the types of progeny in the next (F_3) generation. The one-quarter of the F_2 plants that bore the recessive phenotype were all true-breeding recessives; that is, all their progeny were recessive. However, of the three-quarters that were the dominant phenotype, only one-third were true-breeding dominants, while the other two-thirds were like the F_1 individuals. (To help you understand this, the principles of probability will be discussed later in this chapter.) In other words, upon self-fertilization, three-quarters of the progeny of the plants in the latter category

	F₁		F₂	F₂ Ratio
		Dominant	*Recessive*	*Dominant/Recessive*
Seeds				
Round vs. wrinkled	Round	5,474	1,850	2.96:1
Yellow vs. green	Yellow	6,022	2,001	3.01:1
Pods				
Inflated vs. constricted	Inflated	882	299	2.95:1
Green vs. yellow	Green	428	152	2.82:1
Flowers				
Purple vs. white	Purple	705	224	3.15:1
Axial vs. terminal	Axial	651	207	3.14:1
Stem				
Tall vs. short	Tall	787	277	2.84:1

were dominant and one-quarter were recessive. Overall then, there were three types of F₂ individuals: ¼ true-breeding dominants, ½ dominants that could have both dominant and recessive progeny, and ¼ true-breeding recessives. (See figure 2.4.)

From these experiments, Mendel concluded that particles or factors (which we now call genes) were transmitted intact through the gametes from parents to progeny. He suggested that these factors existed in several alternative forms (which we call **alleles**) that determined the different phenotypes he had observed. He assumed that each individual had two copies of each gene, receiving one copy of the gene (one allele) from the gamete of the male parent and the other copy of the gene (one allele) from the gamete of the female parent. Although this elegant interpretation seems simple to us today, no one in Mendel's time appreciated it. Another term we will also use to indicate a gene is **locus** (pl. **loci**). More precisely, a locus is the place where the gene resides on the chromosome.

MENDEL'S EXPLANATION

Let us examine the hypothesis Mendel developed to explain inheritance for the specific characteristic of round versus wrinkled seeds. We will call the gene determining seed phenotype *R* and say that it has two allelic forms, *R*, the allele associated with the dominant, round phenotype, and *r*, the allele associated with the recessive, wrinkled phenotype. Three **genotypes,** or genetic types, of individuals are possible. (The genotype of an individual is expressed by a pair of letters or symbols because individuals have two copies of each gene.) Two of the genotypes are **homozygotes,** having both their alleles identical. One would have the *RR* genotype for round seeds and the other the *rr* genotype for wrinkled seeds. The third possible type of individual would be a **heterozygote,** *Rr,* having one of each type of allele.

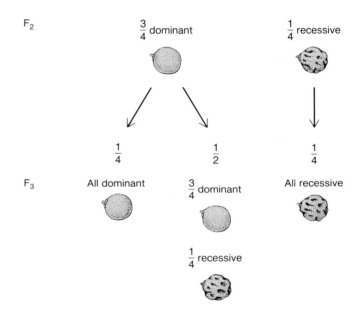

FIGURE 2.4
Proportions of F₃ progeny resulting from selfing F₂ progeny. Note that all F₂ recessive progeny are true-breeding, while only part of the F₂ dominant progeny are true-breeding.

The heterozygous genotype has a round seed phenotype because the round allele is dominant over the wrinkled allele. Sometimes the notation *R*— is used to indicate that the second allele of the genotype may be either *R* or *r;* that is, the individual is either genotype *RR* or *Rr*, both of which have the dominant round phenotype. If we start out with true-breeding round and wrinkled lines as the parents, then they must be *RR* and *rr*, respectively (see figure 2.5). When these parents produce gametes, the *RR* individuals produce gametes with the *R* allele, and the *rr* individuals produce gametes with the *r* allele. Fertilization results in all progeny (the F₁) being heterozygotes, or *Rr*, and having the dominant, round phenotype.

To produce the next generation, Mendel allowed these heterozygotes to self-fertilize. In explaining his results, he assumed that *heterozygotes produced equal numbers of gametes having the two different alleles,* an assumption now known as the **principle of segregation,** or sometimes as **Mendel's first law.** (We will not refer to Mendel's principles as laws in this text because there are important exceptions, some of which will be discussed in later chapters.) As indicated in figure 2.5, this occurs both in males and females, so that half of the pollen from a male heterozygote *Rr* is *R* and half is *r,* and half the eggs from an *Rr* female are *R* and half are *r.* In chapter 4 we will see that the mechanistic basis of the principle of segregation lies in the process of gamete formation and cell division called meiosis.

These gametes can combine in four different, equally frequent ways. First, both the male and female gametes may contain the *R* allele, so that the offspring are homozygous *RR* with a round phenotype. Second, one gamete may be *R* and the other *r,* either with the *R* allele in the male gamete and *r* in the female gamete, or with *R* in the female and *r* in the male. In either case, the offspring will have a heterozygous *Rr* genotype with a round seed phenotype. The other possible combination results when both gametes contain *r,* making all offspring homozygous *rr* with a wrinkled phenotype. Overall then, ¼ of the progeny are *RR,* ½ are *Rr,* and ¼ are *rr,* but because of the dominance of allele *R,* ¾ are the round phenotype (*RR* and *Rr*) and ¼ are the wrinkled phenotype (*rr*). This explanation is consistent with the results of all the crosses listed in table 2.1. In addition, the dominant F₂ plants that were true-breeding in figure 2.4 were genotype *RR,* while those that produced both types of progeny were heterozygous *Rr.*

Mendel confirmed the principle of segregation by conducting several other experiments. In one instance, instead of allowing the F₁ heterozygous individuals to self-fertilize, he crossed them back to the recessive parent. Such a cross to a recessive is called a **testcross** because it is used to test whether a dominant individual is heterozygous or homozygous. In this case, the cross is also a **backcross** because the F₁ is being crossed with an individual having the same genotype as one of its parents. Figure 2.6 gives the genetic basis of this type of cross that leads to equal proportions of the two progeny types because one parent, the homozygous male, contributes only gametes with the recessive *r* allele, while the heterozygous female parent gives equal proportions of gametes with the two types of alleles. In other words, segregation occurs only in the production of gametes from one parent rather than in the production of both types of gametes. The progeny are then equally divided between the dominant phenotype (*Rr*) and the recessive phenotype (*rr*).

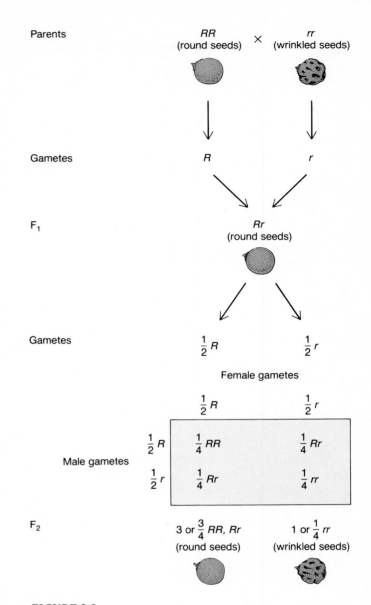

FIGURE 2.5
The genetic basis of the principle of segregation in F₂ progeny in a cross between parents with the round and wrinkled seed phenotypes. Notice that half the gametes from the F₁ individuals have the *R* allele and half have the *r* allele.

After Mendel's genetic principles were rediscovered in 1900, the principle of segregation was reconfirmed by repeating his crosses and carrying out similar crosses in other plants and animals. For example, a number of researchers repeated the cross between yellow and green peas and determined the types of F₂ progeny (see table 2.2). Like Mendel's experiments, the ratio of dominant to recessive in all these experiments was very close to 3:1, and the overall ratio for 175,691 individuals was 3.02:1.

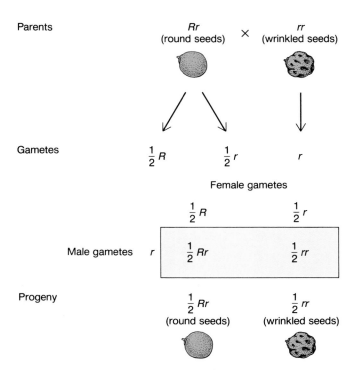

Parents: *Rr* (round seeds) × *rr* (wrinkled seeds)

Gametes: $\frac{1}{2}R$ $\frac{1}{2}r$ r

	Female gametes	
	$\frac{1}{2}R$	$\frac{1}{2}r$
Male gametes r	$\frac{1}{2}Rr$	$\frac{1}{2}rr$

Progeny: $\frac{1}{2}Rr$ (round seeds) $\frac{1}{2}rr$ (wrinkled seeds)

FIGURE 2.6
The genetic basis of the principle of segregation in a testcross. Here segregation occurs only in the female parent, leading to half the progeny having round seeds and half wrinkled seeds.

TABLE 2.2   *Results of Repeat Crosses in Peas*

Researcher	Yellow	Green	F₂ Ratio
			(Dominant/Recessive)
Mendel (1866)	6,022	2,001	3.01:1
Correns (1900)	1,394	453	3.08:1
Tschermak (1900)	3,580	1,190	3.01:1
Bateson (1905)	11,902	3,903	3.05:1
Darbishire (1909)	109,060	36,186	3.01:1
Total	131,958	43,733	3.02:1

TABLE 2.3   *Results of Mendel's Dihybrid Cross*

Generation	Seed Type	Number	F₂ Ratio
Parents	Round, green vs. wrinkled, yellow	—	—
F₁	Round, yellow	—	—
F₂	Round, yellow	315	9.84
	Wrinkled, yellow	101	3.16
	Round, green	108	3.38
	Wrinkled, green	32	1.0
		556	

Because self-fertilization generally does not occur in animals, two different F₁ individuals, or full siblings (brother and sister), had to be crossed in order to examine the segregation ratio in their progeny. The generality of Mendel's principles was confirmed by a number of researchers who crossed full siblings in animals such as mice, pigeons, and *Drosophila*. The first mammalian trait to be analyzed was the albino gene in mice, studied by William Castle and Lucien Cuénot. However, it has been suggested that Mendel actually discovered his principles by secretly crossing mice with different coat color phenotypes, then later validating his findings with peas!

*M*endel was the first scientist to demonstrate that inheritance is based on the segregation of alleles at a gene. From self-fertilization of F₁ garden pea plants obtained from crosses between plants with different characteristics, he observed a 3:1 ratio of dominant to recessive phenotypes in the F₂ progeny. The principle of segregation he developed states that heterozygous individuals produce equal proportions of gametes containing the two alleles.

■

PRINCIPLE OF INDEPENDENT ASSORTMENT

The experiments of Mendel discussed so far have dealt with crosses between lines differing in only one character or in alleles at one gene. What happens if the parents differ in two characters (at two genes)? Mendel examined this situation once he had grasped the pattern of inheritance for a single trait. By performing **dihybrid crosses,** between parents having different parental contributions for two traits, he developed the **principle of independent assortment,** sometimes known as **Mendel's second law.**

In one of Mendel's experiments, he crossed plants that differed in two seed characteristics—round, green seeds in one parent and wrinkled, yellow seeds in the other (see table 2.3). In this case, all of the F₁ progeny had the doubly dominant phenotype of round, yellow seeds. However, when these F₁ individuals were allowed to self-fertilize, four types of progeny were produced in the F₂, with the highest proportion being the dominant type for both traits (round, yellow). The two categories with one dominant type (wrinkled, yellow and round, green) had lower and similar numbers. The fewest had the double recessive phenotype (wrinkled, green). Mendel repeated these experiments using other traits and found that the ratios of the four types were always near 9:3:3:1. For example, if each number in table 2.3 is divided by 32, the number in the double recessive class, the ratio of the types is 9.84 : 3.16 : 3.38 : 1, as shown in the far right column.

Before trying to explain this whole array, let us determine whether the principle of segregation holds in this cross. Examining round versus wrinkled phenotypes in the F_2 reveals 423 round (315 + 108, combining the yellow and green classes) and 133 wrinkled (101 + 32). The ratio of round to wrinkled is then 2.96 : 1, consistent with expectations using the principle of segregation. Likewise, there are 416 yellow (315 + 101) and 140 green (108 + 32), making the yellow to green ratio 2.97 : 1, again close to segregation expectations. From these observations, Mendel concluded that the two genetic systems were independent.

Let us go through the model that Mendel developed to show how he could predict a 9 : 3 : 3 : 1 ratio in the F_2 (see figure 2.7). First, note that the F_1 plants are all heterozygous for both genes, *RrYy*, having received a dominant allele from one parent and a recessive allele from the other. As before, based on the principle of segregation, the frequency of both the *R* and *r* alleles in the F_2 gametes is ½. For the other gene as well, the frequency of both the *Y* and *y* alleles in the gametes is ½. However, the frequency of a gamete that contains both *R* and *Y* is ¼. In a subsequent section of this chapter, we will discuss the use of probability to explain this, but for the time being, we will state that having both *R* and *Y* in a gamete is equal to the product of obtaining an *R* and a *Y*, or (½)(½) = ¼.

If we assume that the same process occurs for all four gametes (*RY, Ry, rY,* and *ry*) and that it occurs in both females and males, we obtain a four-by-four matrix, or grid, as shown in figure 2.7. A matrix like this is often called a **Punnett square** after the British geneticist R. C. Punnett, who employed it for determining the genetic outcome of crosses. In order to determine the proportion of different types of individuals in the F_2 progeny, Mendel assumed that these gametes unite randomly to form genotypes, the same assumption he had made in formulating the principle of segregation.

All the elements (or cells) in the matrix are equally likely and are, therefore, equal in frequency (¼)(¼) = ¹⁄₁₆. In addition, note that the progeny of each genotype is a composite of the two gametes that form it. In other words, the genotype formed from gametes *RY* and *Ry* is *RRYy*. In six cases, two different combinations of gametes yield the same F_2 genotypes. For example, the *RRYy* genotype results from crossing an *Ry* female gamete and an *RY* male gamete as well as from crossing an *RY* female gamete and an *Ry* male gamete.

If we combine the cells having the same combination of traits, the basis of the 9 : 3 : 3 : 1 ratio is apparent. First, only one cell of the sixteen is doubly recessive—the lower right one. Second, there are three cells for each of the two

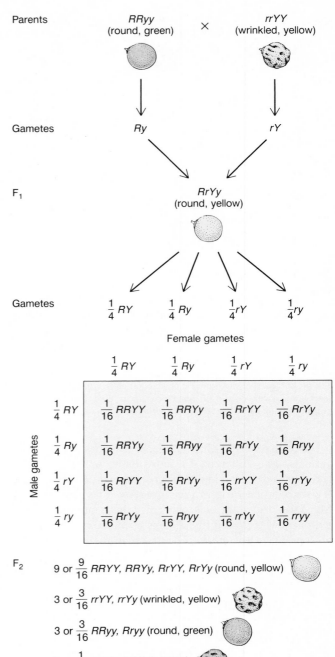

FIGURE 2.7
The genetic basis of the principle of independent assortment in F_2 progeny. Each cell in the 4 × 4 matrix is equally likely. Combine them to find the 9 : 3 : 3 : 1 ratio of F_2 phenotypes.

types of single dominants; for example, the two *Rryy* cells and the one *RRyy* cell are all round, green. The remaining nine cells are double dominants, completing the 9 : 3 : 3 : 1 ratio.

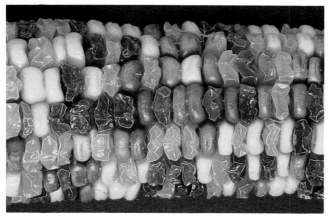

FIGURE 2.8
Each kernel of corn illustrates independent assortment by the color and shape.
Supplied by Carolina Biological Supply Company.

A particularly dramatic example of independent assortment is illustrated in the color and shape of the corn kernels on the cob in figure 2.8. Each kernel is the result of an independent fertilization event so that the array of kernels represents an F_2 progeny array. How many of the four phenotypic classes are observable in this photograph? Is their occurrence close to a $9:3:3:1$ ratio?

The mechanistic basis of independent assortment—that genes on different chromosomes randomly go to different gametes during meiosis—was not known to Mendel. In chapter 4 we will discuss the genetic implications of meiosis in some detail. In addition, not all pairs of genes show independent assortment, a fact Mendel also did not discover. In chapter 5 we will discuss an important exception to Mendel's second principle, called linkage, in which pairs of genes on the same chromosome show association.

To better understand both segregation and independent assortment, it is useful to know some facts about probability and statistics. After we discuss these concepts, we will illustrate how they can be applied to Mendel's garden pea experiments.

> **M**endel's principle of independent assortment states that alleles at different genes assort independently of each other, resulting in a $9:3:3:1$ ratio in F_2 progeny from a dihybrid cross.

■

PROBABILITY

How can we understand Mendel's findings in a general manner? One way to better grasp the principles of segregation and independent assortment is to relate them to some simple theories of probability. These ideas were known to Mendel and were obviously important in his development of a general model of inheritance.

First, the term **probability** is defined as the number of times a particular event occurs, divided by the total number of opportunities for the event to occur. For example, if a coin is flipped ten times and turns up heads four times and tails six times, the probability of heads is $^4/_{10} = 0.4$ and the probability of tails is $^6/_{10} = 0.6$. The sum of these probabilities is unity; that is, $^4/_{10} + ^6/_{10} = 1$. These are the observed probabilities of heads and tails in a given sample of ten coin tosses. If the coin is "fair"—that is, equally likely to land on heads or tails—and if it is flipped many times, the expected probabilities are 0.5 for heads and 0.5 for tails.

SUM RULE

Heads and tails are examples of **mutually exclusive events.** In other words, if a tail is flipped, that excludes the possibility of a head being flipped on the same coin at the same time. *The combined probability of two (or more) mutually exclusive events occurring is the sum of their individual probabilities.* This is known as the **sum rule.** For example, symbolizing the probability of a head as Pr(head) and the probability of a tail as Pr(tail) and assuming a fair coin:

$$Pr(\text{head or tail}) = Pr(\text{head}) + Pr(\text{tail})$$
$$= \frac{1}{2} + \frac{1}{2}$$
$$= 1$$

If there are more than two possible events, the sum of the probabilities of only two of the events should be less than unity, as we will discuss later.

A genetic example of probability occurs when we examine the types of gametes produced by a heterozygote. In this case, because of the genetic consequences of meiosis (see chapter 4), equal numbers of the two gametes having different alleles should be produced. For example, when the parent is heterozygous *Rr,* the probability of an *R* gamete is ½ and the probability of an *r* gamete is also ½. As in the coin example, using the sum rule of probabilities of the two different events, notice that:

$$Pr(R \text{ or } r) = Pr(R) + Pr(r)$$
$$= \frac{1}{2} + \frac{1}{2}$$
$$= 1$$

PRODUCT RULE

Now let us examine the probability of two independent events occurring simultaneously. This probability is expressed by the **product rule,** which states that *the joint probability of both of two independent events occurring is the product of their individual probabilities.* Thus, if we flip two different coins, say a penny and a nickel, what is the probability of obtaining a head from the first and a head from the second? The probability of a head from the penny is ½ and the probability of a head from the nickel is ½. Thus, the joint probability of two heads is:

$$\begin{aligned} \text{Pr(two heads)} &= \text{Pr(head on penny) Pr(head on nickel)} \\ &= (½)(½) \\ &= ¼ \end{aligned}$$

A genetic example of the product rule occurs when examining the progeny of a cross between two heterozygotes. For example, we would like to know the probability of an *rr* homozygote from the cross $Rr \times Rr$ where the first individual is the female. To obtain an *rr* offspring, both gametes must contain an *r* allele. Based on the principle of segregation, the probability of an *r* from the female is ½ and the probability of an *r* from the male is ½. Therefore, the probability of obtaining an *r* simultaneously from the female and male gametes to yield an *rr* offspring is $(½)(½) = ¼$. Mendel used this concept to develop the principle of segregation and the proportions shown in figure 2.5.

We also used the product rule to obtain the probability of different genotypes in a dihybrid cross. First, we assumed that the genes assort independently into F_2 gametes so that we can use the product rule to determine the joint probability of alleles from different genes in the same gamete. For example, the probability of an *R* allele and a *Y* allele in the same gamete is:

$$\text{Pr}(R \text{ and } Y) = \text{Pr}(R)\,\text{Pr}(Y) = (½)(½) = ¼$$

Likewise:

$$\text{Pr}(R \text{ and } y) = \text{Pr}(R)\,\text{Pr}(y) = (½)(½) = ¼$$

$$\text{Pr}(r \text{ and } Y) = \text{Pr}(r)\,\text{Pr}(Y) = (½)(½) = ¼$$

$$\text{Pr}(r \text{ and } y) = \text{Pr}(r)\,\text{Pr}(y) = (½)(½) = ¼$$

Then, in order to determine the probability of F_2 progeny from these gametes, we again used the product rule. For example, the probability of an *RRYY* genotype is:

$$\begin{aligned} \text{Pr}(RY \text{ female gamete and} \\ RY \text{ male gamete)} &= \text{Pr}(RY)\,\text{Pr}(RY) \\ &= (¼)(¼) \\ &= ^1/_{16} \end{aligned}$$

Next, we need to calculate the total probability of two different combined or joint events occurring. To calculate this probability we can again use the sum rule, which states that the probability of either of two mutually exclusive events occurring is the sum of their individual probabilities. Thus, using the coin example again, if we want to know the total probability of obtaining two heads *or* two tails when two coins are flipped:

$$\begin{aligned} \text{Pr(two heads or two tails)} &= \text{Pr(two heads)} + \text{Pr(two tails)} \\ &= \text{Pr(head) Pr(head)} + \text{Pr(tail) Pr(tail)} \\ &= (½)(½) + (½)(½) \\ &= ¼ + ¼ \\ &= ½ \end{aligned}$$

In this case, notice that the events—for example, two heads—are actually themselves composed of two coin flips.

Thinking again of the segregation probabilities of progeny from an F_1 cross, the probability of an *RR* offspring is ¼, that of an *Rr* offspring is ½, and that of an *rr* offspring is ¼. Because both genotypes *RR* and *Rr* are dominant, we may want to know the probability of obtaining either *RR* or *Rr*. Using the sum rule:

$$\begin{aligned} \text{Pr}(RR \text{ or } Rr) &= \text{Pr}(RR) + \text{Pr}(Rr) \\ &= ¼ + ½ \\ &= ¾ \end{aligned}$$

Both the product rule and the sum rule can be extended to more than two events. For example, the probability of simultaneously obtaining heads on each of four coins, using the product rule, is $(½)(½)(½)(½) = (½)^4 = ^1/_{16}$. Likewise, in a standard card deck with thirteen types of cards (aces, twos, threes . . . kings), the probability of drawing a particular type of card (ignoring different suits) is $^1/_{13}$. The probability of drawing a face card (jack, queen, or king) is $^1/_{13} + ^1/_{13} + ^1/_{13} = ^3/_{13}$.

CONDITIONAL PROBABILITY

As we have just discussed, more than two outcomes or events may be possible. For example, a fair, six-sided die should have an equal probability of landing with a 1, 2, 3, 4, 5, or 6 showing. In other words, the probability of each of these outcomes is $^1/_6$. Obviously, half the time the die should land with an even number (2, 4, or 6) showing. We can ask, given that an even number is rolled, what is the probability of its being a 2? *A probability that is contingent on given circumstances* is called a **conditional probability.** In other words, the probability that we will obtain a 2, given that an even number is rolled on the die, is:

$$\begin{aligned} \text{Pr}(2 \mid \text{even number}) &= \text{Pr}(2)/\text{Pr(even)} \\ &= (^1/_6)/(½) \\ &= ^1/_3 \end{aligned}$$

(The vertical line means "given.") In other words, the total probability of a general event, an even number in this case, is subdivided into more specific events.

When Mendel examined his F_2 progeny, he used conditional probability to describe the proportion of heterozygotes among the dominant progeny. In this case, the probability of heterozygotes among dominant progeny is:

$$\text{Pr(heterozygous | dominant)} = \text{Pr(heterozygous)}/\text{Pr(dominant)}$$
$$= (\tfrac{1}{2})/(\tfrac{3}{4})$$
$$= \tfrac{2}{3}$$

In other words, two-thirds of the dominant progeny are expected to be heterozygotes (and one-third are expected to be homozygotes). We will use this same logic later to calculate the probability that normal individuals who have a sibling with a recessive disease are themselves **carriers** of the disease allele; that is, that they are heterozygotes.

BINOMIAL PROBABILITY

Finally, consider the probability that in a group of a given size, say a sibship (a group of brothers and sisters), a certain number of individuals will be of one type and the remainder of another type. To calculate these probabilities we can use an expansion of the product rule. First, we calculate the probability of the occurrence of a particular sequence of events when there are three different events and only two alternatives for each event. As an example, imagine flipping a coin three consecutive times. The following eight sequences are possible:

	HHT	TTH	
HHH	HTH	THT	TTT
	THH	HTT	

There could be three consecutive heads and no tails, as in the first column; two heads and one tail (second column); one head and two tails (third column); or three tails and no heads (fourth column). In other words, if the order is ignored, four outcomes are possible.

Now, assuming again that the probability of heads is $\tfrac{1}{2}$ (and that the probability of tails is $\tfrac{1}{2}$), the probability of their coming up in any one of the orders shown in the second column is, by the product rule:

$$\text{Pr(HTH)} = \text{Pr(H) Pr(T) Pr(H)}$$
$$= (\tfrac{1}{2})(\tfrac{1}{2})(\tfrac{1}{2})$$
$$= \tfrac{1}{8}$$

However, to determine the probability of two heads and one tail, independent of order, we have to use the sum rule to get the total probability of the three sequences that give this constitution, or:

$$\text{Pr(two heads, one tail)} = \text{Pr(HHT)} + \text{Pr(HTH)}$$
$$+ \text{Pr(THH)}$$
$$= \tfrac{1}{8} + \tfrac{1}{8} + \tfrac{1}{8}$$
$$= \tfrac{3}{8}$$

Thus, the probability of three heads is $\tfrac{1}{8}$; the probability of two heads (and one tail) is $\tfrac{3}{8}$; the probability of one head (and two tails) is $\tfrac{3}{8}$; and the probability of no heads (and three tails) is $\tfrac{1}{8}$. The sum of all these probabilities is $\tfrac{1}{8} + \tfrac{3}{8} + \tfrac{3}{8} + \tfrac{1}{8} = 1$, as expected.

Now let us generalize the approach we have just developed. First, assume there are two possible alternatives, say heads or tails, having probabilities p and q, respectively. We know that $p + q = 1$ because there are only two alternatives. If N different events occur (N being equal to three because three coins were flipped), what is the probability of x events of the first type and $N - x$ of the second type? In the previous example, $x = 2$ (two heads) and $N - x = 1$ (one tail). In general, the **binomial formula** (binomial because there are only two different types of events) gives these probabilities:

$$\text{Pr(x type one events}$$
$$\text{out of N total events)} = Cp^x q^{N-x}$$

In this formula, C is the binomial coefficient and indicates the number of sequences that give the same overall constitution. In the current example, $C = 3$.

The binomial coefficient for most of the examples we will consider can be obtained using the following formula:

$$C = \frac{N!}{x!(N - x)!}$$

where the symbol ! indicates a factorial. By definition, $0! = 1$, $1! = 1$, $2! = (2)(1) = 2$, $3! = (3)(2)(1) = 6$, and so on. For example, if $x = 2$ and $N = 3$, then:

$$C = \frac{3!}{2!(3 - 2)!} = \frac{(3)(2)(1)}{(2)(1)(1)} = 3$$

This is the same answer we found previously by writing out all the possibilities.

Another way to obtain the binomial coefficient is to use Pascal's triangle (see figure 2.9). This relationship was named after a seventeenth-century French mathematician and philosopher. It is useful if you forget the formula for the binomial coefficient. Any given number in this array

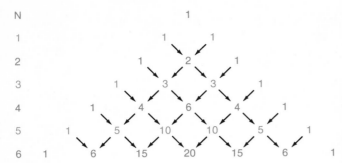

FIGURE 2.9
Pascal's triangle showing the size of the binomial coefficient (C) for different N values, the number of different events. The numbers in a horizontal row indicate C for x = N, N −1, N − 2, . . . , etc.

FIGURE 2.10
A cat with white spotting and short hair. (The cat is also heterozygous for the orange gene, accounting for the black and orange pattern where there is no white spotting.)
© *John Cunningham, Visuals Unlimited.*

is obtained by summing the two numbers directly above. As an example, let us look at N = 3, row 4 in figure 2.9. The first number in this row, C = 1, is the coefficient for all events of one type (for example, three heads), x = 3; the second number is for x = 2 (two heads), giving C = 3 as before; the third number is for x = 1 (one head), giving C = 3; and the fourth number is for x = 0 (no heads), giving C = 1.

We will use the binomial formula to calculate the probability of a particular number of boys and girls in a sibship. Assuming there are four children, what is the probability that four will be girls? In other words, what is the probability that x = 4 and N − x = 0? Assuming that the probability of a girl is ½ and the probability of a boy is also ½, then:

$$\text{Pr(four girls of four children)} = C(\tfrac{1}{2})^4(\tfrac{1}{2})^0$$

From the formula for the binomial coefficient:

$$C = \frac{4!}{4!\ 0!} = 1$$

Or, looking at figure 2.9, where N = 4 and x = 4, then C = 1. Given that any number to the zero power is 1:

$$\text{Pr(four girls of four children)} = (1)(\tfrac{1}{2})^4(1)$$
$$= \tfrac{1}{16}$$

Now let us look at a more complicated situation and calculate the probability of two girls in a sibship of size four. Again, using the formula where x = 2 and N − x = 2:

$$\text{Pr(two girls of four children)} = C(\tfrac{1}{2})^2(\tfrac{1}{2})^2$$

In this case:

$$C = \frac{4!}{2!\ 2!} = 6$$

so that:

$$\text{Pr(two girls of four children)} = 6(\tfrac{1}{2})^2(\tfrac{1}{2})^2$$
$$= \tfrac{3}{8}$$

C = 6 here because there are six orders, all of which have two girls and two boys: BBGG, BGBG, BGGB, GBBG, GBGB, and GGBB.

We can apply these probability concepts to other situations. For example, a number of genes affect the coats of domestic cats. One gene causes white spotting, primarily on the chest of the cat, where *S*, the white-spotting allele is dominant over *s* for no white spotting. Another gene affects hair length, where *L*, the short-hair allele, is dominant over *l*, the long-hair allele. Figure 2.10 shows a cat with white spotting (either genotype *SS* or *Ss*) and short hair (genotype *LL* or *Ll*). Assuming that two cats, both heterozygous at these genes, mate, what proportion of their progeny will exhibit the different phenotypes? Figure 2.11 gives the expected frequency of the progeny after segregation and independent assortment. Most of the time (⁹⁄₁₆), the offspring will have the same phenotype as their parents—white spotting and short hair—but ¹⁄₁₆ of the time, the offspring will be different from their parents for both characters; that is, they will have no white spotting and long hair.

We can apply probability concepts to the following example: Given these same two cats as parents and a sibship of four kittens, what is the probability that all the kittens will have long hair (*ll*) and white spotting (*S*−)?

Chapter 2

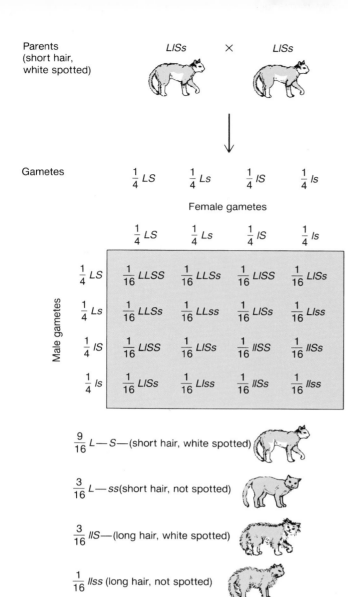

Parents (short hair, white spotted) LlSs × LlSs

Gametes $\frac{1}{4}$ LS $\frac{1}{4}$ Ls $\frac{1}{4}$ lS $\frac{1}{4}$ ls

Female gametes

	$\frac{1}{4}$ LS	$\frac{1}{4}$ Ls	$\frac{1}{4}$ lS	$\frac{1}{4}$ ls
$\frac{1}{4}$ LS	$\frac{1}{16}$ LLSS	$\frac{1}{16}$ LLSs	$\frac{1}{16}$ LlSS	$\frac{1}{16}$ LlSs
$\frac{1}{4}$ Ls	$\frac{1}{16}$ LLSs	$\frac{1}{16}$ LLss	$\frac{1}{16}$ LlSs	$\frac{1}{16}$ Llss
$\frac{1}{4}$ lS	$\frac{1}{16}$ LlSS	$\frac{1}{16}$ LlSs	$\frac{1}{16}$ llSS	$\frac{1}{16}$ llSs
$\frac{1}{4}$ ls	$\frac{1}{16}$ LlSs	$\frac{1}{16}$ Llss	$\frac{1}{16}$ llSs	$\frac{1}{16}$ llss

$\frac{9}{16}$ L—S—(short hair, white spotted)

$\frac{3}{16}$ L—ss(short hair, not spotted)

$\frac{3}{16}$ llS—(long hair, white spotted)

$\frac{1}{16}$ llss (long hair, not spotted)

FIGURE 2.11
A cross between two cats doubly heterozygous at the long-hair and white-spotted genes, giving the expected proportions of progeny.

Because there is only one order in which this can occur and the coat patterns of the different kittens are independent (given these same parental genotypes), then:

Pr(4 of 4 long-hair,
white-spotted kittens) = [Pr(*llS*—)]⁴
= (³⁄₁₆)⁴

In this case, the probability is quite low, with such a litter occurring about one in a thousand times. Perhaps the male parent was misidentified and actually had long hair!

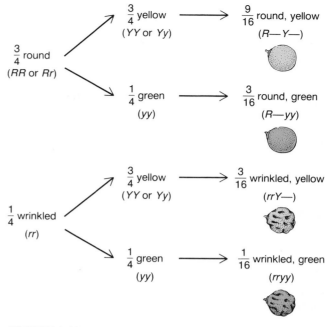

FIGURE 2.12
The forked-line approach for progeny of a dihybrid cross in the garden pea.

S everal basic theories of probability are useful in understanding Mendelian genetics: (1) the product rule, which states that the joint probability of two events occurring is the product of their individual probabilities; (2) the sum rule, which states that the probability of either one of two mutually exclusive events occurring is the sum of their individual probabilities; (3) the principle of conditional probability, which is the probability of a specific event divided by the probability of a more general event; and (4) the binomial probability formula, which states the probability of x events of a particular type out of a total N.

FORKED-LINE APPROACH

The approach just explained is quite cumbersome when used to calculate the probability of different types of progeny for two or more genes. As a result, the **forked-line approach,** in which the probabilities for each category are determined and combined one gene at a time, is quite useful.

The forked-line approach is based on the assumption of independent assortment of all the genes involved. An example is given in figure 2.12 for the garden pea dihybrid cross discussed previously. First, the progeny are separated into proportions as to round or wrinkled (¾ to ¼) at the first gene. Within each of these categories, they are

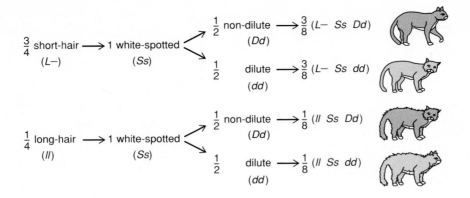

FIGURE 2.13
The four progeny categories and their probabilities for a cross between cats of genotypes *LlssDd* and *LlSSdd*.

then separated into proportions at the second gene (¾ yellow to ¼ green). To determine the number in any combined phenotypic class, the probabilities are multiplied over the genes. For example, the expected proportion of wrinkled, yellow offspring is (¼)(¾) = ³⁄₁₆.

This same approach can be used to calculate proportions for more than two genes. Let us consider cat coat color again and introduce a third gene, *D*, where *dd* homozygotes have dilute (or light) color and *D*— is non-dilute. (The difference between gray and black cats is caused by dilute and non-dilute.) Given a cross between two cats that are *LlssDd* and *LlSSdd*, we can write out all the progeny types using the forked-line approach (see figure 2.13). We can then ask, what is the probability of a long-hair, white-spotted, non-dilute offspring from this mating? This probability, given in the third row of figure 2.13, is ⅛. Or, assuming that these genes assort independently, we can use the product rule to write:

Pr(long-hair, white-spotted,
  non-dilute offspring = Pr(*ll S*— *D*—)
                       = Pr(*ll*) Pr(*S*—) Pr(*D*—)
                       = (¼)(1)(½)
                       = ⅛

We could also ask, what is the probability of an offspring of genotype *LlSsDd*? In this case:

Pr(*LlSsDd*) = Pr(*Ll*) Pr(*Ss*) Pr(*Dd*)
            = (½)(1)(½)
            = ¼

The forked-line approach can be extended to calculate the probability of genotypes or phenotypes in offspring for any number of genes as long as they assort independently.

Often the alleles at a gene are symbolized by a capital letter for the dominant allele and a lowercase letter for the recessive allele, as in alleles *R* and *r*. However, to describe all the mutants that have been discovered, the twenty-six letters of the alphabet are not enough. For example, the recessive mutant in *Drosophila melanogaster* that has small, underdeveloped wings is called *vestigial* and is symbolized by *vg*. The normal allele at this gene, instead of being indicated by a capital letter, is symbolized either by vg^+ or just by +, both of which mean the **wild-type,** or nonmutant allele. Heterozygotes at this locus having normal wings are indicated as either vg^+vg or +*vg*. Mutants themselves may be dominant, such as the curly-wing mutant *Cy* in *Drosophila*, where the first letter of the symbol is capitalized to indicate its dominance. Heterozygotes *Cy*+ have curly wings because *Cy* is dominant over the wild-type allele +.

Let us use these symbols to illustrate the forked-line approach. (Notice that to avoid confusion when using these symbols we will leave a space between the letters indicating the genotypes for different genes.) If we have the cross +*vg Cy*+ *sese* × +*vg* + + +*se* (*se* indicates the eye color mutant *sepia*), what proportion of progeny would we expect to be curly-winged—that is, mutant at the second locus and wild-type at the other two genes? Figure 2.14 shows the eight possible genotypes and their probabilities in the progeny from this cross. The particular category we are interested in is shown in row three of the far right column. To calculate this:

Pr(+— *Cy*+ +—) = Pr(+—) Pr(*Cy*+)
                  Pr(+—)
                = (¾)(½)(½)
                = ³⁄₁₆

Chapter 2

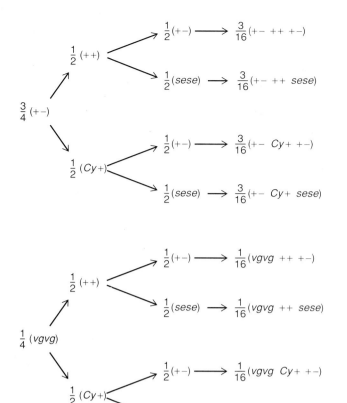

FIGURE 2.14
The eight progeny categories and their probabilities for the cross between *Drosophila* with genotypes $+vg$ $Cy+$ $sese$ and $+vg$ $++$ $+se$.

T he forked-line approach, which is based on the product rule, can be used to determine the probability of various offspring types for any number of genes, assuming they assort independently.

■

GOODNESS OF FIT

How can we use predictions based on probabilities to understand the transmission of genes from one generation to the next? Remember that Mendel's F_2 data given in table 2.1 very closely fit 3:1 expectations. But what if the numbers had not been so close to the expected ratio? Would this have been due to chance or because the underlying hypothesis, the principle of segregation, was incorrect? It is useful to measure the "goodness of fit" to find out whether the observed numbers are adequately close to the

numbers expected according to a particular hypothesis. Mendel himself did not use any statistical tests (they weren't developed until decades later), but such tests have helped identify exceptions to Mendel's principles in other studies.

A test commonly used to measure goodness of fit is the **chi-square test.** In this statistical test, the *observed numbers* in each category, say dominant and recessive progeny, are compared to the *expected numbers* under a certain hypothesis, say Mendel's principle of segregation. Notice that in this test we are not using the observed and expected frequencies but the observed and expected numbers. If the observed number in one category is larger than the expected number, the opposite will be true in another category because the total number over all classes for both observed and expected numbers is the same. To obtain a test that measures the difference between observed and expected numbers, irrespective of the sign of this difference, we use the square of the difference between observed and expected numbers. This value is then standardized by dividing it by the expected number in a category, making the formula for chi-square:

$$\chi^2 = \Sigma \frac{(O - E)^2}{E}$$

where Σ is the Greek letter sigma indicating the sum over all categories, χ is the Greek letter chi, and O and E are the observed and expected numbers.

This statistical technique can be illustrated using the data from one of Mendel's backcross experiments. In this experiment, he crossed F_1 violet-flowered plants (Vv) with white-flowered plants (vv). Of 166 progeny, 85 had violet flowers and 81 had white flowers. According to the principle of segregation, 0.5 should be violet and 0.5 should be white. Or, using numbers, $166(0.5) = 83$ would be expected in each class. Notice that the sum of the expected numbers must always equal the sum of the observed numbers; the sum here is 166. With these observed and expected numbers, we can calculate the chi-square value as in table 2.4. For the violet and white categories, the contributions to chi-square are $(2)^2/83$ and $(-2)^2/83$, respectively, both of which equal 0.05. Notice that if we had not squared -2, this class would exactly cancel out the other class. Summing over both the values, the chi-square is 0.10. Obviously, this is a small value, indicating the similarity of the observed and expected numbers, but how do we know how large or small it really is?

The significance of a chi-square value can be determined by comparing it to a theoretical value from a chi-square table that has the same degrees of freedom. The term **degrees of freedom** refers to an integer that is generally equal to the number of categories in the data, minus

TABLE 2.4   *Values of Mendel's F_2 Data*

Flower Color	Observed Numbers	Expected Numbers	$\dfrac{(O - E)^2}{E}$
Violet	85	166(0.5) = 83	$\dfrac{2^2}{83} = 0.05$
White	81	166(0.5) = 83	$\dfrac{(-2)^2}{83} = 0.05$
	166	166	$\chi^2 = 0.10$

TABLE 2.5   *Probabilities of Different Theoretical Chi-square Values for Given Degrees of Freedom[1]*

Degrees of Freedom	Probability				
	0.9	0.5	0.1	0.05*	0.01**
1	0.02	0.46	2.71	3.84	6.64
2	0.21	1.39	4.60	5.99	9.21
3	0.58	2.37	6.25	7.82	11.34
4	1.06	3.86	7.78	9.49	13.28
5	1.61	4.35	9.24	11.07	15.09
6	2.20	5.35	10.64	12.59	16.81
7	2.83	6.35	12.02	14.07	18.48
8	3.49	7.34	13.36	15.51	20.09
9	4.17	8.34	14.68	16.92	21.07
10	4.86	9.34	15.98	18.31	23.21

[1]Probabilities of 0.05 and 0.01 are known as the statistically significant and highly significant levels and are indicated by * and **, respectively.

one. In other words, the degrees of freedom indicate the number of categories that are independent of each other. For example, if there are two categories, there is only one degree of freedom. This constraint occurs because, given the total number (say 100) and the number in one class (say 22), the number in the other class (78) is fixed.

Assuming that the Mendelian segregation hypothesis we are examining is correct, if we drew many, many samples and calculated a chi-square value for each, we would obtain a theoretical distribution of chi-square values. Due to chance, the chi-square would be large in a small proportion of the cases because the observed numbers would be quite different from the expected numbers. However, most of the time, the observed numbers would be close to those expected, making the chi-square small.

Instead of giving the complete theoretical distribution of chi-square values, we will provide a few important values (table 2.5). These critical chi-square values are exceeded by chance only rarely, given that the underlying hypothesis is correct. There are two different critical theoretical values, 0.05 and 0.01, to which the observed (or calculated) chi-square is generally compared. If the calculated chi-square value exceeds the theoretical values given under 0.05 and 0.01 (for given degrees of freedom), the observations are said to be *statistically significant* or *highly statistically significant,* respectively. Such high chi-square values suggest that the hypothesis used to determine the expected numbers is probably incorrect. The values labeled 0.05 are equalled or exceeded by chance only one in twenty times (0.05) even though the underlying hypothesis is correct. In other words, by chance, the observed numbers could be quite different from those expected one in twenty times, making the chi-square value this large or larger. If the observed and expected numbers are very different, the chi-square may equal or exceed the numbers in the last column, a value that occurs by chance only one in one hundred times, given that the hypothesis is correct.

In the example of violet and white flowers, there are two categories, so there is only one degree of freedom. The calculated or observed chi-square value, 0.10, is substantially less than the critical value needed for statistical significance. (With one degree of freedom, it is 3.84.) Therefore, we can say that the observed chi-square value is not significantly different from the expected chi-square value. In other words, the observed numbers are *consistent* with the numbers expected under the principle of segregation. Notice that we did not say this proves the principle of segregation. One cannot prove a hypothesis by such a statistical test. However, because many such crosses have been evaluated and are consistent with the principle of segregation, we can feel confident that this process is responsible for the inheritance of these genes.

Let us go back to the data from Mendel's dihybrid cross and ask whether the observed phenotypic numbers are consistent with the hypothesis of independent assortment. We can use the chi-square test, but first we need to calculate the expected numbers in each category. There were 556 progeny, so if we multiply 556 by $9/16$, $3/16$, $3/16$, and $1/16$, we obtain the expected numbers in the four classes (see table 2.6). Using the chi-square formula, and then summing over the four categories, the chi-square value is 0.48. Again, like the single-gene example, this is a very small chi-square value. However, in this case, there are four categories, making the degrees of freedom (number of categories minus one) equal to 3. Looking at table 2.5, we find that the calculated chi-square value would have to be equal to, or greater than 7.82 for the observed values

Chapter 2

		TABLE 2.6	*Chi-square of a Dihybrid Cross*	
Seed Type	Observed Numbers	Expected Numbers	$\dfrac{(O-E)^2}{E}$	
Round, yellow	315	$556\,(^9\!/_{16}) = 312.8$	$\dfrac{(2.2)^2}{312.8} = 0.02$	
Wrinkled, yellow	101	$556\,(^3\!/_{16}) = 104.2$	$\dfrac{(-3.2)^2}{104.2} = 0.10$	
Round, green	108	$556\,(^3\!/_{16}) = 104.2$	$\dfrac{(3.8)^2}{104.2} = 0.14$	
Wrinkled, green	32	$556\,(^1\!/_{16}) = 34.8$	$\dfrac{(-2.8)^2}{34.8} = 0.22$	
	556	556	$\chi^2 = 0.48$	

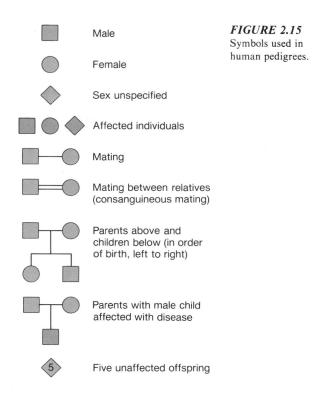

FIGURE 2.15
Symbols used in human pedigrees.

- Male
- Female
- Sex unspecified
- Affected individuals
- Mating
- Mating between relatives (consanguineous mating)
- Parents above and children below (in order of birth, left to right)
- Parents with male child affected with disease
- Five unaffected offspring

in order to be significantly different from expectations with 3 degrees of freedom. Obviously, 0.48 is much less than 7.82, so we can conclude that the observed numbers are consistent with the expected numbers under the hypothesis of independent assortment.

Remember that when calculating chi-square values, observed and expected numbers, not proportions, are used. Also, as a rule of thumb for using a chi-square test, the expected numbers in a category should be five or greater. When the expected number is less than five, a modification of the formula should be used, or else the observed chi-square value may be somewhat inflated.

An interesting historical note is that all of Mendel's results fit expectations very closely and consequently give very low chi-square values. For instance, in the previous example, given that the hypothesis is correct, the chi-square value should be less than 0.46 only 50% of the time. However, the observed value was 0.10, much lower than 0.46. Several statisticians have suggested that Mendel (or some of his student helpers) biased the results to conform to expectations of the principles of segregation and independent assortment. We will probably never know whether or not this is true, but it should not detract from the genius of Mendel's insights into inheritance. We should note too that the workers who repeated Mendel's experiments also obtained results extremely close to expectations (see table 2.2).

A useful statistical test of the goodness of fit between observed and expected numbers is the chi-square. Chi-square values larger than a critical value indicate that the underlying hypothesis used to generate the expected numbers may be incorrect. All of Mendel's results fit expectations very closely and thus yield very low chi-square values.

MENDELIAN INHERITANCE IN HUMANS

The same patterns of inheritance occur in humans as in peas, *Drosophila,* or cats, except that many more traits are involved. In fact, 2,208 single traits and diseases for humans are categorized in the 1988 edition of Victor McKusick's *Mendelian Inheritance in Man;* another 2,001 loci are not yet fully identified or validated. Furthermore, determining the mode of inheritance is more difficult in humans because the sibship sizes are relatively small; that is, there are generally fewer than ten children in a family versus hundreds of seeds produced from a given cross in peas. However, using Mendel's ratios, the same inheritance patterns can be seen even in the pedigrees of small families.

Before looking at Mendelian inheritance in humans, we must introduce the basic symbols used in human pedigree analysis. As shown in figure 2.15, a female or a male in a pedigree chart is indicated by a circle or a square, respectively. Diamonds are used for individuals whose sex has not been determined or in cases where sex is unimportant for genetic analysis. When an individual has a given trait or genetic disease, the symbol is filled in. Mating individuals are joined by a single horizontal line. A mating between relatives, called **consanguineous** (Latin: *consanguinitas* = relationship by blood), is indicated by a double horizontal line. Siblings are joined by a line below their parents and are listed in order of birth, from left to right. A pedigree chart begins with the oldest generation at the top; successive generations are identified by Roman numerals (see figure 2.16).

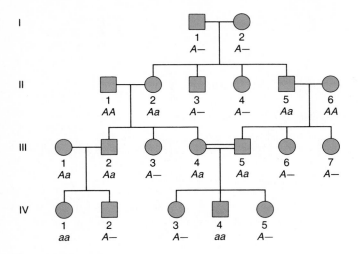

FIGURE 2.16
A human pedigree for a recessive trait, showing the inferred genotypes, where *A* and *a* are the dominant and recessive alleles, respectively. *A*— indicates that the second allele is not known.

From AN INTRODUCTION TO GENETIC ANALYSIS 3/E. By David T. Suzuki et al. Copyright©1976,1981,1986 by W. H. Freeman and Company. Reprinted with permission.

FIGURE 2.17
A Hopi Indian girl with albinism.
Field Museum of Natural History, negative number 118.

RECESSIVE TRAITS

Let us examine a pedigree exemplifying the inheritance pattern for a recessive trait (figure 2.16). Two recessive diseases in humans are **albinism,** in which individuals lack the pigment melanin (see figure 2.17), and **cystic fibrosis,** a severe disorder that affects the production of mucus in the pancreas and lungs. (McKusick lists a total of 626 recessive loci.) Cystic fibrosis is the most common recessive disease among Caucasians, occurring in about one in 2,500 births. In the pedigree in figure 2.16, two individuals, IV-1 and IV-4, have the recessive trait or disease. Notice that the parents of the affected individuals, III-1, III-2, III-4, and III-5, are themselves unaffected, a typical observation in pedigrees for recessive traits. Obviously, the trait appeared because the parents are heterozygotes whose recessive alleles were masked by the dominant alleles. Notice also that either males or females can have the recessive phenotype.

In a large proportion of recessive disease cases, the parents of the affected individual are relatives. For example, the parents of individual IV-4 are first cousins; that is, III-4 and III-5 have the same grandparents, I-1 and I-2. (We will discuss the basis for the increase in recessive diseases from consanguineous matings in chapters 16 and 17.) The other grandparents of III-4 and III-5, as well as some other relatives, are omitted in order to simplify the pedigree.

Several other aspects of recessive pedigrees are important. First, if both parents are affected or have the recessive genotype (not shown in figure 2.16), all progeny should be affected. Second, among the sibs of an affected individual, given that both parents are unaffected (heterozygotes), approximately ¼ should be affected, according to the principle of Mendelian segregation. Finally, the probability that an additional child born to this family would have the trait is also ¼. The attributes of pedigrees for recessive traits are summarized in table 2.7. However, because human families are usually small, these proportions are generally seen only when there are a large number of sibships.

In any given sibship, the numbers of different types of offspring may not be close to Mendelian ratios. To understand this, we can use the expression for binomial probabilities given previously. For example, let us assume we have a sibship of size six in which two individuals are affected. We will designate one affected individual as the **proband** (or index case), the individual who was first identified as having the recessive trait or disease. What is the probability that one of the five remaining sibs will also be affected? Given the Mendelian probabilities of segregation, let us call ¾ dominant (p) and ¼ recessive (q). Thus:

$$Pr(4 \text{ unaffected}, 1 \text{ affected offspring}) = C(¾)^4(¼)^1$$

In this case, C = 5, thus:

$$Pr(4A-, 1aa) = 5(¾)^4(¼) = 0.396$$

This makes the probability of such a sibship quite large. In fact, it is the most likely type of sibship containing five sibs. The next most probable sibships are $Pr(5A-, 0aa) = 0.237$ and $Pr(3A-, 2aa) = 0.264$.

TABLE 2.7 Patterns for Human Traits

Recessive Traits

Parents are generally unaffected.

Approximately one-quarter of the siblings of an affected individual are affected.

The probability that an additional child will be affected is ¼.

Recessive traits often result from consanguineous mating.

Two affected parents cannot have unaffected offspring.

Dominant Traits

Trait occurs every generation; (at least one parent is affected).

When one parent is affected, approximately one-half of progeny are affected.

The probability that an additional child will be affected is ½.

Unaffected individuals do not produce affected offspring.

Two affected parents can produce unaffected offspring.

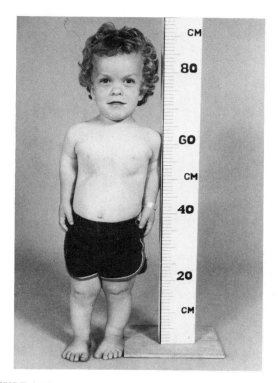

FIGURE 2.18

A boy with achondroplasia, a common type of dwarfism.
Dr. Judith Hall.

Two other points should be made about recessive trait pedigrees before we continue. First, note that unrelated individuals who do not have affected offspring, such as individuals II-1 and II-6 in figure 2.16, are assumed to be homozygous for the dominant allele. Such individuals could possibly be heterozygotes, but because most recessive alleles are rare, it is assumed that unrelated individuals with the dominant phenotype are homozygotes.

Second, given an affected offspring in a family, we know that the unaffected parents are heterozygous (barring mutation, which is extremely rare). Using segregation proportions, we can then calculate the conditional probability that an unaffected sib in such a family is heterozygous. The probability that a sib is heterozygous is ½ and the probability that a sib is unaffected is ¾, so:

$$Pr(heterozygous \mid unaffected) = Pr(heterozygous) / Pr(unaffected)$$
$$= (½)/(¾)$$
$$= ⅔$$

In other words, two-thirds of the time, unaffected sibs of individuals with a recessive disease are carriers (heterozygotes) for the disease allele.

DOMINANT TRAITS

Now let us examine a pedigree that illustrates the inheritance pattern for a dominant trait. **Achondroplasia,** a common type of dwarfism (see figure 2.18), and **Huntington's disease,** a degenerative neurological disorder, are examples of dominant disorders. (McKusick lists a total of 1,443 dominant loci.) For dominant traits, at least one of the parents of an affected individual is also affected, and the trait is present in every generation, as shown in the pedigree in figure 2.19. This pattern results because an offspring must obtain the dominant allele from one of its parents. As a result, unaffected individuals do not produce affected offspring (unless there is a rare mutation). Notice also that either males or females can have the dominant phenotype. Unlike a recessive trait, two affected parents can produce an unaffected offspring, because both parents may be heterozygotes and consequently have recessive alleles. As expected according to Mendelian segregation, when one parent is affected, half the sibs of the proband are expected to be affected. Given such a family, the probability that an additional child will be affected is ½ (see table 2.7 for a summary about dominant traits in pedigrees). Again, because of the small sibship size in humans, sibships in which the proportion of affected individuals is not ½ occur with probabilities that can be calculated using the binomial formula.

The pedigree patterns listed in table 2.7 are not completely true for some diseases, and there are other types of inheritance patterns as well. We will discuss several such situations in chapter 3 and illustrate how they may affect pedigrees. Furthermore, the pedigrees given in figures 2.16 and 2.19 were large ones, allowing the mode of inheritance to be determined with reasonable ease. In many pedigrees having fewer individuals, the mode of inheritance may not be clear.

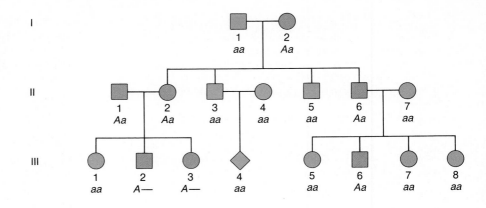

FIGURE 2.19
A human pedigree for a dominant trait, showing the inferred genotypes, where A and a are the dominant and recessive alleles, respectively. $A-$ indicates that the second allele is not known.

From AN INTRODUCTION TO GENETIC ANALYSIS 3/E. *By David T. Suzuki et al. Copyright©1976, 1981,1986 by W. H. Freeman and Company. Reprinted with permission.*

M endelian principles apply to traits in humans, but because human families are relatively small, determining modes of inheritance can be difficult. However, some general patterns have been discerned regarding the passing on of recessive and dominant traits.

■

Summary

M endel first demonstrated that inheritance was based on the segregation of alleles at a gene. Through self-fertilization of F_1 garden pea plants obtained from crosses between plants with different characteristics, he observed a 3:1 ratio of dominant to recessive phenotypes in the F_2 progeny. The principle of segregation he developed from these experiments states that heterozygous individuals produce equal proportions of gametes containing the two alleles. Mendel's second principle, that of independent assortment, states that alleles at different genes assort independently of each other, resulting in a 9:3:3:1 ratio in F_2 progeny from a dihybrid cross.

Several basic theories of probability are useful in understanding Mendelian genetics. These are: (1) the sum rule, which states that the probability of either of two mutually exclusive events occurring is the sum of their individual probabilities; (2) the product rule, which states that the joint probability of two events occurring is the product of their individual probabilities; (3) the theory of conditional probability, which is the probability of a specific event divided by the probability of a more general event; and (4) the binomial probability formula, which figures the probability of x events of a particular type out of a total N.

A useful statistical test of the goodness of fit between the numbers observed and the numbers expected according to Mendel's principles is the chi-square. Chi-square values larger than a critical value indicate that the underlying hypothesis used to generate the expected numbers may be incorrect.

Mendelian principles apply to traits in humans. Human pedigrees show certain patterns that allow recessive and dominant traits to be differentiated.

■

Problems and Questions

1. The F_2 numbers of the dominant and recessive phenotypes for Mendel's seven different traits are given in table 2.1. Calculate the chi-square values for axial versus terminal flowers and for tall versus short stems. Are these values consistent with Mendelian expectations using the chi-square test?

2. If the F_1 yellow-seeded plants from Mendel's experiments (see table 2.1) were crossed to green-seeded plants, what proportion of the progeny would be expected to be yellow? If F_2 yellow-seeded plants from selfed F_1 plants (table 2.1) were crossed to green plants, what proportion of the plants would produce only yellow progeny?

3. A student has a penny, a nickel, a dime, and a quarter. She flips them all simultaneously and checks for heads or tails. What is the probability that all four coins will come up heads? She again flips all four coins. What is the probability that she will get four heads both times? What probability rule did you use to determine your answer?

4. The first child of two normally pigmented parents has albinism, a recessive trait that results from a lack of the pigment melanin. (*a*) Given that the normal allele is A and the albino allele is a, draw this pedigree and label both the phenotypes and the genotypes. (*b*) What is the probability that a second child from these parents will be an albino? What is the probability that the second child will be a carrier of the albino allele? Given that the second child is unaffected, what is the probability that he or she is a carrier?

5. Assuming that the probability of producing a female child is 0.5, what is the probability of having three girls and one boy in a family of four? Calculate this value both by writing out all the possible birth orders that can lead to such a family and by using the binomial formula. The actual proportion of girls at birth is about 0.49 in several human populations. Given this value, what is the probability of a family having three girls and one boy?

6. Mendel crossed purebred wrinkled, green-seeded plants with purebred round, yellow-seeded plants. Then he crossed the F_1 progeny to purebred wrinkled, green-seeded plants and observed 31 plants with round, yellow

seeds, 26 with round, green seeds, 27 with wrinkled, yellow seeds, and 26 with wrinkled, green seeds. (a) Assuming independent assortment, what proportion of these types would be expected? (b) Calculate the appropriate chi-square and determine whether it is consistent with Mendelian expectations.

7. Using the forked-line approach and given the cross between two cats with the genotypes *llSsdd* and *LlSsDd,* what is the probability of a cat having the genotype *llssdd*? What is the probability of a cat having the short-hair, white-spotted, non-dilute phenotype?

8. Using the forked-line approach in a trihybrid cross involving three traits, where the parents are both *AaBbCc,* what is the probability of their producing an offspring recessive for all three traits? What is the probability of producing an offspring with the phenotype $A-bbC-$?

9. Given the following pedigree for a rare genetic disease, would you conclude that the trait is recessive or dominant? Why? What do you expect the genotypes of individuals II-3, II-4, and III-2 to be, assuming A and a are the dominant and recessive alleles, respectively? Given only this information, what is the probability that I-2 is heterozygous? What is the probability that III-1 will have the trait?

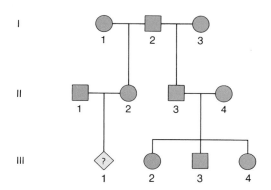

10. Given the following pedigree for a rare genetic disease, would you conclude that the trait is recessive or dominant? Why? What do you expect the genotypes of II-1, II-2, and III-3 to be, assuming that B and b are the dominant and recessive alleles, respectively? What is the probability that a child of II-3 will be affected? Why?

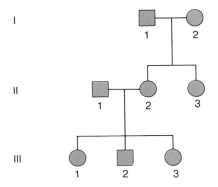

11. A woman with the rare recessive disease phenylketonuria (PKU), who had been treated with a diet having low levels of the amino acid phenylalanine, was told that it was unlikely her children would inherit PKU because her husband did not have it. However, her first child had PKU. What is the most likely explanation? Assuming this explanation is true, what would be the probability of her second child having PKU?

12. In an examination of Mendel's principles, a strain of light brown mice was crossed to a strain of dark brown mice. All F_1 were dark brown. In the F_2, 42 were dark brown and 15 were light brown. Is this consistent with the principle of segregation? (Use a chi-square to check.)

13. The 42 dark brown F_2 mice from question 12 were crossed to light brown mice. Thirty of these mice produced litters composed of both dark and light brown mice, while twelve produced only dark brown mice. Is this consistent with the principle of segregation? (Use a chi-square to check.)

14. What is the probability that, if three identical coins were flipped, all would end up heads? What is the probability that the three coins would *not* be either all heads or all tails?

15. Assume that a cross was made between fruit flies of genotype *AAbb* and those of genotype *aaBB*. (a) Give the Punnett square for the expected F_2 progeny types. (b) What proportions of $A-B-$, $A-bb$, $aaB-$, and *aabb* progeny do you expect in the F_2?

16. In the recessive trait pedigree in figure 2.16, what is the probability that individual IV-2 is a carrier (heterozygote)? What is the probability that a third child of III-1 and III-2 would have the recessive disease?

17. In the dominant trait pedigree in figure 2.19, what is the probability that individual III-2 would be *AA?* If III-1 and III-6 produce an offspring, what is the probability that he or she will have the dominant trait?

18. In a common kind of congenital deaf-mutism, one or both parents are usually affected. Do you think it is a recessive or a dominant trait? What other criteria would you use to distinguish between these two types of deaf-mutism?

19. In the F_2 progeny from two purebred lines of rats, 50 were doubly dominant, $A-B-$. If these were tested by crossing to *aabb* rats, how many would you expect to be double heterozygotes?

20. Assume that one-twelfth of all individuals are born each month. What is the probability that two particular individuals were both born in April? What is the probability that two particular individuals were both born in the same month?

Answers appear at end of book.

Extensions and Applications of Mendelian Genetics

*T*he phenotype of an individual is what is perceived by observation: the organism's structures and functions—in short, what a living being appears to be to our sense organs. . . . The genotype is the sum total of the hereditary materials received by an individual from its parents and other ancestors.

■

Theodosius Dobzhansky
American geneticist

Learning Objectives

In this chapter you will learn:

1. The differences between complete dominance, incomplete dominance, and codominance.

2. The effects of pleiotropy and penetrance.

3. How there can be multiple alleles at a gene.

4. The pattern of inheritance of sex-linked genes.

5. The patterns of inheritance when multiple genes affect a trait.

6. The genetic basis of quantitative traits.

7. How to estimate heritability.

8. The basis of genetic counseling.

9. The basis of paternity exclusion.

*T*he elegance of Mendel's experiments was partly due to the complete consistency between his observations and the hypotheses he developed. However, after Mendel's work was rediscovered, it became clear that the simple Mendelian model did not adequately predict experimental observations in all situations. In this chapter, we will examine some of the more complex phenomena that make it necessary to adopt an expanded view of simple transmission, or Mendelian, genetics. In addition, we will introduce the topics of genetic counseling and paternity analysis, which utilize the principles of Mendelian genetics. In chapter 5 we will discuss genetic linkage, which can also be viewed as an extension of Mendelian principles.

DOMINANCE

All seven traits that Mendel selected had one allele dominant over the other, so that when he examined the F_2 progeny, he found a 3:1 dominant to recessive ratio of the phenotypes. (As in chapter 2, we will use the terms dominant and recessive to refer to alleles as well as to traits.) In fact, Mendel suggested that dominance was a general characteristic of traits, similar to the principles of segregation and independent assortment. However, shortly after 1900, several different patterns of dominance were documented.

COMPLETE DOMINANCE

In most situations, the normal, or wild-type allele is completely dominant over mutant alleles. The basis for this will be more clear when we later examine the biochemical products of genes and find that one functioning copy of a gene is generally enough to produce sufficient gene product for a wild-type phenotype. For example, the albino allele in humans does not produce functioning gene product, and when an individual is homozygous for this allele, the biochemical pathway for the pigment melanin is blocked. However, one copy of the wild-type allele, as is present in heterozygotes, allows sufficient production of the pigment melanin so that normal pigmentation results.

In some instances, the mutant allele is dominant over the wild-type; that is, the wild-type is recessive. There are 1,443 confirmed loci resulting in dominant traits listed in McKusick's 1988 catalog of human traits, many of them rare diseases. For example, the most common type of dwarfism, achondroplasia, is dominant, so that heterozygotes exhibit the mutant phenotype. In a subsequent section of this chapter, we will discuss some aspects of Huntington's disease, a fatal neurological disorder that is also dominant.

INCOMPLETE DOMINANCE

When Carl Correns, one of the rediscoverers of Mendel's principles, experimentally crossed red-flowered four-o'clocks with white-flowered ones, the F_1 plants were unlike either parent, being an intermediate pink instead (figure 3.1). When he allowed pink F_1 plants to self-fertilize, the F_2 ratio of phenotypes was 1 red: 2 pink: 1 white, or ¼, ½, and ¼. Although at first glance these results appear to deviate from Mendel's, what Correns actually observed in the F_2 was the 1:2:1 genotypic ratio, not the 3:1 phenotypic ratio. Because neither parental phenotype was fully expressed in the F_1, the trait is said to exhibit **incomplete dominance** at this gene; that is, all genotypes have different phenotypes, with the heterozygote's phenotype intermediate between the two homozygotes. The notation in figure 3.1 indicates this lack of dominance by using capital letters with different subscripts, R_1 for the red allele and R_2 for the white.

In many instances of incomplete dominance, the observable phenotype of the heterozygote is the same as that of the homozygote (complete dominance) but intermediate on a biochemical level. For example, RR homozygotes and Rr heterozygotes at the gene affecting seed shape in garden peas both have the same round seed phenotype. However, these genotypes differ biochemically—the heterozygote has starch grains that are intermediate between the two homozygotes in type and amount.

CODOMINANCE

In some cases, the traits associated with both alleles are observable in the heterozygote. Such alleles are said to exhibit **codominance.** For example, individuals heterozygous for **red blood cell antigens,** genetically determined chemicals on the membranes of red blood cells, often exhibit properties of both alleles. A well-known case is type AB in the ABO blood group system, which has the antigens from two different alleles (see discussion of multiple alleles later in this chapter for more about ABO). Another blood group system, called the MN, has two alleles, generally denoted as L^M and L^N, and three genotypes, $L^M L^M$, $L^M L^N$, and $L^N L^N$. (Sometimes these genotypes are called *MM, MN,* and *NN,* respectively.) Table 3.1 gives the six possible mating combinations between these genotypes, assuming reciprocal matings are equivalent. The expected proportions of genotypes in the offspring are also shown, and it can be seen that they follow the Mendelian segregation pattern. In chapters 16 and 17, we will discuss naturally occurring allelic forms that determine some biochemical properties of proteins, most of which are also codominant.

Although the example of flower color in four-o'clocks illustrates incomplete dominance on a gross phenotypic scale, close examination of the pink flowers of heterozygotes shows that both red plastids and white plastids are

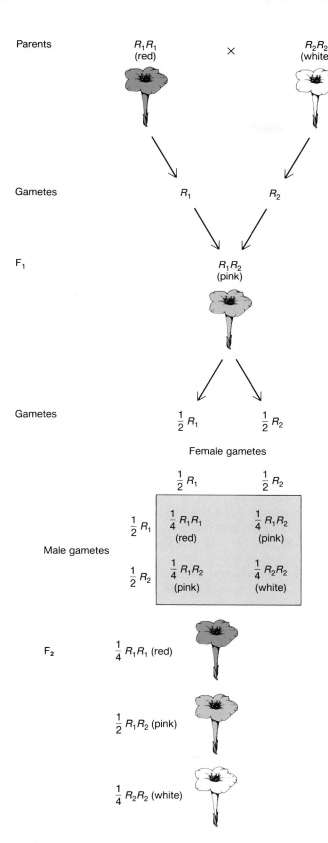

Parents: R_1R_1 (red) × R_2R_2 (white)

Gametes: R_1 R_2

F₁: R_1R_2 (pink)

Gametes: $\frac{1}{2} R_1$ $\frac{1}{2} R_2$

Female gametes

	$\frac{1}{2} R_1$	$\frac{1}{2} R_2$
$\frac{1}{2} R_1$	$\frac{1}{4} R_1R_1$ (red)	$\frac{1}{4} R_1R_2$ (pink)
$\frac{1}{2} R_2$	$\frac{1}{4} R_1R_2$ (pink)	$\frac{1}{4} R_2R_2$ (white)

Male gametes

F₂: $\frac{1}{4} R_1R_1$ (red)

$\frac{1}{2} R_1R_2$ (pink)

$\frac{1}{4} R_2R_2$ (white)

FIGURE 3.1

A cross between red and white four-o'clocks, with the F₂ progeny showing incomplete dominance.

Parents	Offspring		
	L^ML^M	L^ML^N	L^NL^N
$L^ML^M \times L^ML^M$	1	—	—
$L^ML^M \times L^ML^N$	½	½	—
$L^ML^M \times L^NL^N$	—	1	—
$L^ML^N \times L^ML^N$	¼	½	¼
$L^ML^N \times L^NL^N$	—	½	½
$L^NL^N \times L^NL^N$	—	—	1

present. (Plastids are a type of cell organelle.) In other words, we find incomplete dominance on the gross phenotypic level and codominance on a subcellular level.

D ominance occurs in different forms. It may be complete, as found by Mendel, where either the mutant or the wild-type allele is dominant. It may be incomplete, where the heterozygote has a phenotype in between the two homozygotes. Or two traits may be codominant, where both alleles are expressed in the heterozygote.

■

LETHALS

The expectations of segregation given in chapter 2 are true for zygotes (newly fertilized eggs). However, for many traits, the zygotic ratio is modified by differential mortality (that is, survival) as the progeny age. For example, the initial 3:1 dominant to recessive ratio expected in progeny from two heterozygotes for the sickle-cell allele, an allele that causes sickle-cell disease when homozygous, changes to near 0% anemics in adults because of the higher mortality for the sickle-cell homozygotes.

Humans who suffer from genetic diseases have higher mortality than normal individuals in nearly all cases. In other organisms as well, genetic conditions may result in high mortality; for example, genes can cause a lack of chlorophyll in plants or lead to metabolic problems in both plants and animals. When a genetic defect causes 100% mortality, it is termed a **lethal allele.** Lethals are generally recessive, resulting in the loss of the recessive homozygote. However, some lethals are dominant, as in Huntington's disease, where the heterozygote is affected, although generally not until later in life.

In 1904, shortly after the rediscovery of Mendel's principles, the French geneticist Lucien Cuénot, carrying out experimental crosses on coat color in mice, found a gene that was not consistent with Mendelian predictions.

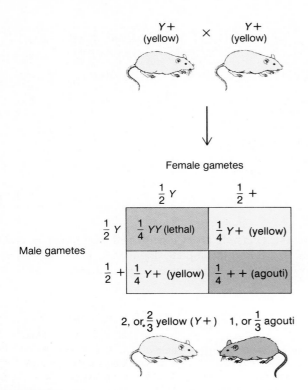

FIGURE 3.2
A cross between two yellow mice, yielding a 2:1 ratio of yellow to agouti-colored mice in the offspring because mice with genotype *YY* die.

FIGURE 3.3
A tailless Manx cat.
© *Robert Pearcy/Animals Animals.*

He observed from his experiments that the yellow body color allele was dominant, but that crosses between two yellow mice yielded approximately a 2:1 ratio of yellow to wild-type, rather than the expected 3:1. Furthermore, when he crossed yellow individuals to the recessive wild-type (agouti-colored mice), Cuénot found that all yellow mice produced some wild-type progeny. He concluded that all yellow mice were heterozygotes and that there were no yellow homozygotes as would have been expected. Later it was suggested that homozygosity for yellow causes lethality, and that these individuals died in utero, a fact eventually validated by the finding that approximately ¼ of the embryos from yellow × yellow crosses failed to develop.

The explanation for this situation is given in figure 3.2, where *Y* indicates the dominant, yellow mutant, and + indicates the recessive, wild-type allele. If one were to calculate a chi-square using the principle of segregation expectations for such data, it would give a statistically significant value, suggesting that the observations were not consistent with the Mendelian hypothesis. However, the principle of segregation obviously works properly in this case. The observed ratio of phenotypes differs from expectations because yellow homozygotes die before they can be counted. Many dominant disease alleles in humans, such as achondroplastic dwarfism, also appear to be lethals when they are homozygous.

The lack of a tail (taillessness) in the Manx cat (figure 3.3) is another trait caused by an allele that has a dominant effect in heterozygotes and is a lethal in homozygotes. This trait is thought to have first occurred as a mutation among domestic cats on the Isle of Man in 1935. The Manx and normal alleles are denoted by *M* and *m,* respectively, so that Manx individuals are the heterozygous genotype *Mm*. True-breeding Manx breeds are not possible because matings between Manx cats, *Mm × Mm,* produce about ⅓ normal-tailed progeny.

Many genes have alleles that affect the rate of mortality but are not lethals. In general, these are termed **deleterious** or **detrimental alleles** and may result in various levels of mortality, ranging from a few percent more than the wild-type genotype to nearly lethal (almost 100% mortality). Natural selection (selection that does not result from human intervention), which we will discuss in chapter 16, results in fewer progeny genotypes (and their alleles) with higher mortality than with lower mortality.

PLEIOTROPY

The genes we have been discussing are generally well-known because they substantially affect an obvious trait such as flower color or seed shape. However, many of these genes may also have secondary, or related effects. For example, the yellow coat color in mice discussed previously is an allele that affects more than one character; that is, it produces yellow body color in heterozygotes, and it also affects survival, causing lethality in homozygotes. Another example of multiple effects is the gene affecting seed shape in garden peas; this gene also affects starch grain morphology. In fact, many genes affect more than one trait.

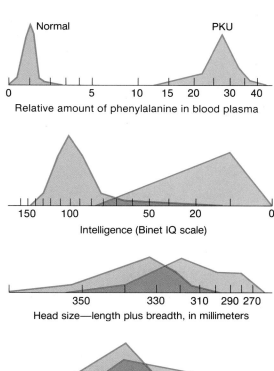

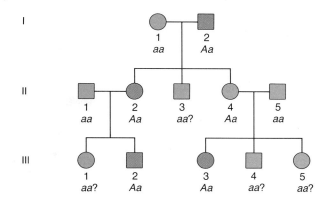

FIGURE 3.5

A pedigree illustrating the pattern of phenotypes that can occur with an incompletely penetrant dominant allele. Individual II-4 has genotype *Aa* but is not affected by the disease.

FIGURE 3.4

The values of four different characteristics in PKU and normal individuals, illustrating the pleiotropic effects of this disease. The purple regions show the overlap in the two distributions.

Reprinted with permission of Jones and Bartlett Publishers, Inc., Boston, MA, from Population Biology, *1984, written by Philip Hedrick.*

Mendel himself noticed that the gene causing the flower colors violet and white also influenced seed color and caused the presence or absence of colored areas on the leaves. This phenomenon where a single gene affects two or more characteristics is called **pleiotropy.**

Many genetic diseases in humans are caused by genes that have pleiotropic effects or a syndrome of diagnostic characteristics. One such disease is phenylketonuria (PKU), which occurs in individuals homozygous for a defective recessive allele. These people lack the enzyme necessary for the normal metabolism of the amino acid phenylalanine. As a result, when normal and PKU individuals are compared, the level of phenylalanine is much higher in the diseased group (figure 3.4). In addition, this basic biochemical difference results in a suite of other changes in untreated PKU patients, such as lower IQ, smaller head size, and lighter hair. All of these pleiotropic effects in PKU can be understood as a consequence of the basic biochemical defect. For example, in PKU patients, phenylalanine accumulates in the head, causing brain damage and leading to a lower IQ and a smaller head size.

Many mutations of *Drosophila* have multiple effects on the phenotype. New mutants are generally recognized by a gross change in the phenotype, such as white eyes or vestigal wings. Invariably, after detailed inspection, less extreme variations in body morphology or other characteristics are found. For example, the *w* allele causes many changes in the fly's internal organs, and the *vg* allele alters its halteres (balancers) and modifies its bristles.

PENETRANCE AND EXPRESSIVITY

All of the genes we have considered to this point have a definite genotype-phenotype relationship; for example, a pea plant with genotype *rr* is always wrinkled. In fact, all of Mendel's genes, all blood group types, and many human diseases show such an absolute pattern. However, for some genes, a given genotype may or may not show a given phenotype. This phenomenon is described as the **penetrance** of a gene. The level of penetrance can be calculated as the proportion of individuals with a given genotype who exhibit a particular phenotype. For example, if there are eight individuals of a particular genotype and five of them express a diseased phenotype, the level of penetrance is $\frac{5}{8} = 0.625$. When all individuals of a particular genotype have the same phenotype, the gene shows complete penetrance and the level of penetrance is 1.0. Otherwise, the gene is termed incompletely penetrant and the level of penetrance can be calculated.

The presence of incomplete penetrance at a gene may cause a trait to skip a generation in a pedigree; that is, a dominant trait present in a given generation may skip the first progeny generation but appear again in the second generation. Such a case is illustrated in figure 3.5, a human pedigree where unaffected individual II-4 is both the daughter and mother of affected individuals. This indicates that she has the genotype *Aa,* but because of incomplete penetrance, she does not show the dominant

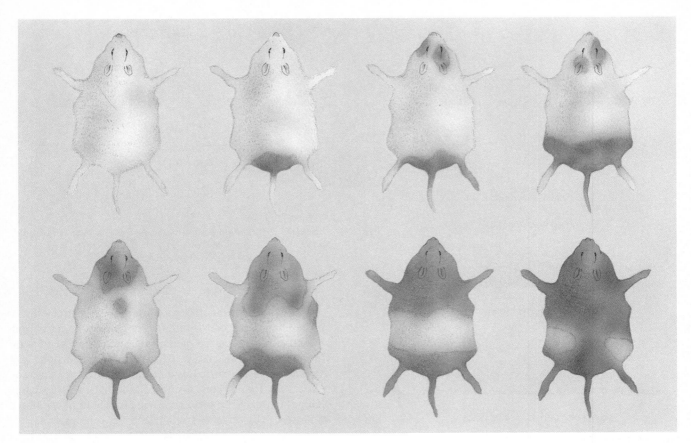

FIGURE 3.6
Variations in spotting of mice homozygous for the *s* allele. All of these mice are genotype *ss,* but their phenotype varies greatly because of variable expressivity.

phenotype. For example, the dominant form of retinoblastoma, a disease that causes malignant eye tumors in children, is only about 90% penetrant.

In addition, given that a particular genotype exhibits the expected phenotype, the level of expression, or **expressivity,** may vary. For example, although a gene causes a detectable disease in most individuals with a given genotype, some may be much more severely affected than others. In other words, individuals of a given genotype may be unaffected, may have intermediate symptoms, or may be severely affected. Figure 3.6 gives the variation seen in spotting in the coats of mice homozygous for the *s* allele. Notice that the phenotype may range from virtually no spotting to completely white.

Huntington's disease in humans is a trait that shows variable expressivity in its age of onset. The disease generally occurs in middle age, but can occasionally occur very early (at younger than ten years) or in an older individual (see figure 3.7). Presumably, some people with the disease genotype die of other causes before they express the disease, resulting in incomplete penetrance as well as variable expressivity for this gene.

What are the causes of incomplete penetrance and variable expressivity? In particular instances, it is often impossible to determine causation, but both environmental factors and other genes are known to influence the penetrance or expressivity of a gene. For example, alleles of a number of genes in the fruit fly *Drosophila melanogaster* affect viability and may be lethal at elevated temperatures (say, greater than 28° C) but have little or no effect at lower temperatures. Such genes, known as **conditional lethals,** generally show their lethal effects in extreme environmental situations.

Chapter 3

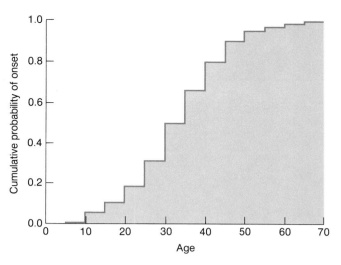

FIGURE 3.7
The cumulative age of onset of Huntington's disease in a human population. All individuals are the diseased genotype *Hh,* but some show symptoms of the disease much earlier than others.

FIGURE 3.8
The color of Siamese cats is darkened at the paws, ears, and nose because of a lower temperature at these extremities.
© *Mary Eleanor Browning/National Audubon Society/Photo Researchers, Inc.*

Temperature can also have a dramatic effect on the phenotypes of plants. For example, primroses have red flowers when grown at 24° C but white flowers when grown above 32° C. The coats of Siamese cats and Himalayan rabbits are darker in color at the paws, ears, and nose because of the lower body temperature at these extremities (figure 3.8). If the animal is placed in a warmer temperature, the fur at these points grows back light in color.

In addition, alleles at other genes may influence the penetrance and expressivity of a gene. For example, the level of expression of a trait is generally more similar among relatives than among unrelated individuals, given that the relatives and unrelated individuals are raised in fairly similar environments. Such genes that have a secondary effect on a trait are called **modifier genes** and can substantially influence a phenotype. The dilute allele that reduces the intensity of pigmentation when homozygous in many animals (such as from black to gray in house cats) is an example of a modifier gene. The extent of taillessness in Manx cats varies from having no tail vertebrae to having a small number of fused tail vertebrae, apparently due to the influence of modifying genes on the expression of the dominant Manx mutant. Mutants of *D. melanogaster* that have been kept in laboratory culture for many years sometimes do not have as extreme a phenotype as when first observed, presumably because of the increased frequency of alleles that modify the mutant phenotype.

A related phenomenon, called a **phenocopy,** occurs when environmental factors induce a particular abnormal phenotype that would usually be genetically determined.

In this case, an extreme environment may disrupt normal development in a pattern similar to that of a mutant gene. One tragic example occurred in western Europe in the late 1950s and early 1960s when a number of babies were born with severely shortened limbs, an abnormality that resembled a rare recessive genetic disorder. Only after an intensive investigation was it shown that the defect had been environmentally induced by the sleeping pill thalidomide, which the mothers of the affected individuals had taken during early pregnancy.

S ome alleles of genes may be lethal. Pleiotropy occurs when one gene affects several traits. Some genotypes may not always have the same phenotype, a phenomenon resulting from incomplete penetrance or variable expressivity of the genotype.

■

MULTIPLE ALLELES

All the genes we have discussed so far (and all the traits Mendel studied) have had only two segregating alleles. However, when detailed analysis of a gene is undertaken in a population, generally more than two alleles are discovered. Although some of these alleles may be rare, the existence of more than two alleles per gene (**multiple alleles**) is relatively frequent in a population.

One commonly known genetic variation resulting from multiple alleles is the ABO blood group system in humans. (There are closely related systems in other mammals such

Female gametes

	I^A	I^B	I^O
I^A	I^AI^A (A)	I^AI^B (AB)	I^AI^O (A)
I^B	I^AI^B (AB)	I^BI^B (B)	I^BI^O (B)
I^O	I^AI^O (A)	I^BI^O (B)	I^OI^O (O)

Male gametes (rows): I^A, I^B, I^O

FIGURE 3.9
The six genotypes produced by the three alleles in the ABO blood group system. The four corresponding phenotypes are in parentheses.

Donor	Phenotype	Recipient A (anti-B)	B (anti-A)	AB	O (anti-A and anti-B)
I^AI^A, I^AI^O	A	+	A, anti-A	+	A, anti-A
I^BI^B, I^BI^O	B	B, anti-B	+	+	B, anti-B
I^AI^B	AB	B, anti-B	A, anti-A	+	A, anti-A B, anti-B
I^OI^O	O	+	+	+	+

FIGURE 3.10
The potential for blood transfusions, indicated by a + for different ABO donor-recipient combinations. A, anti-A, and B, anti-B, indicate a reaction between A antigens and anti-A antibodies and between B antigens and anti-B antibodies.

as chimpanzees and cattle.) The different ABO blood group types were discovered in 1900, although their genetic basis was not elucidated from pedigree examination until 1925. In the ABO system, there are three major types of alleles (these classes themselves being heterogeneous), which are generally identified as I^A, I^B, and I^O (or sometimes as A, B, and O). If we assume that female and male gametes can each be of the three allelic types, then the different genotypes and phenotypes shown in figure 3.9 are possible. Notice that there are six different genotypes—that is, six different ways the three alleles can be combined two at a time.

In general, the number of genotypes possible in a diploid organism with "n" different alleles is found by adding $1 + 2 + 3 + \ldots + n$, where . . . indicates all the integers between 3 and n, or by using the formula $n(n + 1)/2$. Both of these expressions give the number of different ways n objects (or alleles) can be combined two at a time when order is not important. For the ABO system with three alleles, the number of genotypes is either $1 + 2 + 3 = 6$ or $(3)(4)/2 = 6$, as shown in figure 3.9. If there were four alleles, there would be $(4)(5)/2 = 10$ possible genotypes. In addition, the number of homozygotes is equal to the number of alleles (n), and the number of heterozygotes is equal to $n(n - 1)/2$. For example, the ten genotypes resulting from four alleles are composed of four homozygotes and $(4)(3)/2 = 6$ heterozygotes.

Because allele I^O of the ABO system is recessive to both alleles I^A and I^B, there are only four different phenotypes: type A (either genotype I^AI^A or I^AI^O), type B (I^BI^B or I^BI^O), type AB (I^AI^B), and type O (I^OI^O). Alleles I^A and I^B are codominant, illustrating that different dominance relationships are possible among the alleles at a given gene.

Differences among ABO blood group types occur because of the presence of different antigens and antibodies. **Antigens** of the ABO blood group system are molecules composed of a protein-sugar combination that occur on the surface of red blood cells. Individuals with alleles I^A, I^B, and I^O have A, B, and neither A nor B antigens, respectively, on these cells. The A and B antigens differ in the type of terminal sugar on the molecules. The O antigen actually lacks this terminal sugar because when the I^O allele is present, no enzyme to add the terminal sugar is produced.

Antibodies are substances that can be produced by the immune system in large amounts as a response to specific external antigens. (Antibodies are not produced against self antigens.) For example, if cells from a type A individual are injected into rabbits, the immune system of the rabbits will produce anti-A antibodies. In humans, individuals of type A have anti-B antibodies; type B have anti-A antibodies; type AB have neither of these antibodies; and type O have both kinds of antibodies. The presence of antibodies that recognize A and B antigens in individuals lacking these antigens appears to be the result of immune responses to substances produced by bacteria that are similar to these antigens.

In order to transfuse blood successfully from one individual to another, differences between individuals in the ABO blood group system must be taken into account. As an example of transfusion problems that can result from the ABO system, if red blood cells with B antigens (from type B individuals) are transfused to type A individuals, the type B cells will be attacked and eliminated by the anti-B antibodies present in the type A individuals. All combinations of donor and recipient phenotypes are given in figure 3.10. Notice that type O individuals are universal donors; that is, type O blood, because it has no antigens, is accepted by all individuals. On the other hand, type AB individuals are called universal recipients, because they do not have anti-A or anti-B antibodies and can therefore accept A, B, AB, or O blood. Type O recipients given any blood but O will suffer dire consequences because O individuals have both anti-A and anti-B antibodies. Type AB individuals can only donate blood to type AB individuals because other blood types have anti-A, anti-B, or both antibodies.

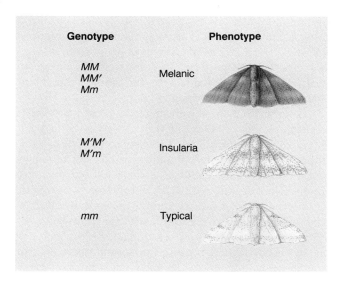

Genotype	Phenotype	
MM MM' Mm	Melanic	
M'M' M'm	Insularia	
mm	Typical	

FIGURE 3.11
The six possible genotypes of the peppered moth from all diploid combinations of three alleles are listed in the left column. The resulting phenotypes are shown in the far right column.

TABLE 3.2 *Numbers of Homozygotes, Heterozygotes, and Genotypes, Given Multiple Alleles*

Alleles	Homozygotes	Heterozygotes	Genotypes
1	1	0	1
2	2	1	3
3	3	3	6
4	4	6	10
5	5	10	15
6	6	15	21
.	.	.	.
.	.	.	.
.	.	.	.
n	n	$\dfrac{n(n-1)}{2}$	$\dfrac{n(n+1)}{2}$

Some genes may have a series of alleles showing sequential dominance; for instance, a number of alleles can affect color in mammals, insects, and mollusks. In some parts of England, the peppered moth, *Biston betularia,* has at least three such alleles. In chapter 16 we will discuss the evolution of these color types, a phenomenon that has been associated with industrial pollution in England. At this gene, the allele that was characteristic of the species before industrial pollution, the pale typical allele (m), is recessive. A second allele, the darkly mottled insularia (M'), is dominant to the typical. The third allele, the nearly black melanic (M), is dominant to both of the other alleles. In other words, there are six genotypes but only three phenotypes: (1) the melanic phenotype, having three genotypes—homozygous melanic (MM) and heterozygous between melanic and the other two alleles (MM' and Mm); (2) the insularia phenotype, composed of the insularia homozygote ($M'M'$) and the insularia-typical heterozygote ($M'm$); and (3) the typical phenotype, all homozygous recessive typical (mm) (see figure 3.11). As an example of the types of mating possible, assume that two heterozygotes, a melanic Mm and an insularia $M'm$, were crossed. The expected progeny genotypes are one-quarter each MM', Mm, $M'm$, and mm. However, because of dominance, half the progeny should be the melanic phenotype (MM' and Mm genotypes), one-quarter insularia ($M'm$), and one-quarter typical (mm).

Some genes, such as HLA (human leucocyte antigen, which determines antigens that are on the surfaces of nearly all cells), may have many alleles. For example, gene *HLA-B* has more than thirty different alleles in some populations. The alleles are denoted by A1, A2, A3, A11, etc.,

	Recipient		
Donor	A1 A28	A2 A10	A1 A9
A2 A2	A2	+	A2
A1 A28	+	A1, A28	A28
A9 A11	A9, A11	A9, A11	A11
A9 A9	A9	A9	+

FIGURE 3.12
The potential for an organ transplant from four donor individuals to three recipients based on HLA-A. The + indicates a match. *A*2, *A*9, etc., indicate the antigens in the donor that would cause rejection.

for HLA-A, and by B5, B7, B8, B12, etc., for HLA-B. As a result of this diversity, an extraordinary number of genotypes generally occur at this gene in a population. To illustrate this, table 3.2 gives the number of possible genotypes for genes with different numbers of alleles. *HLA-B*, with thirty alleles, has 30 different homozygotes, 435 different heterozygotes [(30)(29)/2], and a total of 465 different genotypes [(30)(31)/2], truly an extensive array of genotypic variation.

HLA genes are important in determining the potential success of organ transplants and are related to susceptibility to a number of autoimmune diseases. For these genes, all the alleles are codominant, so the number of genotypes equals the number of phenotypes. Matching the genotypes of *HLA-B* (and the *HLA-A* gene, which has nearly twenty different alleles) in the donor and recipient of a kidney increases the likelihood of a successful transplant. Figure 3.12 gives the HLA-A genotype for three potential recipients and four potential donors. Only where there is a + is the donor compatible with the recipient. For all other combinations, one or more antigens present in the donor are not present in the recipient, and the likelihood of a successful transplant is much lower. Because there are so many different genotypes, donors and recipients generally do not match. Unless a relative has a matching HLA type and is willing to donate an organ, it is necessary to use medicine to reduce the organ recipient's immune response to the foreign antigens and thus achieve a successful transplant.

> **G**enes may have more than two alleles. In diploid organisms, many different genotypes can exist when there are multiple alleles.

■

SEX-LINKED GENES

In chapter 1 we mentioned that most genes are on autosomes—the chromosomes for which each diploid individual has a homologous pair. However, shortly after the rediscovery of Mendel's principles, some experiments with *Drosophila* yielded progeny that differed from Mendelian predictions in the phenotypic proportions of males and females. These initially perplexing results were explained by assuming that the genes in question were not on autosomes but on the X chromosome, which is present in two copies in the female *Drosophila* and in one copy in the male. (These experiments will be discussed in chapter 4.)

Mammals, some insects, and a few plants have this kind of sex chromosome system in which one sex has two homologous chromosomes and the other has two different kinds of chromosomes. (In chapter 4 we will also discuss sex determination.) In mammals, females have two X chromosomes and are designated XX. Males have one X chromosome and one Y chromosome and are designated XY. If we examine the inheritance of genes on these chromosomes, we find patterns different from the autosomal genes examined by Mendel, but the basis of this inheritance is consistent with Mendelian principles.

Let us first examine traits determined by genes on the X chromosome, or **X-linked traits.** McKusick lists 124 identified loci on the human X chromosome. One example is the inheritance of the common type of color blindness, which is caused by a recessive allele located on the X chromosome. We will indicate X chromosomes having the two alternative alleles as X^C and X^c, the normal and color-blind alleles, respectively. In females there are three different genotypes ($X^C X^C$, $X^C X^c$, and $X^c X^c$) in which only the last is color-blind because the normal allele is dominant in heterozygotes. Males have only one X and a Y. Because Y chromosomes do not carry the color blindness gene (the Y does not have any genes allelic to those on the X), the two types of males are $X^C Y$ and $X^c Y$—normal and color-blind, respectively. (Another way to indicate X-linked inheritance is to leave out the X symbol and write the X-linked alleles as we symbolized alleles previously. In other words, a female heterozygous for color blindness would be *Cc,* and a color-blind male would be *c*Y.)

Figure 3.13 shows a test for color blindness. People who have good color vision will see large numbers in both diagrams, while color-blind individuals will not be able to detect the numbers in part (*b*) of the figure. (We should caution that this color vision test is a strict one and that people with some color vision may not be able to distinguish the numbers in figure 3.13*b*.)

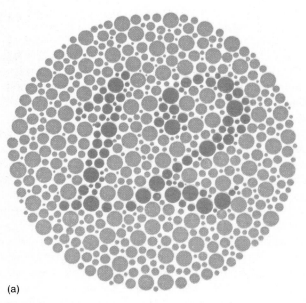

(a)

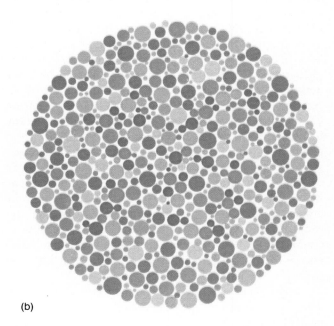

(b)

FIGURE 3.13
Plates to test for color blindness. All individuals with normal sight will read (*a*) as "12" and (*b*) as "16." Color-blind people will not be able to read (*b*) at all.

The above has been reproduced from Ishihara's Tests for Colour Blindness published by KANEHARA & CO., LTD, Tokyo, Japan, but tests for color blindness cannot be conducted with this material. For accurate testing, the original plates should be used.

In females, segregation for X-linked genes is the same as that for autosomal genes, while in males, half of the gametes carry the X and half carry the Y. To illustrate this, let us examine the progeny from a cross between a woman heterozygous for color blindness and a man with normal sight (see figure 3.14). We would expect half the progeny to be female, all normal-sighted (although half of them are homozygotes and half heterozygotes), and half the progeny to be male (half color-blind and half normal-sighted). The ratio of normal to affected progeny is still 3:1, but unlike the F_2 progeny of a monohybrid cross for an autosomal gene, all the recessive types are males. Obviously, the number of progeny from any one mating in humans is small, but if a number of such matings are lumped together, these expectations are appropriate.

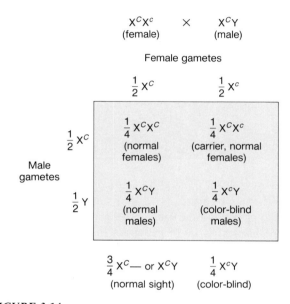

FIGURE 3.14
Probable outcome of a mating between a female heterozygous for color blindness and a male with normal sight.

Pedigrees for X-linked traits have particular patterns that generally allow them to be distinguished from pedigrees for autosomal traits. Figure 3.15 shows a pedigree for an X-linked recessive. First, the X-linked recessive affects males more than females because males need only one copy of the defective allele to express the trait, while females need two. When the allele is rare, it is unlikely that homozygous females will be encountered. Second, the X-linked trait skips generations because males can only receive an X-linked allele from their mother; remember that males receive a Y chromosome from their father, and as just suggested, very few females exhibit X-linked recessive traits.

Several common types of the recessive genetic disease hemophilia, in which blood does not clot normally, are also determined by genes on the X chromosome. As a result, most individuals with X-linked forms of hemophilia are males who have received the hemophilia allele from their unaffected carrier mothers. Hemophilia afflicted a number of the descendants of Queen Victoria of England, suggesting that she was heterozygous for this allele. Figure 3.16 shows the family of Czar Nicholas II of Russia, whose wife Alexandra was a granddaughter of Victoria. Their son Alexis, pictured in front, had hemophilia and was executed, along with the rest of the family, at age 14. Although Alexis did not die from hemophilia, some historians believe his disease was a contributing factor to the whole family's eventual fate.

X-linked dominant traits have a different pattern of inheritance from X-linked recessive traits, as shown in figure 3.17. Notice in this pedigree that both males and females are affected by the trait (sometimes slightly more females are affected), but that the patterns of affected offspring produced by affected individuals differ between the two sexes. Irrespective of sex, approximately half the offspring of affected females, such as I-1 in figure 3.17, are affected, while for affected males, such as II-2, all female offspring and no male offspring are affected.

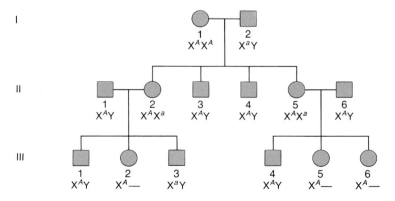

FIGURE 3.15
A pedigree illustrating the pattern of inheritance expected for an X-linked recessive gene.

Extensions and Applications of Mendelian Genetics

FIGURE 3.16
A photograph of Czar Nicholas II, the Czarina Alexandra, their four daughters, and their son, Alexis, who suffered from hemophilia.
The Bettmann Archive.

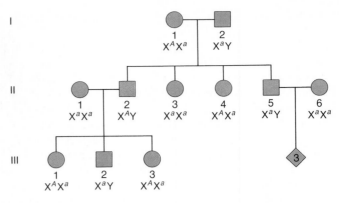

FIGURE 3.17
A pedigree illustrating the pattern of inheritance expected for an X-linked dominant gene.
From AN INTRODUCTION TO GENETIC ANALYSIS *3/E. By David T. Suzuki et al. Copyright©1976,1981,1986 by W. H. Freeman and Company. Reprinted with permission.*

TABLE 3.3   *General Patterns of Sex-linked Inheritance*

X-linked Recessive

Usually more males than females are affected.

No offspring of an affected male are affected, making the trait skip generations in the pedigree, always with an unaffected female in the intermediate generation. (An exception to this pattern occurs in the rare instance when the affected male mates with a female carrier, producing affected female offspring.)

X-linked Dominant

Affected males produce all affected female offspring and no affected male offspring.

Approximately half the offspring of affected females are affected, regardless of their sex.

Y-linked Inheritance

Trait is always passed from father to son.

Only males are affected.

TABLE 3.4   *Example of a Sex-influenced Trait in Sheep*

	HH (Suffolk)	*Hh* (Cross)	*hh* (Dorset Horn)
Males	Polled	Horned	Horned
Females	Polled	Polled	Horned

The last type of sex-linked inheritance we will mention is that expected for genes on the Y chromosome. Although there is little evidence for Y-linked inheritance in humans, except for a gene or genes that determine maleness, other organisms do carry genes on the Y chromosome. One putative Y-linked gene in humans results in hairy ear rims (see figure 3.18). However, determining the precise mode of inheritance of this trait is complicated by a variable age of expression. The characteristics of Y-linked traits are straightforward: (1) They are only expressed in males, and (2) they are always passed from father to son. An interesting analogy to Y-linked inheritance is the traditional transmission of last names in English-speaking (and many other) societies. A summary of the characteristics of pedigrees for the different types of sex-linked traits is given in table 3.3.

Some traits are expressed differently in the two sexes but are not sex linked. First, there are **sex-limited** traits, characters expressed only in one sex. Many such traits are related to reproductive characters, such as milk yield in dairy cattle, egg laying in chickens, or oviposition behavior in insects.

Second, other traits are **sex influenced;** that is, the genotypes determined by autosomal genes are expressed differently in the two sexes. For example, the genes determining the presence of horns in sheep are expressed differently in males and females. In some sheep breeds, such as the Suffolk, neither sex has horns (they are called **polled**), while in other breeds, such as the Dorset Horn, both sexes have horns (see table 3.4). When these two breeds are crossed, the hybrids, which are heterozygous for the autosomal *H* gene, are horned if male but polled if female. In the heterozygotes, the presence of male or female hormones determines whether or not horns develop. It is thought that premature pattern baldness (occurring before age thirty-five) follows this same pattern of inheritance; that is, baldness is dominant in male heterozygotes and recessive in female heterozygotes.

FIGURE 3.18
Men with hairy ear rims, a putative Y-linked trait.
Courtesy of Dr. S. D. Sigamony.

G enes on the X and Y chromosomes have different patterns of inheritance from autosomal genes. X-linked recessive traits occur almost entirely in males and generally skip generations in families.

∎

MULTIPLE GENES AND EPISTASIS

In our discussion of penetrance and expressivity, we mentioned that the phenotype resulting from the genotype at one gene can be altered by alleles at other genes called modifier genes. In fact, many traits are affected by several genes, as was observed by a number of scientists shortly after the rediscovery of Mendel's principles. (Remember that when Mendel examined two genes in his dihybrid crosses, he used genes that affected different traits, not the same trait.)

For a number of traits, different genes affecting the same trait interact so that the phenotype expressed is a function of the particular combination of alleles present at the different genes. The general term for interactions of alleles at different genes to influence a trait is **epistasis** (Greek: *epi* + *stasis* = standing upon). Note that epistasis concerns two (or more) genes and differs from dominance, which occurs when the phenotype is determined by different alleles at the same gene.

One of the first observations of epistasis resulted from a cross between two varieties of white-flowered sweet peas, which was done by early geneticists William Bateson and R. C. Punnett. When they crossed the two white-flowered varieties, the F_1 flowers were unexpectedly purple (see figure 3.19). When the F_1 individuals were allowed to self-fertilize, 382 purple and 269 white flowers resulted, which

is close to a ratio of 9 purple to 7 white in the F_2 progeny. The explanation for this ratio, confirmed by other crosses, was that homozygosity of recessive alleles at two different genes could result in white flowers. When both genes had one or more dominant alleles, the phenotype was purple. Because nine of the sixteen categories in figure 3.19 are like this, there is a 9:7 ratio in the F_2 progeny.

It is useful to understand the mechanistic basis of these phenotypes. Pigmentation in the flowers of sweet peas results from chemicals called anthocyanins. Production of anthocyanins occurs through a series of metabolic steps catalyzed by enzymes that are themselves products of genes. Whenever a step in this synthetic process is blocked by the absence of a functional enzyme, pigmentation does not occur. Although the exact details of pigmentation synthesis in sweet peas are not known, the general pattern is illustrated in figure 3.20, where the enzyme products of gene *C* and gene *P* are both necessary for the production of anthocyanins. When the genotype is *cc,* the first enzyme is not produced; consequently, reaction to the intermediate product is stopped. Likewise, if the genotype is *pp,* the lack of the second enzyme halts the second metabolic step. In other words, if the genotype is either *cc* or *pp,* the synthetic pathway is blocked and no pigments are produced, resulting in a white flower. Such an interaction of genes to jointly produce a specific gene product is a type of epistasis called **complementary gene action.**

As we will discuss later, many recessive diseases in humans are caused by metabolic errors. If two individuals with metabolic errors for the same trait (but in different steps of a biochemical pathway) mate, their offspring should have a normal phenotype. This happens when two

FIGURE 3.19
A cross between two types of white
sweet peas, showing the 9:7 ratio
that results in the F₂ generation.

Parents

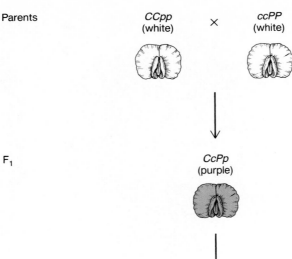

$$\frac{9}{16} C\text{--}P\text{--} : \frac{7}{16} C\text{--}pp, ccP\text{--}, \text{ or } ccpp$$
(purple)   (white)

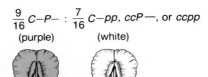

FIGURE 3.20
A diagram illustrating the
relationship between genes *C* and
P and the production of
anthocyanins in sweet peas.

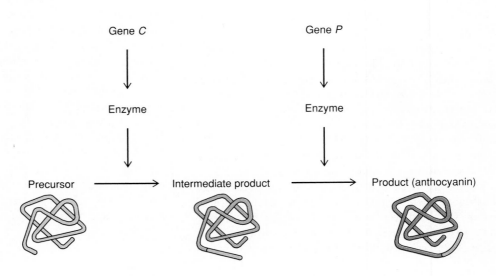

individuals with deaf-mutism caused by different recessive genes mate and have offspring. In the pedigree in figure 3.21, II-3 and II-4 are affected because they are recessive at different genes. Note that the pedigree is different from that of a single recessive gene because no progeny, instead of all progeny, from a mating of two affected individuals are affected.

The 9:7 ratio is only one type of F$_2$ pattern caused by epistasis. In other situations, the F$_2$ ratios may be 9:3:4, 13:3, or 15:1. Notice that the different categories of a 9:3:3:1 ratio can be combined to obtain these ratios. The biochemical basis for the 15:1 ratio can be understood by examining one step of a pathway in which a dominant allele at either of two genes is enough to produce enzymes for catalysis of a given reaction (see figure 3.22). Only when there is a double recessive, say *aabb,* is the pathway blocked and the aberrant phenotype expressed. When either of two genes can function to produce the dominant phenotype, this type of epistasis is called **duplicate gene action.** Presumably, the two different genes produce similar gene products, and one of them may have arisen by duplication from the other gene (see chapter 4). A classic example of duplicate gene action occurs in shepherd's purse, a widespread weed in which round seed pods result when dominant alleles are present at either of two genes, while narrow seed pods are produced only in the absence of dominant alleles at both loci.

Let us diagrammatically summarize the types of epistasis we have discussed. Where there is dominance at both loci and no epistasis, there are four phenotypes, with a 9:3:3:1 F$_2$ ratio as shown in row 1 of figure 3.23. The two types of epistasis we have mentioned, complementary and duplicate gene action, are given next, with their 9:7 and 15:1 ratios, respectively. Two other types of epistasis, in which there are three phenotypic classes, are given in the last rows. Can you find any other phenotypic ratios that are possible by combining the different classes—for example, 12:4 or 9:6:1?

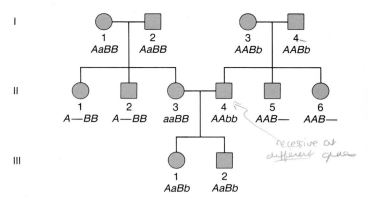

FIGURE 3.21
A pedigree illustrating a recessive disease caused by two genes.

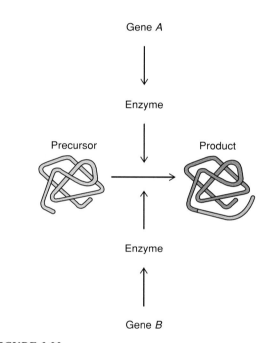

FIGURE 3.22
A diagram illustrating duplicate gene action, in which the enzyme from either gene *A* or gene *B* can catalyze the biochemical reaction.

	A– B–	A–bb	aaB–	aabb
Dominance with no epistasis	9	3	3	1
Complementary gene action	9	7		
Duplicate gene action	15			1
Dominance with epistasis of *aa* over *B*–	9	3	4	
Dominance with epistasis of *A*– over *bb*	12		3	1

FIGURE 3.23
Examples of epistatic F$_2$ ratios from the cross *AaBb* × *AaBb*.

FIGURE 3.24
Mice showing black and brown phenotypes.
© Runk/Schoenberger/Grant Heilman Photography, Inc.

TABLE 3.5	*Genotypes and Resulting Coat Color in Mice*[1]		
A and B Loci		**B and C Loci**	
Genotype	*Phenotype*	*Genotype*	*Phenotype*
$A-B-$	Agouti	$B-C-$	Black
$A-bb$	Cinnamon	$bbC-$	Brown
$aaB-$	Black	$B-cc$	Albino
$aabb$	Brown	$bbcc$	Albino

[1]A = agouti; B = brown; C = albino

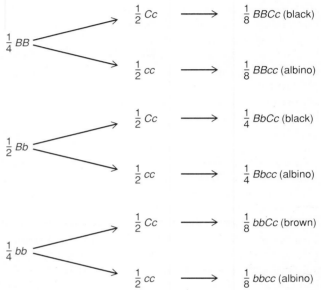

FIGURE 3.25
The expected results of a cross between mouse coat-color genotypes *BbCc* and *Bbcc*.

*M*ore than one gene can affect a single trait. The interaction of two genes to determine a phenotype is called epistasis. Epistatic interaction can result in F_2 ratios different from 9:3:3:1.

■

Coat color in mammals is a trait that is affected by a number of different genes. In fact, at least 63 different genes with 148 different alleles affect coat color in mice. Some of the variations in mouse coat color are shown in figure 3.24. We have already mentioned two coat-color genes, the yellow gene in mice and the dilute gene in house cats. Other coat-color genes interact in a number of different ways, three of which we will discuss here to illustrate epistatic interaction.

First, the dominant, agouti color, controlled by the *A* allele at the *A* gene, results in an overall gray, or "salt and pepper" appearance if you look closely, and it is the common color of many mammals in natural populations. In actual fact, the individual hairs of agouti animals have a band of yellow, giving an overall brindled grayish color to the coat. When the genotype is the recessive non-agouti (*aa*), the hair shaft is all one solid color.

Second, the alleles at the *B,* or brown coat-color gene result in different colors, depending upon the genotype at the *A* gene (see table 3.5). When the *B* allele is present, the phenotype is agouti if the genotype is *AA* or *Aa,* but black if the genotype is *aa.* On the other hand, when the genotype at the *B* gene is *bb,* the phenotype is cinnamon-colored for genotype *A*—and brown for *aa.*

Finally, a third gene, called *C,* allows the formation of coat pigments. (The symbol *C* is used for this albino gene because *A* already represents the agouti gene.) For the *C* gene, the recessive genotype, *cc,* prevents the formation of color from other genes (see table 3.5). For example, the genotype *bbcc* is albino (no pigment) because the *cc* genotype inhibits the formation of the brown color. If two mice of genotype *BbCc* are crossed, the 9:3:4 phenotypic ratio in the fourth row of figure 3.23 results. What phenotypic proportions would you expect if *BbCc* and *Bbcc* were crossed? Using the forked-line approach (figure 3.25), you can see that the expected ratio of phenotypes in the progeny is ⅜ black, ⅛ brown, and ½ albino.

GENOTYPE-PHENOTYPE RELATIONSHIPS

For the traits that Mendel examined, a given genotype always resulted in a certain phenotype. However, in other situations, the phenotype of a given genotype varies, depending upon the penetrance and expressivity of that genotype. Often such variable expression results from environmental factors, and the exact phenotype caused by a given genotype-environment combination is not easily predicted.

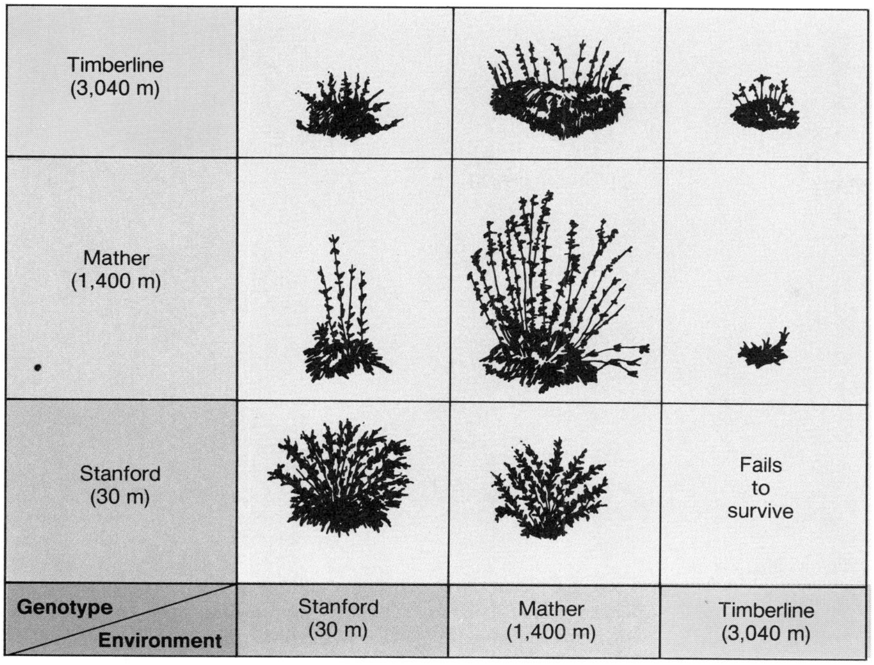

FIGURE 3.26
The phenotypes of *Potentilla* plants, each cloned and grown at three different altitudes.

Genotype / Environment	Stanford (30 m)	Mather (1,400 m)	Timberline (3,040 m)
Timberline (3,040 m)			
Mather (1,400 m)			
Stanford (30 m)			Fails to survive

In an ideal situation, the extent of the effect of the environment on a given genotype can be determined by looking at the phenotype of genetically identical individuals raised in different environments. In humans, identical, or monozygotic, twins separated at birth provide such a test situation. However, the best indications of environmental effects come from studies of species that can be raised in large numbers in controlled environments.

A classic example of such a study was provided for several plant species in the 1940s by J. Clausen and his coworkers. To determine the effect of the environment on phenotypes, these researchers grew **clones** (genetically identical individuals) from cuttings at different altitudes. For example, they collected *Potentilla glandulosa,* a member of the rose family, from three locations: the Pacific coast at an altitude of 100 meters; midway up the Sierras at 1,400 meters; and high in the Sierras at 3,040 meters. From these samples, they generated clones by vegetative propagation and grew them at the three different locations. In other words, genotypes collected from the three locations were grown in each of the three environmental conditions (100 m, 1,400 m, and 3,040 m).

Figure 3.26 shows the results of these experiments. Each row represents a genotype from the different areas, and each column indicates the environment in which these plants were grown. From this, it appears that both genetic and environmental factors are important in determining plant growth. For example, the Stanford (coastal) genotypes grow well at both low and intermediate altitudes but cannot survive at the high altitude (bottom row), indicating the importance of environmental factors. On the other hand, the timberline genotype (top row) grows much better than the other genotypes at the high altitude (far right column), indicating the importance of genetic factors.

The way the phenotype for a given genotype changes according to its environment is called the **norm of reaction.** Figure 3.27 gives norms of reaction similar to those of the low- and middle-altitude genotypes in figure 3.26. Genotype A, like the low-altitude clone, does best at the low end of the environmental spectrum and worst at the high end. But genotype B, like the middle-altitude clone, does best at intermediate altitude levels, falling off in phenotypic value in both the low- and high-altitude environments.

Norms of reaction may differ greatly among genotypes for traits involving behavior. Although we have not discussed behavior as a phenotype until now, a number of mutant genes do affect behavior. For example, individuals with untreated PKU have lower IQs, a behavioral trait that can be measured using IQ tests. Recent reports indicate that some cases of psychological disorders, such as manic-depression and schizophrenia, may result from single-gene defects.

The norm of reaction for a behavioral trait can be shown by comparing two strains of rats and their ability to complete a maze in three different environments: restricted, normal, and enriched. In this experiment, the environments are ranked according to the amount of visual stimulus present during the maturation of the rats. The two strains of rats are selected through a number of generations for their increased or decreased maze-learning

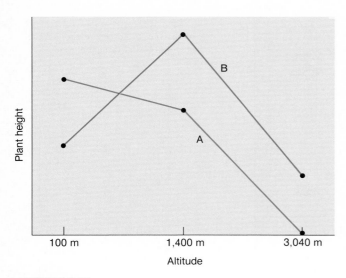

FIGURE 3.27
The norms of reaction for two different genotypes: *A,* the low-altitude clone, and *B,* the middle-altitude clone.

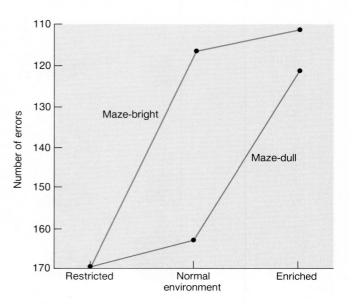

FIGURE 3.28
The numbers of errors in running a maze for two strains of rats, maze-bright and maze-dull, when the two strains are raised in three different environments.

From Cooper and Zubek, in Canadian Journal of Psychology, *Vol. 12, 159–64, 1958. Copyright © 1958 Societe Canadienne d'Orientation, Ottawa, Ontario, Canada.*

ability in a normal environment. The lines are therefore called maze-bright and maze-dull, respectively. In the normal environment, animals from the maze-bright strain make an average of fifty fewer errors than animals from the maze-dull strain, reflecting the effectiveness of selection and the genetic differences between the strains (see figure 3.28). However, if the two strains of rats are raised in either restricted or enriched environments, the two strains make a similar number of errors, primarily because the maze-bright strain cannot run the maze well under restricted conditions and the maze-dull strain can perform well under enriched conditions. The implications of these observations are intriguing; they suggest that some differences in behavior may only be present in a narrow range of environments and may not be generalizable to a diversity of environments.

> *T* he relationship of genotype to phenotype may be complicated. The norm of reaction over different environments is one way to quantify this relationship.

QUANTITATIVE TRAITS

For many traits, the genetic basis is not known precisely, and it may not be feasible, or even possible, to determine the number of genes and their individual effects on a particular trait. The phenotypic value of traits such as size, weight, or shape can generally be measured on a quantitative scale; hence, these are called **quantitative traits.** Although this phenotypic value may result from some general underlying genetic determination, for most quantitative traits, there is no precise relationship between the phe-

notypic value and a particular genotype. In general, quantitative traits have these characteristics: (1) They have a continuous distribution. (2) They appear to be affected by many genes; that is, they are **polygenic.** (3) They are affected by environmental factors.

One of the first good documentations of a polygenic trait was by Swedish geneticist Herman Nilsson-Ehle who showed in 1909 that the color of wheat kernels was affected by several genes. To understand his findings, we will examine the genetic basis of his crosses, where color differences are determined by two genes. Figure 3.29 gives the F_2 results of a cross between wheat plants with red kernels and those with white kernels. The F_1 plants are all intermediate, having pink kernels, as expected with incomplete dominance. However, instead of just three categories in the F_2, there are five different phenotypes because there are two extra classes, light red and light pink, intermediate between the parents and the F_1. The explanation for the ratios of color phenotypes seen in the F_2 (1:4:6:4:1) is that two genes equally affect this trait, so that red occurs with four red alleles present ($A_1A_1B_1B_1$); light red occurs with three red alleles present at either the *A* or the *B* gene ($A_1A_1B_1B_2$ and $A_1A_2B_1B_1$); and so on. Overall, three different genes can affect the color of the wheat kernel, and they do so independently of each other; that is, the phenotype is the sum of the individual effects of the different genes.

To illustrate in general the polygenic nature of quantitative traits, let us assume that we can visualize the influence of multiple genes affecting a trait by comparing

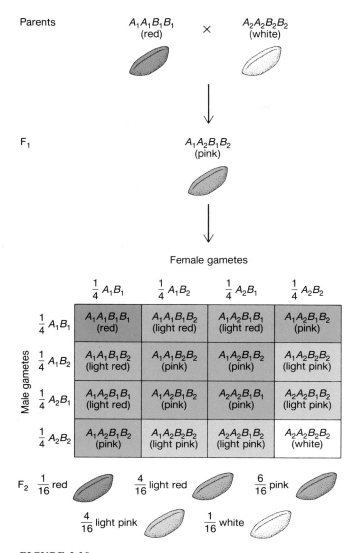

FIGURE 3.29

A cross of red- and white-kerneled wheat differing at two genes, showing the five types of F₂ progeny.

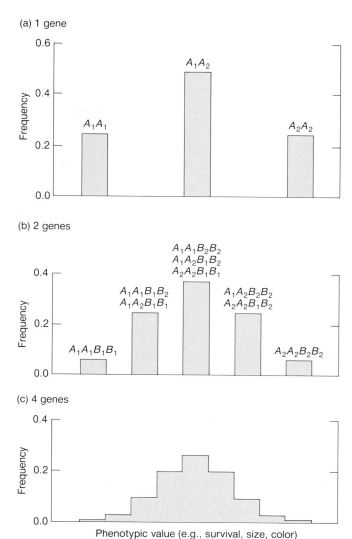

FIGURE 3.30

The determination of a quantitative trait by (*a*) one, (*b*) two, and (*c*) four genes, and the subsequent phenotypic distribution.

Reprinted with permission of Jones and Bartlett Publishers, Inc., Boston, MA, from Population Biology, *1984, written by Philip Hedrick.*

the phenotypic distribution that results when one, two, or four genes, each with two alleles, determine the phenotypic value of a trait (figure 3.30). For simplicity, we assume that heterozygotes are exactly intermediate, that the two alleles at each locus are equal in frequency, and that the environment has no effect on the trait. The frequency of the polygenic genotypes is calculated as the product of the frequencies of the single-locus genotypes. In fact, we have already shown examples of one-gene and two-gene cases in figures 3.1 and 3.29, respectively. From figure 3.30, it is obvious that as the number of genes affecting the trait increases from one to four, the distribution changes from one of discrete classes to one in which there is no discontinuity. In fact, the exact genotype-phenotype correspondence for all classes disappears when there are only two genes. That is, when two genes affect

a trait, the second lowest phenotypic value can result from either genotype $A_1A_1B_1B_2$ or genotype $A_1A_2B_1B_1$.

Most quantitative traits are influenced by the environment. For example, it is obvious that adult size in most organisms is affected by the amount of available resources. Thus, food availability will determine size in an inbred line of mice, just as light and soil differences will result in phenotypic variation among the clones of a fruit tree.

Typically, quantitative traits such as size and weight have a continuous distribution in a population, given that the method of measurement is sufficiently accurate. The distribution of such traits often closely approximates a bell-shaped, or normal, distribution, with a high proportion of

Number of individuals		1	0	0	1	5	7	7	22	25	26	27	17	11	17	4	4	1
Height in inches		58	59	60	61	62	63	64	65	66	67	68	69	70	71	72	73	74

FIGURE 3.31
Frequency distribution of heights of male students at Connecticut Agricultural College.
Library of Congress.

the individuals having an intermediate phenotypic value. Both the polygenic character of such traits and the influence of the environment help determine the shape of such a distribution. A classic example of a normal distribution curve is shown in figure 3.31, depicting a group of 175 male students arranged by height.

MEAN AND VARIANCE

How can we evaluate the phenotypic variation for height in the group of students (or the phenotypic variation in any organism)? To assess the attributes of populations, it is necessary to use descriptive statistics. Here we will briefly discuss how to calculate the **mean** ($\bar{x}$), a measure of the average of a group of values, and the **variance,** a measure of the variability around that central value.

Let us consider the heights of a sample of n different plants. If we symbolize the height of plant i as x_i, we can calculate the mean plant height as:

$$\bar{x} = \frac{1}{n} \sum_{i=1}^{n} x_i$$

where Σ (Greek letter sigma) indicates the summation of all n plant heights in the sample. For example, if there are five plants in the sample and their heights in centimeters are $x_1 = 28$, $x_2 = 30$, $x_3 = 27$, $x_4 = 32$, and $x_5 = 33$, then:

$$\begin{aligned} \bar{x} &= \frac{1}{5}(x_1 + x_2 + x_3 + x_4 + x_5) \\ &= \frac{1}{5}(28 + 30 + 27 + 32 + 33) \\ &= 30 \end{aligned}$$

In other words, the mean plant height in this sample of individuals is 30 centimeters.

Where a sample of individuals has a given mean, different amounts of dispersion, or variation, may exist around that mean. Thus, there may be no dispersion if all plants are exactly the same height, or there may be an extreme amount of dispersion if there are equal numbers of small and large individuals. The variance (V_x) is a measure of the dispersion around the mean and is calculated as:

$$V_x = \frac{1}{n-1} \sum_{i=1}^{n} (x_i - \bar{x})^2$$

In other words, the variance of x is the sum of the squared deviations of individual values from the mean, divided by n − 1. The variance for the data given here is:

$$\begin{aligned} V_x &= \frac{1}{4}[(x_1 - \bar{x})^2 + (x_2 - \bar{x})^2 + (x_3 - \bar{x})^2 \\ &\quad + (x_4 - \bar{x})^2 + (x_5 - x)^2] \\ &= \frac{1}{4}[(28 - 30)^2 + (30 - 30)^2 + (27 - 30)^2 \\ &\quad + (32 - 30)^2 + (33 - 30)^2] \\ &= 6.5 \end{aligned}$$

For the group of students, the mean height is 67.3 inches and the variance is 7.3.

The variance is not on the same scale as the mean because the variance is a squared value. Thus, the square root of the variance, or the **standard deviation,** is often used to measure the dispersion around the mean. For example, the standard deviation for the sample of students is:

$$(V)^{1/2} = (7.3)^{1/2} = 2.7$$

(The exponent ½ indicates the square root.) If the distribution of observations is bell-shaped, or normal, approximately 95% of the sample should be within two standard deviations of the mean. In this case, two standard deviations is 5.4; therefore, 67.3 − 5.4 = 61.9 is two standard deviations less than the mean, and 67.3 + 5.4 = 72.7 is

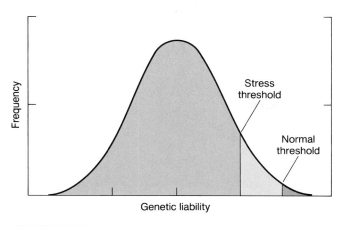

FIGURE 3.32
The genetic liability for a trait, with thresholds indicating the value above which individuals exhibit the disease phenotype.
Reprinted with permission of Jones and Bartlett Publishers, Inc. Boston, MA, from Population Biology, *1984, written by Philip Hedrick.*

two standard deviations greater. Looking at figure 3.31, we can see that a very high proportion of the students are within these limits.

However, some polygenic traits have only discrete classes, such as the number of bird eggs in a clutch, the number of scales on a reptile, the number of petals or leaves on a plant, or the number of vibrissae (whiskers) on the face of a mouse. Such discontinuous traits, sometimes called **countable** or **meristic traits,** are also assumed to be controlled by many genes and affected by environmental factors. Even more extreme, there may be only two phenotypic categories for a discontinuous trait, such as the presence/absence situation that occurs for certain disease states (diseased/not diseased) or for reproductive states (fertile/sterile). In such cases, the presence of one state or the other is the result of an underlying genetic determination of liability or susceptibility that has a continuous distribution.

The presence of a genetic disease in a particular individual is determined by a combination of the individual's underlying genetic liability and his or her particular environment. Figure 3.32 shows the hypothetical threshold in a normal environment for a disease such as diabetes. Here, a relatively small percentage of the population exceeds the threshold, and thus exhibits diabetes (red region). Given a different environment, in which obesity, alcoholism, or other factors increase the likelihood of diabetes, the threshold is moved to the left (indicated as the stress threshold). This shift results in a higher incidence of the disease, because some individuals, represented by the pink region, would not contract the disease in a normal environment but become afflicted under stress conditions.

Hip dysplasia in dogs, a condition that eventually leads to degenerative joint disease, appears to be determined by a combination of underlying genetic liability (that is, susceptibility) and environmental factors. This trait occurs frequently in many large dog breeds, such as German shepherds and Labrador retrievers, despite substantial efforts to eliminate it. The lack of simple inheritance as well as ambiguity in diagnosis (some dogs considered dysplastic by radiographs are clinically unaffected) has made it difficult to reduce the incidence of this condition.

It is important to realize that the genetic determination of quantitative traits may vary considerably among populations. Such a varied response has occurred in populations of insects exposed to various pesticides and may be primarily due to a single gene (or to a number of genes) that confer resistance to a particular insecticide. In addition, the single genes and the method by which they confer resistance may differ among populations. For example, resistance to the insecticide DDT may occur because particular genes influence either detoxification of the insecticide, storage of DDT in fat bodies, or resistance to its diffusion into the circulatory system. (A more detailed discussion of pesticide resistance is given in chapter 17.)

It has been generally believed, and recent molecular genetics evidence has verified, that the genes affecting quantitative traits are essentially like the genes (or are the same genes) that determine single-gene qualitative traits such as color, enzyme, or shape differences. In other words, there is no reason to assume that a different type of gene accounts for quantitative genetic variation; in fact, some genes known for their qualitative effects also affect quantitative traits. Genes affecting quantitative traits follow Mendelian patterns of inheritance, may have multiple alleles, and show various patterns of dominance and epistasis.

Q uantitative traits are affected by many genes as well as by the environment. Polygenic traits generally have a continuous phenotypic distribution in a population, but they may occur in discrete classes, such as diseased or not diseased.

MODEL OF QUANTITATIVE TRAITS

In order to understand and examine the importance of quantitative traits, we need to construct a model that will allow us to break down phenotypic values into genetic and environmental components. We can accomplish this in a simple manner by symbolizing the phenotypic value (P) for a genotype (i) in a particular environment (j) as:

$$P_{ij} = G_i + E_j$$

G_i is the genetic contribution of genotype i, and E_j is the environmental deviation resulting from environment j. E_j may be either positive or negative, depending upon the effect of environment j. This simple model illustrates the basic genetic-environment or nature-nurture dichotomy often discussed in relationship to human intelligence and other traits.

In many cases, different populations have different mean phenotypic values. It is generally difficult to determine whether such a difference is the result of genetic factors, environmental factors, or a combination of both. However, if the populations can be grown in the same environment, called a **common garden experiment** for plants, we can make some inferences. For example, if individuals from different populations retain phenotypic differences in a common environment, this is consistent with the hypothesis that the populations are genetically different. But if the phenotypic differences disappear in the common environment, this indicates that environmental factors are important in determining phenotypic differences. For the *Potentilla* plants shown in figure 3.26, both genetic and environmental factors were important.

A particular genotype may do well in a specific environment. If such specific interactions occur between genotypes and environments, the basic model just presented can be expanded to include a term for **genotype-environment interaction,** with the phenotypic value becoming:

$$P_{ij} = G_i + E_j + GE_{ij}$$

where GE_{ij} measures the interaction between genotype i and environment j. As with the environmental deviation, the genotype-environment interaction may be either positive or negative.

Maze-learning ability in the rat experiment (figure 3.28) is an example of genotype-environment interaction. To further illustrate the complexity introduced by genotype-environment interaction, consider the simple situation of two genotypes in two environments. It is possible that a reversal of the order of the phenotypic values of two genotypes may occur in two different environments, as with the domesticated strain and wild progenitor of a particular species, say pigs or sheep, raised either in association with humans or in the natural environment. The wild progenitor will survive and reproduce better in its natural situation, while the domesticated strain will survive and reproduce better in association with humans (see figure 3.33). As a result, the phenotypic value of a trait like survival is primarily the result of the interaction between the specific genotype and the specific environment and cannot be predicted without information about both the genotype and the environment.

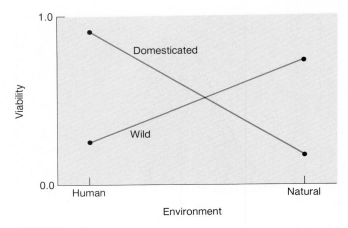

FIGURE 3.33
The viability of domesticated and wild animals in human and natural environments, illustrating the effects of genotype-environment interaction.

The genetic model can be more precisely defined by subdividing the genotypic value into two components, with the value of genotype i being:

$$G_i = A_i + D_i$$

where A_i and D_i are the additive and dominant components, respectively. If the heterozygote is exactly intermediate between the homozygotes, the contribution of the dominant component is zero; that is, all of the genetic component is additive. In other words, the phenotypic value can be symbolized as:

$$P_{ij} = A_i + D_i + E_j$$

assuming that the genotype-environment interaction is negligible.

As pointed out earlier, there is no exact genotype-to-phenotype correspondence for quantitative traits. Therefore, a more practical way to examine quantitative traits is to partition the variation in a population for a particular phenotype into that caused by genetic factors and that caused by environmental factors. If the basic model is used to examine the phenotypic variance, then:

$$V_P = V_G + V_E$$

where V_P, V_G, and V_E are the phenotypic, genetic, and environmental variances, respectively.

Adopting the same approach as before, if we use the expression with separate additive and dominant components to examine the phenotypic variance, it becomes:

$$V_P = V_A + V_D + V_E$$

Chapter 3

where V_A and V_D are the genetic variances due to additive and dominant effects, respectively. If this expression is divided by V_P, we can obtain the proportion of phenotypic variance due to the different components. In particular, the ratio of the additive genetic variance to the phenotypic variance is known as the **heritability** (h^2):

$$h^2 = \frac{V_A}{V_P}$$

The magnitude of the heritability is important in determining the rate and amount of response to directional selection (that is, selecting for an extreme phenotype) and is often estimated by plant and animal breeders prior to initiating a selection program to improve yield or other traits. We will discuss how to estimate heritability and give some examples and applications of this approach in a subsequent section of this chapter.

*Q*uantitative traits can be understood by partitioning out the genetic, environmental, and genotype-environment effects. The heritability is the ratio of the additive genetic variance to the phenotypic variance.

■

ESTIMATING GENETIC VARIANCE AND HERITABILITY

Several different approaches can be used to estimate the amount of genetic variance and heritability. In some organisms, experimental manipulation, or breeding, can either reduce or eliminate the environmental or genetic variance components and make it possible to estimate the complement variance component. In other words, if $V_G = 0$, then $V_P = V_E$, or if $V_E = 0$, then $V_P = V_G$. For example, the genetic variation is zero or near zero in inbred lines, clones, identical twins, hybrids between lines, and repeated measurements on the same individual. Environmental variance may approach zero if all important environmental factors, such as food, moisture, and temperature, are known and carefully controlled.

Identical (monozygotic) twins have the same genotype, thus any differences between the twins should be the result of environmental factors. In fact, identical twins raised together share similar environments as well as identical genotypes. One way to determine the extent of similarity is to measure the **correlation** between individuals for different traits. The highest correlation is 1.0 where all the individuals in pairs have the same trait value. If the traits are uncorrelated, the correlation is 0.0. The first column of table 3.6 gives the correlation (or similarity) for four different traits in monozygotic twins raised together. The second column gives the correlation for the

*T*ABLE 3.6   *Comparisons of Human Twins*

	Correlations		Heritability
	Monozygotic Twins (r_M)	*Dizygotic Twins (r_D)*	
Fingerprint-ridge count	0.96	0.47	0.98
Height	0.90	0.57	0.66
IQ score	0.83	0.66	0.34
Social maturity score	0.97	0.89	0.16

same traits in dizygotic (nonidentical) twins of the same sex. The difference between these values should be the result of the lower genetic similarity of dizygotic twins, who share only half their genes in common. One way to estimate the heritability is from the expression:

$$h^2 = 2(r_M - r_D)$$

where r_M and r_D are the correlations between monozygotic and dizygotic twins, respectively. Using this formula, the heritability for the four traits in table 3.6 varies from 0.98 to 0.16. In other words, most of the phenotypic variation in IQ and social maturity scores appears to be environmental, while most of the variation for fingerprint-ridge count appears to be genetic.

To change a phenotype, the rate and amount of response to selection depend upon the heritability of the trait under selection. This is quite obvious when h^2 is either 0 or 1. When $h^2 = 0$, none of the phenotypic variation is from the additive genetic variance component; as a result, selecting even extreme individuals as parents would not change the phenotypic mean. On the other hand, if $h^2 = 1$, all the phenotypic variation is from the additive genetic variance component, and phenotypically extreme individuals are that way because of their genotypes. The difference in the parental mean and the offspring mean is called the **response** (**R**) and is expressed as:

$$R = h^2 S$$

where S is the **selectional differential,** the difference in the mean of the selected parents and the mean of all the individuals in the parental population. This equation may be rearranged so that an estimate of heritability is:

$$h^2 = \frac{R}{S}$$

This expression is also called the **realized heritability** because it is the ratio of observed response over total response possible.

Separated at birth, the Mallifert twins meet accidentally.

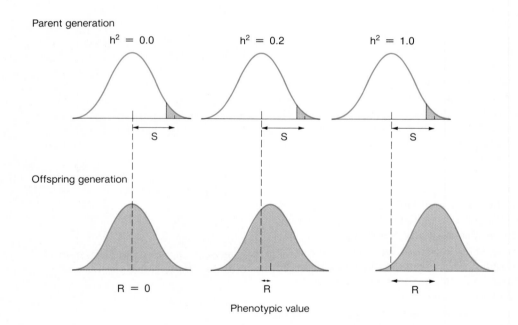

FIGURE 3.34

The change in phenotypic mean for populations with different heritabilities.

Reprinted with permission of Jones and Bartlett Publishers, Inc. Boston, MA, from Population Biology, *1984, written by Philip Hedrick.*

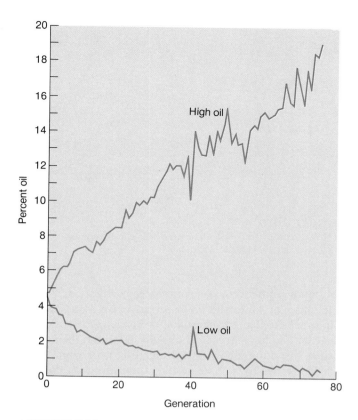

FIGURE 3.35
The percent of high and low oil content in long-term selection experiments in corn.

Reprinted by permission from Proceedings of the International Conference on Quantitative Genetics, August 16–21, 1976, *edited by W. Pollak, O. Kempthorne, and T. B. Bailey, Jr. © 1977 by Iowa State University Press, Ames, IA.*

Perhaps the best way to appreciate this effect is to examine the expected change in phenotypic mean from one generation of selection for traits with different heritabilities (figure 3.34). Notice that even when the selected individuals are quite extreme (blue-shaded region at top of figure), if the heritability is low, the mean of the offspring is similar to that of all the parents (response is low). On the other hand, if the heritability is high, the offspring are much more like the selected parents than the distribution of all parents.

Many directional (or artificial) selection experiments give evidence of a change in phenotypic values over ten to forty generations, often reaching a maximum, or plateau, before the termination of the experiment. Evidence indicates that selection response may continue for many generations, given an adequate initial sample of genetic variation, a reasonable population size during the selection process, and an effective selection scheme.

One of the longest selection experiments ever conducted is still running at the University of Illinois for oil and protein percentage in corn. High and low directional selection has been carried out for seventy-six plant generations, and progress is still continuing, although large changes in oil and protein percentage have already been made. The data for the high oil and low oil lines are given

TABLE 3.7	Approximate Values of Heritability of Various Characters in Several Species	
Organism	**Trait**	**Heritability**
Swine	Back fat thickness	0.55
	Body weight	0.3
	Litter size	0.15
Poultry	Egg weight	0.6
	Body weight	0.2
	Viability	0.1
Mice	Tail length	0.6
	Body weight	0.35
	Litter size	0.15
Drosophila melanogaster	Bristle number	0.5
	Body size	0.4
	Egg production	0.2
Corn	Plant height	0.70
	Yield	0.25
	Ear length	0.17

Reprinted with permission of Jones and Bartlett Publishers, Inc., Boston, MA, from *Population Biology,* 1984, written by Philip Hedrick.

in figure 3.35. The percentage of oil has increased nearly 4-fold in the high line and has been reduced to below 1% in the low line.

Perhaps the most important approach used to measure components of genetic variation and heritability is based on the phenotypic resemblance between relatives. If genetic factors are important, one would expect the offspring of parents with high phenotypic values to have high phenotypic values and the offspring of parents with low phenotypic values to have low phenotypic values. Obviously, close relatives share more genes with each other than with nonrelatives or distant relatives, and thus should have greater phenotypic resemblance. For example, monozygotic twins share all of their genes; siblings share one-half of their genes (on the average); and first cousins share one-eighth of their genes. As a result, phenotypic values between close relatives should be more highly correlated than those between distant relatives. Using this approach, estimates of heritability for a number of traits in different organisms are given in table 3.7.

Fingerprints have been widely used for identification purposes because the fingerprint patterns of an individual are unique. A general way of classifying fingerprints is to count the number of ridges, the raised areas of skin on the fingertips. The three basic configurations of ridges are known as the arch, the loop, and the whorl (figure 3.36). Fingerprint-ridge count seems to be almost entirely genetic, with little noticeable influence from the environment.

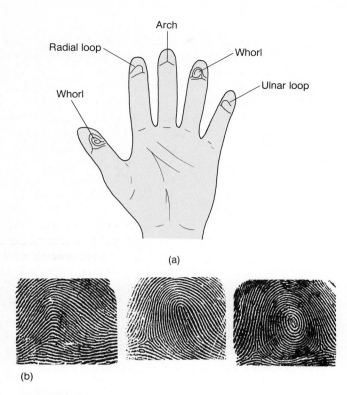

(a)

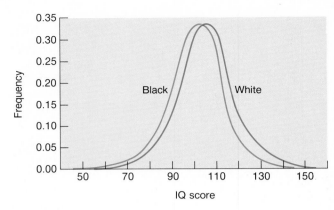

(b)

FIGURE 3.36
(*a*) The position of fingerprints on a hand and (*b*) examples of the three basic fingerprint types. Fingerprint-ridge count for the loop and whorl is the number of ridges crossed by the drawn lines. Arches have a count of zero.
(b) *American Institute of Applied Science.*

FIGURE 3.37
The IQ scores of black and white seven-year-olds from high socioeconomic backgrounds in Boston.

RACE AND INTELLIGENCE

In the late nineteenth century, some British scientists suggested that the high reproductive rate among the "lower" classes was leading to an overall deterioration of the average intelligence of the British population. The discussion that followed became known as the nature-nurture controversy; that is, given real differences in intellectual ability among social classes, were these differences due to genetic or environmental factors? The controversy was revived again in the late 1960s when it was claimed that the difference in IQ scores among racial groups was genetically determined, and in particular, that lower IQ scores among American blacks were the result of genetic factors.

Many researchers have suggested that environmental factors have a substantial effect on IQ scores, citing a range of factors that includes nutrition (prenatal, during childhood, and at the time of testing), socioeconomic status of parents, and educational and testing experience. One concern in comparing the IQ scores of blacks and whites is that these groups differ not only ethnically but also in education, upbringing (urban versus rural), and other socioeconomic factors. An obvious way to evaluate the effect of socioeconomic factors is to compare blacks and whites

within the same socioeconomic category. Figure 3.37 gives the scores of white and black seven-year-olds, all from a high socioeconomic background in Boston. Obviously, the distributions have large amounts of overlap and their mean values are quite close—104.2 and 100.0 for whites and blacks, respectively. As a result, it appears that the differences seen between whites and blacks are probably due in large part to socioeconomic differences between the two groups.

> *G*enetic variance can be estimated by examining identical twins. Realized heritability can be estimated by determining the rate of response from selection. Intelligence is affected by a number of environmental factors, thus any racial differences in IQ are generally confounded with environmental differences.

GENETIC COUNSELING

The importance of genetic disease has increased substantially in the last several decades as the incidence of infectious disease has declined. For example, infectious diseases such as smallpox and diphtheria used to be causes of infant mortality, but genetic disease is now a leading factor. Some of these diseases follow a single-gene pattern of inheritance; others are polygenic, with significant environmental components; and still others are the result of aberrant chromosomal numbers or types (see chapter 4).

After a child is born with a genetic disease, the parents may seek advice concerning the probability of a second child having the same disease. Such **genetic counseling** is offered at a number of medical centers. Once the disease is properly diagnosed and its mode of inheritance known, the **recurrence risk,** or probability that the condition will recur in a given family, can be calculated.

TABLE 3.8 *Probability of Affected Offspring for an Autosomal Recessive[1]*

Parents	Second Child Affected
Both unaffected	¼
One affected	½
Both unaffected; one has affected sib	¼
Both affected	1

[1]Given that one child is already affected.
Reprinted with permission of Jones and Bartlett Publishers, Inc., Boston, MA, from *Population Biology,* 1984, written by Philip Hedrick.

TABLE 3.9 *Empirical Risks for Congenital Pyloric Stenosis in an English Population*

Number of Affected Sibs	Neither Parent Affected	Father Affected	Mother Affected	Both Parents Affected
0	0.003	0.037	0.051	0.298
1	0.030	0.115	0.135	0.365
2	0.119	0.216	0.231	0.431

Reprinted with permission of Jones and Bartlett Publishers, Inc., Boston, MA, from *Population Biology,* 1984, written by Philip Hedrick.

A large proportion of known genetic diseases are inherited as autosomal recessives. Many of these are **inborn errors of metabolism**—that is, enzymatic defects that result in the buildup or absence of an important biochemical product. Most of these diseases are quite rare, generally occurring in fewer than 1 in 10,000 individuals. However, because there are a large number of such diseases, their cumulative incidence is substantial.

The probability of a genetic disease occurring in a child depends on whether the parents or other relatives are affected or unaffected, and on whether other offspring have already been affected. Table 3.8 shows the risk of producing affected children for parents with different disease states when a previous child has been affected. These probabilities follow the pattern of inheritance for recessive traits discussed in chapter 2.

Genetic counseling can also be useful when parents strongly wish to have children but there is a possibility that their children will inherit a genetic disease. For example, if both parents are known carriers of a recessive disease allele (heterozygotes can be detected for sickle-cell disease, Tay-Sachs, and several other recessive diseases), the probability of a sibship of size N with no affected progeny is $(\frac{3}{4})^N$. With a small family of one or two children, the chances of having no affected offspring are 0.75 and 0.5625, respectively, a risk that some people may find acceptable.

A number of diseases, such as diabetes, cleft palate, and spina bifida, have a strong genetic component but do not have a single-gene mode of inheritance. For such diseases, it is assumed that the genetic liability to the disease is polygenically controlled, and that consequently, close relatives of affected individuals are more likely to be affected. Obviously, the closer the affected relative and the more relatives are affected, the higher the genetic liability in an individual.

For such diseases, **empirical recurrence risks** are often calculated from a large number of case studies. These values may be figured where both parents are affected, where one parent is affected, where one sib is affected, or in other situations. Empirical risk values can only be used

as a guideline, since specific genetic or environmental factors may either increase or reduce the risk in a particular family.

Pyloric stenosis is a disease in which the pyloric valve, the sphincter valve between the stomach and the small intestine, is not completely opened. Before corrective surgery was introduced in the 1920s, most affected individuals died in infancy because little or no food could pass on to the small intestine. Nowadays, after surgery, individuals born with pyloric stenosis have normal digestion and generally no other related problems. Risk values for this disease are given in table 3.9 with respect to the disease state of the parents and the number of affected sibs. The risk, as intuitively expected, increases as either the number of affected sibs or the number of affected parents increases. For example, with no affected relatives, the risk is only 0.3%, while with two affected sibs and two affected parents, the risk is greater than 40%.

*G*enetic counseling is used to determine the risks of producing genetically defective infants. Mendelian inheritance probabilities are used to predict single-gene traits, while empirical risk values are used for quantitative traits.

■

PATERNITY EXCLUSION

In cases where the father of a child is unknown or disputed, blood groups, enzymes, or HLA types have been used to determine the probability of parentage. In the past, such data were admitted in United States courts only when they ruled out an individual as a parent. But in recent years, because of better genetic analysis, courts will often accept information that indicates the probability of parenthood.

When a locus has only a few allelic forms in a population, it is difficult to exclude an individual as a parent, because many individuals have the same alleles, making

Mother	Child	Excluded Paternity Types
L^ML^M	L^ML^M	L^NL^N
	L^ML^N	L^ML^M
L^ML^N	L^ML^M	L^NL^N
	L^ML^N	—
	L^NL^N	L^ML^M
L^NL^N	L^ML^N	L^NL^N
	L^NL^N	L^ML^M

TABLE 3.10 *Excluded Paternity Types, Given Genotypes of Mother and Child*

Reprinted with permission of Jones and Bartlett Publishers, Inc., Boston, MA, from *Population Biology,* 1984, written by Philip Hedrick.

TABLE 3.11 *Genotypes of a Mother, Child, and Two Putative Fathers*

Mother	Putative Father		Child	
A3 BW44 / A1 B8	(1)	A1 B17 / A3 B14	—	
	(2)	A26 B7 / A11 B12	A11 B12 / A1 B8	

Reprinted with permission of Jones and Bartlett Publishers, Inc., Boston, MA, from *Population Biology,* 1984, written by Philip Hedrick.

TABLE 3.12 *Genotypes of a Mother, Dizygotic Twins, and the Putative Father of Each Twin*

Mother	Putative Father		Twin	
A11 B27 / A2 BW44	(1)	A2 B15 / A3 —	(1)	A2 B15 / A2 BW44
	(2)	A24 BW54 / A2 B7	(2)	A24 BW54 / A2 BW44

them all possible parents. As an example, mother-child genotype combinations and the excluded paternal genotypes for the MN blood group locus are listed in table 3.10. Given the genotype of the mother and child, one may exclude certain genotypes as the father. For example, if the genotypes of both mother and child are L^ML^M, the father could be either L^ML^M or L^ML^N, and only the paternal genotype L^NL^N can be excluded. Notice that only one genotype at most can be excluded and that when both the mother and child are heterozygotes, no paternal genotypes can be excluded.

Even with a large number of loci, the probability of exclusion is not as high as one might wish. The use of HLA **haplotypes** (all the HLA alleles on a single chromosome) and DNA markers, both of which can have many alleles at a locus, may resolve this difficulty. To illustrate, table 3.11 gives the HLA genotypes of a mother, a child, and two possible fathers. From this information, it is apparent that the mother, and therefore the father, produced an A11–B12 gamete. Putative father 1 does not have this haplotype, so he can be excluded. Putative father 2 does, and if there is no other candidate, this would indicate that he is the father. If there were other possible fathers, one of them could have been the father, so it cannot be stated with complete assurance that putative father 2 is the father.

In an actual case that proved the usefulness of the HLA system, two different men were identified as the fathers of a set of dizygotic twins. The genotypes of the individuals are given in table 3.12. The twins both received an A2–BW44 gamete from their mother, but twin 1 received an A2–B15 gamete from putative father 1 and twin 2 received an A24–BW54 gamete from putative father 2. This was not detectable by looking at the ABO blood group system because all of the individuals were type A.

Recently, the identification of a number of loci, each with many alleles, has allowed the development of **DNA fingerprints.** The principles involved are the same as discussed before, except that nearly every individual has a different DNA pattern for these genes. We will discuss some details of the molecular basis of this procedure in chapter 15.

Paternity exclusion based on genetic variants is used when the father of a given child is unknown. Generally, blood group or HLA alleles are used, but recently, DNA fingerprints have improved the power of this technique.

■

Grandpaternity Exclusion

*G*enetic information has long been used to attempt to identify parents. Recently, the same approach has been extended to grandparents and other relatives. A major impetus for this work was the disappearance of at least nine thousand Argentinians between 1975 and 1983. Among the missing were a number of children who had been abducted by the military and police or who had been born to kidnapped women in captivity. For example, Liliana Pereyra was abducted when she was five months pregnant and kept alive in a torture center until she gave birth to a boy in February 1978. Her son has not been found and may have been kidnapped by the military.

It appears that many of the Argentinian children are now living with military couples who claim to be their biological parents but may actually be the kidnappers. If all four grandparents are alive, and if the child has an allele that none of them have, grandpaternity can be excluded. But if the alleles in the child are also present in the grandparents, the probability of grandpaternity can be calculated. For example, if the shared alleles are rare in the population, the probability of grandpaternity is quite high.

Figure B3.1 is a pedigree in which a child's HLA types were found in the putative grandparents. In this case, Paula Eva Logares, an Argentinian kidnap victim, was raised by a policeman and his wife. One of her HLA haplotypes, *A2–B5*, was present in her paternal grandfather. The reconstructed genotype of her maternal

Liliana Pereyra, an Argentinian kidnap victim.
From Dr. Jared M. Diamond. Reprinted by permission from Nature *327:552. Copyright © 1987 Macmillan Magazines Limited.*

grandparent (from his other children) contains the other haplotype, *A1–B5*. It is very likely that these haplotypes were passed on to her from her deceased parents. These data and probability principles make it 99.9% certain that Paula is the grandchild of the Logares and Grinspon families.

■

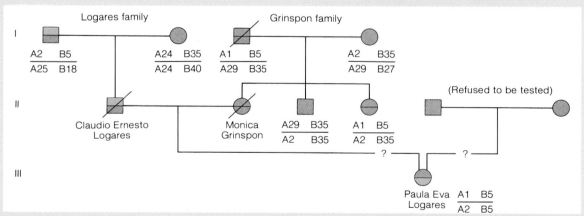

FIGURE B3.1
A pedigree illustrating how probable grandparenthood can be determined if the HLA haplotypes of some of the individuals involved are known. Slashes indicate deceased individuals.

D ominance can occur in different forms: (1) complete, as found by Mendel, where either the mutant or the wild-type allele is dominant; (2) incomplete, with the heterozygote having a phenotype in between the two homozygotes; or (3) codominant, in which both alleles are expressed in the heterozygote.

Many variations on Mendel's original findings have been discovered. Some genes may be lethal. Pleiotropy occurs when one gene affects several traits. Some genotypes may not always have the same phenotype, a phenomenon resulting from incomplete penetrance or variable expressivity of the genotype. Genes may have more than two alleles. In diploid organisms, many different genotypes can exist when there are multiple alleles. Genes on the X and Y chromosomes have different patterns of inheritance than do autosomal genes. X-linked recessive traits occur almost entirely in males and generally skip generations in pedigrees.

More than one gene can affect a single trait. When two genes interact to determine a phenotype, this phenomenon is called epistasis. Epistatic interaction can result in F_2 ratios different from $9:3:3:1$. The relationship of the genotype to the phenotype may be complicated. The norm of reaction over different environments is one way to quantify this relationship. Quantitative traits are affected by many genes and are influenced by the environment. Polygenic traits may occur in a continuous phenotypic distribution in a population or in discrete classes, such as diseased or not diseased.

Quantitative traits can be understood by partitioning out the genetic, environmental, and genotype-environment effects. Heritability is the ratio of the additive genetic variance to the phenotypic variance. Genetic variance can be estimated by examining identical twins. Realized heritability can be estimated by determining the rate of response from selection. Intelligence is affected by a number of environmental factors, and any racial differences in IQ are generally confounded with environmental differences.

Genetic counseling is employed to determine the risks of producing genetically defective infants. The probabilities from Mendelian inheritance are used for single-gene traits, while empirical risk values are used for quantitative traits. Paternity exclusion, based on genetic variants, is applied when the father of a given child is unknown. Generally, blood group or HLA alleles are used, but recently, DNA fingerprints have improved the power of this technique. ∎

1. One experiment to elucidate the inheritance of flower color in four-o'clocks crossed two plants with pink flowers. In the progeny from this cross, there were 42 plants with red flowers, 86 with pink flowers, and 39 with white flowers. Using a chi-square test, determine whether those numbers are consistent with Mendelian expectations.

2. A series of matings between individuals where one parent was blood group phenotype N and the other parent was phenotype MN yielded 84 type MN progeny and 91 type N progeny. It this consistent with expectations? (Use chi-square to test.)

3. Write out all the possible mating types for the ABO blood group system. (Given four phenotypes, there are ten different mating types if it is assumed that reciprocal matings are equivalent.) Specifying genotypes, write out the four mating types designated phenotypically as A × B and the expected proportion of progeny types from each.

4. In trying to understand the inheritance of the dominant yellow gene in mice, two yellow heterozygous mice were mated. A typical result was 56 yellow progeny to 31 wild-type. Use a chi-square test to determine if this is consistent with $3:1$ Mendelian predictions for a dominant gene. Are these data consistent statistically with the lethal model of Cuénot?

5. Some researchers feel that most genes are pleiotropic for many traits. If you discovered a new genetic disease, how would you go about documenting its pleiotropic effects?

6. (a) Discuss the difference between the penetrance and expressivity of a trait. (b) Draw a pedigree for a trait such as Huntington's disease, illustrating variable expressivity. Assuming H and h are the dominant and recessive alleles, respectively, give the genotypes and phenotypes for three generations (one set of grandparents in the first generation, at least three progeny, and at least five grandchildren).

7. (a) Describe how you would document the importance of environmental factors, such as temperature and diet, on the expression of a recessive mutation in *Drosophila*. (b) How would you document the effect of modifier genes on the expression of the same mutation?

Chapter 3

8. Which blood group type of the ABO system is known as the universal donor? The universal recipient? Explain your answers.

9. A female *Biston betularia* moth having the typical phenotype is mated to a male melanic. If half the progeny are melanic and half are insularia, what are the genotypes of the two parents?

10. (*a*) Given that two plants with genotypes B_2B_4 and B_1B_5 are mated, what types of progeny (and in what proportions) would you expect? (*b*) For the same gene, if a progeny array from a single mating has equal numbers of B_1B_2 and B_2B_4 individuals (and no other progeny), what are the parents' genotypes?

11. (*a*) If female *Drosophila* homozygous for the white allele (*ww*) affecting eye color are mated to red-eyed males (w^+Y), what types of F_1 progeny could you expect? (Give both phenotype and sex.) (*b*) F_1 males and females were mated, and in the F_2 progeny there were 40 white-eyed females, 42 red-eyed females, 39 white-eyed males, and 44 red-eyed males. Using a chi-square test, determine whether these results are consistent with the expectations for X-linked genes.

12. Using the forked-line approach, calculate the proportions of the five phenotypes in a cross between two wheat plants with pink kernels and genotype $A_1A_2B_1B_2$. (See figure 3.29 for information on this cross and another way of calculating the proportions.)

13. Assume there is a third gene affecting color in wheat so that the F_1 plants are $A_1A_2B_1B_2C_1C_2$. (*a*) What are expected proportions of the F_2 genotypes using the forked-line approach? (*b*) Assume that gene *C* contributes equally to the other genes for color and that the color of the F_2 can be divided into seven categories: 0 (no "2" alleles, $A_1A_1B_1B_1C_1C_1$); 1 (one "2" allele, $A_1A_2B_1B_1C_1C_1$ or $A_1A_1B_1B_1C_1C_2$); on through to 6 (no "1" alleles, $A_2A_2B_2B_2C_2C_2$). Draw a graph showing the seven phenotypic categories equally spaced on the horizontal axis and their frequencies as obtained from the F_2 progeny on the vertical axis. (*c*) How does this compare to the distribution expected if only one gene or two genes affect color?

14. What is the difference between dominance and epistasis? Why didn't any pairs of genes used by Mendel show epistasis?

15. Bateson and Punnett observed 382 purple-flowered and 269 white-flowered sweet peas in an F_2. Are these numbers consistent with the 9:7 ratio they predicted, using a chi-square test?

16. If the purple F_1 in question 15 were crossed to one of the parental varieties, what proportion of white-flowered progeny would you expect?

17. Two types of epistasis discussed were duplicate and complementary gene action. How would you distinguish between these types of epistasis in determining purple and white flower color if you could set up any crosses necessary?

18. Graph the effect of the environmental temperature on the phenotype for plant height when the environment affects two genotypes, a short and a tall, equally. Assume that as temperature increases, plant height increases. Next, assume that at high temperature, the tall genotype plant is inhibited in growth so that it has a lower height than the short genotype. Draw the norms of reaction of the two genotypes in this case.

19. Compare the terms polygenic trait and quantitative trait. Are traits that have only two phenotypes, such as diseased or healthy, due to single genes or are they polygenic? How would you experimentally determine the genetic basis of such a trait?

20. In figure 3.21, what is the probability that a child from an incestuous mating between III-1 and III-2 would be affected with the recessive disease? If individual II-1 was planning to have children with II-5, an unrelated individual, what is the probability that their first child would be affected?

Answers appear at end of book.

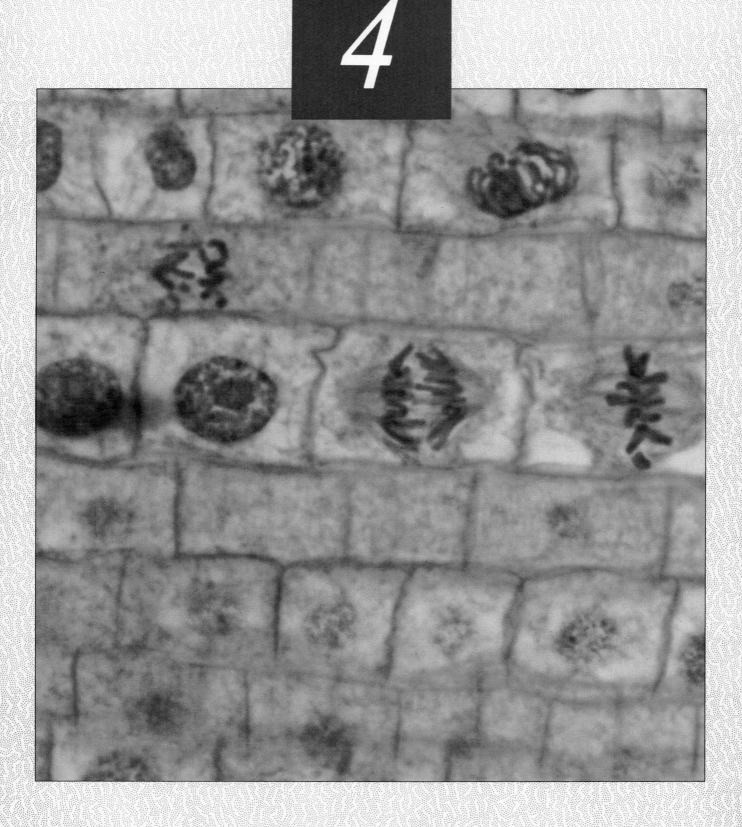

Chromosomes and Heredity

I may finally call attention to the probability that the association of paternal and maternal chromosomes in pairs and their subsequent separation during the reducing division . . . constitute the physical basis of the Mendelian law of heredity.

■

Walter Sutton
American geneticist

In this chapter you will learn:

1. How different chromosomes are characterized.

2. How chromosome number is maintained during the process of mitosis.

3. How chromosome number is halved during the process of meiosis.

4. That genes and chromosomes exhibit parallel behavior.

5. The four types of changes in chromosomal structure: duplication, deletion, inversion, and translocation.

6. The two ways in which chromosome number may vary: polyploidy and aneuploidy.

7. That sex determination is generally caused by the chromosomes, although the mechanism varies with the species.

Photomicrograph of mitosis in the *Allium* (onion) root tip in late metaphase.
© *Arthur M. Siegelman.*

A lthough chromosomes had been noticed as cell structures in the nineteenth century, it was only after the rediscovery of Mendel's principles that their fundamental role in heredity was established. In 1902, Walter Sutton in the United States and Theodor Boveri in Germany both suggested that genes might be on chromosomes. They based their hypotheses on the analogous behavior of genes and chromosomes. In fact, the behavior of chromosomes in meiosis, as we will see later, provides a mechanism for the Mendelian principles of segregation and independent assortment.

SEXUAL REPRODUCTION

Before we describe the behavior of chromosomes during cell division, let us briefly discuss sexual reproduction and the overall structure of chromosomes. Reproduction in eukaryotes in general is either asexual or sexual. **Asexual reproduction** occurs when a single individual produces new individuals identical to itself. This is a common mode of reproduction in plants and simple animals. Asexual reproduction using cuttings is the common method of propagating many horticultural plants. **Parthenogenetic** female organisms (such as some aphids) can also produce offspring without fertilization. These progeny are generally identical to their parent.

Sexual reproduction occurs when separate individuals produce sex cells, or gametes, that in turn unite to form a **zygote,** a single cell from which a new individual develops. Sexual reproduction occurs in nearly all types of organisms, including simple animals (see figure 4.1), plants, and even bacteria (see chapter 13). Usually the male and female gametes come from different individuals, ensuring that the progeny are different from either parent. An important exception occurs in self-fertilization, as in Mendel's garden peas, where the same individual produces both male and female gametes. In the following discussion, we will concentrate on sexual reproduction and its genetic consequences.

The general scheme of reproduction and growth in a sexual organism is shown in figure 4.2. Let us examine two zygotes, one of which will mature into a female and the other into a male. Each zygote has received a complement of chromosomes of number N in the gamete from each parent, making the total number of chromosomes in each zygote 2N. The chromosomal number in the gamete is called the **haploid** (N) number, and the chromosomal number in the zygote is called the **diploid** (2N) number. The N different chromosomes within each haploid set each have different genes. Each haploid set has one copy of each type of chromosome. Two chromosomes of a particular type present in a diploid set are called **homologous chromosomes.**

FIGURE 4.1
Sexual reproduction in barnacles, as shown by two adjacent individuals mating.
© Heather Angel/Biofotos.

· The zygote increases in cell number by the process of cell division called **mitosis.** In this process, the chromosomes are duplicated so that they remain at the 2N number in the daughter cells. This type of cell division continues until reproductive maturity, at which time some cells undergo **meiosis.** Meiosis is a second type of cell division that produces female and male gametes, each having a chromosomal number of N. This process is known as **gametogenesis** in animals and as **sporogenesis** in plants. The female gametes (eggs) and the male gametes (sperm in animals or pollen in plants) then unite to form progeny zygotes by the process of **fertilization.** These new zygotes mature into either males or females, and the whole cycle starts over.

Each species has a characteristic number of chromosomes; that is, virtually all individuals of a given species have the same chromosome number. Generally, this number is expressed in the diploid, or 2N, form. Table 4.1 lists a number of organisms used for genetic research, including corn, with a diploid number of 20 chromosomes, and pink bread mold, with a haploid number of 7 (for this organism, the haploid phase is the major one). The fruit fly has a 2N number of only 8, while carp (and some ferns) have over 100 chromosomes. Notice that closely related species, such as humans and chimpanzees or horses and donkeys, have similar chromosome numbers. In Hymenoptera (bees and wasps), the females are diploid (2N) and the males are haploid (N), resulting in the term **haplodiploid** to describe such organisms. Certain species have some very small chromosomes, called **supernumerary** or **B chromosomes,** that can vary in number among individuals. These chromosomes are generally thought to be

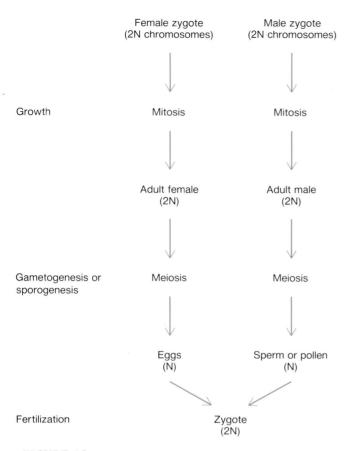

FIGURE 4.2
Outline of growth and reproduction in a sexual organism, indicating the number of chromosomes, diploid (2N) or haploid (N), at various stages.

Labels in figure:
Female zygote (2N chromosomes)   Male zygote (2N chromosomes)
Growth — Mitosis — Mitosis
Adult female (2N)   Adult male (2N)
Gametogenesis or sporogenesis — Meiosis — Meiosis
Eggs (N)   Sperm or pollen (N)
Fertilization — Zygote (2N)

Common Name (Scientific Name)	Diploid Chromosome Number
Mammals	
Human (*Homo sapiens*)	46
Chimpanzee (*Pan troglodytes*)	48
Dog (*Canis familiaris*)	78
Cat (*Felis catus*)	38
Horse (*Equus caballus*)	64
Donkey (*Equus asinus*)	62
House mouse (*Mus musculus*)	40
Other Animals	
Chicken (*Gallus domesticus*)	~78
Frog (*Rana pipiens*)	26
Carp (*Cyprinus carpio*)	104
Fruit fly (*Drosophila melanogaster*)	8
Plants	
Corn (*Zea mays*)	20
Tobacco (*Nicotiana tabacum*)	48
Pine (*Pinus* species)	24
Garden pea (*Pisum sativum*)	14
Fungi	
Yeast (*Saccharomyes cerevisiae*)	17*
Black bread mold (*Aspergillus nidulans*)	8*
Pink bread mold (*Neurospora crassa*)	7*

TABLE 4.1 *Organisms and Their Diploid Chromosome Numbers*

*Haploid number

devoid of genes, although in some instances, supernumerary chromosomes affect fertility. Other species, such as many birds, have a number of very small chromosomes called **microchromosomes.** Although microchromosomes are hard to count exactly because of their number and small size, they are thought to carry genes and to be constant in number, just as the larger chromosomes are.

CHROMOSOMAL MORPHOLOGY

In order to further understand chromosomes and their function, we need to be able to discriminate among different chromosomes. First, chromosomes differ greatly in size. Between organisms, the size difference can be over 100-fold, while within a species, some chromosomes are often ten times (or more) as large as others. In a species **karyotype,** a pictorial or photographic representation of all the different chromosomes in an individual, chromosomes are usually ordered by size and numbered from largest to smallest.

Second, chromosomes may differ in the position of the **centromere,** the place on the chromosome where spindle fibers are attached during cell division. The centromere is identified as a large constriction where the chromosome appears to be pinched. (The term **kinetochore** is used to designate a structure within the centromeric region.) In general, if the centromere is near the middle, the chromosome is **metacentric;** if the centromere is toward one end, the chromosome is **acrocentric;** and if the centromere is very near the end, the chromosome is **telocentric** (see figure 4.3). The centromere divides the chromosome into two **arms,** so that, for example, an acrocentric chromosome has one short and one long arm, while a metacentric chromosome has arms of equal length. All mouse chromosomes are telocentric, while human chromosomes include both metacentrics and acrocentrics, but no telocentrics. Table 4.2 describes the karyotypes of eight common domestic animals to illustrate the variation in gross chromosomal morphology that occurs among these species. In addition to a centromere, there may also be a **secondary constriction** on particular chromosomes that, if located near the end, results in a chromosomal **satellite.**

		Autosomal Pairs		Sex Chromosomes	
	Diploid number (2N)	Metacentrics	Acrocentrics or telocentrics	X	Y
Cat (*Felis catus*)	38	16	2	M	M
Dog (*Canis familiaris*)	78	0	38	M	A
Pig (*Sus scrofa*)	38	12	6	M	M
Goat (*Capra bircus*)	60	0	29	A	M
Sheep (*Ovis aries*)	54	3	23	A	M
Cattle (*Bos taurus*)	60	0	29	M	M
Horse (*Equus caballus*)	64	13	18	M	A
Donkey (*Equus asinus*)	62	24	6	M	A

M = metacentric
A = acrocentric
From F. Nicholas, *Veterinary Genetics*. Copyright © Oxford University Press, Oxford, England. Reprinted by permission.

The secondary constrictions usually represent nucleolus-organizing regions. The ends of a chromosome are called **telomeres** and generally appear to have DNA sequences different from the rest of the chromosome.

Third, chromosomes may be identified by regions that stain in a particular manner when treated with various chemicals. In fact, the term chromosome literally means "colored body." Several different chemical techniques are used to identify certain chromosomal regions by staining them so that they form chromosomal **bands.** For example, darker bands are generally found near the centromeres or on the ends (telomeres) of the chromosomes, while other regions do not stain as strongly. The positions of the dark-staining, or **heterochromatic** regions (also called heterochromatin) and the light-staining, or **euchromatic** regions (called euchromatin) generally remain constant in different cells or individuals of a given species and are therefore useful in identifying particular chromosomes.

Euchromatic regions often undergo a regular cycle of contraction and extension. The two types of heterochromatin are **constitutive** and **facultative.** Constitutive heterochromatin is a permanent part of the genome and is not convertible to euchromatin. On the other hand, facultative heterochromatin consists of euchromatin that takes on the staining and compactness characteristics of heterochromatin during some phase of development. Evidence suggests that the constitutive heterochromatin is genetically inactive and that the euchromatin contains most of the genes. We should note that the G, Q, and R bands (Giemsa, quinacrine, and reverse bands, respectively), which are chromosomal bands that are observed using different chemical techniques, are not associated with centromeric heterochromatin.

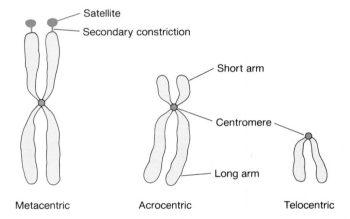

FIGURE 4.3
The three major types of chromosomes as they appear in human karyotypes: metacentric, acrocentric, and telocentric. Satellites may be present on any of the chromosome types.

In humans, a system is used for identifying chromosomes based on chromosomal size, position of the centromere, and banding patterns (see figure 4.4). First, the nonsex chromosomes (autosomes) are numbered 1 to 22 on the basis of length, with the X and Y chromosomes identified separately. Second, the short arm of a chromosome is indicated by p and the long arm by q. Finally, each arm is first divided into regions and then identified by bands. For example, 9q34 indicates the long arm of chromosome 9, region 3, and band 4. This is, in fact, the location of the gene for ABO blood group types.

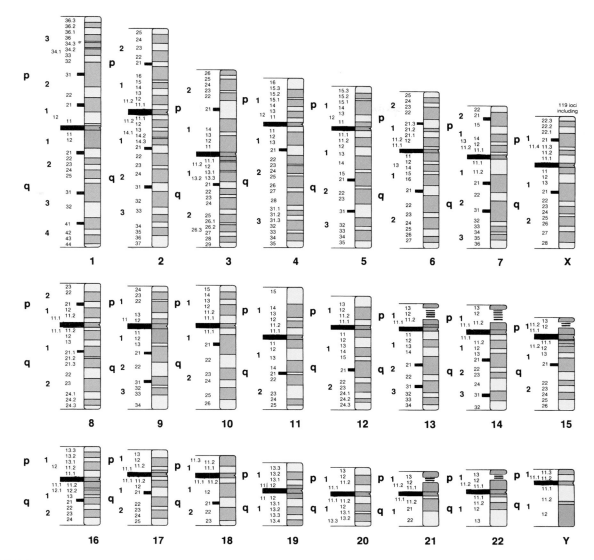

FIGURE 4.4
A schematic representation of human chromosomes, in which the
heavy black bar indicates the centromeric region, and p and q indicate
the short and long arms, respectively. Heterochromatic regions are
shown in blue, while yellow and green indicate other banded regions.

In 1956, Joe-Hin Tjio and Albert Levan determined
the correct diploid number of human chromosomes (46).
(The number had been erroneously thought to be 48 for
over thirty years.) Since then, knowledge of human cytogenetics has increased remarkably. These advances resulted mainly from the development of new techniques—
ways to culture dividing cells, to swell cells using low
osmotic solutions, and to break open cells and spread out
the chromosomes. Various new staining techniques have
allowed each chromosome, and even small parts of each
chromosome, to be identified. Using such biochemical
techniques, it is now possible to locate particular genes to
certain positions on chromosomes.

Figure 4.5 gives a schematic representation of the X
chromosome in *Drosophila melanogaster*, an organism
with a detailed banding pattern. The chromosomes in
the salivary glands of *Drosophila* and other diptera are
extremely large, thus easier to study than other chromosomes. These giant chromosomes, called **polytene chromosomes,** are extended over one hundred times the length
of regular chromosomes and include many parallel replicates of the same chromosome held tightly together. As
a result, more than five thousand bands have been identified in *D. melanogaster*, and more than one thousand of
them are on the X chromosome alone. Notice in figure 4.5

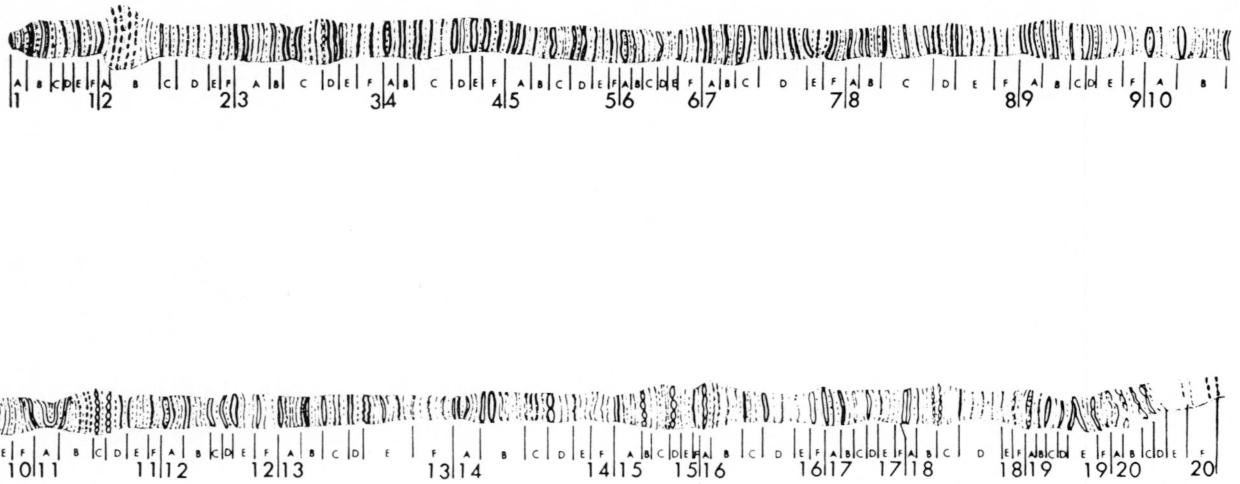

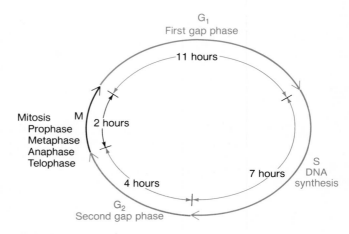

FIGURE 4.5
The chromosome map of the X chromosome of *D. melanogaster*. It has been arbitrarily divided into twenty numbered regions, each beginning with a dark band.

From Modern Genetics *by Ayala and Kiger, p. 141, Benjamin/Cummings Publishers, 1980. (From* Handbook of Genetics, *Vol. 3, by R. C. King, Plenum Press, New York, 1975.)*

the variety in the size of the bands and in their position in relation to each other, allowing identification of particular chromosomal regions. This chromosome is telocentric, having the centromere in region 20, on the far right of the drawing. The gene that causes the white-eyed phenotype, which we will discuss later, is at the very left tip of region 1, at the other end from the centromere. This was the first chromosome on which the location of genes was mapped, as we will discuss in the next chapter.

> **T**he diploid number of chromosomes varies for different species. Specific chromosomes can be identified according to size, position of the centromere, and banding patterns.

■

MITOSIS

In order to understand genetics, we must understand the process by which hereditary information is passed from cell to cell and from parents to offspring. Although nineteenth century researchers documented the behavior of chromosomes in cell division, they did not realize the genetic implications of chromosomal duplication and movement. Only after the rediscovery of Mendel's principles was the connection made between the earlier chromosomal observations and the principles of segregation and independent assortment.

We will first trace the steps of mitosis, the process of cell division in which daughter cells are produced that have chromosomal numbers identical to the parental cell. Mitosis is part of the total **cell cycle** for cells undergoing division. In figure 4.6 notice that the period of mitosis

FIGURE 4.6
A schematic representation of the cell cycle for human white blood cells, showing the time involved for each phase.

(designated M) is a relatively small part of the total cell cycle. (To lend some perspective, the times given in each stage are for dividing human white blood cells. The duration of the different stages varies considerably, depending upon the organism, cell type, temperature, and other factors.)

The rest of the cell cycle, called **interphase,** is divided into three parts, termed G_1, S, and G_2. Interphase is generally a time of high metabolic activity, during which DNA, the hereditary material, is synthesized and replicated. Some mature cell types, such as nerve cells, remain in interphase and never divide further. After a given mitosis is completed, there is an initial time gap, called G_1, in which cells grow, metabolize, and perform their designated functions for the organism. Following the G_1 is a

Chapter 4

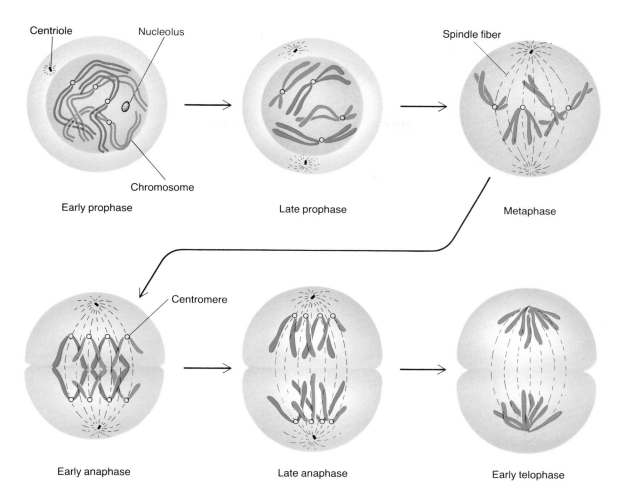

FIGURE 4.7
The behavior of two pairs of chromosomes in mitosis, where pair one
is acrocentric (red) and pair two is metacentric (blue).

period of DNA synthesis, designated S, in which the DNA
is replicated (we will discuss this process in detail in
chapter 8). After synthesis, another gap phase, G_2, occurs,
followed by the next mitotic division. When an organism
is growing, this cycle repeats itself many times, eventually
resulting in an individual with billions of cells.

Mitosis itself can be divided into four different stages,
listed in chronological sequence as: **prophase, metaphase,
anaphase,** and **telophase.** These stages actually merge one
into another, but each has definite characteristics, particularly in relation to chromosomal behavior, that we can
use to identify them. Let us go through these phases to
understand the elements of mitosis and how cells produce
identical daughter cells (see figure 4.7).

PROPHASE

At the beginning of mitosis (the end of G_2 interphase),
the chromosomes coil and condense, thus becoming visible
under the light microscope. The chromosome has two
copies, or **chromatids,** at this stage. The two sister chromatids are attached to each other at the centromere. The
chromatids are identical and are the result of the DNA
replication in the earlier S phase of the cell cycle. Because
the two sister chromatids are attached at the centromeric
region, they are considered one chromosome at this stage.

Also during prophase, the **nucleolus,** the largest cell
organelle besides the nucleus, disappears and the nuclear
membrane begins to break down. A new structure, the
spindle apparatus, which is composed of a series of **spindle
fibers** that stretch to the poles of the cell, also appears (see
figure 4.8). The spindle structure is fundamental to the
proper movement of the chromosomes to the two poles in
the latter part of mitosis.

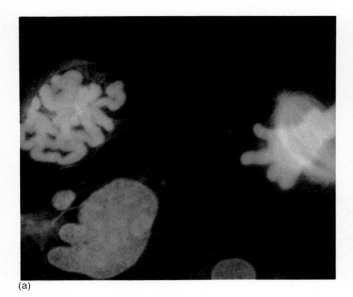 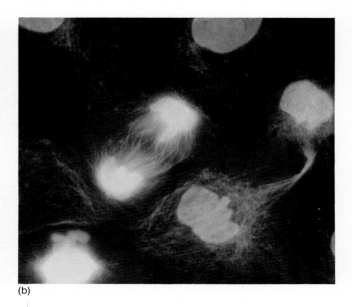

(a)   (b)

FIGURE 4.8

Photo (*a*) shows two cells, one in prophase and the other in metaphase. Photo (*b*) shows one cell in anaphase and another in very late telophase. The images in both slides are stained for immunofluorescence microscopy. The green is from an antibody directed against microtubules, which compose the spindle fibers, and the red is a chemical called propidium iodide, which specifically stains chromosomes.

Courtesy of Mark S. Radinsky and J. Richard McIntosh, University of Colorado, Boulder.

METAPHASE

The nuclear membrane completely disappears during metaphase, leaving the chromosomes free in the cytoplasm. The spindle fibers attach to the centromeres of the chromosomes, and the chromosomes gradually change their orientation, moving to the **metaphase** (or **equatorial) plate,** an imaginary plane midway between the two poles. Generally, the chromosomes are most condensed at this stage of mitosis, making it a good stage for preparation of karyotypes. Because sister chromatids are attached at the centromeres, a chromosome at metaphase appears either as an X or a V, respectively, depending upon whether it is metacentric or telocentric.

ANAPHASE

Anaphase, nearly always the shortest period of mitosis, begins with the separation of the centromeres. As a result, each sister chromatid is now called a daughter chromosome. The spindle fibers contract and cause the movement of the centromeres to the poles, with the two identical daughter chromosomes moving to opposite poles. A chromosome appears as a V, J, or I at this stage, depending upon whether it is metacentric, acrocentric, or telocentric, respectively.

TELOPHASE

In telophase, the last stage of mitosis, the two sets of daughter chromosomes are pulled together at opposite poles where they begin to uncoil. The nuclear membrane reappears and encloses the chromosomes, and the nucleoli are re-formed.

Next, in animals, the cell membrane develops a furrow; in plants, a cell plate develops that divides the cell in two. This splits the two sets of daughter chromosomes and the cytoplasm between the two daughter cells. The daughter cells have chromosome numbers identical to the original cell.

M itosis, cell division that results in the production of two identical daughter nuclei, is divided into four stages: prophase, metaphase, anaphase, and telophase.

■

MEIOSIS

Before describing meiosis, we will summarize why it is fundamental for maintaining the genetic state of an organism and for generating new genetic types. First, meiosis allows the conservation of chromosome number in sexually reproducing species. If not reduced via meiosis prior

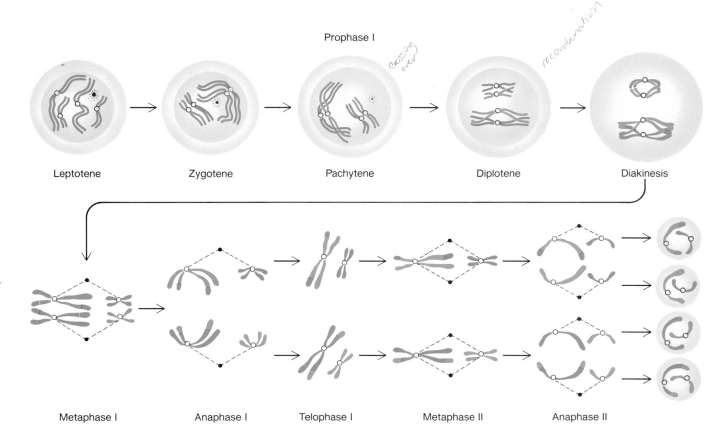

Prophase I

Leptotene   Zygotene   Pachytene   Diplotene   Diakinesis

Metaphase I   Anaphase I   Telophase I   Metaphase II   Anaphase II

FIGURE 4.9

A schematic representation of meiosis for two chromosomes. Meiosis I results in separation of homologous chromosomes, and meiosis II results in separation of duplicated chromosomes.

From F. J. Ayala and J. A. Kiger, Jr., Modern Genetics, *2d ed. Copyright © 1984 Benjamin/Cummings Publishing Co., Menlo Park, CA. Reprinted by permission.*

to fertilization, the chromosome number would be doubled. Second, meiosis permits the different maternal and paternal chromosomes to combine in each gamete. And finally, the presence of crossing over in meiosis I allows the exchange of genetic material on homologous chromosomes and can result in new associations of alleles at different genes on a chromosome. (See chapter 5 for a discussion of crossing over.)

Meiosis produces cells with half the parental number of chromosomes through a process involving two cell divisions, meiosis I and meiosis II. **Meiosis I** is also called **reductional division,** because the number of chromosomes is reduced to the haploid number. In meiosis I, the sister chromatids remain attached to each other. **Meiosis II** is called **equational division** and is much like mitosis in that the centromeres on sister chromatids separate and the chromosome number remains unchanged (see figure 4.9).

MEIOSIS I

As with mitosis, both meiotic divisions can be divided into four stages, based on the position of the chromosomes and other characteristics. The interphase period before meiosis I is similar to that in mitosis, having an S period for DNA synthesis.

Prophase I

The most complex of the whole process of meiosis, **prophase I,** is itself broken down into five different parts (figure 4.9). The first part, called **leptotene** (thin-thread stage), is marked by the appearance of the chromosomes as long threads when seen through a light microscope and is much like the start of prophase in mitosis.

However, in the next stage, **zygotene** (yolk-thread stage), unlike in mitosis, homologous chromosomes pair side by side and gene by gene with each other. The process of the lateral association of homologues, called **synapsis,**

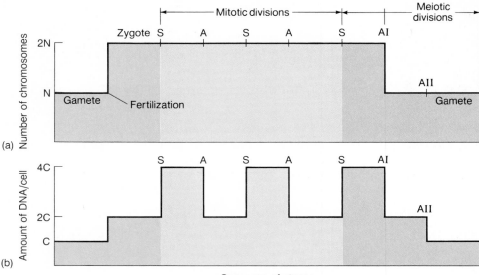

FIGURE 4.10
(a) The numbers of chromosomes per cell in different stages of cell division. (b) The amount of DNA per cell at the various stages of cell division. S indicates DNA synthesis and A indicates anaphase.

occurs during zygotene and is a key difference between meiosis and mitosis. When the two homologous chromosomes (consisting of four chromatids) are paired, the structure is called a **bivalent.** The next, or third stage of prophase I, **pachytene** (thick-thread stage), consists of a shortening and thickening of the bivalents and is the stage during which synapsis is complete. During this stage, portions of homologous chromosomes may be exchanged, a phenomenon called **crossing over** or **recombination.** The site of crossing over appears to be the **synaptonemal complex,** a structure made of protein and DNA that is found between the synapsed homologues. We will discuss the consequences of this phenomenon in chapter 5.

In the fourth stage, **diplotene** (double-thread stage), the homologues begin to separate, particularly in the region surrounding the centromere. The sister chromatids remain attached at the centromeric region. In addition, the homologous chromosomes generally have several areas, called **chiasmata,** in which they are still close or in contact. Chiasmata are the physical evidence that recombination occurred earlier when the homologues were synapsed. Generally, there is at least one chiasma per chromosome arm, but in long chromosomes, there may be several.

The last stage of prophase I, **diakinesis,** is characterized by shortened chromosomes and the terminalization of the chiasmata; that is, the chiasmata appear to move to the ends of the chromosomes. As in mitosis, at the end of prophase, the nucleolus dissolves and the nuclear membrane disappears.

Metaphase I

In metaphase I, the chromosomes move to the equatorial plate and become attached to the spindle fibers. The bivalents orient so that one centromere is on each side of the metaphase plate, with the chiasmata lined up along it. As

a result, metaphase I is different from mitotic metaphase, in which the homologous chromosomes were not paired and all centromeres were along the metaphase plate.

Anaphase I

At the point known as anaphase I, the two centromeres in each bivalent move apart, so that one of each homologous chromosome pair goes toward each pole. As the homologues move apart, the chiasmata completely terminalize and then disappear. In anaphase I, unlike mitotic anaphase, the sister chromatids stay together.

Telophase I

In some organisms, after the chromosomes have migrated to the poles, the nuclear membrane forms around them and the cell divides into two daughter cells, just as in mitosis. However, the exact cytological details of telophase I vary, particularly in plants.

MEIOSIS II

Between meiosis I and meiosis II there is usually only a short interphase called **interkinesis.** No DNA synthesis takes place during this period; at this point, each cell contains only a haploid complement (N) of chromosomes. (Note that each chromosome contains two sister chromatids.) The next stage is **prophase II,** which unlike prophase I, is quick, uncomplicated, and generally similar to mitotic prophase. In **metaphase II** (see figure 4.9), the centromeres attach to spindle fibers and migrate to the metaphase plate. At the start of **anaphase II,** as in mitotic anaphase, the centromeres separate and one centromere begins to move with one daughter chromosome to one pole,

Chapter 4

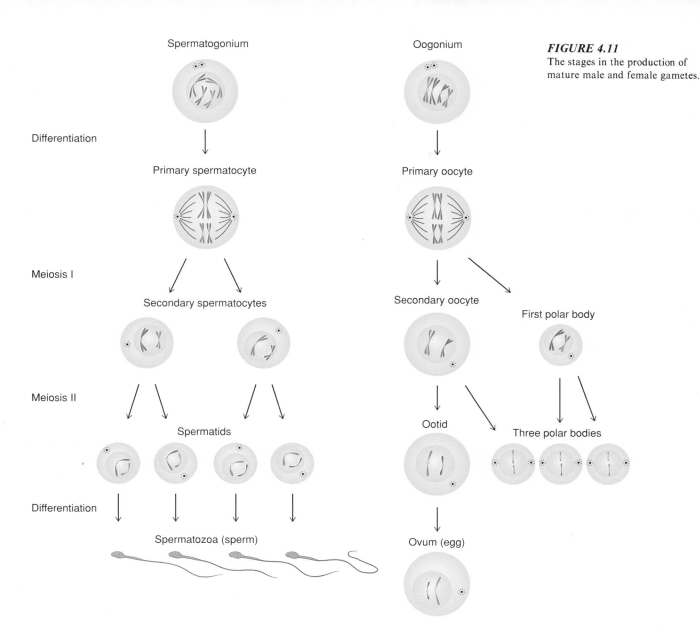

Spermatogonium

Differentiation

Primary spermatocyte

Meiosis I

Secondary spermatocytes

Meiosis II

Spermatids

Differentiation

Spermatozoa (sperm)

Oogonium

Primary oocyte

Secondary oocyte

First polar body

Ootid

Three polar bodies

Ovum (egg)

while the other centromere and chromosome go to the opposite pole. There are N chromosomes at each pole (each chromosome is composed of a single chromatid) when the nuclear membrane forms around them in **telophase II.**

Before we continue, it is useful to compare the changes in chromosome number and amount of DNA that occur during the mitotic and meiotic cycles. First, as we have already discussed, homologous chromosomes synapse in meiosis I, allowing recombination and precise separation of homologues, but this does not occur in mitosis. Second, at the beginning of both mitosis and meiosis, there are 2N chromosomes in each cell (see figure 4.10a). Each chromosome is composed of two sister chromatids, so that the amount of DNA is four times the haploid amount (indicated as 4C in figure 4.10b). At the end of mitosis, there are still 2N chromosomes, but the amount of DNA has

been reduced to 2C per cell. At the end of meiosis I (indicated by AI, or anaphase I, in figure 4.10), the number of chromosomes has been reduced to N and the amount of DNA to 2C. As in mitosis, meiosis II does not change the number of chromosomes, but the amount of DNA is halved to C as the sister chromatids separate.

SPERMATOGENESIS AND OOGENESIS

The sequence of events in meiosis that we have described is similar in the production of both male and female gametes; however, there are some differences (see figure 4.11). The production of male gametes in animals, termed **spermatogenesis,** takes place in the **testes,** the male reproductive organs. The process begins with the growth of an undifferentiated diploid cell called a spermatogonium, which then differentiates into a primary spermatocyte.

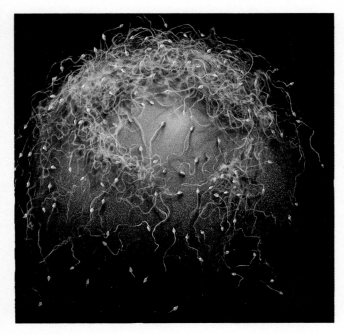

FIGURE 4.12
Sea urchin sperm on a sea urchin egg. Only one sperm will fertilize this egg.
© *Francis Leroy, Biocosmos/Science Photo Library/Photo Researchers, Inc.*

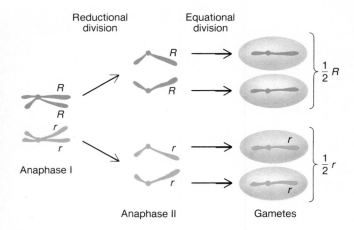

FIGURE 4.13
The chromosomal basis of Mendel's principle of segregation operating in an *Rr* heterozygote. As the diagram shows, ½ *R* gametes and ½ *r* gametes are produced.

This cell, in turn, undergoes the first meiotic division (the reductional division), becoming two haploid secondary spermatocytes. After meiosis II, these cells each divide into two haploid spermatids. The last step in spermatogenesis is the differentiation of the spermatids into new sperm cells with long tails (figure 4.12).

Oogenesis, the production of female gametes in animals, occurs in the **ovaries,** the female reproductive organs. The process is parallel to that of spermatogenesis in that it also begins with a diploid cell, or oogonium. However, the cytoplasm is concentrated into one of the cell products, the primary and secondary oocytes, while the other cell products (first and second polar bodies) receive only a small amount of cytoplasm. Meiosis takes place near the cell membrane, allowing the formation of polar bodies with little cytoplasm. The polar bodies remain on the surface of the egg where they eventually disintegrate. Therefore, only one mature haploid egg is produced in oogenesis, while four mature haploid sperm are produced in spermatogenesis. The concentration of cytoplasm in one egg cell provides nourishment for the developing embryo after fertilization.

Spermatogenesis is continuous in adult animals that reproduce all year round, such as humans, and seasonal in animals that breed seasonally. The capacity for sperm production is extremely high in most animals, with adult males producing many millions of sperm in their lifetimes. On the other hand, egg production in females is generally much lower; for example, a human female produces around five hundred mature eggs in her lifetime. All these eggs develop into primary oocytes before birth and are retained as such in a resting stage until puberty, when they begin maturing into eggs one by one and are ovulated each month. Many other eggs never mature and are reabsorbed.

In plants, the formation of male gametes, or pollen, is known as **microsporogenesis.** It is similar to spermatogenesis in animals in that meiosis gives rise to four haploid cells, all of an equal size and all potentially functional. **Megasporogenesis,** the formation of eggs in plants, is similar to the meiotic process described for animals. However, the products vary considerably among species and are complicated by double fertilization. (It is best to consult a botany text for more details.)

> *M*eiosis is composed of one duplication of DNA and two cell divisions. Meiosis I, the reductional division, results in the separation of homologous chromosomes and the halving of chromosome number. Meiosis II is similar to mitosis and results in the separation of the sister chromatids, with no reduction in chromosome number. In animals, mature male and female gametes are produced by spermatogenesis and oogenesis, respectively.

■

GENES AND CHROMOSOMES

As we mentioned before, several early researchers pointed out the parallelism in the behavior of genes and chromosomes. Now that we have discussed meiosis, we can more precisely show how the behavior of genes and chromosomes is related. For example, figure 4.13 illustrates the principle of segregation in a heterozygote, *Rr*. In reductional division (meiosis I), the two homologues containing different alleles separate into two different cells.

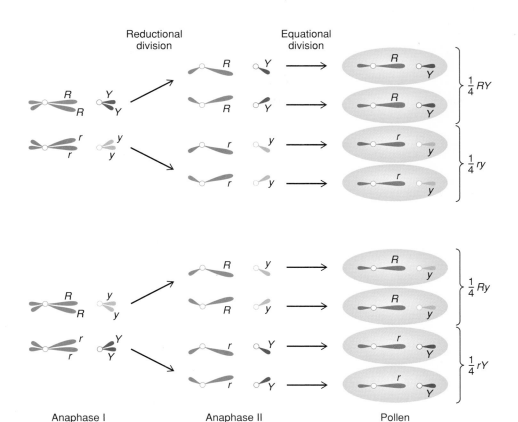

$\frac{1}{4}RY$

$\frac{1}{4}ry$

$\frac{1}{4}Ry$

$\frac{1}{4}rY$

FIGURE 4.14
The chromosomal basis of Mendel's principle of independent assortment in an *RrYy* double heterozygote. Equal proportions of the four types of gametes are produced. The top and bottom halves of the figure are two equally likely outcomes of independent assortment.

Each of these cells then divides into two cells, which contain the same alleles as the cell produced from meiosis I. As a result, half the gametes are *R* and half are *r*. In other words, equal segregation of alleles into gametes is really the result of reductional division in meiosis.

The chromosomal explanation for independent assortment is given in figure 4.14. Let us examine gamete production in the F_1 double heterozygote *RrYy,* where the two genes are on different chromosomes. Two outcomes are equally likely: Either the chromosomes containing *R* and *Y* go to the same pole and the ones containing *r* and *y* go to the other pole, or the *R* and *y* go to one pole and the *r* and *Y* go to the other pole. These two alternatives are equally possible because they result from the chance placement of the centromeres on one side or the other of the metaphase I plate. Each of the four types of cells divides in meiosis II, so that overall, the four gametic types occur in equal proportions. Therefore, the principle of independent assortment also generally results from the behavior of chromosomes in meiosis I. (We should note that this assumes that no recombination takes place between the loci and their centromeres. See chapter 5 for a discussion of recombination.)

The parallel behavior of genes and chromosomes was strong support that genes actually resided in chromosomes, but it was not direct or definitive evidence that specific genes and specific chromosomes were connected. The first evidence of this sort was provided by the experiments of Thomas H. Morgan who in 1910 described X-linked inheritance in *Drosophila.* Morgan had begun to maintain *Drosophila* cultures shortly after the turn of the century. A momentous event occurred when a white-eyed male appeared in a culture of all wild-type (red-eyed) flies. Using this mutant, Morgan experimentally confirmed that genes are on chromosomes.

Morgan carried out several experiments designed to understand the observed association between the white-eyed mutant and its sex. For example, he crossed red-eyed females and white-eyed males to produce F_1 progeny that were all red-eyed. However, when he crossed the F_1 individuals, the F_2 had the expected 3:1 ratio, but all the mutants were males (see figure 4.15). Furthermore, the reciprocal cross, white-eyed females and red-eyed males, gave different results (figure 4.16), with half the F_1 being white-eyed (all of which were males) and half red-eyed (all females). The F_2 progeny from a cross of these F_1 individuals were half white-eyed and half wild-type.

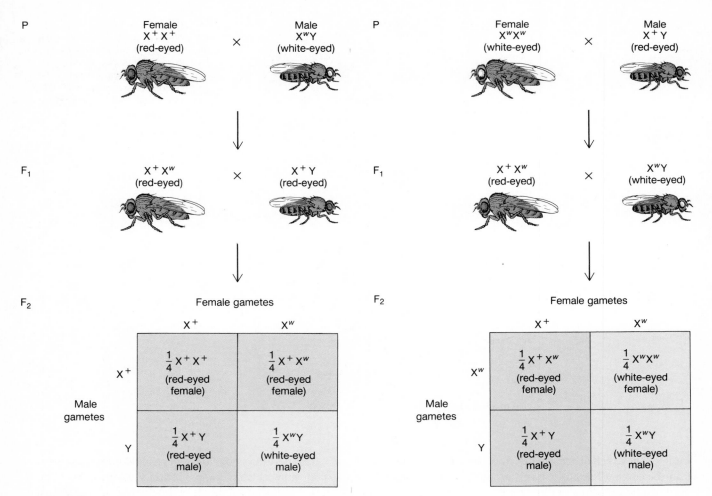

FIGURE 4.15
Results of a cross between a red-eyed female and a white-eyed male to produce an F₂ phenotypic ratio that is sex dependent.

FIGURE 4.16
Results of a cross between a white-eyed female and a red-eyed male to produce an F₂ that is different from the reciprocal cross in figure 4.15.

It was known that male and female *Drosophila* differ with respect to their chromosomes; females have two X chromosomes and males have one X and one Y. Morgan suggested that the results of his crosses could be explained if the eye color gene was on the X chromosome and not on the Y chromosome. Obviously, the explanation that the white-eyed mutant was X-linked was consistent with the observations from his experimental crosses.

Although the association of the *white* alleles with the X chromosome appeared to be convincing proof that genes were on chromosomes, another experiment was necessary to exclude all other explanations. Calvin Bridges, a student of Morgan, provided the crucial evidence in 1916, also using the *white* gene. When white-eyed females and red-eyed males were crossed, as in figure 4.17, approximately one in two thousand F₁ offspring had unexpected eye color; that is, they were either white-eyed females or

red-eyed males. Bridges suggested that these exceptional flies were the product of an abnormal separation of chromosomes in meiosis called **nondisjunction.** In this case, nondisjunction occurred in females such that the two X chromosomes did not separate properly; instead, both went to the same pole, resulting in eggs with either two X chromosomes or no X chromosomes (see figure 4.17).

If nondisjunction did occur, and assuming normal meiosis in males, some offspring would be X^wX^wY, with three sex chromosomes. These would be the unexpected white-eyed females, because there is no wild-type allele. Other flies (the red-eyed males) would be X^+, with only one sex chromosome. (We will discuss the chromosomal basis of sex determination in *Drosophila* later in this chapter.) In this explanation, the white-eyed females inherit both of their X chromosomes from their mother, and the red-eyed males inherit their X chromosome from their father. Bridges corroborated his hypothesis by examining

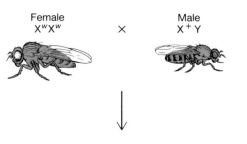

Female
X^w X^w   ×   Male
X^+ Y

FIGURE 4.17
The results of a cross between white-eyed females and red-eyed males. The * indicates gametes produced by rare nondisjunction events.

Female gametes

	X^w	X^w X^w *	— *
$\frac{1}{2}$ X^+	X^+ X^w (red-eyed female)	X^+ X^w X^w (dies)	X^+ (red-eyed male)
$\frac{1}{2}$ Y	X^w Y (white-eyed male)	X^w X^w Y (white-eyed female)	Y (dies)

Male gametes

the chromosomes of the unexpected flies. He found that the exceptional females were XXY and the exceptional males were only X. In other words, Bridges's discovery that an abnormal genetic condition and an abnormal chromosomal condition directly corresponded gave convincing proof that genes were indeed on chromosomes.

*T*he parallel behavior of genes and chromosomes provides a mechanistic explanation for the principles of segregation and independent assortment. The observation of this parallel behavior was also the first indication that genes were on chromosomes. Definitive proof resulted from experiments with X-linked genes in *Drosophila* and from the confirmation that individuals with exceptional eye color were the result of meiotic mistakes that caused abnormal chromosomal numbers.

∎

CHROMOSOMAL CHANGES

Until now, we have been discussing individuals with a normal set of chromosomes. However, an organism's chromosomal constitution may change, or mutate, resulting in differences among individuals of a given species. Chromosomes can change in two basic ways—by altering their structure or by altering their number (see table 4.3). Both types of alterations may have other consequences besides their immediate effect on the chromosomes. For example, individuals heterozygous for chromosomes with different structures often have lowered fertility, and individuals with altered numbers of chromosomes may be inviable or sterile.

*T*ABLE 4.3   *Types of Chromosomal Changes*

Changes in Structure

Duplications
Deletions (or deficiencies)
Inversions
Translocations

Changes in Number

Euploidy level
  Autopolyploids
  Allopolyploids
Aneuploidy
  Monosomies (2N − 1)
  Trisomies (2N + 1)

CHANGES IN CHROMOSOMAL STRUCTURE

The four possible types of changes in chromosomal structure are duplications, deletions (or deficiencies), inversions, and translocations. When breaks occur in chromosomes, any two broken chromosomal ends may reunite. Chromosomal changes are often the result of chromosomal breaks in which the same broken ends do not reunite. Generally, such chromosomal mutations occur infrequently, but some researchers have estimated that more than one in a thousand new gametes may have some type of chromosomal mutation. Note that these chromosomal changes result from chromosomal breaks and abnormal

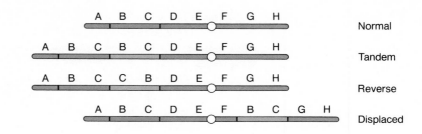

FIGURE 4.18
A normal chromosome (top), followed by three types of duplication that can result, depending on the position and order of the duplicated segment (shown in green).

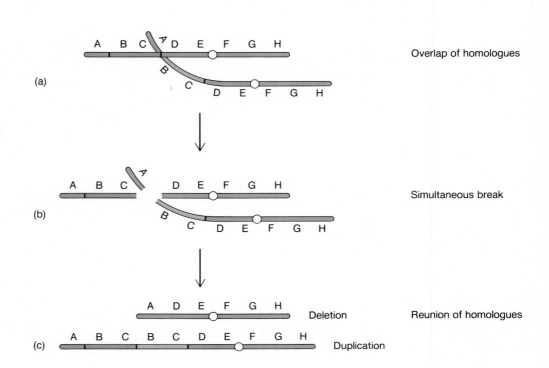

FIGURE 4.19
(*a*) Two overlapping homologues (*b*) break simultaneously, (*c*) producing a deletion and a duplication.

rejoining. The ensuing consequences, which we will discuss in the following section, are the result of the gene-for-gene synapsis of normal and abnormal chromosomes and the subsequent problems that occur in anaphases I and II.

Duplications

When a chromosomal segment is represented twice, it is called a **duplication.** We can categorize duplications by the position and order of the duplicated region (see figure 4.18). First, the duplication may be adjacent to the original chromosomal region. When this occurs, the order may either be the same as the original order, called a **tandem duplication,** or the opposite order, called a **reverse duplication.** Second, the duplicated region may not be adjacent to the original segment, resulting in a **displaced duplication.** In this case, the displaced duplication may still be on the same chromosome, as illustrated in figure 4.18, or it may be on another chromosome.

One possible way a tandem duplication can be generated is illustrated in figure 4.19. It is assumed that the homologues overlap and that there are simultaneous breaks in the two homologous chromosomes at *different* points. If the different homologues reunite, one chromosome will have a tandem duplication and the other a deletion of the duplicated region (figure 4.19*c*). In this case, the duplication and the deletion are reciprocals of each other. In chapter 5, we will discuss another mechanism (unequal crossing over) by which small duplications (and deletions) can be generated.

When an individual is heterozygous for a duplication and a normal chromosome, the duplicated region does not have a homologous segment to pair with in meiosis. As a result, a loop of the duplicated region may develop (see figure 4.20*a*). In some cases, part of the chromosome may bend back and join with the duplicated sequence on the same chromosome. Individuals having two chromosomal types such as this are called **heterokaryotypic.**

Chapter 4

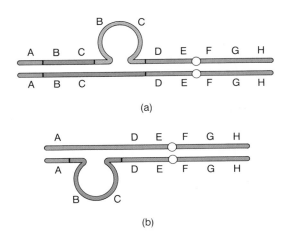

(a)

(b)

FIGURE 4.20
Pairing of chromosomes in individuals heterokaryotypic for (a) duplication or (b) deletion.

Individuals that are either heterozygous or homozygous for small duplicated segments may be viable, although they often exhibit some phenotypic effects (see the discussion of the *Bar* eye mutations in *Drosophila* in chapter 5). If individuals are viable, there is a potential for further evolutionary change in these extra genes. In fact, it is thought that this happened with the different globin genes, the genes that code for the components of the protein hemoglobin. These genes may have descended from an ancestral gene that was duplicated, and then the duplicate copies diverged in their function.

Deletions

A missing chromosomal segment is termed a **deletion** or a **deficiency.** We showed one way a deletion can occur in figure 4.19. This type of deletion, where an internal part of the chromosome is missing, is called an **interstitial deletion.** If there is only one break and the homologue fails to rejoin, a **terminal deletion** can occur. In this case, the tip of the chromosome is usually lost in cell division because it does not have a centromere. Interstitial and terminal deletions are illustrated in figure 4.21.

When deletions are homozygous, they are often lethal, because essential genes are missing. Even when heterozygous, lethals can cause abnormal development. A well-known example in humans is the deletion of one-half of the short arm of chromosome 5 (5p), which when heterozygous, causes the *cri du chat* (cry-of-the-cat) *syndrome.* Infants with this syndrome generally have a characteristic high-pitched, catlike cry as well as microcephaly (small heads) and severe mental retardation. Generally, they die in infancy or early childhood.

In addition, deletion heterozygotes usually show abnormal chromosomal pairing in meiosis. Because the normal chromosome does not have a homologous region

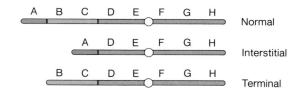

FIGURE 4.21
A normal chromosome (top), followed by two types of deletions that may result, depending on the positions of the deleted segment.

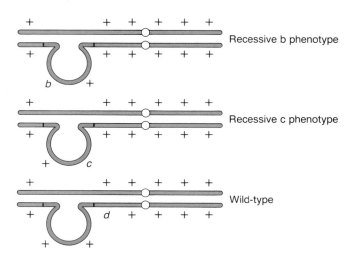

FIGURE 4.22
Deletions in heterozygotes that have recessive mutants at the *B, C,* or *D* gene on the normal chromosome.

to pair with, a deletion loop is formed (figure 4.20*b*). This phenomenon may be seen in meiotic chromosomes or in the polytene chromosomes of *Drosophila* and a few other organisms.

Several other characteristics are useful in identifying deletions. First, deletions, unlike other mutations, generally do not revert, or mutate back, to the wild-type chromosome. Second, in deletion heterozygotes, recessive alleles on the normal chromosome are expressed because the deletion chromosome is missing the homologous region. Expression of recessive alleles in such cases, called **pseudodominance,** is useful in defining the length of the deleted segment. For example, let us assume that genes *B* and *C* were deleted on one chromosome as in figure 4.20*b*. If we have wild-type chromosomes with recessive mutants at different genes, these should be expressed if they are in the deleted region. Figure 4.22 illustrates that the deletion covers genes *B* and *C* but not gene *D.* As will be discussed in chapter 5, deletions can be used to map the sequence of genes on the chromosome.

FIGURE 4.23
A mechanism for the generation and repair of inversions by two chromosomal breaks.

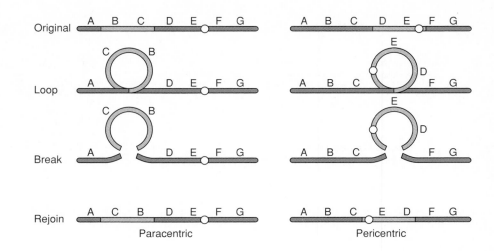

FIGURE 4.24
Loop formed in chromosomal heterozygotes for (*a*) a paracentric inversion and (*b*) a pericentric inversion.

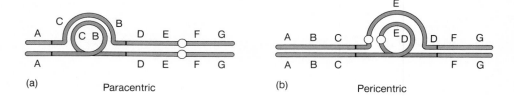

D uplications occur when a chromosomal segment is present more than once; deletions occur when a segment is missing. Both result in phenotypic changes or lethality, although deletions have greater effects. In meiotic chromosomes, duplication and deletion heterozygotes often form loops.

■

Inversions

Most of the homologous chromosomes in a population have genes in the same sequence. However, in some instances, the sequence may differ on different chromosomes. Alterations in the sequence of genes, called **inversions,** may be of two different kinds relative to the position of the centromere. If the inverted segment does not contain the centromere, it is called a **paracentric inversion** (Greek: *para* = next to), while if the inversion spans the centromere, it is called a **pericentric inversion** (Greek: *peri* = around).

Inversions can be generated by a simultaneous break at two points in a chromosome, followed by an incorrect reunion. For example, figure 4.23 illustrates how a looped chromosome may break and rejoin to generate either a paracentric or a pericentric inversion. Note in this example that the paracentric inversion results in an acrocentric chromosome just like the noninverted chromosome, but that the pericentric inversion results in a metacentric chromosome, because the position of the centromere has been changed.

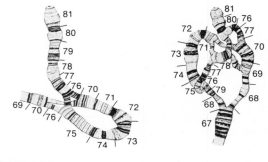

FIGURE 4.25
Inversion heterokaryotypes in *Drosophila pseudoobscura.* The Arrowhead-Standard heterokaryotype (left) illustrates a single inversion, while the Chiricahua-Standard heterokaryotype (right) shows a multiple inversion.
From Genes in Populations *by Eliot B. Spiess, p. 71, John Wiley and Sons, Inc., 1977. (From Genetics 23:28, 1938.)*

Individuals heterozygous for an inversion can be recognized by the presence of inversion loops in meiotic pachytene chromosomes. These structures occur because of the affinity of the two homologues, and the only way the two homologues can pair is if one twists on itself and makes a loop while the other makes a loop without a twist. Examples of such loops for the two types of inversions are given in figure 4.24. These loops can best be seen in the polytene chromosomes of organisms such as *Drosophila pseudoobscura.* Both a single inversion example, Arrowhead-Standard, and a multiple inversion example, Chiricahua-Standard, are shown in figure 4.25. We will mention these sequences again when we discuss population genetics in chapter 16.

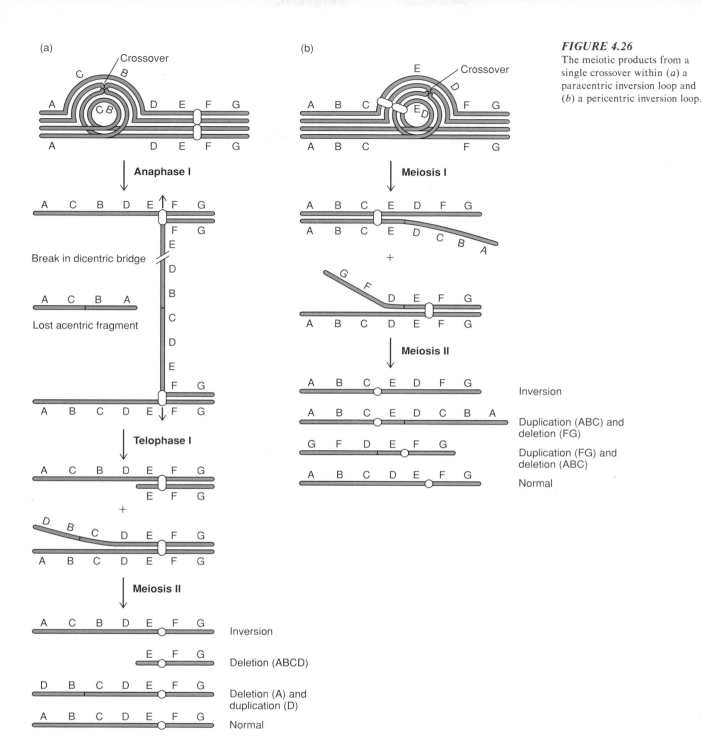

FIGURE 4.26
The meiotic products from a single crossover within (*a*) a paracentric inversion loop and (*b*) a pericentric inversion loop.

If recombination occurs only outside the inversion loop in an inversion heterokaryotype, the products of meiosis have the same genes and the same gene sequence as the parental chromosomes. However, if there is crossing over within the inversion loop, the chromosomes may be abnormal. Figure 4.26*a* shows the products that result from recombination between genes *B* and *C* within a paracentric inversion loop. At anaphase I, the homologous chromosomes pull apart so that one chromatid involved in recombination is **acentric** (having no centromere) and the other is **dicentric** (having two centromeres). The acentric fragment usually does not go to a pole and becomes lost, while the dicentric fragment breaks more or less at random between the two centromeres. When the centromere splits in meiosis II, the result is two balanced chromosomes— one with the normal sequence and one with the inverted sequence—and two chromosomes with either a deletion or

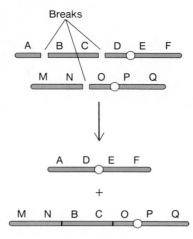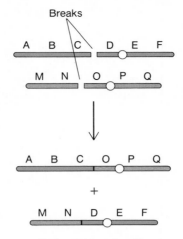

Interstitial translocation   Reciprocal translocation

FIGURE 4.27

The most common types of events leading to two types of translocations. An interstitial translocation involves the one-way movement of a segment, while a reciprocal translocation involves a two-way exchange of chromosomal segments.

a deletion and a duplication. In this example, one abnormal chromosome (*EFG*) is missing more than half the total chromosome, while the other is missing only the *A* region but has a duplication of the *D* region. Zygotes formed from gametes containing these chromosomes would in all probability be lethal.

Recombination within a pericentric inversion loop can also result in abnormal chromosomes. As an example, assume that recombination takes place between genes *D* and *E* (figure 4.26*b*). Again there are two balanced chromosomes, but in this case, the other pair has both a duplication and a deficiency. Zygotes with these unbalanced chromosomes would also likely be lethal.

As we have just seen, when recombination takes place within an inversion, the resulting chromosomes have either a deletion or a deletion and a duplication, and they are consequently often lethal. As a result, there appears to be a suppression of recombination in inverted regions, because no recombinant products are observed. Furthermore, in inversion heterokaryotypes, pairing difficulties often reduce recombination near and in the inverted region.

> *I* nversions occur when the sequence of genes on a chromosome is reversed. When recombination takes place in the inverted region of inversion heterokaryotypes, duplications and deficiencies show up in the meiotic products.
>
> ∎

Translocations

A **translocation** is the movement of a chromosomal segment from one chromosome to another, nonhomologous chromosome. Figure 4.27 illustrates two types of translocations, an **interstitial translocation,** involving the one-way movement of a segment, and the more common **reciprocal translocation,** involving a two-way exchange of chromosomal segments. If two of the segments that join together in a reciprocal translocation are large and the other two are small, the smaller translocated chromosomes may often be lost. In this case, the number of chromosomes is reduced by the chromosomal exchange. Obviously, translocations can change both the size of chromosomes and the position of the centromere.

Even though chromosomal segments have been exchanged between chromosomes in a reciprocal translocation, the affinity of the homologous regions results in pairing during meiosis I. If nearly equal parts of chromosomes are exchanged or not exchanged, the paired chromosomes in a translocation heterozygote have a cross appearance in metaphase I (see figure 4.28). During anaphase I, two major types of segregation occur, one in which alternate centromeres go to the same pole and one in which adjacent centromeres go to the same pole. The third type of segregation pictured in figure 4.28 is rare, because it requires that homologous centromeres go to the same pole.

When alternate centromeres go to the same pole, the chromosomes often form a figure eight shape in early anaphase I. (Remember that the chiasmata terminalize so that the homologous chromosome ends remain together.) The

Chapter 4

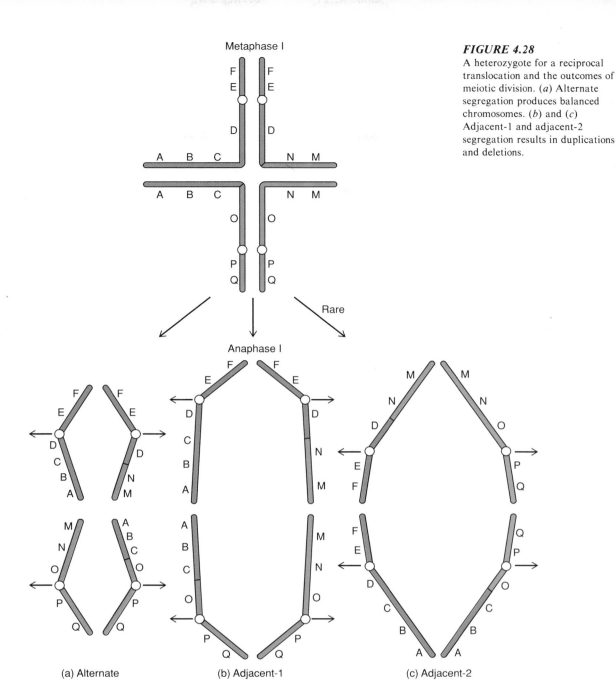

FIGURE 4.28
A heterozygote for a reciprocal translocation and the outcomes of meiotic division. (*a*) Alternate segregation produces balanced chromosomes. (*b*) and (*c*) Adjacent-1 and adjacent-2 segregation results in duplications and deletions.

Metaphase I

Anaphase I

(a) Alternate   (b) Adjacent-1   (c) Adjacent-2

products of this event, which is known as **alternate segregation,** are balanced so that each gamete has a full complement of chromosomes, either two untranslocated or two balanced translocated. On the other hand, when adjacent centromeres segregate together (**adjacent segregation**), the chromosomes appear as a ring at metaphase I. When this occurs, the products are unbalanced, resulting in duplications and deletions in the gametes.

Some plants, and also a few animals, have a series of reciprocal translocations, so that chromosomal heterozygotes have nearly all the chromosomes associated in a large ring (or rings) in meiosis (see figure 4.29). However, at anaphase, these chromosomes may undergo an orderly alternate segregation, producing only zygotes with a balanced chromosomal complement.

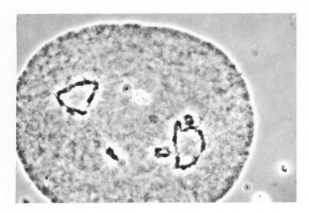

FIGURE 4.29
Cell of *Clarkia,* an annual plant, with two rings of six and three bivalents.
Courtesy of William L. Bloom, University of Kansas.

Although translocations can result in normal chromosomes, they can also cause several human diseases. For example, about 5% of individuals with **Down syndrome** have one parent who is heterozygous for a translocation (see figure 4.30). In this instance, chromosome 14 is translocated onto chromosome 21. Half of the time, the heterozygote produces either the normal set or a balanced translocated set of chromosomes, making the progeny either normal or translocation heterokaryotypes, respectively. The other half of the time, unbalanced chromosomes are produced, either a 14 without the translocated 21 segment or a translocated 14 with the attached 21 plus a normal 21. In the first case, offspring get only one 21 chromosome, a lethal chromosomal complement. In the second instance, three 21 chromosomes are received, resulting in Down syndrome. Overall then, approximately

FIGURE 4.30
Production of individuals with Down syndrome (trisomy 21) from a translocation heterozygote. Chromosome 14 is indicated in blue and chromosome 21 in red.

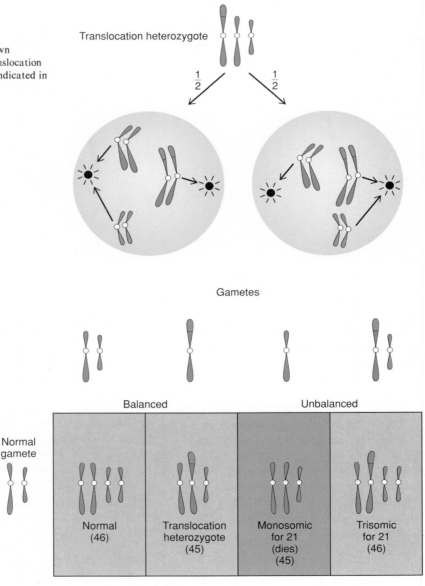

Chapter 4

one-third of the live births from such a translocation heterokaryotype can be expected to have Down syndrome. In actual fact, the proportion is less than this, primarily because some Down individuals do not survive gestation.

Note that this cause of Down syndrome has implications for genetic counseling. First, Down syndrome could recur in children of a translocation heterokaryotype, whereas normally it does not recur in sibs (see discussion of trisomy later in this chapter). Second, half of the phenotypically normal sibs of Down individuals are themselves translocation heterokaryotypes, and therefore, could produce Down progeny.

> *T*ranslocations are the result of the movement of a chromosomal segment to a nonhomologous chromosome. Normal meiotic products result when there is alternate segregation in translocation heterokaryotypes, but otherwise, duplications and deficiencies occur.

■

CHANGES IN CHROMOSOMAL NUMBER

The numbers of chromosomes may vary in two basic ways: **euploid** variants, in which the number of chromosomal sets differ, and **aneuploid** variants, in which the number of a particular chromosome is not diploid. As one might expect, changes in chromosome number, either euploid or aneuploid, generally have an even greater effect on survival than do changes in chromosome structure. In fact, in humans, more than half of the spontaneous abortions that occur in the first three months of pregnancy involve fetuses with aneuploidy, polyploidy, or other large chromosomal aberrations.

Before we discuss variants in chromosome number caused by increased or decreased amounts of genetic material, let us consider the other mechanisms by which chromosome number can change. As already mentioned, if reciprocal translocations occur between two acrocentric chromosomes such that the large segments reattach, the result is a large metacentric chromosome and a small chromosome that may be lost during cell division. In other words, two acrocentric chromosomes may be combined into one metacentric chromosome. This is a mechanism for

chromosomal fusion, in which two nonhomologous chromosomes are joined at their centromeres to form a single metacentric (see figure 4.31). On the other hand, a chromosome may split at its centromere perpendicular to the length of the chromosome, resulting in two smaller chromosomes, a change termed **chromosomal fission.** It is generally thought that chromosomal fusions are much more common than chromosomal fissions. Later in this chapter, we will give an example of a fusion that resulted in the 46 human chromosomes.

Polyploidy (Euploidy Variation)

Diploid organisms with three or more sets of chromosomes are called **polyploids.** If we let the haploid number of chromosomes be x, then organisms with three chromosome sets have 3x chromosomes and are called triploids; those with 4x chromosomes are tetraploids; those with 6x chromosomes are hexaploids; and so on. Earlier in this chapter, we used the symbol N to refer to the haploid number of chromosomes; here, x will stand for the haploid number. However, for organisms that are regularly polyploid, such as many plants, x usually refers to the number of chromosomes in a set and N to the number in a gamete. Thus, in a hexaploid organism with 60 chromosomes, $6x = 2N = 60$, so that $x = 10$ and $N = 30$.

Polyploidy is relatively common in plants but rare in most animals, occurring only in certain beetles, earthworms, salamanders, and a few other organisms. On the other hand, nearly half of all flowering plants are polyploids, as are many important crops. For example, potatoes are tetraploid ($4x = 48$), bread wheat is hexaploid ($6x = 42$), and strawberries are octoploid ($8x = 56$).

Polyploidy is less frequent in animals than in plants for several reasons. First, sex determination is often more sensitive to polyploidy in animals than in plants. Second, plants can often self-fertilize, so a single new polyploid plant with an even number of chromosomal sets (tetraploid, hexaploid, etc.) can still reproduce. Finally, plants generally hybridize more easily with other related species, an important attribute, because the different sets of chromosomes in a polyploid often have different origins.

We can distinguish two types of polyploids: those that receive all their chromosomal sets from the same species (**autopolyploids**) and those that obtain their chromosomal

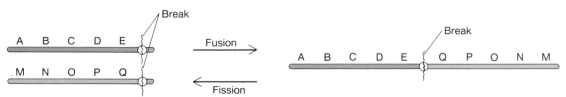

FIGURE 4.31
Centric fusion occurs when two nonhomologous acrocentric chromosomes join to form a metacentric chromosome. Centric fission occurs when the metacentric breaks to form two telocentric chromosomes.

sets from different species (**allopolyploids**). For example, if an unreduced, or diploid pollen grain from a diploid organism fertilizes a diploid ovule of the same species, the offspring are **autotetraploids,** or *AAAA,* where *A* indicates a complete chromosomal set, or **genome,** of type *A.* On the other hand, if diploid pollen of one species fertilizes a diploid ovule of another, related species, the offspring are **allotetraploids,** or *AABB,* where *B* indicates a genome from the second species. All the chromosomal sets in an autopolyploid are homologous, just as they are in a diploid. But in allopolyploids, the different chromosomal sets generally vary somewhat and are called **homeologous,** or partially homologous.

Polyploids occur naturally, but in very low frequency, when a cell undergoes abnormal mitosis or meiosis. For example, if during mitosis all chromosomes go to one pole, that cell will have an autotetraploid chromosome number. If abnormal meiosis takes place, an unreduced gamete may result, having 2N chromosomes. However, in most situations, this diploid gamete would combine with a normal haploid gamete and produce a triploid. Polyploids can also be produced artificially using colchicine, a chemical that interferes with the formation of spindle fibers. As a result of colchicine application, chromosomes do not move to the poles, and autotetraploids are often formed.

Autopolyploids   Triploid organisms are usually autopolyploids (*AAA*) that result from fertilization involving a haploid and a diploid gamete. Triploids are normally sterile, because the probability of producing balanced gametes is quite low. In meiosis, the three homologues may pair and form a trivalent, or two may pair as a bivalent, leaving the third chromosome unpaired. Gametes are equally likely to have one or two homologues of a given chromosome (see figure 4.32). However, because the behavior of nonhomologous chromosomes is independent, the probability of a gamete having exactly N chromosomes is ($\frac{1}{2}$)^N (using the product rule), and the probability of a gamete having exactly 2N chromosomes is also ($\frac{1}{2}$)^N. All other gametes will be unbalanced and generally nonfunctional, as will the zygotes containing them. For example, most bananas are triploids; they produce unbalanced gametes, and as a result, are seedless (they are propagated by cuttings). Other polyploids with odd numbers of chromosome sets (for example, pentaploids) are also usually sterile.

Polyploids are often larger (that is, they produce larger fruit) than related diploids; thus, many food crops are autotetraploids or other types of polyploids. Autotetraploids may have normal meiosis if they form either bivalents or quadrivalents only. However, if the four homologues form a trivalent and a univalent, gametes will generally have either too many or too few chromosomes (see figure 4.33).

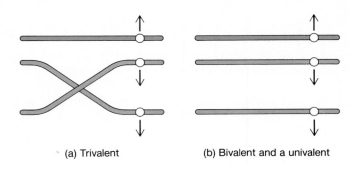

(a) Trivalent   (b) Bivalent and a univalent

FIGURE 4.32
Meiotic pairing and the direction of homologue movement in triploids when (*a*) a trivalent or (*b*) a bivalent and a univalent are formed.

Allopolyploids   Most naturally occurring polyploids are allopolyploids, and they may result in a new species. For example, the bread wheat *Triticum aestivum* is an allohexaploid with 42 chromosomes. By examining wild related species, it appears that bread wheat is descended from three different diploid ancestors, each of which contributed two sets of chromosomes (in this case designated as *AABBDD*). Pairing only occurs between the homologous sets, so that meiosis is normal and results in balanced gametes of N = 21.

A particularly interesting allotetraploid was developed by a Russian, G. Karpechenko, in 1928, when he crossed the cabbage and the radish, both of which have a diploid chromosome number of 18. He wanted to produce a hybrid having the leaves of a cabbage and the root of a radish. When he finally recovered seeds from an artificial hybrid, he planted them and found that they had 36 chromosomes. However, instead of the traits he had hoped for, this hybrid had the leaves of a radish and the roots of a cabbage!

A haploid pollen grain with genome *A* may pollinate a flower of a species with genome *B,* resulting in a sterile hybrid of genome constitution *AB.* If mitotic failure subsequently takes place in one branch, *AABB* cells may be produced. If these are self-fertile, an allopolyploid has been formed. Plant breeders use colchicine on sterile hybrids to produce allopolyploids in much the same manner.

P olyploids, species with more than two chromosomal sets, are most common in plants. They may be either autopolyploids, whose chromosomal sets come from the same species, or allopolyploids, whose chromosomal sets come from different species.

■

Chapter 4

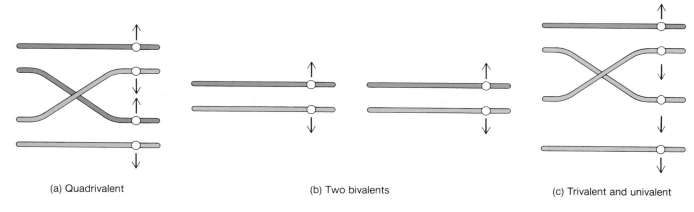

(a) Quadrivalent   (b) Two bivalents   (c) Trivalent and univalent

FIGURE 4.33

Meiotic pairing and the direction of homologue movement in
tetraploids for (a) a quadrivalent and (b) two bivalents. One possible
outcome for a trivalent and a univalent is shown in (c).

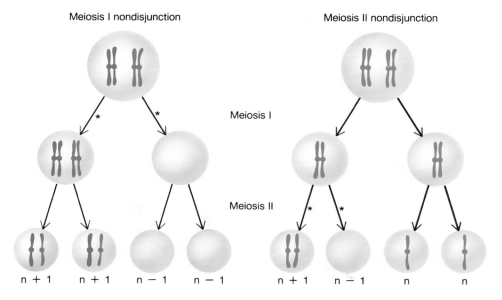

FIGURE 4.34

Nondisjunction occurring in meiosis I and in meiosis II, with the
resulting gametes.

Aneuploidy

We have already discussed the observation by Bridges that
Drosophila occasionally produce individuals having either
one sex chromosome (for example, an X) or three sex
chromosomes (for example, XXY). The cause of such
aneuploidy is nondisjunction; that is, two homologous
chromosomes fail to separate properly during meiosis.
Nondisjunction in meiosis itself is thought to result from
improper pairing of homologues in early meiosis so that
the centromeres are not on opposite sides of the metaphase plate. As a result, both chromosomes may go to the
same pole, leaving one daughter cell with an extra chromosome and the other daughter cell with no chromosome.

When these gametes are fertilized by a normal gamete,
they either have an extra chromosome, $2N + 1$, termed
trisomy, or are missing a chromosome, $2N - 1$, termed
monosomy. Nondisjunction is most common in meiosis I,
but can occur in meiosis II as well (see figure 4.34). Nondisjunction can also take place in mitosis, resulting in
aneuploidy. Other combinations of extra chromosomes are
possible, the most important being a tetrasomic with
$2N + 2$ chromosomes and a nullisomic with $2N - 2$
chromosomes, in which no copies of a particular homologue exist.

FIGURE 4.35
The normal fruit from *Datura* (top) and the twelve different types of trisomies, each having a different appearance.

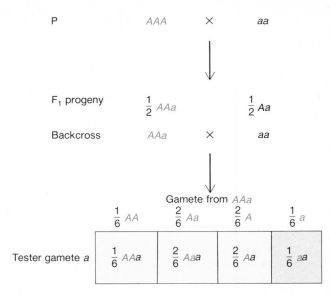

FIGURE 4.36
A cross between an F₁ trisomy and a recessive, giving ⅙ recessive progeny.

Trisomics are known in many different species. They are viable in many plants, but less frequently so in animals. For example, among the aneuploids that have been most thoroughly studied are those in the Jimson weed, or thorn apple, *Datura stramonium*. A series of *Datura* mutants with strange properties, studied by Alfred Blakeslee around 1920, turned out to be trisomics for different chromosomes. In fact, a trisomic for each of the twelve different chromosomes was found, and each had a particular phenotype. The effects on the appearance of the seed capsule were quite different for trisomies of the different chromosomes, suggesting that different chromosomes have different hereditary effects on this trait (see figure 4.35).

Trisomics have been investigated in crop plants such as corn, rice, and wheat in an effort to identify the chromosomes carrying different genes. Crosses involving genes on trisomic chromosomes give unusual segregation ratios. For example, if a homozygous dominant trisomy, *AAA* (the *A* symbol again indicates a dominant allele), is crossed to

a recessive diploid, *aa*, half the progeny are trisomic *AAa* and half are diploid *Aa*. When the trisomic progeny are backcrossed to *aa* individuals, approximately one-sixth of the progeny are recessive *aa* (see figure 4.36). If the gene had been on a chromosome that was not trisomic, the F₁ would be *Aa*, and one-half, not one-sixth, of the backcross progeny would be homozygous recessive (*aa*).

In animals, trisomies and other aneuploid chromosomal complements are more unusual. From analysis of the chromosomal constitution of spontaneous abortions, it appears that nearly all monosomics and many trisomies are fetal lethals. However, several trisomies that sometimes come to full term compose a substantial part of congenitally abnormal births. One of the most common is Down syndrome, trisomy of chromosome 21, with a frequency of one in seven hundred live births (see table 4.4). Down syndrome, first described nearly 150 years ago, is generally characterized by mental retardation, distinctive palm prints, and a common facial appearance. In general, mortality is higher than normal; the average life span is the middle teens to the forties, depending upon the country, but some individuals live much longer. People with Down syndrome generally have a positive disposition, and some are able to be partially independent (see figure 4.37).

The chromosomal basis of Down syndrome was first discovered in 1959, shortly after the correct human diploid number was determined. Detailed banding of human chromosomes has shown that Down syndrome actually results from a trisomy of the smallest chromosome, chromosome 22; however, because Down is known so prevalently as trisomy 21, this association was not changed, and chromosome 22 is still called chromosome 21. We discussed previously how this condition can be generated from

TABLE 4.4   Frequency and Effects of the Most Common Aneuploidies in Humans

Syndrome	Sex	Chromosomes	Frequency		Expected Viability and Fertility
			Abortuses	*Births*	
Down	M or F	Trisomy 21	1/40	1/700	15 years, rarely reproductive
Patau	M or F	Trisomy 13	1/33	1/15,000	< 6 months
Edward	M or F	Trisomy 18	1/200	1/5,000	< 1 year
Turner	F	XO	1/18	1/5,000	Sterile
Metafemale	F	XXX (or XXXX)	0	1/700	Sterile
Klinefelter	M	XXY (or XXXY)	0	1/2,000	Sterile
XYY	M	XYY	?	1/2,000	—

TABLE 4.5   Gametes and Zygotes Resulting from Normal Meiosis and Nondisjunction

Sperm		Egg		
		Normal Meiosis	*Nondisjunction (Meiosis I or Meiosis II)*	
		X	XX	O
Normal Meiosis	X	XX	XXX	XO
	Y	XY	XXY	YO*
Nondisjunction in Meiosis I	XY	XXY	(XXXY)	(XY)
	O	XO	(XXO)	(OO*)
Nondisjunction in Meiosis II	XX	XXX	(XXXX)	(XX)
	YY	XYY	(XXYY)	(YY*)
	O	XO	(XX)	(OO*)

*Inviable zygotes. The zygotes in parentheses are extremely rare because they require nondisjunction in the formation of both gametes.

reciprocal translocation heterokaryotypes, but most commonly (in about 95% of cases), it results from meiotic nondisjunction. The other autosomal trisomies are much rarer, but several of the relatively common ones will be discussed here.

Nondisjunction of the sex chromosomes in humans is the source of several abnormal maladies. Table 4.5 shows the gametes and zygotes resulting from normal meiosis and from nondisjunction. Notice that four common viable, but abnormal chromosomal types, XO, XXX, XXY, and XYY, are produced through nondisjunction. The symbol O here indicates the lack of a sex chromosome in a gamete or zygote.

Klinefelter syndrome, XXY, occurs fairly frequently and generally results in a relatively mild abnormality. These individuals are sterile males with some female characteristics. Individuals with **Turner syndrome** (XO) are sterile females, short in stature with some neck webbing. The frequency of XYY is about one per one thousand males, but they do not appear to have any congenital problems. At one point, it was suggested that XYY individuals had criminal tendencies, but further study indicates minimal correlation, if any. The frequency of XYY

FIGURE 4.37
An individual with Down syndrome.
© *M. Coleman/Visuals Unlimited.*

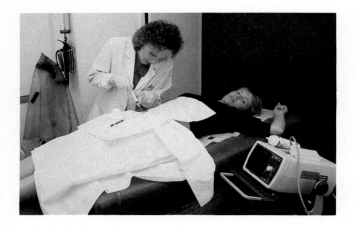

FIGURE 4.38
The procedure for amniocentesis.
© Eric Kroll/Taurus Photos, Inc.

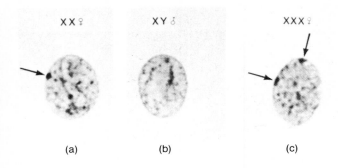

FIGURE 4.39
Nuclei from cells of (*a*) a normal female (XX) with one Barr body, (*b*) a normal male (XY) with no Barr bodies, and (*c*) an XXX female with two Barr bodies.
From M. M. Grumbach and M. L. Barr, Recent Progress in Hormone Research *14:255–324, 1958, fig. 1A, 3A, 4A. Reprinted by permission of Academic Press.*

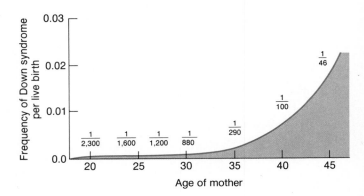

FIGURE 4.40
The frequency of Down syndrome in children as it relates to mothers of different ages.

A neuploidy occurs when there are extra or missing chromosomes compared to the normal complement. Cases of trisomy, one extra chromosome, are well-known in plants and cause a number of human disorders.

■

CHROMOSOMES IN DIFFERENT SPECIES

As we discussed earlier in this chapter, variation in chromosomal structure or number may occur within a species. When the structures of chromosomes in related species are compared, they are often quite similar. Early studies of this sort compared the banding patterns of different related *Drosophila* species and found a high degree of homology. For example, species such as *D. melanogaster* and *D. simulans,* which are exactly the same except in the morphology of the male genitalia, have nearly identical banding sequences, differing only in a few inversions.

individuals in prisons is significantly elevated over that of the general population; however, less than 5% of all XYY individuals are actually institutionalized.

Abnormal chromosome numbers in a fetus can be diagnosed using **amniocentesis** (see figure 4.38). In this procedure, a sample of fluid is withdrawn from the amniotic sac with a needle. The fetal cells contained in this fluid are cultured for two to three weeks. Dividing cells are then stained, and the chromosomes are counted to check for chromosomal abnormalities.

The X chromosome is different from the other chromosomes in that only one is active in a given cell. Normal males have only one X, which is active in all cells. In normal females, only one X is active in a given cell and the other X is heterochromatinized, or inactivated. The inactivated X forms a structure called a **Barr body** (named after its discoverer, Murray Barr) that can be identified in a cell (see figure 4.39). Therefore, normal males and XO individuals have no Barr bodies; normal females and XXY individuals have one; XXX individuals have two; and so on. In other words, by counting the number of Barr bodies in a cell, chromosomal abnormalities involving the X chromosome can be determined.

The incidence of Down syndrome, and to some extent other aneuploidies, is dependent upon the age of the mother (see figure 4.40). The increase for Down is nearly 50-fold for mothers of age forty-five as compared to teenage mothers. Although the exact mechanism for this increase is unknown, it appears to be related to the difference in gametogenesis between females and males. In females, oocytes are formed before birth and held in a resting stage (actually prophase of meiosis I) until just before ovulation. In older mothers, an oocyte may remain at this stage for over forty years, during which time it may be affected by environmental factors.

Chapter 4

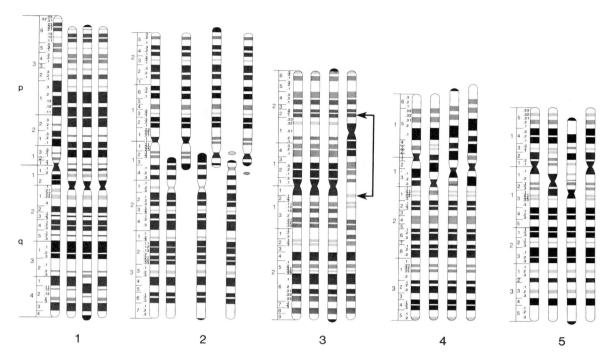

FIGURE 4.41

Comparison of late-prophase banding sequences of the five largest chromosomes (1–5) in humans, chimpanzees, gorillas, and orangutans (left to right).

From J. J. Yunis and O. Prakash, "The Origin of Man: A Chromosomal Pictorial Legacy" in Science, *Vol. 215, March 19, 1982. Copyright © 1982 American Association for the Advancement of Science, Washington, D.C.*

Recently developed high resolution techniques for chromosomal banding permit the comparison of chromosomes in many species. Much research has focused on human chromosomes and on those of the related great apes. Humans have a diploid chromosome complement of 46, while chimpanzees, gorillas, and orangutans have 48. The difference in number is explained by a fusion of one chromosome pair in the great apes to form human chromosome 2 (see figure 4.41). Notice that the two acrocentric chromosomes in the three great apes that are homologous to the metacentric human chromosome 2 have identical banding patterns, except in the fusion region and in the terminal heterochromatin.

A number of other structural differences exist in the chromosomes among these four species. For example, chromosome 3 in humans, chimpanzees, and gorillas has virtually identical banding patterns for each, while the orangutans differ by two inversions. The break points of the larger of these inversions are indicated by the arrows in figure 4.41, showing that the inversion is pericentric and changes the chromosome from a metacentric to an acrocentric. Evidence indicates only one translocation difference among these species—that occurring between chromosomes 5 and 17 in the gorilla and not present in the other species. Overall, the banding patterns for humans are nearly identical to chimpanzees for 13 chromosomes, to gorillas for 9 chromosomes, and to orangutans for 8

chromosomes. Almost all the same bands are present in these species, suggesting that the primary chromosomal differences between them are in the order of the genes, not in the presence or absence of genes.

CHROMOSOMES AND SEX DETERMINATION

In the past, literally hundreds of unscientific theories have attempted to explain sex determination by relating it to phases of the moon, the time of day of fertilization, and other factors. Actually, there is no universal mode of sex determination, but rather many different modes, some of which are environmental. For example, certain reptile eggs that are exposed to high temperatures produce mostly males in particular species and mostly females in others. The sex of some fish is affected by social dominance, and certain plants produce different sexes depending on day length or other factors that affect growth rate.

However, in most organisms, sex is determined by chromosomal differences (see table 4.6). The first example of chromosomal determination of sex was documented in 1905 for the bug *Protenor*, in which it was discovered that females have two X chromosomes and males only one. Male gametes were found to be equally likely to have an X or no X chromosome, and female gametes were found to always carry one X, so that equal numbers of progeny are XX (female) and XO (male).

TABLE 4.6	Different Types of Chromosomal Sex Determination	
Organisms	**Female**	**Male**
Most mammals, some insects, some plants	XX	XY
Protenor, some other insects	XX	XO
Birds, most reptiles, moths	ZW	ZZ
Hymenoptera	Diploid	Haploid

Shortly thereafter, several investigators established the presence of a second sex chromosome, the Y chromosome, in other organisms. In these species, the presence of a Y along with an X results in a male, while two copies of the X produce a female. Again, male gametes are of two types, so that there are equal numbers of X-carrying and Y-carrying sperm or pollen. The sex having both types of sex chromosomes, the XY, is termed the **heterogametic sex,** while the XX is called the **homogametic sex.** In some organisms, such as birds and some reptiles, males are homogametic and females are heterogametic. To avoid confusion, these chromosomes are often termed Z and W; thus, homogametic males are ZZ and heterogametic females are ZW. As a result, the females produce two types of gametes, while the male gametes all contain one Z chromosome.

In chickens, it is important to identify the sex of young birds so that, for example, males will not be raised in an egg laying operation. It is difficult and expensive to determine the sex of young birds, and several systems utilizing the sex chromosome difference between males and females are often used. For example, one gene on the Z chromosome has a recessive allele that determines gold feather color (either Z^sZ^s or Z^sW), and a dominant allele that determines silver feather color (Z^SZ^S, Z^SZ^s, or Z^SW). As a result, the cross given in figure 4.42 produces all silver males and all gold females, allowing easy sex determination of day-old chicks. Could the opposite cross, $Z^SZ^S \times Z^sW$, yield chicks of different colors that could be used to determine their sex?

In nearly all mammals, the presence of a Y chromosome is necessary for the development of the male phenotype. For example, humans that are XO (Turner syndrome) are phenotypically female. Furthermore, individuals that are XXY (Klinefelter syndrome) are phenotypically male, even though they have two X chromosomes. Thus, it appears that the Y chromosome produces a gene product that acts as a necessary switch to begin development toward maleness. Although this switch gene is on the Y chromosome, apparently the rest of the genes involved in the production of sexual phenotypes are on other chromosomes.

Recently, the location of this switch gene on the Y has been identified by examining rare, sex-reversed individuals—that is, XX males and XY females. The XX males actually have a piece of the Y that contains the male-determining gene, and the XY females lack the same region of the Y. In *Drosophila,* there is an autosomal mutant, *tra* (transformer), which when homozygous in XX individuals, causes them to be males, not females as expected. The opposite situation is known in humans where individuals with a mutation on the X chromosome have a condition called testicular feminization. In this case, the XY individuals are phenotypically female, having breasts and a vagina, but they are sterile. This mutation is apparently related to a hormone receptor that prevents these individuals from responding to the presence of testosterone; consequently, they do not develop secondary male characteristics.

Aside from this switch function to maleness, there appear to be few active genes on the Y chromosome in humans according to the following evidence. First, much of the Y chromosome is heterochromatic. Second, the length of the long arm of the Y is quite variable. (Interestingly, the long arm is substantially longer in Japanese males than in males of most other populations, but this variation appears to have no effect on sterility or other phenotypic characteristics.) And finally, there is also extensive variation in the size and intensity of banding on the long arm of the Y in humans without apparent phenotypic effect. These facts support the hypothesis that much of the Y chromosome is inert and does not produce functional gene products.

Although XX individuals in *Drosophila* are female and XY individuals are male, the presence of the Y is not necessary for the male phenotype. For example, the exceptional phenotypes observed by Bridges (see figure 4.17) were red-eyed XO males and white-eyed XXY females. Later, Bridges developed triploid strains of *Drosophila.* When he examined individuals with different numbers of X chromosomes, he found that the sexual phenotype in *Drosophila* was a function of the ratio of the number of X chromosomes (X) to the number of sets of autosomes (A). For example, normal females have two X chromosomes and two autosomal sets, making the ratio $X/A = 2/2 = 1$ (see table 4.7). Normal males have one X and two autosomal sets, so that $X/A = 1/2 = 0.5$. Likewise, his exceptional males and females still had ratios of 0.5 and 1.0, respectively.

When Bridges examined flies with three X chromosomes and two autosomal sets, $X/A = 3/2 = 1.5$, he found that they were sterile females (sometimes called **metafemales**), while flies with one X and three autosomal sets, $X/A = 1/3 = 0.33$, were sterile males (**metamales**).

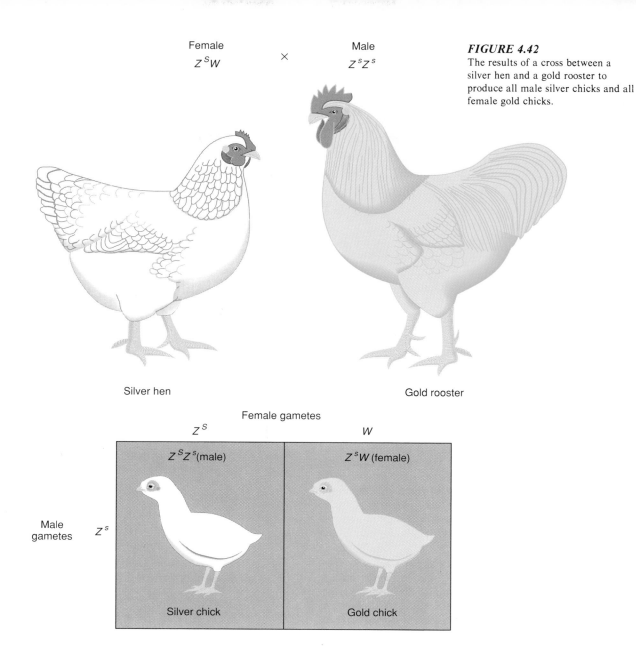

Female
$Z^S W$　×　Male
$Z^s Z^s$

Silver hen

Gold rooster

Female gametes

Z^S　　　　　W

Male gametes Z^s

$Z^S Z^s$ (male)	$Z^S W$ (female)
Silver chick	Gold chick

Finally, flies with two X chromosomes and three autosomal sets, $X/A = 2/3 = 0.67$, had phenotypic characteristics of both sexes and were termed **intersexes.** Based on these findings, Bridges proposed that overall, the X chromosomes were female-determining and the autosomes were male-determining. In other words, when the X/A ratio was 1 or greater, flies would have the female phenotype, and when the X/A ratio was 0.5 or less, flies would have the male phenotype. However, we should note that even though the Y is not necessary for the male phenotype in *Drosophila,* it is necessary for fertility; that is, XO males and metamales are sterile.

TABLE 4.7　*Sexual Phenotypes of* D. melanogaster[1]

Number X Chromosomes	Number Autosomal Sets (A)	X/A Ratio	Sexual Phenotype
3	2	1.5	Metafemale
2	2	1.0	Normal female
2	3	0.67	Intersex
1	2	0.5	Normal male
1	3	0.33	Metamale

[1]Expressed as a function of the ratio of the number of X chromosomes to sets of autosomes.

C hromosomes in related species are generally quite similar. In most organisms, chromosomal differences are involved in sex determination. The Y chromosome determines maleness in nearly all mammals. In *Drosophila*, the relative proportion of X chromosome to autosomes determines sex.

∎

Summary

T he diploid number of chromosomes varies for different species. Specific chromosomes can be identified by size, the position of the centromere, and banding patterns. Mitosis, cell division that results in the production of two identical daughter nuclei, is divided into four stages: prophase, metaphase, anaphase, and telophase. Meiosis is composed of one duplication of DNA and two cell divisions. Meiosis I, the reductional division, results in the separation of homologous chromosomes or a halving of chromosome number. Meiosis II is similar to mitosis and results in the separation of the sister chromatids with no reduction in chromosome number. In animals, mature male and female gametes are produced by spermatogenesis and oogenesis, respectively.

The parallel behavior of genes and chromosomes provides a mechanistic explanation of the principles of segregation and independent assortment. In addition, the first indication that genes were on chromosomes came from the parallel behavior of genes and chromosomes. Definitive proof resulted from experiments using X-linked genes in *Drosophila* and the confirmation that individuals with exceptional eye color were the result of meiotic mistakes that produced abnormal chromosomal numbers.

Duplications and deletions occur when a chromosomal segment is present more than once or is missing. Both result in phenotypic changes or lethality, although deletions have larger effects. In meiotic chromosomes, duplication and deletion heterozygotes often form loops. Inversions occur when the sequence of genes on a chromosome is reversed. When recombination takes place in the inverted region of inversion heterokaryotypes, duplications and deficiencies result in the meiotic products. Translocations are the result of the movement of a chromosomal segment to a nonhomologous chromosome. Normal meiotic products result when there is alternate segregation in translocation heterokaryotypes, but otherwise, duplications and deficiencies occur.

Polyploids, species with more than two chromosomal sets, are most common in plants. They may be either autopolyploids, all chromosome sets from the same species, or allopolyploids, in which the chromosomal sets come from different species. Aneuploidy occurs when there are extra or missing chromosomes compared to the normal complement. Cases of trisomy, one extra chromosome, are well-known in plants and cause a number of human disorders.

Chromosomes in related species are generally quite similar. In most organisms, chromosomal differences are involved in sex determination. In nearly all mammals, the Y chromosome determines maleness. In *Drosophila*, the relative dose of X chromosome to autosomes determines sex.

∎

Problems and Questions

1. A cross between red-eyed F_1 (heterozygous) female fruit flies and red-eyed male fruit flies yielded 94 red-eyed and 30 white-eyed progeny. Does this fit with expectations? (Use a chi-square test.) Of the red-eyed flies, 56 were females. Does this fit with expectations (again, using a chi-square test)?

2. In the cross shown in figure 4.17 involving a group of 4,240 flies, 3 red-eyed males and 2 white-eyed females were observed. From these data, what would be the estimated rate of nondisjunction?

3. Explain why the number of chromosomes per cell (N level) and the amount of DNA per cell (C level) are not synchronous in every stage of cell division.

4. Differentiate between the behavior of chromosomes in metaphase of mitosis, meiosis I, and meiosis II.

5. (*a*) Given a heterozygote for albinism, *Aa*, give the proportions of the sperm genotypes that you would expect after meiosis. (*b*) Diagram the chromosomal behavior as in figure 4.13. (*c*) Extend this analysis to two genes, *Aa*, and another recessive disease heterozygote, *Bb*. Illustrate how chromosomal behavior explains independent assortment by drawing a diagram as in figure 4.14.

6. Figure 4.19 shows how a tandem duplication may be generated. (*a*) Using the same logic, illustrate how a reverse duplication may be generated. (*b*) Show how a displaced duplication may be generated.

7. Let us assume that we are trying to identify the genes involved in a deletion. Crosses to lines homozygous for recessive mutants at genes *e* and *f* gave offspring that were mutant, while crosses to lines homozygous for recessive mutants at genes *g* and *h* yielded wild-type progeny. Which genes are missing in the deletion?

8. What is the difference between pericentric and paracentric inversions? What are the differences in meiosis and in the meiotic products from heterokaryotypes for paracentric and pericentric inversions?

9. Assume that the breaks for the reciprocal translocation in figure 4.27 were between E and F and between P and Q. (*a*) What do the chromosomes with the reciprocal translocation look like? (*b*) Diagram metaphase I and alternate segregation for a heterokaryotype for this reciprocal translocation.

10. Some Down syndrome cases are the result of a translocation of chromosome 21 onto chromosome 14. Approximately one-third of the offspring of a translocation heterozygote are expected to have Down syndrome. Explain why. In fact, only about one-sixth of the offspring of translocation heterozygotes actually do have Down. Explain why.

11. (*a*) Differentiate between the meiotic products expected from nondisjunction in meiosis I and meiosis II. (*b*) Assuming nondisjunction in meiosis I for the sex chromosomes in a male, what progeny genotypes would you expect? (*c*) What progeny genotypes would you expect from nondisjunction in meiosis II in males or in females?

12. In an effort to determine on which chromosome a gene is located, a homozygous recessive mutant was crossed to trisomies of two different chromosomes. A backcross of the trisomy F_1 to the recessive homozygote gave 42 wild-type and 38 mutant progeny for chromosome 5 and 56 wild-type and 12 mutant progeny for chromosome 6. On which chromosome do you think the gene was located?

13. How many copies of a gene does an autotetraploid have? Assuming that there are two types of alleles at a gene, *A* and *a*, give the five possible genotypes. If *A* is dominant, how many possible phenotypes would you expect?

14. Given an autotriploid with $3x = 12$, what is the probability of producing a gamete having exactly one copy of each of the four homologues? Having exactly two copies of each homologue?

15. In figure 4.41, identify the smaller inversion on chromosome 3 that differs between orangutans and the other great apes. Identify the pericentric inversion that differentiates chromosome 4 for humans and chimpanzees.

16. In chickens, males are the homogametic sex, *ZZ*, and females are the heterogametic sex, *ZW*. In a cross between $Z^A W$ and $Z^A Z^a$, where the superscripts indicate the dominant and recessive alleles, what proportions of progeny would you expect? Give both genotype and sex.

17. Given the rules developed by Bridges to determine sex in *Drosophila*, what would be the sex of an individual with three X chromosomes and three autosomal sets? Of an individual with two X chromosomes and four autosomal sets? Explain why.

18. Using diagrams, show the differences between prophase in mitosis and in meiosis I.

19. Why might allotetraploids have higher rates of growth than either of the parental diploids? If the two sets of chromosomes from different species in an allotetraploid do not pair, would you expect balanced gametes?

20. Two related species of mice have different numbers of chromosomes. Species 1 has twenty pairs of metacentrics, and species 2 has twenty-three pairs of chromosomes, seventeen of which are metacentric and six acrocentric. How many chromosomal fissions would be necessary to produce a karyotype of species 2 from species 1?

Answers appear at end of book.

C·H·A·P·T·E·R

5

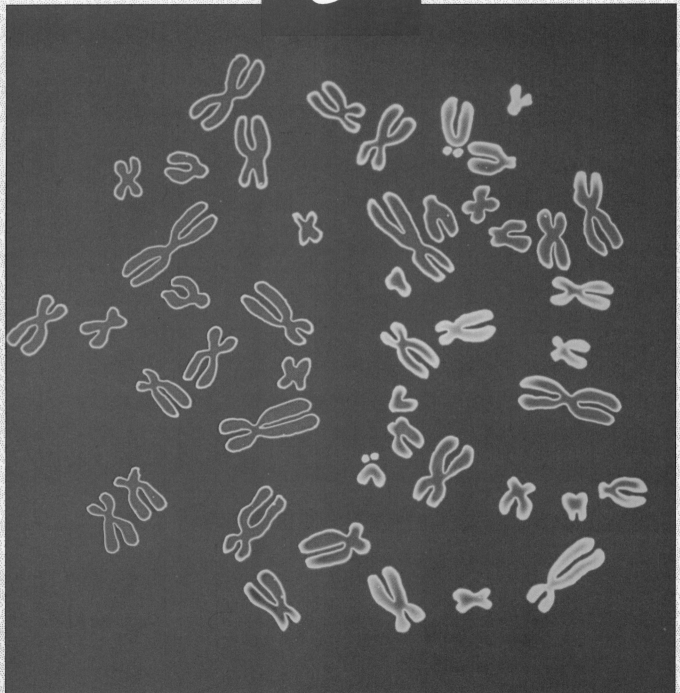

Genetic Linkage

*T*he results are a simple mechanical result of the location of the materials in the chromosomes, and of the method of union of homologous chromosomes, and the proportions that result are not so much the expression of a numerical system as the relative location of the factors in the chromosomes.

■
_____

Thomas Hunt Morgan
American geneticist

Learning Objectives

In this chapter you will learn:

1. That genes on the same chromosome violate Mendel's principle of independent assortment.

2. That by estimating the rate of recombination between linked loci, a linkage map of genes on the same chromosome can be constructed.

3. How crosses of three linked genes can be used to order genes on a chromosome.

4. That the recombination frequency is generally depressed in a region where there has already been recombination.

5. That linkage in humans has traditionally been determined from particular matings and somatic cell hybridization—and that recently, DNA markers have made mapping in humans more feasible.

6. That crossing over takes place in the four-stranded stage of meiosis I.

7. That there is generally a good correspondence between genetic map distance and physical DNA distance, with important exceptions.

Light micrograph of human chromosomes, artistically enhanced in coloration.
© *Briolle/Rapho/Photo Researchers, Inc.*

s we discussed in chapter 4, the principles of segregation and independent assortment are consistent with the fact that genes are on chromosomes and with the behavior of chromosomes in meiosis. In that discussion, we assumed that the different genes we followed were on different chromosomes. However, shortly after the rediscovery of Mendel's principles, Thomas Hunt Morgan (figure 5.1) realized there were more genes than chromosomes in *Drosophila*. In fact, we know today that hundreds of genes exist on each of the three major *Drosophila* chromosomes. As a result, the pattern of inheritance predicted by the principle of independent assortment does not apply to many pairs of genes. In this chapter, we will examine the pattern of inheritance that is present when two or more genes are on the same chromosome. We will also describe how genes can be organized into genetic maps showing their positions on the various chromosomes.

LINKAGE

In the first decade of this century, William Bateson and R. C. Punnett investigated the inheritance patterns for various traits in the sweet pea. In one study, they examined the simultaneous inheritance of flower color, dominant purple (P) versus recessive red (p), and pollen shape, dominant long (L) versus recessive round (l). They carried out a dihybrid cross as Mendel had done for traits in the garden pea; that is, they crossed $PPLL$ with $ppll$, self-fertilized the F_1 ($PpLl$), and counted the numbers of different types of F_2 plants. But instead of the expected 9:3:3:1 ratio, they obtained the results given in table 5.1. Comparing the observed numbers to those expected under a 9:3:3:1 ratio, they found many more phenotypes like the original parents and many fewer single dominant or nonparental types. In this case, there are 3 degrees of freedom, making the chi-square value of 222.0 for the data in table 5.1 highly significant (see table 2.5).

Bateson and Punnett knew that sometimes F_2 ratios were a modification of the 9:3:3:1 expectations because of epistasis, but the F_2 numbers they observed in this experiment did not fit a modified Mendelian ratio either (although they did try to fit it to a 7:1:1:7 ratio). Besides, the two genes affected quite different traits, unlike the epistatic examples with modified F_2 ratios. What could be the cause of the strange ratios they had found in F_2 progeny?

Bateson and Punnett suggested that because the two parental phenotypes were in excess in the F_2 progeny, there might be a physical connection between the parental alleles—the dominant alleles from one parent and the recessive alleles from the other parent. This phenomenon, which they termed **coupling,** interfered with independent assortment. On the other hand, they attributed the low

FIGURE 5.1
Thomas Hunt Morgan, a leading geneticist of the early twentieth century, in his fly lab.
Lowe Memorial Library, Columbia University.

TABLE 5.1 *F₂ Progeny of a Dihybrid Cross in Sweet Peas*

Phenotype	Numbers		
	Observed	*Expected*	$(O - E)^2/E$
Purple, long	296	240.2	13.1
Purple, round	19	80.1	46.5
Red, long	27	80.1	35.1
Red, round	85	26.7	127.3
	427	427	$\chi^2 = 222.0$

numbers of the F_2 types, those having a dominant phenotype at one gene and a recessive at the other, to some innate negative affinity of dominant and recessive alleles, which they termed **repulsion.**

A MECHANISTIC EXPLANATION

A mechanistic explanation for Bateson and Punnett's observations came later from the research of Morgan, using two genes in *Drosophila*. One gene affected eye color (recessive purple, *pr,* and dominant wild-type red, *pr+*, alleles), and the other gene affected wing development (recessive vestigial, *vg,* and dominant normal, *vg+*, alleles). Instead of examining the F_2 progeny of a dihybrid cross, he used a testcross for both genes simultaneously, as outlined in figure 5.2. Note that both wild-type alleles were originally in one parent of the F_1 and that the recessives were in the other parent. The value of such a cross

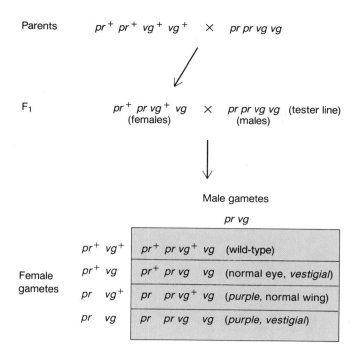

Parents $pr^+\ pr^+\ vg^+\ vg^+$ $\times$ $pr\ pr\ vg\ vg$

F$_1$ $pr^+\ pr\ vg^+\ vg$ $\times$ $pr\ pr\ vg\ vg$ (tester line)
(females)   (males)

Male gametes

$pr\ vg$

Female gametes		
$pr^+\ vg^+$	$pr^+\ pr\ vg^+\ vg$	(wild-type)
$pr^+\ vg$	$pr^+\ pr\ vg\ vg$	(normal eye, *vestigial*)
$pr\ vg^+$	$pr\ pr\ vg^+\ vg$	(*purple*, normal wing)
$pr\ vg$	$pr\ pr\ vg\ vg$	(*purple*, *vestigial*)

FIGURE 5.2
The genotypes and phenotypes from a two-gene testcross using the *vestigial* and *purple* mutants.

TABLE 5.2   *Results of Morgan's First Testcross*

Phenotype	Gamete	Numbers		
		Observed	Expected	$(O-E)^2/E$
Wild-type	pr^+vg^+	1,339	709.75	558
Normal, vestigial	pr^+vg	151	709.75	440
Purple, normal	$pr\ vg^+$	154	709.75	435
Purple, vestigial	$pr\ vg$	1,195	709.75	332
		2,839	2,839.00	$\chi^2 = 1,765$

TABLE 5.3   *Results of Morgan's Second Testcross*

Phenotype	Gamete	Numbers		
		Observed	Expected	$(O-E)^2/E$
Wild-type	pr^+vg^+	157	583.75	312
Normal, vestigial	pr^+vg	965	583.75	249
Purple, normal	$pr\ vg^+$	1,067	583.75	400
Purple, vestigial	$pr\ vg$	146	583.75	328
		2,335	2,335.00	$\chi^2 = 1,289$

is that only one parent, in this case the female, is doubly heterozygous, and therefore, segregation and independent assortment would only be expected in the production of the gametes from that sex.

If the two genes assorted independently in Morgan's testcross, then the four types of female gametes, and consequently the four different phenotypes, should be equally frequent. However, as shown in table 5.2, the numbers of the parental phenotypes were many times those of the nonparental phenotypes. The chi-square value for these data is very large (with 3 degrees of freedom, a value greater than 11.34 is highly significant), indicating that the independent assortment hypothesis does not hold up. Although the four gametic numbers were not those expected from independent assortment, the two parental gametes, pr^+vg^+ and $prvg$, were nearly equal in number, and the two nonparental gametes were also nearly equal in number, but much less frequent than the parental gametes. Today we generally consider gametes with either two wild-type or two mutant alleles *coupling gametes* and those with one wild-type and one mutant allele *repulsion gametes*. In other words, the two parental gametes in table 5.2 were in coupling, and the two nonparental gametes were in repulsion.

In a second experiment, Morgan crossed two different genotypes, each of which was homozygous for a wild-type allele at one gene and homozygous for a recessive allele

at another gene ($pr^+pr^+vgvg \times prprvg^+vg^+$). The F$_1$ flies, double heterozygotes, were then crossed to the tester line $prprvgvg$. The results of this cross were quite different from the first cross (see table 5.3). In this case, the progeny numbers in the F$_2$ of the single dominants were much larger than expected, while those of the double dominants and double recessives were much lower than expected. In other words, there were more repulsion types than coupling types. Note that in this cross, the parental gametes were in repulsion and the nonparental gametes were in coupling.

To explain his observations, Morgan suggested that the genes having the alleles for these traits are on the same chromosome. As a result, alleles at these genes tend to remain associated between generations because they are physically linked to each other. A schematic representation of this explanation is given in figure 5.3, where the wild-type alleles are on the blue chromosomes and the recessive alleles are on the red chromosomes. The F$_1$ flies have one of each type of chromosome. However, instead of these chromosomes always staying intact during the production of gametes, in some chromosomes, recombination occurs between the chromosomes. As a consequence, these gametes contain nonparental chromosomes composed of the pr^+ and vg alleles or the pr and vg^+ alleles. A similar explanation is true for the cross discussed

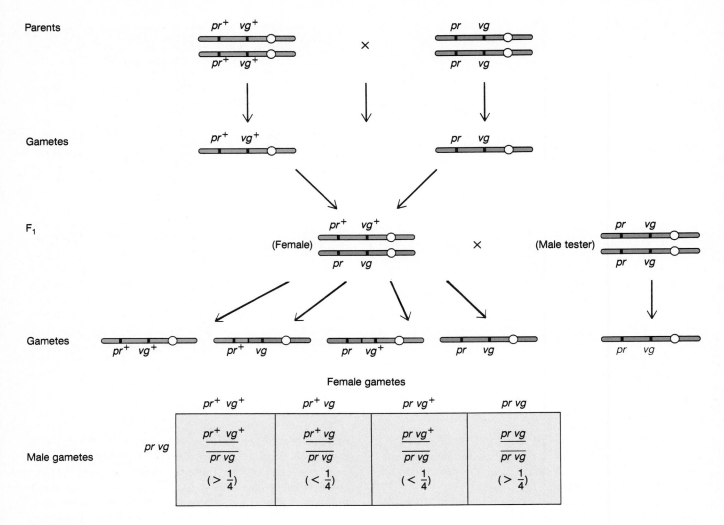

Parents

pr^+ vg^+ × pr vg
pr^+ vg^+ pr vg

Gametes

pr^+ vg^+ pr vg

F₁

pr^+ vg^+
(Female) × (Male tester) pr vg
pr vg pr vg

Gametes

pr^+ vg^+ pr^+ vg pr vg^+ pr vg pr vg

Female gametes

	pr^+ vg^+	pr^+ vg	pr vg^+	pr vg
Male gametes pr vg	$\dfrac{pr^+\ vg^+}{pr\ vg}$ $(> \frac{1}{4})$	$\dfrac{pr^+\ vg}{pr\ vg}$ $(< \frac{1}{4})$	$\dfrac{pr\ vg^+}{pr\ vg}$ $(< \frac{1}{4})$	$\dfrac{pr\ vg}{pr\ vg}$ $(> \frac{1}{4})$

FIGURE 5.3

A schematic representation of the gametes and progeny produced in a testcross, with coupling in one parent.

in table 5.3, except that the parental chromosomes have one dominant and one recessive allele, rather than two dominants or two recessives.

The intact chromosomes are often called **parental chromosomal types,** while the new combinations are called **nonparental chromosomal types.** As we mentioned in chapter 4, the physical exchange through which new chromosomal types are formed is termed **crossing over** or **recombination** and is often indicated in diagrams by an X joining homologous chromosomes. The other common names for parental and nonparental chromosomal types are **nonrecombinant** and **recombinant,** respectively, reflecting the process involved. We will discuss the biochemical mechanisms of recombination in a subsequent chapter.

At this time, we can envision that the homologous chromosomes sometimes break but then reattach to the other member of the homologous pair. The farther apart two genes are on a chromosome, the more likely it is that such breakage-reunion between them will occur.

If a breakage and reunion were to take place between homologous chromosomes, one might expect some cytological manifestation of the event. In fact, it was observed in meiosis that sometimes nonsister chromatids appear to be attached to each other, forming a cross configuration. As we mentioned in chapter 4, this structure (termed a **chiasma**) is a cytological observation resulting from recombination.

Genes that are on the same chromosome are said to be **linked,** and the general phenomenon in which genes occur on the same chromosome is called **linkage.** We can refer to alleles that are linked on the same chromosome as being in either the coupling phase—that is, both dominants of pr^+vg^+ or both recessives as $prvg$—or in the repulsion phase, pr^+vg or $prvg^+$. Genotypes containing the two different coupling chromosomes are called coupling double heterozygotes and can be indicated by $pr^+vg^+/prvg$. Genotypes with two different repulsion chromosomes are called repulsion double heterozygotes and can be expressed as $pr^+vg/prvg^+$. Note that the terms coupling and repulsion initially came from traits with complete dominance in which the dominant alleles were on the same chromosome. In other words, the definition is based on the phase in these initial crosses, an artifact that becomes apparent when we examine traits having intermediate dominance or codominance.

S hortly after the rediscovery of Mendel's principles, exceptions to the principle of independent assortment were discovered. The tendency for the alleles at different genes on parental chromosomes to remain together is explained by linkage of genes on a chromosome.

■

RATE OF RECOMBINATION

When two genes are on different chromosomes, all four gametic types (or phenotypes) are represented equally in the progeny of a testcross. In other words, independent assortment results in equal numbers of parental and nonparental gametes or offspring. However, when two genes are linked on a chromosome, the proportion of recombinant gametes is less than 50%, the specific amount depending upon the physical proximity of the two genes on the chromosome. As an extreme, if the two genes are very, very close to each other, the number of recombinant gametes or offspring should be close to zero. We can quantify the **rate of recombination** (r) as the proportion of recombinant offspring observed in a testcross or in other crosses. For example, in table 5.2, 305 (151 + 154) of 2,839 progeny were recombinant, making the rate of recombination 305/2,839 = 0.107. This value suggests that the genes are relatively close to each other but not exactly at the same location on the chromosome.

A general way to express the proportions of recombinant and nonrecombinant gametes is given in table 5.4. It is assumed that the parent is doubly heterozygous, either in coupling (second column) or in repulsion (third column).

TABLE 5.4 *Proportions of Gametes*[1]

Gamete	Parental Genotype	
	Coupling AB/ab	*Repulsion Ab/aB*
AB	½(1 − r)	½r
Ab	½r	½(1 − r)
aB	½r	½(1 − r)
ab	½(1 − r)	½r
	1	1

[1]r indicates the frequency of recombinations.

For a coupling parent, a proportion (r) of the gametes *Ab* and *aB* are recombinant. Because we expect equal proportions of these two recombinant classes, the probability of each class is ½r. Likewise, the proportion of nonrecombinant gametes, 1 − r, is equally divided between *AB* and *ab*. The same logic applies in determining the expected proportion of gametes for the repulsion genotype, although the recombinant and nonrecombinant gamete categories are reversed.

THE LINKAGE MAP

Alfred Sturtevant, a student of Morgan's, was assigned the project of consolidating the early information on crossing over in *D. melanogaster.* He conceived an approach using the recombination data to describe the physical relationship of genes on a chromosome in a linear arrangement called a **linkage map** or **genetic map.** A genetic map uses the frequency of recombinant gametes (or offspring) from a testcross, or from other crosses, as a measure of the distance between two genes. For example, 10.7% of the offspring from the cross in table 5.2 were recombinants between genes *pr* and *vg*, making the **map distance** between them 10.7 **map units.** Morgan and Sturtevant suggested that distance along the chromosome is measured in units determined by the percentage of recombination. In fact, later studies have shown that genetic distance measured by this statistical approach is generally similar to cytologically or biochemically measured distances on a chromosome.

Morgan and his students discovered a number of mutants in *D. melanogaster* that showed X-linked inheritance. When they crossed strains differing in two of these mutants and examined their progeny, they found substantial variation in the frequency of recombination for different gene pairs (table 5.5). For example, out of 21,736 potentially recombinant gametes between the genes for

Genes	Recombination Frequency
yellow (y)—white (w)	$\dfrac{214}{21{,}736} = 0.010$
yellow (y)—vermilion (v)	$\dfrac{1{,}464}{4{,}551} = 0.322$
white (w)—vermilion (v)	$\dfrac{471}{1{,}584} = 0.297$
vermilion (v)—miniature (m)	$\dfrac{17}{573} = 0.030$
white (w)—miniature (m)	$\dfrac{2{,}062}{6{,}116} = 0.337$
white (w)—rudimentary (r)	$\dfrac{406}{898} = 0.452$
vermilion (v)—rudimentary (r)	$\dfrac{109}{405} = 0.269$

TABLE 5.6   *Chromosomal Locations for the Seven Traits in Mendel's Peas*

Character	Chromosome
Seed color (yellow—green)	1
Flowers (violet—white)	1
Pods (smooth—wrinkled)	4
Flowers (axial—terminal)	4
Stem (tall—short)	4
Pods (green—yellow)	5
Seed form (round—wrinkled)	7

DROSOPHILA RESEARCH LAB

Cartoon by J. Chase.

yellow and white, only 214, or 1%, were recombinants. On the other hand, 32.2% of the gametes for the genes for yellow and vermilion were recombinants.

Using the data in table 5.5, Sturtevant determined the physical relationship of these five genes and suggested that they formed a linear array. First, *y* and *w* had very few recombinant gametes, indicating that they were close to each other. Second, the next lowest level of recombination between *w* and another gene was with *v*, which had 0.297 recombinants. But because the frequency of recombination between *y* and *v* is greater than that between *w* and *v*, the sequence of these three genes must be *y–w–v*, with *w* in between *y* and *v*. Next, the gene closest to *v* is *m*, only 3 map units away. It must be to the right of *v*, because the distance *w–m* is approximately the sum of *v–m* and *w–v*. Finally, *r* is loosely linked to *v* but shows no evidence of linkage to *w* in genetic crosses. As a result, *r* must be to the right of *v*, making the array of five genes *y–w–v–m–r*.

Sturtevant constructed a map beginning with *y* at 0.0 on the left side and using the frequencies of recombination between adjacent genes (figure 5.4). That is, he put *w* at 1.0, *v* at 30.7 (1.0 + 29.7), *m* at 33.7 (30.7 + 3.0), and *r* at 57.6 (30.7 + 26.9). Notice in the figure that the positions for *w, v, m,* and *r* today are slightly different from those given by Sturtevant because many intermediate genes have since been used to more accurately map them.

As indicated here, a genetic map can be constructed by estimating the recombination frequency of all pairs of genes. However, a more efficient approach, as we will discuss later, is to use three genes simultaneously. Using more marker genes also allows identification of cases with more than one crossover.

Notice that although both the *white* and *rudimentary* loci are X-linked, the data in table 5.5 indicate a very high recombination frequency between them, nearly that which would be found if the genes were on different chromosomes. This suggests that genes far apart on the same chromosome would not show evidence of linkage in genetic crosses. Genes on the same chromosome, whether or not they show linkage in genetic crosses, are called **syntenic genes**.

People have often wondered why Mendel did not observe linkage in his dihybrid crosses. It turns out that, although he picked seven traits and there are seven chromosomes in garden peas, two of the traits he studied are determined by genes on chromosome 1 and three by genes on chromosome 4 (see table 5.6). Thus, out of the twenty-one different possible pairs of genes (the number of combinations of seven genes taken two at a time), four pairs of genes are syntenic. However, three of these pairs are far apart on the chromosome and show no indication of linkage in hundreds of crosses. The remaining pair of genes, containing the alleles that determine pod morphology and stem length, should have shown linkage in genetic crosses but were not investigated by Mendel. In

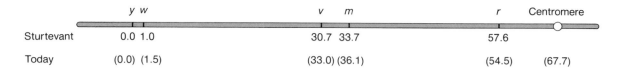

y w
v m
r
Centromere
Sturtevant
0.0 1.0
30.7 33.7
57.6
Today
(0.0) (1.5)
(33.0) (36.1)
(54.5)
(67.7)

FIGURE 5.4
The five X-linked genes studied by Sturtevant, showing the positions he gave and those recognized today.

Parents: + + + / + + + × y w m / Y

↓

+ + + / y w m (F$_1$ females) × y w m / Y (tester line males)

↓

	Gametes	Numbers	Frequency
Parental	+ + +	3,501	0.664
	y w m	3,471	
Recombination of *m* with *y* and *w*	+ + m	1,754	0.329
	y w +	1,700	
Recombination of *y* with *w* and *m*	y + +	28	0.0057
	+ w m	32	
Recombination of *w* with *y* and *m*	+ w +	6	0.00086
	y + m	3	
		10,495	1.0

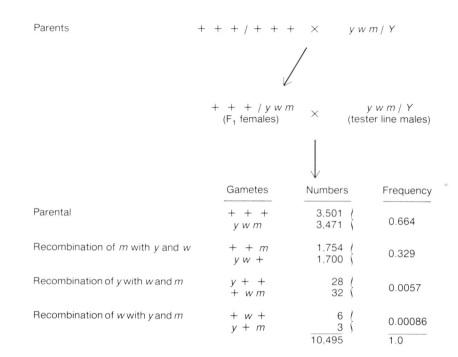

FIGURE 5.5
A testcross of females heterozygous for three X-linked genes, giving the number and frequency of different progeny types.

other words, because only one of twenty-one pairs of traits would have shown linkage and Mendel examined only a few pairs, it is not surprising (and perhaps fortunate) that he did not observe patterns differing from the expectations of independent assortment.

The rate of recombination estimates the linkage of genes and ranges from 0 for extremely tightly linked loci to ½ for very loosely linked loci (or those on different chromosomes). The relationship of genes on a chromosome can be shown on a genetic map.

THREE-POINT CROSSES

Although the linearity of the chromosome was apparent from the data for pairs of X-linked genes, complete substantiation of this model came by simultaneously crossing three different genes. One test again used the three X-linked genes in *Drosophila*—*y, w,* and *m*. In this experiment, a backcross of a female in coupling for all three genes was done (figure 5.5), and the gametes in the progeny were scored, receiving either an X chromosome with *y, w,* and *m* or a Y chromosome from the male tester parent. In the F_2 gametes from the F_1 triple heterozygote female, eight types are possible, two of which are parental and six of which are recombinant.

Assuming the three genes involved are arranged in a linear fashion, there are three possible orders: *y–w–m, w–y–m,* and *y–m–w,* each having a different gene in the center. Two of the recombined classes can be generated by a single recombination event, but the third must be the product of two crossovers involving the same chromatids. Because the likelihood of such a double crossover is small (on the order of the product of the probability of two different crossover events), the smallest recombined class should represent double recombinants. The smallest observed recombinant class in figure 5.5 is that composed of

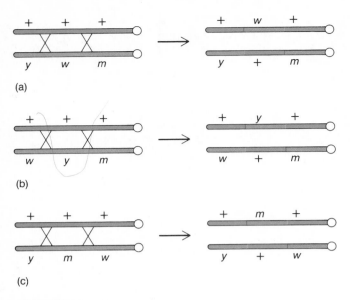

(a)

(b)

(c)

FIGURE 5.6
The three possible orders of genes *y, w,* and *m* with gametes produced
from double recombination. Order (*a*) with gene *w* in the center is
consistent with the data in figure 5.5.

the complementary gametes + *w* + and *y* + *m*. Using the
three different gene orders, double recombination can only
produce these rare gametes if the order of the genes is
y–w–m (figure 5.6). Double recombination for the other
two gene orders does not result in + *w* + and *y* + *m* as
the rarest class but produces the other, more common recombinant classes.

With three linked genes, there are three different recombination frequencies between pairs of loci: between *w*
and *m,* between *y* and *w,* and between *y* and *m*. From figure
5.5, the total frequency of recombination between *w* and
m is the sum of the two recombinant classes that have recombined in this region, or 0.329 + 0.00086 = 0.330.
(Double crossover gametes must have had recombination
between *w* and *m* as well.) Likewise, for *y* and *w* the frequency of recombination is the sum 0.0057 + 0.00086 =
0.0066. Between *y* and *m,* the frequency is 0.329 + 0.0057
= 0.335. (Another way to ascertain the order of the genes
is to note that the largest of these recombination values
must be between the outside genes, *y* and *m* in this case.)

Given that the order of the genes is *y–w–m,* a double
recombinant must involve both an exchange between loci
y and *w* and one between *w* and *m*. Using the recombination frequencies just calculated, the expected frequency
of a double recombination event (using the product rule)
should be (0.007)(0.330) = 0.00231. The observed frequency of double recombination is 0.00086, which is
somewhat lower than that expected (see further discussion later in this chapter). The map distance between *y*
and *m* is then calculated as the sum of the recombination
between *y* and *w* plus that for *w* and *m,* or 0.007 + 0.330
= 0.337. Notice that the same information is obtained

from one three-point cross as from three crosses involving
different pairs of genes. In addition, three-point crosses
allow us to identify gametes produced by double recombination.

Using such crosses and related techniques, researchers have constructed a genetic map of *Drosophila
melanogaster* that is one of the most complete for any eukaryotic organism. Some of this information is given in
figure 5.7, listing the positions of some of the important
genes on the four different chromosomes. Besides the X
chromosome, which has 70 map units, *Drosophila* has
three autosomes. The two large autosomes have 108 and
106 map units, while the small dot chromosome has only
3 map units. The total length of these chromosomes has
been determined by the summation of empirically mapped
intervals as discussed previously. Overall then, the *Drosophila* genome has 287 map units. The linkage maps of
other organisms used in genetic research, such as *Neurospora,* corn, and mice, are also quite well-known.

> *T* hree-point crosses are used to determine the order
> of genes on a chromosome.

∎

INTERFERENCE

Recombination frequency is defined as the proportion (or
probability) of exchange between two linked genes. As we
have suggested, more than one recombinational exchange
may occur on a chromosome. Therefore, it is important to
know whether these recombinations are independent of
each other or whether they somehow influence each other.
If different recombinant events are independent, the probability of two such events occurring in the same gamete
should be equal to the product of the probability of the
separate events. For the *y, w,* and *m* chromosomes discussed previously, the observed frequency of double crossovers was 0.00086, somewhat lower than the 0.00231
proportion expected. Fewer double crossovers than expected are generally found in other organisms as well.

A standard way of describing the difference between
the observed and expected numbers of double crossovers
was first used by H. J. Muller. The method takes the ratio
of the observed to expected numbers (or proportions) of
double crossovers and subtracts it from unity. This value
gives a measure of the **interference** (I) of one crossover on
another. This expression is:

$$I = 1 - \frac{\text{Observed double recombinants}}{\text{Expected double recombinants}}$$

Thus, for the *y, w,* and *m* data, I = 1 − 0.00086/0.00231
= 1 − 0.374 = 0.626. (Sometimes the ratio of observed

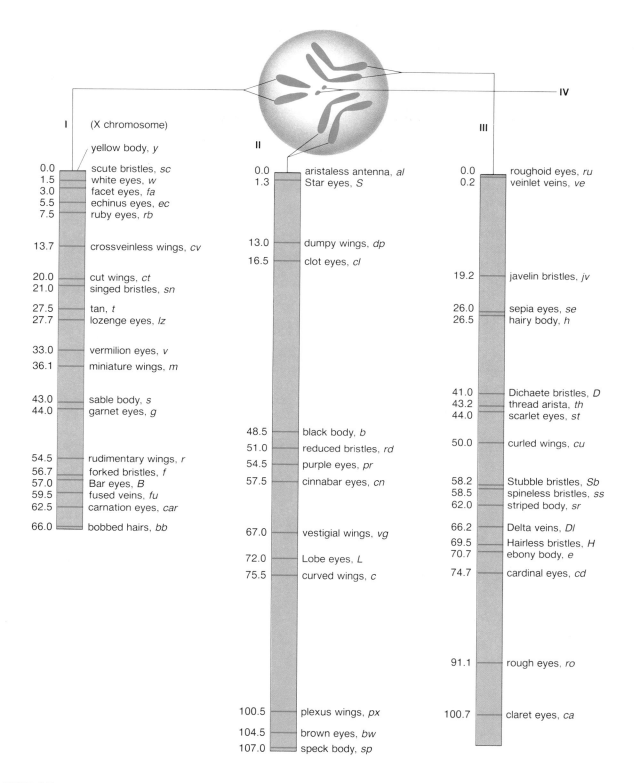

FIGURE 5.7

The genetic map of *D. melanogaster,* showing some of the important
genes on the three large chromosomes.

From W. E. Sinnot, et al., Principles of Genetics, *5th ed. Copyright © McGraw-Hill Book Company, New York, NY. Reprinted by permission of McGraw-Hill, Inc.*

to expected doubles is used, a value known as the **coefficient of coincidence.**) Positive values such as this one indicate interference between recombination events, while values near zero suggest no interference—that is, that different recombinant events are independent of each other. For closely linked genes, the measure of interference may often be near unity, while for widely separated genes on the same chromosome, I is often near zero.

One theory to explain the infrequency of double crossovers is that the high values of interference for closely linked genes appear to be associated with a physical inflexibility of the chromatids that somewhat inhibits them from bending enough for two crossovers to occur in a very short region. However, interference may occur in genes up to 30 map units apart, indicating that other factors must be important as well. Interestingly, the centromere for metacentric chromosomes in *Drosophila* (and perhaps other organisms) acts as a barrier for interference. In other words, crossing over in one arm of a metacentric chromosome has no inhibitory effect on recombination in the other arm.

We should also note that when genes are far apart on a chromosome, the observed frequency of recombination may be less than that expected from knowledge of map distances. For example, we know that genes w and r are 53 map units apart on the X chromosome. First, the maximum proportion of recombinant gametes when the genes are on different chromosomes is 0.5, based on the principle of independent assortment, so we would not expect 0.53

recombinant gametes. Second, multiple even numbers of recombination will reduce the observed recombination frequency. Figure 5.8 illustrates that both double and quadruple crossovers appear as nonrecombinant gametes when there are no intermediate marker genes. As a result, the observed proportion of recombinant gametes for genes 50 map units apart is generally less than 0.5. Remember that the proportion of recombinants between w and r observed by Sturtevant was 0.452. Only when genes are 70 to 80 map units apart does the observed frequency of recombinants approach 0.5.

> *T*he frequency of double recombination between closely linked genes is lower than that expected, a phenomenon called interference.

■

LINKAGE IN HUMANS

Obviously, planned genetic crosses are not possible in humans, and the number of progeny from any one mating is small. As a result, until the recent introduction of biochemical techniques, linkage had been determined for only a few human genes, mainly genes on the X chromosome. By studying informative matings (particular pedigrees segregating for the two genes of interest), tight linkage was only established between the ABO blood group system

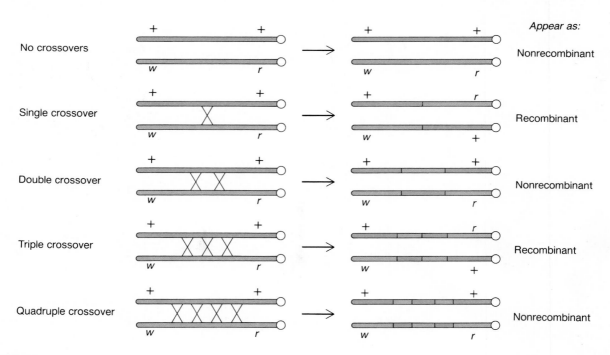

FIGURE 5.8
The types of gametes produced by different numbers of crossovers between two genes.

and the nail-patella syndrome (a condition causing abnormal and underdeveloped fingernails and small or absent kneecaps), between the β- and δ-globin genes, and between a few other gene pairs.

Mapping the X chromosome is made easier because recombination between X chromosomes only takes place in females (remember that males have only one X chromosome), and potential recombinant gametes can be directly identified in the male progeny. In other words, if the female parent is a double heterozygote with known gametic types, and if a sizeable number of male offspring are available for study, the recombination frequency can be estimated.

An example of linkage is given in figure 5.9 for the X-linked traits color blindness and hemophilia. In this case, a woman (II-1) was unaffected by either recessive trait but produced three sons who had both hemophilia and color blindness, one son who was just color-blind, and two who were normal. Because her father (I-2) was unaffected by both traits, the woman must have inherited an X chromosome with normal alleles from him and an X with both recessive alleles from her mother. In other words, she was a double heterozygote in coupling phase ($H\ C/h\ c$). Therefore, the five sons who were either normal ($H\ C/Y$) or affected with both traits ($h\ c/Y$) received a nonrecombinant gamete from her, while the sixth son (III-5) received a recombinant gamete, $H\ c$. The frequency of recombination in this pedigree is then $\frac{1}{6} = 0.167$. Of course, to obtain a good estimate of the map distance between these genes, many such families

must be examined, and the results from different families must then be combined using various statistical techniques. Generally, pedigrees segregating for these two traits do not have recombinants because the genes are closely linked at the end of the long arm of the X chromosome. These genes and a number of other X-linked genes, along with their relative map positions, are shown in figure 5.10.

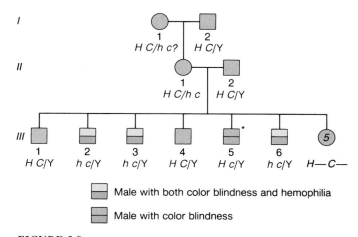

FIGURE 5.9
In this pedigree, a female (II-1) heterozygous for two X-linked genes has five nonrecombinant sons (1, 2, 3, 4, and 6) and one recombinant son (5, denoted by an asterisk). In addition, there are five unaffected daughters.

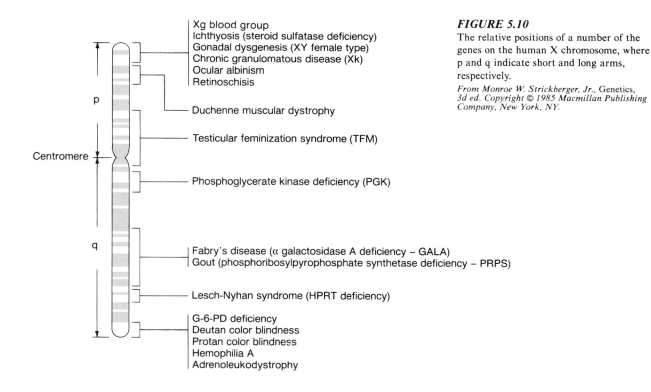

FIGURE 5.10
The relative positions of a number of the genes on the human X chromosome, where p and q indicate short and long arms, respectively.

From Monroe W. Strickberger, Jr., Genetics, 3d ed. Copyright © 1985 Macmillan Publishing Company, New York, NY.

Surprisingly, the recombination rate for autosomal genes in males and females is often different. In fact, the British evolutionary biologist J. B. S Haldane suggested that in organisms with a chromosomal mechanism of sex determination, recombination is generally higher in the homogametic sex. In mammals, the homogametic sex is usually the female. And in both humans and mice, the recombination rate is higher in females than in males—about 50% to 100% higher for the human gene pairs examined. In *Drosophila,* which also has homogametic females, no recombination occurs in males; but in the silk moth *Bombyx mori,* which has heterogametic females, the females show no recombination.

The human genome is not yet mapped in the same detail as the *Drosophila* genome. However, a rough estimate of the total map length for humans can be obtained by observing the average number of chiasmata in meiosis I. The estimated mean number of chiasmata in male meiosis is about fifty per cell, or slightly more than two per bivalent. Assuming that a chiasma occurs approximately every 50 map units, the total map distance is around 2,000 to 2,500 map units, spread over the twenty-three human chromosomes—nearly ten times the map length of the *Drosophila* genome.

For many human genes, it is not known on which chromosome a gene is located, much less what linkage may exist within the chromosome. In some cases, chromosomal abnormalities can give a clue to the location of genes. For example, trisomies are used to locate genes in a number of crops (see figure 4.36), and in theory, trisomy 21 (Down syndrome) individuals should have a different level of enzymes produced by genes on the affected chromosome. However, Down individuals exhibit many abnormalities, so it is not clear whether any biochemical differences are due to three doses of a gene or to the abnormal physiological condition of these individuals. Likewise, individuals with deletions might also have lower levels of proteins produced by genes in the deleted region.

In both *Drosophila* and prokaryotes, **deletion mapping** is a technique through which genes can be located to a particular chromosome segment (see also figure 4.22). For example, assume an individual is heterozygous for a deletion on a given chromosome that has a number of codominant genes (blood group, enzyme, or DNA variants). If different alleles are present at each gene, each genotype should be recognizable as a heterozygote (figure 5.11*a*). However, when a gene is in the deleted region, a heterozygote should always appear as a homozygote (figure 5.11*b*). To distinguish deletion heterozygotes from true homozygotes, some pedigree information is necessary. For example, if a deletion heterozygote B_2- was mated with a B_1B_2, there could be B_1- offspring appearing as B_1B_1 homozygotes. But if the parent really was a B_2B_2 homozygote and was mated with a B_1B_2, the offspring should be half B_1B_2 and half B_2B_2, with no individuals that are B_1- (and appear as B_1B_1 homozygotes).

SOMATIC CELL HYBRIDIZATION

Until recently, it was extremely difficult to locate human genes to specific chromosomes. One ingenious approach is **somatic cell hybridization.** In this technique, cells from two different species are fused in the laboratory, using a fusion agent such as Sendai virus. The Sendai virus has several points of attachment to its host cell, a characteristic that allows it to serve as a vehicle to fuse two quite different host cells. The chromosomes of the two species in the fused cells and the genetic trait (or traits) of interest are both monitored over time; human cells are usually hybridized with mouse or hamster cells. Eventually, the human chromosomes are randomly lost, so that different cell lines lose different human chromosomes. The chromosome that contains the gene of interest is determined by noting which cell lines still carry the trait and finding a human chromosome that they have in common.

The first gene to be located in this way codes for thymidine kinase, an enzyme important in DNA replication. Mouse cells deficient for this enzyme (symbolized by Tk^-) were hybridized with human cells having the gene for the enzyme (Tk^+). These hybrid cells were then grown on a medium that required the cells to produce thymidine kinase in order to survive. Figure 5.12 shows the chromosomes found for five cell lines that were able to produce thymidine kinase. Notice that in these cell lines, human chromosome 17 is the only chromosome present in all lines. Therefore, these data indicate that the thymidine kinase gene must be on chromosome 17.

This experimental technique can be extended in several ways. For example, some cell lines also have deletions of a particular chromosomal segment. Hybrid cells with a deleted long arm of human chromosome 17 are Tk^-, indicating that the *TK* gene must be on the long arm of chromosome 17. Furthermore, if several genetic traits are

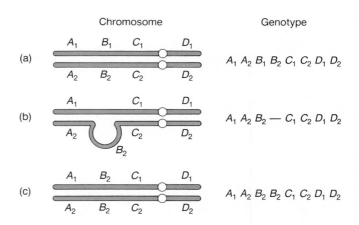

FIGURE 5.11

(*a*) A normal heterozygote for codominant alleles; (*b*) a heterozygote missing the gene *B* region; and (*c*) an individual homozygous for B_2.

monitored simultaneously, the linkage relationship of the genes can be determined by the joint occurrence or absence of their protein products.

Studies locating genes to particular chromosomes in different organisms have shown that genes in related species are often linked in the same pattern. For example, the same genes appear on the X chromosome among related mammals. Other genes, such as *TK,* are on the same chromosome in all the great apes, supporting the evidence from chromosomal banding that indicates a high genetic homology for these species (see chapter 4). In other words, if either the linkage relationship between two genes or the location of a gene on a chromosome is established in one species, similar species may have the same relationship. Knowledge of this sort could help shortcut the task of determining linkage patterns in each species. For example, the chromosomal location of the albino gene in humans is not known. However, the homologous albino genes in mice and cats are on regions with banding patterns like those on human chromosome 11p.

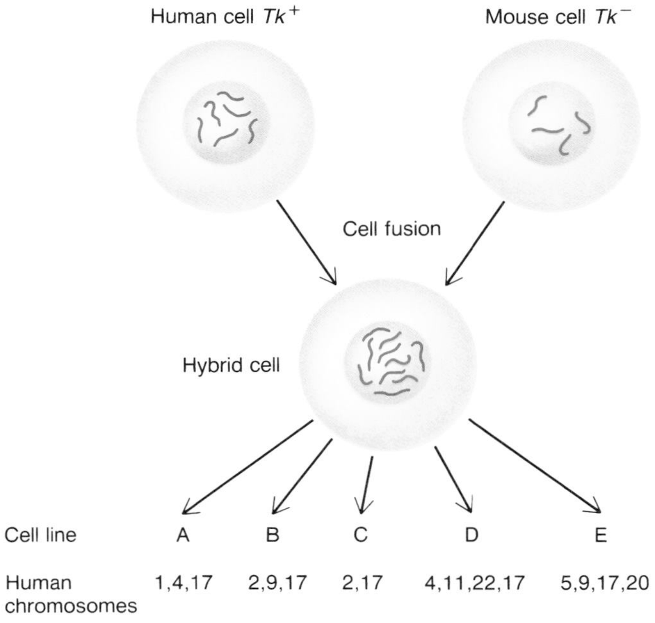

*I*n the past, linkage of genes in humans generally was determined by accumulating information from informative matings. However, the technique of somatic cell hybridization with mouse cells has allowed many more human genes to be assigned to particular chromosomes.

FOUR-STRANDED CROSSING OVER

Morgan's theories about the linkage of genes and crossing over were supported by extensive data. However, it was not clear initially whether recombination took place between chromosomes before they replicated (at the two-stranded stage) or after replication (at the four-stranded stage). If crossing over took place before replication, all four chromatids would be recombinants, while if it took place after replication, only two of the four chromatids would be recombinants (figure 5.13).

FIGURE 5.12

Fusion of human and mouse cells as a technique to identify the chromosomal location of the gene for thymidine kinase. The only human chromosome the five cell lines have in common is 17, demonstrating that the *Tk* gene must be on this chromosome.

Cell line	A	B	C	D	E
Human chromosomes	1,4,17	2,9,17	2,17	4,11,22,17	5,9,17,20

Human cell *Tk*⁺   Mouse cell *Tk*⁻

Cell fusion

Hybrid cell

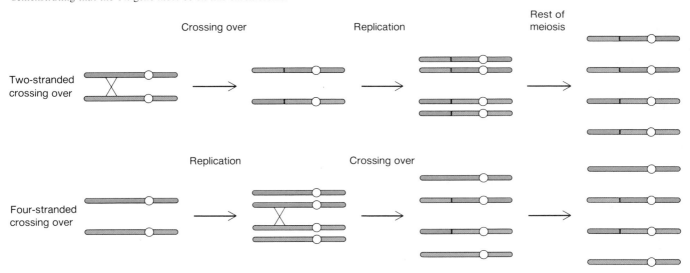

FIGURE 5.13

The different consequences of two-stranded and four-stranded crossing over. Four-stranded crossing over produces both parental and recombinant gametes, while two-stranded crossing over produces only recombinant gametes.

From AN INTRODUCTION TO GENETIC ANALYSIS *3/E. By David T. Suzuki et al. Copyright©1976,1981,1986 by W. H. Freeman and Company. Reprinted with permission.*

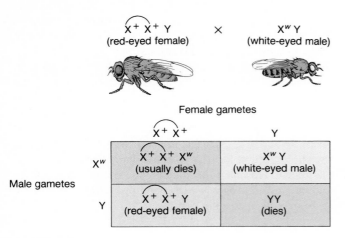

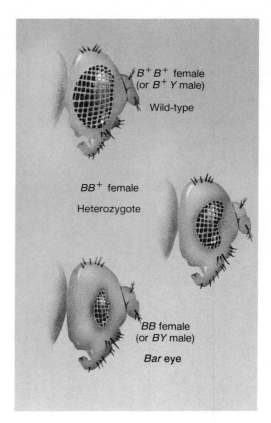

FIGURE 5.14

A cross between an attached X wild-type female and a white-eyed male. Notice that the surviving progeny are exactly like their parents.

FIGURE 5.15

The phenotypes of wild-type, heterozygous *Bar* eye, and homozygous *Bar* eye.

To distinguish between these alternatives, a genetic system was necessary in which all of the genetic products from meiosis could be identified. The first such system was developed in *Drosophila,* utilizing the **attached X,** a complex chromosome that has two X chromosomes attached at its centromere. As a result of this attachment, gametes from females having the attached X chromosome and a separate Y chromosome ($\overline{XX}Y$) contain either two X chromosomes or no X chromosomes (just a Y). An example of a cross in which both X chromosomes carry the wild-type allele at the white locus is given in figure 5.14. $\overline{XX}Y$ females with the wild-type allele on both X chromosomes are crossed to normal white-eyed males. The viable offspring are identical to the parents, because the YY flies always die and the $\overline{XXX}$ flies usually die.

When the alleles are different on the two attached X chromosomes, two-stranded and four-stranded crossing over can be differentiated. It is useful to use a mutant that is identifiable as a heterozygote in this experiment. For example, the *Bar* mutant reduces the number of eye facets from the normal number of approximately 780 to about 360 in the heterozygotes and about 70 in the homozygote *BB* (see figure 5.15).

The top row of figure 5.16 indicates the types of chromosomes obtained when no crossing over occurs. In this case, all the gametes with X chromosomes have the heterozygous attached X and therefore the heterozygous *Bar* phenotype. Next, if crossing over takes place at the two-stranded stage (before replication), it will involve both strands, even though there is exchange between the two X chromosomes. Therefore, when replication occurs later, all gametes are again heterozygous. In other words, the outcomes for no crossing over and two-stranded crossing over are the same.

Four-stranded crossing over yields different results in the progeny. In this instance, only two of the four strands participate in recombination, so that new types are produced. With recombination at the four-stranded stage, both wild-type and *BB* homozygotes should result. In fact, this is the observation from actual experiments, indicating that recombination takes place after replication—that is, at the four-stranded stage.

TETRAD ANALYSIS

Confirmation of four-stranded crossing over was also obtained in *Neurospora crassa,* a fungus with the useful genetic attribute of having its meiotic products arranged in linear fashion. The life cycle of *Neurospora* is given in figure 5.17, illustrating mating and the production of a cell with the diploid chromosome number, which then undergoes meiosis. The meiotic products, called ascospores, are aligned in an ascus (or sac) so that the four on the left half descend from one type of segregants in meiosis I, while the four on the right half descend from the complementary segregants. In addition, within each half of

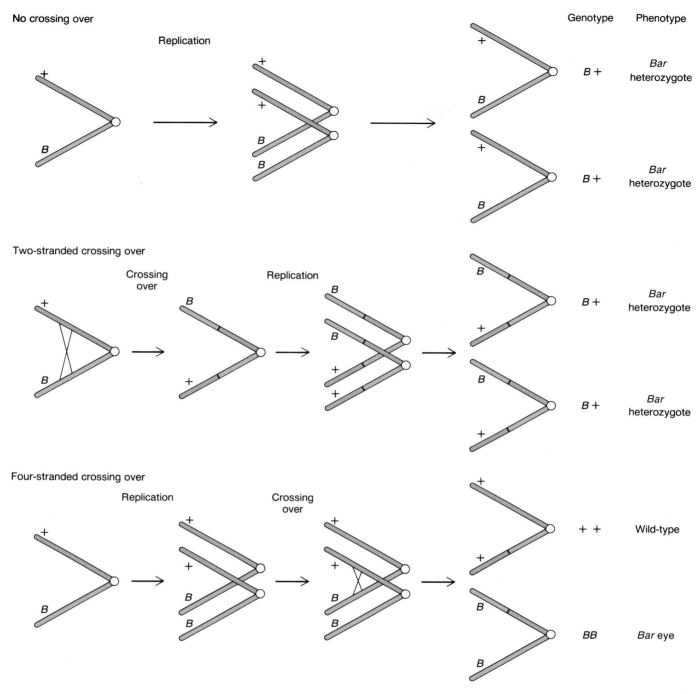

No crossing over

Replication

Genotype   Phenotype

B +   *Bar* heterozygote

B +   *Bar* heterozygote

Two-stranded crossing over

Crossing over   Replication

B +   *Bar* heterozygote

B +   *Bar* heterozygote

Four-stranded crossing over

Replication   Crossing over

+ +   Wild-type

BB   *Bar* eye

FIGURE 5.16
The gametes produced from an attached X heterozygous for *Bar* with
no crossing over (top), followed by the results expected from two- and
four-stranded crossing over. In reality, wild-type and *Bar* eye progeny
are observed, indicating that crossing over takes place at the four-
stranded stage.

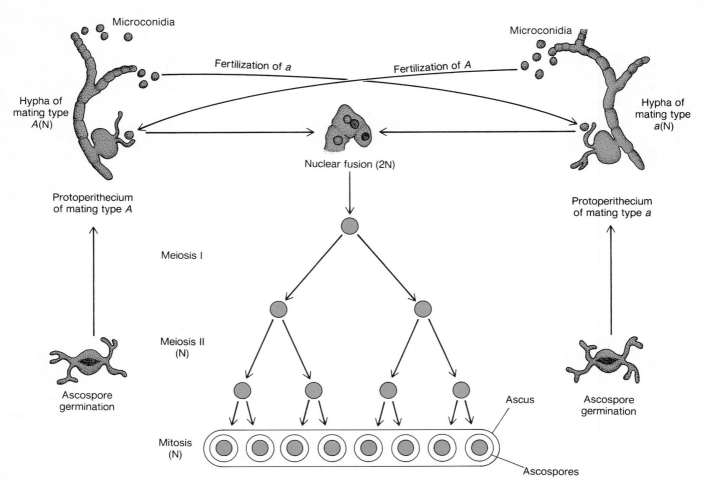

FIGURE 5.17
The production of ascospores following meiosis in *Neurospora crassa*.
The ascospores are in a linear array such that each half is descended
from the segregants at meiosis I. Each quarter is descended from
chromatids associated with a particular centromere.

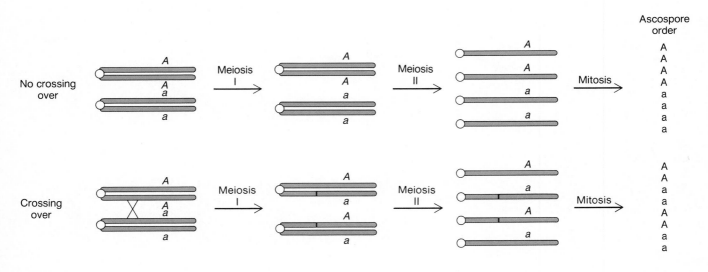

FIGURE 5.18
The ascospore order, illustrating the difference between no crossing
over and a single crossover.

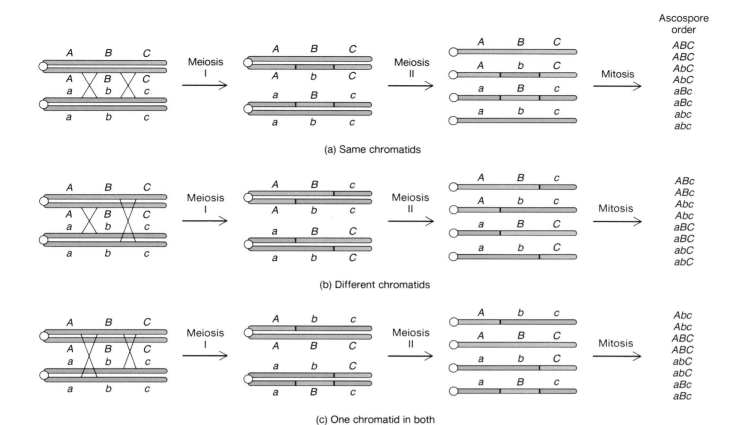

FIGURE 5.19
The ascospore order that results from the three different types of double crossovers: (*a*) Involving the same chromatids; (*b*) involving different chromatids; and (*c*) involving one chromatid in both events.

the ascospore, one-half are products from half of meiosis II and the other half from the complementary half. After meiosis is completed, one mitotic division occurs, so that each pair of ascospores is descended from one of the products of meiosis II. These four pairs of ascospores, linearly arranged, are called a **tetrad.**

When there is no crossing over in a heterozygote, one-half of the ascus has one allele and the other half has the second allele (figure 5.18). However, when crossing over does occur between the gene and the centromere, the pairs of the ascospores alternate as to which alleles they contain. This observation could only be true if recombination took place at the four-stranded stage. Two-stranded recombination would give results indistinguishable from no crossing over, assuming there were no other marker genes.

Other experiments in *Neurospora* have increased our understanding of the fundamental features of meiotic recombination. For example, the three basic ways that nonsister chromatids can have double crossing over are distinguishable in *Neurospora* (figure 5.19). First, if there are three genes on the chromosome and both crossovers involve the same chromatids, let us say the center ones in

figure 5.19*a*, then four types of ascospores are produced, with the parental ones on the outside. This is the type of double recombination event we assumed in our earlier discussion.

Second, different nonsister chromatids may be involved in the two recombination events (see figure 5.19*b*). And finally, one chromatid could be involved in both events, while the second chromatid is different in the two crossovers (see figure 5.19*c*). Notice that in this instance, one parental gametic type (*ab*) is on the outside of the ascus and the other (*AB*) is on the inside. An important observation from these results is that the products of recombination are reciprocal.

T he fact that crossing over takes place in the four-stranded stage of meiosis I was first demonstrated using attached X chromosomes in *Drosophila*. Confirmation of this theory was obtained from studies of the fungus *Neurospora*.

■

FIGURE 5.20

Top: The results of normal crossing over. Bottom: The results of mispairing and unequal crossing over that produce both wild-type and *Ultrabar* chromosomes. The number of progeny are those few abnormal progeny observed out of 20,000.

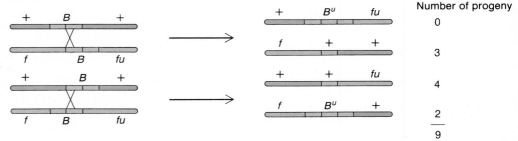

Normal pairing and normal crossing over

Mispairing and unequal crossing over

	Number of progeny
	0
	3
	4
	2
	9

FIGURE 5.21

The banding patterns of homozygous *Bar* eye mutants. Note that wild-type, *Bar*, and *Ultrabar* have one, two, and three copies of band 16A, respectively.

From Monroe W. Strickberger, Jr., Genetics, 3d ed. Copyright © 1985 Macmillan Publishing Company, New York, NY.

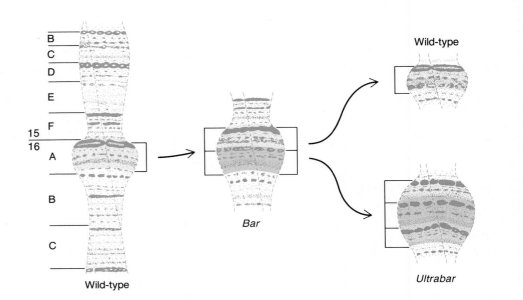

UNEQUAL CROSSING OVER

Duplication and deletion of a relatively small amount of genetic material can result from a mistake related to recombination. Chromatids are strongly attracted to their homologous counterparts in prophase I, and they pair gene by gene, as discussed earlier. Occasionally, however, pairing does not occur precisely as it should. In particular, when regions already have duplications or repeated sequences, homologous chromosomes sometimes mispair. If recombination occurs within this mispaired region, duplications or deletions can result.

The classic demonstration of this phenomenon is for the *Bar* mutant in *Drosophila melanogaster*. In stocks homozygous for *Bar*, approximately one of 1,600 new chromosomes is either wild-type or a more extreme variant having only twenty-five eye facets, called *Ultrabar* (B^U).

Instead of these being new mutants, it was shown by Sturtevant and Morgan that they are the products of recombination.

To understand the origin of *Ultrabar,* Sturtevant and Morgan constructed marked chromosomes that had the *Bar* mutations, as well as other mutations, on either side of the *Bar* region. In particular, these chromosomes had the recessive mutation for forked bristles (*f*) on one side and the recessive mutation causing fusion of two wing veins (*fu*) on the other side (see figure 5.20). The map positions of these genes are 56.7 for *f,* 57.0 for *B,* and 59.5 for *fu*; thus, a relatively small region of about 3 map units on the X chromosome can be examined. Females homozygous at *Bar,* but heterozygous (and in coupling) at these other genes, were crossed with males having the triply mutant chromosome, so that any recombinant chromosomes could

Multigene Families

Many genes are now known to exist as multigene families. Most of these families have probably arisen as a result of unequal crossing over and subsequent tandem duplication. Generally, the individual genes in a multigene family retain identical or similar functions. However, if the duplication occurred far back in the evolutionary past, the different individual genes in the family may have developed somewhat different functions.

A particularly striking multigene family is the cluster of linked genes comprising the **homeotic** genes in the flour beetle *Tribolium*. Different homeotic genes direct cells in different body segments of insects to differentiate into the appropriate anatomical structures (see figure B5.1). Five of these genes have been mapped to a chromosomal segment of 2.6 map units. In addition, the order of the genes is identical to the order of the body segments they affect (see figure B5.2). ∎

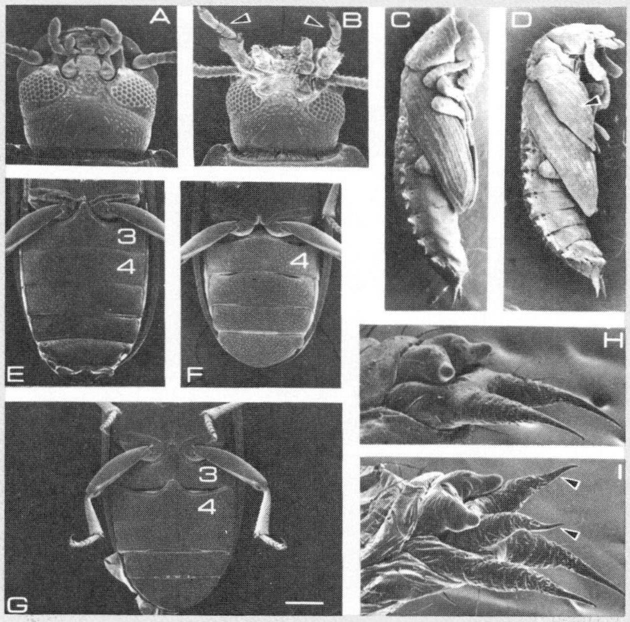

FIGURE B5.1
Normal phenotypes for five closely linked homeotic loci in a flour beetle are shown in (*a*), (*c*), (*e*), and (*h*); mutant phenotypes are shown in (*b*), (*d*), (*f*), (*g*), and (*i*). The five genes that cause the mutant phenotypes are shown in figure B5.2.
Courtesy of Richard W. Beeman, USDA.

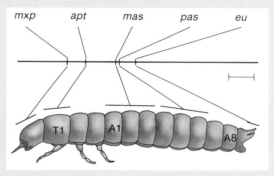

FIGURE B5.2
Five genes in *Tribolium*, showing the body segments they affect.
($\vdash\!\!\!-\!\!\!\dashv$ = one map unit length.)
Courtesy of Richard W. Beeman, USDA.

be identified. Of the approximately 20,000 progeny examined, most chromosomes were nonrecombinant; that is, their chromosomes were either $+ B +$ or $f B fu$, because the marker loci were so tightly linked to B. Of the recombinant chromosomes, nearly all had *Bar* eyes, but seven had the wild-type allele and two had *Ultrabar*. In fact, all of the chromosomes that did not have the *Bar* allele were recombinant, three being $f + +$, four $+ + fu$, and two $f B^u +$.

Confirmation that duplication of the region was responsible for the *Bar* phenotype was obtained by examining in detail the banding pattern of the X chromosome. *Bar* chromosomes have a duplication of region 16A (figure 5.21), and the subsequent unequal recombinants have either a single copy of the genetic material in this region (wild-type) or three copies (B^u).

In some cases, related genes may be closely linked, such as the different globin genes or the genes of the major histocompatibility complex (MHC) in mammals. In a

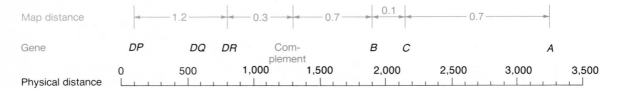

Map distance									
	←——1.2——→		←—0.3—→	←——0.7——→	←0.1→		←———0.7———→		
Gene	DP		DQ	DR	Com-plement	B	C		A
Physical distance	0	500		1,000	1,500	2,000		2,500	3,000	3,500

FIGURE 5.22
Comparison of map distance (given in map units) and physical distance (given in kilobases) for the HLA region in humans.

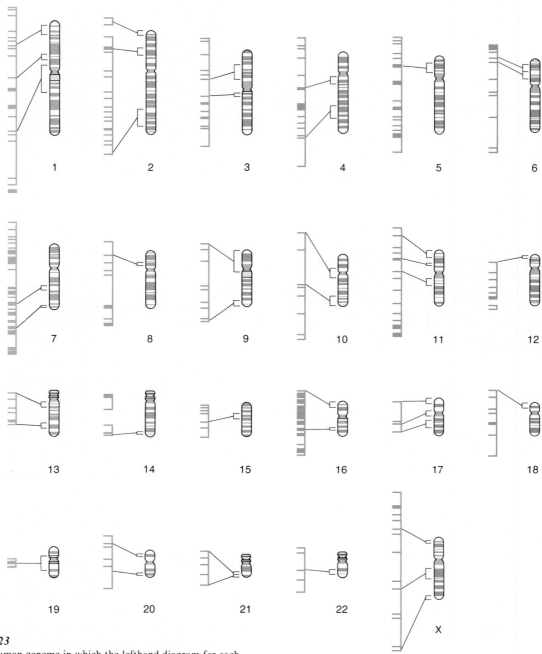

FIGURE 5.23
Map of the human genome in which the lefthand diagram for each chromosome indicates the relative map distance between different RFLPs. The righthand chromosome gives the major banding patterns along with the general location of some RFLP sites.

subsequent section of this chapter, we will discuss linkage of the genes in the MHC in humans, usually called the HLA region. It is thought that linked genes of similar structure and function are originally generated by duplication resulting from unequal crossing over. A group of related genes is called a **multigene family.**

> *I*n some cases, unequal crossing over occurs, and this can increase the amount of DNA by generating duplications.

MAP DISTANCE AND PHYSICAL DISTANCE

As we have mentioned before, the map distance between two genes estimated by genetic crosses generally corresponds with the physical distance, or length of DNA, that separates the genes. One major exception is the centromeric region, which has no recombination but can contain a sizeable length of DNA. Furthermore, it appears that recombination may often be localized into particular regions, and that it is not completely random over the length of the chromosome.

Comparing genetic and physical maps, figure 5.22 gives both the estimated map distance and the number of kilobases in the DNA (physical distance) for the major human HLA genes. For example, genes *B* and *C* are close physically and are also only 0.1 map unit apart. On the other hand, genes *DP* and *DR* are physically closer than genes *C* and *A,* but the map distance is greater. This suggests that there may be higher than expected recombination between *DP* and *DR.*

Recently, a substantial effort has been focused toward obtaining a genetic map of the human genome (see figure 5.23). Most of the landmarks on this map are DNA variants called **restriction fragment length polymorphisms (RFLPs);** we will discuss RFLPs more fully in chapter 15. Notice that some chromosomes, such as 1 and 7, are well covered by these genetic variants while others, such as 14 and 19, have very few markers. Adjacent to the genetic map of each chromosome is the physical map based on chromosome banding studies. This new map has great potential for determining the location of genes that cause particular diseases and for subsequently examining the DNA to see what has caused the defect.

> *T*he genetic map distance and the physical DNA distance generally correspond quite well. Recently, DNA variants have been employed to map the human genome.

Summary

*I*n the early 1900s, some exceptions to Mendel's principle of independent assortment were discovered—namely, that alleles at different genes on parental chromosomes tend to remain together because of linkage of genes on a chromosome. The rate of recombination estimates the linkage of genes and ranges from 0 for extremely tightly linked loci to ½ for very loosely linked loci (or for those on different chromosomes). A genetic map illustrates the relationship of genes on a chromosome. Three-point crosses are used to determine the order of genes on a chromosome. In a phenomenon known as interference, the frequency of double recombination between closely linked genes is depressed compared to that expected.

Linkage of genes in humans has been difficult to determine. Originally, it was done by accumulating information from informative matings, but a more recent technique using somatic cell hybridization with mouse cells has located many more human genes at particular chromosomes. The fact that crossing over takes place in the four-stranded stage of meiosis I was first demonstrated in a study of attached X chromosomes in *Drosophila.* The theory was later confirmed using the fungus *Neurospora.*

In some cases, crossing over occurs unequally, thus generating duplications and increasing the amount of DNA. Generally, there is a close correspondence between the genetic map distance and the physical DNA distance. DNA variants are now being used to map the human genome.

Problems and Questions

1. In a cross between two F_1 garden peas having genotype *PpLl,* the progeny types were 142 purple, long; 8 purple, round; 11 red, long; and 47 red, round. (This is a progeny array similar to that shown in table 5.1.) Does this fit 9:3:3:1 expectations, using a chi-square test?

2. Using a testcross, a recombination rate of 0.114 was found for the genes for purple and long. In the cross by Bateson and Punnett, what proportions of the four gametes would you expect from F_1 parents?

3. Using the recombination rate in question 2, what proportions of the four phenotypes would you expect in the F_2 progeny? Use a chi-square test to see if the observed data in table 5.1 are consistent with this explanation.

4. Calculate the rate of recombination between genes *pr* and *vg* using the data from Morgan's second testcross (table 5.3). Compare your results to the value calculated from the first testcross in which the parents were in coupling.

5. Show diagrammatically (using chromosomes as in figure 5.3) how gametes for the F_2 are produced when the original parents are $prprvg^+vg^+$ and $pr^+pr^+vgvg.$

6. In a series of testcrosses, the rate of recombination between genes *a* and *c* is 0.1; between *a* and *b,* 0.27; and between *b* and *c,* 0.2. Determine the order of the three genes. Explain why the sum of rates of recombination for the two closest pairs of genes is not equal to the rate between the two genes farthest apart.

7. Another gene, *d,* had a rate of recombination of 0.15 with gene *b* and a rate of 0.14 with gene *a*. Where would you place this gene on the map constructed for question 6? What cross would you make to confirm your map?

8. Assume a three-point cross in which the F_1 *kk+ll+mm+* was crossed to *kkllmm* and the offspring had the following genotypes:

kk+ll+mm+	621
kk+ll+mm	3
kk+llmm+	64
kkll+mm+	103
kk+llmm	109
kkll+mm	57
kkllmm+	7
kkllmm	608

Construct a genetic map of these three loci, including their order and the map distance between them.

9. Using the data in problem 8, calculate the expected proportion of double recombinants. What is the value of interference (I) estimated from these data? Explain what this level of I means.

10. Diagram the types of gametes you would expect from a *y+/+w* female, given the following situations between these loci: no crossovers, one crossover, two crossovers, three crossovers, and four crossovers.

11. In a large family, the mother is heterozygous for both hemophilia and color blindness with repulsion gametes, or genotype *H c/h C*. Two of her sons have hemophilia, two other sons are color-blind, and the fifth son is normal. Give the genotypes for the sons. What is the estimated rate of recombination between the genes in this family?

12. Several families having a female parent heterozygous for both hemophilia and color blindness were gathered. In these families, the following numbers of recombinants and sons were observed: 1 of 5 sons, 2 of 8, 1 of 4, 1 of 5, and 1 of 6. What is the estimated rate of recombination, using all these data? In fact, the actual rate is much less. Can you give a possible explanation?

13. Assume that a deletion heterozygote (A_1-) was mated with an A_2A_2 individual. What type of progeny, and in what proportions, would you expect? Would this differ from the results expected if the first parent were A_1A_1?

14. Using somatic cell hybridization, a number of cells having a particular human enzyme were generated. Five of these cell lines had human chromosomes: (A) 4, 9, 11; (B) 2, 7, 11, 21; (C) 4, 9, 10, 11, 21; (D) 1, 2, 11, 18; and (E) 4, 11, 15. On which chromosome is the gene for this enzyme?

15. Give the progeny genotypes and phenotypes of a cross between $\overline{X^wX^+}Y$ and X^+Y where the two X chromosomes in the female parent are attached. What types of progeny survive? Why can't this cross be used to differentiate between two- and four-stranded crossing over?

16. Assume that *Neurospora* with allele D are crossed to those with allele d. If there were no crossing over between the centromere and this gene, what order of ascospores would you expect? If there was one crossover, what order would you expect?

17. Assume that we have two genes, E and F, which we know are on a particular *Neurospora* chromosome. If we cross EF with ef and examine the resulting ascospores, most of them are *EF, EF, EF, EF, ef, ef, ef, ef*. However, some are *EF, EF, eF, eF, Ef, Ef, ef, ef,* or *eF, eF, EF, EF, ef, ef, Ef, Ef*. What is the order of these genes and the centromere?

18. Sturtevant and Morgan used genes f (56.7) and fu (59.5) as markers in a cross to examine *Ultrabar*. What proportion of recombinant progeny would you expect between these markers? Sturtevant and Morgan found 9 out of approximately 20,000 progeny that were recombinant for the markers and had either wild-type or *Ultrabar* alleles. Assuming that mispairing takes place one out of ten times for *Bar* homozygotes, what is the estimated size of the region that can produce duplications or deficiencies?

19. In this chapter, we have been focusing on genetic or map distance between genes. Generally, map distance is similar to physical distance on the chromosome, but in some cases, it is not. Suggest reasons why map and physical distance may not be exactly concordant.

20. Assume that two of the genes Mendel picked for his studies in garden peas were tightly linked. How might this have influenced the principle of independent assortment?

Answers appear at end of book.

P·A·R·T

3

Molecular Genetics

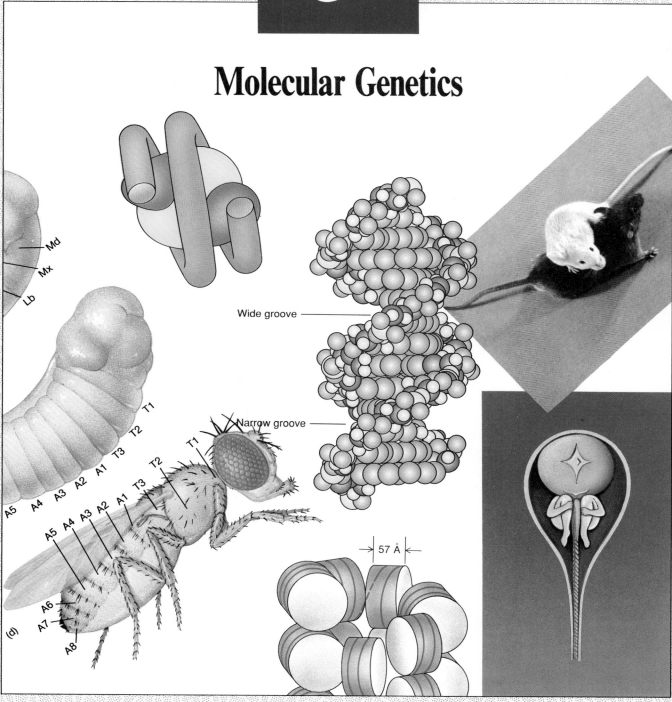

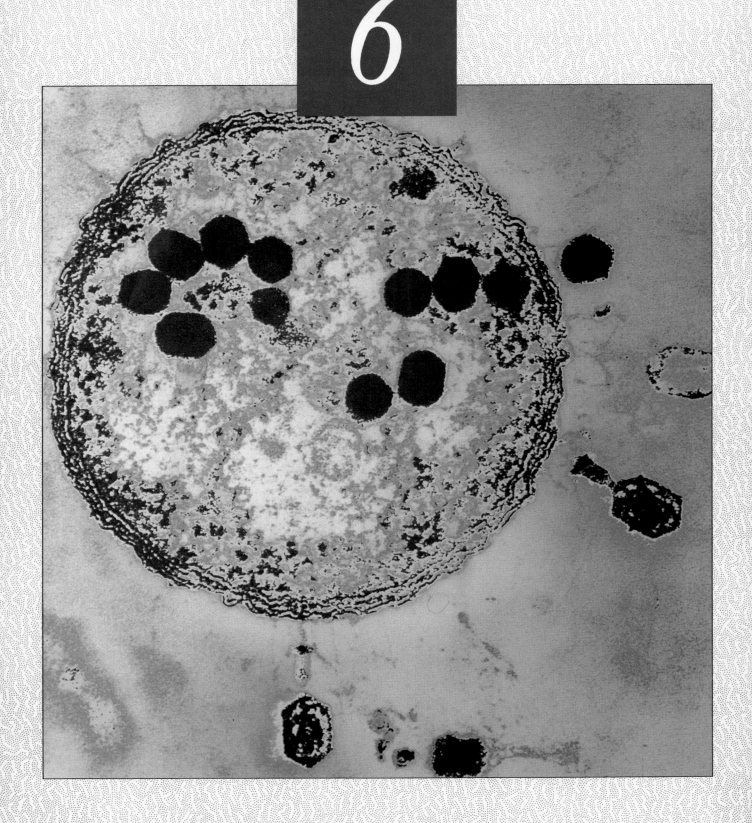

Chemistry of the Gene

*T*he meeting of two personalities is like the contact of two chemical substances. If there is any reaction, both are transformed.

■

Carl Gustav Jung
Swiss psychiatrist and psychologist

Learning Objectives

In this chapter you will learn:

1. That most genes are made of a chainlike substance called DNA, and that the others are composed of a similar substance called RNA.

2. The chemical natures of DNA and RNA.

3. That most genes contain two strands of DNA wound around each other in a double helix.

4. How the two strands in a DNA double helix can be separated and put back together again.

5. That DNA comes in a variety of sizes and shapes.

enetics, as practiced for most of its history, has been a science of whole organisms. Living things are mated, and their genetic makeup is inferred from the characteristics of their offspring. But genetics also has a chemical side. In fact, many modern geneticists work primarily at the molecular level.

This does not mean that there is no longer a place for classical genetics research. Molecular analysis of a gene would make little sense if its function were unknown, and our understanding of a gene's function usually comes from observing the effects of mutations in whole organisms. Indeed, the very existence of most genes has been discovered this way. So, modern genetics is a highly fruitful blend of traditional and molecular approaches. This chapter will introduce the molecular aspects.

THE NATURE OF GENETIC MATERIAL

If we are going to discuss molecular genetics, we must first get to know the molecules. The three molecules at the heart of genetics—DNA, RNA, and protein—have already been mentioned in chapter 1. We will examine DNA and RNA in more detail later in this chapter, but first it will help to introduce a few very basic chemical structures that recur over and over in these molecules.

Substances derived from living things are termed "organic," and chemists had noted even before Mendel's time that organic molecules are rich in carbon. This led to the division of the discipline of chemistry into two parts—organic chemistry, the study of carbon-containing compounds, and inorganic chemistry, the study of all other compounds. This is not to imply that all organic molecules come from living things. Plastics such as polyethylene and nylon are surely organic substances, but they are just as surely synthetic.

Thus, the molecules we will be talking about in our study of genetics are organic, or carbon-containing, molecules. The simplest organic compound is the natural biological product natural gas, or methane. Its structure:

$$
\begin{array}{c}
\text{H} \\
| \\
\text{H} - \text{C} - \text{H} \\
| \\
\text{H}
\end{array}
$$

contains only hydrogen and carbon, so it is termed a hydrocarbon.

Carbon can bond to a variety of other atoms. In genetics, the most important of these are nitrogen, oxygen, hydrogen, phosphorus, and other carbons. Frequently, these other atoms are present in groups, sometimes called *functional groups*, as illustrated in table 6.1. Notice that

a shorthand convention is used to represent most of these groups. Thus, the methyl group, with a carbon and three hydrogens, is not usually shown with its three hydrogens projecting out as in the structure of methane, but simply as CH_3.

Note that phosphate is written with two negative charges. At very low pH, both of the negatively charged oxygens would be linked to hydrogen atoms. However, at neutral pH, such as we find in living cells, the positively charged hydrogen nuclei (protons) dissociate, leaving negative ions behind. The same principle applies to most carboxyl groups; for example, acetic acid is mostly deprotonated (ionized) at neutral pH, yielding acetate ($CH_3 - COO^-$). These deprotonation reactions are illustrated in figure 6.1.

We will encounter these functional groups repeatedly in this and other chapters. We will also see how some of these groups can link together to form different kinds of compounds.

The studies that have led to our understanding of the gene's chemistry began in Tübingen, Germany, in 1869. There, Friedrich Miescher isolated nuclei from pus cells in waste surgical bandages. He found that these nuclei contained a novel phosphorus-bearing compound that he named *nuclein*. Nuclein is mostly **chromatin,** a complex of deoxyribonucleic acid (DNA) and chromosomal proteins.

By the end of the nineteenth century, both DNA and ribonucleic acid (RNA) had been separated from the protein that clings to them in the cell. This allowed more detailed chemical analysis of these **nucleic acids.** (Notice that the term nucleic acid and its derivatives, DNA and RNA, come directly from Miescher's term "nuclein.") By the beginning of the 1930s, P. Levene, W. Jacobs, and others had demonstrated that RNA is composed of a sugar (ribose) plus four nitrogen-containing bases, and that DNA contains a different sugar (deoxyribose) plus four bases. They discovered that each base is coupled with a sugar-phosphate to form a nucleotide. Furthermore, the four bases in DNA or RNA seemed to be present in roughly equal quantities.

TRANSFORMATION IN BACTERIA

Frederick Griffith laid the foundation for the identification of DNA as the genetic material in 1928 with his experiments on **transformation** in the bacterium *Pneumococcus,* now known as *Streptococcus pneumoniae.* The wild-type organism is a spherical cell surrounded by a mucous coat called a capsule. The cells form large, glistening colonies, characterized as smooth (S) (see figure 6.2). These cells are *virulent*—capable of causing lethal

TABLE 6.1   *Functional Groups Found in Organic Molecules*

Group Name	Group Structure	Name	Structure (Example)
Hydroxyl	$-OH$	Ethanol (grain alcohol)	CH_3-CH_2-OH, or H-C-C-OH (with H's on each carbon)
Methyl	$-CH_3$, or $-C-H$ (with H above and below)	Methanol (wood alcohol)	CH_3-OH, or H-C-OH (with H above and below)
Carbonyl	$\backslash C = O$	Formaldehyde	$H_2C=O$, or $C=O$ (with H above and below)
Carboxyl	$-C-OH$ (with $=O$)	Acetic acid	CH_3-C-OH (with $=O$), or H-C-C-OH (with H's)
Amino	$-NH_2$	Methylamine	CH_3-NH_2, or H-C-NH$_2$ (with H above and below)
Phosphate	$-O-P-O^-$ (with $=O$ above and O^- below)	Methyl phosphate	$CH_3-O-P-O^-$, or H-C-O-P-O$^-$

(a)

$CH_3-O-P-OH \xrightarrow{-H^+} CH_3-O-P-O^- \xrightarrow{-H^+} CH_3-O-P-O^-$

Methyl phosphoric acid (fully protonated)   Methyl phosphate (half ionized)   Methyl phosphate (fully ionized)

(b)

$CH_3-C-OH \xrightarrow{-H^+} CH_3-C-O^-$

Acetic acid (protonated)   Acetate (ionized)

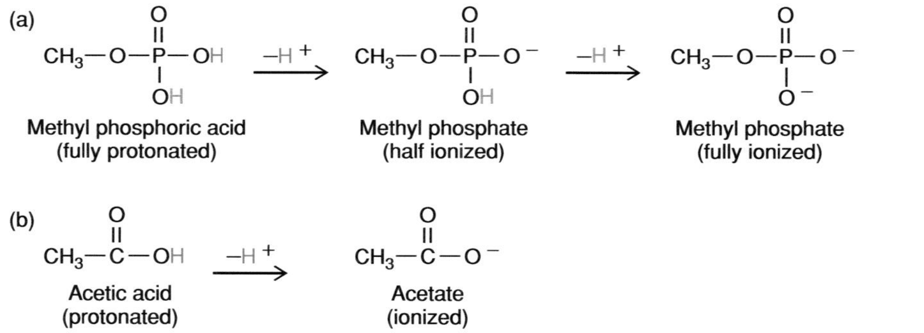

FIGURE 6.1
Deprotonation reactions. (*a*) Methyl phosphoric acid loses two protons (red and green) in successive deprotonations. (*b*) Acetic acid loses a proton (red) in its conversion to acetate.

– deprotonated @ neutral PH.

(a)

(b)

FIGURE 6.2
Variants of *Streptococcus pneumoniae.* (*a*) The large, glossy colonies
contain smooth (S) virulent bacteria; (*b*) the small, mottled colonies
are composed of rough (R) avirulent bacteria.
Photographs by Harriett Ephrussi-Taylor.

infections upon injection into mice. A mutant strain of
S. pneumoniae has lost the ability to form a capsule. As
a result, it grows as small, rough (R) colonies. More importantly, it is *avirulent;* it has no protective coat, so it is
engulfed by the host's white blood cells before it can proliferate enough to do any damage.

The key finding of Griffith's work was that heat-killed
virulent colonies of *S. pneumoniae* could **transform** avirulent cells to virulent ones. Neither the heat-killed virulent bacteria nor the live avirulent ones by themselves
could cause a lethal infection. Together, however, they were
deadly. Somehow the virulent trait passed from the dead
cells to the live, avirulent ones. This transformation phenomenon is illustrated in figure 6.3. Transformation was
not transient; the ability to make a capsule and therefore
to kill host animals, once conferred upon the avirulent
bacteria, was passed to their descendants as a heritable
trait. In other words, the gene for virulence, missing in the
avirulent cells, was somehow restored during transformation. This made it very likely that the transforming
principle in the heat-killed bacteria was the gene for virulence itself. The missing piece of the puzzle was the
chemical nature of the transforming principle. Whoever
discovered this would reveal the nature of genes.

DNA: The Transforming Principle

Oswald Avery, Colin MacLeod, and Maclyn McCarty
supplied the missing piece in 1944. They used a variation
of the transformation assay that Griffith had introduced,
and they took pains to define the transforming substance
from virulent cells. First, they destroyed the protein in the
extract with organic solvents and found that it still transformed. Next, they subjected it to digestion with various
enzymes. Trypsin and chymotrypsin, which destroy protein, had no effect on transformation. Neither did ribonuclease, which degrades RNA. These experiments ruled
out protein or RNA as the transforming principle. On the
other hand, Avery and his coworkers found that deoxyribonuclease (DNase), which breaks down DNA, destroyed the transforming ability of the virulent cell extract.

Finally, direct physical-chemical analysis showed the
purified transforming principle to be DNA. The analyt-
ical tools Avery and his colleagues used were:

1. *Ultracentrifugation* (spinning the transforming
    substance in an ultracentrifuge to estimate its
    size). The material with transforming activity
    sedimented rapidly (moved rapidly toward the
    bottom of the centrifuge tube), suggesting a very
    high molecular weight, characteristic of DNA.

2. *Electrophoresis* (placing the transforming
    substance in an electric field to see how rapidly it
    moved). The transforming activity had a high
    mobility, also characteristic of DNA.

3. *Ultraviolet absorption spectrophotometry.* A
    solution of the transforming substance was placed
    in a spectrophotometer to see what kind of
    ultraviolet light it absorbed most strongly. Its
    absorption spectrum matched that of DNA. That
    is, the light it absorbed most strongly had a
    wavelength of about 260 nanometers (nm).

4. *Elementary chemical analysis.* This yielded an
    average nitrogen/phosphorus ratio of 1.67, about
    what one would expect for DNA, which is rich in
    both elements, but vastly lower than the value
    expected for protein, which is rich in nitrogen but
    poor in phosphorus. Even a slight protein
    contamination would have revealed itself by
    raising the nitrogen/phosphorus ratio.

Further Confirmation

These findings should have settled the issue, but they had
little immediate impact. The mistaken notion that DNA
was a monotonous repeat of a four-nucleotide sequence,
such as ACTG, persuaded many geneticists that it could

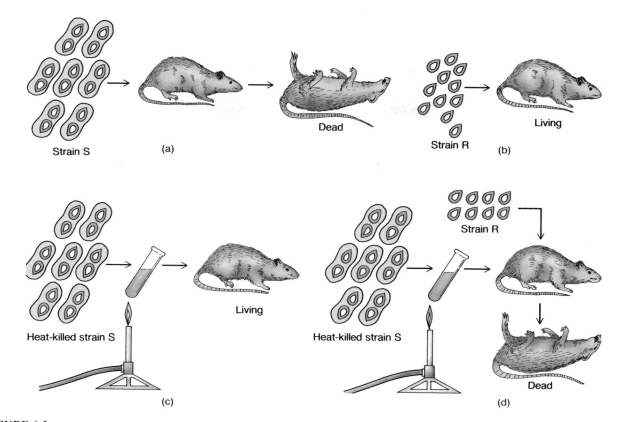

FIGURE 6.3
Griffith's transformation experiments. (*a*) Virulent strain S
Pneumococci kill their host; (*b*) avirulent strain R bacteria cannot
infect successfully, so the mouse survives; (*c*) strain S bacteria that are
heat-killed can no longer infect; (*d*) a mixture of strain R and heat-
killed strain S bacteria kills the mouse. The killed virulent (S)
bacteria have transformed the avirulent (R) bacteria to virulent (S).
From R. Sager and J. Ryan, Cell Heredity. *Copyright © 1961 John Wiley &
Sons, Inc., New York, NY.*

not be the genetic material. Furthermore, controversy
persisted about possible protein contamination in the
transforming principle, whether transformation would
work with other characteristics besides R and S, and even
whether bacterial genes were like the genes of higher organisms.

Yet, by 1953, when James Watson and Francis Crick
published the double-helical model of DNA structure,
most geneticists agreed that genes were made of DNA.
What had changed? For one thing, Erwin Chargaff had
shown in 1950 that the bases were not really found in equal
proportions in DNA as some geneticists believed, and that
the base composition of DNA varied from one species to
another. This is in fact exactly what one would expect for
genes, which also vary from one species to another. Furthermore, Rollin Hotchkiss had refined and extended
Avery's findings. He purified the transforming principle
to the point where it contained only 0.02% protein, and
showed that it could change other genetic characteristics
besides R and S.

Finally, in 1952, A. D. Hershey and Martha Chase
performed another experiment that was not definitive by
itself, but added to the weight of evidence that genes were
made of DNA. This experiment involved a **phage** (bacterial virus) called T2 that infects the bacterium *Escherichia coli* (figure 6.4). During infection, the phage genes
enter the host cell and direct the synthesis of new phage
particles. The phage is composed of protein and DNA only.
The question is: Do the genes reside in the protein or in
the DNA? The Hershey-Chase experiment answered this
question by showing that most of the DNA entered the
bacterium, along with only a little protein; the bulk of the
protein stayed on the outside (figure 6.5). Since the DNA
is the major component that gets into the host cells, it is
likely that it contains the genes. Of course, this conclusion
is not unequivocal; the small amount of protein that entered along with the DNA could conceivably carry the
genes. But taken together with the more definitive work
that had gone before, this study helped convince geneticists about the chemical nature of genes.

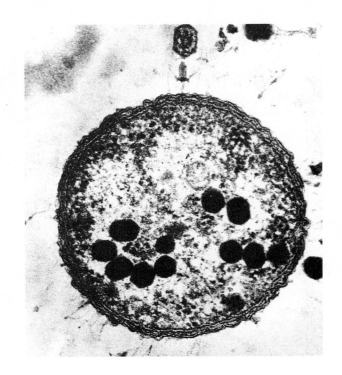

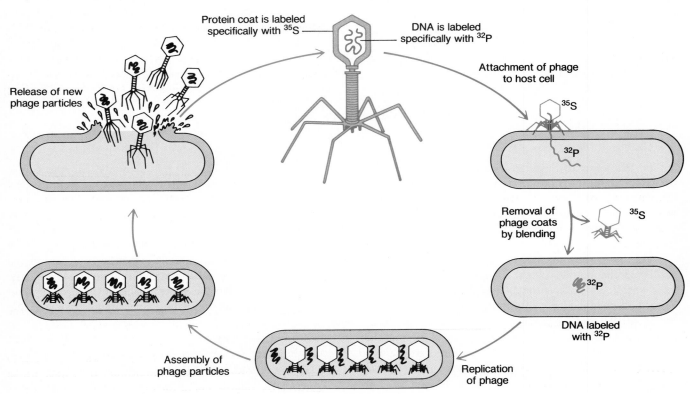

FIGURE 6.5
The Hershey-Chase experiment. Phage T2 contains genes that allow it to replicate in *E. coli*. Since the phage is composed of DNA and protein only, its genes must be made of one of these substances. To discover which, Hershey and Chase labeled phage protein with ^{35}S (green) and phage DNA with ^{32}P (red), then infected bacteria with these labeled phages in separate experiments. In one experiment, the protein was labeled; in the other, the DNA was labeled. Here, both are shown labeled together, for simplicity's sake. Since the phage genes must enter the cell, the experimenters reasoned that the type of label found in the infected cells would indicate the nature of the genes. Most of the labeled protein remained on the outside and was stripped off the cells by blending, while most of the labeled DNA entered the infected cells. The conclusion was that the genes of this phage are made of DNA. The left side of the figure shows the completion of the phage replication cycle, which is irrelevant to this experiment.

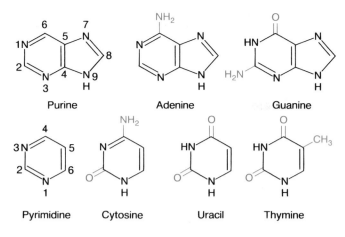

FIGURE 6.6
The bases of DNA and RNA. The parent bases, purine and pyrimidine, on the left, are not found in DNA and RNA. They are shown for comparison with the other five bases.

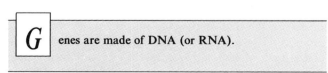

FIGURE 6.7
The sugars of nucleic acids. Note the OH in the 2-position of ribose.

be nitrogenous **bases, phosphoric acid,** and the sugar **deoxyribose.** Similarly, RNA yielded bases and phosphoric acid, plus a different sugar, **ribose.** The four main bases found in DNA are **adenine (A), cytosine (C), guanine (G),** and **thymine (T).** RNA contains the same bases, except that **uracil (U)** replaces thymine. The structures of these bases, shown in figure 6.6 reveal that adenine and guanine are derived from the parent molecule, purine. Therefore, we refer to these compounds as **purines.** The other bases resemble pyrimidine, so they are called **pyrimidines.** These structures constitute the alphabet of genetics.

The sugars found in nucleic acids are depicted in figure 6.7. Notice that they differ only in one place. Where ribose contains a hydroxyl (OH) group in the 2-position, deoxyribose lacks the oxygen and simply has a hydrogen (H) represented by the vertical line. Hence the name *deoxy-*ribose. The bases and sugars in RNA and DNA are joined together into units called **nucleosides** (figure 6.8). The names of the nucleosides derive from the corresponding bases:

This experiment depended on radioactive labels on the DNA and protein—a different label for each. Such labeling became possible as a by-product of nuclear research before and during World War II. The labels used were phosphorus-32 (^{32}P) for DNA and sulfur-35 (^{35}S) for protein. These choices make sense, considering that DNA is rich in phosphorus while phage protein has none, and that protein contains sulfur but DNA does not.

The labeled phages were allowed to attach by their tails to bacteria and inject their genes into their hosts. Then the empty phage coats were removed by homogenizing in a Waring blender. (Yes, they had those back then.) Since we know that the genes must go into the cell, our question is: What went in, the ^{32}P-labeled DNA or the ^{35}S-labeled protein? Of course it was the DNA. In general, then, genes are made of DNA. On the other hand, as we will see in chapter 7, later experiments showed that some genes consist of RNA. Such genes are found in some viruses and other simple genetic systems.

Base	Nucleoside (RNA)	Deoxynucleoside (DNA)
Adenine	Adenosine	Deoxyadenosine
Guanine	Guanosine	Deoxyguanosine
Cytosine	Cytidine	Deoxycytidine
Uracil	Uridine	Not usually found
Thymine	Not usually found	(Deoxy)thymidine

G enes are made of DNA (or RNA).

Because thymine is not usually found in RNA, the "deoxy" designation for its nucleoside is frequently assumed, and the deoxynucleoside is simply called thymidine. There does not seem to be much rhyme or reason to these names, but they are important. It is also worth learning the numbering of the carbon atoms in the sugars of the nucleosides (figure 6.8). Note that the ordinary numbers are used in the bases, so the carbons in the sugars

THE CHEMICAL NATURE OF POLYNUCLEOTIDES

By the mid-1940s, the fundamental chemical structures of DNA and RNA were known. When DNA was broken into its component parts, these constituents were found to

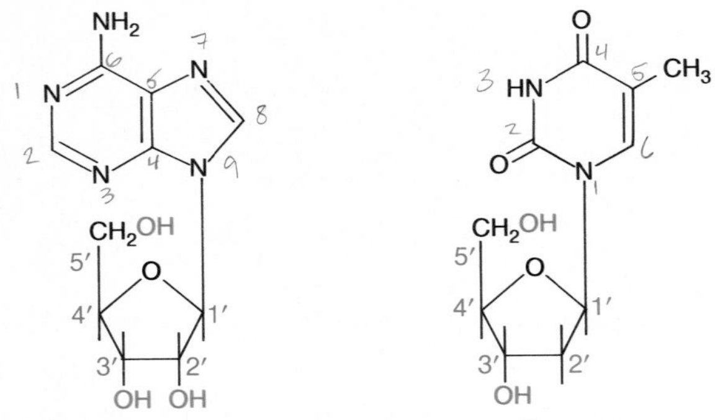

FIGURE 6.8

Adenosine

2'-deoxythymidine

FIGURE 6.8
Two examples of nucleosides.

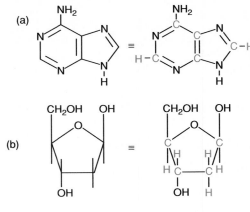

FIGURE 6.9
The structures of (*a*) adenine and (*b*) deoxyribose. Note that the structures on the left omit most or all of the carbons and some of the hydrogens. These are included in the structures on the right, in red and green, respectively.

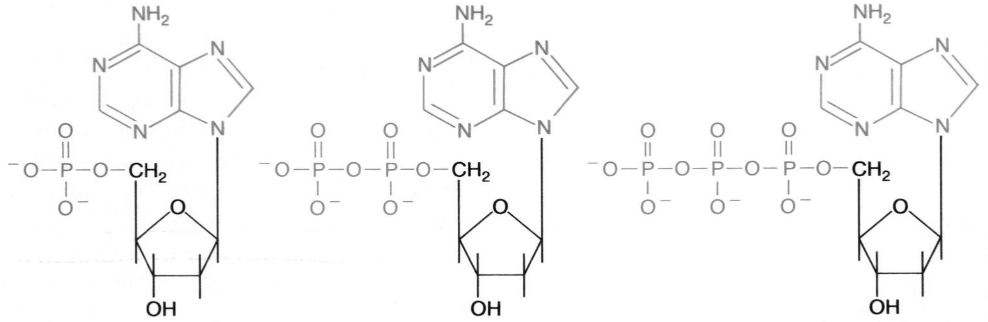

Deoxyadenosine-5'-
monophosphate (dAMP)

Deoxyadenosine-5'-
diphosphate (dADP)

Deoxyadenosine-5'-
triphosphate (dATP)

FIGURE 6.10
Three nucleotides. The 5'-nucleotides of deoxyadenosine are formed by phosphorylating the 5'-hydroxyl group. The addition of one phosphate results in deoxyadenosine-5'-monophosphate (dAMP). One more phosphate yields deoxyadenosine-5'-diphosphate (dADP). Three phosphates give deoxyadenosine-5'-triphosphate (dATP).

are called by primed numbers. Thus, for example, the base is linked to the 1'-position of the sugar, and the 2'-position is deoxy in deoxynucleosides.

The structures in figure 6.6 were drawn using an organic chemistry shorthand that leaves out certain atoms for simplicity's sake. Figures 6.7 and 6.8 use a slightly different convention in which a straight line with a free end denotes a C–H bond. Figure 6.9 shows the structures of adenine and deoxyribose, first in shorthand, then with every atom included.

The subunits of DNA and RNA are **nucleotides,** which are nucleosides with a phosphate group attached through a phospho-ester bond (figure 6.10). An ester is an organic

compound formed from an alcohol (bearing a hydroxyl group) and an acid. In the case of a nucleotide, the alcohol group is the 5'-hydroxyl of the sugar, and the acid is phosphoric acid, which is why we call the ester a phospho-ester. The structure of one of the four precursors of DNA synthesis, deoxyadenosine-5'-triphosphate (dATP), is also shown in figure 6.10. Upon incorporation of dATP into DNA, two phosphate groups are removed, leaving deoxyadenosine-5'-monophosphate (dAMP). The other three nucleotides in DNA (dCMP, dGMP, and dTMP) have analogous structures and names.

Chapter 6

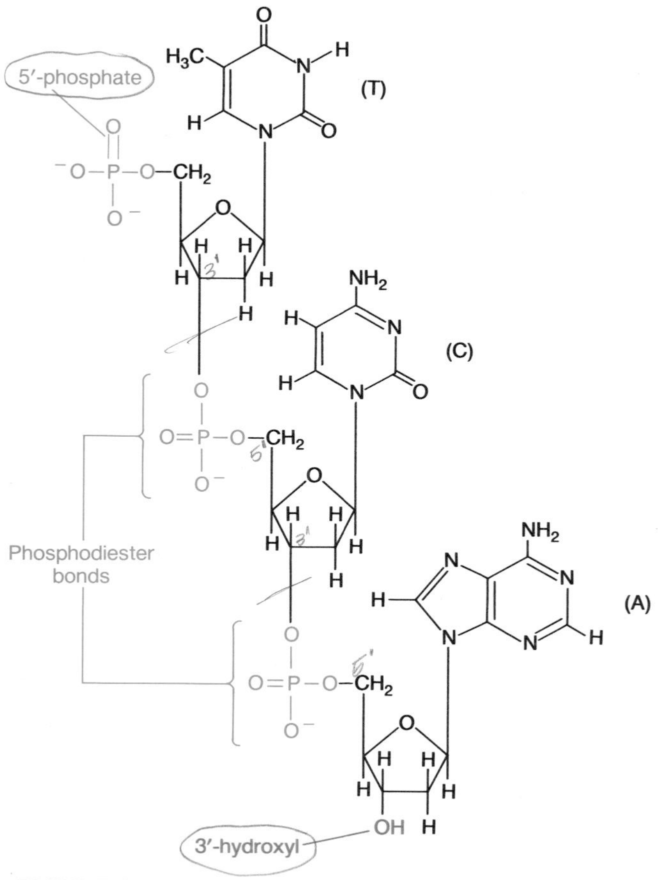

FIGURE 6.11
A trinucleotide. This little piece of DNA contains only three nucleotides linked together by phosphodiester bonds (red) between the 5'- and 3'-hydroxyl groups of the sugars. The 5'-end of this DNA is at the top, where there is a free 5'-group (blue); the 3'-end is at the bottom, where there is a free 3'-group (green). The sequence of this DNA could be read as 5'pdTpdCpdA3'. This would usually be simplified to TCA. All of the hydrogens are shown in this illustration.

D NA and RNA are chainlike molecules composed of subunits called nucleotides. The nucleotides, in turn, contain a base linked to the 1'-position of a sugar (ribose in RNA or deoxyribose in DNA) and a phosphate group. The phosphate joins the nucleotides in a DNA or RNA chain through their 5'- and 3'-hydroxyl groups by phosphodiester bonds.

∎

DNA STRUCTURE

All the facts about DNA and RNA just mentioned were known by the end of the 1940s, and by that time, it was also becoming clear that DNA was the genetic material and that it therefore stood at the very center of the study of life. Yet the detailed structure of DNA was unknown. For these reasons, several researchers dedicated themselves to solving the riddle of DNA.

EXPERIMENTAL BACKGROUND

One of the scientists interested in DNA structure was Linus Pauling, a theoretical chemist at the California Institute of Technology. He was already famous for his studies on chemical bonding and for his elucidation of the α-helix, an important feature of protein structure. Indeed, the α-helix, held together by hydrogen bonds, laid the intellectual groundwork for the double helix model of Watson and Crick. Another group toiling on the DNA problem was composed of Maurice Wilkins, Rosalind Franklin, and their colleagues at King's College in London. They were using X-ray diffraction to reveal the three-dimensional structure of DNA. Finally, there were James Watson and Francis Crick. Watson, in his early twenties and already holding a Ph.D. degree from Indiana University, had come to the Cavendish Laboratories in Cambridge, England, to learn about DNA. There he met Crick, a physicist who at age thirty-five, was retraining as a molecular biologist. Watson and Crick performed no experiments themselves. Their tactic was to use other groups' data to build a DNA model.

Erwin Chargaff was another very important contributor. We have already seen how his 1950 paper helped identify DNA as the genetic material, but the paper contained another piece of information that was even more significant. Chargaff's studies of the base compositions of DNAs from various sources revealed that the content of purines always equalled the content of pyrimidines. Furthermore, the amounts of adenine and thymine were always equal, as were the amounts of guanine and cytosine. These data, known as Chargaff's rules, provided a valuable confirmation of Watson and Crick's model.

We will discuss the synthesis of DNA in detail in chapters 7 and 8. For now, it will be sufficient to point out that the nucleotides are joined together in DNA or RNA by **phosphodiester bonds.** These bonds are called phosphodiesters because they involve phosphoric acid linked to *two* sugars—one through the 5'-group, the other through the 3'-group (figure 6.11). You will notice that the bases have been rotated in this picture, relative to their positions in previous figures. This is more like their geometry in DNA or RNA. Note also that this *trinucleotide,* or string of three nucleotides, has polarity: The top of the molecule bears a free 5'-group, so it is called the **5'-end.** The bottom, with a free 3'-group, is called the **3'-end.**

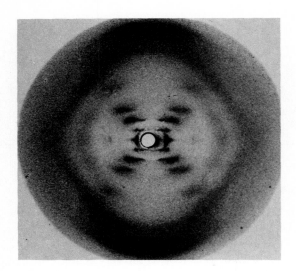

FIGURE 6.12
Franklin's X-ray picture of DNA. The regularity of this pattern
indicated that DNA is a helix. The spacing between the strong bands
at the top and bottom gave the spacing between elements of the helix
(base pairs) as 3.4 angstroms. The spacing between neighboring lines
in the pattern gave the overall repeat of the helix (the length of one
helical turn) as 34 angstroms.
© *M. H. F. Wilkins, Biophysics Department, King's College, London.*

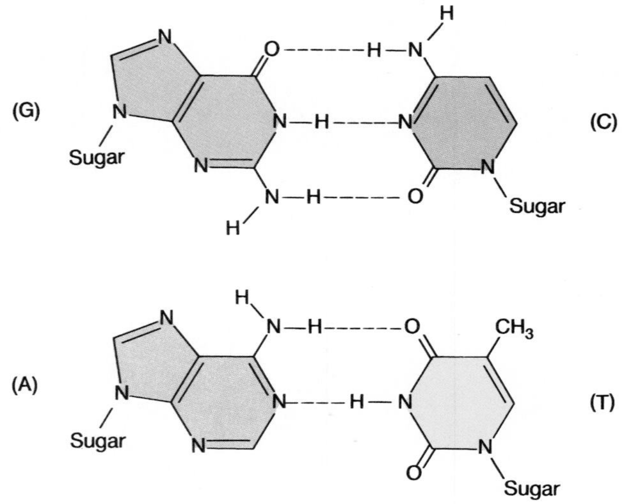

FIGURE 6.13
The base pairs of DNA. A guanine-cytosine pair (G–C), held together
by three hydrogen bonds, has almost exactly the same shape as an
adenine-thymine pair (A–T), held together by two hydrogen bonds.

Perhaps the most crucial piece of the puzzle came
from an X-ray diffraction picture of DNA taken by
Franklin in 1952—a picture that Wilkins shared with
James Watson over dinner in London on the night of January 30, 1953. The X-ray technique worked as follows:
The experimenter made a very concentrated, viscous solution of DNA, then reached in with a needle and pulled
out a fiber. This was not a single molecule, but a whole
batch of DNA molecules, forced into a side-by-side arrangement by the pulling action. Given the right relative
humidity, this fiber was enough like a crystal that it dif-
fracted X rays in an interpretable way. In fact, the X-ray
diffraction pattern in Franklin's picture (figure 6.12) was
so simple—a series of spots arranged in an X shape—that
it required the DNA structure itself to be very simple. By
contrast, a complex, irregular molecule like a protein gives
a complex X-ray diffraction pattern with many spots,
rather like a surface peppered by a shotgun blast. Since
DNA is very large, it can only be simple if it has a regular,
repeating structure. And the simplest repeating shape that
a long, thin molecule can assume is a spiral, or helix.

THE DOUBLE HELIX

Franklin's X-ray work strongly suggested that DNA was
a helix. Not only that, it gave some important information
about the size and shape of the helix. In particular, the
spacing between adjacent spots in an arm of the X is related to the overall repeat distance in the helix (34Å), and
the spacing from the top of the X to the bottom is related

to the spacing (3.4Å) between the repeated elements (**base
pairs**) in the helix. However, even though the Franklin
picture told much about DNA, it presented a paradox:
DNA was a helix with a regular, repeating structure, but
for DNA to serve its genetic function, it must have an *irregular* sequence of bases.

Watson and Crick saw a way to resolve this contra-
diction and satisfy Chargaff's rules at the same time: DNA
must be a **double helix** with its sugar-phosphate backbones on the outside and its bases on the inside. Moreover,
the bases must be paired, with a purine always across from
a pyrimidine. This way the helix would be regular; it would
not have bulges where two large purines are paired or constrictions where two small pyrimidines are paired. Watson
has joked about the reason he seized on a double helix: "I
had decided to build two-chain models. Francis would have
to agree. Even though he was a physicist, he knew that
important biological objects come in pairs."

But Chargaff's rules went further than this. They de-
creed that the amounts of adenine and thymine were equal
and so were the amounts of guanine and cytosine. This fit
very neatly with Watson and Crick's observation that an
adenine-thymine base pair held together by hydrogen
bonds has almost exactly the same shape as a guanine-
cytosine base pair (figure 6.13). So Watson and Crick postulated that adenine must always pair with thymine, and
guanine with cytosine. This way, the double-stranded DNA
will be uniform, composed of very similarly shaped base
pairs, regardless of the unpredictable sequence of either
DNA strand by itself. This was their crucial insight, the
key to the structure of DNA.

The double helix, often likened to a spiral ladder, is
presented in several ways in figure 6.14. The twisted sides
of the ladder represent the sugar-phosphate backbones of

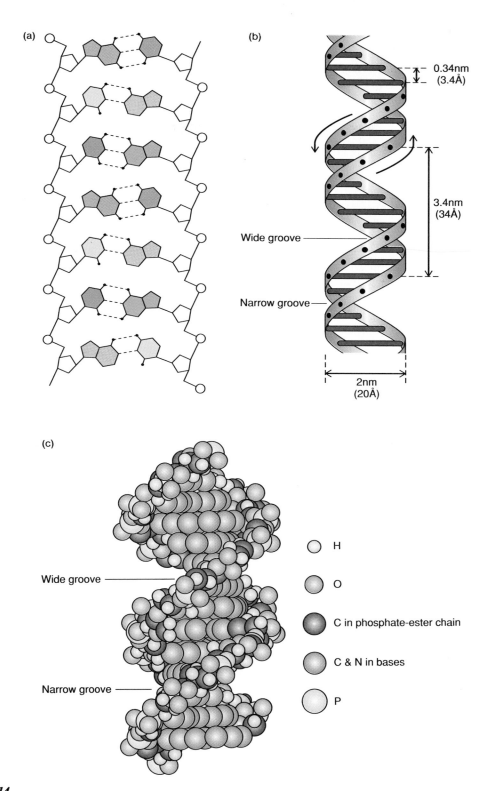

0.34nm
(3.4Å)

3.4nm
(34Å)

Wide groove

Narrow groove

2nm
(20Å)

(c)

Wide groove

Narrow groove

○ H

○ O

● C in phosphate-ester chain

● C & N in bases

○ P

FIGURE 6.14

Three models of DNA structure. (*a*) The helix is straightened out to show the base pairing in the middle. Each type of base is represented by a different color, with the sugar-phosphate backbones in black. Note the three hydrogen bonds in the G–C pairs and the two in the A–T pairs. (A = green, T = yellow, G = red, C = blue.) (*b*) The DNA double helix is represented as a spiral ladder whose sides represent the sugar-phosphate backbones of the two strands and whose rungs represent base pairs. The two strands are antiparallel; as suggested by the curved arrows, one strand (red) runs 5′ → 3′ top to

bottom and the other strand (blue) runs 5′ → 3′ bottom to top. (*c*) A space-filling model. The sugar-phosphate backbones appear as strings of mostly black, red, white, and yellow spheres, while the base pairs are rendered as horizontal flat plates composed of blue spheres. Note the wide and narrow grooves in the helix depicted in (*b*) and (*c*).

(a) *Adapted from "The Synthesis of DNA" by Arthur Kornberg. Copyright © 1968 by SCIENTIFIC AMERICAN, Inc. All rights reserved. (b) Reprinted by permission from Nature, vol. 171, p. 737. Copyright © 1953 Macmillan Magazines, Ltd. (c) Reprinted by permission from Nature, vol. 175, p. 834. Copyright © 1955 Macmillan Magazines, Ltd.*

the two DNA strands; the rungs are the base pairs. The spacing between base pairs is 3.4 angstroms, and the overall helix repeat distance is about 34 angstroms, meaning that there are about ten base pairs per turn of the helix. (One angstrom is one ten-billionth of a meter or one-tenth of a nanometer.) The arrows indicate that the two strands are **antiparallel.** If one has $5' \rightarrow 3'$ polarity from top to bottom, the other must have $3' \rightarrow 5'$ polarity. In solution, DNA has a structure very similar to the one just described, but the helix contains about 10.4 base pairs per turn.

Watson and Crick published the outline of their model in *Nature,* back-to-back with papers by Wilkins and Franklin and their coworkers showing the X-ray data. The Watson-Crick paper is a classic of simplicity—only nine hundred words, barely over a page long. It was published very rapidly, less than a month after it was submitted. Actually, Crick wanted to spell out the biological implications of the model, but Watson was uncomfortable doing that. They compromised on a sentence that is one of the great understatements in scientific literature: "It has not escaped our notice that the specific base pairing we have proposed immediately suggests a possible copying mechanism for the genetic material."

As this cryptic sentence indicates, Watson and Crick's model does indeed suggest a copying mechanism for DNA. Since one strand is the **complement** of the other, the two strands can be separated, and each can then serve as the template for building a new partner. Figure 6.15 shows schematically how this could be accomplished.

> T he DNA molecule is double-helical, with sugar-phosphate backbones on the outside and base pairs on the inside. The bases pair in a specific way: A with T and G with C.

GENES MADE OF RNA

The genetic system explored by Hershey and Chase was a phage—a bacterial virus. A virus particle by itself is essentially just a bag of genes. It has no life of its own, no metabolic activity; it is inert. But when the virus infects a host cell, it seems to come to life. Suddenly the host cell begins making viral proteins. Finally, the viral genes are replicated and the newly made genes, together with viral coat proteins, assemble into progeny virus particles. Because of their behavior as inert particles outside, but life-like activity inside their hosts, viruses resist classification. Some scientists refer to them as "living things" or even "organisms." We prefer a label that, although more cumbersome, is also more descriptive of a virus's less-than-living status; thus, we will call a virus a "genetic system." This term applies to any entity—be it a true organism, a virus, or even bare circles of infectious RNA called *viroids*—that contains genes capable of replicating.

Most genetic systems studied to date contain genes made of DNA. But some viruses, including several bacterial phages, plant and animal viruses (e.g., influenza virus), and all viroids, have RNA genes. Sometimes viral RNA genes are double-stranded, but usually they are single-stranded.

We have already encountered one famous example of the use of viruses in genetics research. We will see many more in subsequent chapters. In fact, without viruses, the field of genetics would be immeasurably poorer.

> C ertain viruses and viroids, alone among genetic systems, contain genes made of RNA instead of DNA.

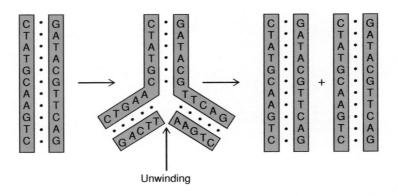

FIGURE 6.15
Replication of DNA. For simplicity, the two parental DNA strands (blue) are represented as parallel lines. During replication these parental strands separate, or unwind, and new strands (red) are built with bases complementary to those of the parental strands. The end result is two double-stranded molecules of DNA identical to the original. Each daughter DNA duplex therefore ends up with one parental strand (blue) and one new strand (red). Since only one parental strand is conserved in each of the daughter duplexes, this mechanism of replication is called semiconservative.

PHYSICAL CHEMISTRY OF NUCLEIC ACIDS

We conclude this chapter by briefly discussing the variety of structures available to DNA and RNA molecules and the behavior of DNA under conditions that encourage the two strands to separate and then come together again.

A VARIETY OF DNA STRUCTURES

The DNA structure proposed by Watson and Crick is right-handed; the helix runs clockwise away from you whether you look at it from the top or the bottom. But Alexander Rich and his colleagues discovered in 1979 that DNA does not always have to be right-handed. They showed that double-stranded DNA containing strands of alternating purines and pyrimidines (e.g., poly[dG–dC] · poly[dG–dC]) can exist in an extended left-handed helical form. Because of the zigzag look of this DNA's backbone when viewed from the side, it is often called **Z-DNA**. Figure 6.16 presents a picture of Z-DNA along with B-DNA for comparison's sake. Although Rich discovered Z-DNA in studies of model compounds like poly[dG-dC] · poly[dG-dC], this structure seems to be more than just a laboratory curiosity. There is evidence that living cells contain a small component of Z-DNA. Figure 6.17 presents computer graphic models of the A and Z forms of DNA.

> *I*n the cell, DNA may exist in the common B form, with base pairs horizontal, or a small fraction of the DNA may assume an extended left-handed helical form called Z-DNA.

■

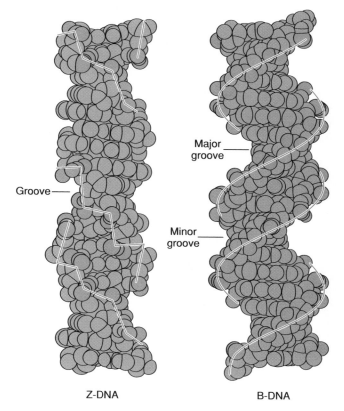

Z-DNA   B-DNA

FIGURE 6.16
Structures of Z- and B-DNA. Note the left-handed nature of the Z-DNA helix, in which the backbone (red line connecting red phosphate groups) goes up from right to left. By contrast, the B-DNA is clearly right-handed. Note also the contrast between the zigzag appearance of the Z-DNA backbone and the smooth B-DNA backbone.

From A. H. J. Wang, et al., Nature 282:684, 1979. Copyright © 1979 Macmillan Magazines Limited, London, England.

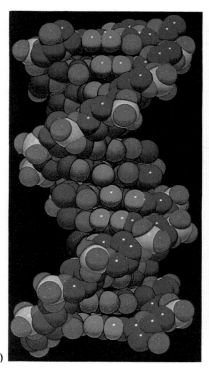
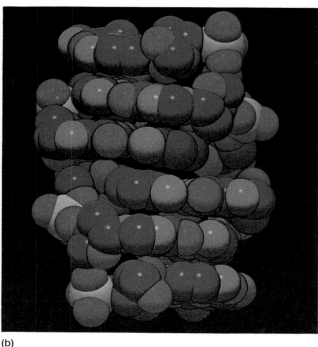

(a)   (b)

FIGURE 6.17
Computer graphic models of B- and Z-DNA. (*a*) B-DNA. Note the familiar right-handed helix, with horizontal base pairs. (*b*) Z-DNA. Note that the base pairs are grouped in twos in an alternating side-to-side pattern. This gives Z-DNA its characteristic zigzag shape. Carbons are dark blue; nitrogens, light blue; oxygens, red; and phosphorous atoms, yellow. Green represents a bromine atom added to aid the X-ray crystallography.
© Dr. Nelson Max, Lawrence Radiation Laboratory, University of California, Berkeley.

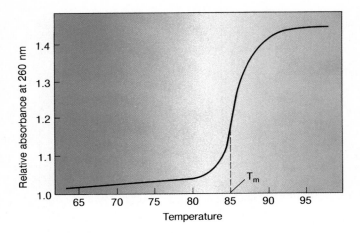

FIGURE 6.18

Melting curve of *Streptococcus pneumoniae* DNA. The DNA was heated, and its melting was measured by the increase in absorbance at 260 nm. The point at which the melting is half complete is the melting temperature, or T_m. The T_m for this DNA under these conditions is about 85° C.

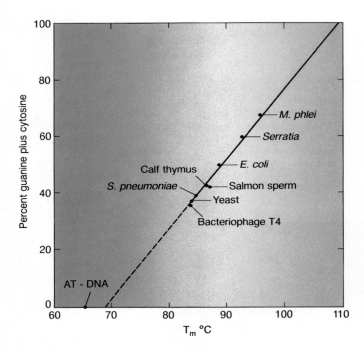

FIGURE 6.19

Relationship between DNA melting temperature and GC content. AT-DNA refers to synthetic DNAs composed exclusively of A and T (GC content = 0).

From Paul Doty, Harvey Lectures 55:121, 1961. Copyright © 1961 Academic Press, Orlando, FL. Reprinted by permission of the author and publisher.

SEPARATING THE TWO STRANDS OF A DNA DOUBLE HELIX

When the temperature of a DNA solution rises high enough, the forces that hold the two strands together weaken and finally break. When this happens, the two strands come apart in a process known as **DNA denaturation** or **DNA melting.** The temperature at which the DNA strands are half denatured is called the melting temperature, or T_m. Figure 6.18 contains a melting curve for DNA from *Streptococcus pneumoniae*. The amount of strand separation, or melting, is measured by the absorbance of the DNA solution at 260 nm. Nucleic acids absorb light at this wavelength because of the electronic structure in their bases, but when two strands of DNA come together, the close proximity of the bases in the two strands quenches some of this absorbance. When the two strands separate, this quenching disappears and the absorbance rises 20% to 30%. The precipitous rise in the curve shows that the strands hold fast until the temperature approaches the T_m and then rapidly let go.

The GC content of a DNA has a significant effect on its T_m. In fact, as figure 6.19 shows, the higher a DNA's GC content, the higher its T_m. Why should this be? Recall that one of the forces holding the two strands of DNA together is hydrogen bonding. Remember also that G–C pairs form three hydrogen bonds, whereas A–T pairs have only two. It stands to reason, then, that two strands of DNA rich in G and C will hold to one another more tightly than those of AT-rich DNA. Consider two pairs of embracing centipedes. One pair has two hundred legs each, the other three hundred. Naturally, the latter pair will be harder to separate.

Heating is not the only way to denature DNA. Organic solvents such as dimethyl sulfoxide and formamide disrupt the hydrogen bonding between DNA strands and thereby promote denaturation. Lowering the salt concentration of the DNA solution also aids denaturation by removing the ions that shield the negative charges on the two strands from one another. At low ionic strength, the mutually repulsive forces of these negative charges are strong enough to denature the DNA at a relatively low temperature.

> *T* he GC content of a DNA can have a strong effect on the physical properties of the DNA, in particular on its melting temperature, which increases linearly with GC content. The melting temperature (T_m) of a DNA is the temperature at which the two strands are half-dissociated, or denatured. Low ionic strength and organic solvents also promote DNA denaturation.

Chapter 6

REJOINING THE SEPARATED DNA STRANDS

Once the two strands of DNA separate, they can, under the proper conditions, come back together again. This is called **annealing** or **renaturation**. Several factors contribute to renaturation efficiency. Three of the most important are:

1. *Temperature.* The best temperature for renaturation of a DNA is about 25° C below its T_m. This temperature is low enough that it does not promote denaturation, but high enough to allow rapid diffusion and weaken the transient bonding between mismatched sequences and short intrastrand base-paired regions. This suggests that rapid cooling following denaturation would frustrate renaturation. Indeed, a common procedure to ensure that denatured DNA stays denatured is to plunge the hot DNA solution into ice. This is called *quenching*.

2. *DNA concentration.* The concentration of DNA in the solution is also important. Within reasonable limits, the higher the concentration, the more likely it is that two complementary strands will encounter each other within a given time. In other words, the higher the concentration, the faster the annealing.

3. *Renaturation time.* Obviously, the longer the time allowed for annealing, the more annealing will occur.

HYBRIDIZATION OF TWO DIFFERENT POLYNUCLEOTIDE CHAINS

Consider a piece of single-stranded DNA forming a double-stranded structure with a single-stranded RNA of complementary sequence (figure 6.20). Because we are putting together a **hybrid** of two different kinds of polynucleotide chains, we call this **hybridization.** The two chains do not have to be as different as DNA and RNA. If we put together two different strands of DNA having complementary or nearly complementary sequences we can call it hybridization—as long as the strands are of different origin. The difference between the two complementary strands may be very subtle; for example, one may be radioactive, while the other is not. As we will see later, especially in chapter 15, molecular hybridization is an extremely valuable technique. In fact, it would be difficult to overestimate the importance of such hybridization in genetics.

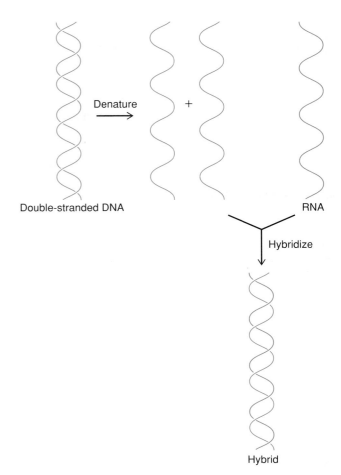

FIGURE 6.20
Hybridizing DNA and RNA. First, the DNA at upper left is denatured to separate the two DNA strands (blue). Then the DNA strands are mixed with a strand of RNA (red) that is complementary to one of the DNA strands. If this hybridization reaction is carried out under appropriate conditions (relatively high temperature, for example), an RNA-DNA hybrid will form. This duplex has one DNA strand (blue) and one RNA strand (red).

DNAS OF VARIOUS SIZES AND SHAPES

Table 6.2 shows the sizes of the DNAs of several organisms and viruses. The sizes are expressed three ways: molecular weight, number of base pairs, and length. These are all related, of course. We already know how to convert number of base pairs to length, since there are ten base pairs per helical turn, which is 34 angstroms long. To convert base pairs to molecular weight, we simply need to multiply by 660, the approximate molecular weight of one average nucleotide pair.

Source	Molecular Weight	Base Pairs	Length
Subcellular Genetic Systems:			
SV40 (mammalian tumor virus)	3.5×10^6	5,226	1.7 μm
Bacteriophage ϕX174 (double-stranded form)	3.2×10^6	5,386	1.8 μm
Bacteriophage λ	33×10^6	5×10^4	13 μm
Bacteriophage T2 or T4	1.3×10^8	2×10^5	50 μm
Mouse mitochondria	9.5×10^6	1.4×10^4	5 μm
Prokaryotes:			
Hemophilus influenzae	8×10^8	1.2×10^6	300 μm
Escherichia coli	2.8×10^9	4.2×10^6	1.4 mm
Salmonella typhimurium	8×10^9	1.1×10^7	3.8 mm
Eukaryotes (content per haploid nucleus):			
Saccharomyces cerevisiae (yeast)	1.2×10^{10}	1.8×10^7	6.0 mm
Neurospora crassa (pink bread mold)	1.9×10^{10}	2.7×10^7	9.2 mm
Drosophila melanogaster (fruit fly)	1.2×10^{11}	1.8×10^8	6.0 cm
Frog	1.4×10^{13}	2.3×10^{10}	7.7 m
Mouse	1.5×10^{12}	2.2×10^9	75 cm
Human	1.9×10^{12}	2.8×10^9	94 cm
Zea mays (corn, or maize)	4.4×10^{12}	6.6×10^9	2.2 m
Lilium longiflorum (lily)	2×10^{14}	3×10^{11}	100 m

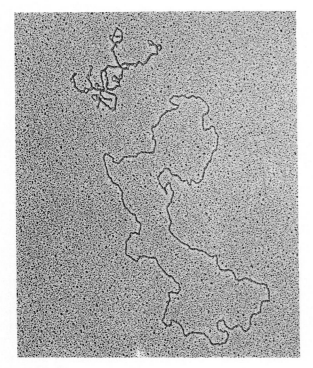

FIGURE 6.21
Autoradiograph of phage PM2 DNA in open circular and supercoiled forms.
© *Dr. Jack Griffith.*

How do we measure these sizes? For small DNAs, this is fairly easy. For example, consider phage PM2 DNA, which contains a double-stranded, circular DNA. How do we know it is circular? The most straightforward way to find out is simply by looking at it. We can do this using an electron microscope, but first we have to treat the DNA so that it stops electrons and will show up in a micrograph just as bones stop X rays and therefore show up in an X-ray picture. The most common way of doing this is by *shadowing* the DNA with a heavy metal such as platinum. We place the DNA on an electron microscope grid and bombard it with minute droplets of metal from a shallow angle. This makes the metal pile up beside the DNA like snow behind a fence. We rotate the DNA on the grid so that it becomes shadowed all around. Now the metal will stop the electrons in the electron microscope and make the DNA appear as light strings against a darker background. Printing reverses this image to give a picture such as figure 6.21, which is an electron micrograph of PM2 DNA in two forms: an open circle (lower left) and a **supercoil** (upper right), in which the DNA coils around itself rather like a twisted rubber band. You can also use pictures like these to measure the length of the DNA. This is more accurate if you include a DNA of known length in the same picture.

The size of a very small DNA can be estimated by **gel electrophoresis.** In this technique, we make a gel with slots in it, as shown in figure 6.22. We put a little DNA

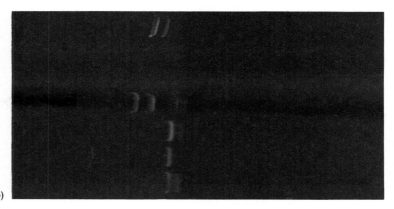

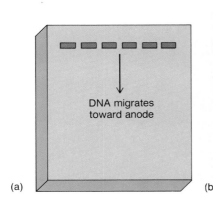

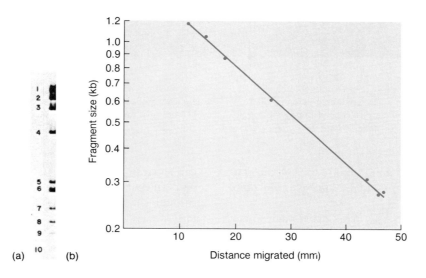

FIGURE 6.22

DNA gel electrophoresis. (*a*) Scheme of the method: This is a horizontal gel made of agarose (a substance derived from seaweed, and the main component of agar). The agarose melts at high temperature, then gels as it cools. A "comb" is inserted into the molten agarose; after the gel cools, the comb is removed, leaving slots, or wells. The DNA is then placed in the wells, and an electric current is run through the gel. The DNA electrophoreses, or migrates, toward the positive pole, or anode. (*b*) A close-up of gels that reveal DNA as pink bands when placed under UV light.

Photo (b) © Hank Morgan/Photo Researchers, Inc.

FIGURE 6.23

Gel electrophoresis of φX174 phage DNA fragments. (*a*) Autoradiograph of radioactive fragments. The fragments are numbered from 1 to 10 in order of decreasing size. (*b*) Plot of migration of the DNA fragments versus the log of their sizes in kilobase pairs (kb).

Photo (a) from Lee and Robert Sinsheimer, Proceedings of the National Academy of Science, vol. 71, p. 2884, 1974.

in a slot and run an electric current through the gel. The negatively charged DNA migrates toward the positive pole (the *anode*) at the end of the gel. The secret of the gel's ability to distinguish DNAs of different sizes lies in its porosity. Small DNA molecules can fit easily through the minute holes in the gel, so they migrate rapidly. Large DNAs, by contrast, move with difficulty through the gel's pores, so their mobility is retarded. The result is that the electric current will distribute the DNA fragments according to their sizes: the largest near the top, the smallest near the bottom. Finally, we stain the DNA with a fluorescent dye and look at the gel under ultraviolet illumination. Figure 6.23 depicts the results of such analysis on fragments of phage DNA of known size. The mobilities of these fragments are plotted versus the log of their molecular weights (or number of base pairs). Now any unknown DNA can be electrophoresed in parallel with the standard fragments, and its size can be determined if it falls within the range of the standards. For example, a

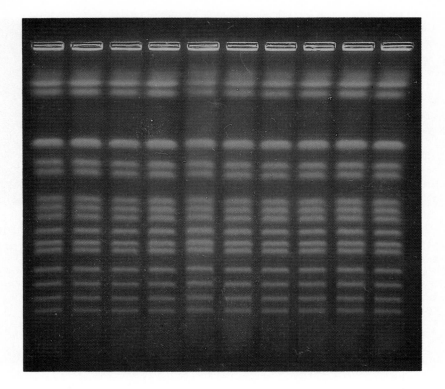

FIGURE 6.24
Pulsed-field gel electrophoresis of yeast chromosomes. Identical samples of yeast chromosomes were electrophoresed in ten parallel lanes and stained with ethidium bromide. The bands represent chromosomes having sizes ranging from 0.2 megabases (at bottom) to 2.2 megabases (at top). Original gel is 5-¼″ wide by 5″ long.
Bio-Rad Labs, CHEF-DR®II pulsed-field electrophoresis system.

DNA with a mobility of 20 mm in figure 6.23 would contain about 820 base pairs (bp), or 0.82 **kilobase pairs (kbp, or simply kb).**

The size of a larger DNA is not as easy to determine. One reason is that double-stranded DNA is a relatively rigid rod—very long and thin. The longer it is, the more fragile it is. In fact, large DNAs break very easily; even seemingly mild manipulations, like swirling in a beaker or pipetting, create shearing forces sufficient to fracture them. To visualize this, think of DNA as a piece of uncooked spaghetti. If it is short—say an inch or two—you can treat it roughly without harming it, but if it is long, breakage becomes almost inevitable.

In spite of these difficulties, molecular geneticists have recently developed a new kind of gel electrophoresis that can separate DNA molecules up to several million base pairs (**megabases, mb**) long. Instead of a constant current through the gel, this method uses pulses of current. This *pulsed-field electrophoresis* is valuable for measuring the sizes of DNAs even as large as some of the chromosomes found in yeast. Figure 6.24 presents the results of pulsed-field gel electrophoresis on yeast chromosomes. The sixteen visible bands represent chromosomes containing 0.2 to 2.2 mb.

N atural DNAs come in sizes ranging from a few kilobases to hundreds of megabases. Circular DNAs frequently twist around themselves, forming supercoils.

■

Summary

G enes of all true organisms are made of DNA; certain viruses and all viroids have genes made of RNA. DNA and RNA are chainlike molecules composed of subunits called nucleotides. DNA in most genetic systems is a double-helical molecule with sugar-phosphate backbones on the outside and base pairs on the inside. The bases pair in a specific way: A with T and G with C. When DNA replicates, the parental strands separate; each then serves as the template for making a new, complementary strand.

The $G+C$ content of a DNA can vary from about 25% to about 75%, and this can have a strong effect on the physical properties of DNA, particularly its melting temperature. The melting temperature (T_m) of a DNA is the temperature at which the two strands are half-dissociated, or denatured. Separated DNA strands can be induced to renature, or anneal. Complementary strands of polynucleotides (either RNA or DNA) from different sources can form a double helix in a process called hybridization. Natural DNAs vary widely in length. The size of a DNA can be estimated by a variety of methods, including electron microscopy and gel electrophoresis. ∎

Problems and Questions

1. What evidence suggested that the transforming principle in Griffith's experiments was really a gene or genes?

2. One possible explanation for Griffith's transformation results was that avirulent (R) bacteria somehow restored the dead virulent (S) cells to life. How did the experiments of Avery et al. eliminate this unlikely possibility?

3. Which experimental approach more clearly points to DNA as the genetic material—that of Avery et al. or that of Hershey and Chase? *Hint:* Which experiment is easier to criticize on the basis of possible impurities in the genetic material that enters the test cell?

4. Would ^{14}C and ^{3}H have been suitable radioactive tracers for the Hershey-Chase experiment? Why, or why not?

5. If one strand of double-stranded DNA has the following sequence: 5'-AGCATTCG-3', what is the sequence of the opposite strand in *its* 5' to 3' direction? In its 3' to 5' direction?

6. If bovine DNA has an adenine content of 30% and a guanine content of 20%, what is its content of the other two bases?

7. An animal virus contains a circular double-stranded DNA with 120,000 base pairs. (*a*) Approximately how many helical turns does this DNA contain? (*b*) How long is the DNA in microns (1 micron $= 10^4$ angstroms)? (*c*) What is the approximate molecular weight of this DNA? (*d*) How many phosphorous atoms does the DNA contain?

8. A certain DNA fragment has a mobility of 30 mm in the electrophoresis experiment depicted in figure 6.23. What is its size in kilobase pairs?

Answers appear at end of book.

7

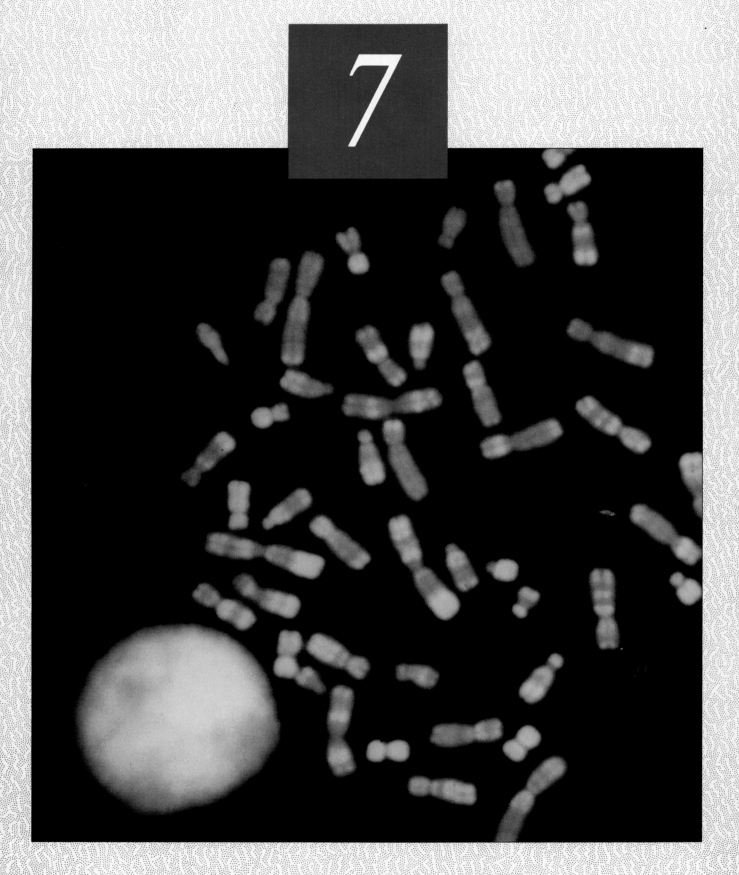

An Introduction to Gene Function

M

an is only man at the surface. Remove his skin, dissect, and immediately you come to machinery.

■

Paul Valéry
French poet and critic

Learning Objectives

In this chapter you will learn:

1. The three main functions of genes.

2. The outline of how cells duplicate their own genes.

3. The outline of how genes serve as blueprints for making important cellular components: RNAs and proteins.

4. How a change in a human gene can have serious consequences for the individual bearing the altered gene.

Human chromosomes. Magnification: ×1,250.
© *Dr. Ram Verma/Phototake.*

typical gene has three main functions: (1) It is capable of faithful replication so that its information is passed essentially unchanged from one generation to the next. (2) It holds the information for making one of the cell's key constituents—a protein. (3) It is able to accept occasional change so that life forms can evolve. In other words, it is susceptible to mutation. This chapter outlines how genes perform their functions. Subsequent chapters will treat each function in detail.

REPLICATION

The Watson-Crick model for DNA replication introduced in chapter 6 assumes that as new strands of DNA are made, they follow the usual base-pairing rules of A with T and G with C. This is essential because the DNA-replicating machinery must be capable of discerning a good pair from a bad one, and the Watson-Crick base pairs give the best fit. The model also presupposes that the two parental strands separate and that each then serves as a template for a new progeny strand. This is called **semiconservative** replication because each daughter double helix has one parental strand and one new strand (figure 7.1*a*). In other words, one of the parental strands is "conserved" in each daughter double helix. This is not the only possibility. Another potential mechanism (figure 7.1*b*) is **conservative** replication, in which the two parental strands stay together and somehow produce another daughter helix with two completely new strands. Yet another possibility is **dispersive** replication, in which the DNA becomes fragmented so that new and old DNA coexist in the same strand after replication (figure 7.1*c*).

Fortunately, the three mechanisms can be distinguished experimentally. Matthew Meselson and Franklin Stahl worked out a clever procedure to do just that, using a nonradioactive heavy isotope of nitrogen. Ordinary nitrogen, the most abundant isotope, has an atomic weight of 14, so it is called ^{14}N. A relatively rare isotope (^{15}N) has an atomic weight of 15. Meselson and Stahl found that if bacteria grow in a medium enriched in ^{15}N, they incorporate the heavy isotope into their DNA, which becomes denser than normal. This labeled DNA clearly separates from ordinary DNA in a gradient of cesium chloride (CsCl) spun in an ultracentrifuge. CsCl is used because it is a very dense salt and therefore makes a dense enough solution that DNA (a dense material itself) will form a band, rather than sinking to the bottom. Figure 7.2 shows a schematic diagram of this experimental procedure. Figure 7.3, from Meselson and Stahl's 1958 paper, demonstrates the clean separation of normal ^{14}N- and heavy ^{15}N-tagged DNAs.

The crux of the experiment was to grow ^{15}N-labeled bacteria in ^{14}N-medium for one or more generations and then to look at the density of the DNA. If replication is conservative, then after one generation, the two heavy parental strands will stay together and two newly made light

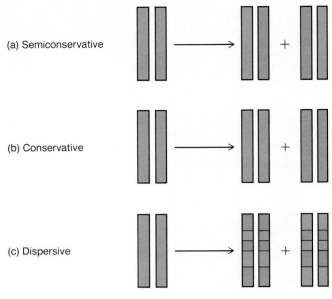

(a) Semiconservative

(b) Conservative

(c) Dispersive

FIGURE 7.1

Three hypotheses for DNA replication. (*a*) Semiconservative replication (see also figure 6.15) gives two daughter duplex DNAs, each of which contains one old strand (blue) and one new strand (red). (*b*) Conservative replication yields two daughter duplexes, one of which has two old strands (blue) and one of which has two new strands (red). (*c*) Dispersive replication gives two daughter duplexes, each of which contains strands that are a mixture of old and new DNA.

strands will appear as a separate band in the CsCl gradient (figure 7.4). On the other hand, if replication is semiconservative, each heavy parental strand will take a new, light partner. These heavy/light hybrid duplexes will have a density halfway between the heavy/heavy parental DNA and light/light ordinary DNA. This is exactly what happened; after the first DNA doubling, a single band appeared midway between the labeled heavy/heavy DNA and a normal light/light DNA. This ruled out conservative replication, but was still consistent with either semiconservative or dispersive replication.

The results of one more round of DNA replication destroyed the dispersive hypothesis. Dispersive replication would give a product with one-fourth ^{15}N and three-fourths ^{14}N after two rounds of replication in ^{14}N-medium. Semiconservative replication would yield half of the products as heavy/light and half as light/light (figure 7.4). In other words, the hybrid heavy/light products of the first round of replication would each split and take new, light partners, giving the 1:1 ratio of heavy/light to light/light DNAs. Again, this is precisely what occurred. Therefore, the data supported only the semiconservative mechanism.

D NA replicates in a semiconservative manner. When the parental strands separate, each serves as the template for making a new, complementary strand.

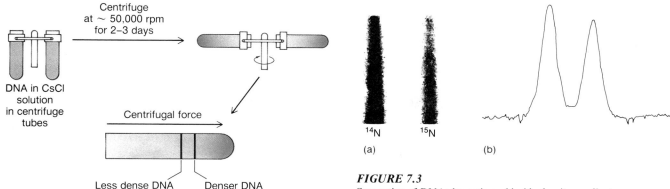

FIGURE 7.2

Principle of cesium chloride density gradient centrifugation. As a solution of cesium chloride spins in the ultracentrifuge, the dense cesium chloride sediments toward the bottom until equilibrium is established between the centrifugal force and diffusion forces. At equilibrium there is a gradient of cesium chloride, and therefore of density, from high density at the bottom of the tube to relatively low density at the top. If DNA is included in the solution, it will also seek an equilibrium position—at the point in the tube where its density exactly matches that of the cesium chloride solution. Thus, two DNAs of different density will form two separate bands in the tube.

FIGURE 7.3

Separation of DNAs by cesium chloride density gradient centrifugation. DNA containing the normal isotope of nitrogen (^{14}N) was mixed with DNA labeled with a heavy isotope of nitrogen (^{15}N) and subjected to cesium chloride density gradient centrifugation. The two bands have different densities, so they separate cleanly. (*a*) A photograph of the spinning tube under ultraviolet illumination. The two dark bands correspond to the two different DNAs that absorb ultraviolet light. (*b*) A graph of the darkness of each band, which gives an idea of the relative amounts of the two kinds of DNA.

From Meselson and Stahl, in Proceedings of the National Academy of Science *44:671–82, 1958. Copyright © 1958 National Academy Press.*

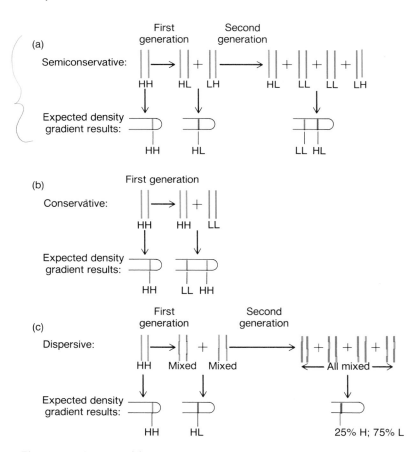

FIGURE 7.4

Three replication hypotheses. The conservative model (*b*) predicts that after one generation there will be equal amounts of two different DNAs (light/light and heavy/heavy). Both the semiconservative (*a*) and dispersive (*c*) models predict a single band of DNA with a density halfway between the light/light and heavy/heavy densities. Meselson and Stahl's results confirmed the latter prediction, so the conservative mechanism was ruled out. The dispersive model predicts that the

DNA after the second generation will have a single density, corresponding to a molecule that is 25% heavy and 75% light. This should give one band of DNA halfway between the light/light band and the heavy/light band. The semiconservative model predicts equal amounts of two different DNAs (light/light and heavy/light). Again, the latter prediction matched the experimental results, confirming the semiconservative model.

An Introduction to Gene Function

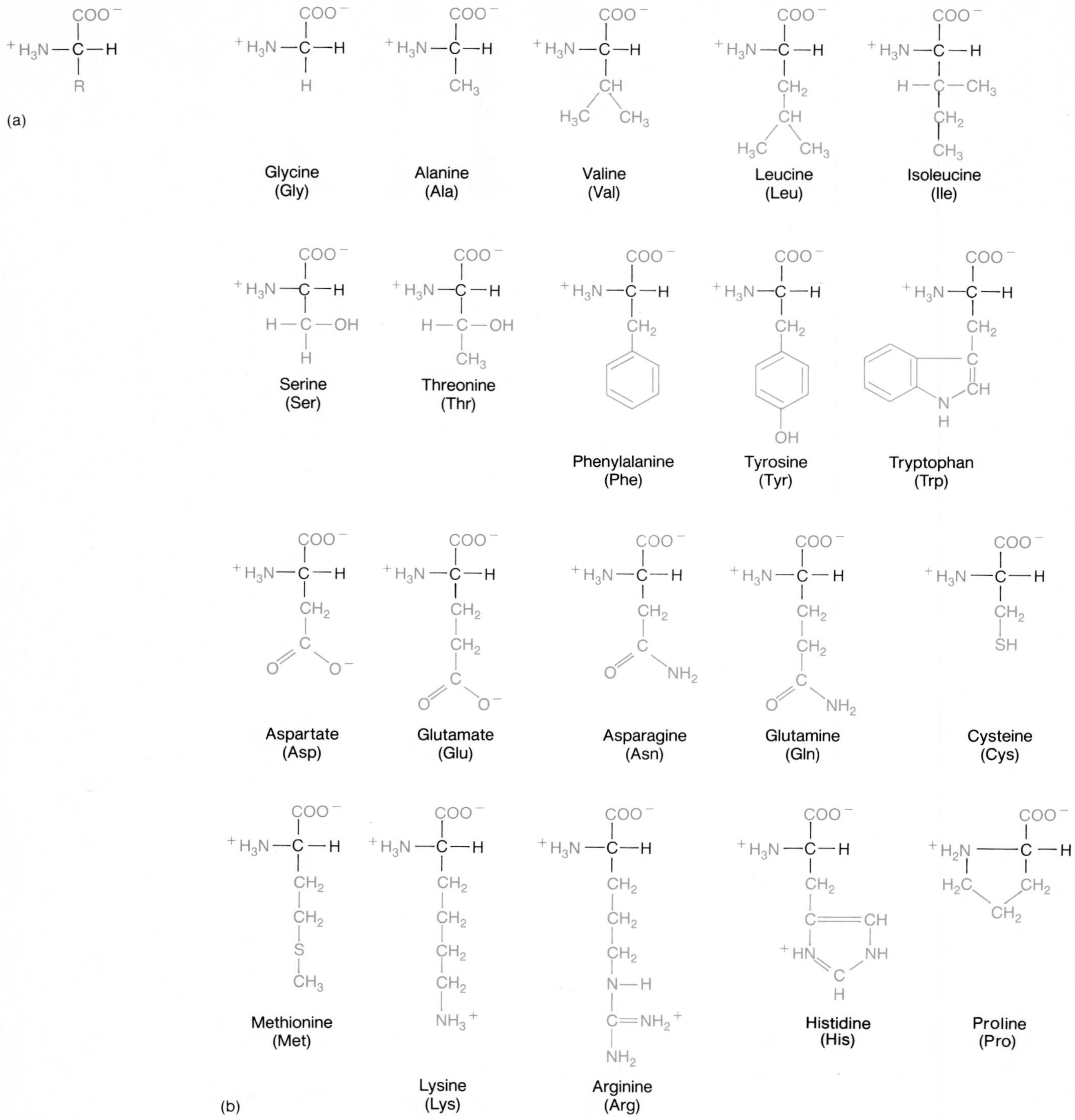

(a)

Glycine
(Gly)

Alanine
(Ala)

Valine
(Val)

Leucine
(Leu)

Isoleucine
(Ile)

Serine
(Ser)

Threonine
(Thr)

Phenylalanine
(Phe)

Tyrosine
(Tyr)

Tryptophan
(Trp)

Aspartate
(Asp)

Glutamate
(Glu)

Asparagine
(Asn)

Glutamine
(Gln)

Cysteine
(Cys)

Methionine
(Met)

Lysine
(Lys)

Arginine
(Arg)

Histidine
(His)

Proline
(Pro)

(b)

FIGURE 7.5

Amino acid structure. (a) The general structure of an amino acid. It has both an amino group (NH_3^+; red) and an acid group (COO^-; blue); hence the name. Its other two positions are occupied by a proton (H) and a side chain (R, green). (b) There are twenty different side chains on the twenty different amino acids, all of which are illustrated here.

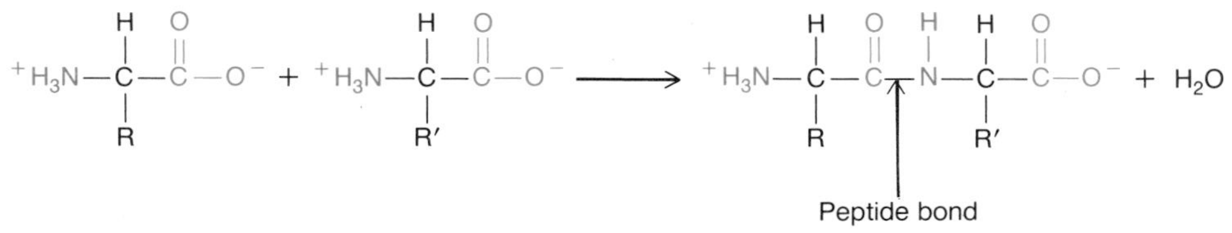

$$^+H_3N-\underset{R}{\overset{H}{\underset{|}{\overset{|}{C}}}}-\overset{O}{\overset{\|}{C}}-O^- + {}^+H_3N-\underset{R'}{\overset{H}{\underset{|}{\overset{|}{C}}}}-\overset{O}{\overset{\|}{C}}-O^- \longrightarrow {}^+H_3N-\underset{R}{\overset{H}{\underset{|}{\overset{|}{C}}}}-\overset{O}{\overset{\|}{C}}-\overset{H}{\overset{|}{N}}-\underset{R'}{\overset{H}{\underset{|}{\overset{|}{C}}}}-\overset{O}{\overset{\|}{C}}-O^- + H_2O$$

Peptide bond

FIGURE 7.6

Formation of a peptide bond. Two amino acids with side chains R and R' combine through the acid group of the first and the amino group of the second to form a dipeptide, two amino acids linked by a peptide bond. One molecule of water also forms as a by-product.

STORING INFORMATION

Most genes are blueprints for **proteins;** that is, each gene carries the coded information for making one protein. In order to understand this relationship, we first need to explore the nature of proteins.

PROTEIN STRUCTURE

Proteins, like nucleic acids, are chainlike polymers of small subunits. In the case of DNA and RNA, the links in the chain are nucleotides. The chain links of proteins are **amino acids.** Whereas DNA contains only four different nucleotides, proteins contain twenty different amino acids. The structures of these compounds are shown in figure 7.5. The amino acid subunits join together in proteins via **peptide bonds,** as shown in figure 7.6. Another name for a chain of amino acids is **polypeptide.**

The linear order of amino acids constitutes a protein's **primary structure.** The way these amino acids interact with their neighbors gives a protein its **secondary structure.** The α-*helix* is a common form of secondary structure. It results from hydrogen bonding among near-neighbor amino acids, as shown in figure 7.7. Another common secondary structure found in proteins is the β-pleated sheet. This involves extended protein chains, packed side-by-side, that interact by hydrogen bonding. The packing of the chains next to each other creates the sheet appearance. Silk is a protein very rich in β-pleated sheets. A third example of secondary structure is simply a turn. Such turns connect the α-helices and β-pleated sheet elements in a protein.

The total three-dimensional shape of a polypeptide is its **tertiary structure.** Figure 7.8 illustrates how the protein myoglobin folds up into its tertiary structure. Elements of secondary structure are apparent, especially the several α-helices of the molecule and the turns that connect them. Note the overall roughly spherical shape of myoglobin. Most polypeptides take this form, which we call *globular.* Many proteins are large enough that they contain more than one compact structural region. Each of

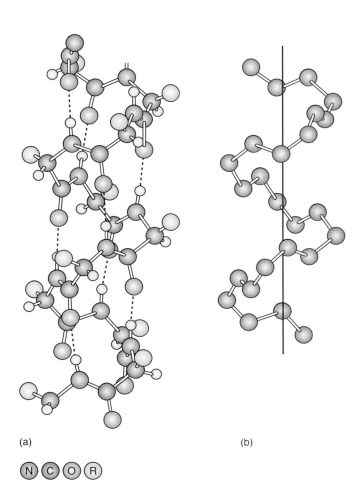

(a)   (b)

Ⓝ Ⓒ Ⓞ Ⓡ

FIGURE 7.7

An example of protein secondary structure: the α-helix. (a) The positions of the amino acids in the helix are shown, with the helical backbone in black and red. The dotted lines represent hydrogen bonds between hydrogen and oxygen atoms on neighboring amino acids. The small white circles represent hydrogen atoms. (b) A simplified rendition of the α-helix, showing only the atoms in the helical backbone.

(a) Reprinted from Linus Pauling: The Nature of the Chemical Bond, *3rd edition. Copyright 1939 and 1940, third edition © 1960 by Cornell University. Used by permission of the publisher, Cornell University Press.*

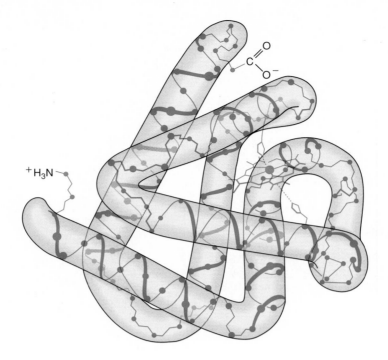

$C = O$
O^-

^+H_3N

FIGURE 7.8

Tertiary structure of myoglobin. The several α-helical regions of this protein are represented by spirals. The overall molecule seems to resemble a sausage, twisted into a roughly spherical or globular shape. The heme group is shown in red, bound to two histidines (green) in the protein.

From R. E. Dickerson, The Proteins, *2d ed. Copyright © 1964 Academic Press, Orlando, FL. Reprinted by permission of the publisher and author.*

these regions is called a *domain.* Antibodies (the proteins our white blood cells make to repel invaders) provide a good example of domains. Each of the four polypeptides in the IgG-type antibody contains globular domains, as shown in figure 7.9. This figure also illustrates the highest level of protein structure: **quaternary structure.** This is the way two or more individual polypeptides fit together in a complex protein. It has long been assumed that a protein's amino acid sequence determines all of its higher levels of structure, much as the linear sequence of letters in this book determines word, sentence, and paragraph structure. However, this is an oversimplification. Many proteins cannot fold properly by themselves outside their normal cellular environment. Some cellular factors besides the protein itself seem to be required in these cases.

P roteins are polymers of amino acids linked through peptide bonds. The sequence of amino acids in a polypeptide (primary structure) gives rise to that molecule's local shape (secondary structure), overall shape (tertiary structure), and interaction with other polypeptides (quaternary structure).

■

PROTEIN FUNCTION

Why are proteins so important that most genes are dedicated to their creation? Some proteins provide the structure that helps give cells integrity and shape. Others serve as *hormones* to carry signals from one cell to another. For example, the pancreas secretes the hormone insulin that signals liver and muscle cells to take up the sugar glucose from the blood. Proteins can bind and carry substances. The protein hemoglobin carries oxygen from the lungs to remote areas of the body; myoglobin stores oxygen in muscle tissue until it is used. Proteins also bind to genes and control their activity, as we will see in chapters 9, 10, and 14. Perhaps most important of all, proteins serve as enzymes that catalyze the myriad chemical reactions necessary for life. Left to themselves, these reactions would take place far too slowly at ordinary body temperatures; without enzymes, life as we know it would be impossible.

Making Proteins

Our knowledge of the gene-protein link dates back as far as 1902, when Archibald Garrod, a physician, noticed that a human disease, *alcaptonuria,* behaved as if it was caused

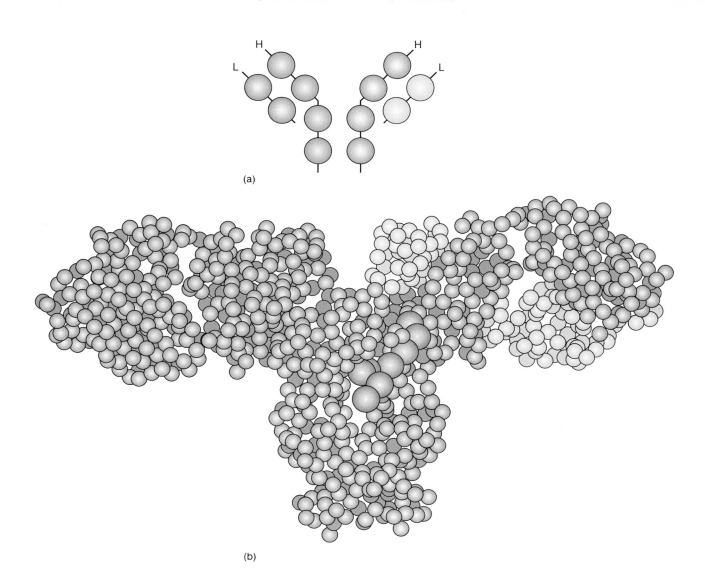

(a)

(b)

FIGURE 7.9

The globular domains of immunoglobulin. (*a*) Schematic diagram, showing the four polypeptides that comprise the immunoglobulin: two light chains (L) and two heavy chains (H). The light chains each contain two globular regions, while the heavy chains have four globular domains apiece. (*b*) Space-filling model of immunoglobulin. The colors correspond to those in (*a*). Thus, the two H-chains are in red and blue, while the L-chains are in green and yellow. A complex sugar attached to the protein is shown in gray. Note the globular domains in each of the polypeptides. Also note how the four polypeptides fit together to form the quaternary structure of the protein.

by a single recessive gene. Fortunately, Mendel's work had been rediscovered two years earlier and provided the theoretical background for Garrod's observation. Patients with alcaptonuria excrete copious amounts of *homogentisic acid,* which has the startling effect of coloring their urine black. Garrod reasoned that the abnormal buildup of this compound resulted from a defective metabolic pathway. Somehow, a blockage somewhere in the pathway was causing the intermediate, homogentisic acid, to accumulate to abnormally high levels, much as a dam causes water to accumulate behind it. Several years later, Garrod postulated that the problem came from a defect in the pathway that degrades the amino acid phenylalanine (figure 7.10).

By that time, metabolic pathways had been studied for years and were known to be controlled by enzymes— one enzyme catalyzing each step. Thus, it seemed that alcaptonuria patients carried a defective enzyme. And since the disease was inherited in a simple Mendelian fashion, Garrod concluded that a gene must control the enzyme's

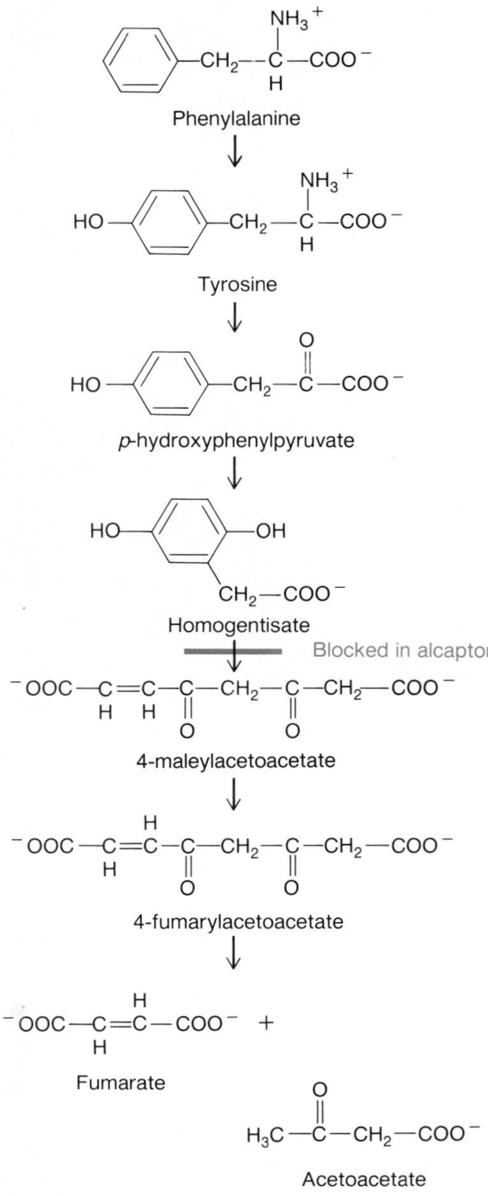

FIGURE 7.10

Phenylalanine

Tyrosine

p-hydroxyphenylpyruvate

Homogentisate

Blocked in alcaptonuria

4-maleylacetoacetate

4-fumarylacetoacetate

Fumarate +

Acetoacetate

FIGURE 7.10

Pathway of phenylalanine breakdown. Alcaptonuria patients are defective in the enzyme that converts homogentisic acid (homogentisate) to 4-maleylacetoacetate.

production. When that gene is defective, it gives rise to a defective enzyme. This established the crucial link between genes and proteins.

George Beadle and E. L. Tatum carried this argument a step further with their studies of a common bread mold, *Neurospora,* in the 1940s. They performed their experiments as follows: First, they bombarded the peritheca (spore-forming parts) of *Neurospora* with X rays or ultraviolet radiation to cause mutations. Then, they collected the spores from the irradiated mold and germinated

them separately to give pure strains of mold. They screened many thousands of strains to find a few mutants. The mutants revealed themselves by their inability to grow on minimal medium composed only of sugar, salts, inorganic nitrogen, and the vitamin biotin. Wild-type *Neurospora* grows readily on such a medium; the mutants had to be fed something extra—a vitamin, for example—in order to survive.

Next, Beadle and Tatum performed biochemical and genetic analyses on their mutants. By carefully adding substances, one at a time, to the mutant cultures, they pinpointed the biochemical defect. For example, the last step in the synthesis of the vitamin pantothenate involves putting together the two halves of the molecule: pantoic acid and β-alanine. One "pantothenateless" mutant would grow on pantothenate, but not on the two halves of the vitamin. This demonstrated that the last step in the biochemical pathway leading to pantothenate was blocked, so the enzyme (enzyme 3 in figure 7.11) that carries out that step must have been defective.

The genetic analysis was just as straightforward. *Neurospora* is an *ascomycete,* in which nuclei of two different mating types fuse and undergo meiosis to give eight haploid *ascospores,* borne in a fruiting body called an *ascus* (see chapter 5). A mutant can therefore be crossed with a wild-type strain of the opposite mating type to give eight spores (figure 7.12). If the mutant phenotype is due to a mutation in a single gene, then four of the eight spores should be mutant and four should be wild-type. Beadle and Tatum collected the spores, germinated them separately, and checked the phenotypes of the resulting molds. Sure enough, they found that four of the eight spores gave rise to mutant molds, demonstrating that the mutant phenotype was controlled by a single gene. This happened over and over again, leading these investigators to the conclusion that each enzyme in a biochemical pathway is controlled by one gene. Subsequent work has shown that many enzymes contain more than one polypeptide chain and that each polypeptide is usually encoded in one gene. This is the **one gene-one polypeptide hypothesis.**

*M*ost genes contain the information for making one polypeptide.

GENE EXPRESSION

Gene expression is the process by which gene products are made. The information in the gene is decoded and expressed as a product. As we have seen, this product is usually a protein, but sometimes it is an RNA. We will discuss these RNA-specifying genes later in this chapter; for now, let us consider genes whose products are proteins—**structural genes.** The term "code" and its derivatives, aside from

H₃C—CH—C—COOH →(1, HCHO) H₂C—C—C—COOH →(2, 2H⁺) ... (chemical pathway figure)

The pathway figure shows:

Top row:
H_3C\
\ CH—C—COOH →(1)(HCHO) H_2C—C—C—COOH →(2)($2H^+$)\
H_3C/   ‖ | | ‖\
              O   OH CH₃ O

Second row:
        CH₃   CH₃   H
H_2C—C—CH—COOH →(3)(ATP) H_2C—C—CH—C—N—CH₂—CH₂—COOH\
    | | | | | | ‖\
    OH CH₃ OH   OH CH₃ OH O\
                                                    Pantothenic acid\
                                                    + AMP + PPi

Pantoic acid H_2N—CH₂—CH₂—COOH\
                              β-alanine

FIGURE 7.11
Pathway of pantothenic acid synthesis. The last step (step 3), formation of pantothenic acid from the two half-molecules, pantoic acid (blue) and β-alanine (red), was blocked in one of Beadle and Tatum's mutants. The enzyme that carries out this step must have been defective.

providing a useful metaphor to describe the language of the gene, are legitimate scientific terms. Thus, when a gene contains the information for making a certain protein, we say that it **encodes** or **codes for** that protein. These structural genes are expressed in two major steps: **transcription** and **translation.** These steps will be discussed in detail in chapters 9 through 11. We will give a brief introduction here.

Transcription means making an RNA copy of a gene. Translation means that the information in this RNA, still "written" in genetic code, is used to make protein. Again, the terms are apt. In everyday language, transcription refers to making a close copy of an original, just as RNA is very similar to DNA. By contrast, translation means rendering something in a different language, which is exactly what happens when the genetic code in RNA is translated to the amino acid code of protein.

The RNA copy, or **transcript,** of a structural gene is called a **messenger RNA (mRNA).** It carries the information, or **message,** for making the protein from the gene to the cell's protein factories, the **ribosomes.**

DISCOVERY OF MESSENGER RNA

The concept of a messenger RNA carrying information from gene to ribosome developed in stages during the years following the publication of Watson and Crick's DNA model. In 1958, Crick himself proposed that RNA serves as an intermediate carrier of genetic information. He based his hypothesis in part on the fact that the DNA resides in the nucleus of eukaryotes, whereas proteins are made in the cytoplasm. This means that something must carry the information from one place to the other. Crick noted that ribosomes contain RNA and suggested that this ribosomal RNA (rRNA) is the information bearer. But rRNA is an integral part of ribosomes; it cannot escape. Therefore, Crick's hypothesis implied that each ribosome, with its own rRNA, would produce the same kind of protein over and over. This turned out to be wrong.

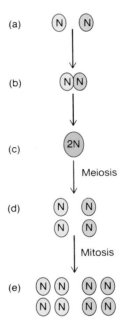

FIGURE 7.12
Sporulation in the mold *Neurospora crassa.* (*a*) Two haploid nuclei, one wild-type (yellow) and one mutant (blue), have come together in the immature fruiting body of the mold. (*b*) The two nuclei begin to fuse. (*c*) Fusion is complete, and a diploid nucleus (green) has formed. One haploid set of chromosomes is from the wild-type nucleus and one set is from the mutant nucleus. (*d*) Meiosis occurs, producing four haploid nuclei. If the mutant phenotype is controlled by one gene, two of these nuclei (blue) should have the mutant allele and two (yellow) should have the wild-type allele. (*e*) Finally, mitosis occurs, producing eight haploid nuclei, each of which will go to one ascospore. Four of these nuclei (blue) should have the mutant allele and four (yellow) should have the wild-type allele. If the mutant phenotype is controlled by more than one gene, the results will be more complex.

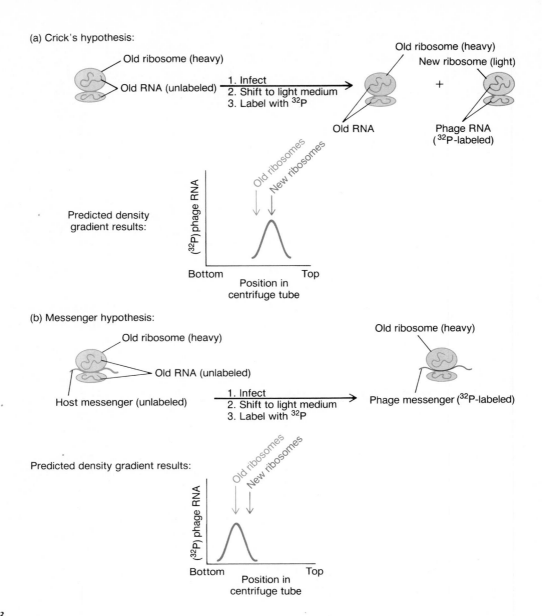

(a) Crick's hypothesis:

Old ribosome (heavy)

Old RNA (unlabeled)

1. Infect
2. Shift to light medium
3. Label with ^{32}P

Old ribosome (heavy)
New ribosome (light)

+

Old RNA

Phage RNA
(^{32}P-labeled)

Predicted density
gradient results:

(^{32}P) phage RNA

Old ribosomes
New ribosomes

Bottom   Top
Position in
centrifuge tube

(b) Messenger hypothesis:

Old ribosome (heavy)

Old RNA (unlabeled)

Host messenger (unlabeled)

1. Infect
2. Shift to light medium
3. Label with ^{32}P

Old ribosome (heavy)

Phage messenger (^{32}P-labeled)

Predicted density gradient results:

(^{32}P) phage RNA

Old ribosomes
New ribosomes

Bottom   Top
Position in
centrifuge tube

FIGURE 7.13

Experimental test of the messenger hypothesis. *E. coli* ribosomes were made heavy by labeling the bacterial cells with heavy isotopes of carbon and nitrogen. The bacteria were then infected with T2 or T4 phage and simultaneously shifted to "light" medium containing the normal isotopes of carbon and nitrogen, plus some ^{32}P to make the phage RNA radioactive. (*a*) Crick had proposed that ribosomal RNA carried the message for making proteins. If this were so, then whole new ribosomes with phage-specific ribosomal RNA would have been made after phage infection. In that case, the new, ^{32}P-labeled RNA (red) should have moved together with the new, light ribosomes (red). (*b*) Jacob and Monod had proposed that a messenger RNA carried genetic information to the ribosomes. According to this hypothesis, phage infection would cause the synthesis of new, phage-specific messenger RNAs that would be ^{32}P-labeled (red). These would associate with old, heavy ribosomes (blue). Therefore, the radioactive label would move together with the old, heavy ribosomes in the density gradient. This was indeed what happened.

François Jacob and Jacques Monod proposed an alternate hypothesis calling for nonspecialized ribosomes that translate unstable RNAs called **messengers.** The messengers are independent RNAs that bring genetic information from the genes to the ribosomes. In 1961, Jacob, along with Brenner and Meselson, probed the validity of the messenger hypothesis. This study used the same bacteriophage (T2) that Hershey and Chase had employed

almost a decade earlier to show that the genes were made of DNA. The premise of the experiments was this: When phage T2 infects *E. coli,* it subverts its host from making bacterial proteins to making phage proteins. If Crick's hypothesis was correct, this switch to phage protein synthesis should be accompanied by the production of new ribosomes equipped with phage-specific RNAs.

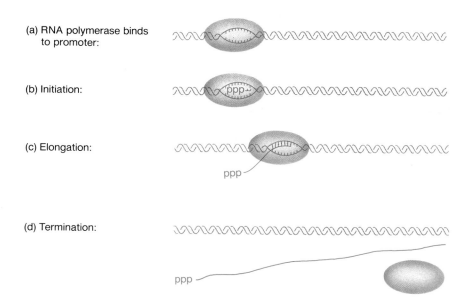

(a) RNA polymerase binds to promoter:

(b) Initiation:

(c) Elongation:

ppp

(d) Termination:

ppp

FIGURE 7.14
Transcription. (*a*) RNA polymerase (blue) binds tightly to the promoter and "melts" a short stretch of DNA. (*b*) In the initiation step, the polymerase joins the first two nucleotides of the nascent RNA (red) through a phosphodiester bond. The first nucleotide retains its triphosphate group. (*c*) During elongation, the melted bubble of DNA moves with the polymerase, allowing the enzyme to "read" the bases of one strand and make a complementary RNA. (*d*) Termination occurs when the polymerase reaches a termination signal at the end of a gene, causing the RNA and the polymerase to fall off the DNA template.

To distinguish new ribosomes from old, these investigators labeled the ribosomes in uninfected cells with heavy isotopes of nitrogen (^{15}N) and carbon (^{13}C). This made "old" ribosomes heavy. Then they infected these cells with phage T2 (or the related phage T4) and simultaneously transferred them to medium containing light nitrogen (^{14}N) and carbon (^{12}C). Any "new" ribosomes made after phage infection would therefore be light and would separate from the old, heavy ribosomes upon density gradient centrifugation. Brenner et al. also labeled the infected cells with ^{32}P to tag any phage RNA as it was made. Now our question is: Was the radioactively labeled phage RNA associated with new or old ribosomes? Figure 7.13 shows that the phage RNA was found on old ribosomes whose rRNA was made before infection even began. Clearly, this old rRNA cannot carry phage genetic information; by extension, it was very unlikely that it could carry host genetic information either.

Other workers had already identified a better candidate for the messenger: a class of unstable RNAs that associate transiently with ribosomes. Interestingly enough, in phage T2-infected cells, this RNA had a base composition very similar to that of phage DNA—and quite different from that of bacterial DNA and RNA. This is exactly what we would expect of phage messenger RNA, and that is exactly what it is. On the other hand, host mRNA, unlike host rRNA, has a base composition similar to that of host DNA. This lends further weight to the hypothesis that it is the mRNA, not the rRNA, that is the "instructional" molecule.

*M*ost genes are expressed in a two-step process: transcription, or synthesis of an mRNA copy of the gene, followed by translation of this message to protein.

■

TRANSCRIPTION

As you might expect, transcription follows the same base-pairing rules as DNA replication: T,G,C, and A in the DNA pair with A,C,G, and U, respectively, in the RNA product. (Notice that uracil appears in RNA in place of thymine in DNA.) This base-pairing pattern ensures that an RNA transcript is a faithful copy of the gene.

Of course, highly directed chemical reactions such as transcription do not happen by themselves—they are enzyme-catalyzed; the enzyme that directs transcription is called **RNA polymerase.** Figure 7.14 presents a schematic diagram of *E. coli* RNA polymerase at work. There are three phases of transcription: initiation, elongation, and termination.

1. *Initiation.* First, the enzyme recognizes a region called a **promoter,** which lies just "upstream" from the gene. The polymerase binds tightly to the promoter and causes localized melting, or separation of the two DNA strands within the promoter. Approximately ten base pairs are melted. Next, the polymerase starts building the RNA chain. The substrates, or building blocks, it uses for this job are the four **ribonucleoside triphosphates:** ATP, GTP, CTP, and UTP. The first, or initiating substrate is usually a purine nucleotide. After the first nucleotide is in place, the polymerase binds the second nucleotide and joins it to the first nucleotide, forming the initial phosphodiester bond in the RNA chain. At this point, the initiation process is complete.

2. *Elongation.* During the elongation phase of transcription, RNA polymerase directs the sequential binding of ribonucleotides to the

growing RNA chain in the 5′ → 3′ direction (from the 5′-end toward the 3′-end of the RNA). As it does so, it moves along the DNA **template,** and the "bubble" of melted DNA moves with it. This melted region exposes the bases of the template DNA one-by-one so that they can pair with the bases of the incoming ribonucleotides. As soon as the transcription machinery passes, the two DNA strands wind around one another again, re-forming the double helix. This points to two fundamental differences between transcription and DNA replication: (a) RNA polymerase makes only one RNA strand during transcription, which means that it copies only one DNA strand. Transcription is therefore said to be **asymmetric.** This contrasts with replication, in which both DNA strands are copied. (b) In transcription, DNA melting is limited and transient. Only enough strand separation occurs to allow the polymerase to "read" the DNA template strand. However, during replication, the two parental DNA strands separate permanently.

3. *Termination.* Just as promoters serve as initiation signals for transcription, other regions at the ends of genes, called **terminators,** signal termination. These work in conjunction with RNA polymerase, sometimes aided by another protein, to loosen the association between RNA product and DNA template. The result is that the RNA dissociates from the RNA polymerase and DNA; transcription is thereby terminated.

A final, important note about conventions: RNA sequences are usually written 5′ to 3′, left to right. This feels natural to a molecular geneticist because RNA is made in a 5′ to 3′ direction, and as we shall see, mRNA is also translated 5′ to 3′. Thus, since ribosomes read the message 5′ to 3′, it is appropriate to write it 5′ to 3′ so that *we* can read it like a sentence.

Genes are also usually written so that their transcription proceeds in a left to right direction. This "flow" of transcription from one end to the other gives rise to the term *upstream,* which refers to the DNA close to the start of transcription (near the left end when the gene is written conventionally). Thus, we can describe most promoters as lying just upstream from their respective genes.

T ranscription takes place in three stages: initiation, elongation, and termination. Initiation involves binding RNA polymerase to the promoter, local melting, and forming the first phosphodiester bond; during elongation, the RNA polymerase links together ribonucleotides in the 5′ → 3′ direction to make the rest of the RNA; finally, in termination, polymerase and RNA product dissociate from the DNA template.

TRANSLATION

The mechanism of translation is considerably more complex than that of transcription, as the relative complexities of the ribosome and RNA polymerase suggest. The details of translation will concern us in chapter 11; for now, let us look briefly at two substances that play key roles in translation: ribosomes and transfer RNA.

Ribosomes: Protein Synthesizing Machines

Figure 7.15 shows the probable shapes of the *E. coli* **ribosome** and its two subunits, the *50S* and *30S* particles. The numbers 50S and 30S refer to the **sedimentation coefficients** of the two subunits. These coefficients are a measure of the speed with which the particles sediment in an ultracentrifuge. The 50S particle, with a larger sedimentation coefficient, migrates more rapidly to the bottom of the centrifuge tube under the influence of a centrifugal force. The coefficients are functions of the mass and shape of the particles. Heavy particles sediment more rapidly than light ones; spherical particles migrate faster than extended or flattened ones—just as a skydiver falls more rapidly in a tuck position than with arms and legs extended. The 50S particle is actually about twice as massive as the 30S. Together, the 50S and 30S subunits comprise a 70S ribosome. Notice that the numbers do not add up. This is because the sedimentation coefficients are not proportional to the particle weight; in fact, they are roughly proportional to the ⅔-power of the particle weight.

Each ribosomal subunit contains RNA and protein. The 30S particle includes one molecule of **ribosomal RNA (rRNA)** with a sedimentation coefficient of 16S, plus twenty-one ribosomal proteins. The 50S particle is composed of two rRNAs (23S + 5S) and thirty-four proteins (figure 7.16). All these ribosomal proteins are of course gene products themselves. Thus, a ribosome is produced by dozens of different genes.

Note that rRNAs participate in protein synthesis but do not code for proteins. Transcription is the only step in expression of the genes for rRNA, aside from some trimming of the transcripts (chapter 10). No translation occurs.

R ibosomes are the cell's protein factories. Bacteria contain 70S ribosomes with two subunits, called 50S and 30S. Each of these contains ribosomal RNA and many proteins.

∎

Transfer RNA: The Adapter Molecule

The transcription mechanism was easy for molecular geneticists to predict. RNA resembles DNA so closely that it follows the same base-pairing rules. By following these rules, RNA polymerase produces replicas of the genes it

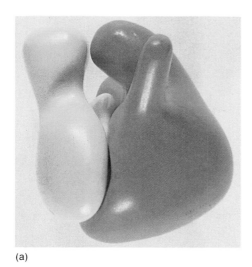

(a)

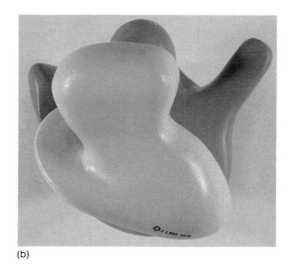

(b)

FIGURE 7.15

E. coli ribosome structure. (*a*) The 70S ribosome is shown from the "side" with the 30S particle (yellow) and the 50S particle (red) fitting together. (*b*) The 70S ribosome is shown rotated 90° relative to the view in (*a*). The 30S particle (yellow) is in front, with the 50S particle (red) behind.

James A. Lake, Journal of Molecular Biology *105(1976):131–59. Reproduced by permission of Academic Press.*

transcribes. But what rules govern the ribosome's translation of mRNA to protein? This is a true translation problem. A nucleic acid language must be translated to a protein language. Francis Crick suggested the answer to this problem in a 1958 paper before much experimental evidence was available to back it up. What is needed, Crick reasoned, is some kind of adapter molecule that can recognize the nucleotides in the RNA language as well as the amino acids in the protein language. He was right. He even noted that there was a type of small, soluble RNA of unknown function that might play the adapter role. Again, he guessed right. Of course there were some bad guesses in this paper as well, but even they were important. By their very creativity, Crick's ideas stimulated the research (some from Crick's own laboratory) that led to solutions to the puzzle of translation.

The adapter molecule in translation is indeed a small RNA that recognizes both RNA and amino acids; it is called **transfer RNA (tRNA).** Figure 7.17 shows a schematic diagram of a tRNA that recognizes the amino acid phenylalanine (Phe). In chapter 11 we will discuss the structure and function of tRNA in detail. For the present, the *cloverleaf* model, though it bears scant resemblance to the real shape of tRNA, will serve to point out the fact that the molecule has two "business ends." One end (the top of the model) attaches to an amino acid. Since this is a tRNA specific for phenylalanine (tRNA^{Phe}), only phenylalanine will bind. An enzyme called *phenylalanine-tRNA synthetase* catalyzes this binding. The generic name for such enzymes is *aminoacyl-tRNA synthetase*.

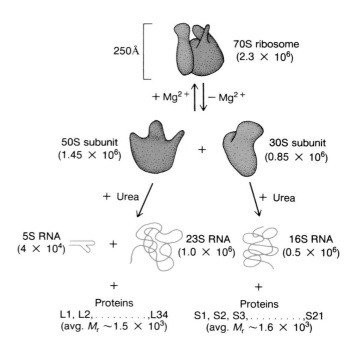

FIGURE 7.16

Composition of the *E. coli* ribosome. The arrows at the top denote the dissociation of the 70S ribosome into its two subunits when magnesium ions are withdrawn. The lower arrows show the dissociation of each subunit into RNA and protein components in response to the protein denaturant urea. The masses (M_r, in daltons) of the ribosome and its components are given in parentheses.

Reprinted by permission from Nature, *vol. 331, p. 225. Copyright © 1988 Macmillan Magazines, Ltd.*

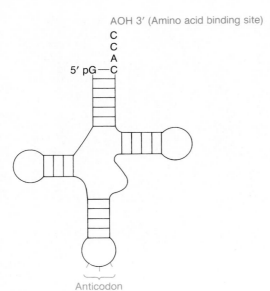

C
C
A
5′ pG—C

Anticodon

FIGURE 7.17

Transfer RNA. A schematic representation of the yeast phenlylanine tRNA (tRNA^{Phe}), the tRNA that functions in yeast to insert the amino acid phenylalanine into proteins. The tRNA does not really look like this in three dimensions; this diagram simply points out the secondary structure of the molecule—how base pairing occurs within the RNA. The four ladderlike stems show this base pairing. For example, observe the G–C pair between the 5′-terminal G and the C that is five bases from the 3′-end. The 3′-end of a tRNA always contains the same three bases: CCA. The terminal A (top, blue) is the site of attachment of the amino acid—phenylalanine in this case. The other business end of the tRNA is at the bottom of the diagram. This is the anticodon (green), which pairs with a three-base phenylalanine codon in a messenger RNA.

The other end (the bottom of the model) contains a three-base sequence that pairs with a complementary three-base sequence in an mRNA. Such a triplet in mRNA is called a **codon;** naturally enough, its complement in a tRNA is called an **anticodon.** The codon in question here has attracted the anticodon of a tRNA bearing a phenylalanine. That means that this codon says: "Please insert one phenylalanine into the protein at this point." A schematic of codon-anticodon recognition appears in figure 7.18. The recognition between codon and anticodon, mediated by the ribosome, obeys the same Watson-Crick rules as any other double-stranded polynucleotide, at least in the case of the first two base pairs. The third pair is allowed somewhat more freedom, as we will see in chapter 11.

It is apparent from figure 7.18 that UUC is a codon for phenylalanine. This implies that the **genetic code** contains three-letter words, as indeed it does. We can predict the number of possible three-letter code words as follows: The number of permutations of four different bases taken three at a time is 4^3, which is 64. But there are only twenty amino acids. Are some code words not used? Actually, three of the possible codons (UAG, UAA, and UGA) code for termination; all of the other codons specify amino acids.

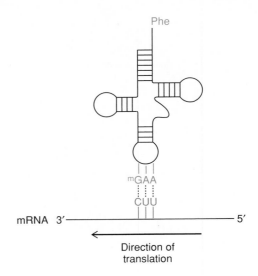

Phe

^{m}GAA
CUU

mRNA 3′——————————————5′

Direction of
translation

FIGURE 7.18

Codon-anticodon recognition. The recognition between a codon in an mRNA and a corresponding anticodon in a tRNA obeys essentially the same Watson-Crick rules as apply to other polynucleotides. Here, a 5′^{m}GAA3′ anticodon (green) on a tRNA^{Phe} is recognizing a 5′UUC3′ codon (red) for phenylalanine in a mRNA. The ^{m}G denotes a methylated G, which base-pairs like an ordinary G. Notice that the mRNA is pictured backwards (3′ → 5′) relative to normal convention, which is 5′→ 3′, left to right. That was done to put its codon in the proper orientation (3′ → 5′, left to right) to base-pair with the tRNA, shown conventionally with its anticodon reading 5′ → 3′, left to right. Remember that the two strands of DNA are antiparallel; this applies to any double-stranded polynucleotide, including one as small as the three-base codon-anticodon pair.

This means that most amino acids have more than one codon; the genetic code is therefore said to be **degenerate.** Chapter 11 presents a fuller description of the code and how it was broken.

*T*he two business ends of tRNAs allow them to recognize both amino acids and nucleic acids. They are therefore capable of serving the adapter role postulated by Crick and are the key to the mechanism of translation.

■

CODING AND ANTICODING STRANDS OF DNA

Figure 7.19 summarizes the transcription-translation process and introduces the coding-anticoding nomenclature we apply to the strands of DNA. Notice that the mRNA has the same sequence (except that Us substitute for Ts) as the top strand (blue) of the DNA. In turn, the mRNA codes for protein, in this case the tetrapeptide fMet-Ser-Asn-Ala. The term fMet refers to N-formyl methionine, a special amino acid used in initiation of protein synthesis. Since the top DNA strand codes for protein in the same way that mRNA does, we call it the **coding strand.** The opposite DNA strand is therefore the **anticoding strand.**

Gene: ATGAGTAACGCG Coding strand
TACTCATTGCGC Anticoding strand

| Transcription

mRNA: AUGAGUAACGCG

| Translation

Protein: fMetSerAsnAla

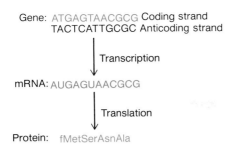

FIGURE 7.19

Outline of gene expression. In the first step, transcription, the anticoding strand is transcribed into mRNA. Note that the coding strand (blue) of the DNA has the same sequence (except for the T–U change) as the mRNA (red). In the second step, the mRNA is translated into protein (green). This little "gene" is only twelve base pairs long and codes for only four amino acids (a tetrapeptide). Real genes are much larger.

A possible source of confusion is the fact that the anti-coding DNA strand is the one that is copied, or transcribed, into mRNA. Just remember that the transcribed strand is complementary, or opposite, to the mRNA, and is therefore the anticoding strand.

MUTATIONS

The third function of a gene is to accumulate changes—to mutate. By this process, life itself can change. As we will see in chapter 16, mutation is the driving force of evolution. Now that we know most genes are strings of nucleotides that code for proteins, which in turn are strings of amino acids, it is easy to see the consequences of changes in DNA. If a nucleotide in a gene changes, it is likely that there will be a corresponding change in an amino acid in that gene's protein product. Sometimes, because of the degeneracy of the genetic code, a nucleotide change will not affect the protein. For example, changing the codon AAA to AAG is a mutation, but it would probably not be detected because both AAA and AAG code for the same amino acid: lysine. More often, a changed nucleotide in a gene results in an altered amino acid in the protein, which frequently impairs or destroys the function of that protein.

SICKLE-CELL DISEASE

An excellent example of such a defect is *sickle-cell disease,* a true genetic disease. People who are homozygous for this condition have normal looking red blood cells when their blood is rich in oxygen. The shape of normal cells is a *biconcave disc;* that is, the disc is concave viewed from both the top and bottom. However, when these people exercise, or otherwise deplete the oxygen in their blood, their red blood cells change dramatically to a sickle, or crescent shape (figure 7.20). This has dire consequences. The sickle cells cannot fit through tiny capillaries, so they clog and

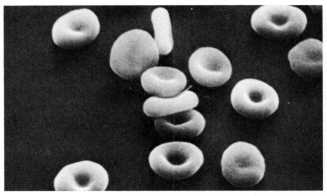

(a)

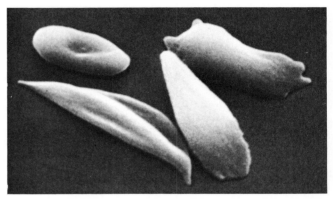

(b)

FIGURE 7.20

Normal red blood cells and a sickle cell. These scanning electron micrographs contrast (*a*) the regular, biconcave shape of normal cells with (*b*) the distorted shape of red blood cells from a sickle-cell disease patient. Magnifications: (*a*) ×3,333; (*b*) ×5,555.
(a) © Jeroboam/Photo Researchers, Inc.; (b) © Omikron/Photo Researchers, Inc.

rupture them, starving parts of the body for blood and causing internal bleeding and pain. Furthermore, the sickle cells are so fragile that they burst, leaving the patient anemic. Without medical attention, patients undergoing a sickling crisis are in mortal danger.

What causes this sickling of red blood cells? The problem is in *hemoglobin,* the red, oxygen-carrying protein in the red blood cells. Normal hemoglobin remains soluble under ordinary physiological conditions, but the hemoglobin in sickle cells precipitates when the blood oxygen level falls, forming long, fibrous aggregates that distort the blood cells into the sickle shape.

What is the difference between normal hemoglobin (HbA) and sickle-cell hemoglobin (HbS)? Vernon Ingram answered this question in 1957 by determining the amino acid sequences of parts of the two proteins using a process that was invented by Frederick Sanger and is known as **protein sequencing.** Ingram focused on the β-globins of the two proteins. β-globin is one of the two different polypeptide chains found in the tetrameric (four-chain) hemoglobin protein. First, Ingram cut the two polypeptides into pieces with an enzyme that breaks selected peptide bonds.

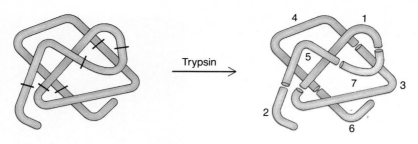

(a) Cutting protein to peptides

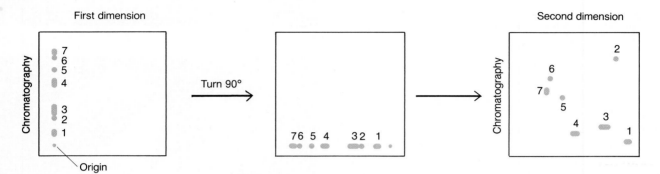

(b) Two-dimensional separation of peptides

FIGURE 7.21

Fingerprinting a protein. (*a*) A hypothetical protein, with its six trypsin-sensitive sites indicated by slashes. After digestion with trypsin, seven peptides are released. (*b*) These tryptic peptides separate partially during chromatography in the first dimension, then fully after the paper is turned 90 degrees and rechromatographed in the second dimension with another solvent.

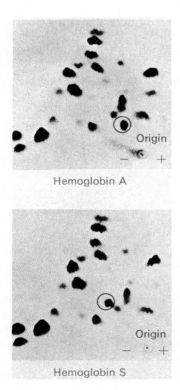

FIGURE 7.22

Fingerprints of hemoglobin A and hemoglobin S. The fingerprints are identical except for one peptide (circled), which shifts up and to the left in hemoglobin S.

Stryer, Biochemistry, *W. H. Freeman and Company Publishers, 1975, p. 92.*

These pieces, called peptides, can be separated by paper chromatography run in two dimensions (figure 7.21). One solvent is used to cause partial resolution of the peptides in the first dimension; the paper is then turned 90 degrees and rechromatographed with another solvent to separate the peptides still further. The peptides usually appear as spots on the paper; different proteins, because of their different amino acid compositions, give different patterns of spots. These patterns are aptly named **fingerprints.**

When Ingram compared the fingerprints of HbA and HbS, he found that all the spots matched except for one (figure 7.22). This spot had a different mobility in the HbS fingerprint than in the normal HbA fingerprint, which indicated that it had an altered amino acid composition. Ingram checked the amino acid sequences of the two peptides in these spots. He found that they were the amino-terminal peptides located at the very beginning of both proteins. And he found that they differed in only one amino acid. The glutamic acid in the sixth position of HbA becomes a valine in HbS (figure 7.23). This is the only difference in the two proteins, yet it is enough to cause a profound distortion of the protein's behavior.

Knowing the genetic code, we can ask: What change in the β-globin gene caused the change Ingram detected in its protein product? The two codons for glutamic acid (Glu) are GAA and GAG; two of the four codons for valine

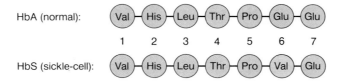

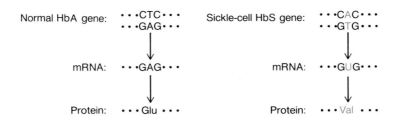

FIGURE 7.23

Sequences of amino-terminal peptides from normal and sickle-cell β-globin. The numbers indicate the positions of the corresponding amino acids in the mature protein. The only difference is in position 6, where a valine (Val) in HbS replaces a glutamic acid (Glu) in HbA.

FIGURE 7.24

The sickle-cell mutation and its consequences. The GAG in the sixth codon of the coding strand of the normal gene changes to GTG. This leads to a change from GAG to GUG in the sixth codon of the β-globin mRNA of sickle cells. This, in turn, results in insertion of a valine in the sixth amino acid position of sickle-cell β-globin, where a glutamic acid ought to go.

(Val) are GUA and GUG. If the glutamic acid codon in the HbA gene is GAG, a single base change to GTG would alter the mRNA to GUG, and the amino acid inserted into HbS would be valine instead of glutamic acid. A similar argument can be made for a GAA → GTA change. Notice that, by convention, we are presenting the DNA strand that has the same sense as the mRNA (the coding strand). Actually, the opposite strand (the anticoding strand), reading CAC, is transcribed to give a GUG sequence in the mRNA. Figure 7.24 presents a summary of the mutation and its consequences. We can see how changing the blueprint does indeed change the product.

Sickle-cell disease is a very common problem among black people of central African descent. The question naturally arises: Why has this deleterious mutation spread so successfully through the population? The answer seems to be that although the homozygous condition can be lethal, heterozygotes have little if any difficulty because their normal genes make enough product to keep their blood cells from sickling. Moreover, heterozygotes are at an advantage in central Africa, where malaria is rampant, because HbS helps protect against replication of the malarial parasite in their blood cells. In chapter 15 we will see how genetic engineering techniques offer hope of combating sickle-cell disease and related genetic disorders.

S ickle-cell disease is a human genetic disorder. It results from a single base change in the gene for β-globin. The altered base causes insertion of the wrong amino acid into one position of the β-globin. This altered protein results in distortion of red blood cells under low-oxygen conditions.

∎

COLINEARITY OF GENE AND PROTEIN

The idea of strict correspondence between the linear arrangement of nucleotides in a gene and of amino acids in its protein product is called the **colinearity** of gene and protein. It was proposed by Crick in his very fertile 1958 paper "On Protein Synthesis." In 1964, Charles Yanofsky and his colleagues proved that the idea was right, at least for bacteria. They studied the A protein of *E. coli* tryptophan synthetase and its gene. First, they mapped the gene by techniques described in chapter 5. This allowed them to place sixteen mutations in order on the gene and to establish the relative distances between any pair of these mutations. Next, they obtained the amino acid sequences of the mutant proteins to locate the altered amino acids.

		ABLE 7.1	Relations Between Map Distances and Residue Distances	

Mutant Pair	Map Distance	Amino Acid Residue Distance	Map Distance / Residue Distance
A58–A78	< 0.01	0	—
A46–A23	0.002	0	—
A58–A169	0.01	1	0.01
A78–A169	0.015	1	0.015
A46–A187	0.08	2	0.04
A23–A187	0.04	2	0.02
A446–A487	0.04	2	0.02
A487–A223	0.3	6	0.05
A446–A223	0.19	8	0.02
A46–A58	0.44	23	0.02
A46–A78	0.52	23	0.02
A23–A58	0.78	23	0.03
A23–A78	0.85	23	0.04
A46–A169	0.48	24	0.02
A23–A169	0.8	24	0.03
A223–A46	0.44	28	0.02
A487–A46	0.48	34	0.01
A446–A46	0.28	36	0.01
A78–A223	0.61	51	0.01
A169–A487	0.75	58	0.01

From Charles Yanofsky, et al., "On the Colinearity of Gene Structure and Protein Structure," in *Proceedings of the National Academy of Science*, Vol. 51, 1964. Copyright © 1964 Charles Yanofsky. Reprinted by permission.

The two sets of data compiled by Yanofsky (table 7.1) show quite clearly that the gene and its product are colinear. The farther apart the two mutations (that is, the longer the genetic map distance between them), the farther apart were the resulting altered amino acids. In fact, the data show an almost constant ratio of genetic map distance to amino acid residue distance. This colinearity is a common feature of bacterial genes and their products, but as we will see in chapter 10, the genes of higher organisms are frequently interrupted; in these genes, colinearity breaks down. This does not mean that the coding regions of eukaryotic genes are scrambled, just that they are sometimes interrupted with noncoding sequences.

The Relationship Between DNA Size and Genetic Capacity

How many genes are in a given DNA? It is impossible to tell just from the sizes of the DNAs, because we do not know how much of a given DNA is devoted to genes and how much is *intergenic space,* or even intervening sequences within genes. However, we can estimate an upper limit on the number of genes a DNA can hold. We start with the assumption that an average protein has a molecular weight of about 40,000. How many amino acids does this represent? The molecular weights of amino acids vary,

but they average a bit over 100. To simplify our calculation, let us assume that it is 100. That means our average protein contains 40,000/100, or 400 amino acids. Since each amino acid requires three base pairs to code for it, a protein containing 400 amino acids needs a gene of about 1,200 base pairs.

Consider a few of the DNAs listed in table 6.2. The *E. coli* chromosome contains 4.2×10^6 base pairs, so it could encode about 3,500 average proteins. The phages T2 and T4, which infect *E. coli,* have only 2×10^5 base pairs, so they can code for only about 160 proteins. The smallest double-stranded DNA on the list, belonging to the mammalian tumor virus SV40, has a mere 5,226 base pairs. In principle, that is only enough for four proteins, but the virus squeezes in some extra information by overlapping its genes. We will look at that phenomenon in more detail in chapter 13.

DNA Content and the C-value Paradox

You would probably predict that complex organisms like vertebrates would need more genes than simple organisms like yeast. Therefore, they should have higher **C-values,** or DNA content per haploid cell. In general, your prediction would be right; mouse and human haploid cells contain more than one hundred times more DNA than yeast cells. In addition, yeast cells have about five times more DNA than *E. coli* cells, which are even simpler. However, this correspondence between an organism's physical complexity and the DNA content of its cells is not perfect. Consider, for example, the frog. There is no reason to suspect that an amphibian has a great many more genes than a human, yet the frog has seven times more DNA per cell. Even more dramatic is the fact that the lily has one hundred times more DNA per cell than a human.

This perplexing situation is called the **C-value paradox.** It becomes even more difficult to explain when we look at organisms within a group. For example, some amphibian species have a C-value one hundred times higher than that of others, and the C-values of flowering plants vary even more widely. Does this mean that one kind of higher plant has one hundred times more genes than another? That is simply unbelievable. It would raise the questions of what all those extra genes are good for and why we do not notice tremendous differences in physical complexity among these organisms. The more plausible explanation of the C-value paradox is that organisms with extraordinarily high C-values simply have a great deal of extra, noncoding DNA. The function, if any, of this extra DNA is still mysterious.

In fact, even mammals have a lot more DNA than they need for genes. Applying our simple rule (dividing the number of base pairs by 1,200) to the human genome yields an estimate of about two million for the maximum number of genes. We will see in chapter 10 that most genes in higher eukaryotes actually have within them one or more noncoding DNA regions known as introns. However, even assuming five to ten times as much intron DNA as coding DNA leaves us with an estimate of about 200,000 to 400,000 genes, which is still almost certainly too high.

If we cannot use the C-value as a reliable guide to the number of genes in higher organisms, how can we make a reasonable estimate? One approach has been to count the number of different mRNAs in a given mammalian cell type, which gives an estimate of about 10,000. Of course, mammals have over one hundred different cell types, but they all produce many common mRNAs. Therefore, we cannot simply multiply 10,000 times 100. In fact, the overlap in genes expressed in different mammalian cells is so extensive that we probably need to multiply 10,000 by a number less than 10. By this reasoning, the upper limit on the number of functional mammalian genes is about 100,000.

*T*here is a rough correlation between the DNA content and the number of genes in a cell or virus. However, this correlation breaks down in several cases of closely related organisms where the DNA content per haploid cell (C-value) varies widely. This C-value paradox is probably explained, not by extra genes, but by extra noncoding DNA in some organisms.

■

Summary

*T*he three main functions of genes are replication, information storing, and mutation. DNA replicates in a semiconservative manner: When the parental strands separate, each serves as the template for making a new, complementary strand.

Proteins, or polypeptides, are polymers of amino acids linked through peptide bonds. Most genes contain the information for making one polypeptide, and are expressed in a two-step process: transcription or synthesis of an mRNA copy of the gene, followed by translation of this message to protein. Translation takes place in three stages: initiation, elongation, and termination, on complex structures called ribosomes, the cell's protein factories. Translation also requires adapter molecules that can recognize both the genetic code in mRNA and the amino acids the mRNA encodes. Transfer RNAs (tRNAs), with their two business ends, fill this role.

A change, or mutation, in a structural gene usually causes a change at a corresponding position in the gene's protein product. Sickle-cell disease is an example of the deleterious effect of most such mutations.

There is a rough correlation between the DNA content and the number of genes in a cell or virus. However, this correlation does not hold in several cases of closely related organisms where the DNA content per haploid cell (C-value) varies widely. This C-value paradox is probably explained by extra noncoding DNA in some organisms.

■

Problems and Questions

1. An animal virus contains a circular double-stranded DNA with 120,000 base pairs. How many proteins of average size (400 amino acids) can the virus code for?

2. Draw schematic diagrams of the DNA molecules in a Meselson-Stahl experiment after *three* generations in ^{14}N medium assuming: (*a*) dispersive, (*b*) conservative, and (*c*) semiconservative replication. Use solid lines for ^{15}N-labeled DNA and dotted lines for ^{14}N-labeled DNA.

3. Draw diagrams of the bands of DNA in CsCl gradients that would arise from the three types of replication in question 2. Show the positions of fully ^{15}N-labeled DNA and fully ^{14}N-labeled DNA.

4. You have isolated a *Neurospora* mutant that cannot make amino acid *z*. Normal *Neurospora* cells make *z* from a common cellular substance *x* by a pathway involving three intermediates, *a, b,* and *c:* $x \rightarrow a \rightarrow b \rightarrow c \rightarrow z$. How would you establish that your mutant contains a defective gene for the enzyme that catalyzes the $b \rightarrow c$ reaction?

5. The G + C content of phage T3 DNA is 53%. What would you expect the G + C content of T3 mRNA to be? Why?

6. Given the G + A content of phage T3 DNA, and not knowing which strand is transcribed, could you predict the G + A content of T3 mRNA? Why, or why not?

7. Here is a schematic diagram of a double-stranded DNA being transcribed.
   1. _____
   2. _____
   3. - - - - - - - - - - - - - - - - - - - →
   Strand 1 is the coding strand. Strand 3 is the nascent RNA. (*a*) Label the 5′ and 3′ ends of the RNA. (*b*) Label the 5′ and 3′ ends of the coding strand. (*c*) Label the 5′ and 3′ ends of the anticoding strand.

8. If a mutant *E. coli* cell had an aminoacyl-tRNA synthetase that should attach the amino acid phenylalanine to its tRNA (tRNA^{Phe}), but at the elevated temperature of 42°C attached the amino acid arginine instead, what consequence would this have for the proteins that cell made at the elevated temperature? Would this affect the function of these proteins? Why, or why not?

9. Here is the sequence of the anticoding strand of a DNA fragment:

   5′GTCAATCGTAGCGGCCATAT3′

   (*a*) Write the sequence of this DNA in double-stranded form. (*b*) Assuming that transcription of this DNA begins with the first nucleotide and ends with the last, write the sequence of the transcript of this DNA in conventional form (5′ to 3′ left to right).

Answers appear at end of book.

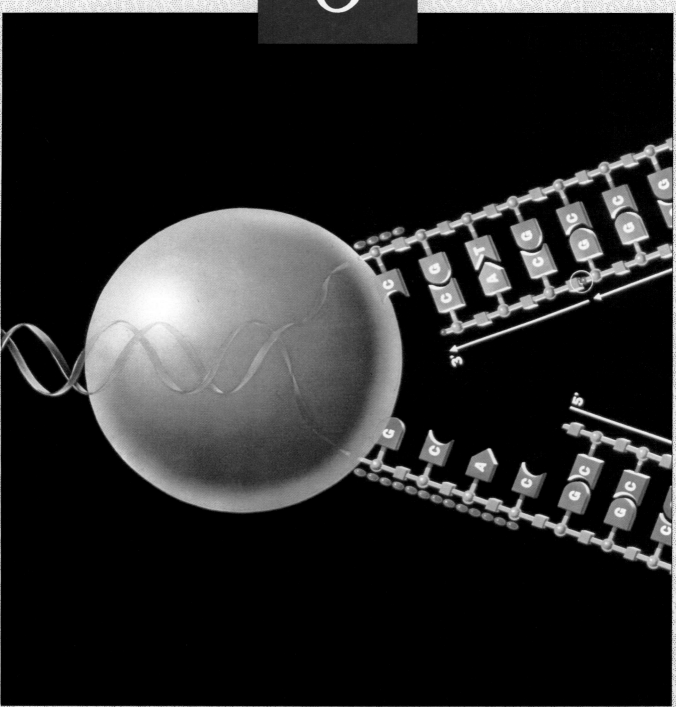

Replication and Recombination of Genes

*M*y own suspicion is that the universe is not only queerer than we suppose, but queerer than we can suppose.

■

_____

J. B. S. Haldane
British geneticist

In this chapter you will learn:

1. The mechanism of DNA replication.

2. The enzymes involved in DNA replication.

3. The way cells ensure that their genes are replicated faithfully.

4. The way DNA molecules are thought to recombine.

Artwork illustrating a mechanism for replication of DNA, the substance of genetic inheritance. Conservation of information during cell replication occurs through each parental DNA strand acting as a template for the formation of a complementary daughter strand.
© *Francis Leroy, Biocosmos/Science Photo Library/Photo Researchers, Inc.*

*I*n chapter 7 we learned one fundamental fact about DNA synthesis: It is semiconservative. But this basic statement serves only to introduce the complex topic of DNA replication. Our job in this chapter will be to fill in some of the considerable gaps we have left.

MECHANISM OF DNA REPLICATION

In the years since Matthew Meselson and Franklin Stahl's classic experiment, other investigators have uncovered the following interesting and unexpected general features of DNA replication: (1) It is usually bidirectional—that is, it proceeds in both directions away from a common starting point; (2) it is at least half discontinuous (made in short pieces that are later stitched together); and (3) it requires an RNA primer.

BIDIRECTIONAL REPLICATION

One direct approach to investigating DNA replication is to catch DNA molecules in the act of replicating and take pictures of them. This has been done, and we will see examples of this kind of experiment later in this chapter, but the early studies used a related technique: **autoradiography.** Autoradiography involves incorporating a radioactive label into a substance and then placing the radioactive substance in contact with a photographic emulsion. Since the radioactive emissions expose the emulsion, the radioactive substance "takes a picture of itself." That is why we call this process autoradiography.

John Cairns exploited this technique in the early 1960s to investigate the replication of the *E. coli* chromosome. Figure 8.1*a* shows an autoradiograph of replicating *E. coli* DNA, along with Cairns's interpretation, which is amplified in figure 8.1*b* Figures 8.1*a* and *b* also attempt to explain the difference in labeling of the three loops (A,B, and C) of the DNA as follows: The DNA underwent one round of replication and part of a second. That means all the DNA should have at least one labeled strand. During the second round of replication, the unlabeled parental strand will serve as the template to make a labeled partner and this duplex will therefore become singly labeled. The labeled parental strand will obtain a labeled partner, making this duplex doubly labeled. This doubly labeled part (loop B) should expose the film more and therefore appear darker than the rest of the DNA (loops A and C).

The structure represented in figure 8.1*a* is a so-called theta structure because of its resemblance to the Greek letter θ. Since it may not be immediately obvious that the DNA in figure 8.1*a* looks like a theta, figure 8.1*c* provides a schematic diagram of the events in the second round of

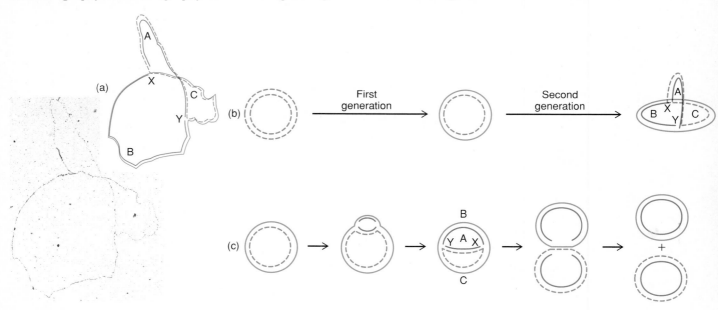

FIGURE 8.1

The theta mode of DNA replication in *Escherichia coli*. (*a*) An autoradiograph of replicating *E. coli* DNA with an interpretive diagram. The DNA was allowed to replicate for one whole generation and part of a second in the presence of radioactive nucleotides. That part of the DNA that has replicated only once will have one labeled strand (blue) and one "cold" strand (dotted, green). The DNA that has replicated twice will have one part that is doubly labeled (loop B, blue and red) and one part with only one labeled strand (loop A, red

and green). Cairns uses a dotted line to show unlabeled DNA and a solid line for labeled DNA. Note that loop B contains two solid lines, indicating that it is the part of the DNA that has become doubly labeled. The fact that loop B is darker than loops A and C fits with this interpretation. (*b*) Interpretation of Cairns's autoradiograph, using the same dotted and solid line and color conventions as in (*a*). (*c*) More detailed description of the theta mode of DNA replication. The dotted and solid lines and colors have the same meaning as in (*a*) and (*b*).

(a) *From John Cairns*, Symposia on Quantitative Biology *(Cold Spring Harbor) 28:44, 1963.*

replication that led to the autoradiogram. This drawing shows that DNA replication begins with the creation of a "bubble" or small region where the parental strands have separated and progeny DNA has been synthesized. As the bubble expands, the replicating DNA begins to take on the theta shape. We can now recognize the autoradiograph as representing a structure shown in the middle of figure 8.1c, where the crossbar of the theta has grown long enough to extend above the circular part.

The theta structure contains two **forks,** marked X and Y in figure 8.1. This raises an important question: Does one of these forks, or do both, represent sites of active DNA replication? In other words, is DNA replication **unidirectional,** with one fork moving away from the other, which remains fixed at the **origin of replication?** Or is it **bidirectional,** with two replicating forks moving in opposite directions away from the origin? Cairns's autoradiographs were not designed to answer this question, and in fact, they led to the erroneous conclusion that replication in *E. coli* is unidirectional. A subsequent study performed by Elizabeth Gyurasits and R. B. Wake showed clearly that DNA replication in *Bacillus subtilis* is bidirectional.

These investigators' strategy was to allow *B. subtilis* cells to grow for a short time in the presence of a weakly radioactive DNA precursor, then for a short time with a more strongly radioactive precursor. These short bursts of labeling with a radioactive substance are called **pulses** of label. The labeled precursor was the same in both cases: ^{3}H-thymidine. Tritium (^{3}H) is the radioactive isotope of hydrogen. It is especially useful for this type of autoradiography because its radioactive emissions are so weak that they do not travel far from their point of origin before they stop in the photographic emulsion and create silver grains. This means that the pattern of silver grains in the autoradiograph will bear a close relationship to the shape of the radioactive DNA. It is important to note that autoradiography reveals the shape of radioactive substances only; unlabeled DNA will not show up. The pulses of label in this experiment were short enough that only the replicating bubbles are visible (figure 8.2a). You should not mistake these for whole bacterial chromosomes such as in figure 8.1.

If you look carefully at figure 8.2a, you will notice that the pattern of silver grains is not uniform. There are concentrations of silver grains near both forks in the bubble. This extra labeling identifies the regions of DNA that were replicating during the "hot," or high-radioactivity pulse period. Both forks incorporated extra label, telling us that they were both operating during the

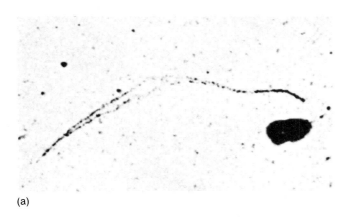

(a)

FIGURE 8.2

Experimental demonstration of bidirectional DNA replication. (*a*) Autoradiograph of replicating *Bacillus subtilis* DNA. Dormant bacterial spores were germinated in the presence of a low radioactivity DNA precursor, so the newly formed replicating bubbles immediately became slightly labeled. After the bubbles had grown somewhat, a more radioactive DNA precursor was added to label the DNA for a short period. (*b*) Interpretation of the autoradiograph. The purple color represents the slightly labeled DNA strands produced during the low radioactivity pulse. The orange color represents the more highly labeled DNA strands produced during the later, high radioactivity pulse. Since both forks picked up the high radioactivity label, both must have been functioning during the high radioactivity pulse. Therefore, DNA replication in *B. subtilis* is bidirectional.

Photo (a) from E. B. Gyurasits and R. G Wake. Journal of Molecular Biology, Vol. 73, 1973, facing pg. 58. Reprinted by permission of Academic Press.

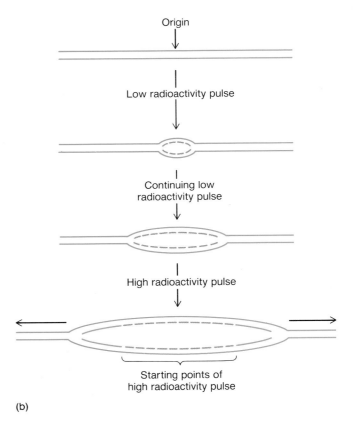

(b)

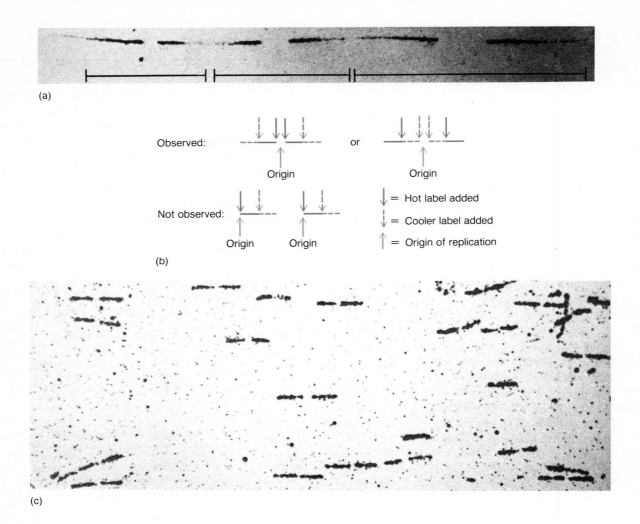

(a)

Observed:

Origin   or   Origin

Not observed:

Origin   Origin

↓ = Hot label added

⇣ = Cooler label added

↑ = Origin of replication

(b)

(c)

FIGURE 8.3

Bidirectional DNA replication in eukaryotes. (*a*) Autoradiograph of replicating *Drosophila melanogaster* DNA, pulse-labeled first with high radioactivity DNA precursor, then with low. Note the pairs of streaks (denoted by brackets) tapering away from the middle. This reflects the pattern of labeling of a replicon with a central origin and two replicating forks. (*b*) Idealized diagram showing the patterns observed with hot-then-cold or cold-then-hot labeling, respectively, and the pattern expected if the pairs of streaks represent two independent unidirectional replicons whose replicating forks move in the same direction. The latter pattern was not observed. (*c*) Autoradiograph of replicating embryonic *Triturus vulgaris* DNA. Note the constant size and shape of the pairs of streaks, suggesting that all the corresponding replicons are replicating at the same time.

Photo (a) from J. A. Huberman and A. Tsai, Journal of Molecular Biology, *Vol. 75, 1973, facing pg. 8(A). Reprinted by permission of Academic Press. Photo (c) from H. G. Callan,* Symposia on Quantitative Biology *(Cold Spring Harbor), vol. 38, p. 196, 1973.*

hot pulse. Therefore, DNA replication in *B. subtilis* is bidirectional; two forks arise at a fixed starting point—the origin—and move in opposite directions around the circle until they meet on the other side. Later experiments employing this and other techniques have shown that the *E. coli* chromosome also replicates bidirectionally.

J. Huberman and A. Tsai have performed the same kind of autoradiography experiments in a eukaryote, the fruit fly *Drosophila melanogaster*. Here, the experimenters gave a pulse of strongly radioactive (high specific activity) DNA precursor, followed by a pulse of weakly radioactive (low specific activity) precursor. Alternatively, they reversed the procedure and gave the low specific activity label first, followed by the high. Then they autoradiographed the labeled insect DNA. The spreading of DNA in these experiments did not allow the replicating bubbles to remain open; instead, they collapsed and appear on the autoradiographs as simple streaks of silver grains.

One end of a streak marks where labeling began, the other shows where it ended. But the point of this experiment is that the streaks always appear in pairs (figure 8.3*a*). The pairs of streaks represent the two replicating forks that have moved apart from a common starting point. Why doesn't the labeling start in the middle, at the origin of replication, the way it did in the experiment with *B. subtilis* DNA? In the *B. subtilis* experiment, the investigators were able to synchronize their cells by allowing them to germinate from spores, all starting at the same time. That way they could get label into the cells before any of them had started making DNA (i.e., before germination). That was not tried in the *Drosophila* experiments, where it would have been much more difficult. As a result, replication usually began before the label was added, so there is a blank area in the middle where replication was occurring but no label could be incorporated.

Notice the shape of the pairs of streaks in figure 8.3*a*. They taper to a point, moving outward, rather like a waxed mustache. That means the DNA incorporated highly radioactive label first, then more weakly radioactive label, leading to a tapering off of radioactivity as we move outward in both directions from the origin of replication. The opposite experiment—"cooler" label first, followed by "hotter" label—gives a reverse mustache, with points on the inside. It is possible, of course, that closely spaced, independent origins of replication gave rise to these pairs of streaks. But we would not expect that such origins would always give replication in opposite directions. Surely some would lead to replication in the same direction, producing asymmetric autoradiographs such as the hypothetical one in figure 8.3*b*. But these are not seen. Thus, these autoradiography experiments confirm that each pair of streaks we see really represents *one* origin of replication, rather than two that are close together. It therefore appears that replication of *Drosophila* DNA is bidirectional.

These experiments were done with *Drosophila* cells originally derived from mature fruit flies and then cultured in vitro. H. G. Callan and his colleagues performed the same type of experiment using highly radioactive label and embryonic amphibian cells. These experiments (with embryonic cells of the newt) gave the striking results shown in figure 8.3*c*. In contrast to the pattern in adult insect cells, the pairs of streaks here are all the same. They are all approximately the same length and they all have the same size space in the middle. This tells us that replication at all these origins began simultaneously. This must be so, because the addition of label caught them all at the same point—the same distance away from their respective origins. This phenomenon probably helps explain how embryonic newt cells complete their DNA replication so rapidly (in as little as one hour, compared to forty hours in adult cells). Replication at all origins begins simultaneously, rather than in a staggered fashion.

This discussion of origins of replication helps us define an important term: **replicon.** The DNA under the control of one origin is called a replicon, even though it may require two replicating forks. The *E. coli* chromosome comprises a single replicon because it replicates from a single origin. Obviously, eukaryotic chromosomes have many replicons; otherwise, it would take far too long to replicate a whole chromosome. This would be true even at the breakneck speed of DNA synthesis in prokaryotes—about a thousand nucleotides per second. In fact, eukaryotic DNA replicates much more slowly than this, which makes the need for multiple replicons even more obvious.

UNIDIRECTIONAL REPLICATION

Do all genetic systems replicate bidirectionally? Apparently not. For example, consider the replication of a *plasmid* called colicin El. A plasmid is a circular piece of DNA that replicates independently of the cell's chromosome; *E. coli* harbors colicin El, which explains how this plasmid got the "coli" part of its name. Lovett et al. used an electron microscope to examine replicating molecules of colicin El and found that only one fork moves. Thus, this plasmid replicates unidirectionally.

*I*n most genetic systems, DNA replication is bidirectional. The replication machinery starts at a fixed point, the origin of replication, and two replicating forks move in opposite directions away from this point. The DNA that replicates under the control of a single origin of replication is called a replicon. The chromosomes of simple genetic systems like plasmids, viruses, and even bacteria contain just one replicon. More complex eukaryotic chromosomes have multiple replicons. Some genetic systems, colicin El, for example, replicate unidirectionally.

■

SEMIDISCONTINUOUS REPLICATION

If we were charged with the task of designing a DNA replicating machine, we might come up with a system such as the one pictured in figure 8.4*a*. DNA would replicate continuously on both strands, following in the same direction as the moving fork. However, there is one fatal flaw in this scheme. It demands that the replicating machine be able to make DNA in both the $5' \rightarrow 3'$ and $3' \rightarrow 5'$ directions. That is because of the antiparallel nature of the two strands of DNA; if one goes $5' \rightarrow 3'$, the other must go $3' \rightarrow 5'$ (recall chapter 6). But the DNA-synthesizing part of all natural replicating machines (**DNA polymerase**) can only go one way: $5' \rightarrow 3'$. That is, it inserts the $5'$-most nucleotide first and extends the chain toward the $3'$-end.

Following this line of reasoning, Reiji Okazaki concluded that both strands could not replicate continuously. DNA polymerase could theoretically make one strand (the **leading strand**) continuously in the $5' \rightarrow 3'$ direction, but the other strand (the **trailing strand**), in order to be made $5' \rightarrow 3'$, would have to be made **discontinuously** as shown in figure 8.4*b* and *c*. The discontinuity of synthesis of the trailing strand comes about because its direction of synthesis is opposite to that of the replicating fork. Therefore, as the fork opens up and exposes a new region of DNA to replicate, the trailing strand is growing in the wrong direction, away from the fork. The only way to replicate this newly exposed region is to restart DNA synthesis at the fork, behind the piece of DNA that has already been made.

(a) Continuous:

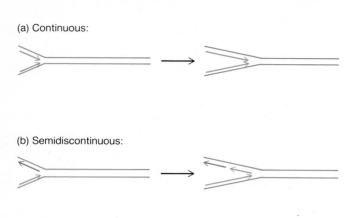

(b) Semidiscontinuous:

(c) Discontinuous:

FIGURE 8.4

Continuous and discontinuous models of DNA replication. (*a*) Continuous model. As the replicating fork moves to the right, both strands are replicated continuously in the same direction, left to right. (*b*) Semidiscontinuous model. Synthesis of one of the new strands (the leading strand, bottom) is continuous, as in the model in (*a*); synthesis of the other (the trailing strand, top) is discontinuous, with the DNA being made in short pieces. The red arrow indicates the first piece of trailing strand DNA made. (*c*) Discontinuous model. Both leading and trailing strands are made in short pieces (i.e., discontinuously). The first piece of DNA made on both strands is rendered in red.

This starting and restarting of DNA synthesis occurs over and over again. The short pieces of DNA thus created would of course have to be joined together somehow to produce the continuous strand that is the final product of DNA replication.

This model makes two predictions that Okazaki's research team tested experimentally: (1) Since at least half of the newly synthesized DNA appears first as short pieces, we ought to be able to label and catch these before they are stitched together by giving very short pulses of radioactive DNA precursor. (2) If we eliminate the enzyme (**DNA ligase**) that is responsible for stitching together the short pieces of DNA, we ought to be able to see these pieces even with long pulses of DNA precursor.

For his model system, Okazaki chose replication of bacteriophage T4 DNA. This had the advantage of simplicity, as well as the availability of mutants of T4 DNA ligase. To test the first prediction, Okazaki gave shorter and shorter pulses of tritium-labeled thymidine to *E. coli* cells that were replicating T4 DNA. To be sure of catching short pieces of DNA before they could be joined together, he even administered pulses as short as two seconds. Finally, he measured the approximate sizes of the newly synthesized DNAs by ultracentrifugation. Just as in the

case of ribosomes (chapter 7), large DNA species sediment faster toward the bottom of the centrifuge tube than smaller species do.

Figure 8.5*a* shows the results. Already at two seconds, some labeled DNA is visible in the gradient; within the limits of detection, it appears that all of the label is in very small DNA pieces, 1,000–2,000 bases long, which remain near the top of the tube. With increasing pulse time, a new peak of labeled DNA appears much nearer the bottom of the centrifuge tube. This is the result of attaching the small, newly formed pieces of labeled DNA to much larger, preformed pieces of DNA that were made before labeling began. These large pieces, because they were unlabeled before the experiment began, do not show up until enough time has elapsed for DNA ligase to join the smaller, labeled pieces to them; this takes only a few seconds. Appropriately enough, the small pieces of DNA that are the initial products of replication have come to be known as **Okazaki fragments.** This serves as a fitting memorial to Okazaki, a survivor of Hiroshima, who died of leukemia at the pinnacle of his career in cancer research.

The discovery of Okazaki fragments provided powerful evidence for at least partially discontinuous replication of T4 DNA. The clincher was the demonstration that these small DNA fragments accumulate to very high levels when the stitching enzyme, DNA ligase, does not operate. Okazaki's group performed this experiment with the T4 mutant containing a defective DNA ligase gene. Figure 8.5*b* shows that the peak of Okazaki fragments grows to a tremendous size in this mutant. Even after a full minute of labeling, this is virtually the only species of labeled DNA.

The latter experiment suggests that Okazaki fragments are an authentic initial product of DNA synthesis and not just an artifact of very short labeling times. Indeed, the fact that *only* small pieces of DNA were labeled either with short pulses of radioactive precursor or when DNA ligase did not operate suggested that replication proceeded discontinuously on *both* strands (figure 8.4*c*). Subsequent experiments have shown that half the Okazaki fragments seen in the early studies were probably artifacts caused by breakage of the continuously replicating strand. Thus, replication is really **semidiscontinuous:** One strand replicates continuously, the other discontinuously.

*I*n prokaryotes and eukaryotes, DNA replication is semidiscontinuous. The leading strand grows continuously, 5' → 3', in the direction of motion of the replicating fork. The other, trailing strand grows discontinuously, in pieces that are made, 5' → 3', in a direction opposite to the motion of the replicating fork. These pieces are later stitched together by DNA ligase.

■

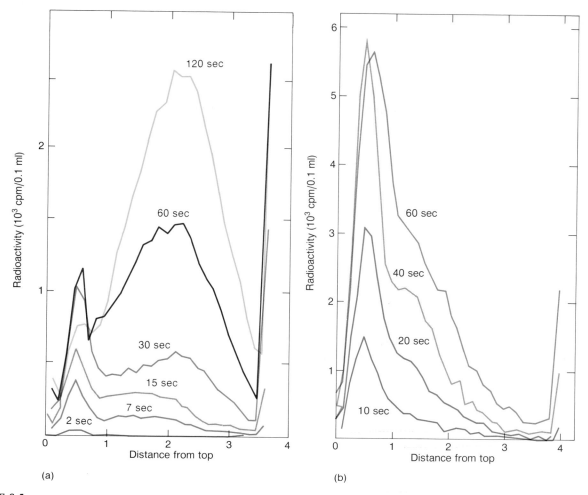

FIGURE 8.5
Experimental demonstration of at least semidiscontinuous DNA replication. (*a*) Okazaki and his colleagues labeled replicating phage T4 DNA with very short pulses of radioactive DNA precursor and separated the product DNAs according to size by ultracentrifugation. At the shortest times, the label went primarily into short DNA pieces (found near the top of the tube), as the discontinuous model predicted. (*b*) When a mutant phage with a defective DNA ligase gene was used, short DNA pieces accumulated even after relatively long labeling times (one minute in the results shown here).

PRIMING OF DNA SYNTHESIS

We saw in chapter 7 that RNA polymerase initiates transcription simply by starting a new RNA chain; it puts the first nucleotide in place and then joins the next to it. Therefore, it may come as something of a surprise to learn that DNA polymerase cannot perform the same trick with initiation of DNA synthesis. If we supply a DNA polymerase with all the nucleotides and other small molecules it needs to make DNA, then add either single-stranded or double-stranded DNA with no strand breaks, the polymerase will make no DNA. What is missing?

We now know that the missing component here is a **primer,** a piece of nucleic acid that the polymerase can "grab onto" and extend by adding nucleotides to its 3'-end. Furthermore, this primer is not DNA at all, but a little piece of RNA. Notice that priming with RNA neatly sidesteps the inability of DNA polymerase to get started. The first line of evidence supporting RNA priming was the finding that replication of M13 phage DNA by an *E. coli* extract is inhibited by the antibiotic rifampicin. This was a surprise because rifampicin inhibits RNA polymerase, not DNA polymerase. The explanation is that M13 uses *E. coli* RNA polymerase to make the RNA that primes its DNA synthesis. However, this is not a general phenomenon. Even *E. coli* does not use its own RNA polymerase for priming; it has a special enzyme system, tailor-made for that purpose.

Perhaps the best evidence for RNA priming is the discovery that the DNA-degrading enzyme DNase cannot completely destroy Okazaki fragments. It leaves little pieces of RNA 10–12 bases long. Other work has shown that these RNA **oligonucleotides** are covalently linked, not just hydrogen-bonded, to the 5'-end of Okazaki fragments.

> **B**ecause DNA polymerase is incapable of initiating DNA synthesis by itself, it needs a primer to supply a free 3'-end upon which it can build the nascent DNA. Very short pieces of RNA serve this priming function.

■

"STITCHING TOGETHER" NASCENT DNA FRAGMENTS

The mechanism we have described so far for DNA replication leaves us with nascent DNA fragments of the lagging strand that need to be linked together. We have already seen that DNA ligase serves this linking function, but two chores must be accomplished before ligase can operate. First, the RNA primers have to be removed so that the final DNA product is not interrupted with patches of RNA. Second, the gaps left by removing the primers need to be filled in.

Figure 8.6 demonstrates how a single enzyme performs both these tasks in *E. coli.* This multipurpose enzyme is **DNA polymerase I,** the first DNA-synthesizing enzyme discovered by Arthur Kornberg in his pioneering work on DNA replication. DNA polymerase I, in addition to its 5' → 3' polymerizing activity, has two nuclease activities. One degrades polynucleotides in the 5' → 3' direction, the other in the 3' → 5' direction. The 5' → 3' nuclease activity destroys the RNA primers, then the DNA polymerizing activity fills in the gap. In fact, these two activities may even work in concert, with primer-destruction and gap-filling going on simultaneously. In this respect, DNA polymerase would resemble a highway paving machine that tears up old pavement at its front end and lays down new at its rear.

After DNA polymerase has done its job, we have the situation depicted in figure 8.6c. All the nucleotides have been replaced, but a single-stranded break called a **nick** remains in the DNA. DNA polymerase, for all its versatility, cannot repair nicks. That task is performed by DNA ligase, which forms a new phosphodiester bond to seal the DNA. This process is repeated as many times as necessary to stitch together all the Okazaki fragments into a single, covalently closed DNA duplex.

> **A**fter Okazaki fragments are made, their primers are removed, the resulting gaps are filled in, and the remaining nicks are sealed.

■

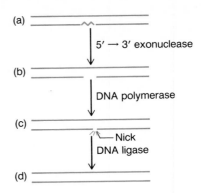

FIGURE 8.6
Joining nascent DNA fragments. (*a*) We start with two adjacent fragments, the right-hand one containing a primer (blue) at its 5'-end. (*b*) First, the 5' → 3' exonuclease activity of DNA polymerase I removes the primer. (*c*) Next, the polymerizing activity of the same enzyme fills in the gap, leaving a nick between the two DNA fragments. (*d*) Finally, another enzyme, DNA ligase, removes the nick by forming a phosphodiester bond between the two fragments.

ENZYMOLOGY OF DNA REPLICATION

Over thirty distinct polypeptides cooperate in replicating *E. coli* DNA. In the following section, we will examine the activities of some of these proteins.

STRAND SEPARATION

In the foregoing discussion of the general features of DNA replication, we have been assuming that the two DNA strands at the fork must somehow come unwound. This does not just happen automatically as DNA polymerase does its job; the two parental strands hold tightly to one another, and it takes energy and enzyme action to separate them.

Helicases

The enzyme that harnesses the chemical energy of ATP to separate the two parental DNA strands at the replicating fork is called a **helicase.** We have already seen in chapter 7 that ATP is used in RNA synthesis; it is also the primary source of chemical energy in a cell. When ATP breaks down to ADP and phosphate, considerable energy is released. Many enzymes are able to trap this energy and use it to perform work. Helicases contain an *ATPase* activity that breaks down two molecules of ATP for every base pair separated.

Single-Strand-Binding Protein

Once the parental DNA strands are locally separated by one or more helicases, another kind of protein binds to the newly exposed single-stranded regions and prevents their

FIGURE 8.7
Cairns's swivel concept. As the closed circular DNA replicates, the two strands must separate at the fork (F). The strain of this unwinding would be released by a swivel mechanism. Cairns actually envisioned the swivel as a machine that rotated actively and thus drove the unwinding of DNA at the fork.

reassociation. This protein is called a **single-strand-binding protein** or **SSB.** Actually the SSB of *E. coli* also binds to double-stranded DNA, but the mode of binding is different. When SSB binds to double-stranded DNA, only a single molecule binds. When it encounters single-stranded DNA, it binds cooperatively. In other words, once one molecule binds, it helps more molecules attach. The result is much greater binding of SSB to single- than to double-stranded DNA. Eukaryotic SSBs appear similar to the *E. coli* protein with one important exception: They do not bind cooperatively to single-stranded DNA. Thus, they must bind preferentially to single strands by another mechanism.

Topoisomerases

Sometimes we refer to the separation of DNA strands as "unzipping." We should not forget, when using this term, that DNA is not like a zipper with straight, parallel sides. It is a helix. Therefore, when the two strands of DNA separate, they must rotate around one another. Helicase could handle this task alone if the DNA were linear and unnaturally short, but closed circular DNAs, such as the *E. coli* chromosome, present a special problem. As the DNA unwinds at the replicating fork, a compensating winding up of DNA will occur elsewhere in the circle. This

tightening of the helix will create intolerable strain unless it is relieved somehow. Cairns recognized this problem when he first observed circular DNA molecules in *E. coli* in 1963, and he proposed a "swivel" in the DNA duplex that would allow the DNA strands on either side to rotate to relieve the strain (figure 8.7). We now know that an enzyme known as **DNA gyrase** serves the swivel function. DNA gyrase belongs to a class of enzymes called **topoisomerases** that introduce transient single- or doublestranded breaks into DNA and thereby allow it to change its form or topology.

To understand how the topoisomerases work, we need to look more closely at the phenomenon of supercoiled, or superhelical DNA mentioned in chapter 6. All naturally occurring, closed circular, double-stranded DNAs studied so far exist as supercoils. Closed circular DNAs are those with no single-strand breaks, or nicks. When a cell makes such a DNA, it causes some unwinding of the double helix; the DNA is then said to be "underwound." As long as both strands are intact, there can be no free rotation about the bonds in either strand's backbone, so there is no way the DNA can relieve the strain of underwinding except by supercoiling. The supercoils introduced by underwinding are called negative, by convention. This is the only kind of supercoiling found so far in nature; positive supercoils exist only in the laboratory.

You can visualize the supercoiling process with a Mr. Wizard-type experiment (figure 8.8). Take a mediumsized rubber band, hold it at the top with one hand, and cut through one of the sides about halfway down. With your other hand, twist one free end of the rubber band one full turn and hold it next to the other free end. You should notice that the rubber band resists the turning as strain is introduced, then relieves the strain by forming a supercoil. The more you twist, the more supercoiling you will observe: one superhelical turn for every full twist you introduce. Reverse the twist and you will see supercoiling of the opposite handedness or sign. DNA works the same way. Of course, the rubber band is not quite a perfect model. For one thing, you have to hold the severed ends together; in a circular DNA, chemical bonds join the two ends. For another, the rubber band is not a double helix.

If you release your grip on the free ends of the rubber band in figure 8.8, of course the superhelix will relax. In DNA, it is only necessary to cut one strand to relax a supercoil because the other strand can rotate freely, as demonstrated in figure 8.9.

Underwinding can occur as follows: A nicked DNA is partially unwound, then the nick is sealed to give a closed circular DNA. However, unwinding DNA at the replicating fork would have the opposite effect if there were no way to relax the strain. The reason is that here we are *permanently* unwinding one region of the DNA without nicking it; this means that the rest of the DNA would become *over*wound, and therefore positively supercoiled,

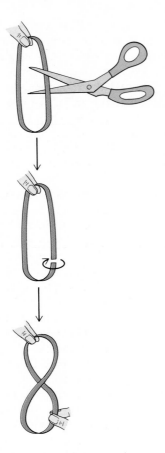

FIGURE 8.8
Rubber band model of supercoiling in DNA. If you cut the rubber band and twist one free end through one complete turn while preventing the other end from rotating, the rubber band will relieve the strain by forming one supercoil.

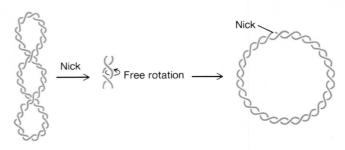

FIGURE 8.9
Nicking one strand relaxes supercoiled DNA. A nick in one strand of a supercoiled DNA (left) allows free rotation around a phosphodiester bond in the opposite strand (middle). This releases the strain that caused the supercoiling in the first place and so allows the DNA to relax to the open circular form (right).

to compensate. To appreciate this, look at the circular arrow ahead of the replicating fork in figure 8.7. Notice how twisting the DNA in the direction of the arrow causes unwinding behind the arrow but overwinding ahead of it. Therefore, unwinding the DNA at the replicating fork introduces *positive* superhelical strain that must be constantly relaxed so that replication will not be retarded. You can appreciate this when you think of how the rubber band increasingly resisted your twisting as it became more tightly wound. In principle, any enzyme that creates transient breaks (either single- or double-stranded) in DNA should be able to relax this strain and thereby serve as a swivel. In fact, of all the topoisomerases in an *E. coli* cell, only one, DNA gyrase, appears to perform this function.

Topoisomerases are classified according to whether they operate by causing single- or double-stranded breaks in DNA. Those in the first class (type I topoisomerases) introduce temporary single-stranded breaks. Enzymes in the second class (type II topoisomerases) break and reseal both DNA strands. DNA gyrase is in this latter class. Why are type I topoisomerases incapable of providing the swivel

function needed in DNA replication? It is because they can only relax *negative* supercoils, not the positive ones that would form in replicating DNA in the absence of a swivel. Obviously, the nicks created by these enzymes do not allow free rotation in either direction. But DNA gyrase pumps negative supercoils into closed circular DNA and therefore counteracts the tendency to form positive ones. Hence, it can operate as a swivel.

There is direct evidence that DNA gyrase is crucial to the DNA replication process. First of all, mutations in the genes for the two polypeptides of DNA gyrase are lethal and they block DNA replication. Second, antibiotics such as novobiocin, coumermycin, and nalidixic acid attack DNA gyrase and thereby inhibit replication. However, these findings do not yet make clear the exact role of DNA gyrase in replication. In addition to operating as a swivel, this enzyme could: (1) aid initiation by making it easier for the two parental strands to separate and bind the replicating machinery; (2) aid elongation by favoring strand separation; and (3) aid termination by breaking bonds and allowing the two intertwined daughter DNAs to separate. These processes will be examined more fully later in this chapter. At any rate, in the absence of any evidence for another enzyme capable of acting as a swivel, the working hypothesis is that DNA gyrase fills this role.

*O*ne or more enzymes called helicases use ATP energy to separate the two parental DNA strands at the replicating fork. As helicase unwinds the two parental strands of a closed circular DNA, it introduces a compensating positive supercoiling force into the DNA. The stress of this force must be overcome or it will resist progression of the replicating fork. The name given to this stress-release mechanism is the swivel. DNA gyrase is the leading candidate for this role. Once the two strands are separated, single-strand-binding protein (SSB) keeps them that way.

■

Biochemists Like to Grind Things Up

T he tendency of biochemists to break things down to their component parts and then see how they go back together again has given rise to a joke about three scientists who were asked to figure out how a wristwatch worked. The physiologist stuck electrodes into it, the geneticist bombarded it with mutagenic radiation, and the biochemist ground it up in a Waring blender. In general, though, the separating and reconstituting strategy has proven very powerful. Arthur Kornberg summed up his support for it with an aphorism: "I remain faithful to the conviction that anything a cell can do, a biochemist should be able to do. He should do it even better. . . . Put another way, one can be creative more easily with a reconstituted system." In the investigation of DNA replication in *E. coli,* the marriage of this approach with genetics has been especially fruitful.

∎

INITIATION OF REPLICATION

We have already seen that a crucial aspect of initiation in DNA replication is synthesis of a short RNA primer. (There are a few exceptions to this rule, notably adenovirus, which uses a special protein to prime its DNA synthesis.) A rich variety of mechanisms exist to form these RNA primers, and each genetic system uses one of them. A large aggregate of about twenty polypeptides combine to make the primers in *E. coli* DNA replication. This **primosome** is so complex that the identification of its components would have taken much longer were it not for the existence of single-stranded DNA phages that use only parts of the primosome. These have been used as models for unraveling the mechanism of primer formation in *E. coli* itself.

Primer Synthesis

Two different experimental approaches have been used to identify the components of the DNA replication machinery, including the **primases,** or primer synthesizers, in these phage systems. The first is a combination genetic-biochemical approach, the strategy of which is to isolate mutants with defects in the ability to replicate phage DNA, then to complement extracts from these mutants with proteins from wild-type cells. The mutant extracts will be incapable of replicating the phage DNA in vitro unless the right wild-type protein is added. Using this system as an assay, the protein can be highly purified and then characterized. The second approach is the classical biochemical one: purify all of the components needed and then add them all back together to reconstitute the replication system in vitro.

As luck would have it, the first phage DNA replication system to be investigated was the simplest: the filamentous phage M13. As mentioned before, one of the striking early findings of this study was that the replication of this phage DNA is inhibited by rifampicin, an antibiotic that inhibits *E. coli* RNA polymerase. This means that the phage uses the host RNA polymerase as its primase. Another single-stranded DNA phage, G4, uses another host protein, the product of the *dnaG* gene, to make its primers. This is the same primase that *E. coli* uses. Still another well-studied phage, φX174, uses a whole host of *E. coli* proteins, including the dnaG protein (primase) for primer synthesis. These seem to be the three examples that all other single-stranded *E. coli* phages follow. The last has been the most help in understanding the host primosome because it resembles it most closely.

A combination of biochemical and genetic approaches has been used to identify the components of the primer-synthesizing systems of three classes of single-stranded DNA phages, all of which infect *E. coli.* The first class is represented by phage M13, which uses host RNA polymerase to make its primers; the second, represented by phage G4, uses the host primase (the dnaG protein); the third class, represented by the phage φX174, uses a complex primosome composed of the dnaG primase and about twenty other host polypeptides. This seems to be the same primosome the host employs.

∎

Making Primers in *E. coli*

More than twenty-five elements combine to make up the DNA replicating machinery of *E. coli,* a system rivaling the ribosome in complexity. For the moment, let us concentrate on some of the proteins involved in the synthesis of primers.

We have already introduced the primase, the enzyme that makes the primer, but we have not said much about how it operates. The difference between putting together a primer only twelve nucleotides long and ordinary transcription of long genes is striking. It suggests that the primase works in a fundamentally different way from RNA polymerase involved in transcription.

This supposition is borne out by other characteristics of primase activity. During replication of double-stranded DNA, the primosome moves *processively* along the template for the lagging strand in the direction of motion of the replication fork. "Processively" means that it moves continuously along the DNA without dissociating from it. It is advantageous for the primosome to bind to the lagging strand template since that is where DNA is being made discontinuously and new primers are needed for each new Okazaki fragment. However, this presents a problem for the primase, because the direction the replicating fork is moving on the lagging strand template is $5' \rightarrow 3'$, which means a primer synthesized complementary to that strand would be made in the $3' \rightarrow 5'$ direction. And RNAs, even primers, have to be made in the $5' \rightarrow 3'$ direction. Actually, the primer *is* made in the $5' \rightarrow 3'$ direction, even though the primosome moves in the opposite direction (figure 8.10). Just how this is accomplished is not known.

Kornberg likens the primosome to a locomotive moving along the railroad track of DNA in the direction of the replicating fork. The **n'** protein provides the energy for this process by cleaving ATP and thus plays the engine role in the locomotive. This protein also serves as a "cowcatcher" by bumping SSB molecules off the DNA "track" in advance of the moving primosome. The n' protein also binds first to single-stranded ϕX174 DNA and initiates the assembly of the remaining elements of the primosome. On the other hand, experiments with the *E. coli* origin of replication have shown that n' protein alone cannot bind, so some other factors may be required.

According to Kornberg's metaphor, if the n' protein is the power plant, then the dnaB protein is the engineer. It has binding sites for many of the other components involved in the primosome, including double- and single-stranded DNA, ATP, the dnaC protein, and the primase. Thus, it can act as a primosome organizer. Furthermore, it has the ability to harness the energy in ATP to change the conformation of DNA so that it is receptive to primase activity.

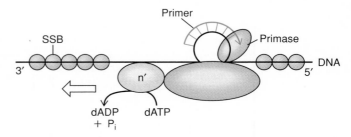

FIGURE 8.10

Primosome forming a primer on ϕX174 DNA. The complex primosome is moving in the $5' \rightarrow 3'$ direction on the template strand (open arrow), while making a primer in the opposite direction (red arrow). The actual synthesis of the primer is performed by the primase (orange). The n' protein (yellow) provides energy for this process by cleaving energy-rich ATP or (as shown here) dATP. The rest of the primosome is represented by a green oval.

*T*he primosome is a complex structure containing about twenty polypeptides that makes primers for *E. coli* and phage ϕX174 DNA replication. Some of the proteins involved in the primosome are: the n' protein, which uses energy from ATP to pull the primosome along behind the replicating fork, removing SSB as it goes; the dnaB protein, which organizes some of the other components of the primosome and prepares the DNA strand for primase activity; and the primase itself, which forms the primer. In *E. coli* DNA replication, the primosome moves along the template for the lagging strand and makes primers in the opposite direction.

ELONGATION

Once the primosome has made a primer, it is time for elongation to begin. The key enzyme involved in the elongation process is **DNA polymerase III holoenzyme,** a complex of seven different polypeptides. The term holoenzyme simply means "the whole thing," including the active center (for DNA synthesis in this case) as well as any attendant factors that participate in the reaction. We do not know for sure, but we suspect that the DNA polymerase III holoenzyme joins with the primosome to make a **replisome** that then carries out elongation of the DNA chain.

Aid from the *E. coli* Replisome

A major question immediately arises concerning the process of elongation: How does the replisome coordinate the synthesis of the leading and lagging strands? It appears

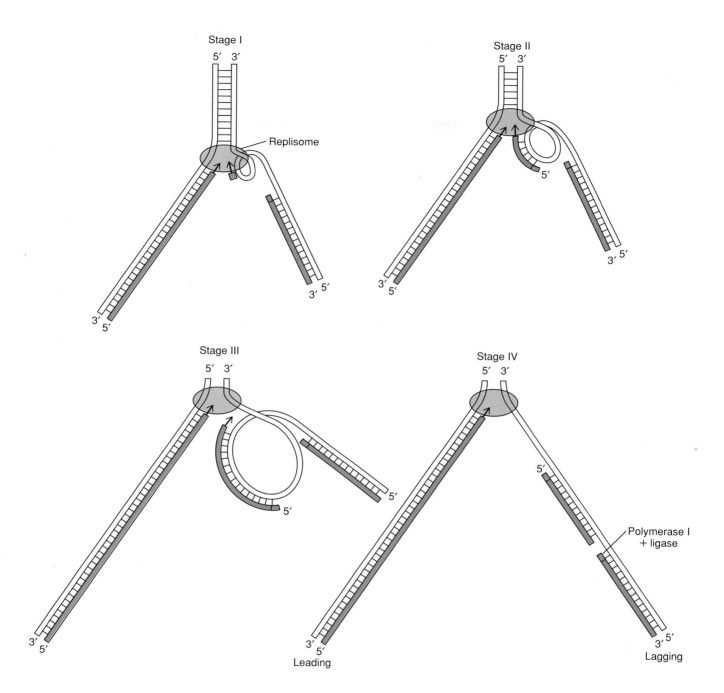

Stage I

Repliome

Stage II

Stage III

Leading

Stage IV

Polymerase I
+ ligase

Lagging

at first that each replicating fork would either need two replisomes, one for each strand, or that the replisome would have to be very flexible to carry out DNA synthesis in opposite directions on the two strands.

However, single DNA strands are considerably more flexible than proteins, so it is not unreasonable to expect that the lagging strand template curves 180 degrees around the replisome as shown in figure 8.11. If this happens, as

FIGURE 8.11

A model for simultaneous synthesis of both DNA strands. At stage I, the primosome has made a primer (red square) for the lagging strand. The lagging strand template is looped through one full turn. At stage II, the two DNA polymerase molecules in the replisome (blue) are elongating both leading and lagging strands (green). The loop in the lagging strand template allows the polymerase making the lagging strand to move in the same direction as the one making the leading strand, even though the two strands are antiparallel. In stage III, the new lagging strand fragment has grown still further. It will eventually reach the point where it adjoins the previously made fragment. In stage IV, the loop is released. Now DNA polymerase I and ligase can remove the old primer, fill in, and join the fragments together, and a new cycle can begin (back to stage I).

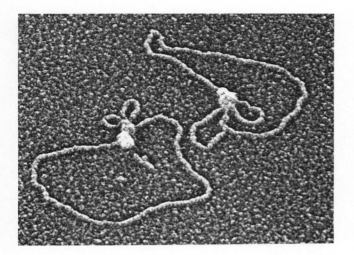

FIGURE 8.12
Replisomes on φX174 DNAs. Each double-stranded DNA molecule contains one replisome and at least one loop structure. At least one loop is predicted by the model of DNA elongation presented in figure 8.11.
© *Dr. Jack Griffith.*

Kornberg suggests, then the replisome could read both strands simultaneously because both would feed into the DNA polymerase center in the same orientation ($5' \rightarrow 3'$, top to bottom).

This model has the great advantage that it permits one replisome to replicate both strands processively, without having to leave the replicating fork. This is important because once dissociated from the DNA, the replisome would waste much time rebinding. To put this in perspective, consider that the rate of elongation in *E. coli* is about 1,000 nucleotides per second and the rate of formation of a complex capable of initiation is one or two minutes. This means that while the replisome paused to rebind to the DNA, it could have been elongating the DNA chain by at least 60,000 nucleotides. Some time will be lost even in the looping and releasing process described here, but the intramolecular transfer involved only requires two to five seconds—quite a saving compared to one to two minutes.

Of course, this model requires that the replisome have *two* DNA polymerase centers, and Kornberg lists the following evidence suggesting that this is so: (1) Each replicating center contains two molecules of at least two of the subunits. (2) It is possible to isolate a form of polymerase called DNA polymerase III′ whose molecular weight suggests that it contains two each of four of its subunits. (3) The anticipated looped structures are actually observed by electron microscopy of replicating DNA (figure 8.12). (4) It appears that the eukaryotic enzyme analogous to DNA polymerase III holoenzyme is a dimer.

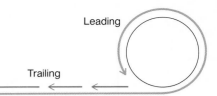

FIGURE 8.13
Rolling circle model for phage λ DNA replication. As the circle rolls to the right, the leading strand (orange) elongates continuously. The trailing strand (blue) elongates discontinuously, using the unrolled leading strand as template. The progeny double-stranded DNA thus produced grows to many genomes in length (a concatemer) before one genome's worth is clipped off and packaged into a phage head.

E longation during DNA synthesis in *E. coli* apparently requires a replisome made up of the primosome plus a DNA polymerase III holoenzyme composed of at least ten different polypeptides. Kornberg's model for simultaneous replication of both strands calls for a double-headed replisome with two DNA polymerase centers. One would make the leading strand continuously; the other would synthesize the lagging strand discontinuously, but without having to leave the replicating fork.

Rolling Circle Replication

Certain circular DNAs replicate, not by the theta mode we have already examined, but by a mechanism called **rolling circle** replication. Lambda (λ) phage actually uses both means of replication. During the early phase of λ DNA replication, the phage follows the theta mode of replication to produce several copies of circular DNA. These circular DNAs are not packaged into phage particles; they serve as templates for rolling circle synthesis of linear λ DNA molecules that *are* packaged.

Figure 8.13 shows how this rolling circle operates. Here, the replicating fork resembles that in *E. coli* DNA replication, with continuous synthesis on the leading strand (the one going around the circle) and discontinuous synthesis on the trailing strand. It is not difficult to see how the name "rolling circle" arose. In fact, you can almost think of the replicating DNA as a roll of toilet paper unrolling as it speeds across the floor. The unrolled part represents the growing double-stranded progeny DNA. In λ, this progeny DNA reaches lengths that are several genomes long before it is packaged. These multiple-length DNAs are called **concatemers.** The packaging mechanism is designed to accept only one genome's worth of linear DNA into each phage head, so the concatemer must be cut during packaging. This rolling circle mechanism of DNA replication is also sometimes called the sigma mode because the intermediate resembles the lowercase Greek letter sigma (σ).

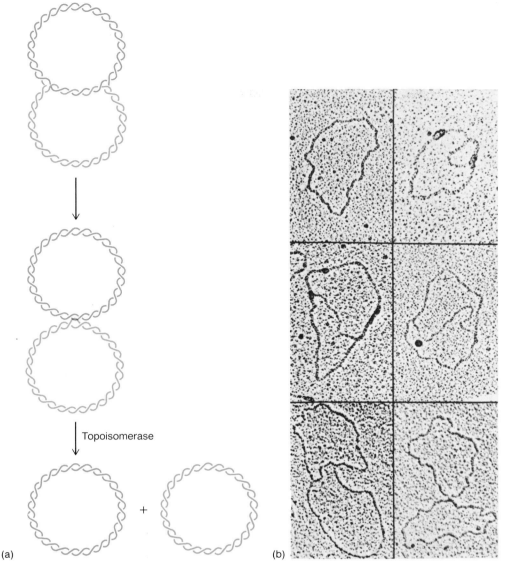

FIGURE 8.14
A role for a topoisomerase in termination of DNA replication. (*a*) When a circular DNA reaches the end of a cycle of theta-style replication (top), the two daughter duplexes are still linked to one another (middle). A topoisomerase probably nicks at least one of the molecules to separate them (bottom). (*b*) Electron micrographs of replicating polyoma virus DNA molecules show the two daughters still linked at bottom right. A topoisomerase would disentangle them to generate two independent molecules such as the one at upper left. Magnification: ×60,000.

(b) From William Meinke and David Goldstein, Journal of Molecular Biology, *Vol. 61, 1971, facing pg. 556. Reprinted by permission of Academic Press.*

(a)

(b)

C ircular DNA can replicate by a rolling circle, or sigma mechanism. One strand of a double-stranded DNA is nicked and the 3′-end is extended, using the intact DNA strand as template. This displaces the 5′-end. In phage λ, the displaced strand serves as the template for discontinuous, trailing strand synthesis.

TERMINATION

Termination of replication is not a problem for λ and other phages that produce a long, linear concatemer. The concatemer simply continues to grow as genome-sized parts of it are snipped off and packaged into phage heads. But for *E. coli* and other systems where replication has a definite end as well as a beginning, we are interested in the mechanism of termination. Relatively little is known about this process, but one aspect of it that has at least encouraged speculation is the problem of separating the two daughter duplexes at the end of replicating double-stranded genomes. Figure 8.14 illustrates this problem. As the replicating machinery approaches the end, the two daughter duplexes are still entwined. Finally, they are separated except for a simple linkage of one strand of each duplex with one of the other. A logical way this impasse could be broken would be to use a topoisomerase, DNA gyrase, for example, to nick one or both of the strands of one duplex so the two daughters can separate completely.

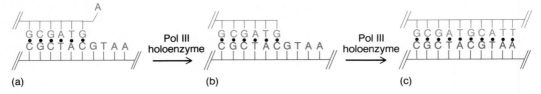

FIGURE 8.15

Proofreading in DNA replication. (*a*) An adenine nucleotide (red) has been mistakenly incorporated across from a guanine. This destroys the perfect base pairing required at the 3′-end of the primer, so the replicating machinery stalls. (*b*) This then allows polymerase III holoenzyme to use its 3′ → 5′ exonuclease function to remove the offending nucleotide. (*c*) With the appropriate base pairing restored, polymerase III holoenzyme is free to continue replication.

Gyrase is an attractive candidate for this chore, since it is needed to generate negative supercoils in the daughter molecules in any event.

> *W*hen replication of a double-stranded circular DNA approaches completion, the two daughter molecules are still topologically entwined. A topoisomerase is presumably needed to separate them.

■

FIDELITY OF REPLICATION

By this time, you may be asking yourself: "Why is DNA replication so complex? Why go to all the trouble of semidiscontinuous replication with RNA priming?" The answer seems to be that both of these complex mechanisms contribute to something very important to life: faithful DNA replication. Consider the alternative to this fidelity. What if every time our chromosomes replicated, the DNA synthesizing machinery made many mistakes? These mistakes are mutations, which would accumulate with every new cell division. We have already seen in chapter 7 an example of the very deleterious effects of most mutations; it is clear that life cannot tolerate too many of them. In other words, if our replicating machinery were much less faithful, our genome would have to be much smaller in order to minimize the number of mutations occurring each generation. Complex life forms, relying on many genes, would be impossible.

How good are DNA polymerases at pairing up the right bases during replication? The *E. coli* DNA polymerase III holoenzyme, operating in vitro, makes about one pairing mistake in one hundred thousand—not a very good record, considering that even the *E. coli* genome contains about three million base pairs. At this rate, replication would introduce errors into a significant percentage of genes. Fortunately, the DNA polymerase III holoenzyme includes its own **proofreading** activity— a 3′ → 5′ exonuclease—that allows it to synthesize DNA with only one error in at least ten million nucleotides polymerized.

Figure 8.15 shows the proofreading process. The DNA polymerase III holoenzyme in the replisome can only add a nucleotide to the growing strand if the previously incorporated nucleotide is properly base-paired. This is a critical requirement. It means that the polymerase cannot add to a mismatched nucleotide. Therefore, if the replisome adds the wrong nucleotide, it will stall until the mismatch is repaired. The 3′ → 5′ exonuclease activity of DNA polymerase III holoenzyme can then remove the mismatched nucleotide, perhaps along with several others. With the proper base-paired primer restored, the replisome can continue.

Now consider the implications of this proofreading mechanism. It immediately explains the need for a primer. If DNA polymerase needs a base-paired nucleotide to add to, it clearly cannot start a new DNA chain unless a primer is there. But why primers made of RNA? The reason seems to be the following: Primers are made with more errors, since their synthesis is presumably not subject to proofreading. Making primers out of RNA guarantees that they will be recognized, removed, and replaced with DNA by extending the neighboring Okazaki fragment. The latter process is, of course, relatively error-free. Besides these theoretical arguments, there is the simple fact that RNA synthesizing enzymes can start without primers, whereas DNA polymerase cannot.

The proofreading mechanism also helps rationalize semidiscontinuous replication. If there were a polymerase capable of synthesizing DNA in the 3′ → 5′ direction, thus making fully continuous replication possible, we would need a whole new 5′ → 3′ proofreading function, with the ability to restore the 5′-triphosphates that would inevitably be destroyed during proofreading. Semidiscontinuous replication, as cumbersome as it seems, is probably simpler.

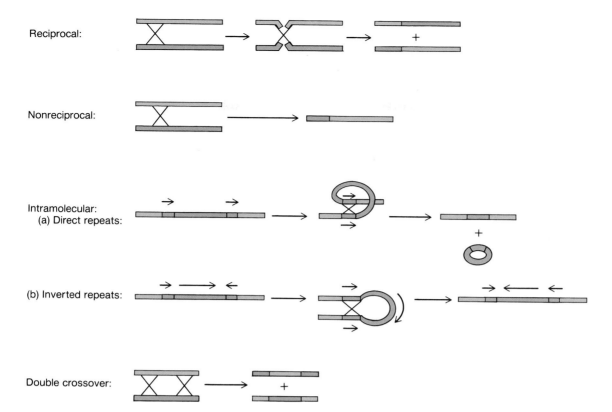

Reciprocal:

Nonreciprocal:

Intramolecular:
(a) Direct repeats:

(b) Inverted repeats:

Double crossover:

FIGURE 8.16

Examples of recombination. The X's represent crossovers between the two chromosomes or parts of the same chromosome. To visualize how these crossovers work, look at the intermediate form of the reciprocal recombination on the top line. Imagine the DNAs breaking and forming new, interstrand bonds as indicated by the arms of the X. This same principle applies to all the examples shown; in the nonreciprocal case, one of the recombinants is lost.

> T he DNA replicating machinery has a built-in proofreading system that requires priming. Only a base-paired nucleotide can serve as a primer for the replicating enzyme; therefore, if the wrong nucleotide is incorporated by accident, replication stalls until a nuclease erases the error by removing the most recently added nucleotide. The fact that the primers are made of RNA may mark them for degradation.

■

THE MECHANISM OF RECOMBINATION

As we have seen in our discussion of meiosis in chapter 4, DNA replication in meiotic cells is usually followed by recombination between homologous chromosomes. This process scrambles the genes of maternal and paternal chromosomes so that nonparental combinations occur in the progeny. Thus, recombination is a central feature of the science of genetics and will be a recurring theme throughout this book. Now that we have become well acquainted with DNA by studying its replication, we can appreciate the molecular mechanism of recombination. We should pause at this point to ask: How does recombination work?

FORMS OF RECOMBINATION

The examples of recombination shown in figure 8.16 always involve a **crossover** event that joins DNA segments previously separated. This does not mean the two segments must start out on separate DNA molecules. Recombination can be intramolecular, in which case a segment of DNA is removed as two flanking segments

FIGURE 8.17

Mechanism of recombination. (*a*) Nicks occur in the same places in homologous chromosomes (blue and red). (*b*) Strands of the two chromosomes cross over. (*c*) Nicks are sealed, permanently joining the two chromosomes through two of their four strands and yielding a chi structure. (*d*) Branch migration (optional) occurs by breaking some base pairs and re-forming others. (*e*) and (*f*) These are simply bends and twists in the chi structure to make the subsequent events easier to understand. (*g*) One kind of resolution of the chi structure. The same strands are nicked as were nicked in (*a*). (*h*) Sealing these nicks leaves two DNA duplexes with short stretches of "heteroduplex," containing one strand from each of the recombining partners. No true recombination has occurred. (*i*) The alternative resolution of the chi structure. The strands opposite those nicked originally are nicked. (*j*) When the nicks are sealed, two recombinant DNA duplexes emerge.

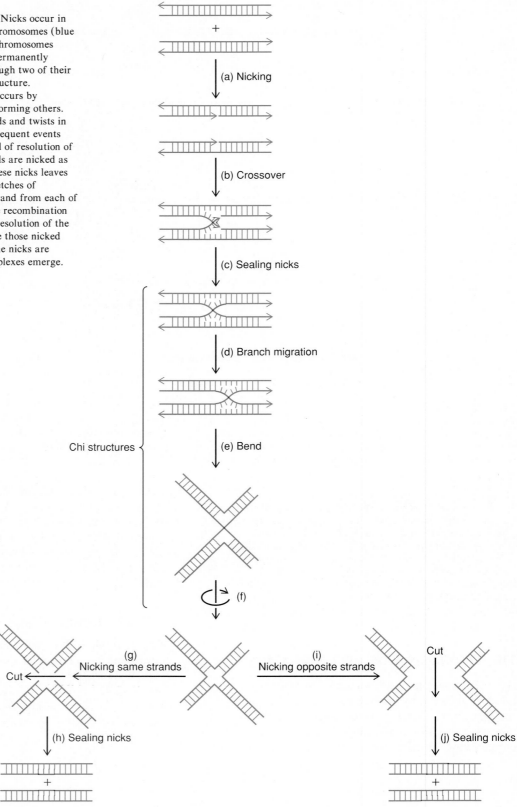

come together. On the other hand, bimolecular recombination involves crossover events between two independent DNA molecules. Ordinarily, recombination is **reciprocal**—a "two-way street" in which the two participants trade DNA segments.

What does recombination look like at the molecular level? First, consider that there are three fundamentally different kinds of recombination: (1) **Site-specific recombination** depends on limited sequence similarity between the recombining DNAs and always involves the same DNA regions, which explains why it is called site-specific. A prominent example of this kind of recombination is the insertion of λ phage DNA into the host DNA (chapter 9). (2) A related form is **illegitimate recombination.** This also requires little if any sequence similarity between the recombining DNAs, but it is not site-specific. Transposable elements (chapter 12) usually use the illegitimate recombination pathway. (3) By far the most common form of recombination depends on extensive sequence similarity between the participating DNA molecules. This is the kind of recombination that occurs in meiosis. It is called **homologous** or **generalized recombination.**

A MODEL FOR GENERALIZED RECOMBINATION

Our current models for the mechanism of generalized recombination share the following essential steps: (1) pairing of two homologous DNA duplexes; (2) breaking two of the homologous DNA strands; (3) re-forming phosphodiester bonds to join the two homologous strands (or regions of the same strand during recombination within one strand); (4) breaking the other two strands and joining them. The main point of difference among the models lies in the means by which the second strand break occurs in step 2. The basic mechanism, originally proposed by Robin Holliday, is presented in figure 8.17.

The first step in the Holliday model is the pairing of two homologous DNA duplexes. The mechanism of this pairing is still not well understood. The second step is creation of nicks in corresponding positions in two homologous DNA strands. These nicks could theoretically come from a variety of sources. They could be leftover breaks that occurred during discontinuous DNA replication; they could be the result of random breaking of the DNA strands; they could be introduced by the recombination machinery itself; or a random break in one strand could provoke a break in the corresponding position of the homologous strand by a "strand invasion" process that we

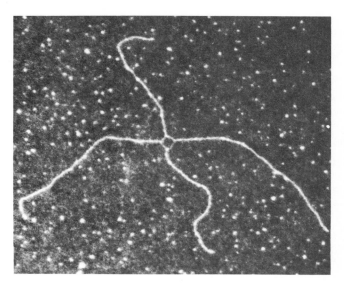

FIGURE 8.18
Chi structure generated during recombination. This corresponds to the intermediate generated in figure 8.17*f*, after some base pairs have opened up around the central "diamond" to produce short stretches of single-stranded DNA.
Courtesy of Dr. Huntington Potter and Dr. David Dressler.

will explain later in this chapter. Once both homologous strands are broken, the free ends can cross over and join with ends in the other duplex, forming intermolecular bonds, rather than rejoining to form the old intramolecular bonds.

When the two strands have crossed over and DNA ligase has sealed the new intermolecular phosphodiester bonds, we have a **half chiasma,** or **chi structure,** a cross-shaped structure pictured in several ways in figure 8.17. The branch in the half chiasma can migrate in either direction simply by breaking old base pairs and forming new ones, but this **branch migration** is not essential to the recombination process. The chi structure is first shown with two of its strands crossed, but a 180-degree rotation of either the top or bottom half uncrosses these strands to give an intermediate with a diamond shape at the center where all the strands meet. If a few base pairs break at the junction point, the diamond, which is composed of single-stranded DNA, will expand. Figure 8.18 is an electron micrograph of recombining plasmid DNAs, which demonstrates beautifully that such structures really do form.

Of course, the two duplexes that are participating in the chi structure cannot remain linked indefinitely. The other two DNA strands must also break and rejoin to resolve the chi structure into two independent DNA duplexes. This can happen in two different ways. If the same strands that broke the first time break again and cross back, no true recombination results. What we get instead are two DNA duplexes with patches of **heteroduplex.** These patches contain one strand from each of the DNAs involved in the chi structure. On the other hand, if the strands that did *not* break the first time are involved in the second break, the chi structure resolves into two recombinant molecules.

INDUCTION OF DNA STRAND BREAKS

One problem with Holliday's model is that it assumes two breaks in exactly corresponding positions in homologous DNA strands. Single-strand breaks occur frequently, so there is no problem in imagining one break appearing eventually in any given position in a DNA molecule. Nevertheless, it seems unlikely that two breaks, conveniently located right across from one another in two paired molecules, would occur spontaneously. Other investigators have therefore proposed alternate models to explain the origin of the second break. One of these is the Meselson-Radding model (figure 8.19), which suggests that the free end created by the first break "invades" the other duplex, base-pairing with one of the strands and displacing the other. The displaced strand forms a loop called a *D-loop,* where D stands for "displacement." According to this model, the act of forming the D-loop breaks the looping strand, allowing formation of the chi structure. Hotchkiss proposed a similar model, which includes the notion that the breakage results, not from D-loop formation itself, but from extension of the loop by DNA replication.

I n generalized recombination, the two homologous DNA duplexes first align themselves. Then single-stranded breaks occur in two homologous strands, allowing the two free ends to cross over and form new, interstrand bonds. This yields a chi structure, which is resolved when the two other strands break and cross over.

■

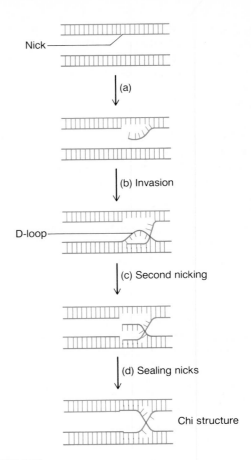

FIGURE 8.19
The Meselson-Radding model of nicking the second strand during recombination. (*a*) The first strand breaks. (*b*) A loose end generated by this first break invades the second DNA duplex. (*c*) This invasion creates a nick in the second duplex. (*d*) The two nicks are sealed, forming the chi structure.

Summary

T he DNA that replicates under the control of a single origin of replication is called a replicon. The chromosomes of simple genetic systems like plasmids, viruses, and even bacteria contain just one replicon. DNA replication in most genetic systems is bidirectional. The replication machinery starts at a fixed point—the origin of replication—and two replicating forks move in opposite directions away from this point. DNA replication in prokaryotes and eukaryotes is semidiscontinuous. The leading strand grows continuously, 5′ → 3′, in the direction of motion of the replicating fork. The other, trailing strand grows discontinuously, in pieces (Okazaki fragments) that are made, 5′ → 3′, in a direction opposite to the motion of the replicating fork. DNA polymerase is incapable of initiating DNA synthesis by itself. It needs a primer to supply a free 3'-end upon which it can build the nascent DNA. Very short pieces of RNA serve this priming function. After Okazaki fragments are made, their primers are removed, the resulting gaps are filled in, and the remaining nicks are sealed.

As helicase unwinds the two parental strands at the replicating fork of a closed circular DNA, it introduces superhelical strain into the DNA. DNA gyrase releases this strain, permitting replication to continue. When replication of a double-stranded circular DNA approaches completion, the two daughter molecules are still entwined. A topoisomerase is presumably needed to separate them.

Circular DNA can replicate by a rolling circle mechanism. One strand of a double-stranded DNA is nicked, and the 3'-end is extended using the intact DNA strand as template. This displaces the 5'-end. In phage λ, the displaced strand serves as the template for discontinuous, lagging strand synthesis.

DNA polymerase, like all enzymes, is not perfect. If left to itself, it would make about one mistake per one hundred thousand nucleotides of DNA. To improve on this intolerably poor performance, the DNA replicating machinery has a built-in proofreading system that depends on primers. Only a base-paired nucleotide can serve as a primer for the replicating enzyme; therefore, if the wrong nucleotide is incorporated by accident, replication stalls until a nuclease erases the error by degrading the most recently made stretch of DNA. Thus, the error can be repaired.

DNA replication in meiotic cells is usually followed by recombination, using the generalized recombination mechanism. First, homologous regions of two DNA duplexes line up, then single-strand breaks occur in two homologous strands. The free ends of the broken strands cross over and form new, interstrand bonds, yielding a chi structure. This cross-shaped structure is resolved in two recombinant DNA products when the other two strands break and cross over. ∎

Problems and Questions
---

1. (a) Draw a schematic diagram of a replicating bubble in *Drosophila* DNA, showing both parental and progeny strands after all Okazaki fragments have been ligated together. (b) Assuming that replication began before labeled thymidine was added, show on the schematic where labeled and unlabeled DNA will be found.

2. Assume the following sequence of events: Replication begins; DNA is pulse-labeled with low specific activity thymidine; DNA is pulsed with high specific activity thymidine. Then show what the resulting autoradiogram will actually look like if the two daughter duplexes twist around one another.

3. What could we conclude from an autoradiographic pattern in which all pairs of silver streaks fall into two categories: _____ _____ or _____ _____ ?

4. Colicin El is a small, double-stranded circular DNA that replicates unidirectionally. Consider an autoradiography experiment in which you label colicin El DNA uniformly with low specific activity thymidine. Then, *after* a round of replication begins, you label with high specific activity thymidine for one-fourth of a replication cycle. Draw a picture of the expected autoradiogram.

5. An extraterrestrial being brings you a vial of bacteria from its planet and informs you that there all living things replicate their DNA fully continuously. (a) Describe the results of a pulse labeling experiment you would perform that verifies this assertion. (b) What unearthly characteristic must the DNA polymerase of these bacteria have?

6. What role do helicases play in DNA replication? Why is an ATPase a necessary part of this role?

7. Contrast the function of a helicase with that of a topoisomerase.

8. Why does the single-strand-binding protein (SSB) of *E. coli* bind so much better to a single-stranded than to double-stranded DNA?

9. Would a swivel be required in replicating a linear chromosome? Why, or why not?

10. If there were no swivel, what kind of supercoils would develop in a replicating closed circular DNA?

11. Unwinding DNA will relax what kind of supercoils? Why?

12. Unwinding is easier in which kind of DNA: positively supercoiled, negatively supercoiled, or relaxed circular? Why?

13. The *E. coli* primosome binds to the template for which strand? Why does it seem more useful for the primosome to bind to this strand?

14. What advantage would a double-headed replisome with two DNA polymerase centers offer? What structural shift would the DNA have to make in order to accommodate such a double-headed replisome?

15. Why is primer synthesis inherently more error-prone than DNA elongation?

Answers appear at end of book.

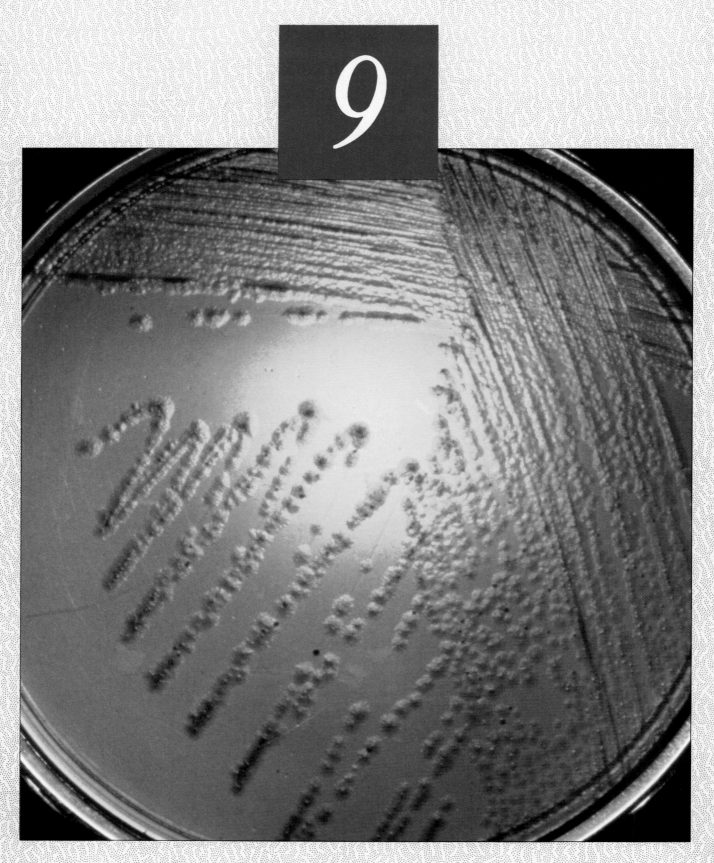

Transcription and Its Control in Prokaryotes

*T*here are more microbes *per person than the entire population of the world. Imagine that. Per person.*

■

——————————————

Alan Bennett

Learning Objectives

In this chapter you will learn:

1. How the enzyme RNA polymerase transcribes a gene, producing an RNA.

2. How bacterial genes of related function are arranged in functional units called operons, each controlled by a binding site for RNA polymerase called a promoter.

3. How expression of bacterial genes is controlled primarily by controlling which promoter the RNA polymerase will recognize.

4. How different programs of gene expression can be activated by altering the RNA polymerase.

n chapter 7 we learned that RNA polymerase transcribes one of the strands of its DNA template, following the Watson-Crick base-pairing rules. We also looked briefly at the mechanism of transcription, divided into three phases: initiation, elongation, and termination. In this chapter we will explore in greater detail the process of transcription in prokaryotes. We will see that bacteria and their phages have evolved intricate and fascinating methods of controlling their gene expression by regulating transcription.

THE MECHANISM OF TRANSCRIPTION

Let us begin by examining a little more closely the mechanism of transcription—the making of an RNA strand. Figure 9.1 demonstrates the essential step in transcription, the formation of a phosphodiester bond between nucleotides. The nucleotides that participate in this process can be any of the four RNA nucleotides: ATP, GTP, CTP, or UTP. In this case, ATP and GTP come together to form the shortest possible RNA, a dinucleotide. In transcription, phosphodiester bond formation occurs over and over again; each time, a new nucleotide link joins the growing RNA chain. Figure 9.2 shows this process repeating three times, producing a short RNA containing only four nucleotides (a tetranucleotide). The participants in this process are: (1) the DNA template, shown in pink; (2) RNA nucleotides, shown in blue; and (3) the enzyme, RNA polymerase, which catalyzes the reaction and makes it specific, so that the RNA product is a faithful copy of the DNA template.

STRUCTURE OF RNA POLYMERASE

RNA polymerase stands at the very center of the study of transcription; after all, this is the enzyme that does the transcribing. Therefore, we will continue our investigation of transcription in prokaryotes with a study of this key enzyme.

As early as 1960–61, RNA polymerases were discovered in animals, plants, and bacteria. And, as you might anticipate, the bacterial enzyme was the first to be studied in great detail. By 1969, the polypeptides that make up the *E. coli* RNA polymerase had been identified by gel electrophoresis. In chapter 6 we saw how gel electrophoresis can be used to separate DNA fragments according to their sizes. The same technique is often applied to proteins, in which case the gel is usually made of polyacrylamide. We therefore call it **polyacrylamide gel electrophoresis,** or **PAGE.** Many enzymes are composed of multiple polypeptides, or **subunits.** To determine the polypeptide makeup of such an enzyme, the experimenter must treat the protein so that the enzyme subunits will electrophorese independently. This is usually done by mixing the enzyme with a detergent (sodium dodecyl sulfate, or SDS) to **denature** the subunits so that they no longer bind to one another. The SDS has two added advantages: (1) It coats all the polypeptides with negative charges, so they all electrophorese toward the anode. (2) It masks the natural charges of the subunits, so they all electrophorese according to their molecular weights. Small polypeptides fit easily through the pores in the gel, thus they migrate rapidly. Larger polypeptides migrate more slowly.

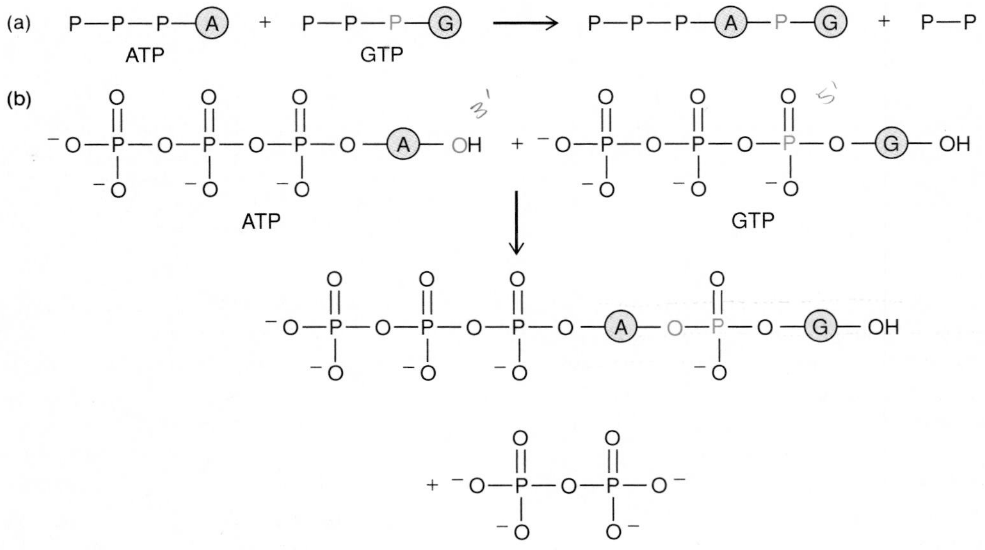

FIGURE 9.1
Phosphodiester bond formation in RNA synthesis. (*a*) Simplified reaction showing only the phosphorous atoms. Note that the phosphorous atom (red) closest to the guanosine is retained in the phosphodiester bond. The other two phosphorous atoms (black) are removed as a by-product. (*b*) More complete structures showing all the atoms participating in phosphodiester bond formation. Note the phosphorous atom (red) on the GTP and the oxygen atom (green) in the ATP hydroxyl group, both of which are retained in the phosphodiester bond.

Chapter 9

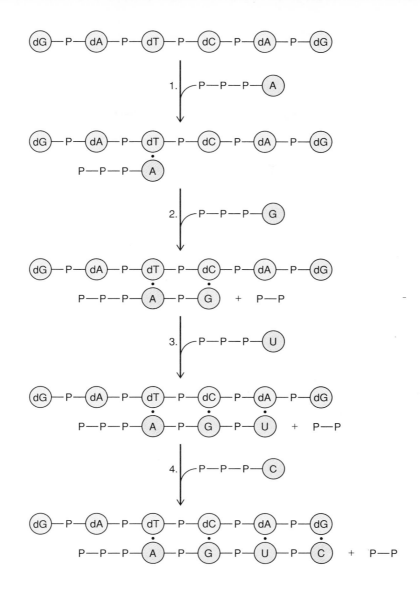

FIGURE 9.2

Synthesis of RNA on a DNA template. Step 1: RNA polymerase initiates RNA synthesis by matching an RNA nucleotide, an ATP (blue) in this case, with a dT (pink) in the DNA template. Transcription in vivo never starts right at the beginning of the DNA, which is why we do not initiate with a CTP to match the dG at left.

Step 2: RNA polymerase completes the initiation step by forming the first phosphodiester bond—between the ATP and an incoming GTP. Step 3: RNA polymerase elongates the chain to a trinucleotide by adding a UTP. Step 4: The RNA chain (a tetranucleotide) is completed by the addition of a CTP.

Figure 9.3 gives the results of a PAGE experiment on *E. coli* RNA polymerase holoenzyme. Polymerase is remarkable in that it contains two very large subunits: **beta** (β) and **beta-prime** (β'), with molecular weights of 150,000 and 160,000 respectively. The other RNA polymerase subunits are called **sigma** (σ) and **alpha** (α), with molecular weights of 70,000 and 40,000, respectively. The subunit content of an RNA polymerase holoenzyme is: β', β, σ, α_2; in other words, there are two molecules of α and one of all the others. A schematic diagram of this subunit structure is presented in figure 9.4.

When Richard Burgess, Andrew Travers, and their colleagues subjected the RNA polymerase holoenzyme to ion exchange chromatography on a negatively charged resin called phosphocellulose, they discovered that they had separated the σ subunit from the remainder of the enzyme, called the **core enzyme.** In the process, they had caused a profound change in the enzyme's activity (table 9.1). Whereas the holoenzyme could transcribe intact phage T4 DNA in vitro quite actively, the core enzyme had little ability to do this. On the other hand, core polymerase retained its basic RNA polymerizing function,

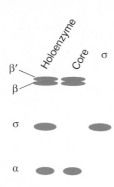

FIGURE 9.3
Separation of the subunits of *E. coli* RNA polymerase by PAGE. Lane 1, holoenzyme; lane 2, core enzyme after σ is removed by phosphocellulose chromatography; lane 3, σ.

T	ABLE 9.1	*Ability of Core and Holoenzyme to Transcribe DNAs*

	Transcription Activity	
DNA	*Core*	*Holoenzyme*
T4 (native, intact)	0.5	33.0
Calf thymus (native, nicked)	14.2	32.8

since it could still transcribe highly nicked templates very well. (As will be explained, transcription of nicked DNA is a laboratory artifact and has no biological significance.)

Adding σ back to the core reconstituted the enzyme's ability to transcribe un-nicked T4 DNA. Even more significantly, the holoenzyme transcribed only a certain class of T4 genes (called immediate early genes), but the core enzyme showed no such specificity; it transcribed the whole T4 genome, though weakly, in vitro. Furthermore, transcription of T4 DNA in vitro by the holoenzyme was asymmetric—just as it is in vivo—but the weak transcription by the core enzyme was symmetric. In other words, not only did the core enzyme read the whole genome, it read both strands. Clearly, depriving the holoenzyme of its σ subunit leaves a core enzyme with basic RNA synthesizing capability but lacking specificity. Adding σ back restores specificity. In fact, σ was named only after this characteristic came to light, and the σ, or Greek letter s, was chosen to stand for "specificity."

> **T**he key player in the transcription process is RNA polymerase. The *E. coli* enzyme is composed of a core, which contains the basic transcription machinery, and a σ factor, which directs the core to transcribe specific genes.

■

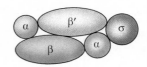

FIGURE 9.4
Subunit structure of *E. coli* RNA polymerase. This schematic drawing is intended to show the numbers and relative sizes of the subunits, not necessarily their orientations to one another.

BINDING RNA POLYMERASE TO PROMOTERS

Why was core enzyme still capable of transcribing nicked DNA, but not intact DNA? Nicks and gaps in DNA provide ideal initiation sites for RNA polymerase, even core polymerase, but this kind of initiation is necessarily nonspecific. There were few nicks or gaps on the intact T4 DNA, so the core polymerase encountered only a few such artificial initiation sites. On the other hand, when σ was present, the holoenzyme could recognize the authentic RNA polymerase binding sites—the **promoters**—on the intact T4 DNA and begin transcription there. This initiation is specific and mirrors the initiation that would occur in vivo. Thus, σ operates by directing the polymerase to initiate at specific promoters.

And how does σ change the way the core polymerase behaves toward promoters? Michael Chamberlin and his coworkers used binding studies to help answer this question. They measured the binding between phage DNA and RNA polymerase core or holoenzyme and found two different classes of binding sites. There were just a few sites that bound RNA polymerase holoenzyme extremely tightly. Once bound, it took 30–60 hours for half of the holoenzymes to dissociate from these tight binding sites, which are, of course, promoters. There were also many loose binding sites that were available to core polymerase as well as to holoenzyme.

The interpretation of these and other experiments is that the RNA polymerase can exist in one of two states: one appropriate for tight binding to promoters (the initiation state) and one appropriate for loose binding (the elongation state). The tight binding, or initiating state can only occur in the presence of σ; after initiation is achieved, σ can dissociate and leave the core enzyme to carry out elongation. Furthermore, tight binding of holoenzyme to promoters involves local melting of the DNA to form a so-called **open promoter complex**. This conversion of a loosely bound polymerase in a **closed promoter complex** to the tightly bound polymerase in the open complex requires σ, and this is what allows transcription to begin (figure 9.5). We can now appreciate how σ fulfills its role in determining specificity of transcription. It selects the promoters to which RNA polymerase will bind tightly. The genes adjacent to these promoters will then be transcribed.

> *T*he σ factor allows initiation of transcription by causing the RNA polymerase holoenzyme to bind tightly to a promoter. This tight binding depends on local melting of the DNA to form an open promoter complex, and it is possible only in the presence of σ. σ can therefore select which genes will be transcribed. After σ has participated in initiation, it dissociates from the core polymerase, leaving the core to carry on with the elongation process.

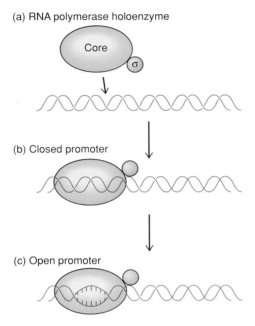

(a) RNA polymerase holoenzyme

Core

σ

(b) Closed promoter

(c) Open promoter

FIGURE 9.5
RNA polymerase-promoter binding. (*a*) The polymerase holoenzyme searches for a binding site on the DNA. (*b*) The holoenzyme has found a promoter and has bound loosely, forming a closed promoter-polymerase complex. (*c*) The holoenzyme has bound tightly, melting a local region of DNA and forming an open promoter-polymerase complex.

COMMON BASE SEQUENCES IN *E. COLI* PROMOTERS

What is the special nature of a prokaryotic promoter that attracts RNA polymerase? David Pribnow compared several *E. coli* and phage promoters, and discerned two regions they held in common. The first was a sequence of six or seven bases centered approximately ten bases upstream from the start of transcription. This was originally dubbed the *Pribnow box,* but is now frequently referred to as the **−10 box.** The second was another short sequence centered approximately thirty-five bases upstream from the transcription start; it is usually called the **−35 box.** More than one hundred promoters have now been examined, and a "typical" sequence for each of these boxes has emerged (figure 9.6).

These so-called **consensus sequences** represent probabilities. The capital letters in figure 9.6 denote bases that have a high probability of being found in the given position. The small letters correspond to bases that are usually found in the given position, but at a lower frequency than those denoted by capital letters. The probabilities are such that you rarely find −10 or −35 boxes that match the consensus sequences perfectly. However, when such perfect matches are found, they tend to occur in very strong promoters that initiate transcription unusually actively. In fact, mutations that destroy matches with the consensus sequences tend to be **down mutations.** That is, they make the promoter weaker, resulting in less transcription. For example, mutation of a −10 box sequence from TATAAT to TGTAAT would usually be a down mutation because it would weaken the promoter. By contrast, mutations that make promoter sequences more like the consensus sequences usually make the promoters stronger; these are called **up mutations.** For instance, mutation of a −10 box sequence from TATCTT to TATATT would probably be an up mutation. In chapter 10 we will see that eukaryotic promoters have their own consensus sequences, one of which resembles the −10 box quite closely.

−35 box   −10 box   Transcription

TTGACa   TAtAaT
AACTGt   ATaTtA

Unwound region

FIGURE 9.6
A prokaryotic promoter. The positions of the −10 and −35 boxes and the unwound region are shown relative to the start of transcription for a typical *E. coli* promoter. Capital letters denote bases found in those positions in more than 50% of promoters examined; small letters denote bases found in those positions in 50% or fewer of promoters examined.

*P*rokaryotic promoters contain two regions centered at −10 and −35 bases upstream from the transcription start site. In *E. coli,* these bear a greater or lesser resemblance to two consensus sequences: TATAAT and TTGACA, respectively. In general, the more closely regions within a promoter resemble these consensus sequences, the stronger that promoter will be.

■

TERMINATION OF TRANSCRIPTION

The *E. coli trp* operon contains a DNA sequence called an attenuator that causes premature termination of transcription. Terry Platt and colleagues used attenuation as an experimental probe of the mechanism of termination. They manipulated the attenuator region by introducing mutations and then looking to see what the effect on attenuation would be. The rationale is that if a certain sequence of the attenuator is important in causing termination, mutations in this sequence should block attenuation. Two features of the attenuator turned out to be important. These are (1) an inverted repeat that gives the transcript the ability to form a "hairpin" and (2) a string of T's in the coding strand.

Inverted Repeats and Hairpins

Before we go on, we should understand how an inverted repeat predisposes a transcript to form a hairpin. Consider the inverted repeat:

TACGAAGTTCGTA
ATGCTTCAAGCAT

Such a sequence is symmetrical about its center, indicated by the dot; it would read the same if rotated 180 degrees in the plane of the paper. Now observe that a transcript of this sequence:

UACGAAGUUCGUA

because of *its* symmetry, is self-complementary about its center. That means that the self-complementary bases can pair to form a hairpin as follows:

U · A
A · U
C · G
G · C
A · U
A   U
  G

The inverted repeat in the *trp* attenuator is not perfect, but eight base pairs are still possible, and seven of these are strong G–C pairs, held together by three hydrogen bonds, compared to just two in A–U pairs. The hairpin looks like this:

A · U
G · C
C · G
C · G
C · G
G · C
C · G A
C · G A
U   U
  A — A

Notice that there is a small loop at the end of this hairpin because of the U–U and A–A combinations that cannot base-pair. Furthermore, one A has to be "looped out" to allow eight base pairs instead of just seven. Still, the hairpin should form and be relatively stable.

A Model for Termination

The discovery that the inverted repeat and string of T's are key features of the transcription termination signal supports the termination hypothesis illustrated in figure 9.7. The RNA polymerase transcribes the inverted repeat, producing a transcript that tends to form a hairpin (figure 9.7*a*). When the hairpin forms (figure 9.7*b*), the RNA polymerase pauses for some unexplained reason. This leaves the base pairs between the dA's in the DNA and the rU's in the RNA as the only force holding the transcript and template together. Such rU–dA pairs are exceptionally weak, which leads to the hypothesis that the RNA can easily fall off the template, terminating transcription (figure 9.7*c*). However, things may not be that simple. There are other terminators that bear a sequence just downstream from the inverted repeat that is not especially AT rich. Mutating them to make them more AT rich actually reduces their termination efficiency. Thus, the string of T's may simply be the second part of the signal recognized by RNA polymerase in the *trp* attenuator— but not necessarily because of any inherent weakness it causes between template and transcript.

The Termination Factor ρ

The model of termination discussed previously really applies only to termination that is carried out by the *E. coli* RNA polymerase itself. There are many termination events in *E. coli* that involve not only the polymerase but another factor called **rho** (ρ). If σ is the initiation, or specificity factor, ρ is the termination, or release factor. Jeffrey

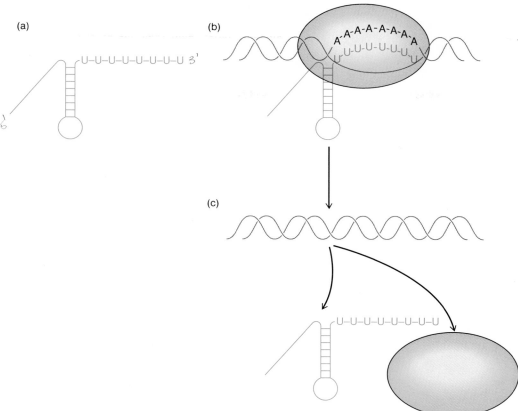

(a)

U–U–U–U–U–U–U–U 3'

(b)

A-A-A-A-A-A-A-A
U-U-U-U-U-U

(c)

U–U–U–U–U–U–U–U

FIGURE 9.7
Illustration of ρ-independent termination. (*a*) The 3'-end of the transcript shows the two elements necessary for ρ-independent termination: a base-paired hairpin loop, followed by a second motif, in this case a string of U's. (*b*) When the hairpin forms, the polymerase (red) pauses, leaving only the A–U bonds holding the transcript to the template. (*c*) The bonds break, releasing the RNA (blue) and terminating transcription.

Roberts discovered ρ as a protein that caused an apparent depression of the ability of RNA polymerase to transcribe certain phage DNAs in vitro. This depression is simply the result of termination. Whenever ρ causes a termination event, the polymerase has to reinitiate to begin transcribing again. And, since initiation is a time-consuming event, less net transcription can occur.

Rho-dependent termination follows somewhat different rules than ρ-independent termination. The most obvious difference is that ρ-dependent terminators have only one part—the inverted repeat. There is no string of T's or other second sequence motif. Therefore, the polymerase presumably pauses at the hairpin but cannot release the transcript by itself. ρ performs that function.

How does ρ do its job? Since we have seen that σ operates through RNA polymerase by binding tightly to the core enzyme, we might envision a similar role for ρ. But ρ seems not to have affinity for RNA polymerase. Whereas σ is a critical part of the polymerase holoenzyme, ρ is not found there. An alternative hypothesis is that ρ binds to DNA at terminators and waits for polymerase to come along, at which point it can perform its function. A more attractive possibility (figure 9.8) is that ρ binds to the transcript and follows the RNA polymerase. This chase would continue until the polymerase stalled in the terminator region just after making the hairpin. Then ρ could catch up and release the transcript. There is some evidence to support this idea. For example, it can be demonstrated that ρ probably does bind to RNA, at least under certain circumstances.

*T*ermination signals (terminators) recognized by RNA polymerase alone consist of an inverted repeat followed by another important base sequence, which is sometimes a string of T's. The transcript of the inverted repeat can form a base-paired structure called a hairpin, which somehow stalls the polymerase. The second part of the terminator then ensures that the RNA will fall off, terminating transcription. Other terminators require an ancillary factor, ρ, in addition to RNA polymerase. These terminators contain only an inverted repeat. ρ may loosen transcript-template binding after RNA polymerase pauses at the hairpin.

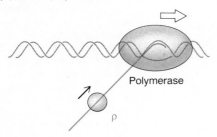

(a) ρ pursues polymerase

Polymerase

ρ

(b) Hairpin forms; polymerase pauses; ρ catches up

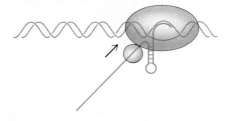

(c) ρ causes termination

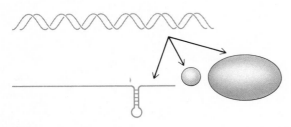

FIGURE 9.8
A model of ρ-dependent termination. (*a*) As polymerase (red) makes RNA, ρ (blue) binds to the transcript (green) and pursues the polymerase. (*b*) When the hairpin forms in the transcript, the polymerase pauses, giving ρ a chance to catch up. (*c*) ρ destabilizes the RNA-DNA bonds and releases the transcript.

OPERONS

In chapter 7 we discussed biochemical pathways that are catalyzed by groups of enzymes. One example was the pathway that breaks down the amino acid phenylalanine. Obviously, all of the enzymes in the pathway will be needed in roughly equal amounts, since one enzyme deficiency would stall the whole process. Therefore, if four enzymes are needed for a pathway to operate, it would be wasteful to turn on the genes for three of them and leave the other turned off. In other words, for greatest efficiency, the genes for proteins that function together should be controlled together. One mechanism that allows coordinate control is to group functionally related genes together so they can be regulated together easily. Such a group of contiguous, coordinately controlled genes is called an **operon.**

THE *lac* OPERON

The first operon to be discovered has become the paradigm of the operon concept. It contains three **structural genes**—genes that code for proteins. These three code for the enzymes that allow *E. coli* cells to use the sugar **lactose,** hence the name *lac* **operon.** Consider a flask of *E. coli* cells growing on the sugar **glucose.** We take away the glucose and replace it with lactose. Can the cells adjust to the new nutrient source? For several minutes it appears that they cannot; they stop growing as soon as the glucose disappears and behave as if the lactose is no help at all. Then, after the lag period, growth resumes. During the lag, the cells have been turning on the *lac* operon and beginning to accumulate the enzymes they need to metabolize lactose.

What are these enzymes? First, *E. coli* needs an enzyme to transport the lactose into the cells. The name of this enzyme is **galactoside permease.** Next, the cells need an enzyme to break the lactose down into its two component sugars, galactose and glucose. Figure 9.9 shows this reaction. Since lactose is composed of two simple sugars, we call it a *disaccharide.* As mentioned previously, these are the six-carbon sugars galactose and glucose; they are joined together by a linkage called a *β-galactosidic bond.* Lactose is therefore called a *β-galactoside,* and the enzyme that cuts it in half is called *β*-**galactosidase.** The genes for these two enzymes, galactoside permease and *β*-galactosidase, are found side-by-side in the *lac* operon, along with another structural gene—for **galactoside transacetylase**—whose function in lactose metabolism is still not clear. These names give some insight into enzyme nomenclature. The "ase" suffix is an almost universal feature of enzyme names; the rest of the name usually gives a clue to the reaction the enzyme catalyzes.

The three genes coding for enzymes that carry out lactose metabolism are grouped together in the order: *β*-galactosidase (*lacZ*), galactoside permease (*lacY*), galactoside transacetylase (*lacA*). They are all transcribed together on one messenger RNA, called a **polycistronic message,** starting from a single promoter. Thus, they can all be controlled together simply by controlling that promoter. The term "polycistronic" comes from "cistron," which is a synonym for "gene." A polycistronic message is therefore simply a message with information from more than one gene.

L actose metabolism in *E. coli* is carried out by two enzymes, with possible involvement by a third. The genes for all three enzymes are clustered together and transcribed together from one promoter, yielding a polycistronic message. Therefore, these three genes, linked in function, are also linked in expression. They are turned off and on together.

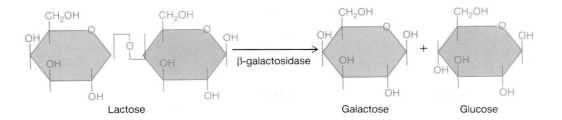

FIGURE 9.9
The β-galactosidase reaction. The enzyme breaks the β-galactosidic bond (green) between the two sugars, galactose (red) and glucose (blue), that compose lactose.

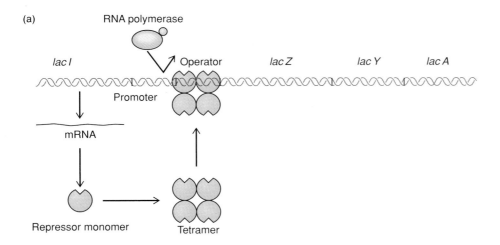

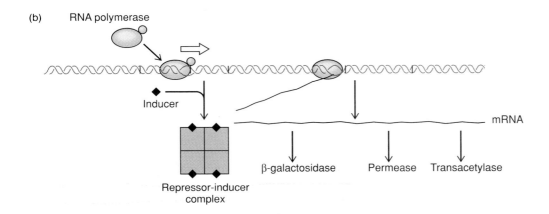

FIGURE 9.10
Negative control of the *lac* operon. (*a*) Repression. The *lacI* gene produces repressor (green), which binds to the operator and prevents RNA polymerase (red and blue) from binding to the promoter. (*b*) Derepression. Inducer (black) binds to repressor, changing it to a form (green squares) that no longer binds well to the operator. Now polymerase can bind successfully to the promoter and transcribe the operon.

Negative Control of the *lac* Operon

Figure 9.10 illustrates one aspect of *lac* operon regulation: negative control. The term **negative control** implies that the operon is turned on unless something intervenes to stop it. The "something" that can turn off the *lac* operon is the **lac repressor.** This repressor, the product of the ***lacI* gene** shown at the extreme left in figure 9.10, is a tetramer of four identical polypeptides; it binds to the **operator** just to the right of the promoter. When the repressor is bound to the operator, the operon is **repressed.** That is because the operator and promoter are contiguous, and when the repressor occupies the operator, it blocks RNA polymerase's access to the promoter. If RNA polymerase cannot get to the promoter, it cannot transcribe the three structural genes of the operon, so the operon is off, or repressed.

FIGURE 9.11
Conversion of lactose to allolactose. A side reaction carried out by β-galactosidase rearranges lactose to the inducer, allolactose. Note the change in the galactosidic bond from β-1,4 (green) to β-1,6 (gray).

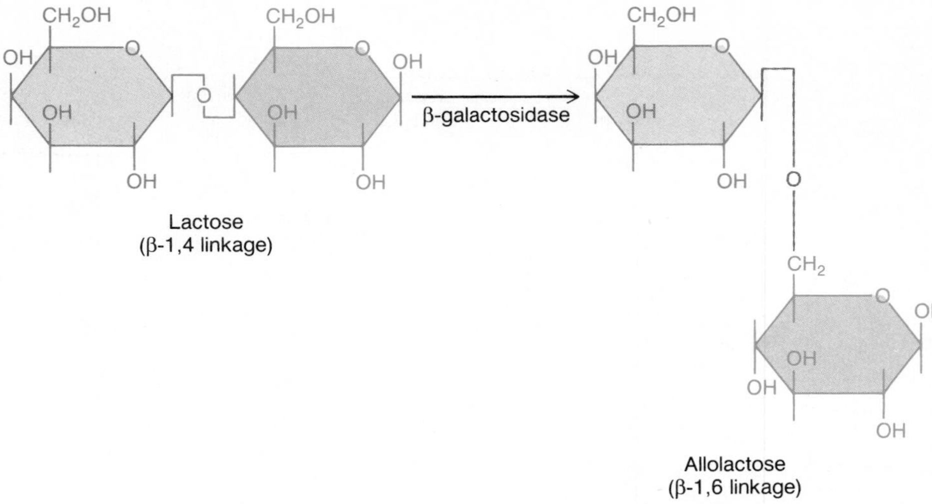

Lactose
(β-1,4 linkage)

β-galactosidase

Allolactose
(β-1,6 linkage)

The *lac* operon is repressed as long as glucose is present and no lactose is available. That is economical; it would be wasteful for the cell to produce enzymes needed to use an absent sugar, especially when an alternate sugar, glucose, is available. In fact, if all of an *E. coli* cell's genes were turned on all the time, this would drain the cell of so much energy that it could not survive. Thus, control of gene expression is essential to life.

On the other hand, when all the glucose is gone and lactose is at hand, there should be a mechanism for removing the repressor so the operon can turn on and take advantage of the new nutrient. How does this work? The repressor is a so-called **allosteric protein,** which means that it changes its conformation, and therefore its function, when it binds to a certain small molecule. The small molecule in this case is called the **inducer** of the *lac* operon, because when it binds to the repressor, it causes the protein to change to a conformation that favors dissociation from the operator, thus inducing the operon (figure 9.10).

What is the nature of this inducer? It is actually an alternate form of lactose called **allolactose** (Greek: *allo*=other). When β-galactosidase cleaves lactose to galactose plus glucose, it rearranges a small fraction of the lactose to allolactose. Figure 9.11 shows that allolactose is just galactose linked to glucose in a different way than in lactose. (In lactose, the linkage is through a β-1,4 bond; in allolactose, the linkage is β-1,6.)

You may be asking yourself: How can lactose be metabolized to allolactose if there is no permease to get it into the cell and no β-galactosidase to perform the metabolizing, since the *lac* operon is repressed? The answer is that nothing in life is perfect; even repression is somewhat leaky, and a low basal level of the *lac* operon products is always present. This is enough to get the ball rolling

by producing a little inducer. It does not take much inducer to do the job, since there are only about ten tetramers of repressor per cell. Furthermore, the derepression of the operon will snowball as more and more operon products are available to produce more and more inducer.

The Operon Hypothesis

The development of the operon concept by François Jacob and Jacques Monod and their colleagues was one of the classic triumphs of the combination of genetic and biochemical analysis. The story begins in the 1940s, when Monod began studying the inducibility of lactose metabolism in *E. coli*. He chose this inducible system over others, such as the maltose system, because of the availability of a variety of galactosides that could serve as inducers without themselves being metabolized. This allowed the two key phenomena of induction and metabolism to be analyzed separately.

Monod knew from the outset that a key feature of lactose metabolism was β-galactosidase, and that this enzyme was inducible by lactose and by other galactosides. Furthermore, he and Melvin Cohn had used an anti-β-galactosidase antibody to measure the amount of β-galactosidase protein, and they showed that this increased upon induction. Therefore, since more gene product appeared in response to lactose, it appeared that the β-galactosidase gene itself was being induced.

To complicate matters, certain mutants were found that could make β-galactosidase but still could not grow on lactose. What was missing in these *cryptic* mutants? To answer this question, Monod and his coworkers added a radioactive galactoside to wild-type and cryptic bacteria. They found that uninduced wild-type cells did not

Chapter 9

Genotype	Inducer	Accumulation of Galactoside
Z^+Y^+	−	−
Z^+Y^+	+	+
Z^+Y^- (cryptic)	−	−
Z^+Y^- (cryptic)	+	−

take up the galactoside, and neither did the cryptic mutants, even if they were induced. Of course, previously induced wild-type cells did accumulate the galactoside. This revealed two things: First, a substance (galactoside permease) is induced along with β-galactosidase in wild-type cells and is responsible for transporting galactosides into the cells; second, the cryptic mutants seem to have a defective gene (Y^-) for this substance (see table 9.2).

Monod named this substance "galactoside permease," and then endured criticism from his colleagues for naming a protein before it had been isolated. He has since remarked, "This attitude reminded me of that of two traditional English gentlemen who, even if they know each other well by name and by reputation, will not speak to each other before having been formally introduced." In their efforts to purify galactoside permease, Monod and his colleagues identified another protein, galactoside transacetylase, which is induced along with β-galactosidase and galactoside permease, and whose function in the *lac* operon remains somewhat obscure to this day.

By the late 1950s, Monod knew that three enzyme activities (and therefore at least three genes) were induced together by galactosides. He had also found some mutants, called **constitutive mutants,** that needed no induction. They produced the three gene products all the time. Monod realized that further progress would be greatly accelerated by genetic analysis, so he teamed up with François Jacob, who was working just down the hall at the Pasteur Institute.

In collaboration with Arthur Pardee, Jacob and Monod created merodiploids (partial diploids) carrying both the wild-type (inducible) and constitutive alleles. The inducible allele proved to be dominant, demonstrating that wild-type cells produce some substance that keeps the *lac* genes turned off. Because this substance turned off the genes from the constitutive as well as the inducible parent, it made the merodiploids inducible. Of course, this substance is the *lac* repressor, and Jacob and Monod recognized it quickly because the λ phage repressor, which acts similarly, was already under study in André Lwoff's laboratory at the other end of the corridor.

The existence of a repressor required that there be some specific receptor to which the repressor would bind. Jacob and Monod called this the operator. The specificity of this interaction suggested that it should be subject to genetic mutation; that is, some mutations in the operator should abolish its interaction with the repressor. Of course, these would also be constitutive mutations, so how are we to distinguish them from constitutive mutations in the repressor gene?

Jacob and Monod realized that they could do this by determining whether the mutation was dominant or recessive. Jacob explained this with an analogy. He likened the two operators in a merodiploid bacterium to two radio receivers, each of which controls a separate door to a house. The repressor genes in a merodiploid are like little radio transmitters, each sending out an identical signal to keep the doors closed. A mutation in one of the repressor genes would be like a breakdown in one of the transmitters; the other transmitter would still function and the doors would both remain closed. In other words, both *lac* operons in the merodiploid would still be repressible. Thus, such a mutation should be recessive (figure 9.12*a*), as we have already observed it is.

On the other hand, a mutation in one of the operators would be like a breakdown in one of the receivers. Since it could no longer receive the radio signal, its door would swing open; the other, with a functioning receiver, would remain closed. In other words, one of the *lac* operons in the merodiploid would be constantly derepressed, while the other would remain repressible. Therefore, this mutation would be dominant, but only with respect to the *lac* operon under the control of the mutant operator (figure 9.12*b*). We call such a mutation ***cis*-dominant** because it is only dominant with respect to genes on the same piece of DNA (*in cis*). Jacob and Monod did indeed find such *cis*-dominant mutations, and they defined the operator. These mutations are called O^c, for *operator constitutive*.

What about mutations in the repressor gene that render the repressor unable to respond to inducer? Such mutations should make the *lac* operon uninducible and should be dominant both *in cis* and *in trans* because the mutant repressor will remain bound to both operators even in the presence of inducer (figure 9.12*c*). We might liken this to a radio signal that cannot be turned off. Monod and his colleagues found two such mutants, and Suzanne Bourgeois later found many others. These are named I^s to distinguish them from constitutive repressor mutants (I^-), which make a repressor that cannot recognize the operator.

Both kinds of constitutive mutants (I^- and O^c) affected all three of the *lac* genes in the same way. The genes had already been mapped and were found to be adjacent on the chromosome. These findings strongly suggested that

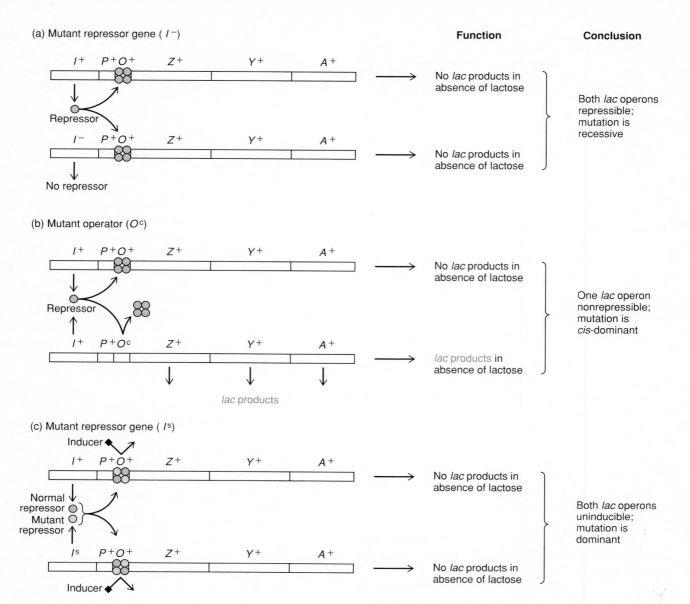

(a) Mutant repressor gene (I^-)

I^+ P^+O^+ Z^+ Y^+ A^+

Repressor

No repressor

I^- P^+O^+ Z^+ Y^+ A^+

Function

→ No *lac* products in absence of lactose

→ No *lac* products in absence of lactose

Conclusion

Both *lac* operons repressible; mutation is recessive

(b) Mutant operator (O^c)

I^+ P^+O^+ Z^+ Y^+ A^+

Repressor

I^+ P^+O^c Z^+ Y^+ A^+

lac products

→ No *lac* products in absence of lactose

→ *lac* products in absence of lactose

One *lac* operon nonrepressible; mutation is *cis*-dominant

(c) Mutant repressor gene (I^s)

Inducer

I^+ P^+O^+ Z^+ Y^+ A^+

Normal repressor
Mutant repressor

I^s P^+O^+ Z^+ Y^+ A^+

Inducer

→ No *lac* products in absence of lactose

→ No *lac* products in absence of lactose

Both *lac* operons uninducible; mutation is dominant

FIGURE 9.12

Effects of regulatory mutations in the *lac* operon in merodiploids. (*a*) This merodiploid has one wild-type operon (top) and one operon (bottom) with a mutation in the repressor gene (I^-). The wild-type repressor gene (I^+) makes enough normal repressor (green) to repress both operons, so the I^- mutation is recessive. (*b*) This merodiploid has one wild-type operon (top) and one operon (bottom) with a mutation in the operator (O^c) that makes it incapable of binding repressor (green). The wild-type operon remains repressible, but the mutant operon is not; it makes *lac* products even in the absence of lactose.

Since only the operon connected to the mutant operator is affected, this mutation is *cis*-dominant. (*c*) This merodiploid has one wild-type operon (top) and one operon (bottom) with a mutant repressor gene (I^s) whose product (yellow) cannot bind inducer. The mutant repressor therefore binds irreversibly to both operators and renders both operons uninducible. This mutation is therefore dominant. Notice that these repressor tetramers containing some mutant and some wild-type subunits behave as mutant proteins. That is, they remain bound to the operator even in the presence of inducer.

the operator lay near the three structural genes, although it was still formally possible that the repressor acted on a polycistronic mRNA rather than on the DNA itself.

Notice that Jacob and Monod, by skillful genetic analysis, were able to develop the operon concept. They predicted the existence of two key control elements: the

repressor gene and the operator. Deletion mutations revealed a third element (the promoter) that was necessary for expression of all three structural genes. Furthermore, they could conclude that all three genes were clustered into a single control unit: the *lac* operon. Subsequent biochemical studies have amply confirmed Jacob and Monod's beautiful hypothesis.

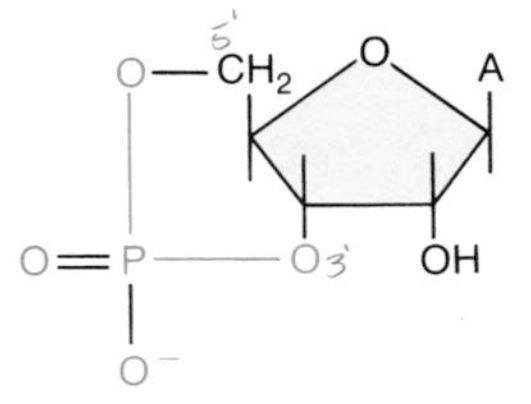

FIGURE 9.13
Cyclic AMP. Note the cyclic 5'-3' phosphodiester bond (blue).

Negative control of the *lac* operon occurs as follows: The operon is turned off as long as repressor binds to the operator, because the repressor interferes with RNA polymerase's binding to the adjacent promoter. When the supply of glucose is exhausted and lactose is available, the few molecules of *lac* operon enzymes produce a few molecules of allolactose from the lactose. The allolactose acts as an inducer by binding to the repressor and causing a conformational shift that encourages dissociation from the operator. With the repressor removed, RNA polymerase is free to bind to the *lac* promoter and transcribe the three structural genes.

■

Positive Control

As you might predict, positive control is the opposite of negative control; that is, the operon is turned off unless something intervenes to turn it on. If negative control is like the brake of a car, positive control is the ignition switch. In the case of the *lac* operon, this means that removing the repressor from the operator (releasing the brake) is not enough to turn on the operon. An additional, positive factor (turning on the ignition) is needed.

What is the selective advantage of this positive control system? Isn't negative control enough? Negative control alone can only respond to the presence of lactose. If there were no positive control, the presence of lactose alone would suffice to activate the operon. But that would be inappropriate if glucose were still available, because *E. coli* cells metabolize glucose more easily than lactose; it would therefore be wasteful for them to turn on the *lac* operon in the presence of glucose. In fact, *E. coli* cells keep the *lac* operon turned off as long as glucose is present. This selection in favor of glucose metabolism and against use of other energy sources was long supposed to be due to the influence of some breakdown product, or *catabolite,* of glucose. It is therefore known as **catabolite repression.**

The ideal positive controller of the *lac* operon would be a substance that sensed the lack of glucose and responded by activating the *lac* promoter so that RNA polymerase could bind and transcribe the structural genes (assuming, of course, that lactose is present and repressor is therefore gone). One substance that responds to glucose concentration is a nucleotide called **cyclic AMP (cAMP)** (figure 9.13). The lower the level of glucose drops, the higher the concentration of cAMP rises.

Is cAMP the positive effector, then? Not exactly. The positive controller of the *lac* operon is a complex composed of two parts: cAMP and a binding protein called **CAP,** for **catabolite activator protein** by its discoverers, Geoffrey Zubay and his colleagues. The protein binds cAMP, and the resulting complex binds to the *lac* promoter and turns it on by helping RNA polymerase bind.

Figure 9.14 illustrates this mechanism. Notice that the *lac* promoter is divided into two parts, the CAP binding site on the left and the RNA polymerase binding site on the right.

How does binding CAP plus cAMP to one side of the promoter facilitate RNA polymerase binding to the other? X-ray diffraction studies of the CAP/cAMP/promoter complex have shown that when CAP/cAMP binds, it causes the promoter to bend very noticeably. This bending might make it easier for the RNA polymerase to separate the two DNA strands, forming an open promoter complex.

CAP and cAMP also stimulate transcription of other inducible operons, including the well-studied *ara* and *gal* operons. Just as the *lac* operon makes possible the metabolism of lactose, these other operons code for enzymes that break down the alternative sugars arabinose and galactose. Thus, for greatest efficiency, all three operons should remain turned off as long as the cell has glucose available. Since cAMP responds to glucose concentration, it is not surprising that all three operons share a common regulatory mechanism involving cAMP. The CAP-cAMP complex binds at or near each operon's promoter and facilitates the binding of RNA polymerase. However, the exact location and character of the CAP binding site varies from one operon to the next.

Positive control (catabolite repression) of the *lac* operon (or the *ara* or *gal* operon) works as follows: A complex composed of cAMP plus a protein known as catabolite activator protein (CAP) binds to the upstream part of the promoter and facilitates binding of RNA polymerase to the downstream part. This greatly enhances transcription of the operon. The physiological significance of this positive control mechanism is that it can only operate when cAMP concentration is elevated; this occurs when glucose concentration is low and there is a corresponding need to metabolize an alternate energy source.

■

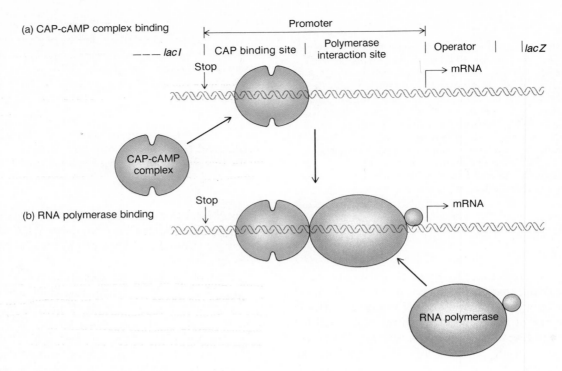

(a) CAP-cAMP complex binding

Promoter

___ lacI | CAP binding site | Polymerase interaction site | Operator | lacZ

Stop

CAP-cAMP complex

→ mRNA

(b) RNA polymerase binding

Stop

→ mRNA

RNA polymerase

FIGURE 9.14
Positive control of the *lac* operon. (*a*) CAP-cAMP complex (green) binds to the CAP binding site part of the promoter. (*b*) The CAP-cAMP complex facilitates binding of RNA polymerase (red and blue) to the polymerase-interaction-site part of the promoter. Transcription can now begin.

TEMPORAL CONTROL OF TRANSCRIPTION

Up to this point we have been discussing the ways in which bacteria control the transcription of a very limited number of genes at a time. For example, when the *lac* operon switches on, only three structural genes are activated. There are other times in a bacterium's life when more radical shifts in gene expression take place. When a phage infects a bacterium, it usually subverts the host's transcription machinery to its own use; further, it establishes a time-dependent, or temporal program of transcription. In other words, the phage early genes are transcribed first, then the later genes. By the time phage T4 infection of *E. coli* reaches its late phase, there is essentially no more host transcription—only phage transcription. This massive shift in specificity would be hard to explain by the operon mechanisms we have looked at so far. Instead, it is engineered by a fundamental change in the transcription machinery: a change in RNA polymerase itself. Another profound change in gene expression occurs during sporulation in bacteria such as *Bacillus subtilis*. Here, genes that are needed in the vegetative phase of growth switch off, and other, sporulation-specific genes switch on. Again, this switch is accomplished by changes in RNA polymerase. The remainder of this chapter is devoted to this important aspect of gene control.

MODIFICATION OF THE HOST RNA POLYMERASE

What part of RNA polymerase would be the logical candidate to change the specificity of the enzyme? We have already seen that σ is the key player in determining specificity of T4 DNA transcription in vitro, so σ is the obvious answer to our question. And experimentation has shown it to be the correct answer. It is interesting, however, that these experiments were not done first with the *E. coli*-T4 system, which turned out to be very grudging about yielding further secrets. A more favorable experimental climate was found with *B. subtilis* and its phages, especially phage SPO1.

SPO1, like T4, is a large phage with many genes. It has a temporal program of transcription as follows: In the first five minutes or so of infection, the early genes are expressed; next, the middle genes turn on (about five to ten minutes post-infection); from about the ten-minute point until the end of infection, the late genes switch on. Since the phage has a large number of genes, it is not surprising that it uses a fairly elaborate mechanism to control this time program. J. Pero and her colleagues have been the leaders in developing the model illustrated in figure 9.15.

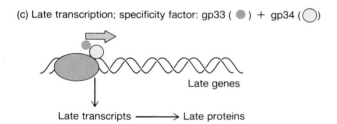

(a) Early transcription; specificity factor: host σ ( )

Early genes

Early transcripts

gene product of gene 28

Early proteins, including gp28 ( )

(b) Middle transcription; specificity factor: gp28 ( )

Middle genes

Middle transcripts

Middle proteins, including gp33 (●) and gp34 (○)

(c) Late transcription; specificity factor: gp33 (●) + gp34 (○)

Late genes

Late transcripts ⟶ Late proteins

FIGURE 9.15
Temporal control of transcription in phage SPO1-infected *B. subtilis*.
(*a*) Early transcription is directed by the host RNA polymerase
holoenzyme, including the host σ factor (blue); one of the early phage
proteins is gp28 (green), a new σ factor. (*b*) Middle transcription is
directed by gp28, in conjunction with the host core polymerase (red);
two middle phage proteins are gp33 and gp34 (purple and yellow,
respectively); together, these constitute yet another σ factor. (*c*) Late
transcription depends on the host core polymerase plus gp33 and 34.

The host RNA polymerase holoenzyme handles transcription of early SPO1 genes, which is analogous to the
T4 model, where the earliest genes are transcribed by the
host holoenzyme. This is necessary because the phage does
not carry its own RNA polymerase. Therefore, when the
phage first infects the cell, the host holoenzyme is the only
RNA polymerase available. The *B. subtilis* holoenzyme
resembles the *E. coli* enzyme closely. Its core consists of

two large β and β′ and two small α polypeptides; its σ factor
has a molecular weight of 43,000, somewhat smaller than
E. coli's σ. One of the genes transcribed in the early phase
of SPO1 infection is called gene 28. Its product, **gp28,** associates with the host core polymerase, displacing the host
σ ($σ^{43}$). With this new, phage-coded polypeptide in place,
the RNA polymerase changes specificity. It begins transcribing the phage middle genes instead of the phage early
genes and host genes. In other words, gp28 is a novel σ
factor that accomplishes two things: It diverts the host's
polymerase from transcribing host genes, and it switches
from early to middle transcription.

The switch from middle to late transcription occurs
in much the same way, except that two polypeptides team
up to bind to the polymerase core and change its specificity. These are **gp33** and **gp34,** the products of two phage
middle genes (genes 33 and 34, respectively). These proteins replace gp28 and direct the altered polymerase to
transcribe the phage late genes in preference to the middle
genes. Note that the polypeptides of the host core polymerase remain constant throughout this process; it is the
progressive substitution of σ factors that changes the specificity of the enzyme and thereby accomplishes the transcriptional program.

How do we know that the σ-switching model is valid?
Two lines of evidence, genetic and biochemical, support
it. First, genetic studies have shown that mutations in gene
28 prevent the early-to-middle switch, just as we would
predict if the gene 28 product is the σ factor that turns on
the middle genes. Similarly, mutations in either gene 33
or 34 prevent the middle-to-late switch, again in accord
with the model. The biochemical studies have relied on
purified RNA polymerases that contain one or another of
the σ factors in question. When polymerase bearing host
σ encounters phage DNA in vitro, it transcribes only the
early genes; when the polymerase bears gp28, it transcribes the middle genes preferentially, and when it is
bound to gp33 and gp34, it is specific for the late genes.

*T*ranscription of phage SPO1 genes in infected
B. subtilis cells proceeds according to a temporal
program in which early genes are transcribed first, then
middle genes, and finally late genes. This switching is
directed by a set of phage-encoded σ factors that associate
with the host core RNA polymerase and change its
specificity from early to middle to late. The host σ is
specific for the phage early genes; the phage gp28 switches
the specificity to the middle genes; and the phage gp33 and
gp34 switch to late specificity.

■

THE RNA POLYMERASE ENCODED IN PHAGE T7

Phage **T7** belongs to a class of relatively simple *E. coli* phages that also includes T3 and φII. These have a considerably smaller genome than SPO1 and, therefore, much fewer genes. In these phages we distinguish three phases of transcription, called classes I, II, and III. (They could just as easily be called early, middle, and late, to conform to SPO1 nomenclature.) One of the five class I genes (gene 1) is necessary for class II and class III gene expression. When it is mutated, only the class I genes are transcribed. Having just learned the SPO1 story, you are probably all primed to hear that gene 1 codes for a σ factor directing the host RNA polymerase to the later phage genes. In fact, this was the conclusion reached by the first workers on T7 transcription, but it was erroneous.

The gene 1 product is actually not a σ factor, but a whole phage-specific RNA polymerase contained in one polypeptide. This polymerase, as you might expect, reads the T7 phage class II and III genes specifically, leaving the class I genes completely alone. Indeed, this enzyme is unusually specific; it will transcribe the class II and III genes of phage T7, and virtually no other natural template. The switching mechanism in this phage is thus quite simple (figure 9.16). When the phage DNA enters the host cell, the *E. coli* holoenzyme transcribes the five class I genes, including gene 1. The gene 1 product—the phage-specific RNA polymerase—then transcribes the phage class II and class III genes.

> *P* hage T7, instead of coding for a new σ factor to change the host polymerase's specificity from early to late, encodes a whole new RNA polymerase with absolute specificity for the later phage genes. This polymerase, composed of a single polypeptide, is a product of one of the earliest phage genes, gene 1. The temporal program in the infection by this phage is simple. The host polymerase reads the earliest (class I) genes, one of whose products is the phage polymerase, which then reads the later (class II and class III) genes.

■

CONTROL OF TRANSCRIPTION DURING SPORULATION

We have already seen how phage SPO1 changes the specificity of its host's RNA polymerase by replacing its σ factor. In the following section, we will show that the same kind of mechanism applies to changes in gene expression in the host itself during the process of sporulation. *E. coli*, the bacterium we have examined in greatest detail, lives its whole life in the **vegetative,** or growth state. If you starve a culture of *E. coli*, the cells stop growing and then begin to die; they have little defense against adverse conditions.

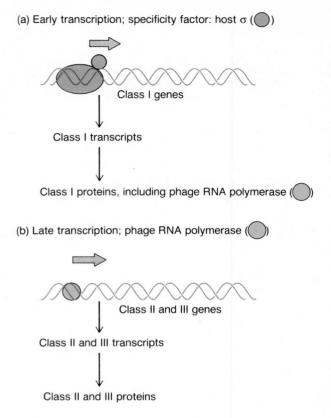

FIGURE 9.16
Temporal control of transcription in phage T7-infected *E. coli*. (*a*) Early (class I) transcription depends on the host RNA polymerase holoenzyme, including the host σ factor (blue); one of the early phage proteins is the T7 RNA polymerase (green). (*b*) Late (class II and III) transcription depends on the T7 RNA polymerase.

But there are other bacteria that sense the onset of hard times and protect themselves by **sporulating.** *B. subtilis*, for example, forms **endospores**—tough, dormant bodies that can survive until favorable conditions return (figure 9.17).

Gene expression must change during sporulation; cells as different in morphology as vegetative and sporulating cells must contain at least some different gene products. In fact, when *B. subtilis* cells sporulate, they activate a whole new set of sporulation-specific genes. The switch from the vegetative to the sporulating state is accomplished by a complex σ-switching scheme that turns off transcription of some vegetative genes and turns on sporulation-specific transcription.

Sporulation is a fundamental change, involving large numbers of old genes turning off and new genes turning on. Furthermore, the change in transcription is not absolute; some genes that are active in the vegetative state remain active during sporulation. How is the transcription-switching mechanism able to cope with these complexities?

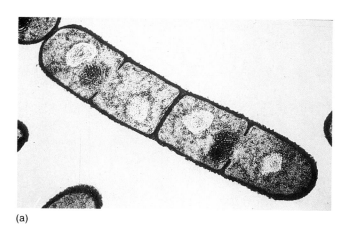

(a)

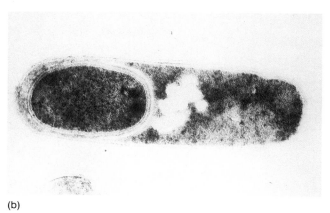

(b)

FIGURE 9.17

(a) *B. subtilis* vegetative cells and (b) sporulating cells.
Dr. Kenneth Bott.

As you might anticipate, more than one new σ factor is involved in sporulation. In fact, there are at least three: σ^{29}, σ^{30}, and σ^{32} (now called σ^E, σ^H, and σ^C, respectively), in addition to the vegetative σ^{43} (now called σ^A). These polypeptides have molecular weights indicated in their names. Each recognizes a different class of promoter, as shown in table 9.3. For example, the vegetative σ^{43} recognizes promoters that are very similar to the promoters recognized by the *E. coli* σ factor, with a −10 (Pribnow) box that looks something like TATAAT and a −35 box having the consensus sequence TTGACA. By contrast, the sporulation-specific factors σ^{32} and σ^{29} have their own recognition sequences.

When the original σ was discovered in *E. coli* in 1969, Burgess and Travers speculated that many different σ's would be found, and that these would direct transcription of different classes of genes. This principle was first verified in *B. subtilis,* as we have seen, but it has also proven true in *E. coli.* For example, a distinct *E. coli* σ (σ^{32}) directs the transcription of a class of genes called *heat-shock genes* that switch on when the bacteria experience high temperature or other environmental insults.

W hen the bacterium *B. subtilis* sporulates, a whole new set of sporulation-specific genes turns on, and many, but not all vegetative genes turn off. This switch takes place largely at the transcription level. It is accomplished by several new σ factors that displace the vegetative σ factor from the core RNA polymerase and direct transcription of sporulation genes instead of vegetative genes. Each σ factor has its own preferred promoter sequence.

■

T ABLE 9.3   σ *Factors and Promoters*

σ Factor	When Used	−35 Box*	−10 Box*
σ^A (σ^{43})	Vegetative	TTGACA	TATAAT
σ^B (σ^{37})	Unknown	AGGNTTT	GGNATTGNT
σ^C (σ^{32})	Sporulation	AAATC	TANTGTTNTA
σ^D (σ^{28})	Late vegetative	CTAAA	CCGATAT
σ^E (σ^{29})	Sporulation	CTNAAA	CATATT
σ^F (SpoIIAC)	Sporulation	(similar to σ^G; see below)	
σ^G (SpoIIIG)	Sporulation	TGaATa	CAtacTA
σ^H (σ^{30})	Sporulation	—	—
σ^K (σ^{27})	Sporulation	—	—
gp^{28}	SPO1 middle phase	AGGAGA	TTTNTTT
gp^{33}–gp^{34}	SPO1 late phase	CGTTAGA	GATATT

*Some of these are not true consensus sequences, because not enough promoters have been examined.

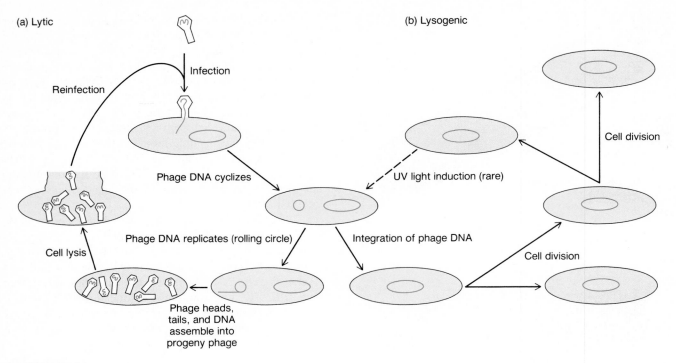

(a) Lytic

Reinfection

Infection

Phage DNA cyclizes

Phage DNA replicates (rolling circle)

Cell lysis

Phage heads, tails, and DNA assemble into progeny phage

(b) Lysogenic

Cell division

UV light induction (rare)

Integration of phage DNA

Cell division

FIGURE 9.18
Lytic versus lysogenic infection by phage λ. (*a*) Lytic infection; (*b*) lysogenic infection.

INFECTION OF *E. COLI* BY PHAGE λ

Many of the phages we have studied so far (T2, T4, T7, and SPO1, for example) are **virulent** phages. When they replicate, they kill their host by **lysing** it, or breaking it open. On the other hand, **lambda (λ)** is a **temperate** phage; when it infects an *E. coli* cell, it does not necessarily kill it. In this respect, λ is more versatile than many phages; it can follow two paths of replication (figure 9.18). The first is the **lytic** mode, in which infection progresses just as it would with a virulent phage. It begins with phage DNA entering the host cell and then serving as the template for transcription by host RNA polymerase. Phage mRNAs are translated to yield phage proteins, the phage DNA replicates, and progeny phages assemble from these DNA and protein components. The infection ends when the host cell lyses to release the progeny phages.

In the **lysogenic** mode, something quite different happens. The phage DNA enters the cell, and its early genes are transcribed and translated, just as in a lytic infection. But then a phage protein (the λ **repressor**) appears and binds to the two phage operator regions, shutting down transcription of all genes except for *cI,* the gene for the λ repressor itself. Under these conditions, with only one phage gene active, it is easy to see why no progeny phages can be produced. Furthermore, when lysogeny is established, the phage DNA integrates into the host genome.

A bacterium harboring this integrated phage DNA is called a **lysogen.** The integrated DNA is called a **prophage.** The lysogenic state can exist indefinitely, and it should not be considered a big disadvantage for the phage, because the phage DNA in the lysogen replicates right along with the host DNA. In this way, the phage genome multiplies without the necessity of making phage particles; thus, it gets a "free ride." Under certain conditions, such as when the lysogen encounters mutagenic chemicals or radiation, lysogeny can be broken and the phage enters the lytic phase.

P hage λ can replicate in either of two ways: lytic or lysogenic. In the lytic mode, almost all of the phage genes are transcribed and translated, and the phage DNA is replicated, leading to production of progeny phages and lysis of the host cells. In the lysogenic mode, the λ DNA is incorporated into the host genome; after that occurs, only one gene is expressed. The product of this gene, the λ repressor, binds to the two early phage operators and prevents transcription of all the rest of the phage genes. However, the incorporated phage DNA (the prophage) still replicates, since it has become part of the host DNA.

■

(a)

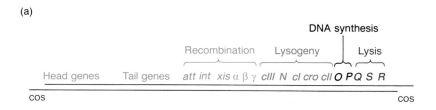

(b)

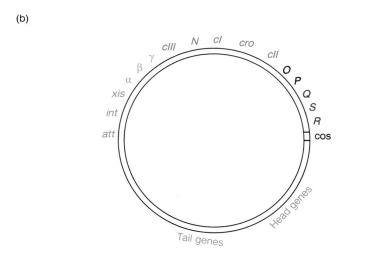

FIGURE 9.19

Genetic map of phage λ. (*a*) The map is shown in linear form, as the DNA exists in the phage particles; the cohesive ends (cos) are at the ends of the map. The genes are grouped primarily according to function. (*b*) The map is shown in circular form, as it exists in the host cell during a lytic infection after annealing of the cohesive ends.

Lytic Replication of Phage λ

The lytic replication cycle of phage λ resembles that of the virulent phages we have studied in that it contains three phases of transcription. These phases are called **immediate early, delayed early,** and **late.** These genes are sequentially arranged on the phage DNA, which helps explain how they are regulated, as we will see. Figure 9.19 shows the λ genetic map in two forms: linear, as the DNA exists in the phage particles, and circular, the shape the DNA assumes shortly after infection begins. The circularization is made possible by 12-base overhangs, or "sticky" ends, at either end of the linear genome. These cohesive ends go by the name **cos.** Note that cyclization brings together all the late genes, which had been separated at the two ends of the linear genome.

As usual, the program of gene expression in this phage is controlled by transcriptional switches, but λ uses a switch we have not seen before: **antitermination.** Figure 9.20 outlines this scheme. Of course, the host RNA polymerase holoenzyme transcribes the immediate early genes first. There are only two of these genes, *cro* and *N,* which lie immediately downstream from the rightward and leftward promoters, P_R and P_L, respectively. At this stage in the lytic cycle, no repressor is bound to the operators that govern these promoters (O_R and O_L, respectively), so transcription proceeds unimpeded. When the polymerase reaches the ends of the immediate early genes, it encounters ρ-dependent terminators and stops short of the delayed early genes.

The products of both immediate early genes are crucial to further expression of the λ program. The *cro* gene product is an **antirepressor** that turns off transcription of the repressor gene, *cI,* and therefore prevents synthesis of repressor protein. This is, of course, necessary if any of the other phage genes are to be expressed. The *N* gene product, **pN,** is an **antiterminator** that permits RNA polymerase to ignore the terminators at the ends of the immediate early genes and continue transcribing right on into the delayed early genes. When this happens, the delayed early phase begins. Note that the same promoters (P_R and P_L) are used for both immediate early and delayed early transcription. The switch does not involve a new σ factor or RNA polymerase that recognizes new promoters and starts new transcripts, as we have seen with other phages; instead, it involves an extension of transcripts controlled by the same promoters.

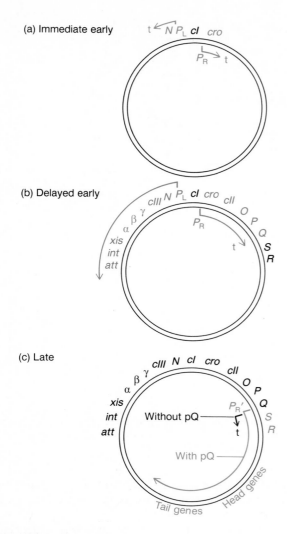

(a) Immediate early

(b) Delayed early

(c) Late

FIGURE 9.20

Temporal control of transcription during lytic infection by phage λ. (*a*) Immediate early transcription (red) starts at the rightward and leftward promoters (P_R and P_L, respectively) that flank the repressor gene (*cI*); transcription stops at the ρ-dependent terminators (t) after the *N* and *cro* genes. (*b*) Delayed early transcription (blue) begins at the same promoters, but bypasses the terminators by virtue of the *N* gene product, pN, which is an antiterminator. (*c*) Late transcription (green) begins at a new promoter ($P_{R'}$); it would stop short at the terminator (t) without the *Q* gene product, pQ, another antiterminator.

The delayed early genes are important not only in continuing the lytic cycle, but also in establishing lysogeny. The delayed early genes that help continue the lytic cycle are all on the rightward side, downstream from *cII*. Genes *O* and *P* code for proteins that are necessary for phage DNA replication, a key part of lytic growth. The *Q* gene product (**pQ**) is another antiterminator, which permits transcription of the late genes.

The late genes are all transcribed in the rightward direction, but not from P_R, as you may be thinking. The late promoter, $P_{R'}$, lies just downstream from *Q*. Transcription from this promoter terminates after only 194 bases, unless pQ intervenes to prevent termination. The *N* gene product will not do; it is specific for antitermination after *cro* and *N*. The late genes code for the proteins that make up the phage head and tail, and for proteins that lyse the host cell so the progeny phages can escape.

*T*he immediate early/delayed early/late transcriptional switching in the lytic cycle of phage λ is controlled by antiterminators. One of the two immediate early genes is *cro*, which codes for an antirepressor that allows the lytic cycle to continue. The other, *N*, codes for an antiterminator, pN, that overrides the terminators after the *N* and *cro* genes. Transcription then continues on into the delayed early genes. One of the delayed early genes, *Q*, codes for another antiterminator (pQ) that permits transcription of the late genes from the late promoter, $P_{R'}$, to continue without premature termination.

■

Establishing Lysogeny

We have mentioned that the delayed early genes are required not only for the lytic cycle, but also for establishing lysogeny. The delayed early genes help establish lysogeny in two ways: (1) Most of the delayed early gene products are needed for integration of the phage DNA into the host genome, a prerequisite for lysogeny. (2) The products of the *cII* and *cIII* genes allow transcription of the *cI* gene and therefore production of the λ repressor, the central component in lysogeny.

There are two promoters for the *cI* gene (figure 9.21): P_{RM} and P_{RE}. The RM in P_{RM} stands for repressor maintenance. This is the promoter that is used *during* lysogeny to ensure a continuing supply of repressor to maintain the lysogenic state. It has the peculiar property of requiring its own product—repressor—for activity. Therefore, this promoter cannot be used to establish lysogeny because at the start of infection, there is no repressor to activate it. Instead, the other promoter, P_{RE}, is used. The RE in P_{RE} stands for repressor establishment. P_{RE} lies to the right of both P_R and *cro*. It directs transcription leftward through *cro* and then through *cI*. Of course, the natural direction of transcription of *cro* is rightward from P_{RE}, so the leftward transcription from P_R gives an anticoding transcript of *cro*. This does not matter, since the purpose of this transcription is to express *cI*, not *cro*.

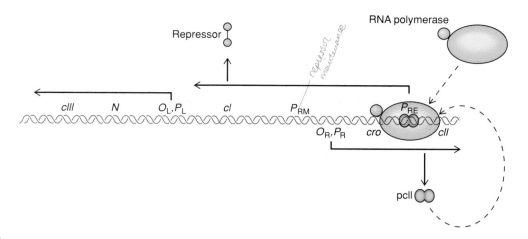

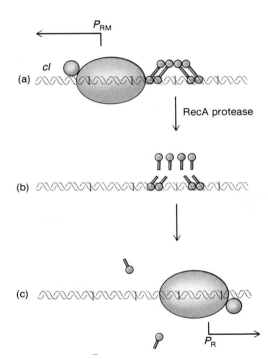

FIGURE 9.21

Establishing lysogeny. Delayed early transcription from P_R gives *cII* mRNA that is translated to pcII (purple). pcII allows RNA polymerase (blue and red) to bind to P_{RE} and transcribe the *cI* gene, yielding repressor (green).

P hage λ establishes lysogeny by causing production of enough repressor to bind to the early operators and prevent further early RNA synthesis. The promoter used for establishment of lysogeny is P_{RE}, which lies to the right of P_R and *cro*. Transcription from this promoter goes leftward through the *cI* gene. Another promoter, P_{RM}, comes into play later, to maintain lysogeny.

■

Lysogen Induction

We mentioned that a lysogen can be induced by treatment with mutagenic chemicals or radiation. The mechanism of this induction is as follows: *E. coli* cells respond to environmental insults, such as mutagens or radiation, by inducing a set of genes whose collective activity is called the **SOS response;** one of these genes, in fact the most important, is ***recA***. The *recA* product participates in recombination repair of DNA damage (which explains part of its usefulness to the SOS response), but environmental insults also induce a new activity, protease, in the RecA protein. One of the targets of this protease, or protein-cleaving activity is none other than the λ repressor. As figure 9.22 shows, RecA cuts the repressor between the two domains and thereby releases it from the operators. As soon as that happens, transcription begins from P_R and P_L. One of the first genes transcribed is *cro*, whose product shuts down any further transcription of the repressor gene. Lysogeny is broken.

FIGURE 9.22

Inducing the λ prophage. (*a*) Lysogeny. Repressor (green) is bound to O_R (and O_L) and *cI* is being actively transcribed. (*b*) The RecA protease (activated by ultraviolet light or other mutagenic influence) cleaves the repressor. (*c*) The repressor falls off the operator, allowing polymerase (red and blue) to bind to P_R and transcribe *cro*. Lysogeny is broken.

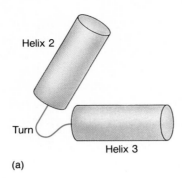

Helix 2

Turn

Helix 3

(a)

(b)

FIGURE 9.23
The helix-turn-helix motif as a DNA-binding element. (*a*) The helix-turn-helix motif of the λ repressor. (*b*) The fit of the helix-turn-helix motif of one repressor monomer with the λ operator. Helix 3 (blue)

lies in the major groove of its DNA target; some of the amino acids on the back of this helix (away from the viewer) are available to make contacts with the DNA.

Surely λ would not have evolved with a repressor that is sensitive to cleavage by RecA unless there were an advantage to the phage. That advantage seems to be this: The SOS response signals that the lysogen is under some kind of DNA-damaging attack. It is expedient under those circumstances for the prophage to get out by inducing the lytic cycle, rather like rats deserting a sinking ship.

hen a lysogen suffers DNA damage, it induces the SOS response. The initial event in this response is the appearance of protease activity in the RecA protein. This protease cuts λ repressors in half, removing them from the λ operators and inducing the lytic cycle. In this way, λ phages can escape the potentially lethal damage that is occurring in their host.

■

SPECIFIC DNA-PROTEIN INTERACTIONS

We have seen several examples of proteins that bind to specific regions of DNA. These include the *trp* and λ repressors and Cro, which bind to their respective operators, and CAP, which binds to the CAP binding site in the *lac* promoter. The binding of these proteins to their targets is very specific. They can locate and bind to one particular short DNA sequence among a vast excess of unrelated DNA.

How do these proteins accomplish such specific binding? All the proteins listed above have a similar structural motif: two alpha helices connected by a short protein "turn." This **helix-turn-helix** motif allows the second helix (the **recognition helix**) to fit snugly into the

wide groove of the target DNA site, as illustrated in figure 9.23. Furthermore, the amino acids of the recognition helix of a DNA-binding protein have chemical groups that fit hand-in-glove with the specific bases protruding into the wide DNA groove at that protein's binding site. Other proteins with helix-turn-helix motifs will not be able to bind as well at that same site, because they do not have the correct amino acids in their recognition helices.

M any DNA-binding proteins in prokaryotes employ a helix-turn-helix binding motif. The second (recognition) helix of each motif fits into the wide groove of the target site on the DNA, and certain key amino acids in this helix make intimate contact with specific bases in the DNA.

■

| Summary |

R NA polymerase is the key player in the transcription process. The *E. coli* enzyme is composed of a core, which contains the basic transcription machinery, and a polypeptide called σ, which directs the core to transcribe specific genes. σ allows initiation of transcription by causing the RNA polymerase holoenzyme to bind tightly to a promoter, melting a local region of DNA.

Termination signals (terminators) recognized by RNA polymerase alone consist of an inverted repeat followed by a second sequence motif—sometimes a string of T's. The transcript of the inverted repeat can form a base-paired structure called a hairpin, which somehow stalls the polymerase. The polymerase then causes the RNA to fall off, terminating transcription. Other terminators require an ancillary factor, ρ, in addition to RNA polymerase.

Bacterial genes that are linked in function are frequently clustered together in operons and are controlled together from a common promoter. For example, the genes for the three enzymes involved in lactose metabolism in *E. coli* are clustered together in the *lac* operon. This operon is controlled both positively (by cAMP plus catabolite activator protein) and negatively (by the *lac* repressor).

Gene expression is controlled primarily at the transcription level in prokaryotes. The temporal control of transcription in *B. subtilis* cells infected by phage SPO1 is directed by a set of phage-encoded σ factors that associate with the host core RNA polymerase and change its specificity from early to middle to late. Phage T7, instead of coding for a new σ factor to change the host polymerase's specificity from early to late, encodes a whole new RNA polymerase with absolute specificity for the later phage genes. Transcriptional switching in the lytic cycle of phage λ is controlled by antiterminators that override ρ-dependent terminators and allow transcription to continue into the next class of genes.

Phage λ can replicate in either of two ways: lytic or lysogenic. In the lytic mode, almost all of the phage genes are transcribed and translated, and the phage DNA is replicated, leading to production of progeny phage and lysis of the host cells. In the lysogenic mode, the λ DNA is incorporated into the host genome; after that occurs, only the λ repressor gene (*cI*) is expressed. The product of this gene, the λ repressor, binds to the two early phage operators and prevents transcription of all the rest of the phage genes. However, the incorporated phage DNA (the prophage) still replicates, since it has become part of the host DNA.

When a lysogen suffers DNA damage, it induces the SOS response. The initial event in this response is the appearance of protease activity in the RecA protein. This protease cuts λ repressors in half, removing them from the λ operators and inducing the lytic cycle. In this way, λ phages can escape the damage that is occurring in their host.

Several prokaryotic DNA-binding proteins, such as the λ repressor, the *trp* repressor, and Cro, bind to their DNA targets via helix-turn-helix motifs that insert specifically into the DNA wide groove. Amino acids in the second (recognition) helix in these motifs make specific contact with bases in the wide groove. ∎

1. Contrast the activities of RNA polymerase holoenzyme and core polymerase on un-nicked phage DNA in vitro.

2. Consider a -10 box with the following sequence: CATAGT. (*a*) Would a C → T mutation in the first position be likely to be an up or a down mutation? (*b*) Would a T → A mutation in the last position be likely to be an up or a down mutation?

3. What structural characteristic of an operon ensures that the genes contained in the operon will be coordinately controlled?

4. Consider an *E. coli* cell with the following mutations. What effect would each mutation have on the function of the *lac* operon (assuming no glucose is present)?
   (*a*) A mutant *lac* operator that cannot bind repressor.
   (*b*) A mutant *lac* repressor that cannot bind to the *lac* operator.
   (*c*) A mutant *lac* repressor that cannot bind to allolactose.
   (*d*) A mutant *lac* promoter that cannot bind CAP plus cAMP.

5. Which phage genes would be transcribed in *B. subtilis* cells infected with SPO1 phages with mutations in the following genes? Why?
   (*a*) Gene 28.
   (*b*) Gene 33.
   (*c*) Gene 34.

6. What would be the effect of a mutation in gene 1 of phage T7? Why?

7. What would be the effect of a mutation in the *B. subtilis* gene for σ^{29} or σ^{32}? Why?

8. Which phage gene would *not* be expressed after the λ lytic cycle is underway?

9. Which phage genes would *not* be expressed in a λ lysogen?

10. pN is an immediate early product that is required for delayed early transcription, yet it is very unstable. Why does this not pose a problem for continued delayed early expression in the lytic cycle?

11. RNA polymerase transcribing from P_{RE} traverses *cro*. Can this result in *cro* expression? Why, or why not?

12. Would you classify the antiterminator activity of pN as positive or negative control? Why?

Answers appear at end of book.

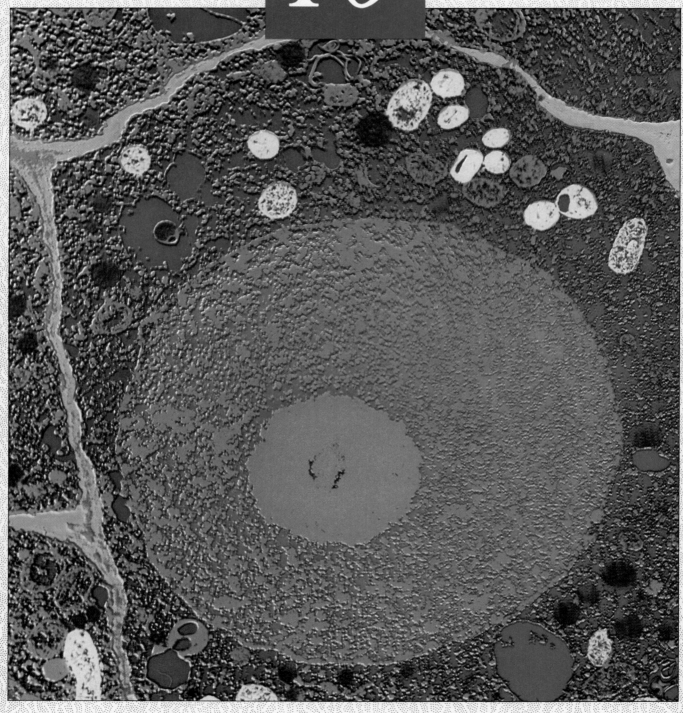

Eukaryotic Gene Structure and Expression

N ature has a short menu, but the items on it are quite reliable.

.

Dagmar Ringe

Learning Objectives

In this chapter you will learn:

1. How eukaryotic DNA is complexed with protein and condensed into chromosomes.

2. The roles of the three RNA polymerases found in eukaryotic cells.

3. The structures of the promoters recognized by the three eukaryotic RNA polymerases.

4. How many eukaryotic genes are interrupted.

5. The extra steps in eukaryotic gene expression that take place between transcription and translation.

False-color transmission electron micrograph of a cell in the root tip of a corn plant.
© *Dr. Jeremy Burgess/Science Photo Library/Photo Researchers, Inc.*

T he picture of prokaryotic gene structure and activity presented in chapter 9 seems rather complex. Nevertheless, compared to the situation in eukaryotes, it is simple. One obvious example of this difference between eukaryotes and prokaryotes is the structure of their chromosomes. The prokaryotic chromosome is a circle of DNA encumbered with relatively little protein, while the eukaryotic chromosome contains a large amount of protein in addition to the DNA. In this chapter we will examine the structure of eukaryotic chromosomes and the genes they contain. We will also learn that expression of these genes involves not only transcription and translation, but a variety of other events that are unique to eukaryotes.

CHROMATIN STRUCTURE

Have you ever dealt with a length of kite string, or fishline, or thread, or even a garden hose? If so, you know that long, thin things tend to get hopelessly tangled. One look at figure 10.1 should convince you that the DNA in a chromosome is longer in relationship to its width than any kite string. In fact, you would have to fly a kite from sea level to an altitude far above Mt. Everest to make the length/width ratio of the string approach that of the DNA in an average human chromosome (more than ten million to one). How do those immensely long molecules stay untangled? As we will see, they are complexed with protein (forming a mixture called **chromatin**) and coiled up in a highly organized way.

When chromatin is lightly treated with a DNase called *micrococcal nuclease*, DNA-protein particles called **nucleosomes** are released (figure 10.2a). These nucleosomes,

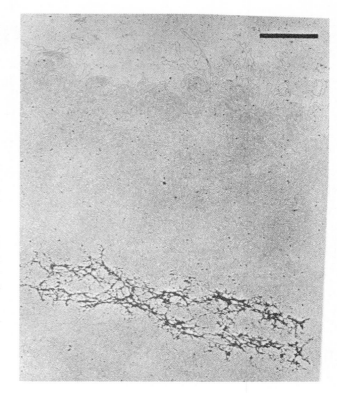

FIGURE 10.1

DNA in a human chromosome. A metaphase chromosome from HeLa (human) cells was extracted to remove most of its protein. This allowed the DNA to uncoil, leaving a scaffolding in the shape of the chromosome behind. Bear in mind that this is a single molecule of DNA. The bar represents one micron.
Courtesy of James R. Paulson. (From James R. Paulson and U. K. Laemmli, Cell 12:820, 1977.) Reprinted with permission from Cell Press.

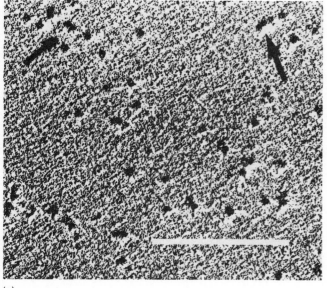

(a)

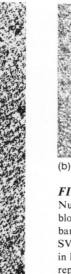

(b)

FIGURE 10.2

Nucleosomes. (*a*) Nucleosomes (arrow) released from chicken red blood cells by treatment of chromatin with micrococcal nuclease. The bar represents 250 nm. (*b*) Nucleosomes in the minichromosome of SV40 virus. The minichromosome has been relaxed by suspending it in low-salt buffer, which reveals the beaded structure. The bar represents 100 nm.
Photo (a) from P. Oudet, M. Gross-Bellard, and P. Chambon, Cell 4:287, 1975. Reprinted with permission from Cell Press. Photo (b) © Dr. Jack Griffith.

ubiquitous in the eukaryotic kingdom, are remarkably uniform in size, about 11 nanometers (nm) in diameter, and they contain about 200 base pairs of DNA. Further trimming with DNase reduces the amount of DNA to 160 base pairs; the resulting particle is called a **chromato-some.** More stringent treatment with DNase removes all loose DNA, leaving a **core particle** with just 146 base pairs. In practice, we usually use the generic term "nucleosome" to refer to all these particles.

The abundance and uniformity of nucleosomes suggests that they are a repeating unit of chromatin structure, and there is direct evidence for this supposition. Consider, for example, the *minichromosome* of the mammalian tumor virus SV40. When this virus infects a host cell, its circular DNA attracts host proteins to form a miniature chromosome. If we lower the salt concentration of a solution of such minichromosomes, we reveal a beaded structure (figure 10.2*b*). The beads are nucleosomes.

NUCLEOSOMES: THE FIRST ORDER OF CHROMATIN FOLDING

Nucleosomes contain, in addition to the 200 base pairs of DNA, a set of basic proteins called **histones.** Most eukaryotic cells contain five different kinds of histones called **H1, H2A, H2B, H3,** and **H4.** Figure 10.3 shows how these five species can be separated by gel electrophoresis. The "bead" of a nucleosome is actually a ball of histones with the DNA wound twice around the outside. Each ball contains exactly eight histone molecules, a pair each of H2A, H2B, H3, and H4. A single molecule of histone H1 binds to

DNA outside the ball. Nucleosomes and chromatosomes contain H1, but when the last few base pairs are removed to generate core particles, H1 is finally lost. Figure 10.4 shows the probable arrangement of histones and DNA in the nucleosome core particle. It appears that the core histones are really not arranged in pairs. H3 and H4 form a tetramer at the center of the particle, and this tetramer is flanked by H2A–H2B dimers at top left and lower right.

What is the function of nucleosomes? Whatever it is, two properties of these particles suggest that their function is fundamentally important. The first is the universality of nucleosomes in eukaryotes; the second is the extreme evolutionary conservation of most of the histones. The classic example of this conservation is that histone H4 from the cow differs from H4 from the pea plant by only two amino acids out of 102, and these are *conservative substitutions* that should not change the protein structure significantly. In other words, in all the eons that have intervened since the cow and pea lines diverged from a common ancestor, only two amino acids in histone H4 have changed. Obviously, the histones, and the nucleosomes of which they are a part, play a vital role, or their structures would have been allowed more evolutionary latitude.

The most likely role for nucleosomes is, of course, a structural one; nucleosomes provide the first order of condensing, or coiling, of the extremely long, thin chromosomal fiber. Figure 10.5 makes this point very clearly. SV40 DNA that has been stripped of its protein is shown next to an SV40 minichromosome at the same magnification. The apparent decrease in length of the DNA in

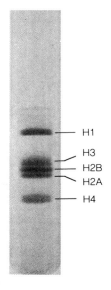

FIGURE 10.3
Separation of histones by electrophoresis. The histones of calf thymus were separated by polyacrylamide gel electrophoresis.
From S. Panyim and R. Chalkley, Archives of Biochemistry and Biophysics *130:343, fig. 6(A), 1969.*

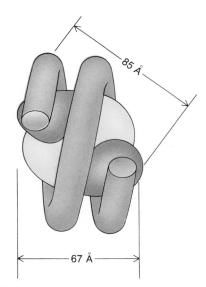

FIGURE 10.4
One interpretation of nucleosome structure, based on X-ray and neutron diffraction studies. The DNA wraps around the core of histones, composed of the $H3_2$–$H4_2$ tetramer in the center (yellow) and an H2A–H2B dimer at each end (purple).

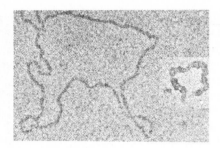

FIGURE 10.5
Condensation of DNA in nucleosomes. Deproteinized SV40 DNA is shown next to an SV40 minichromosome (inset). The condensation of DNA afforded by nucleosome formation is apparent.

Jack Griffith, "Chromatin Structure: Deduced from a Minichromosome." Science, *March 28, 1975, Vol. 187, p. 1202, figure 1. Copyright 1975 by AAAS.*

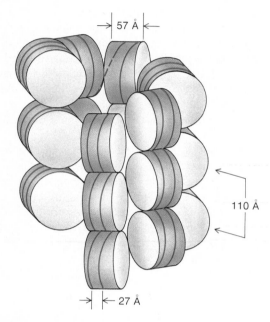

57 Å
110 Å
27 Å

FIGURE 10.6
The solenoid model of chromatin folding. A string of nucleosomes coils into a hollow spiral to form the solenoid. Each nucleosome is represented by a blue cylinder with a wirelike DNA (red) coiled around it.

From Widom and Klug, in Cell *43:210, 1985. Copyright © 1985 Cell Press, Cambridge, MA. Reprinted by permission of the publisher and author.*

the minichromosome is due to the formation of nucleosomes. In fact, the DNA achieves about a 6-fold condensation by coiling up into nucleosomes.

---

*E*ukaryotic DNA combines with basic protein molecules called histones to form structures known as nucleosomes. These structures contain four pairs of histones (H2A, H2B, H3, and H4) in a ball, around which is wrapped a stretch of about 200 base pairs of DNA. Histone H1, bound to DNA outside the ball, completes the nucleosome structure. The first order of chromatin folding is represented by a string of nucleosomes.

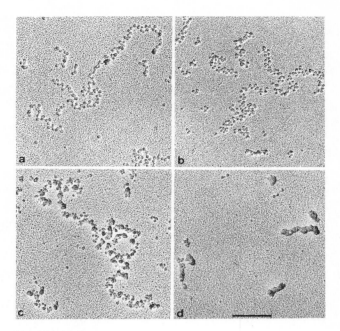

FIGURE 10.7
Formation of the solenoid. (*a*) A string of nucleosomes, extended due to very low ionic strength environment. (*b*) As the salt concentration is raised, the string begins to coil. (*c*) Still higher salt concentration, with more coiling. (*d*) At physiological salt concentration, the strings of nucleosomes have coiled all the way into tightly packed solenoids. The bar represents 200 nm.

F. Thoma, T. Koller, A. Klug. Reproduced from Journal of Cell Biology *83:403–27, Rockefeller University Press, 1979. By copyright permission of the Rockefeller University Press.*

FURTHER FOLDING OF THE NUCLEOSOME STRING

Chromatin in interphase nuclei takes on various thicknesses. One of these, 11 nm in thickness, simply represents a string of nucleosomes, or *nucleosome fiber*. The next thickness, about 25 nm, is due to further winding of the nucleosome string to form a hollow coil called a **solenoid** (figure 10.6). Aaron Klug, who first described the solenoid, showed how it forms by making electron micrographs of chromatin in solutions of increasing salt concentration (figure 10.7). At very low salt concentration, the chromatin appears as a string of nucleosomes, much like the SV40 minichromosome in figure 10.2. As the salt concentration rises, coiling takes place, until the typical solenoid structures appear. The DNA achieves another 6–7-fold condensation in coiling up into the solenoid. We know that histone H1 participates in this coiling, because coiling cannot occur in salt-treated chromatin that lacks H1. Furthermore, it seems that H1-H1 interactions are important in coiling.

Further observations of chromatin suggest at least one more possible order of folding in the chromosome. First of all, it has been observed that eukaryotic chromosomes undergo supercoiling. We learned in chapter 8 that supercoiling can occur in bacterial and certain viral DNAs

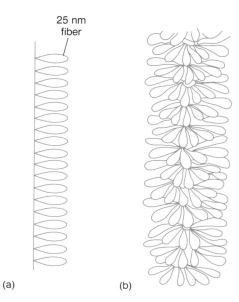

25 nm
fiber

(a)   (b)

FIGURE 10.8
Radial loop models of chromatin folding. (*a*) This is only a partial model, showing some of the loops of chromatin attached to a central scaffold; of course, all the loops are part of the same continuous DNA molecule. (*b*) A more complete model, showing how the loops are arranged in three dimensions around the central scaffold.

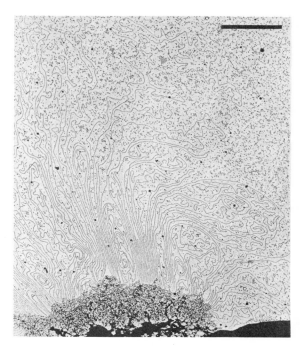

FIGURE 10.9
Deproteinized chromosome showing looped structure. The scaffold is at the bottom, with loops of DNA extending upward. The bar represents 2 microns.
Courtesy of James R. Paulson. (From James R. Paulson and U. K. Laemmli, Cell 12:823, 1977.) Reprinted with permission from Cell Press.

because the circular structure of these molecules prevents the release of strain. But eukaryotic chromosomes are linear, so how can the DNA in them be supercoiled? One way this could happen would be if the chromatin fiber were looped and held fast at the base of each loop (figure 10.8). Each loop would be the functional equivalent of a circle, at least for supercoiling purposes, and the winding of DNA around the nucleosomes would provide the strain necessary for supercoiling.

The direct visualization of a human chromosome in figure 10.1 also supports the loop idea, although the DNA is so tightly bunched in that picture that the looping is difficult to see. Figure 10.9, also depicting a deproteinized chromosome, makes it very clear that loops begin and end at adjacent points on the chromosomal scaffold. Figure 10.10, showing cross sections through several swollen chromosomes, further supports the notion of radial loops. Figure 10.11 summarizes our current thinking about chromatin folding, from the 2-nm bare DNA fiber to the postulated radial looped chromosome.

T he second order of chromatin folding involves a coiling of the string of nucleosomes into a 25-nm-thick fiber called a solenoid. Histone H1 is thought to participate in this folding by interacting with other H1 molecules. The third order of folding is probably looping of the 25-nm fiber into a brushlike structure with the loops anchored to a central matrix.

∎

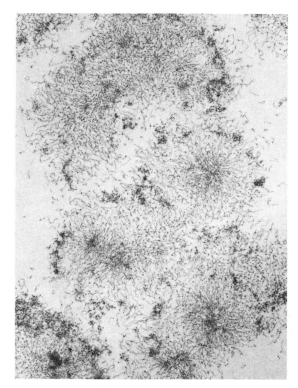

FIGURE 10.10
Swollen chromosomes showing radial loops. The chromatids in the lower portion of the photograph are shown in cross section. The chromatid at the top is cut lengthwise.
M. P. F. Marsden and U. K. Laemmli, Cell 17:851, 1979. Reprinted with permission from Cell Press.

FIGURE 10.11
Orders of chromatin folding.
Structures (*a*), (*b*), and (*c*) rest on
solid experimental ground; (*d*) and
(*e*), also shown in figure 10.8, are
more speculative.

Reproduced, with permission, from the
Annual Review of Genetics, *vol. 12.*
© *1978 by Annual Reviews, Inc.*

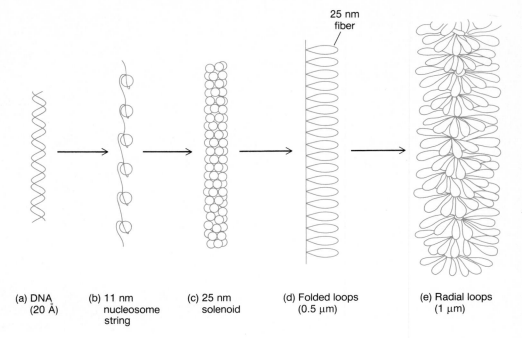

25 nm
fiber

(a) DNA
(20 Å)

(b) 11 nm
nucleosome
string

(c) 25 nm
solenoid

(d) Folded loops
(0.5 μm)

(e) Radial loops
(1 μm)

NONHISTONE PROTEINS

Of course, histones are not the only proteins involved in chromatin. Another, more complex class of proteins also associates with eukaryotic chromosomes. Since little is known about many of these proteins, they are usually called by a name that tells us not what they are, but simply what they are not: **nonhistone proteins.** RNA polymerase is an example of a nonhistone protein about which we know quite a bit. There are many other nonhistones, some that undoubtedly play structural roles in the chromosome and others that help control gene expression. We will focus on some possible regulatory molecules later in the chapter.

RNA POLYMERASES AND THEIR ROLES

In chapter 9 we learned that prokaryotes have just one RNA polymerase (not counting primase), which makes all the different types of RNA the cell needs. True, the σ factor can change to meet the demands of a changing environment, but the core enzyme remains essentially the same. Furthermore, the normal holoenzyme can make all three major types of RNA: messenger RNA, ribosomal RNA, and transfer RNA. Quite a different situation prevails in the eukaryotes. Robert Roeder and William Rutter showed in 1969 that eukaryotic cells have not one but three different RNA polymerases in their nuclei. Furthermore, these three enzymes have distinct roles in the cell.

Roeder and Rutter separated the three enzymes by **chromatography** on an *ion exchange resin* called DEAE-Sephadex. Chromatography is an old term that originally referred to the pattern one sees after separating colored substances on paper. In ion exchange chromatography of proteins, colors are usually not involved, but we continue to use the term because the principle is still to separate different substances. The DEAE groups on DEAE-Sephadex contain positive charges, so they bind negatively charged substances. The greater the negative charge, the tighter the binding.

Robert Roeder and William Rutter named the three peaks of polymerase activity in order of their emergence from the ion exchange column: **polymerase I, polymerase II,** and **polymerase III** (figure 10.12). The three enzymes turned out to have different properties—different responses to salts and divalent metal ions, for example. More importantly, we now know that the three RNA polymerases have distinct roles; each makes a different kind of RNA.

It is intriguing that there are three different kinds of RNA polymerase and three different kinds of RNA to make: mRNA, rRNA, and tRNA. In the fourteenth century, William of Occam developed a maxim now called "Occam's razor," which states in effect that when you are confronted with several possible explanations for a phenomenon, you should choose the simplest. The simplest explanation for the existence of three kinds of polymerase and three kinds of RNA is that each polymerase makes one of the classes of RNA. In fact, there *is* such a one-to-one correspondence, and in simplified form it goes like this: Polymerase I makes ribosomal RNA (the 18S, 28S, and 5.8S varieties); polymerase II makes messenger RNA; polymerase III makes transfer RNA.

Of course, there are some complicating factors. One of these is that the three kinds of RNA are not originally transcribed in mature form; instead, they are made as

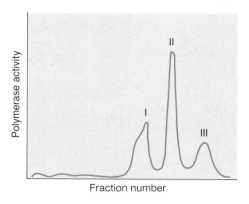

FIGURE 10.12
Separation of three forms of RNA polymerase from eukaryotic cells. Crude RNA polymerase from sea urchin cells was chromatographed on a DEAE-Sephadex column. As buffer of increasing salt concentration flowed through the column, three peaks of RNA polymerase activity emerged: polymerases I, II, and III.

TABLE 10.1	*Roles of Eukaryotic RNA Polymerases*
Polymerase	**RNA Product**
I	rRNA precursor (18S + 28S + 5.8S)
II	mRNA precursor
III	tRNA precursor; 5S rRNA

larger precursors that must later be cut down to size. So it is more correct to say, for example, that polymerase I makes an rRNA *precursor,* not rRNA itself. We have also glossed over the fact that there is a small 5S rRNA, in addition to the varieties made by polymerase I. This 5S rRNA is made in mature form, not as a precursor, by polymerase III. Thus, polymerase III specializes in synthesizing small RNAs. The roles of the three polymerases are summarized in table 10.1.

E ukaryotic cells contain three distinct RNA polymerases, named polymerase I, polymerase II, and polymerase III. The roles of these three enzymes are as follows: Polymerase I makes a large rRNA precursor; polymerase II makes mRNA precursors; and polymerase III makes tRNA precursors, as well as 5S rRNA.

■

PROMOTERS

As we saw in chapter 9, prokaryotic promoters recognized by different σ factors vary quite a bit. For example, the −10 and −35 boxes of σ^{43}-specific promoters in *B. subtilis* are so different from those of σ^{32}-specific promoters that they would not be recognized by an RNA polymerase bearing the latter σ factor. Still, the basic architecture of the prokaryotic polymerases remains the same: core plus σ; and so does the structure of the promoters they recognize: −10 and −35 boxes, lying upstream from the gene(s) they control. No evidence exists for σ factors in eukaryotes, and the subunit structures of the three RNA polymerases in a given cell vary considerably. It should come as no surprise then that these three quite different RNA polymerases recognize quite different promoters.

PROMOTERS RECOGNIZED BY POLYMERASE II

We start with the promoters recognized by polymerase II because they will be most familiar to you; they bear the closest resemblance to prokaryotic promoters. In general, these promoters have more than one element.

The TATA Box

Closest to the transcription start site, approximately at position −25, is an A–T-rich sequence that is an almost constant feature of polymerase II promoters. Its consensus sequence is TATAAAA, although T's frequently replace the A's in the fifth and seventh positions. Its name, **TATA box,** is derived from its first four bases.

You may have noticed the close similarity between the eukaryotic TATA box and the prokaryotic −10 box. In fact, David Pribnow included a TATA box from a eukaryotic virus called SV40 in his original list of consensus sequences, even though it is found farther upstream than its prokaryotic counterparts (−25 compared to −10). As is the case with all consensus sequences, there are exceptions to the rule. Sometimes G's and C's creep in, as in the "TATA" box of the rabbit β-globin gene, which starts with a C, giving CATA. Occasionally there is no recognizable TATA box at all, as in the adenovirus early II gene.

Another conserved region in polymerase II promoters is not as prevalent as the TATA box. It is usually found about fifty base pairs farther upstream, approximately at position −75, and it has the consensus sequence GGCCAATCT. This sequence commonly goes by the name CAAT box or, more often, **CCAAT box** (pronounced "cat box").

It is one thing to find conserved sequences upstream from the transcription start site; it is another to demonstrate that they actually function as parts of the promoter. The latter task has been accomplished for both the TATA box and the CCAAT box, and cloned genes have figured prominently in these studies. The experimental plan has been to clone (reproduce in large quantity) a particular gene transcribed by polymerase II, introduce mutations into its promoter, and then observe the effects of these mutations on transcription of the gene. These mutations differ from those that occur naturally at random in that we can choose just which bases we want to change and how. Thus, we call this process *site-directed mutagenesis.* We will discuss this topic in greater detail in chapter 15.

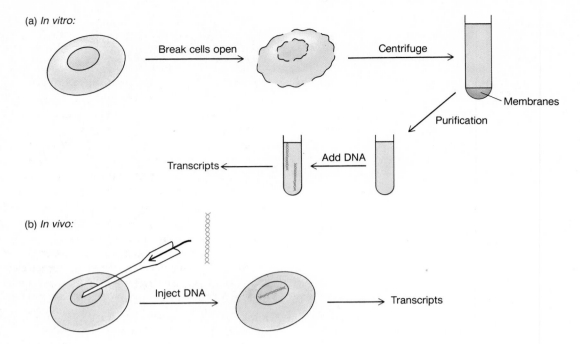

FIGURE 10.13
In vitro and in vivo transcription systems. (*a*) In vitro systems use components from broken cells; (*b*) in vivo systems use whole cells. In both cases, foreign DNA (red) is introduced into the system, and its transcription is measured.

To test for the effect of a mutation in the TATA box, we have to be able to add our cloned genes (mutant or wild-type) to some kind of transcription system so that we can observe the difference in the way these genes are transcribed. The two basic types of transcription systems we use, in vitro and in vivo, are shown in figure 10.13. To make an in vitro system, we start by grinding up the cells, then slightly purifying their contents to yield a cell-free extract. This extract usually contains RNA polymerases and all the other factors needed to transcribe the DNAs that we add, as long as these DNAs contain promoters the extract can recognize. In vivo systems come in several types, but all are living cells to which we can add DNAs. We then check by one means or another to see if our added DNA is transcribed. (Both terms are derived from Latin: *vitro* = glass; *vivo* = life.)

When such tests were applied to genes with mutations in their TATA boxes, something surprising happened. Instead of eliminating transcription, as one might have predicted, TATA box mutations in these genes caused an increase in the number of different transcription start sites. Figure 10.14 shows the results Pierre Chambon and his colleagues got when they tested TATA box mutants of the SV40 early promoter using an in vitro transcription system. Whereas the wild-type DNA gave transcription from three tightly clustered start sites, the TATA-less mutant showed starts at more than twenty different sites.

On the other hand, mutating the TATA box in certain other genes (the rabbit β-globin gene, for example) did

not change the transcription initiation site. Instead, it caused a profound decrease in efficiency of transcription. Thus, the TATA box seems to play different roles in different promoters.

Upstream Promoter Elements

Similar studies have been performed in which sequences upstream from the TATA boxes in a variety of genes have been deleted or changed. The CCAAT box, for example, seems to influence promoter activity. Thus, removing the CCAAT box from the rabbit β-globin promoter destroys most of that promoter's activity. The CCAAT box plays a similar role in the promoter for the thymidine kinase gene of herpes simplex virus (HSV). The SV40 early promoter also contains an important upstream element; instead of a CCAAT box, it is a 21-base-pair sequence that is repeated three times between positions −40 and −103 just upstream from the TATA box. Each of these 21-bp repeats contains two copies of a **GC box** with the sequence GGGCGG on the coding strand. The HSV thymidine kinase promoter also contains GC boxes: one on each side of the CCAAT box. HSVs, by the way, cause cold sores and genital herpes infections in humans.

As a general rule then, promoters recognized by RNA polymerase II contain a TATA box near the transcription start site and at least one other important element farther upstream. We will see later in this chapter that these upstream elements serve as binding sites for transcription factors that activate the attached genes.

Chapter 10

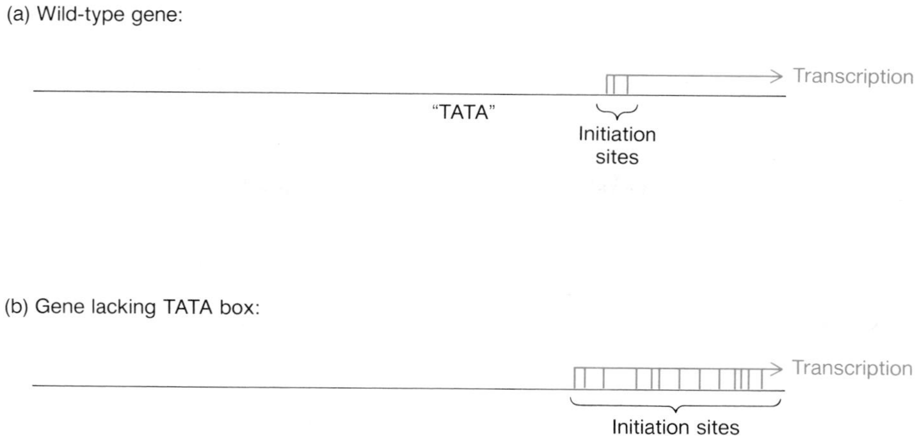

(a) Wild-type gene:

"TATA"

→ Transcription

Initiation
sites

(b) Gene lacking TATA box:

→ Transcription

Initiation sites

FIGURE 10.14
Effect of TATA box on transcription initiation. (*a*) In the wild-type
SV40 early gene that contains a TATA box, transcription starts from
three tightly clustered sites. (*b*) In the mutant that lacks a TATA box,
we observe a multitude of transcription initiation sites.

(a) SV40 early promoter:

| GC | GC | GC | GC | GC | GC | TATA |

21 bp   21 bp   21 bp

(b) Herpes simplex virus thymidine kinase promoter:

| GC | CAAT | GC | TATA |

(c) Typical polymerase II promoter + enhancer:

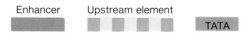

Enhancer   Upstream element

TATA

FIGURE 10.15
Structures of promoters recognized by RNA polymerase II. (*a*) The
SV40 early promoter contains a TATA box (red) and an upstream
element composed of three 21-bp repeats. Each of these repeats
contains two 6-base GC boxes (green). (*b*) The HSV thymidine kinase
promoter contains a TATA box (red) plus two upstream elements. The
one nearest the TATA box is a GC box (green); the other contains a
CCAAT box (yellow) as well as a GC box. (*c*) A typical polymerase
II promoter has a TATA box and at least one upstream element. It is
also usually associated with an enhancer (blue), which generally lies
upstream from the promoter.

Many eukaryotic genes contain sequences farther upstream in the flanking regions (or even within the genes)
that are crucial to active transcription. But these are not
strictly part of the promoters. We call them enhancers,
and we will discuss them in detail later in the chapter.
Figure 10.15*a* and *b* shows the arrangement of the promoter elements for the SV40 early and HSV thymidine

kinase genes. Figure 10.15*c* presents a generalized view
of the control region of a typical gene recognized by RNA
polymerase II, including both promoter and enhancer. This
is the arrangement of these elements in the SV40 early
region—a favorite example of a polymerase II promoter-
enhancer complex.

Mouse: ————————ATCTTT————————————TATTG——————————— TACTGACACGC

Rat: ————————ATCTTT————————————TATTG——————————— TACTGACACGC

Human: — ATCTTTT ——————————————————TTTTGG——————————— TGCTGACACGC

FIGURE 10.16

Some mammalian polymerase I promoters.

Eukaryotic promoters that are recognized by RNA polymerase II have at least one feature in common: a consensus sequence found about twenty-five base pairs upstream from the transcription start site and containing the motif TATA. This TATA box acts either to position the start of transcription about twenty-five base pairs downstream or to increase transcription efficiency. Polymerase II promoters also contain important sequences upstream from the TATA box. Examples are the CCAAT box and the GC box.

PROMOTERS RECOGNIZED
BY POLYMERASE I

Several groups of investigators have searched for conserved sequences and have applied site-directed mutagenesis techniques to elucidate the nature of the promoter recognized by RNA polymerase I. Promoter is written here in the singular because each species should have only one kind of gene recognized by polymerase I: the rRNA precursor gene. It is true that this gene is present in hundreds of copies in each cell, but each copy is virtually the same as the others, and they all have the same promoter sequence. Figure 10.16 shows the regions of DNA thought to be important in polymerase I promoter function in three mammals. These sequences resemble those of polymerase II promoters only in that they appear in the 5'-flanking region of the gene (that is, upstream from the start of transcription); their base sequences do not remind us at all of TATA, GC, or CCAAT boxes. Notice that the end of the gene where transcription begins is called the 5'-end. This refers to the 5'-end of the transcript, or of the coding DNA strand.

Promoters recognized by polymerase I, like polymerase II promoters, are found in the 5'-flanking region of the gene (the rRNA precursor gene). However, there is no resemblance in sequence between the conserved sequences of polymerase I and polymerase II promoters.

PROMOTERS RECOGNIZED
BY POLYMERASE III

Polymerase III ordinarily transcribes genes that are much shorter than those transcribed by the other two polymerases. Should this make a big difference in the nature of the polymerase III promoters? You might not expect so, but the polymerase III promoters are radically different from those recognized by polymerases I and II and by bacterial polymerases. Instead of being located exclusively in the 5'-flanking region of the gene, as all these other promoters are, polymerase III promoters have elements located *within* the genes they control.

Donald Brown and his coworkers performed the first experiments leading to this astonishing conclusion using a cloned 5S rRNA gene from *Xenopus*. The experimental plan was to mutate the 5S rRNA gene and measure the effects of these mutations on transcription of the gene in vitro. Correct transcription was scored by measuring the size of the transcripts by gel electrophoresis. An RNA of approximately 120 bases (the size of 5S RNA) was deemed an accurate transcript, even if it did not have the same sequence as real 5S rRNA. We have to allow for incorrect sequence in the transcript for the following reason: By changing the internal sequence of the gene in order to disrupt the promoter, we automatically change the sequence of the transcript.

The surprising result was that the entire 5'-flanking region of the gene could be removed without affecting transcription very much. Furthermore, big chunks of the gene could be removed, or extra DNA could be inserted into the gene, and a transcript of about 120 bases would still be made. However, there was one sensitive spot in the gene where one could not remove or insert DNA and still retain promoter function. This was a region between bases 50 and 83 of the transcribed sequence. This is the location of the **internal control region (ICR)** of the *Xenopus* 5S gene. In addition, a DNA sequence upstream from the transcription start site helps initiate transcription accurately.

Figure 10.17 summarizes the results of these experiments and similar ones performed on tRNA genes. The promoters for tRNA genes contain an additional surprise: The ICR is divided into two parts; the space in between can be altered without destroying promoter function. There

Gene

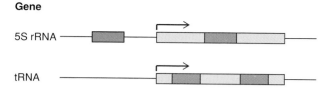

5S rRNA

tRNA

FIGURE 10.17
Polymerase III promoters. The long blocks represent the genes; the blue areas within the blocks are the internal control regions (ICRs). The red block upstream from the 5S rRNA transcription start site (arrow) is necessary for correct transcription initiation.

are limits on such alteration; if one inserts too much DNA between the two ICR halves (called the *A block* and the *B block*), efficiency of transcription suffers.

> R NA polymerase III transcribes a set of short genes, including those for 5S rRNA and tRNA. The promoters for these genes contain internal control regions (ICR's) that lie wholly within the genes. The 5S RNA gene has a one-part ICR lying between bases 50 and 83 of the transcribed region. The tRNA genes have two-part, divided ICR's.

■

REGULATION OF TRANSCRIPTION

In our study of transcription in prokaryotes, we discovered an elegant means of controlling several genes together: the operon. As long as two or more genes belong to the same operon and are transcribed together to produce a polycistronic (multigene) mRNA, they can all be turned off and on together under the control of a common promoter and operator.

Like prokaryotes, eukaryotes control their genes primarily by controlling transcription. However, unlike prokaryotes, eukaryotes have no operons and no polycistronic mRNAs. Does this mean that coordinate gene control is not necessary in higher organisms? No, eukaryotes have just as much need as do lower organisms to control genes together. The blood protein hemoglobin again provides a good example. This protein is composed of two kinds of polypeptides: two molecules of α-globin and two of β-globin. Clearly, since these two molecules are needed in equal quantities, it would be wasteful to make a lot more of one than of the other. And, indeed, they are produced in approximately equal amounts. But the genes for these two polypeptides are not linked. In fact, they are not even on the same chromosome. This situation is a far cry from an operon system, and it challenges us to explain how the expression of these two genes can be coordinated. In the remainder of this chapter, we will consider the elements of eukaryotic gene structure that allow for control; in chapter 14, we will examine in greater detail the ways genes are switched on and off in developing organisms.

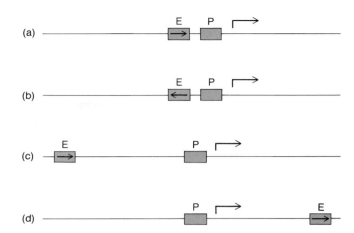

FIGURE 10.18
Enhancers are orientation- and position-independent. (*a*) The wild-type relationship between a typical enhancer (blue) and its associated promoter (red). (*b*) The enhancer has been inverted. (*c*) The enhancer has been moved a few thousand base pairs upstream from the promoter. (*d*) The enhancer has been moved a few thousand base pairs downstream from the promoter. In all four cases, the enhancer functions normally.

ENHANCERS

The early site-directed mutagenesis experiments on polymerase II promoters also pointed to very important areas of the 5'-flanking regions, aside from the TATA and CCAAT boxes we have discussed. The best-studied of these upstream sequences is a **72-base-pair repeat** in the early region of the monkey tumor virus **SV40,** or simian virus number 40. The early region of SV40 codes for a protein called **T antigen** that is needed for viral DNA synthesis and also allows SV40 virus to transform normal cells to malignant ones. In other words, the early region is what makes SV40 a tumor virus. The 72-bp repeat is located just upstream from the three 21-bp repeats in the 5'-flanking region of the SV40 early gene.

The 72-bp repeat is just what its name implies: a side-by-side duplication of a sequence 72 base pairs long. When this element was deleted from the SV40 early region, transcription was greatly depressed. This is exactly the behavior we would expect from part of a promoter, so why did we not include the 72-bp repeat in our discussion of polymerase II promoters? Because this DNA element exhibits two characteristics that are very unpromoterlike: First, as figure 10.18 illustrates, it can be moved to a new location thousands of base pairs away from the SV40 early promoter (either in front of it or behind it) and still enhance early transcription. Second, it can be inverted without diminishing its effect. For this reason, we classify the 72-bp repeat and other elements with the same characteristics as **enhancers** that stimulate transcription from promoters but are not really part of the promoters themselves. The term "promoter" is reserved for elements that have a relatively fixed spatial relationship with the genes they control.

Enhancers have now been found to be associated with a variety of genes—in eukaryotes as well as their viruses and in genes transcribed by all three RNA polymerases. They are usually tissue-specific; that is, a gene with a given enhancer can only function in certain kinds of cells. For example, a mouse immunoglobulin enhancer can function in cells that express the immunoglobulin genes (the cells of the immune system), but it does not function in mouse connective tissue cells. Sometimes enhancers have sequences in common, but we cannot discern real consensus sequences shared by all enhancers.

> *E* nhancers are DNA elements that can greatly stimulate transcription from the eukaryotic promoters with which they associate. They differ from promoter elements in that they work in either orientation and can operate at some distance from the genes they control.

■

TRANSCRIPTION FACTORS

We learned in chapter 9 that prokaryotic RNA polymerase holoenzyme is sufficient for specific transcription in vitro. For example, recall the experiments performed in the late 1960s showing that purified *E. coli* holoenzyme could transcribe the immediate early genes of purified T4 phage DNA but leave the other genes alone. People working on eukaryotic RNA polymerase in that same era were disappointed to discover that their systems did not work that way. Instead, purified eukaryotic RNA polymerases, when presented with purified viral DNA, responded by transcribing the whole genome promiscuously. Indeed, transcription did not even begin at specific sites, but more or less at random. Obviously, something was missing.

Since the usual biochemical trick of separating, purifying, and reconstituting did not work, molecular biologists turned to crude cellular extracts and found that some of these could support specific transcription of added genes. The substances missing from the purified systems were still there in the crude extracts. The first of these eukaryotic in vitro transcription systems was developed for transcription of a 5S rRNA gene by RNA polymerase III. Later, similar systems were found that could support faithful transcription of genes for mRNA—genes recognized by RNA polymerase II.

TFIIIA: An RNA Polymerase III Factor

Once you have such a transcription system, one of the first questions you want to answer is: What factors, missing from polymerase alone, allow this system to transcribe genes accurately? One can answer the question in the traditional biochemical way by putting together RNA polymerase and DNA template along with purified components from the cell-free extract. When the right component is added, it should turn on specific transcription. As you would probably predict by now, these right components turn out to be protein factors.

In the case of the 5S rRNA gene from *Xenopus laevis,* a protein called **TFIIIA** (transcription factor A for RNA polymerase III) was the first factor to be purified. Without the factor, no transcription took place; with it, considerable transcription occurred, and the product electrophoresed as a single band just like 5S rRNA, demonstrating that the transcription was specific. Structural experiments on TFIIIA have shown it contains a repeating domain that is finger-shaped (figure 10.19a). Because each of these nine finger domains binds to a zinc ion, they are known as **zinc fingers** or **metal fingers.** Similar zinc fingers have been found in a variety of DNA-binding proteins; it appears that they assist in DNA binding by inserting into the wide groove of the DNA double helix as shown in figure 10.19b.

A Variety of RNA Polymerase II Transcription Factors

Experiments with polymerase II transcription systems have yielded similar results. For instance, Robert Tjian (pronounced "Tee-jun" by his American colleagues) has purified a protein factor called **Sp1** from human tumor cells (HeLa cells) that greatly stimulates transcription from the SV40 early promoter and from the HSV thymidine kinase promoter. At the same time, this factor actually seems to decrease transcription from the major late promoter of another human virus called **adenovirus.** Thus, this factor has the ability to select certain promoters for transcription while ignoring others.

The key to Sp1's activity is its ability to bind to the GC boxes in the 5'-flanking region of some promoters. As we discussed earlier, the GC boxes in the SV40 early promoter are arranged in the 21-bp repeats between the gene's TATA box and enhancer (see figure 10.15). The spacing between GC boxes is almost exactly two full turns of the DNA double helix, which puts the bound Sp1 molecules all on the same face of the DNA molecule. Sp1 can bind to each of the GC boxes individually, but because some of the sites are a little closer together than others, only five of them are effectively occupied. Like TFIIIA, Sp1 is a finger protein, but it contains three zinc fingers instead of nine.

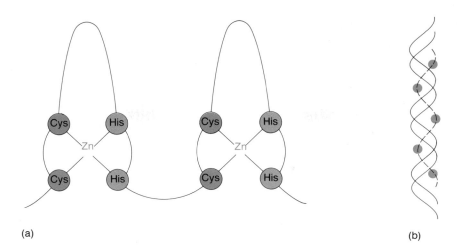

(a)   (b)

FIGURE 10.19

TFIIIA, with metal fingers. (*a*) Postulated structure of two of the nine metal fingers of TFIIIA. The zinc (green) in each finger is bound to four amino acids: two cysteines (blue) and two histidines (red), holding the finger in the proper shape for DNA binding. These four key amino acids are found in the same places in each finger. (*b*) A possible scheme for interaction between the metal fingers of TFIIIA and the 5S internal control region. The protein is shown by the dotted line, with the metal fingers represented in cross section by the green circles.

B · O · X

10.1

HeLa Cells

W hen people see the term "HeLa cells" for the first time, they may think it is a misprint. Shouldn't the "L" be lowercase? No, HeLa is correct; the reason for capitalizing the H and the L is that the term stands for the name of the woman who donated the cells: Henrietta Lacks. In 1952, George Gey was trying to grow human tumor cells in culture at Johns Hopkins University. No one had succeeded in this enterprise before, but Gey managed to get human cervical cancer cells to grow in a serum clot. One of these specimens came from Henrietta Lacks, so Gey called the culture HeLa cells. Ms. Lacks died of her cancer not long afterward, but she has achieved a kind of immortality in that her cells are still growing in laboratories all over the world. They are one of the favorite sources of human cells for molecular genetics investigations. In fact, they grow so easily—almost like weeds—that many researchers have cultured them inadvertently. That is, their cultures of other cells became contaminated with HeLa cells, which then outgrew their neighbors until only HeLa cells remained. This change sometimes went undetected, and several studies, ostensibly on the behavior of other kinds of cells, have actually been investigations of already well-studied HeLa cells.

■

Carl Wu and coworkers have looked at another example of a transcription factor: a protein called **heat-shock activator protein (HAP)** from the fruit fly *Drosophila melanogaster*. Heat-shock genes apparently exist in all species; they turn on in response to heat and other environmental insults. For the most part, their functions are unknown, but they are convenient tools for the investigation of gene control because we can turn them off and on at will simply by adjusting the temperature. HAP can turn on a particular fruit fly heat-shock gene (*hsp*82) in vitro. Yet another example of a transcription factor was discovered by Chambon's group in HeLa cells. (See box 10.1 for a brief history of these popular cells.) This factor has the ability to turn on transcription of a variety of genes in vitro, including the adenovirus late genes and the chicken **conalbumin** gene. Like the ovalbumin gene, the gene for conalbumin switches on in the oviduct in response to estrogen.

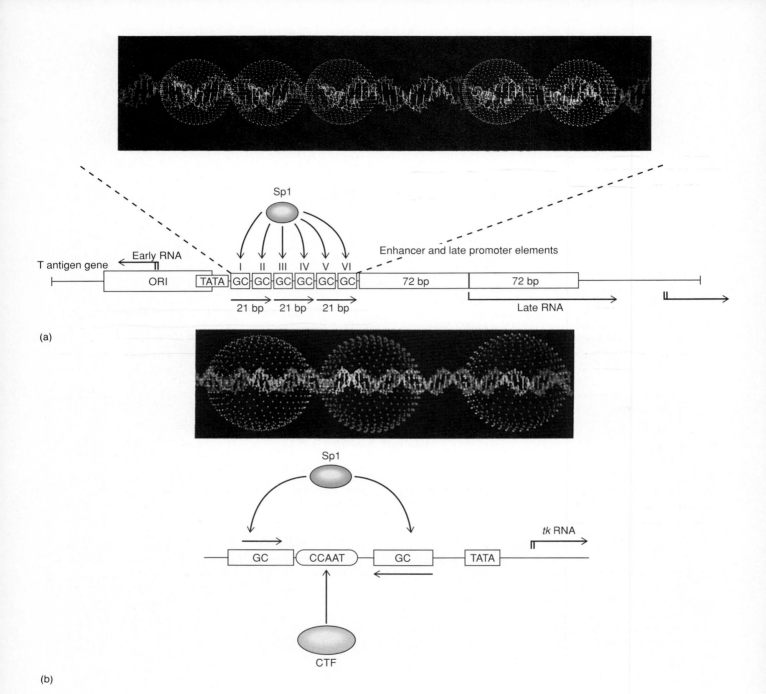

(a)

(b)

FIGURE 10.20

Interactions of transcription factors with two promoters. (*a*) The drawing shows the whole SV40 control region, including both late and early promoters. Note that the early promoter is in the opposite orientation from that in figure 10.15. The drawing also depicts the interaction between the transcription factor Sp1 and the six GC boxes. (Box IV shows little if any interaction under normal conditions.) The color computer graphic above the drawing further illustrates the interaction between Sp1 and five of the GC boxes. Each GC box is rendered in green; bases that are protected by Sp1 are highlighted by small red hemispheres and molecules of Sp1 bound to the GC boxes are represented by large magenta spheres. (*b*) The drawing shows the HSV thymidine kinase promoter elements and their interaction with the transcription factors Sp1 and CTF (CCAAT box transcription factor). The computer graphic shows the GC boxes in green and the CCAAT box in yellow. Sp1 is represented by the two magenta spheres and CTF by the red sphere.

From Steven McKnight and Robert Tjian, Cell 46:796, 799, 1986. Copyright © 1986 Cell Press, Cambridge, MA. Reprinted by permission of the author and publisher.

Chambon's group showed that the HeLa transcription factor binds to the promoter's TATA box. There is also a TATA-binding protein (TAB) required for *Drosophila* heat-shock gene transcription, but it is not the only factor involved. The heat-shock activator protein (HAP) just described is also necessary, and it binds about 50 base pairs upstream from TAB. The CCAAT boxes in the HSV thymidine kinase promoter and in other promoters attract their own transcription factor, called the **CCAAT box transcription factor (CTF)**.

The preceding discussion has perhaps reminded you of our description of σ factors in prokaryotes. In fact, it has been suggested that eukaryotic transcription factors are evolutionary relatives of prokaryotic σ factors. They both share the ability to direct RNA polymerase to transcribe specific genes. Another salient feature of σ factors is that they cause the holoenzyme to bind tightly to promoters. Do eukaryotic transcription factors do this too? We do not have direct evidence for this, although the transcription caused by these factors could not take place without tight polymerase-promoter binding. However, we do know that, unlike σ factors, eukaryotic transcription factors themselves bind to the promoters.

It will probably turn out that most eukaryotic promoters contain binding sites for more than one factor necessary for that gene's transcription. Figure 10.20 presents a summary of the structures of two well-studied promoters and the transcription factors that interact with them. Enhancer-binding proteins have also been found. These specifically turn on genes with certain kinds of enhancers—an arrangement that could be inferred from the tissue-specificity of enhancer function. Presumably, an enhancer only works in a tissue with the right kind of factor.

E ukaryotic genes depend on protein factors, as well as on RNA polymerase, for their transcription. A factor called TFIIIA binds to the internal promoter of the *Xenopus* 5S rRNA gene and helps polymerase III transcribe the gene. Transcription factors for polymerase II genes also bind to their respective promoters. Since these promoters generally contain more than one element, it is likely that they usually associate with multiple transcription factors.

∎

CHROMATIN STRUCTURE AND GENE EXPRESSION

A strong relationship exists between the physical state of the chromatin in the vicinity of a particular gene and the activity of that gene. To begin with, there are two fundamental classes of chromatin structure: **heterochromatin** and **euchromatin.** The latter is the type of chromatin we

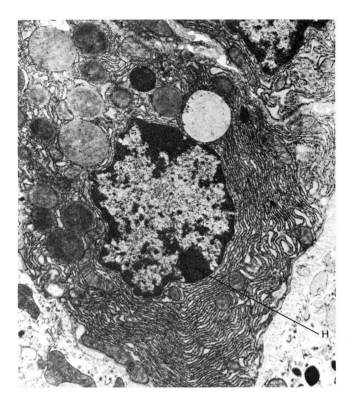

FIGURE 10.21

Interphase nucleus showing heterochromatin. Bat stomach lining cell with nucleus at center. Dark areas around periphery of nucleus are heterochromatin (H).

© *Dr. Keith Porter.*

have been discussing so far—relatively extended and open. By contrast, heterochromatin is very condensed and its DNA is inaccessible. It even appears as clumps when viewed microscopically (figure 10.21). One example of heterochromatin is found at or near the centromeres of chromosomes. The DNA associated with the centromeres belongs to a class of DNA that is repeated many times in the genome, in contrast to structural genes, which are usually found in only one or a few copies each. This *pericentromeric* DNA is complexed with protein and coiled tightly; obviously, DNA in such a state is not available to RNA polymerase and is not transcribed.

This does not imply that the relatively extended euchromatin is automatically transcribed, just because it does not physically exclude RNA polymerase. On the contrary, the majority of the euchromatin in a given cell is inactive. Why is this? That, of course, is an extremely important question; it is the same as asking what controls eukaryotic genes. We have already learned about the importance of transcription factors and enhancer binding proteins. Later in this chapter and in chapter 14, we will examine this question in more detail.

■

THE RELATIONSHIP BETWEEN GENE STRUCTURE AND EXPRESSION

In our discussion of transcription of eukaryotic genes, we have thus far been ignoring a central fact: Many eukaryotic genes are fundamentally different from their prokaryotic counterparts. A prokaryotic structural gene is simple. Once it begins, it continues uninterrupted until it ends. The colinearity of gene and protein is a restatement of this principle. In fact, this seems such an obvious way of doing things that nobody imagined eukaryotes would behave any differently, but they do. Eukaryotic genes are frequently arranged in pieces; in other words, they are interrupted. For example, if we expressed the sequence of the human β-globin gene as a sentence, this is how it might read:

This is bhgty the human β-globin qwtzptlrbn gene.

INTERRUPTED GENES

There are obviously two regions within the β-globin gene that make no sense; they contain sequences totally unrelated to the globin coding sequences surrounding them. These are sometimes called **intervening sequences,** or **IVSs,** but they usually go by the name Walter Gilbert gave them: **introns.** Similarly, the parts of the gene that make sense are sometimes called coding regions, or expressed regions, but Gilbert's name for them is more popular: **exons.** Some genes have no introns at all; others have an abundance. The current record (at least sixty introns and probably more than one hundred) is held by the dystrophin gene, whose failure causes Duchenne muscular dystrophy.

The usefulness of introns remains something of a mystery; fortunately, we can do a lot better at describing them—how we know they are there and what their implications are. Consider the major late locus of adenovirus, a favorite system for studying gene expression and one of the first places introns were found by Phillip Sharp and his colleagues in 1977. Several lines of evidence converged at that time to show that the genes of the adenovirus major late locus are interrupted, but perhaps the easiest to understand comes from studies using a technique called *R-looping.*

In R-looping experiments, we **hybridize** RNA to its DNA template. In other words, we separate the DNA template strands and form a double-stranded hybrid be-

tween one of these strands and the RNA product. Such a hybrid double-stranded polynucleotide is actually a bit more stable than a double-stranded DNA. After we form the hybrid, we examine it by electron microscopy. There are two basic ways to do these experiments: (1) using DNA whose two strands are only separated enough to let the RNA hybridize, or (2) completely separating the two DNA strands before hybridization. Sharp used the latter method, hybridizing single-stranded adenovirus DNA to mature mRNA for one of the viral coat proteins: the hexon protein. Figure 10.22 shows the results.

What would we expect to see if there were no introns in the hexon gene? In that case, there should be a smooth, linear hybrid where the mRNA lined up with its DNA template. But what if there *are* introns in this gene? Clearly, there can be no introns in the mature mRNA, or nonsense would appear in the protein product. Therefore, introns are sequences that occur in the DNA but are missing from mRNA. That means the hexon DNA and hexon mRNA will not be able to form a smooth hybrid. Instead, the intron regions of the DNA will not find partners in the mRNA and so will form unhybridized loops. Of course, that is exactly what happened in the experiment shown in figure 10.22.

The electron micrograph shows an RNA-DNA hybrid interrupted by three single-stranded DNA loops (labeled A, B, and C). These loops represent the introns in the hexon gene. Each loop is preceded by a short hybrid region, and the last loop is followed by a long hybrid region. This means there are four exons in the gene: three short ones near the beginning, followed by one large one. The three short exons are transcribed into **leader** regions that appear at the 5'-end of the hexon mRNA before the coding region; the long exon contains the coding region of the gene.

So far we have only discussed introns in mRNA genes, but tRNA genes also have introns, and even rRNA genes sometimes do. The introns in both these latter types of genes are a bit different from those in mRNA genes. For example, tRNA introns are relatively small, ranging in size from four to about fifty bases long. Not all tRNA genes have introns; those that do have only one, and it is adjacent to the DNA bases corresponding to the anticodon of the tRNA.

■

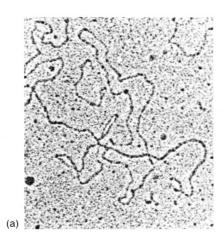

(a)

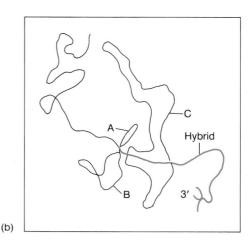

(b)

(c)

FIGURE 10.22

R-looping experiments reveal introns in adenovirus. (*a*) Electron micrograph of a cloned fragment of adenovirus DNA containing the 5′-part of the late hexon gene, hybridized to mature hexon mRNA. The loops represent introns in the gene that cannot hybridize to mRNA. (*b*) Interpretation of the electron micrograph, showing the three intron loops (labeled A, B, and C), the hybrid (heavy red line), and the unhybridized region of DNA upstream from the gene (upper left). The fork at the lower right is due to the 3′-end of the mRNA, which cannot hybridize because the 3′-end of the gene is not included.

Therefore, the mRNA forms intramolecular double-stranded structures that have a forked appearance. (*c*) Linear arrangement of the hexon gene, showing the three short leader exons, the two introns separating them (A and B), and the long intron (C) separating the leaders from the coding exon of the hexon gene. All exons are represented by red boxes.

Photo (a) from Susan M. Berget, Claire Moore, and Philip Sharp. Proceedings of the National Academy of Science, *vol. 74, p. 3173, 1977. (b) and (c) from Susan M. Berget, et al.,* Proceedings of the National Academy of Science, *vol. 74, p. 3173, 1977.*

RNA Splicing

Consider the problem introns pose. They are present in genes but not in mature RNA. How is it that the information in introns does not find its way into the final products of the genes? There are two main possibilities: (1) The introns are never transcribed; the polymerase somehow jumps from one exon to the next and ignores the introns in between. (2) The introns are transcribed, yielding a **primary transcript,** an overlarge gene product that is cut down to size by removing the introns. As wasteful as it seems, the latter possibility is the correct one. The process of cutting introns out of immature RNAs and stitching together the exons to form the final product is called **RNA splicing.** The splicing process is outlined in figure 10.23, although, as we will see later in the chapter, this picture is considerably oversimplified.

How do we know splicing takes place? Actually, at the time introns were discovered, circumstantial evidence to support splicing already existed. A class of large nuclear RNAs called **heterogeneous nuclear RNA (hnRNA),** widely believed to be precursors to mRNA, had been found. These hnRNAs are the right size (larger than mRNAs) and have the right location (nuclear) to be

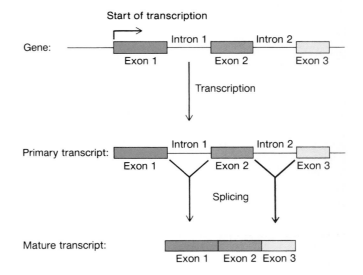

FIGURE 10.23

Outline of splicing. The introns in a gene are transcribed along with the exons (colored boxes) in the primary transcript. Then they are removed as the exons are spliced together.

unspliced mRNA precursors. Furthermore, hnRNA turns over very rapidly, which means it is made and degraded quickly. This too suggested that these RNAs are merely intermediates in the formation of more stable RNAs. However, no direct evidence existed to show that hnRNA could be spliced to yield mRNA.

The mouse β-globin mRNA and its precursor provided an ideal place to look for such evidence. The mouse globin mRNA precursor is a member of the hnRNA population. It is found only in the nucleus, turns over very rapidly, and is about twice as large (1,500 bases) as mature globin mRNA (750 bases). Also, immature red blood cells make so much globin (about 90% of their protein) that globin mRNA is abundant and can be purified relatively easily; even its precursor exists in appreciable quantities. This made experiments feasible. Furthermore, the precursor is the right size to contain both exons and introns. Charles Weissmann and Philip Leder and their coworkers used R-looping to test the hypothesis that the precursor still contained the introns.

The experimental plan was to hybridize mature globin mRNA, or its precursor, to the cloned globin gene, then observe the resulting R-loops (figure 10.24). We know what the results with the mature mRNA should be. Since this RNA has no intron sequences, the introns in the gene will loop out. On the other hand, if the precursor RNA still has all the intron sequences, no such loops will form. Of course, that is what happened. You may have a little difficulty recognizing the structures in figure 10.24 because this R-looping was done with double-stranded DNA instead of single-stranded. Thus, the RNA simply hybridizes to one of the DNA strands, displacing the other. The precursor RNA gave a smooth, uninterrupted R-loop; the mature mRNA gave an R-loop interrupted by an obvious loop of double-stranded DNA, which represents the large intron. For some reason, the small intron escaped detection in this experiment.

M essenger RNA synthesis in eukaryotes occurs in stages. The first stage is synthesis of the primary transcription product, an mRNA precursor that still contains introns copied from the gene, if any were there. This precursor is part of a pool of large nuclear RNAs called hnRNAs. The second stage is mRNA maturation. Part of the maturation of an mRNA precursor is the removal of its introns in a process called splicing. This yields the mature-sized mRNA.

■

Splicing Signals

Pause to consider how important it is that splicing be accurate. If too little RNA is removed from an mRNA precursor, the mature RNA will be interrupted by "garbage." If too much is removed, important sequences may be left out.

Given the importance of accurate splicing, signals must occur in the mRNA precursor that tell the splicing machinery exactly where to "cut and paste." What are these signals? One way to find out is to look at the base sequences of a number of different genes, locate the intron boundaries, and see what sequences they have in common. In principle, these common sequences could be part of the signal for splicing. The most striking observation, first made by Chambon, is that almost all introns in nuclear mRNA precursors begin and end the same way:

exon/GU-intron-AG/exon

In other words, the first two bases in the intron of a transcript are GU and the last two are AG. This kind of conservation does not occur by accident; surely the GU---AG motif is part of the signal that says: "Splice here." However, GU---AG motifs occur too frequently. A typical intron will contain several GU's and AG's within it. Why are these not used as splice sites? The answer is that GU---AG is not the whole story; there is more to it than that. By this time, the sequences of many intron/exon boundaries have been examined and a consensus sequence has emerged:

$$5'\text{-}^C_A\text{AG/GU}^A_G\text{AGU-intron-}Y_N\text{NAG/G-}3'$$

where Y_N denotes a string of about nine pyrimidines (U's and C's) and N is any base.

Again, finding consensus sequences is one thing; showing that they are really important is another. Several research groups have found ample evidence supporting the importance of these splice junction consensus sequences. Their experiments were of two basic types. In one, they mutated the consensus sequences at the splice junctions in cloned genes, then checked whether proper splicing still occurred. In the other, they collected defective genes from human patients with presumed splicing problems and examined the genes for mutations near the splice junctions. Both approaches gave the same answer: Disturbing the consensus sequences usually destroys normal splicing.

For example, consider the β-globin gene in several cases of human β-thalassemia, an often fatal disease in which no functional β-globin mRNA appears in the cytoplasm. This can be for a number of different reasons; in the case cited here, it is because the patient carried a β-globin gene with defective splice junctions. Figure 10.25 shows the sequence of the mutated β-globin gene of this thalassemia patient and the consequences of this mutation. In this case, a single base change destroyed a normal splice site at the beginning of the first intron, so the splicing system used a series of alternate sites (cryptic sites) where splicing does not normally occur. Two of these are upstream from the normal site, within the first exon; the other is downstream, in the intron.

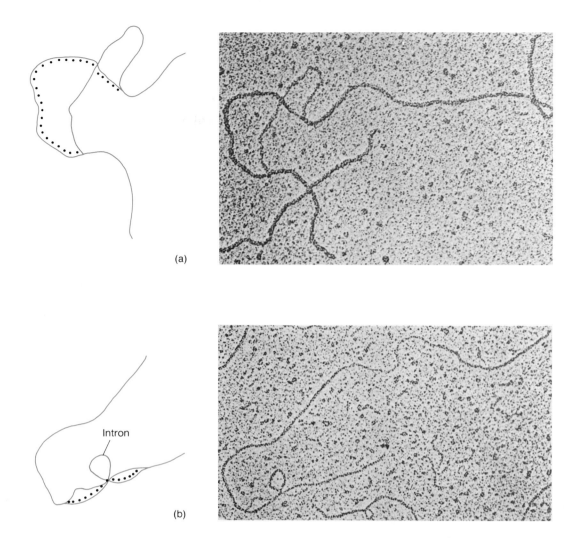

FIGURE 10.24

Introns are transcribed. (*a*) R-looping experiment in which mouse globin mRNA precursor was hybridized to a cloned mouse β-globin gene. A smooth hybrid formed, demonstrating that the introns are represented in the mRNA precursor. (*b*) Similar R-looping experiment in which mature mouse globin mRNA was used. Here, the large intron in the gene looped out, showing that this intron was no longer present in the mRNA. The small intron was not detected in this experiment. In the interpretive drawings at left, the dotted black lines represent RNA and the solid red lines, DNA.

Photos from Shirley Tilgham, Peter Curtis, David Tiemeier, Philip Leder, and Charles Weismann, Proceedings of the National Academy of Science *75:1312, 1978.*

The sequences of the cryptic sites resemble that of the normally used site. In this way, the splicing machinery is like a child who always buys the same candy bar until one day, that particular brand is sold out. For the first time, the child notices other brands that don't look too bad, and buys not one, but three different kinds. These alternate brands had been cryptic, or hidden, to the child before, even though they were in plain sight, because the favorite was there.

Unfortunately for thalassemia patients, these cryptic splice sites are in the wrong places, so the wrong sequences are removed from the mRNA precursors. In the first two cases, parts of exon 1 are spliced out; in the third case, part of intron 1 is left in. In none of these cases is functional mRNA produced. Thus, even though the globin coding sequence may have been perfectly normal in this mutated gene, the fact that they could not be spliced properly rendered them nonfunctional, with life threatening consequences.

*T*he splicing signals in nuclear mRNA precursors are remarkably uniform. The first two bases of the intron are almost always GU, and the last two are almost always AG. In addition, the 5'- and 3'-splice sites have consensus sequences that are important to proper splicing; when they are mutated, abnormal splicing occurs.

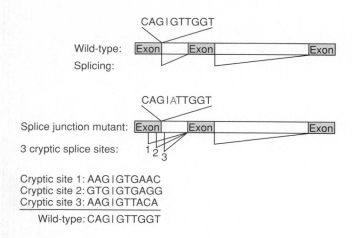

CAG|GTTGGT

Wild-type: Exon   Exon   Exon
Splicing:

CAG|ATTGGT

Splice junction mutant: Exon   Exon   Exon
3 cryptic splice sites: 1 2 3

Cryptic site 1: AAG|GTGAAC
Cryptic site 2: GTG|GTGAGG
Cryptic site 3: AAG|GTTACA
  Wild-type: CAG|GTTGGT

FIGURE 10.25

Importance of consensus sequences at splice junctions. With the wild-type splice junctions, splicing proceeds normally; when the first G of the intron changes to A (blue), splicing no longer occurs at that location, but at three alternate, cryptic sites. The sequences of the three cryptic splice junctions are shown, with the wild-type sequence below for comparison.

Small Nuclear RNAs

Joan Steitz and her coworkers first focused attention on a class of **small nuclear RNAs (snRNAs)** as potential agents in splicing (figure 10.26). These snRNAs exist in the cell coupled with proteins in complexes called **small nuclear ribonucleoproteins (snRNPs),** usually referred to informally as "snurps." Steitz noticed that one snRNA, called U1, has a region whose sequence is almost perfectly complementary to both 5'- and 3'-splice site consensus sequences. In principle, this would allow base pairing between U1 RNA and an mRNA precursor. In practice, such base pairing does occur, but probably only at the 5'-splice site.

How do we know that snRNPs are involved in splicing? One line of evidence comes from in vitro splicing studies using antibodies directed against snRNPs. Antibodies react with a specific substance and frequently interrupt its normal function. Therefore, if snRNPs are needed for splicing, one would expect that antibodies directed against snRNPs would prevent splicing. This is what the experiments showed.

We also know that splicing of nuclear mRNA precursors occurs on a complex particle called a **spliceosome.** The mammalian spliceosome is about the size of a large ribosomal subunit (50–60S). It includes the mRNA precursor as well as a set of snRNPs: U2, U4, U5, and U6. The fact that U1 snRNP is absent from the spliceosome does not mean U1 is not involved in splicing; excellent evidence indicates that it *is* involved. Perhaps it has already served its function by the time the spliceosome forms.

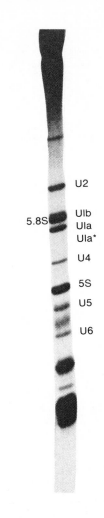

       U2

       U1b
5.8S   U1a
       U1a*

       U4

       5S

       U5

       U6

FIGURE 10.26

Electrophoresis of snRNAs. Small nuclear RNAs from mouse tumor cells are displayed by polyacrylamide gel electrophoresis; 5S and 5.8S rRNAs are also present as size markers.

From Michael R. Lerner, John A. Boyle, Stephen M. Mount, Sandra L. Wolin, and Joan A. Steitz. Reprinted by permission from Nature *283:221. Copyright © 1980 Macmillan Magazines Limited.*

A Lariat-shaped Intermediate

The detailed mechanism by which cells splice mRNA precursors seems at first so illogical that it would be best to present the conclusion first, then the evidence for it. Figure 10.27 illustrates the two-step **lariat model** of mRNA splicing. The first step is the formation of a looped intermediate that looks like a lariat or lasso. This occurs when a nucleotide in the middle of the intron attacks the G at the beginning of the intron, forming the loop, and simultaneously breaking the phosphodiester bond between the first exon and the intron. The splicing is completed when the 3'-end of the first exon attacks the 5'-end of the second exon, forming the exon-exon phosphodiester bond and releasing the intron, in lariat form, at the same time.

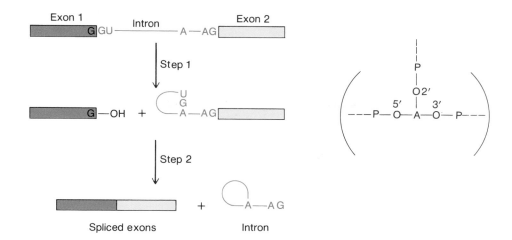

FIGURE 10.27
Lariat splicing model. In reaction 1, the bond between the first exon (blue) and the intron breaks, and the terminal G of the intron bonds to an A within the intron, forming a lariat-shaped intermediate. The branched nucleotide (A) in the intron is shown in parentheses at right. In reaction 2, the first exon (blue) is spliced to the second (yellow), liberating the lariat-shaped intron (red).

This mechanism seemed unlikely enough that rigorous proof had to be presented in order for it to be accepted. There is, in fact, very good evidence for the existence of all the intermediates shown in figure 10.27, most of it collected by Sharp and his research group. Only one piece of this evidence will be presented here. The key to the whole scheme is the branched nucleotide at the "knot" of the lariat. In the experimental system Sharp used, this is an adenine nucleotide. This nucleotide will therefore be unique in participating in three phosphodiester bonds instead of the usual two. (See parentheses in figure 10.27.) The extra bond joins the 2'-hydroxyl group of the adenosine to the 5'-phosphate of the guanine nucleotide at the 5'-end of the intron. Therefore, we ought to be able to digest the lariat with an appropriate RNase and leave only the branch:

$$pA \begin{matrix} \diagup pG \\ \diagdown pG \end{matrix}$$

Sharp found this structure, just as predicted.

A class of small nuclear RNAs (snRNAs) appears to be important in splicing nuclear mRNA precursors. In particular, U1 snRNA has a region that is complementary to the consensus splice site sequences in mRNA precursors. This RNA has also been shown to base-pair with mRNA precursors at the 5'-splice site in vivo. The mechanism of mRNA precursor splicing involves intramolecular attack by a nucleotide within the intron on the G at the 5'-end of the intron, breaking the 5'-exon/intron bond and creating a lariat-shaped intermediate. Then the two exons are joined, releasing the intron, still in its lariat shape.

■

OTHER RNA PROCESSING EVENTS

Splicing is one example of something that happens to RNA between the time it is transcribed and the time it is exported to the cytoplasm for use in protein synthesis. But several other things occur during this **posttranscriptional** period. Some of them are peculiar to mRNA; others pertain to rRNA and tRNA only; and some involve all three types of RNA. All these events can be lumped together under the heading **RNA processing.** Among these processing events are: (1) trimming RNA precursors outside the coding regions; (2) methylating (adding methyl groups to) RNA; (3) adding nucleotides to the 3'-ends of RNAs; and (4) blocking the 5'-ends of mRNAs with special structures called caps.

Trimming Ribosomal RNA Precursors

The ribosomal RNA genes in eukaryotes are repeated several hundred times and clustered together in the nucleolus of the cell. Their arrangement in amphibians has been especially well studied, and as figure 10.28a shows, they are separated by regions called *nontranscribed spacers.* This is to distinguish them from *transcribed spacers,* regions of the gene that are transcribed as part of the rRNA precursor and then removed in the processing of the precursor to mature rRNA species.

This clustering of the reiterated rRNA genes in the nucleolus made them easy to find and therefore provided Oscar Miller and his colleagues with an excellent opportunity to observe genes in action. These workers looked at amphibian nuclei with the electron microscope and uncovered a visually appealing phenomenon, shown in figure 10.28b. The DNA containing the rRNA genes can be seen

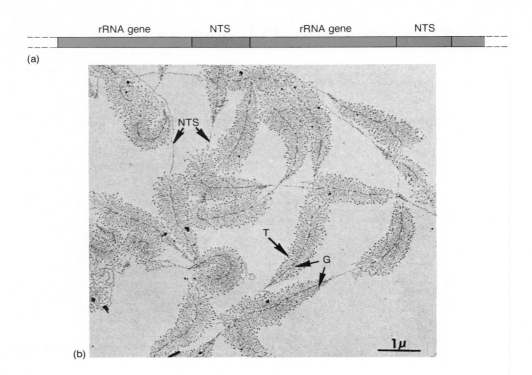

(a) rRNA gene | NTS | rRNA gene | NTS

(b)

FIGURE 10.28

Transcription of rRNA precursor genes. (*a*) A portion of the newt (amphibian) rRNA precursor gene cluster, showing the alternating rRNA genes (blue) and nontranscribed spacers (NTS, red). (*b*) Electron micrograph of part of a newt nucleolus, showing rRNA precursor transcripts (T) being synthesized in a "Christmas tree"

pattern on the tandemly duplicated rRNA precursor genes (G). At the base of each transcript is a polymerase I, not visible in this picture. The genes are separated by nontranscribed spacer DNA.

(b) From O. L. Miller, Barbara R. Beatty, Barbara Hankalo, and C. A. Thomas. Symposia on Quantitative Biology *(Cold Spring Harbor), vol. 35, p. 506, 1970.*

winding through the picture, but the most obvious feature of the micrograph is a series of "Christmas tree" structures. These include the rRNA genes (the trunk of the tree) and growing rRNA transcripts (the branches of the tree). We must remember that these transcripts are actually rRNA precursors, not mature rRNA molecules (more about this later). The spaces between "Christmas trees" are of course the nontranscribed spacers. You can even tell the direction of transcription from the lengths of the transcripts within a given gene; the shorter RNAs are at the beginning of the gene and the longer ones are at the end.

Ribosomal and transfer RNAs are sometimes called *stable RNAs* because they are so much longer-lived than mRNA in the cell. It is sometimes useful to lump rRNA and tRNA together this way, because they behave differently than mRNA in several respects. One of these is in processing. We have seen that mRNA precursors frequently require splicing but no other trimming. On the other hand, rRNAs and tRNAs first appear as precursors that sometimes need splicing, but they also have excess nucleotides at their ends, or even between regions that will

become separate mature RNA sequences (e.g., 5.8S, 18S, and 28S rRNA sequences). These excess regions must also be removed.

The rRNA precursors of vertebrates are more typical than the one found in *Tetrahymena* because they contain no introns. Figure 10.29 depicts the structure and scheme of processing of the 45S human rRNA precursor. It contains the 28S, 18S, and 5.8S sequences, imbedded between transcribed spacer RNA regions. The processing of this precursor occurs in the nucleolus. The first step is to cut off the spacer at the 5′-end, leaving a 41S intermediate. The next step involves cleaving the 41S RNA into two pieces, 32S and 20S, that contain the 28S and 18S sequences, respectively. The 32S precursor also retains the 5.8S sequence. Finally, the 32S precursor is split to yield the mature 28S and 5.8S RNAs, which base-pair with each other, and the 20S precursor is trimmed to mature 18S size.

The details of this processing scheme are not universal; even the mouse does things a little differently, and the frog precursor is only 40S, which is quite a bit smaller than 45S. Still, the basic mechanism of rRNA processing, including the order of mature sequences in the precursor, is preserved throughout the eukaryotic kingdom.

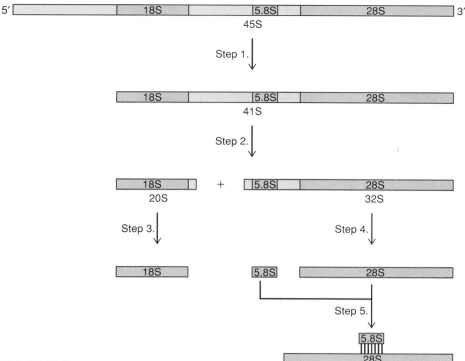

FIGURE 10.29
Processing scheme of 45S human (HeLa) rRNA precursor. Step 1: The 5'-end of the 45S precursor RNA is removed, yielding the 41S precursor. Step 2: The 41S precursor is cut into two parts, the 20S precursor to the 18S rRNA, and the 32S precursor to the 5.8S and 28S rRNAs. Step 3: The 3'-end of the 20S precursor is removed, yielding the mature 18S rRNA. Step 4: The 32S precursor is cut to liberate the 5.8S and 28S rRNAs. Step 5: The 5.8S and 28S rRNAs associate by base pairing.

Poly(A) Tails

We have already seen that hnRNA is a precursor to mRNA. One finding that suggested such a relationship between these two types of RNA was that they shared a unique structure at their 3'-ends: a long chain of AMP residues called **poly(A)**. Neither rRNA nor tRNA has a poly(A) tail. James Darnell and his coworkers performed much of the early work on poly(A) and the process of adding poly(A) to RNA (called **polyadenylation**). They released the poly(A) intact from the rest of the mRNA by treating with two enzymes: ribonuclease A, which cuts RNA after the pyrimidine nucleotides C and U, and ribonuclease T$_1$, which cuts after G nucleotides. In other words, the RNA was cut after every nucleotide except the A's, so only runs of A's were preserved. The poly(A) prepared this way was electrophoresed to determine its size, which turned out to be an average of 150 to 200 nucleotides. This also implies that the poly(A) is not interrupted by other kinds of nucleotides, or it would have been broken up by the RNases.

It is apparent that the poly(A) goes on the 3'-end of the mRNA or hnRNA because it can be released very quickly with an enzyme that degrades RNAs from the 3'-end inward. We also know that poly(A) is added post-transcriptionally, because there are no runs of T's in the DNA long enough to encode it. In fact, there is an enzyme in nuclei called *poly(A) polymerase* that adds AMP residues one at a time to mRNA precursors once they are complete. We know that poly(A) is added to mRNA precursors because it is found on hnRNA. When we look

closely, we can even see that specific unspliced mRNA precursors (the 15S mouse globin mRNA precursor, for example) contain poly(A). Once in the cytoplasm, the poly(A) on an mRNA "turns over"; in other words, it is constantly being broken down by RNases and rebuilt by a cytoplasmic poly(A) polymerase.

Most mRNAs contain poly(A). One famous exception is histone mRNA, which somehow manages to perform its function without a detectable poly(A) tail. The near-universality of poly(A) suggests that it has an important function. What is it? The best guess is that poly(A) helps protect the vulnerable mRNA from degradation by ribonucleases lurking about in the cell. This is not apparent in short-term experiments in which mRNAs are deprived of their poly(A) tails and then translated in vitro. Under these circumstances, there is little difference between RNA containing poly(A) [poly(A)$^+$ RNA] and that lacking it [poly(A)$^-$ RNA].

In long-term experiments, however, the usefulness of poly(A) becomes very apparent. One such experiment tested the ability of globin mRNA with and without poly(A) to support protein synthesis over a two-day period after injection into frog eggs. The mRNA without poly(A) was "dead" after only six hours; it would no longer direct globin synthesis. On the other hand, the poly(A)$^+$ RNA could still be translated at least sixty hours after injection. This suggests that the poly(A) helps protect the mRNA against degradation or at least keeps it in a translatable condition.

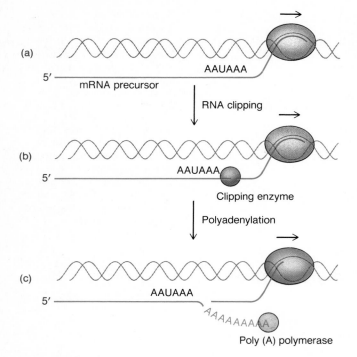

FIGURE 10.30

Polyadenylation. (*a*) The mRNA precursor (red) is being synthesized by RNA polymerase II (blue). (*b*) A clipping enzyme (purple) recognizes the polyadenylation signal (AAUAAA) and cuts the growing RNA about twenty bases downstream. (*c*) The poly(A) polymerase (yellow) adds AMP residues (green) stepwise to the new 3′-end, as transcription continues. The transcript beyond the polyadenylation site is presumably discarded.

> **M**ost eukaryotic mRNAs carry a poly(A) tail about 150–200 bases long on their 3′-ends. Poly(A) polymerase adds this tail to mRNA precursors in the nucleus as soon as transcription is complete, and the poly(A) apparently remains with the message through splicing and transport to the cytoplasm. Poly(A) seems to protect mRNA from degradation.

■

Polyadenylation

It would be logical to assume that poly(A) polymerase simply waits for a transcript to be finished, then adds poly(A) to its 3′-end. It would also be wrong, at least in most cases. Instead, the mechanism of polyadenylation usually involves clipping an mRNA precursor, sometimes even before transcription has terminated, and then adding poly(A) to the newly exposed 3′-end (figure 10.30). This means that, contrary to expectations, polyadenylation and transcription termination are not necessarily linked. RNA polymerase II can still be elongating an RNA chain as it approaches a termination signal, while the polyadenylating apparatus has already located a polyadenylation signal, cut the growing RNA, and polyadenylated it.

What constitutes a polyadenylation signal? One consensus sequence appears in the transcript about twenty bases upstream from the polyadenylation site. This is usually AAUAAA in higher eukaryotes, although deviations, especially AUUAAA, are also found. To establish the importance of this consensus sequence, several experiments have been done. For example, bases have been deleted between the AAUAAA motif and the normal polyadenylation site. The result was simply that the polyadenylation site moved down by the number of bases deleted, preserving the normal spacing between AAUAAA and the poly(A) site. In other experiments, a fragment of the SV40 early gene, containing the polyadenylation site, was cloned and injected into frog eggs. The resulting RNA was then examined for proper polyadenylation. As long as the AAUAAA motif was intact, everything worked normally. However, when it was changed to AAUACA or AAUUAA or AACAAA or AAUGAA, no normal polyadenylation occurred.

This experiment implies that no deviation from the consensus sequence is allowed, but this is not really the case. When the base sequences of 134 mRNAs were compared, the "correct" bases were found in the consensus region with the following frequencies:

$$A_{98}A_{91}U_{100}A_{99}A_{99}A_{98}$$

In other words, some "incorrect" sequences are tolerated, including several that did not work in the experiment. This suggests that the context of the polyadenylation signal is also important. This same conclusion comes from the finding that many genes have multiple AAUAAA motifs, yet only one ordinarily serves as a polyadenylation signal. That one must be in the proper context.

What about the termination signals, or terminators, as opposed to the polyadenylation signals in eukaryotes? Unfortunately, our knowledge of these sequences is still meager. The problem seems to be that termination of most polymerase II transcripts is sloppy; the terminations occur in several different places spread out over hundreds or even thousands of bases. This is apparently not a problem, since eukaryotes have the clipping and polyadenylation system to tidy up the 3′-ends of their mRNAs.

> **P**olyadenylation occurs by clipping a transcript upstream from its 3′-end, sometimes while transcription is still in progress, then adding poly(A) to the newly created 3′-end. The signal for such clipping and polyadenylation, at least in higher eukaryotes, includes the consensus sequence AAUAAA, plus some other information that is not yet clear. The termination signals in genes transcribed by polymerase II have not been identified.

■

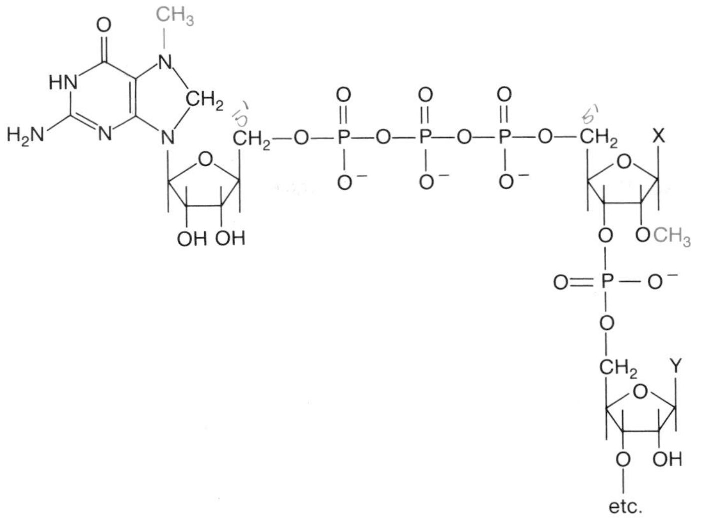

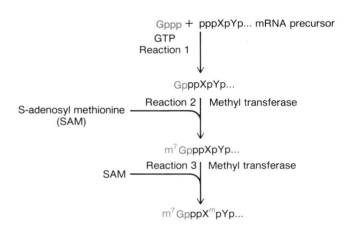

FIGURE 10.31
Cap 1 structure at the 5'-end of an mRNA. The methyl groups of the cap are shown in red. X and Y represent the bases on the second and third nucleotides, respectively. Cap 0 would have no methyl group on the penultimate ribose. Cap 2 (rare) would have both methyl groups of cap 1, plus another on the ribose linked to base Y.

Eukaryotic mRNA Cap Structures

The 3'-end of a eukaryotic messenger is not the only place that differs from its prokaryotic counterpart. The 5'-end also has something different—a **cap.** The structure of a cap can vary somewhat, but its typical architecture is shown in figure 10.31. The cap contains a methylated guanine nucleotide linked by a triphosphate to the *penultimate nucleotide,* or next-to-last nucleotide. The penultimate nucleotide is frequently methylated also, but the methyl group is on the 2'-hydroxyl group of the ribose instead of on the base.

The capping process is illustrated in figure 10.32. The growing RNA has a triphosphate group at its 5'-end, just as a prokaryotic mRNA does. Then, before the RNA has grown to fifty bases long, a capping enzyme adds a GTP to its 5'-end. This reaction is unlike the normal RNA polymerase reaction in several respects: (1) The incoming nucleotide adds to the 5'-end of the RNA chain instead of the 3'-end. (2) Instead of a simple phosphodiester bond forming between the two end nucleotides, a triphosphate linkage forms. (3) The triphosphate does not link the 3'-site of one nucleotide with the 5'-site of the next; instead, the two end nucleotides are linked through their 5'-sites. Thus, the capping nucleotide "backs in" to join the growing RNA chain. Two enzymes, called *methyl transferases,* perform the methylation that finishes capping. They transfer methyl groups from the methyl donor, *S-adenosyl methionine (SAM),* to the appropriate positions on the cap. Note that, strictly speaking, capping cannot be considered a posttranscriptional process, because it occurs before transcription is complete. It might be better to call it a "cotranscriptional" process.

Yasuhiro Furuichi and Aaron Shatkin and their colleagues first discovered caps in viral mRNA in 1974. As frequently happens when such a surprising discovery

FIGURE 10.32
Capping and methylating mechanism. Reaction 1: GTP caps the mRNA precursor; the blue p indicates that only one of GTP's phosphate groups is preserved in the cap. Reaction 2: A methyl transferase transfers a methyl group (red) from SAM to the 7-position of the capping guanine. Reaction 3: Another methyl transferase methylates (green) the 2'-hydroxyl group of the penultimate nucleotide.

occurs, one wondered whether it would be a general phenomenon or an isolated curiosity. It turns out that caps are ubiquitous among eukaryotes. What, then, is the function of these widespread structures? Furuichi showed that caps have two important roles. First, they protect mRNA from degradation, just as poly(A) does. They can be envisioned as tough blocking structures at each end of the mRNA, protecting the tender midsection from attack by ribonucleases. Second, caps are necessary for binding mRNA to the ribosome. Little if any translation of decapped messages occurs.

FIGURE 10.33

Levels of gene expression in eukaryotes.

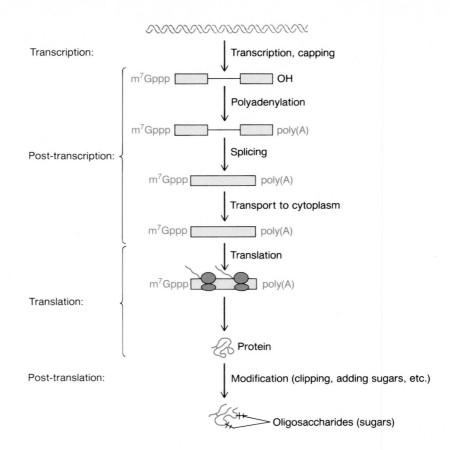

Transcription:

Transcription, capping

m⁷Gppp ☐——☐ OH

Polyadenylation

m⁷Gppp ☐——☐ poly(A)

Post-transcription:

Splicing

m⁷Gppp ☐ poly(A)

Transport to cytoplasm

m⁷Gppp ☐ poly(A)

Translation

Translation:

m⁷Gppp ☐ poly(A)

Protein

Post-translation:

Modification (clipping, adding sugars, etc.)

Oligosaccharides (sugars)

T he 5′-ends of eukaryotic mRNAs are blocked with structures called caps. A cap is added by the capping enzyme, which joins a GTP through a 5′-5′-triphosphate linkage to the penultimate nucleotide. Methyl transferases then add methyl groups to the 7-nitrogen of the terminal guanine and to the 2′-hydroxyl of the penultimate ribose. Caps serve a dual purpose: They protect messages from degradation, and they allow them entry to ribosomes for translation.

■

TYPES OF GENE EXPRESSION CONTROL

By now, you can readily appreciate that gene expression in eukaryotes is much more complex than in prokaryotes. This extra complexity gives a eukaryotic cell more opportunities to control gene expression. In principle, control could occur at the transcriptional, posttranscriptional, or translational level.

TRANSCRIPTIONAL CONTROL

We will see examples of eukaryotic gene control at all three levels (figure 10.33), but it seems that the most common level in the eukaryotes, as in the prokaryotes, is transcription. Why is this? One reason is that transcription is by far the most economical level at which to control genes.

It allows for genes to be transcribed only if their products are needed. Consider the alternatives. If there were no transcriptional control, all genes would be transcribed equally. Then those unneeded transcripts would either fail to be processed properly or, if they were processed, would fail to be translated. In either case, the surplus transcripts would be degraded. This would be grossly wasteful, especially considering the energy it takes to produce RNA. In fact, it is doubtful that an organism could survive by transcribing all its genes all of the time.

However, nature does not always do things in what we would regard as the most sensible way, so we need to ask: What is the evidence that transcription is the primary level of control in eukaryotes? The best evidence comes from direct examination of transcription products in differentiated cells. To begin with, we can look at the levels of mature mRNA in the cytoplasms of various differentiated cells. These RNA levels can be measured in several ways, but most are variations on the theme of **hybridization.** Recall from chapter 6 that a strand of RNA and a complementary strand of DNA can form a double-stranded helix, much like a DNA double helix. Therefore, if we obtain a piece of DNA complementary to an mRNA whose concentration we want to measure, we can hybridize it to radioactive mRNA and measure the amount of radioactivity in the RNA-DNA hybrid (figure 10.34).

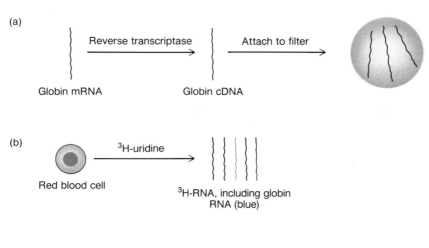

(a)

Reverse transcriptase → Attach to filter →

Globin mRNA   Globin cDNA

(b)

Red blood cell —^{3}H-uridine→ ^{3}H-RNA, including globin RNA (blue)

(c)

Globin-cDNA filter + ^{3}H-RNA —Hybridize→ ^{3}H-globin mRNA hybridized to globin cDNA on filter → Count radioactivity on filter

FIGURE 10.34
Hybridization assay for specific transcripts. (*a*) Reverse transcriptase makes a globin cDNA from purified globin mRNA. The cDNA is attached to a filter. (*b*) Red blood cells make RNA, including globin mRNA (blue), in the presence of a radioactive precursor, so the RNA becomes radioactive. (*c*) This radioactive RNA is hybridized to the globin cDNA on the filter; the more globin mRNA is present, the more radioactivity will stick to the filter.

The key to this procedure is a tumor virus enzyme called *reverse transcriptase*. This enzyme does just what its name implies—the reverse of transcription; it uses RNA as a template and makes a DNA copy. If we supply the enzyme with a little bit of the mRNA we want to detect, it makes a complementary DNA, or copy DNA (**cDNA**). This cDNA can then be cloned or reproduced in very large quantity (see chapter 15 for a fuller explanation).

When we hybridize a globin cDNA to mRNA from the cytoplasms of several different kinds of cells, including red blood cells, we find that the cDNA hybridizes successfully only to mRNA from red blood cells. In fact, we can show that red blood cells contain about 50,000 molecules of globin mRNA, compared to less than one molecule in other types of cells. Does this mean that the globin genes are not turned on in the other types of cells? It would be tempting to say so, but there are other possible explanations. For example, perhaps the globin gene is active in all types of cells, but the globin RNA is degraded before it reaches the cytoplasm of all cells except red blood cells.

In order to prove our thesis, we need to examine newly synthesized RNA before it has a chance to be degraded. That means looking in the nucleus, not in the cytoplasm. Furthermore, we must observe periods of RNA synthesis that are short enough that the RNA has not suffered significant degradation. In other words, we need to look at pulse-labeled RNA. In case after case where this has been done, we find evidence for transcriptional control. For example, red blood cells make vast quantities of globin (about ten million molecules per cell) but no ovalbumin. Accordingly, their nuclei contain many transcripts of the globin

TABLE 10.2 *Synthesis of Ovalbumin mRNA in Various Tissues[1]*

Nuclei From	Ovalbumin mRNA (% of Total)
Oviduct + DES	0.238
Oviduct − DES	0.002
Spleen	<0.001
Liver	<0.001

[1]The limit of detectability was 0.001%.
Data from Swaneck et al., *Proceedings of the National Academy of Science* 76:1049 (1979).

genes, but almost no transcripts of the ovalbumin gene. Conversely, oviduct cells make great amounts of ovalbumin, but no globin; as expected, their nuclei contain abundant ovalbumin transcripts, but almost no globin transcripts.

Table 10.2 shows the data that led to these conclusions. Bert O'Malley and his colleagues isolated nuclei from a variety of chicken cells, then allowed these nuclei to continue RNA synthesis for several minutes in the presence of a radioactive RNA precursor to label the RNA. Isolated nuclei usually cannot initiate RNA chains, so it is assumed that the RNA made in these nuclei represents RNA chains that were started in vivo. Next, these investigators hybridized the labeled RNA to filters containing a cloned ovalbumin gene. Only ovalbumin mRNA will hybridize, so the radioactivity on the filters is a direct measure of ovalbumin mRNA synthesis in the isolated nuclei, and therefore of ovalbumin gene activity in vivo. In table 10.2 we see that spleen and liver nuclei support no detectable ovalbumin mRNA synthesis. On the other hand, oviduct nuclei from chickens exposed to the hormone *diethylstilbestrol (DES)* make a considerable amount of ovalbumin mRNA. Nuclei from chickens that have been withdrawn from DES for three days drop back to a barely detectable level of ovalbumin mRNA synthesis. These results are certainly not compatible with a model in which all genes are transcribed equally in all cells. On the contrary, it appears that DES stimulates transcription of the ovalbumin gene in oviduct cells by at least a factor of 100. In other words, a great deal of transcriptional control occurs.

S pecialized genes of eukaryotes are controlled, at least primarily, at the transcription level. They are significantly active only in those differentiated cells where their protein products are produced.

■

POSTTRANSCRIPTIONAL CONTROL

Cells could, in principle, control their gene expression at the posttranscriptional level in several different ways. For example, they could add caps or poly(A) tails to certain mRNAs, but not to others; or they could transport certain mRNAs, but not others, to the cytoplasm; or they could selectively degrade certain mRNAs or their precursors. In practice, cells do not seem to use all of these potential control mechanisms: (1) Cells do not withhold capping from certain types of messages. If they did, we would expect to find some uncapped messages, or at least message precursors, but virtually all mRNA precursors are capped at a very early time after transcription initiation. (2) With one exception, cells seem not to withhold polyadenylation from certain types of messages. The lone exceptions are the histone mRNAs, which never seem to be polyadenylated in higher eukaryotes.

One way that posttranscriptional control does appear to occur is by selective degradation of mRNAs in the cytoplasm. The response of breast tissue to the hormone *prolactin* provides a good example of this mechanism. When cultured breast tissue is stimulated with prolactin, it responds by producing the milk protein *casein*. One would expect an increase in casein mRNA concentration to accompany this casein buildup, and it does. The number of casein mRNA molecules increases about 20-fold in twenty-four hours following the hormone treatment. But this does not mean the rate of casein mRNA synthesis has increased 20-fold. In fact, it only increases about 2- to 3-fold. The rest of the increase in casein mRNA level depends on an approximately 20-fold increase in stability of the casein mRNA.

T ABLE 10.3 *Effect of Prolactin on Half-life of Casein mRNA[1]*

Species	RNA Half-life (H$_r$)	
	Without Prolactin	*With Prolactin*
rRNA	>90	>90
Poly(A)$^+$ RNA (short-lived)	3.3	12.8
Poly(A)$^+$ RNA (long-lived)	29	39
Casein mRNA	1.1	28.5

[1]Prolactin has relatively little effect on rRNA and total mRNA (poly[A]$^+$ RNA), but causes a great increase in the half-life of casein mRNA. Data from Guyette, et al., *Cell* 17:1013 (1979).

A **pulse-chase** experiment was performed to measure the **half-life** of casein mRNA. The half-life is the time it takes for half the RNA molecules to be degraded. Casein mRNA was radioactively labeled in cultured breast cells for a short time in the presence or absence of prolactin. In other words, cells were given a pulse of radioactive nucleotides, which they incorporated into mRNA. Then the cells were transferred to "cold" medium lacking radioactivity. This chases the radioactivity out of the RNA, as "hot" RNAs break down and are replaced by "cold" ones. After various chase times, the level of labeled casein mRNA remaining was measured by hybridizing it to a cloned casein gene. The faster the labeled casein mRNA disappeared, the shorter its half-life. The surprising conclusion, shown in table 10.3, was that the half-life of casein mRNA increased dramatically, from 1.1 hours to 28.5 hours, in the presence of prolactin. At the same time, the half-life of total polyadenylated mRNA increased only 1.3- to 4-fold in response to the hormone. It appears prolactin causes a selective stabilization of casein mRNA that is largely responsible for the enhanced expression of the casein gene.

O f the possible mechanisms of posttranscriptional control, there is good evidence only for altering mRNA stability. Differential capping and polyadenylation appear not to be significant means of controlling gene expression.

■

TRANSLATIONAL CONTROL

Maternal messages in eggs provide good examples of how genes can be controlled at the translational level. Maternal messages are mRNAs made in great quantity during **oogenesis,** the development of an egg, then stored, complexed with protein, in the egg cytoplasm. These stored mRNAs are abundant enough that they can support the early development of the embryo even in the absence of

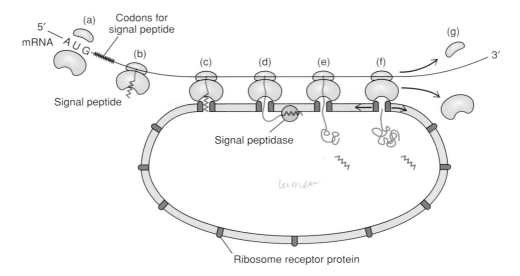

FIGURE 10.35
Signal peptide hypothesis. (*a*) A ribosome (yellow) attaches to the 5'-end of an mRNA. (*b*) The first part of the protein to be made is the signal peptide (blue) at the amino terminus. (*c*) The signal peptide attaches to ribosome receptor proteins (green) on the outside of the endoplasmic reticulum (large oval) and begins to draw the growing protein into the interior. (*d*) The signal peptide is removed by a signal peptidase (orange) within the endoplasmic reticulum. (*e*) The protein (red) continues to grow and fold. (*f*) The completed protein moves completely into the endoplasmic reticulum. (*g*) The ribosome dissociates from the mRNA.

From P. Walter, et al., Cell *38:6, 1984. Copyright © 1984 Cell Press, Cambridge, MA. Reprinted by permission.*

transcription. Maternal messages in sea urchin eggs are translated only poorly until fertilization when their translation increases at least 50-fold. We know that this represents enhanced translation of maternal messages, because the increase occurs even when new RNA synthesis is blocked, or even in cells that lack nuclei.

Surf clam eggs are also interesting in this regard. The translation rate does not change after fertilization, but the spectrum of messages translated does change. We can assay the total population of maternal messages in these eggs before and after fertilization by isolating them, translating them in vitro, and displaying their protein products by gel electrophoresis. The same bands appear before and after fertilization, showing that the same mRNAs are present. Next, we can look at the proteins produced from these mRNAs in vivo before and after fertilization. The spectrum of proteins made in vivo changes after fertilization, even though the mRNA population does not change. In other words, something is controlling which messages are read.

> *T*he expression of some genes, at least in early embryogenesis, can be controlled at the level of translation.

■

POSTTRANSLATIONAL CONTROL

In principle, at least, there is one more level at which gene expression can be controlled. This is the **posttranslational** level, or what happens to a protein after it is released from the ribosome. Cells modify and decorate proteins in a variety of ways. Proteins can be clipped by proteases and thus converted from immature to mature form, or they can be coupled to sugars, phosphate groups, or other compounds.

One type of clipping actually occurs *during* synthesis of certain proteins. It involves the removal of a stretch of about twenty amino acids at the protein's **amino terminus**—the end with the free amino group, where protein synthesis began. This portion of the protein, called the **signal peptide,** anchors the nascent protein and its polysome to membranes called **endoplasmic reticulum (ER),** which simply means "network in the cell." As figure 10.35 demonstrates, the signal peptide also serves as a "needle" to pull the protein "thread" into the interior, or *lumen,* of the ER. Once inside, the signal peptide is removed. The shortened protein moves through the ER to the **Golgi apparatus,** where it is packaged for export. Therefore, this type of processing is characteristic of proteins that are destined for export from a cell.

The classic example of **proteolytic** (protein-cleaving) processing is provided by the hormone insulin, which appears first as a precursor called **preproinsulin.** After the signal peptide of preproinsulin is removed, we call it

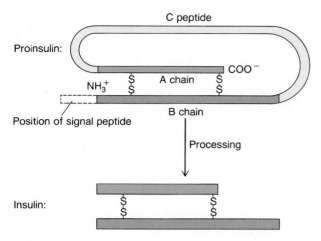

FIGURE 10.36
Converting proinsulin to insulin. Proinsulin consists of the B chain (orange), the C peptide (yellow), plus the A chain (blue). The A and B chains are linked through two disulfide bonds (S–S). The signal peptide (dashed lines) has already been removed during proinsulin synthesis. The conversion of proinsulin to insulin is accomplished by excising the C peptide.

proinsulin, because there is still clipping to be done. During this latter process, an enzyme clips out a part of the proinsulin called the **C peptide,** leaving two pieces, the **A peptides** and **B peptides** that comprise mature insulin (figure 10.36). If these clipping steps did not occur, active insulin would not be produced, and expression of the insulin gene would be blocked as surely as if it were never transcribed. Expression of the insulin gene is apparently not controlled this way, but some other genes do seem to be.

> **M**any proteins experience posttranslational alterations such as proteolytic clipping or addition of sugars. Because these alterations are frequently important for the formation of an active protein, they are potential steps at which gene expression can be controlled.

∎

MECHANISMS OF TRANSCRIPTION CONTROL

We have seen that eukaryotic genes, at least the luxury genes that are expressed actively, are controlled primarily at the transcription level. For this reason, and because we understand much more about transcription control, a few examples of such control will be given here. However, it is important to remember that eukaryotes also use other mechanisms to regulate their genes. Even though a gene is subject to transcriptional control, these other mechanisms may also play an important role.

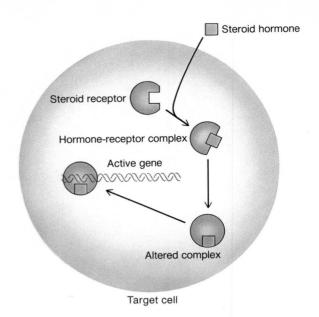

FIGURE 10.37
Steroid hormone action. A steroid hormone enters a cell and binds to a steroid receptor. The hormone-receptor complex changes shape, binds to a target gene, and activates it.

Hormone Activation of Genes

Many hormones, notably the steroids, act by turning on transcription of specific genes in certain target cells. They do this by a complex process. First, the hormone penetrates the cell membrane; then it couples with a receptor protein, forming a hormone-receptor complex and changing the structure of the receptor in the process. Ultimately, the hormone-receptor complex appears in the nucleus where it binds to a set of sites on the chromatin and activates the corresponding genes (figure 10.37). A key element missing from this simplified scheme is the nature of the chromatin sites that recognize the hormone-receptor complex.

Let us consider a well-studied example, the activation by estrogen of the ovalbumin gene in the chick oviduct. Pierre Chambon and his coworkers have identified a site about 250 base pairs upstream from the ovalbumin gene's TATA box that binds the estrogen-receptor complex. They pinned down this binding site by a competition assay that worked as follows (figure 10.38a): First, they made the hormone-receptor complex radioactive, then they mixed it with calf thymus DNA preattached to cellulose. The hormone-receptor complex binds to calf thymus DNA nonspecifically and therefore binds only weakly to the DNA cellulose. The trick is to add competing chicken DNA and try to remove the radio-labeled hormone-receptor complex from the calf thymus DNA. The more

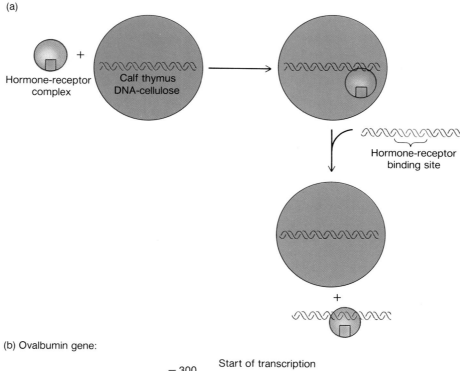

(a)

Hormone-receptor complex

Calf thymus DNA-cellulose

Hormone-receptor binding site

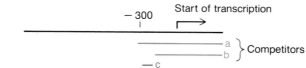

(b) Ovalbumin gene:

−300

Start of transcription

a
b   } Competitors
−c

(c)

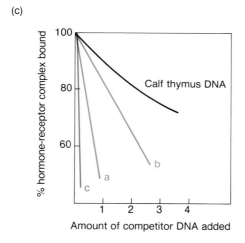

% hormone-receptor complex bound

100

80

60

Calf thymus DNA

c
a
b

1   2   3   4

Amount of competitor DNA added

FIGURE 10.38

The steroid hormone-receptor complex acts on the target gene's 5'-flanking region. (*a*) Outline of the assay. The hormone-receptor complex binds loosely to calf thymus DNA on a filter. The competitor DNA is then added. If it contains a hormone-receptor binding site, it efficiently removes the hormone-receptor from the filter. (*b*) Three of the competitors used. Competitor *a* (red) is a piece of the chicken ovalbumin gene that includes a region about 300 bp upstream from the start of transcription, plus several hundred bases downstream. Competitor *b* (blue) includes about 50 bp less upstream from the start of transcription. Competitor *c* (green) includes only the region between about 250 bp and 300 bp upstream from the start of transcription. (*c*) Results of the assay. Competitor *c* competes extremely well, showing that it includes the hormone-receptor binding site.

avidly the competing DNA binds the complex, the more efficiently it will remove radioactivity from the calf thymus DNA-cellulose. We would anticipate that the chicken DNA containing the authentic binding site for the hormone-receptor complex would compete best of all for the labeled complex.

Figure 10.38b and c shows some of the locations of the competitors Chambon's group tried and the results of the competition experiments. When they used DNA extending at least 300 base pairs upstream from the ovalbumin cap site (the start of transcription), they saw very good competition. Pieces of DNA that did not extend that far did not compete well. Very long pieces did not compete as well either, because their binding sites were diluted out by a mass of nonspecific DNA. The best competition of all came from a very small piece of DNA that included only the region between 250 and 300 base pairs upstream from the ovalbumin cap site. This localized the hormone-receptor binding site to that region in the 5'-flank of the ovalbumin gene (figure 10.38c). This site is in fact an enhancer. Thus, when the hormone-receptor complex binds to it, transcription of the ovalbumin gene is greatly stimulated.

It is important to note that only a few types of cells respond to a given hormone. The key to whether or not a cell will respond is the presence or absence of the specific hormone receptor in that cell. The receptor is, in turn, a protein—a gene product—made only in certain differentiated cells. In other words, the receptor is itself the product of cell differentiation; further differentiation, with activation of new genes, will occur when the cell encounters the hormone.

E strogen stimulates the chicken ovalbumin gene in oviduct cells by first complexing with a cell-specific receptor. The hormone-receptor complex then binds to a site on the chicken chromatin. This activating site lies in the 5'-flanking region of the ovalbumin gene, between 250 and 300 base pairs upstream from the start of transcription.

■

Summary

E ukaryotic chromosomes consist of chromatin: DNA combined with five basic proteins called histones and a large group of nonhistone proteins. This chromatin is organized by several types of folding. The first order of chromatin folding is represented by a string of nucleosomes. These structures contain four pairs of histones (H2A, H2B, H3, and H4) in a ball, around which are wrapped about 200 base pairs of DNA. The fifth histone, H1, binds to DNA outside the ball. The second order of chromatin folding involves a coiling of the string of nucleosomes into a 25-nm-thick fiber called a solenoid. This folding is apparently mediated by H1-H1 interactions. The third order of folding is probably looping of the 25-nm fiber into a brushlike structure with the loops anchored to a central matrix.

Chromatin exists in eukaryotic cells in two forms: heterochromatin, which is condensed and transcriptionally inactive, and euchromatin, which is extended and at least potentially active.

Eukaryotic cells contain three distinct RNA polymerases, named polymerase I, polymerase II, and polymerase III. The roles of these three enzymes are as follows: Polymerase I makes a large rRNA precursor; polymerase II makes mRNA precursors; and polymerase III makes tRNA precursors, as well as 5S rRNA. Eukaryotic promoters that are recognized by RNA polymerase II have at least one feature in common: a consensus sequence found about twenty-five base pairs upstream from the transcription start site and containing the motif TATA. This TATA box acts either to position the start of transcription about twenty-five base pairs downstream or to increase efficiency of transcription. Polymerase II promoters generally contain other important sequences upstream from the TATA box. These sequences are typically recognition sites for specific activating proteins called transcription factors. Many eukaryotic genes also contain, or reside close to, position- and orientation-independent elements called enhancers. These are not promoter elements, but they strongly stimulate transcription of nearby genes. Polymerase I promoters also lie in the 5'-flanking regions of the genes they control, but polymerase III promoters lie wholly within their genes.

Most eukaryotic genes coding for mRNA and tRNA, and a few coding for rRNA, are interrupted by unrelated regions called intervening sequences, or introns. The other parts of the gene, surrounding the introns, are called exons. Part of the maturation of an mRNA precursor is the removal of its introns in a process called splicing.

The splicing signals in eukaryotic mRNA precursors are remarkably uniform. The first two bases of the intron are almost always GU and the last two are almost always AG. In addition, there are consensus sequences at the 5'- and 3'-splice sites.

The rRNA precursor of the protozoan *Tetrahymena* contains an intron in the 28S part of the molecule. This intron can be spliced out in vitro by the RNA precursor itself. Some mitochondrial mRNA precursors are also self-splicing.

Most eukaryotic mRNAs carry a poly(A) tail about 150–200 bases long on their 3'-ends. Poly(A) polymerase adds this tail to mRNA precursors in the nucleus, and the poly(A) apparently remains with the message through splicing and transport to the cytoplasm. Poly(A) seems to protect mRNA from degradation. The 5'-ends of eukaryotic mRNAs are blocked with structures called caps. Caps serve a dual purpose; they protect messages from degradation and they allow them entry to ribosomes for translation.

Specialized genes of eukaryotes are controlled primarily at the transcription level, by interacting with regulatory proteins. Other levels at which gene expression is controlled are post-transcription and translation. In principle, it can also be controlled at the post-translation level.   ∎

Problems and Questions

1. If posttranscriptional processing were suddenly stopped, which type of RNA would not be affected?

2. Why do changes in polymerase III promoters change the sequence of the gene product?

3. Cite two lines of evidence suggesting that nucleosomes are important to life.

4. Without histone H1, would the 11-nm chromatin fiber be possible? The 25-nm fiber? What is the rationale for your predictions?

5. In each somatic cell of a female mammal, one X chromosome is euchromatic and the other is heterochromatic. Which chromosome carries the active genes?

6. What important characteristic do eukaryotic transcription factors have in common with prokaryotic σ factors? In what important respect do they differ?

7. Here is the structure of a gene you want to examine:

   e_1 i_1 e_2 i_2 e_3 i_3 e_4

   The boxes represent exons and the single lines, introns. Draw a picture of the results of an R-looping experiment on this gene, using mRNA and single-stranded DNA. Label the parts of your diagram (e_1, i_1, etc).

8. What would the experimental results look like if the gene in problem 7 were left in double-stranded form and R-looped with mRNA? Assume that all the introns are big enough to produce a visible loop.

9. What would the results be if the experiment in problem 7 were run with an unspliced mRNA precursor?

10. Draw a diagram of the "Christmas tree" pattern of transcription of the rRNA gene cluster in amphibian nucleoli. Point out: (a) an rRNA precursor gene; (b) a nontranscribed spacer; (c) an rRNA precursor being synthesized; (d) an RNA polymerase; (e) the direction of transcription.

11. What would a cap look like in the absence of methylation?

12. If you are interested in demonstrating whether control of expression of a given gene is exerted at the transcription level, why is it not sufficient to sample mRNA levels in the cytoplasm before and after that gene's expression is activated?

13. How would you measure the stability of an mRNA?

14. An mRNA for a certain protein has an AUG initiation codon followed by twenty other codons before the first glutamine codon, yet glutamine is the first amino acid in the mature protein product of this mRNA. How can this be?

15. Gene A is stimulated by estrogen. In order to determine where the hormone-receptor complex binds to this gene, you perform a competition experiment such as the one depicted in figure 10.38. The competition exhibited by various DNA fragments from the 5'-flanking region of gene A is as follows: The 0 to −200 fragment and the −250 to −400 fragment compete weakly; the 0 to −400 fragment competes moderately; the −150 to −250 and the −200 to −300 fragments compete very well. (a) Where is the likely binding site for the hormone-receptor complex? (It helps to draw a diagram of the positions of the competitor fragments.) (b) What experiment would you do to confirm this?

16. You have discovered a new form of life that shares characteristics with both eukaryotes and prokaryotes. You are curious to know whether its mRNAs have poly(A) at their 3'-ends. What experiments would you perform to answer this question?

17. You have discovered a new class of eukaryotic algae growing near thermal vents deep in the ocean. These organisms lack TATA boxes, but have a conserved GAGA motif in about the same position. Outline an experiment to test the importance of this GAGA box. What effect would you expect if the GAGA box behaves like the TATA box in the SV40 early promoter?

18. Here is the sequence of the 5'-end of the coding strand of a hypothetical eukaryotic gene transcribed by RNA polymerase II:

    5'AATTGCGTATAAATTCGAGACTACTTACGGCTAC CCGTTAGCT

Answers appear at end of book.

Translation

In this chapter you will learn:

1. The structures of two important elements of translation: tRNA and the ribosome.

2. The nature of the genetic code.

3. The mechanism of translation.

4. How antibiotics interfere with translation.

Computer graphic image of the protein myoglobin, the oxygen-storage molecule found in muscle.
© *Dr. Arthur Lesk/Science Photo Library/Photo Researchers, Inc.*

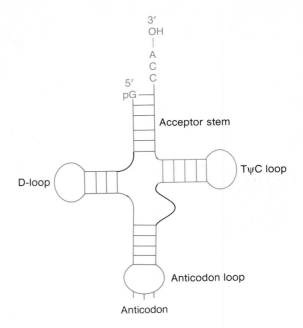

*I*n chapter 7 we learned some of the basics of translation: Messenger RNA carries the information for making proteins, encoded in three-letter codons. Transfer RNAs serve as the adapter, or translator molecules that recognize these codons in mRNA and place the corresponding amino acids into the growing protein chain. The key to this adapter function of tRNA is its double-ended nature. One end of a tRNA is coupled to a specific amino acid; the other contains an anticodon that can pair with the corresponding codon in the mRNA. Ribosomes provide the environment for the process of stringing together amino acids to make a protein; they are the protein synthesis "factories."

TRANSFER RNA

We begin our more detailed examination of protein synthesis with a closer look at the remarkable tRNA molecule.

SECONDARY STRUCTURE OF tRNA

Figure 11.1 shows again the familiar cloverleaf structure of tRNA. Remember that the molecule does not really look like this in three dimensions; the cloverleaf model simply points out the **secondary structure** of tRNA—that is, the base-paired regions of the molecule. For example, at the top of the diagram lie the two ends of the RNA: the 5'-end on the left and the 3'-end on the right. These two ends base-pair with one another to form the so-called **acceptor stem.** Four other base-paired stems occur, defining three loops: the **dihydrouracil loop (D-loop),** named for the modified uracil bases this region always contains; the **anticodon loop,** named for the all-important anticodon near its apex; and the **TψC loop,** which takes its name from a nearly invariant sequence of three bases, TψC. The psi (ψ) stands for a modified nucleoside in tRNA, *pseudouridine.* It is the same as normal uridine, except that the base is linked to the ribose through the 5-carbon of the base instead of the 1-nitrogen. The region between the anticodon loop and the TψC loop in figure 11.1 is called the **variable loop** because it varies in length from four to thirteen nucleotides; some of the longer variable loops contain base-paired stems.

Since tRNAs contain many modified bases in addition to dihydrouracil, a reasonable question would be: Is the tRNA made with modified bases, or are the bases modified after synthesis of the tRNA is finished? The answer is that tRNAs are made with the same four ordinary bases as any other RNA. Then, once transcription is complete, multiple enzyme systems modify the bases. What effects, if any, do these modifications have on tRNA function? We know they are important, because at least two tRNAs have been made in vitro with the four normal, unmodified bases, and they were unable to bind amino acids. On the other hand, D. Wang and a team of over two hundred Chinese colleagues chemically synthesized a fully modified yeast alanine tRNA and found that it was active. This demonstrated that it was not the in vitro synthesis of the inactive tRNAs that was at fault, but the fact that they were unmodified.

*T*he cloverleaf model of tRNA shows its secondary (base-pairing) structure and reveals four base-paired stems that define three loops of RNA: the D-loop, the anticodon loop, and the TψC loop. Transfer RNAs contain many modified bases, at least some of which are vital to tRNA activity.

THREE-DIMENSIONAL STRUCTURE OF tRNA

Alexander Rich and his colleagues used X-ray diffraction techniques to reveal the three-dimensional structure, or **tertiary structure,** of tRNAs. Since all tRNAs have essentially the same secondary structure, represented by the cloverleaf model, it is perhaps not too surprising that they all have essentially the same tertiary structure as well. This L-shaped structure is shown in figure 11.2.

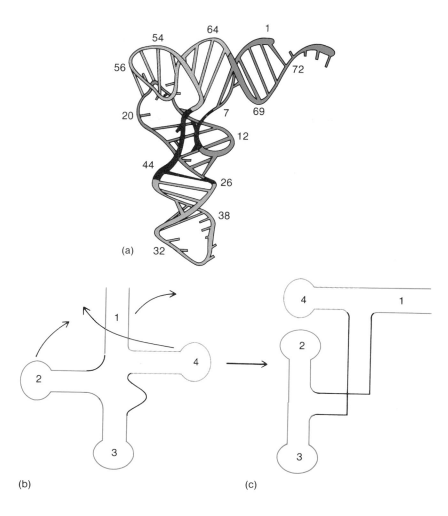

(a)

(b)   (c)

FIGURE 11.2

Three-dimensional structure of tRNA. (*a*) A planar projection of the three-dimensional structure of yeast tRNA^Phe. The various parts of the molecule are color-coded to correspond to the diagrams in figure 11.1 and to (*b*) and (*c*) of this figure. (*b*) Familiar cloverleaf structure of

tRNA with same color scheme as (*a*). Arrows indicate the contortions this cloverleaf would have to go through to achieve the approximate shape of a real tRNA, shown in (*c*).

(a) From Quigley and Rich, Science *185:436, 1974. Copyright © 1974 by the American Association for the Advancement of Science, Washington, D.C.*

The key to understanding the three-dimensional structure of tRNA is to realize that the longer the base-paired regions in a tRNA are, the more stable they will be. The centipede metaphor we used in chapter 6 helps illuminate this point. Each base pair is like a pair of legs on two embracing centipedes; the more legs, the tighter the grip. However, this imagery does not convey the idea that ten legs in a row are much better than two separated sets of five, which may not be the case with centipede legs, but is certainly true of base pairs. Transfer RNA maximizes the length of its base-paired stems by lining them up in two extended regions. One of these lies at the top of the molecule and encompasses the acceptor stem and the TψC stem; the other forms the vertical axis of the molecule and includes the D stem and the anticodon stem. Even though the two parts of each stem are not aligned perfectly and the stems therefore bend slightly, the alignment confers stability.

Figure 11.3 illustrates the double-helical nature of these extended double stems. This is a stereo diagram; you can see it in three dimensions with a stereo viewer. Without a viewer, you can force the two images to merge by staring at a point far behind the page or by crossing your eyes slightly (although the eye-crossing technique interchanges the front and back of the diagram). This may take a little practice, and some people never get the hang of it, but the fun of seeing the helical stems of tRNA in three dimensions is worth some effort. If all else fails, you can get an idea of the extent of the helical regions by observing the base pairs in one of the two-dimensional drawings in figure 11.3. The base pairs are the flat structures seen edge-on near the centers of the two helices. They are more or less parallel over the length of each helix—a feature that is especially noticeable in the horizontal helix at the top, which includes the TψC stem plus the acceptor stem.

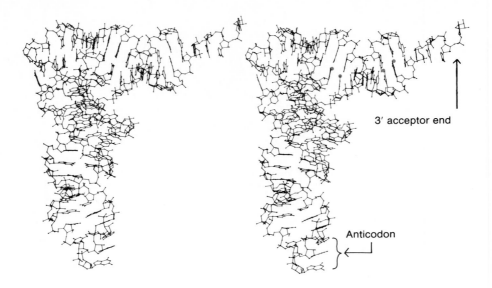

FIGURE 11.3
Stereo view of tRNA. To see the molecule in three dimensions, use a stereo viewer, or force the two images to merge either by relaxing your eyes as if focusing on something in the distance or by crossing your eyes slightly. Each colored dot indicates a base pair in the acceptor stem that is especially easy to see. It may take a little time for the three-dimensional effect to develop.

From Gary J. Quigley and Alexander Rich, "Structural Domains of Transfer RNA Molecules" in Science, *194:796–806, Figure 2, November 19, 1976. Copyright © 1976 by the American Association for the Advancement of Science. Reprinted by permission.*

Even in two dimensions, figure 11.3 demonstrates clearly that the bases of the anticodon do not extend inward to base-pair with other parts of the tRNA molecule. Rather, they protrude outward to the right so they can base-pair with a codon in an mRNA. Notice that the three bases of the anticodon are stacked parallel to one another, suggesting a helical arrangement. In fact, the structure of the tRNA twists the anticodon into a helical shape that greatly facilitates its base pairing with the codon. The ribosome presumably makes this binding even easier.

> *T*he three-dimensional structure of tRNA consists of an L-shaped framework in which the four stems stack upon one another to produce two long, double-helical regions that help stabilize the molecule. The bases of the anticodon protrude to one side and are arranged in a helical shape so they can more easily form a partial double helix by pairing with the three-base codon.

∎

AMINO ACID BINDING TO tRNA

When an amino acid binds to the 3'-end of a tRNA, we say the tRNA is **charged** with that amino acid. All tRNAs have the same three bases (CCA) at their 3'-ends, and the terminal adenosine is the target for charging. Figure 11.4 gives the chemical structure of the end of this terminal CCA and shows an amino acid attached through its acid group to the 2'-hydroxyl group of the adenosine's ribose. As we learned in chapter 6, this kind of linkage between an acid group and a hydroxyl group is called an ester bond.

Charging takes place in two steps, both catalyzed by the enzyme **aminoacyl-tRNA synthetase.** In the first step, the amino acid is activated, using energy from ATP; the product of the reaction is aminoacyl-AMP. The pyrophosphate by-product is simply the two end phosphate groups from the ATP, which were lost in forming AMP:

$$\text{amino acid} + \text{ATP} \rightarrow \text{aminoacyl-AMP} + \text{pyrophosphate}$$

The bonds between phosphate groups in ATP (and the other nucleoside triphosphates) are "high energy bonds." When they are broken, this energy is released. In this case, the energy is trapped in the *aminoacyl-AMP*, which is why we call this an *activated amino acid*. The principle is the same as burning wood to release energy, which is then used to boil water. The boiling water is activated in much the same way as the aminoacyl-AMP. In the second reaction of charging, the energy in the aminoacyl-AMP is used to transfer the amino acid to a tRNA, forming **aminoacyl-tRNA.**

$$\text{aminoacyl-AMP} + \text{tRNA} \rightarrow \text{aminoacyl-tRNA}$$

Just like other enzymes, an aminoacyl-tRNA synthetase plays a dual role. Not only does it catalyze the reaction leading to an aminoacyl-tRNA, but it also determines the

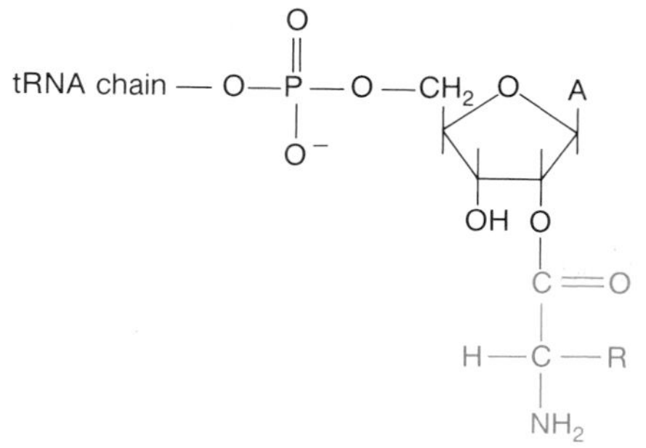

FIGURE 11.4
Linkage between tRNA and an amino acid. Most amino acids are bound by an ester linkage to the 2'-hydroxyl group of the terminal adenosine of the tRNA as shown, but some bind to the 3'-hydroxyl group.

specificity of this reaction. Only twenty synthetases exist, one for each amino acid, and they are very specific. Each will almost always place its amino acid on the right kind of tRNA.

Consider what would happen if these enzymes made many mistakes. For example, if serine-tRNA synthetase, which attaches the amino acid serine to its cognate tRNA (tRNA^Ser), suddenly started attaching lysine instead, the result would be catastrophic. The lysine-tRNA^Ser would still recognize serine codons, because its anticodon would be unchanged and the amino acid plays no role in codon-anticodon recognition. But this aminoacyl-tRNA contains lysine instead of serine, so it would insert lysines into proteins where serines ought to go. Since a protein's structure and function depend absolutely on its primary structure, these changes would in all likelihood destroy the protein's usefulness.

> *E*ach aminoacyl-tRNA synthetase couples its specific amino acid to one of its cognate tRNAs. The reaction, which is very specific for both amino acid and tRNA, requires activation of the amino acid by ATP to form aminoacyl-AMP, then transfer of the activated amino acid to the appropriate tRNA.

THE GENETIC CODE

The term **genetic code** refers to the set of three-base code words (**codons**) in mRNAs that stand for twenty amino acids in proteins. Like any code, this one had to be broken before we knew what the codons stood for. Indeed, before 1960, other more basic questions about the code included: Do the codons overlap? Are there gaps, or "commas," in the code? How many bases make up a codon? These questions were answered in the 1960s by a series of imaginative experiments, which we will describe here.

NONOVERLAPPING CODONS

A. Tsugita and H. Frankel-Conrat reasoned that if the code were nonoverlapping, a change of one base in an mRNA (a **missense mutation**) would be expected to change no more than one amino acid in the resulting protein. For example, consider the following "micromessage":

AUGCUA

Assuming that the code is triplet (three bases per codon) and this message is read from the beginning, the codons will be AUG and CUA if the code is nonoverlapping. A change in the third base (G) would change only one codon (AUG) and therefore at most only one amino acid. On the other hand, if the code were overlapping, base G could be part of three adjacent codons (AUG, UGC, and GCU). Therefore, if the G were changed, up to three adjacent amino acids could be changed in the resulting protein. But when the investigators introduced one-base alternations into mRNA from tobacco mosaic virus (TMV), they found that these caused changes in only one amino acid. Hence, the code must be nonoverlapping.

NO GAPS IN THE CODE

If the code contained untranslated gaps, or commas, mutations that add or subtract a base from a message might change a few codons, but we would expect the ribosome to get back on track after the next comma. In other words, these mutations might frequently be lethal, but there should be many cases where the mutation occurs just before a comma in the message and therefore has little if any effect. If there were no commas to get the ribosome back on track, these mutations would be lethal except when they occur right at the end of a message.

Such mutations do occur, and they are called **frame-shift mutations;** they work as follows. Consider another tiny message:

AUGCAGCCAACG

If translation starts at the beginning, the codons will be AUG, CAG, CCA, and ACG. If we insert an extra base (X) right after base U, we get:

AUXGCAGCCAACG

Now this would be translated from the beginning as AUX, GCA, GCC, AAC. Notice that the extra base changes not only the codon (AUX) in which it appears, but every codon from that point on. The **reading frame** has shifted one base to the left; whereas C was originally the first base of the second codon, G is now in that position. This could be compared to a filmstrip viewing machine that slips a cog, thus showing the last one-third of one picture together with the first two-thirds of the next.

1. Wild-type:	CAT|CAT|CAT|CAT|CAT
2. Add a base:	CAG|TCA|TCA|TCA|TCA
3. Delete a base:	CAT|CTC|ATC|ATC|ATC
4. Cross #2 and #3:	CAG|TCT|CAT|CAT|CAT
5. Add 3 bases:	CAG|GGT|CAT|CAT|CAT

FIGURE 11.5

Frameshift mutations. Line 1: An imaginary mRNA has the same codon, CAT, repeated over and over. The vertical dotted lines show the reading frame, starting from the beginning. Line 2: Adding a base, G (red), in the third position changes the first codon to CAG and shifts the reading frame one base to the left so that every subsequent codon reads TCA. Line 3: Deleting the fifth base, A (marked by the triangle), from the wild-type gene changes the second codon to CTC and shifts the reading frame one base to the right so that every subsequent codon reads ATC. Line 4: Crossing the mutants in lines 2 and 3 sometimes gives pseudo-wild-type revertants with an insertion and a deletion close together. The end result is a DNA with its first two codons altered, but every other one put back into the correct reading frame. Line 5: Adding three bases, GGG (red), after the first two bases disrupts the first two codons, but leaves the reading frame unchanged. The same would be true of deleting three bases.

On the other hand, a code with commas would be one in which each codon is flanked by one or more untranslated bases, represented by Z's in the following message. The commas would serve to set off each codon so the ribosome could recognize it:

AUGZCAGZCCAZACGZ

Deletion or insertion of a base anywhere in this message would change only a single codon. The comma (Z) at the end of the damaged codon would then put the ribosome back on the right track. Thus, addition of an extra base (X) to the first codon would give the message:

AUXGZCAGZCCAZACGZ

The first codon(AUXG) is now wrong, but all the others, still neatly set off by Z's, would be translated normally.

When Francis Crick and his colleagues treated bacteria with acridine dyes that usually cause single-base insertions or deletions, they found that such mutations were very severe; the mutant genes gave no functional product. This is what we would expect of a "comma-less" code with no gaps; base insertions or deletions cause a shift in the reading frame of the message that persists until the end of the message.

Moreover, Crick found that adding a base could cancel the effect of deleting a base, and vice versa. This phenomenon is illustrated in figure 11.5, where we start with an artificial gene composed of the same codon, CAT, repeated over and over. When we add a base, G, in the third position, we change the reading frame so that all codons

thereafter read TCA. When we start with the wild-type gene and delete the fifth base, A, we change the reading frame in the other direction, so that all subsequent codons read ATC. Crossing these two mutants sometimes gives a recombined "pseudo-wild-type" gene like the one on line 4 of the figure. Its first two codons, CAG and TCT, are wrong, but thereafter, the insertion and deletion cancel, and the original reading frame is restored; all codons from that point on read CAT.

Finally, Crick and Leslie Barnett discovered that a set of three adjacent insertions or deletions could produce a wild-type gene (figure 11.5, line 5). This of course demands that a codon consist of three, or a multiple of three, bases. As Crick remarked to Barnett when he saw the experimental result: "We're the only two [who] know it's a triplet code!"

THE TRIPLET CODE

We have been assuming that the codons contain three letters, not a multiple of three as Crick and Barnett's experiment still allowed. How do we know this is true? Marshall Nirenberg and Johann Heinrich Matthaei performed the experiment that laid the groundwork for answering this question and for breaking the genetic code itself. The experiment was deceptively simple; it showed that synthetic RNA could be translated in vitro. In particular, when Nirenberg and Matthaei translated poly(U), a synthetic RNA composed only of U's, they made polyphenylalanine. Of course, that told them that the codon for phenylalanine contained only U's. This finding by itself was important, but the long-range implication was that one could design synthetic mRNAs of defined sequence and analyze the protein products to shed light on the nature of the code. Har Gobind Khorana and his colleagues were the chief practitioners of this strategy.

Here is how Khorana's synthetic messenger experiments demonstrated that the codons contain three bases: First, if the codons contain an odd number of bases, then a repeating dinucleotide [poly(UC) or UCUCUCUC . . .] should contain two alternating codons (UCU and CUC, in this case), no matter where translation starts. The resulting protein would be a repeating dipeptide—two amino acids alternating with each other. If codons have an even number of bases, only one codon (UCUC, for example) should be repeated over and over. Of course, if translation started at the second base, the single repeated codon would be different (CUCU). In either case, the resulting protein would be a homopolypeptide, containing only one amino acid repeated over and over. Khorana found that poly(UC) translated to a repeating dipeptide, poly(serine-leucine) (figure 11.6a), proving that the codons contained an odd number of bases.

Repeating triplets translated to homopolypeptides, as had been expected if the number of bases in a codon was three or a multiple of three. For example, poly(UUC)

FIGURE 11.6
Coding properties of several synthetic mRNAs. (*a*) Poly(UC) contains two alternating codons, UCU and CUC, which code for serine (Ser) and leucine (Leu), respectively. Thus, the product is poly(Ser-Leu). (*b*) Poly(UUC) contains three codons, UUC, UCU, and CUU, which code for phenylalanine (Phe), serine (Ser), and leucine (Leu), respectively. Therefore, the product is poly(Phe), or poly(Ser), or Poly(Leu), depending on which of the three reading frames the ribosome uses. (*c*) Poly(UAUC) contains four codons in a repeating sequence: UAU, CUA, UCU, and AUC, which code for tyrosine (Tyr), leucine (Leu), serine (Ser), and isoleucine (Ile), respectively. The product is therefore poly(Tyr-Leu-Ser-Ile).

translated to polyphenylalanine plus polyserine plus polyleucine (figure 11.6*b*). The reason for three different products is that translation can start at any point in the synthetic message. Therefore, poly(UUC) can be read as UUC, UUC, etc., or UCU, UCU, etc., or CUU, CUU, etc., depending on where translation starts. In all cases, once translation begins, only one codon is encountered, as long as the number of bases in a codon is divisible by three.

The clincher was a repeating tetranucleotide, which translated to a repeating tetrapeptide. For example, poly(UAUC) yielded poly(tyrosine-leucine-serine-isoleucine) (figure 11.6*c*). As an exercise, you can write out the sequence of such a message and satisfy yourself that it is compatible with codons having three bases, or nine, or even more, but not six. Since it not likely that codons would be as cumbersome as nine bases long, three is the best choice. Look at the problem another way: Three is the lowest number that gives enough different codons to specify all twenty amino acids. (The number of permutations of four different bases taken three at a time is 4^3, or 64.) There are only sixteen two-base codons ($4^2 = 16$), not quite enough. But there would be over 200,000 ($4^9 = 262,144$) nine-base codons. Nature is usually more economical than that.

---

*T*he genetic code is a set of three-base code words, or codons, in mRNA that instruct the ribosome to incorporate specific amino acids into a polypeptide. The code is nonoverlapping; that is, each base is part of only one codon. It is also devoid of gaps, or commas; that is, each base in the coding region of an mRNA is part of a codon.

■

BREAKING THE CODE

Obviously, Khorana's synthetic mRNAs gave strong hints about some of the codons. For example, since poly(UC) yields poly(serine-leucine), we know that one of the codons (UCU or CUC) codes for serine and the other codes for leucine. The question remains: Which is which? Nirenberg developed a powerful assay to answer this question. He found that a trinucleotide was usually enough like an mRNA to cause a specific aminoacyl-tRNA to bind to ribosomes. For example, the triplet UUU will cause phenylalanyl-tRNA to bind, but not lysyl-tRNA or any other aminoacyl-tRNA. Therefore, UUU is a codon for phenylalanine. This method was not perfect; some codons did not cause any aminoacyl-tRNA to bind, even though they were authentic codons for amino acids. But it provided a nice complement to Khorana's method, which by itself would not have given all the answers either, at least not easily.

Here is an example of how the two methods could be used together: Translation of the polynucleotide poly(AAG) yielded polylysine plus polyglutamic acid plus polyarginine. There are three different codons in that synthetic message: AAG, AGA, and GAA. Which one codes for lysine? All three were tested by Nirenberg's assay, yielding the results shown in figure 11.7. Clearly, AGA and GAA cause no binding of (^{14}C)lysyl-tRNA to ribosomes, but AAG does. Therefore, AAG is the lysine codon in poly(AAG). Something else to notice about this experiment is that the triplet AAA also causes lysyl-tRNA to bind. Therefore, AAA is another lysine codon. This illustrates a general feature of the code: In most cases, more than one triplet codes for a given amino acid. In other words, the code is **degenerate.**

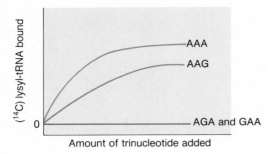

FIGURE 11.7
Binding of lysyl-tRNA to ribosomes in response to various codons. Lysyl-tRNA was labeled with radioactive carbon (^{14}C) and mixed with *E. coli* ribosomes in the presence of the following trinucleotides: AAA, AAG, AGA, and GAA. Lysyl-tRNA-ribosome complex formation was measured by binding to nitrocellulose filters. (Unbound lysyl-tRNA does not stick to these filters, but a lysyl-tRNA-ribosome complex does.) AAA was a known lysine codon, so binding was expected with this trinucleotide.

Figure 11.8 shows the entire genetic code. As predicted, there are sixty-four different codons and only twenty different amino acids, yet all of the codons are used. Three are "stop" codons found at the ends of messages, but all the others specify amino acids, which means that the code is highly degenerate. Leucine, serine, and arginine have six different codons; several others, including proline, threonine, and alanine, have four; isoleucine has three; and many others have two. Just two amino acids, methionine and tryptophan, have only one codon.

*T*he genetic code was broken by using either synthetic messengers or synthetic trinucleotides and observing the polypeptides synthesized or aminoacyl-tRNAs bound to ribosomes, respectively. There are sixty-four codons in all. Three are stop signals, and the rest code for amino acids. This means that the code is highly degenerate.

■

UNUSUAL BASE PAIRS BETWEEN CODON AND ANTICODON

How does an organism cope with multiple codons for the same amino acid? One way would be to have multiple tRNAs (*isoaccepting species*) for the same amino acid, each one specific for a different codon. In fact, this is part of the answer, but there is more to it than that; we can get along with considerably fewer tRNAs than that simple hypothesis would predict. Again, Francis Crick anticipated experimental results with insightful theory. In this case, Crick hypothesized that the first two bases of a codon must pair correctly with the anticodon according to Watson-Crick base-pairing rules, but the last base of the

	Second position				
First position (5′-end)	**U**	**C**	**A**	**G**	**Third position (3′-end)**
U	UUU }Phe UUC } / UUA }Leu UUG }	UCU UCC UCA UCG }Ser	UAU }Tyr UAC } / UAA }STOP UAG }	UGU }Cys UGC } / UGA STOP / UGG Trp	U C A G
C	CUU CUC CUA CUG }Leu	CCU CCC CCA CCG }Pro	CAU }His CAC } / CAA }Gln CAG }	CGU CGC CGA CGG }Arg	U C A G
A	AUU }Ile AUC } AUA } / AUG Met	ACU ACC ACA ACG }Thr	AAU }Asn AAC } / AAA }Lys AAG }	AGU }Ser AGC } / AGA }Arg AGG }	U C A G
G	GUU GUC GUA GUG }Val	GCU GCC GCA GCG }Ala	GAU }Asp GAC } / GAA }Glu GAG }	GGU GGC GGA GGG }Gly	U C A G

FIGURE 11.8
The genetic code. All sixty-four codons are listed, along with the amino acid for which each codes. To find a given codon, ACU for example, we start with the wide horizontal row labeled with the name of the first base of the codon (A) on the left border. Then we move across to the vertical column corresponding to the second base (C). This brings us to a box containing all four codons beginning with AC. It is now a simple matter to find the one among these four we are seeking, ACU. We see that this triplet codes for threonine (Thr), as do all the other codons in the box: ACC, ACA, and ACG. This is an example of the degeneracy of the code. Notice that three codons (red) do not code for amino acids; instead, they are stop signals.

codon can "wobble" from its normal position to form unusual base pairs with the anticodon. This proposal was called the **wobble hypothesis.** In particular, Crick proposed that a G in an anticodon can pair not only with a C in the third position of a codon, but also with a U. This would give the *wobble base pair* shown in figure 11.9*b*. Notice how the U has wobbled from its normal position to form this base pair.

Furthermore, Crick noted that one of the unusual nucleosides found in tRNA is **inosine (I),** which has a structure similar to that of guanosine. This nucleoside can ordinarily pair like guanine, so we would expect it to pair with C (Watson-Crick base pair) or U (wobble base pair) in the third position of a codon. But Crick proposed that inosine could form still another kind of wobble pair, this time with A in the third position of a codon (figure 11.9*c*). That means an anticodon with I in the first position can

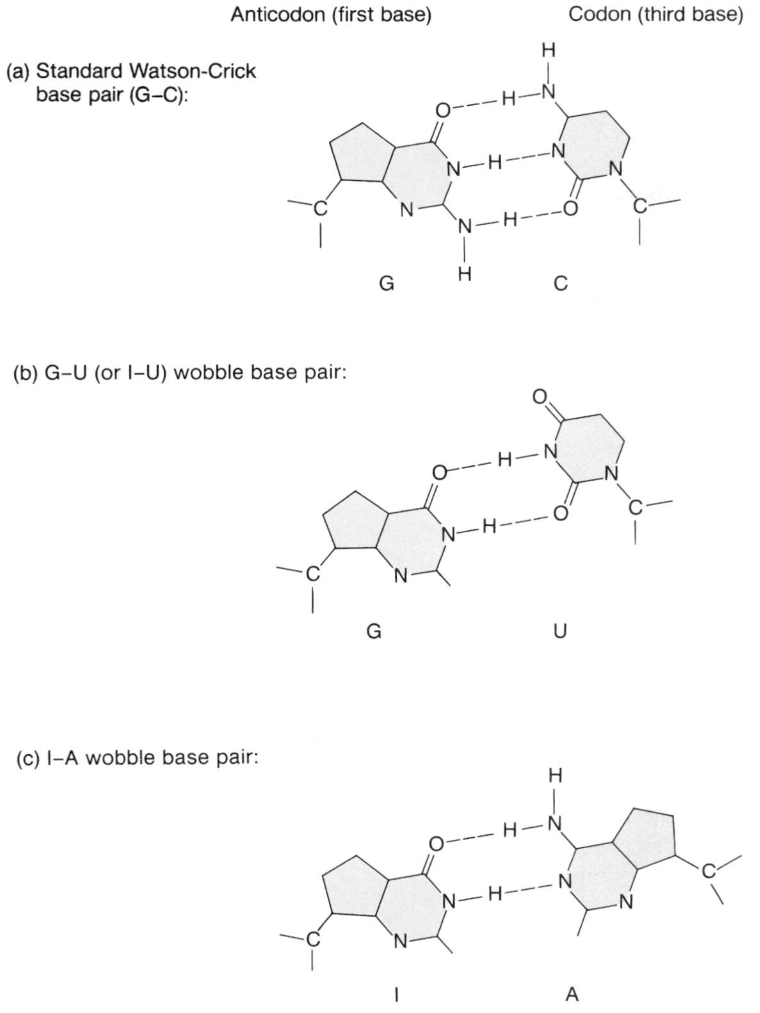

Anticodon (first base)   Codon (third base)

(a) Standard Watson-Crick base pair (G–C):

G   C

(b) G–U (or I–U) wobble base pair:

G   U

(c) I–A wobble base pair:

I   A

FIGURE 11.9

Wobble base pairs. (a) Relative positions of bases in a standard (G–C) base pair. The base on the left in (a) and in the wobble base pairs (b) and (c) is the first base in the anticodon. The base on the right is the third base in the codon. (b) Relative positions of bases in a G–U (or I–U) wobble base pair. Notice that the U has to "wobble" upward to pair with the G (or I). (c) Relative positions of bases in an I–A wobble base pair. The A has to rotate (wobble) counterclockwise in order to form this pair.

potentially pair with three different codons ending with C, U, or A. Such a situation arises in the case of isoleucine, whose three codons are AUC, AUU, and AUA. The wobble phenomenon obviously reduces the number of tRNAs required to translate the genetic code.

---

*P*art of the degeneracy of the genetic code is accommodated by isoaccepting species of tRNA that bind the same amino acid but recognize different codons. The rest is handled by wobble, in which the third base of a codon is allowed to move slightly from its normal position to form a non-Watson-Crick base pair with the anticodon. This allows the same aminoacyl-tRNA to pair with more than one codon. The wobble pairs are G–U (or I–U) and I–A.

■

THE (ALMOST) UNIVERSAL CODE

In the years after the genetic code was broken, all organisms examined, from bacteria to humankind, were shown to share the same code. It was therefore generally assumed (incorrectly, as we will see) that the code was universal, with no deviations whatsoever. This apparent universality led in turn to the notion of a single origin of present life on earth.

The reasoning for this idea goes like this: There is nothing inherently advantageous about each specific codon assignment we see. There is no obvious reason, for example, why UUC should make a good codon for phenylalanine, whereas AAG is a good one for lysine. Rather, in a sense, the genetic code may be an "accident"; it just happened to evolve that way. However, once these codons were established, there is a very good reason why they should not change: A change that fundamental would almost certainly be lethal.

Consider, for instance, a tRNA for the amino acid cysteine and the codon it recognizes, UGU. In order for that relationship to change, the anticodon of the cysteinyl-tRNA would have to change so it can recognize a different codon, say UCU, which is a serine codon. At the same time, all the UCU codons in that organism's genome that code for important serines would have to change to alternate serine codons so they would not be recognized as cysteine codons. The chances of all these things happening together, even over vast evolutionary time, are negligible. That is why the genetic code is sometimes called a "frozen accident"; once it was established, for whatever reasons, it had to stay that way. So a universal code would be powerful evidence for a single origin of life. After all, if life started independently in two places, we would hardly expect the two lines to evolve the same genetic code by accident!

In light of all this, it is remarkable that the genetic code is *not* absolutely universal; there are some exceptions to the rule. The first of these to be discovered were in the genomes of mitochondria. In mitochondria of the fruit fly *D. melanogaster,* UGA is a codon for tryptophan rather than for "stop." Even more remarkably, AGA in these mitochondria codes for serine, whereas it is an arginine codon in the universal code. Mammalian mitochondria show some deviations, too. Both AGA and AGG, though they are arginine codons in the universal code, have a different meaning in human and bovine mitochondria; there they code for "stop." Furthermore, AUA, ordinarily an isoleucine codon, codes for methionine in these mitochondria.

These aberrations might be dismissed as relatively unimportant, occurring as they do in mitochondria, which have very small genomes coding for only a few proteins and therefore more latitude to change than nuclear genomes. But there are exceptional codons in nuclear genomes and prokaryotic genomes as well. In at least three ciliated protozoa, including *Paramecium,* UAA and UAG, which are normally stop codons, code for glutamine. In the prokaryote *Mycoplasma capricolum,* UGA, normally a stop codon, codes for tryptophan. All the known deviations from the standard genetic code are summarized in table 11.1.

Clearly, the so-called universal code is not really universal. Does this mean that the evidence now favors more than one origin of present life on earth? If the deviant codes were radically different from the standard one, this might be an attractive possibility, but they are not. In many cases, the novel codons are stop codons that have been recruited to code for an amino acid: glutamine or tryptophan. There is a well-established mechanism for this sort of occurrence, as we will see later in the chapter. The only examples of codons that have switched their meaning from one amino acid to another occur in mitochondria. Again, mitochondrial genomes, because they code for far fewer proteins than nuclear genomes or even prokaryotic genomes, might be expected to change a codon safely every now and then. In summary, even if there is no universal code, there is a standard code from which the deviant ones almost certainly evolved.

*T*he genetic code is not strictly universal. In certain eukaryotic nuclei and mitochondria and in at least one bacterium, codons that cause termination in the standard genetic code can code for amino acids such as tryptophan and glutamine. In several mitochondrial genomes, the sense of a codon is changed from one amino acid to another. These deviant codes are still closely related to the standard one from which they probably evolved.

■

RIBOSOMES

In chapter 7 we learned that **ribosomes,** complex structures composed of RNA and protein, are the cell's protein factories. All ribosomes contain two subparticles; in prokaryotes these have sedimentation coefficients of 30S and 50S, and the whole ribosome sediments at 70S. The ribosomes of eukaryotes are larger and more complex: Two subunits, having sedimentation coefficients of 40S and 60S, make up an 80S ribosome.

SELF-ASSEMBLY OF RIBOSOMES

As we also learned in chapter 7, each ribosomal particle is in turn composed of at least one RNA molecule and many different proteins. The small subunit of the *E. coli* ribosome contains a 16S rRNA and twenty-one different proteins; the large subunit includes a 23S and a 5S rRNA and thirty-four different proteins. We know exactly how many proteins there are, because several research groups, primarily Masayasu Nomura's, have taken ribosomes apart and put them back together again. This work, in the best biochemical tradition of separating and reconstituting, allows us to identify the components that are necessary to yield an active ribosome.

Figure 11.10 demonstrates the resolution of the proteins from the 30S particle by polyacrylamide gel electrophoresis. The proteins are named S1–S21, the S standing for "small," since they derive from the small ribosomal subunit. The proteins of the large subunit are therefore named L1–L31, where the L, of course, stands for "large." In some cases, two or more proteins electrophoresed together, but further purification separated these proteins completely. Once the proteins were completely resolved, they could be added back to 16S rRNA to produce fully active 30S particles. The test for activity was to combine these reconstituted 30S particles with 50S particles to give 70S ribosomes that could carry out translation in vitro.

TABLE 11.1 Deviations from the "Universal" Genetic Code

Source	Codon	Usual Meaning	New Meaning
Fruit fly mitochondria	UGA	Stop	Tryptophan
	AGA & AGG	Arginine	Serine
	AUA	Isoleucine	Methionine
Mammalian mitochondria	AGA & AGG	Arginine	Stop
	AUA	Isoleucine	Methionine
	UGA	Stop	Tryptophan
Yeast mitochondria	CUN*	Leucine	Threonine
	AUA	Isoleucine	Methionine
	UGA	Stop	Tryptophan
Protozoa nuclei	UAA & UAG	Stop	Glutamine
Higher plants	UGA	Stop	Tryptophan
	CGG	Arginine	Tryptophan
Mycoplasma	UGA	Stop	Tryptophan

*N = Any base

FIGURE 11.10

Resolution of *E. coli* 30S ribosomal proteins. The proteins were electrophoresed on a polyacrylamide gel, which gave only partial resolution (fewer than twenty bands appear, so some contain more than one protein). The proteins that were incompletely resolved here were separated completely by chromatography.

From P. Traub and M. Nomura. Laboratory of Genetics, University of Wisconsin, Madison. Reprinted from Proceedings of the National Academy of Science, *Vol. 59, No. 3, fig. 2, pp. 777–84, March 1968.*

What about the spatial relationship of the ribosomal proteins to each other and to the rRNA? Two techniques, neutron scattering and immune electron microscopy, have been most valuable in providing answers to these questions, at least for the small ribosomal particle of *E. coli*. Neutron scattering works according to the same principle as X-ray crystallography. The pattern of scattering of neutrons passing through a sample of ribosomes is related to the structure of the ribosome. Immune electron microscopy relies on antibodies that bind specifically to one type of ribosomal protein. We can observe these antibodies on the surface of a ribosome by electron microscopy; in effect, the antibody is an arrow that points out the corresponding protein's location. Figure 11.11a shows the arrangement of the 30S ribosomal proteins relative to each other, and figure 11.11b superimposes the location of part of the 16S rRNA on this collection of proteins.

The 50S particle proved much more difficult to reconstitute than the small particle. In fact, the first experiments failed because the temperatures needed for reassociation of the 50S particle were so high that some of the proteins denatured. Later, tricks were discovered that could allow 50S ribosomes from *E. coli* to reassemble at a lower temperature. This means that the whole ribosome is a *self-assembly system;* it can be reconstituted from its component parts, with no intervention from outside enzymes or factors.

We assume that a similar assembly process occurs in vivo in prokaryotes. But what about assembly of eukaryotic ribosomes? Do they also self-assemble, and if so, where? We cannot answer the first part of that question because the complex eukaryotic ribosome has not been self-assembled in vitro. However, we do know the site of assembly of the eukaryotic ribosome: the nucleolus. The large ribosomal RNA genes are located in the nucleolus, and therefore the 5.8S, 18S, and 28S rRNAs are made there. The ribosomal proteins are made on cytoplasmic ribosomes, of course, so they must be imported into the nucleolus. There they join with rRNAs and build new ribosomes.

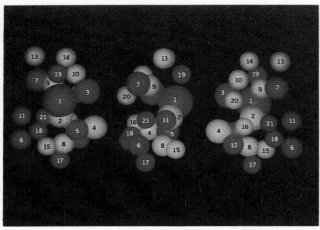

(a)

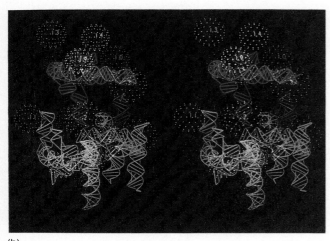

(b)

FIGURE 11.11

Arrangement of components in the *E. coli* 30S ribosomal particle.
(*a*) Positions of proteins. The relative positions of the ribosomal
proteins (numbered 1–21) are shown from three perspectives: the side
facing away from the 50S particle (left); edge-on (middle); and the
side facing the 50S particle (right). (*b*) Relationship of proteins and
rRNA in the *E. coli* 30S ribosomal particle. The positions of the
proteins, indicated by dotted spheres, are the same as on the left in
(*a*). Only the best-studied 60% of the 16S rRNA is shown. The 5′-
part of the RNA is in red, the center in yellow, and the 3′-part in
blue.

(a) From M. S. Capel, M. Kjeldgaard, D. M. Engelman, and P. B. Moore. 1988.
Journal of Molecular Biology. *Reprinted by permission of the Academic Press.*
(b) From Peter B. Moore. Reprinted by permission of Nature, *Vol. 331, p. 224,*
January 1988. Copyright © 1988 Macmillan Magazines Limited.

R ibosomal subunits can be dissociated into their
component parts—RNAs and proteins—and then
reassociated. Thus, the ribosome is a self-assembly system
that needs no help from the outside.

FUNCTIONS OF RIBOSOMAL PROTEINS

The in vitro reassociation of ribosomal particles makes it
possible to examine the role of each ribosomal protein in
protein synthesis. The basic idea of such an experiment is
to create a mutant with a suspected defect in its ribosomes, then reassemble the particles in several different
reactions. In each reaction there is one ribosomal protein
from the mutant bacterium, plus all the other proteins from
a wild-type bacterium. The protein from the mutant is
changed in each reaction until a defective ribosome results. The mutant protein in that reassembly reaction must
be the one that caused the defect.

The classic example of this approach is the identification of the protein responsible for resistance to the
antibiotic **streptomycin.** Streptomycin kills bacterial cells
by causing their ribosomes to misread mRNAs. For
example, ribosomes given poly(U) as mRNA would ordinarily only incorporate phenylalanine into polyphenylalanine. In the presence of streptomycin, however, ribosomes also incorporate isoleucine, serine, and leucine.
Apparently, streptomycin distorts the ribosome in such a
way that it partially destroys fidelity of translation. Where
does streptomycin act? To answer that question, we can
take components from wild-type (streptomycin-sensitive)
ribosomes and from streptomycin-resistant ribosomes, mix
them, and see whether the reconstituted ribosomes are
streptomycin-sensitive or streptomycin-resistant.

For example, when the 30S particle came from a
streptomycin-resistant bacterium and the 50S particle
came from a streptomycin-sensitive bacterium, the reconstituted ribosome was streptomycin-resistant. Therefore,
streptomycin apparently acts on the 30S particle. Does the
streptomycin act on the ribosomal RNA or on protein?
Reconstitution experiments showed that if the proteins
came from a resistant bacterium, the ribosome would be
resistant, regardless of the origin of the rRNA. So streptomycin involves the protein, not the RNA. But which
protein(s) are involved? The same kind of experiment
showed that when the protein S12 came from a resistant
bacterium, the reconstituted ribosome was resistant, regardless of the origin of the other proteins. Thus, S12 is
the apparent site of action of streptomycin.

This does not necessarily mean that S12 is the binding
site for streptomycin. In fact, the antibiotic will still bind
tightly to 30S particles in the absence of S12, but only if
two other proteins, S3 and S5, are present. Furthermore,
an antibody against S12 only faintly inhibits streptomycin
binding to ribosomes. Apparently, therefore, even though
S12 seems to be the locus of action of streptomycin, other
ribosomal proteins define the antibiotic binding site.

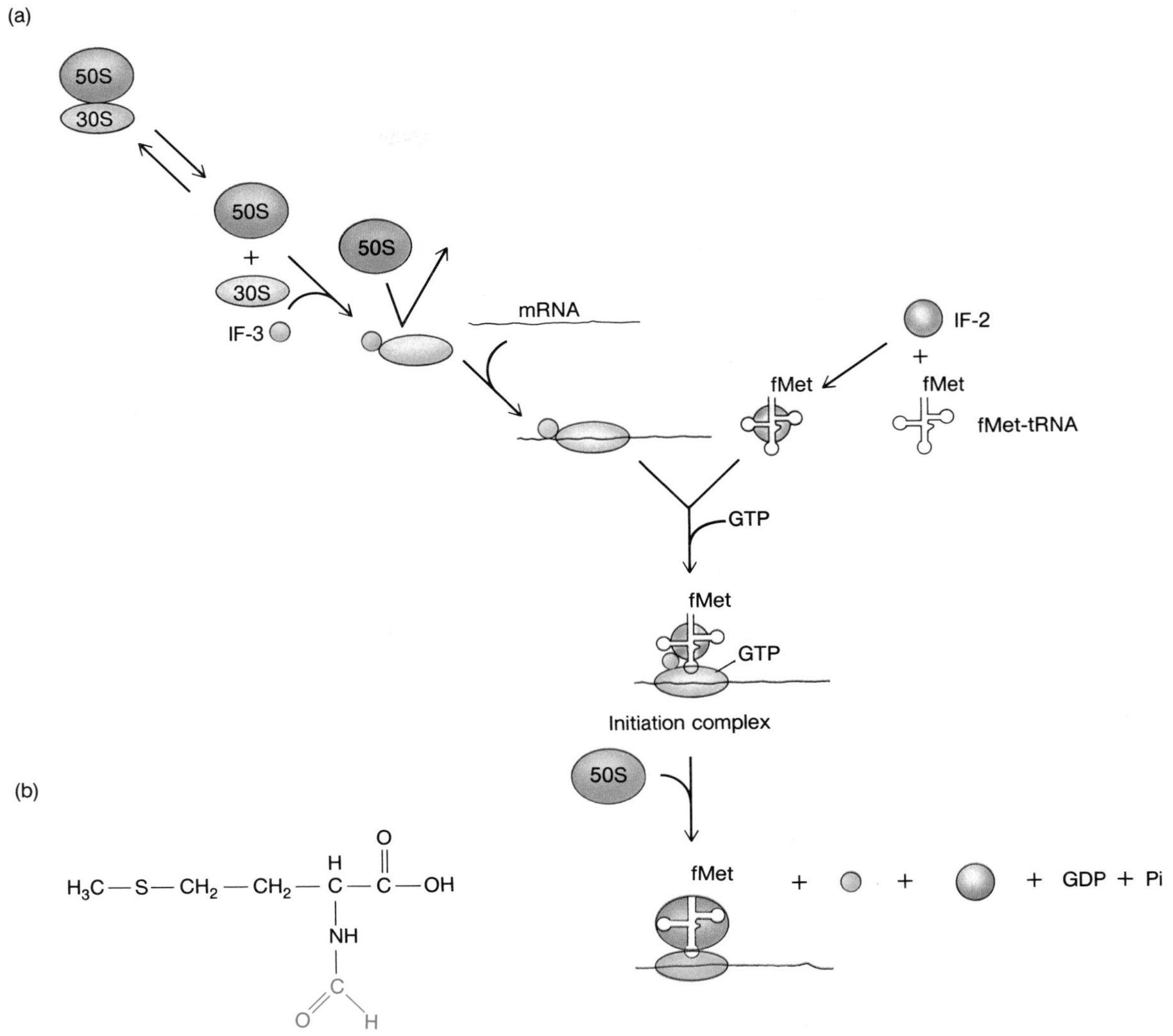

(a)

(b)

FIGURE 11.12

Initiation of translation in *E. coli*. (*a*) Initiation sequence. First, the 70S ribosome dissociates into its component subunits (red and blue). The initiation factor IF-3 (yellow) aids this dissociation by binding to the free 30S particle and preventing its reassociation with the 50S subunit. IF-1, not shown, is also thought to participate at this point. Next, IF-3 promotes binding of mRNA to the 30S particle. In the next step, IF-2 (purple) binds fMet-tRNA and attaches it to the 30S particle; GTP also binds in this reaction, completing the initiation complex. Finally, the 50S particle joins the initiation complex, and the two initiation factors dissociate; at the same time, GTP is broken down to GDP and inorganic phosphate. (*b*) Structure of N-formyl methionine, with formyl group in green.

> *I* n vitro reconstitution experiments with mutant and wild-type ribosomal proteins allow us to pinpoint the defective component by mixing one mutant protein at a time with all other wild-type components and observing the effects on the reconstituted ribosomes. This strategy showed that protein S12 is the site of action of the antibiotic streptomycin.

∎

MECHANISM OF TRANSLATION

The mechanism of translation is considerably more complex than that of transcription, as one might guess from the relative complexities of the ribosome and RNA polymerase. Translation, like transcription, is divided into three phases: initiation, elongation, and termination. We will deal with each in turn.

INITIATION

Initiation of translation in *E. coli* takes place in several steps (figure 11.12*a*). First, a ribosome that has finished translating an mRNA must be separated into its component subunits. An **initiation factor, IF-3,** effects this step, joining the 30S subunit and preventing its reassociation with free 50S subunits. **IF-1** is also thought to participate in this dissociation process, although its role is not as clear.

Next, the 30S subunit binds a new mRNA in a reaction directed by IF-3. **Initiation factor 2 (IF-2)** then joins the growing collection of molecules and attaches the first aminoacyl-tRNA to the 30S subunit. GTP also binds to this *initiation complex*. Next, the 50S subunit joins the complex, and IF-3 dissociates. Finally, GTP is broken down to GDP and phosphate, and IF-2 dissociates, leaving the complete ribosome bound to a fresh mRNA and the first aminoacyl-tRNA. Initiation is now complete.

In prokaryotes, the first amino acid, the one that initiates protein synthesis, is always the same: **N-formyl methionine (fMet).** The structure of fMet is given in figure 11.12*b*. N-formyl methionyl-tRNA (fMet-tRNA) forms in two steps. First, ordinary methionine binds to a special tRNA, then the methionine on the tRNA receives the formyl group. This unique amino acid participates only in initiation; it never goes into the interior of a polypeptide.

A Special tRNA for Initiation

The codon for fMet is AUG (or sometimes GUG). But these codons only cause initiation when they occur near the beginning (the 5'-end) of an mRNA. If they appear in the middle of a message, they simply cause elongation, coding for methionine and valine, respectively. Thus, something special about the context of the codons AUG and GUG at the beginning of a message identifies them as initiation codons. These initiation codons are recognized by a special tRNA called $tRNA_F^{Met}$, which bears fMet. AUG and GUG in the interior of a message attract $tRNA_M^{Met}$ and $tRNA^{Val}$, bearing methionine and valine, respectively.

Eukaryotes perform the initiation step by a mechanism similar to that used by prokaryotes, but ordinary methionine, instead of fMet, plays the initiating role. The methionine may be ordinary, but it is set apart by being bound to a special tRNA, analogous to $tRNA_F^{Met}$, that participates only in initiation.

From this discussion, you might conclude that fMet would always be the first (the N-terminal) amino acid in mature prokaryotic proteins and that methionine would be the first in eukaryotic proteins. But, in fact, fMet is never found in mature prokaryotic proteins, and methionine is the N-terminal amino acid only about half the time. Methionine is rarely, if ever, the first amino acid in eukaryotic proteins. This means that something removes the formyl group from the fMet in prokaryotic proteins during protein synthesis, and that the whole fMet is frequently removed. In eukaryotes, the initiating methionine must usually, if not always, be removed.

I nitiation consists of binding an mRNA and the first aminoacyl-tRNA to the ribosome to form an initiation complex. This complex is built in stages by three initiation factors (IF-1, IF-2, and IF-3). Prokaryotes use a special amino acid (N-formyl methionine, fMet) and tRNA ($tRNA^{Met}$) for initiation. Eukaryotes use a unique tRNA, but unmodified methionine. AUG is ordinarily the initiating codon in prokaryotes and eukaryotes.

■

mRNA Ribosome Binding Sites

We have mentioned that there must be something special about the context of AUG codons at the beginning of a message that identifies them as initiation codons rather than ordinary methionine codons. Presumably, that "something" could be either a special primary structure (sequence of bases) or a characteristic secondary structure of the mRNA, such as a hairpin loop. Whatever the signal, it could be expected to be part of the ribosome binding site, the place on the mRNA where ribosomes first bind to initiate translation.

Using the RNA phage R17 as a model, Joan Steitz and her colleagues showed at the end of the 1960s that there is no secondary structure common to all initiation codons, even in this one phage. Furthermore, relaxing the secondary structure of the phage RNA increases, rather than decreases, its overall efficiency of translation. Therefore, it appears that ribosomes do not recognize a common secondary structure as a binding signal.

The alternative was to look for a common primary structure. J. Shine and L. Dalgarno pointed out that all the initiation regions recognized by *E. coli* ribosomes contained at least part of the sequence 5'-AGGAGGU-3' a few bases upstream from the initiation codon. This sequence is complementary to a sequence contained at the very 3'-end of *E. coli* 16S rRNA: 5'-CACCUCCUUA-3'. Base pairing could therefore occur as follows:

16S rRNA: 3'-AUUCCUCCAC-5'
Initiation site: 5'-AGGAGGU-3'

This would provide a convenient mechanism for binding an mRNA to a ribosome. The 16S rRNA in the ribosome would simply base-pair with the **Shine-Dalgarno sequence (SD sequence)** in the initiation region of the mRNA. Most mRNAs do not have such a good fit with the 16S rRNA sequence. The R17 coat gene, for instance, has only three

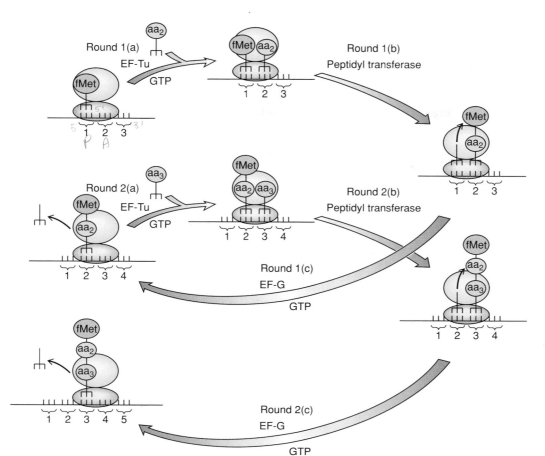

FIGURE 11.13
Elongation in translation. Note first of all that this is a highly schematic view of protein synthesis. For example, tRNAs are represented by forklike structures that show only the two business ends of the molecule. Round 1: (*a*) EF-Tu brings in the second aminoacyl-tRNA (yellow) to the A site on the ribosome. The P site is already occupied by fMet-tRNA (red). (*b*) Peptidyl transferase forms a peptide bond between fMet-tRNA and the second aminoacyl-tRNA. (*c*) In the translocation step, EF-G shifts the message and tRNAs one codon's width to the left. This moves the dipeptidyl-tRNA into the P site, bumps the deacylated tRNA aside, and opens up the A site for a new aminoacyl-tRNA. In round 2, these steps are repeated to add one more amino acid (green) to the growing polypeptide.

bases that can base-pair, yet it is still translated actively. But mRNAs that can form fewer than three base pairs are not translated well.

Eukaryotic mRNAs do not contain a sequence analogous to the Shine-Dalgarno sequence of prokaryotes. Instead, the cap at the 5'-end of a eukaryotic mRNA (chapter 10) seems to provide a "docking site" for ribosomes. First, a **cap binding protein (CBP)** attaches to the cap; then initiation factors bind to CBP and promote 40S ribosomal subunit binding to form an initiation complex.

M ost prokaryotic mRNAs have a sequence of bases (the Shine-Dalgarno sequence) just to the 5'-side of the initiation codon that is more or less complementary to a sequence at the very 3'-end of the 16S rRNA in the ribosome. Base pairing between these two regions of RNA presumably helps the ribosome recognize and bind to the initiation regions of mRNAs. Eukaryotic mRNAs have no Shine-Dalgarno sequences. Instead, the cap ensures binding of mRNA to ribosomes.

ELONGATION: ADDING AMINO ACIDS TO THE GROWING CHAIN

As suggested in the preceding discussion, ribosomes translate an mRNA in the 5' → 3' direction, the same direction in which the RNA was made. We will see that the protein grows in the amino (N) to carboxyl (C) direction. In other words, the N-terminal amino acid is added first, and the C-terminal amino acid is added last. Therefore, the 5'-end of the mRNA encodes the amino end of the protein.

The elongation step in protein synthesis is better visualized than described verbally. Figure 11.13 schematically depicts two rounds of elongation (adding two amino acids to a growing polypeptide chain) in *E. coli*. We start with fMet-tRNA bound to a ribosome. There are two binding sites for aminoacyl-tRNAs on the ribosome, called the *P (peptidyl) site* and the *A (aminoacyl) site*. In our schematic diagram, the P site is on the left and the A site is on the right. The fMet-tRNA is in the P site. Here are the elongation events:

1. To begin elongation, we need another amino acid to join with the first. This second amino acid arrives bound to a tRNA, and the nature of this

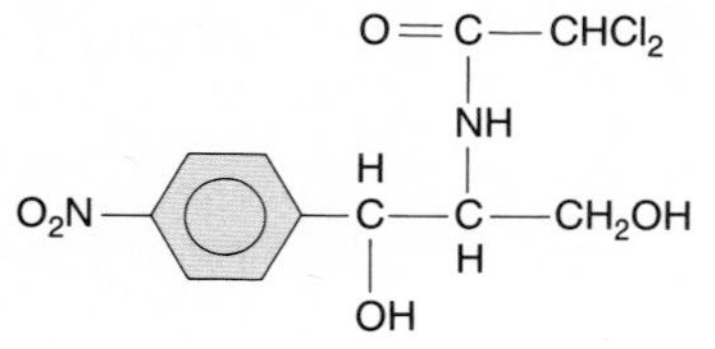

FIGURE 11.14
Chloramphenicol.

aminoacyl-tRNA is dictated by the second codon in the message. The second codon is in the A site, which is otherwise empty, so our second aminoacyl-tRNA will bind to this site. Such binding requires a catalyst and energy. The catalyst is a protein *elongation factor* (EF) known as **EF-Tu.** GTP provides the energy.

2. Next, the first peptide bond forms. An enzyme called **peptidyl transferase**—an integral part of the large ribosomal subunit—transfers the fMet from its tRNA in the P site to the aminoacyl-tRNA in the A site. This forms a two-amino acid unit called a *dipeptide* linked to the tRNA in the A site. The whole assembly is called a *dipeptidyl-tRNA*. The tRNA in the P site, minus its amino acid, is said to be *deacylated*. The peptidyl transferase step in prokaryotes is inhibited by an important antibiotic called **chloramphenicol** (figure 11.14).

   This drug has no effect on most eukaryotic ribosomes, which makes it selective for bacterial invaders in higher organisms. However, the mitochondria of eukaryotes have their own ribosomes, and chloramphenicol does inhibit their peptidyl transferase. Thus, chloramphenicol's selectivity for bacteria is not absolute.

3. In the next step, called **translocation,** the mRNA with its peptidyl-tRNA attached in the A site moves one codon's length to the left. This has the following results: (a) The deacylated tRNA in the P site (the one that lost its amino acid during the peptidyl transferase step when the peptide bond formed) leaves the ribosome. (b) The dipeptidyl-tRNA in the A site, along with its corresponding codon, moves into the P site. (c) The codon that was "waiting in the wings" to the right moves into the A site, ready to interact with an aminoacyl-tRNA. Translocation requires an elongation factor called EF-G plus energy from GTP.

The process then repeats itself to add another amino acid: (1) EF-Tu, in conjunction with GTP, brings the appropriate aminoacyl-tRNA to match the new codons in the A site. (2) Peptidyl transferase brings the dipeptide from the P site and joins it to the aminoacyl-tRNA in the A site, forming a tripeptidyl-tRNA. (3) EF-G, using energy from GTP, translocates the tripeptidyl tRNA, together with its mRNA codon, to the P site.

We have now completed two rounds of peptide chain elongation. We started with an aminoacyl-tRNA (fMet-tRNA) in the P site and we have lengthened the chain by two amino acids to a tripeptidyl-tRNA. This process continues over and over until the ribosome reaches the last codon on the message. The protein is now complete; it is time for chain termination.

Some controversy continues about the two-site model presented here, but there is general agreement about the existence of at least the A and P sites, based on experiments with the antibiotic **puromycin** (figure 11.15). This drug has a structure that resembles an aminoacyl-tRNA; as such, it can be accepted by the A site of the ribosome. Then, since puromycin has an amino group in the same position as an aminoacyl-tRNA would, peptidyl transferase forms a peptide bond between the peptide in the P site and puromycin in the A site, yielding a *peptidyl-puromycin*. At this point, the ruse is over. Puromycin has no association with the mRNA and cannot be translocated to the P site. Therefore, the peptidyl-puromycin dissociates from the ribosome, aborting translation prematurely. This is why puromycin kills bacteria and other cells.

The link between puromycin and the two-site model is this: Before translocation, because the A site is occupied by a peptidyl-tRNA, puromycin cannot bind and release the peptide; after translocation, the peptidyl-tRNA has moved to the P site, and the A site is open. At this point, puromycin can bind and release the peptide. We therefore see two states the ribosome can assume: puromycin-reactive and puromycin-unreactive. Those two states require at least two binding sites on the ribosome for the peptidyl-tRNA.

The model in figure 11.13 is simplified for ease of understanding; for example, it pictures the ribosome as a two-dimensional entity, with the mRNA traveling in only one dimension. This does not accurately reflect the real situation, which must exist in three dimensions. Recent evidence suggests that the mRNA enters and leaves the ribosome at adjacent sites. If so, it must make at least one U-turn within the ribosome.

E longation takes place in three steps: (1) EF-Tu, with energy from GTP, binds an aminoacyl-tRNA to the ribosomal A site. (2) Peptidyl transferase forms a peptide bond between the peptide in the P site and the newly arrived aminoacyl-tRNA in the A site. This lengthens the peptide by one amino acid and shifts it to the A site. (3) EF-G, with energy from GTP, translocates the growing peptidyl-tRNA, with its mRNA codon, to the P site.

■

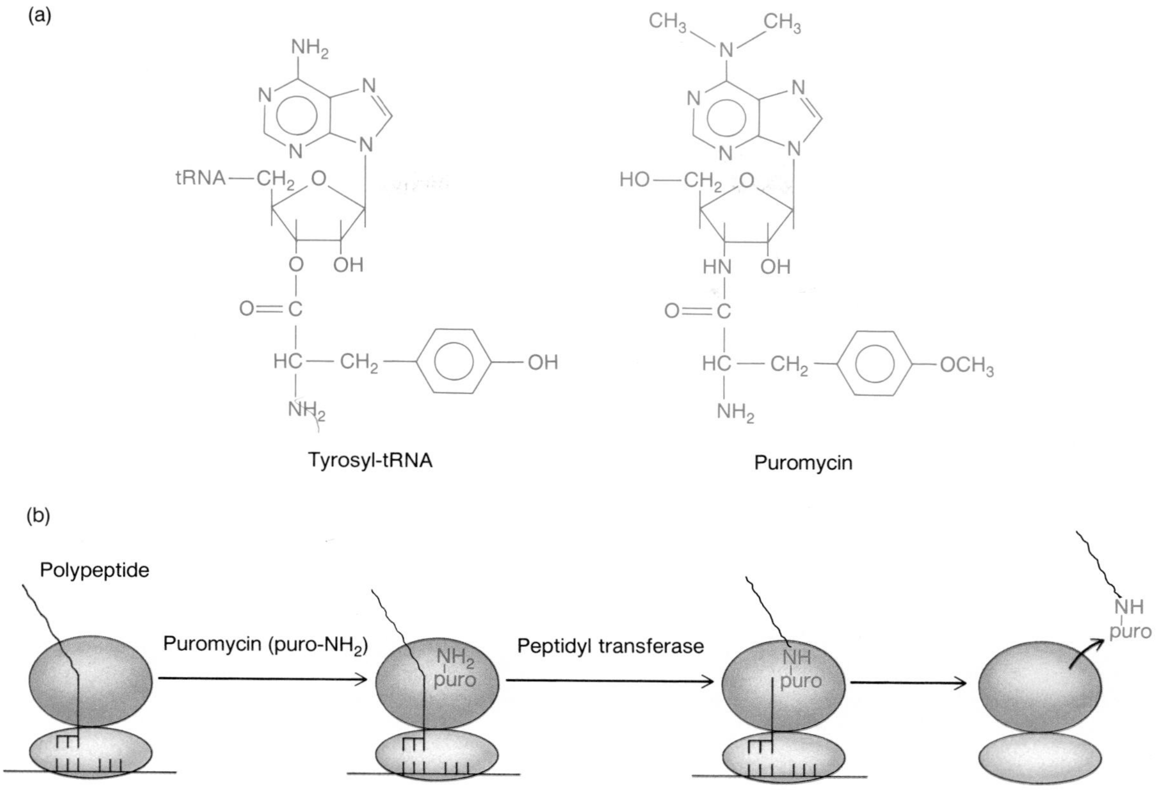

(a)

Tyrosyl-tRNA

Puromycin

(b)

Polypeptide

Puromycin (puro-NH₂)

Peptidyl transferase

FIGURE 11.15

Puromycin structure and activity. (*a*) Comparison of structures of tyrosyl-tRNA and puromycin. Note the rest of the tRNA attached to the 5′-carbon in the aminoacyl-tRNA, where there is only a hydroxyl group in puromycin. (*b*) Mode of action of puromycin. First, puromycin (puro-NH₂) binds to the open A site on the ribosome. (The A site must be open in order for puromycin to bind.) Next, peptidyl transferase joins the peptide in the P site to the amino group of puromycin in the A site. Finally, the peptidyl-puromycin dissociates from the ribosome, terminating translation prematurely.

TERMINATION: RELEASE OF THE FINISHED POLYPEPTIDE

In our discussion of the genetic code, we saw that three codons (UAG, UAA, and UGA), instead of coding for amino acids, cause termination of protein synthesis. In the early 1960s, Seymour Benzer and his colleagues discovered the first stop codon (later determined to be UAG) as a mutation in phage T4 that rendered the phage incapable of growing, except in special **suppressor** strains of *E. coli* that counteract the effect of the mutation. These workers called the mutation **amber,** and that name has stuck to the codon **UAG** ever since (see box 11.1). The reason for the lethality of the amber mutation is that it causes premature termination in the middle of messages. This was first demonstrated with amber mutations in the alkaline phosphatase gene of *E. coli.* Such mutants produce no full-length alkaline phosphatase; instead, they make fragments that begin at the amino terminus of the protein and stop somewhere in the middle.

Later, a new mutation occurred that could not grow in **amber suppressor** strains, but could grow in certain other suppressor strains. In keeping with the colorful name amber, these new mutants were called **ochre;** they involve the codon **UAA,** which also causes premature termination. The third stop codon, **UGA,** has been named in kind: **opal.**

The sequences of the amber and ochre codons were discovered in 1965, not by sequencing mRNA, but by sequencing protein products. The rationale was as follows: Amber mutants regularly *reverted* to a phenotype that no longer required the amber suppressor. But these *revertants* did not usually give a product exactly like that in wild-type cells. Instead, one amino acid, the one corresponding to the position of the amber mutation, was different. Martin Weigert and Alan Garen studied such revertants of an amber mutation at one position in the alkaline phosphatase gene of *E. coli.* The amino acid at this position in wild-type cells was tryptophan, whose sole codon is UGG. Since the amber mutation originated with a one-base change, we already know that the amber codon is related to UGG by a one-base change. To find out what this change was, Weigert and Garen determined the amino acids inserted in this position by several different revertants. The amino acids found there were serine, tyrosine,

How the Amber Mutation Got Its Name

Molecular geneticists frequently invent colorful names for the materials and phenomena they discover, and some of these names actually make it into the normally dry scientific literature. The CCAAT box is one example we have encountered; another, which we have not discussed, is "magic spot," an unusual nucleotide that accumulates when bacteria are starved for amino acids. The amber codon is another well known example, but the origin of its name is somewhat mysterious.

Here is the story, related to us by the man who did the naming, Dr. Harris Bernstein. One evening, when Bernstein was a student at Cal Tech, he went down to Seymour Benzer's lab to try to persuade his friends, Charles Steinberg and Richard Epstein, to go to the movies with him. He was unsuccessful, because Steinberg and Epstein were busy making mutant phages and could not leave. Instead, they persuaded Bernstein to make

himself useful and help them. As they worked, they explained the experiments to Bernstein, who became intrigued with the project and made a prediction about its outcome. The others disagreed, so they made a bet. Whoever guessed correctly about the characteristics of the mutant phage would get to name the mutant. Bernstein guessed that it would grow on a mutant strain of *E. coli,* but not on wild-type. He proposed to name the mutant phage for his mother, Mrs. Bernstein.

Sure enough, the mutant behaved as Bernstein predicted. He chuckles now about the fact that his reasons were faulty and that he managed to kill all the bacteria he plated that night because he heated his wire loop too much. Nevertheless, the honor of naming was his. As it turned out, the mutant was not exactly called Bernstein, but the English equivalent, amber, and the name has stuck. ∎

leucine, glutamic acid, glutamine, and lysine. Even without a microcomputer, it was possible to deduce that UAG is the only codon that can yield at least one codon for each of those amino acids, including tryptophan, by a one-base change (figure 11.16).

By the same logic, including the fact that amber mutants can mutate by single-base changes to ochre mutants, Sydney Brenner and his colleagues reasoned that the ochre codon must be UAA. Severo Ochoa and his group verified that UAA is a stop signal when they showed that the synthetic message AUGUUUUAAA---directed the synthesis and release of the dipeptide fMet-Phe. (AUG codes for fMet; UUU codes for Phe; and UAA codes for stop.) With UAG and UAA assigned to the amber and ochre codons, respectively, UGA must be the opal codon. These three codons, because they code for no amino acid, are frequently referred to as **nonsense codons.** This is really a misnomer. They make perfect sense; they say, "Stop."

How do suppressors overcome the lethal effects of premature termination signals? A clue came from an experiment showing that an amber mutation could be suppressed in vitro by purified tRNA from an amber suppressor strain. Therefore, tRNA is the suppressor molecule. Furthermore, suppressor strains insert amino acids into polypeptides in the position corresponding to the stop codon. Together, these data suggest that a tRNA from the

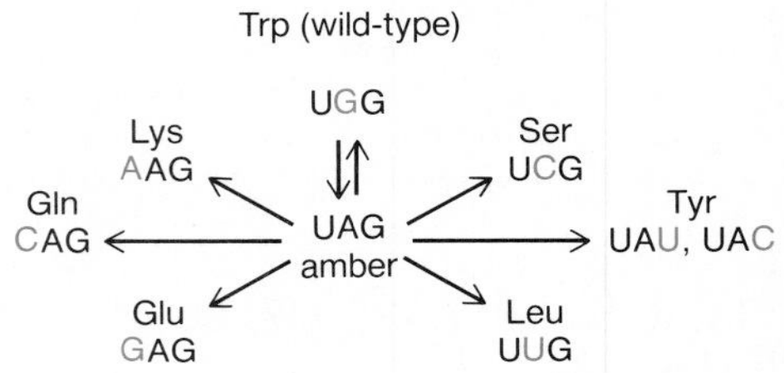

FIGURE 11.16

The amber codon is UAG. The amber codon (middle) came via a one-base change from the tryptophan codon (UGG), and the gene reverts to a functional condition in which one of the following amino acids replaces tryptophan: serine, tyrosine, leucine, glutamic acid, glutamine, or lysine. The red represents the single base that is changed in all these revertants, including the wild-type revertant that codes for tryptophan.

suppressor strain recognizes the stop codon and inserts an amino acid, thus preventing termination. Proof of this mechanism came from experiments on an amber suppressor strain that inserts a tyrosine in the position corresponding to an amber codon. A tRNA^{Tyr} from this strain has an altered anticodon. Instead of GUA, which would respond to the tyrosine codon UAC, it has the anticodon CUA, which responds to the amber codon UAG and inserts a tyrosine (figure 11.17). Thus, the suppressor is itself

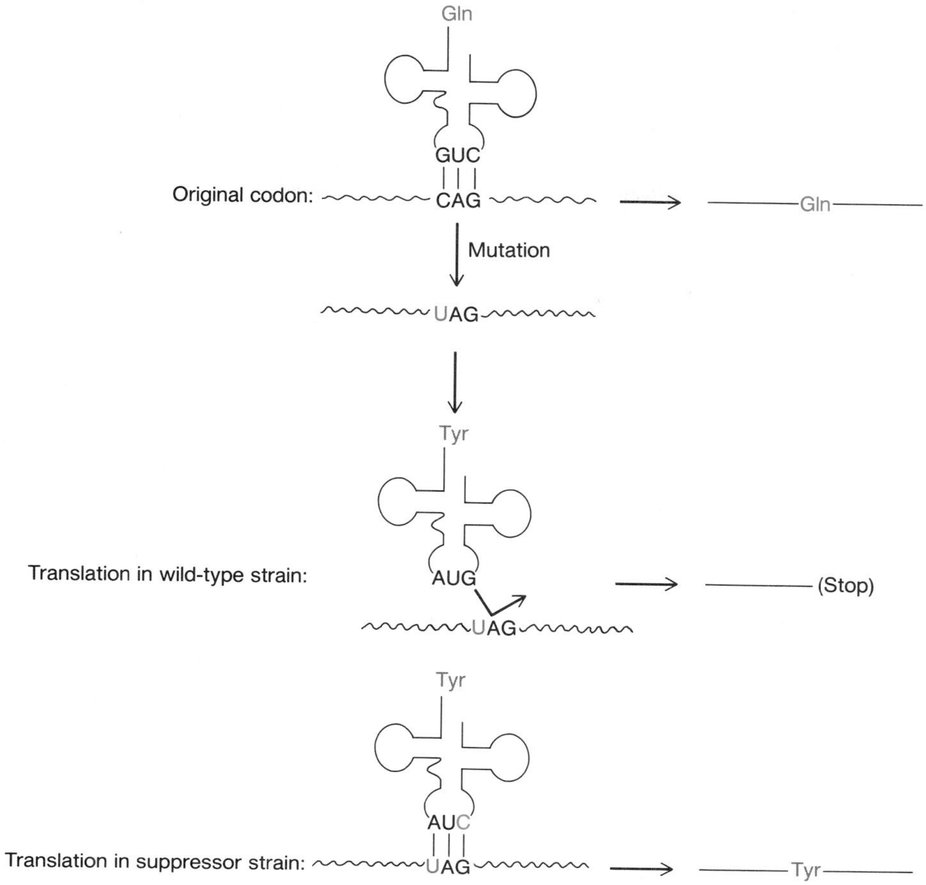

FIGURE 11.17

Mechanism of suppression. Top: The original codon in the wild-type *E. coli* gene was CAG, which was recognized by a glutamine tRNA. Middle: This codon mutated to UAG, which was translated as a stop codon by a wild-type strain of *E. coli*. Notice the tyrosine tRNA, whose anticodon (AUG) cannot translate the amber codon. Bottom: A suppressor strain contains a mutant tyrosine tRNA with the anticodon AUC instead of AUG. This altered anticodon recognizes the amber codon and causes the insertion of tyrosine (gray) instead of allowing termination.

a mutant. Its mutation is in a gene for a tRNA; in particular, in the part of the gene that specifies the anticodon.

We have seen that three codons, UAG, UAA, and UGA, can cause premature termination when they occur in the middle of a message. Are these actually the termination signals at the ends of messages? Every gene for an mRNA sequenced to date (except in certain mitochondria) has at least one of these codons at the end, just after the codon for the last amino acid in the protein. Frequently there are two stop codons in a row, such as the UAAUAG sequence at the end of the coat gene in R17 phage.

Because a special tRNA participates in initiation of translation, it would seem logical to propose a similar mechanism for termination, especially since the termination signals are three-base codons just like all other codons. Indeed, it was suggested that special tRNAs recognize the termination signals. However, when Mario Capecchi devised an assay for termination in 1967, he discovered that when he primed the ribosome with a polynucleotide containing an early stop codon, termination was triggered, not by RNA, but by protein factors. These factors, termed **release factor-1** and **release factor-2 (RF-1** and **RF-2)**, recognize the stop codons as follows: RF-1 reacts to UAA and UAG; RF-2 reacts to UAA and UGA. Both cause breakage of the bond between the polypeptide and its tRNA, releasing the product from the ribosome. Another factor, **RF-3,** appears to cooperate with both of the others. In eukaryotes, a single release factor, **eEF,** causes termination. This reaction, unlike the termination reactions in prokaryotes, requires GTP.

T ranslation release factors, RF-1 and RF-2 recognize the stop signals UAA, UAG, and UGA, and along with RF-3, cause termination in prokaryotes. Along with GTP, eEf performs the same function in eukaryotes. When a mutation creates one of these stop signals in the middle of a message, premature termination occurs. Such mutations can be overridden by suppressors—tRNAs that respond to termination codons by inserting an amino acid, thus preventing termination.

■

Summary

T he cloverleaf model of tRNA shows its secondary (base-pairing) structure and reveals four base-paired stems that define three loops of RNA: the D-loop, the anticodon loop, and the TψC loop. Transfer RNAs contain many modified bases, at least some of which are vital to tRNA activity. The three-dimensional structure of tRNA consists of an L-shaped framework, with the bases of the anticodon protruding to one side and arranged in a helical shape for easy pairing with the three-base codon. Each aminoacyl-tRNA synthetase couples a specific amino acid to one of its cognate tRNAs.

The genetic code is a set of three-base code words, or codons, in mRNA that instruct the ribosome to incorporate specific amino acids into a polypeptide. The code is nonoverlapping; that is, each base is part of only one codon. It is devoid of gaps, or commas; each base in the coding region of an mRNA is part of a codon.

The code is also degenerate. More than one codon can code for one amino acid. Part of the degeneracy of the code is accommodated by isoaccepting species of tRNA that bind the same amino acid but recognize different codons. The rest is handled by wobble, in which the third base of a codon is allowed to move slightly from its normal position to form a non-Watson-Crick base pair with the anticodon.

Ribosomal subunits can be dissociated into their component parts, RNAs and proteins, and then reassociated. Thus, the ribosome is a self-assembly system that needs no help from the outside.

Initiation of translation consists of binding the first aminoacyl-tRNA to the ribosome-mRNA complex. Prokaryotes use a special amino acid (N-formyl methionine, fMet) and tRNA (tRNA$_F^{Met}$) for initiation. A sequence of bases (the Shine-Dalgarno sequence) just to the 5'-side of the initiation codon in prokaryotic mRNAs is more or less complementary to a sequence at the very 3'-end of the 16S rRNA in the ribosome. Base pairing between these two regions of RNA presumably helps the ribosome recognize and bind to the initiation regions of the mRNAs.

Elongation takes place in three steps: (1) EF-Tu, with energy from GTP, binds an aminoacyl-tRNA to the ribosomal A site. (2) Peptidyl transferase forms a peptide bond between the peptide in the P site and the newly arrived aminoacyl-tRNA in the A site. This lengthens the peptide by one amino acid and shifts it to the A site. (3) EF-G, using energy from GTP, translocates the growing peptidyl-tRNA, with its mRNA codon, to the P site.

Translation release factors RF-1 and RF-2 recognize the stop signals UAA, UAG, and UGA, and along with RF-3, cause termination in prokaryotes. Along with GTP, eEF performs the same function in eukaryotes. When a mutation creates one of these stop signals in the middle of a message, premature termination occurs. Such mutations can be overridden by suppressors—tRNAs that respond to termination codons by inserting an amino acid, thus preventing termination.

■

Problems and Questions

1. Would you predict that two tRNAs with complementary anticodons would bind to one another less tightly, more tightly, or the same as two complementary trinucleotides? Why? *Hint:* Free trinucleotides have no special shape, but tRNAs do.

2. If you charged a methionine tRNA with methionine, then chemically modified the amino acid to alanine and used this alanyl-tRNA^{Met} in an in vitro translation reaction, what do you predict would be the result? In other words, would the alanine go into the protein where an alanine is supposed to go or where a methionine is supposed to go? Why?

3. What does the answer to question 2 say about the importance of accurate aminoacyl-tRNA synthetases in faithful translation?

4. What is the product of the reaction catalyzed by aminoacyl-tRNA synthetase?

5. Consider this short message: 5'-AUGGCAGUGCCA-3'. Answer the following questions, assuming first that the code is overlapping and then that it is nonoverlapping. (*a*) How many codons would be represented in this oligonucleotide? (*b*) If the second G were changed to a C, how many codons would be changed?

6. What would be the effect on reading frame and gene function if: (*a*) two bases were inserted into the middle of a message? (*b*) three bases were inserted into the middle of a message? (*c*) one base was inserted in one codon and one subtracted from the next?

7. If the length of a codon were six bases, what kind of product would you expect from a repeating tetranucleotide such as poly(UUCG)?

8. If a mutant *E. coli* cell had an aminoacyl-tRNA synthetase that should attach the amino acid phenylalanine to its tRNA (tRNA^Phe), but at the elevated temperature of 42°C, attached the amino acid arginine instead, what consequence would this have for the proteins that cell made at elevated temperature? Would this affect these proteins' function? Why, or why not?

9. What would be the effect of a mutation in the initiating codon?

10. (*a*) Use the genetic code to predict the amino acid sequence of a peptide encoded in this mini-message: 5-'AUGUUCAAGAUGGUGACUUGGUAAAUC-3'. (*b*) What amino acid sequence would result if the first G in this message were changed to C, assuming that only physiologically meaningful start codons are used? (*c*) What amino acid sequence would result if the first C were changed to a U? (*d*) If the first C were changed to a G? (*e*) If the last G were changed to an A?

11. (*a*) Which three amino acids have codons with 6-fold degeneracy? (*b*) Which two amino acids have only one codon each?

12. How many different codons would exist in a genetic code with four-base codons?

13. Assuming wobble can occur, what two codons could be recognized by the anticodon 3'-GAG-5'? Write the codons in the 5' → 3' direction, and remember that the pairing between codon and anticodon is antiparallel.

14. What anticodon could recognize all three codons for isoleucine, assuming wobble? Write the anticodon in the 3' → 5' direction.

15. A ribonuclease called cloacin DF13 specifically cleaves off 49 bases from the 3'-end of 16S rRNA in the intact ribosome. What effect would you expect this RNase to have on translation? Why?

16. A mutation in the genes for the following proteins would stall elongation before which steps? (*a*) EF-G; (*b*) EF-Tu; (*c*) peptidyl transferase.

17. Of all the aminoacyl-tRNAs, only one enters the P site on the ribosome. Which one is it? Why is it important that this aminoacyl-tRNA not stay in the A site?

18. A bacterium makes a protein that contains 403 amino acids, counting the terminal methionine that is retained. How many peptidyl transferase reactions occur during synthesis of this protein?

19. If every stop signal in phage QT were UAGUGA, what would happen to all the phage QT proteins made in an amber suppressor strain in which the suppressor tRNA is tRNA^Tyr? Why?

20. A certain ochre suppressor inserts glutamine (Gln) in response to the ochre codon. What is the likeliest change in the anticodon of a tRNA^Gln that created this suppressor strain?

Answers appear at the end of book.

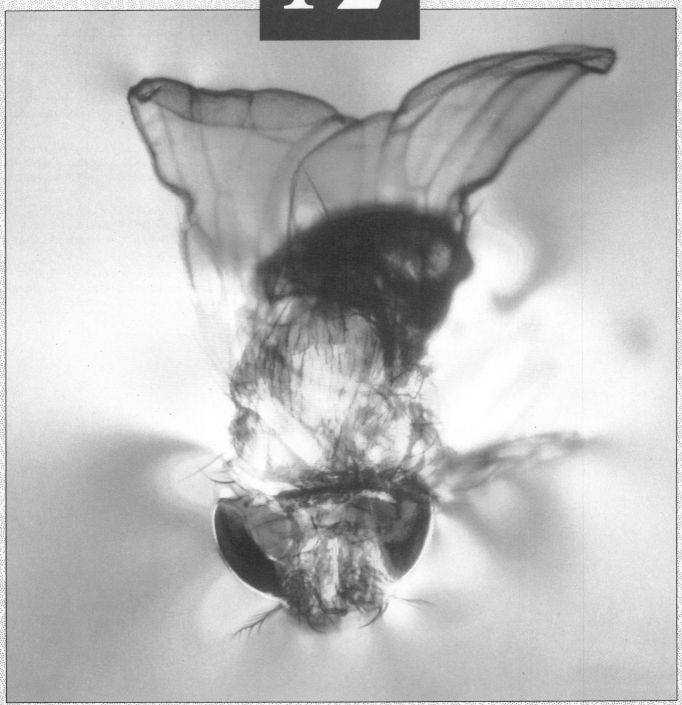

Gene Mutation, Transposable Elements, and Cancer

nd nothin's quite as sure as change.

■

—————————————————

John Phillips
The Mamas and the Papas

Learning Objectives

In this chapter you will learn:

1. The different phenotypic effects of mutations.

2. The different ways in which chemicals and radiation can damage DNA and thereby cause mutations.

3. How transposable elements can foster mutation.

4. How DNA damage can be repaired.

5. That mutagens, including chemicals and radiation, can cause cancer.

6. That retroviruses contain oncogenes capable of causing cancer, and that normal cells have oncogene counterparts called proto-oncogenes.

7. That proto-oncogenes can be converted to oncogenes by mutation.

Drosophila—a curly-winged mutation.
© *Arthur M. Siegelman.*

C hapters 8-11 have treated in detail the first two fundamental characteristics of genes: They replicate faithfully, and they store information for the production of RNAs and proteins. The third characteristic of genes is that they accumulate alterations, or mutations. We have already mentioned mutations several times in this book; indeed, there could hardly be a science of genetics without mutations, since we usually do not notice a gene until a mutation changes it. So far, however, we have not said much about the mechanisms by which genes are changed. That will be the main purpose of this chapter.

Sometimes a mutation goes beyond the alteration of one or a few bases and involves a larger part of a chromosome. Such changes are called chromosome mutations. Genomic mutations, or aneuploidies, are even more noticeable and involve loss or gain of whole chromosomes. Chromosome mutations and aneuploidies have already been discussed in chapter 5 and will not be considered further here. Instead, we will focus on **gene mutations,** alterations that are confined to a single gene.

TYPES OF GENE MUTATIONS

We can look at gene mutations in several different ways. For example, we can classify them according to their effects on the phenotype of the mutant organism or according to their effects on the genetic material itself. Let us first examine the various ways that mutations affect phenotype.

SOMATIC VERSUS GERM-LINE MUTATIONS

Multicellular organisms that reproduce sexually can experience mutations in their sex cells, in which case the gametes can be altered and the mutation passed on to the progeny. Such mutations are called **germ-line,** or **germinal mutations.** A germ-line mutation probably happened to Queen Victoria of Great Britain (chapter 3), because some of her male descendants were afflicted with hemophilia, a sex-linked blood clotting disorder, yet none of her ancestors apparently had this disease.

On the other hand, mutations in the nonsex cells, or **somatic cells,** may change the phenotype of the individual that suffers the mutation, but the mutant trait will stop with that individual. Since the mutation does not affect the gametes, it cannot be passed on.

Examples of such **somatic mutations** abound. Consider, for instance, a kernel of Indian corn such as the one pictured in figure 12.1. A wild-type kernel would be solid purple, but a mutation has inactivated the color gene, so this kernel is mostly white. However, you can easily see that the kernel is not solid white, but speckled. Each spot of color represents a back-mutation, or reversion, that reactivated the color gene in one cell, which then grew

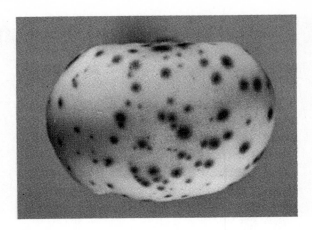

FIGURE 12.1
Somatic mutation. Each colored spot on this speckled kernel of corn represents a group of cells descended from one that experienced a back-mutation in a color gene. Without the back-mutations, the kernel would be all white.

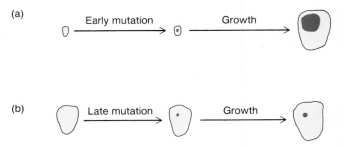

FIGURE 12.2
Influence of timing on the effect of somatic mutation. (*a*) The mutation occurs early in development of the seed, so many colored cells descend from the one with the back-mutation, producing a large colored spot. (*b*) The mutation occurs late in seed development, so not many more cell divisions take place. As a result, the colored spot is small.

into a patch of colored cells. If the mutation takes place early in kernel development (figure 12.2*a*), when there are few cells, the mutant cell will give rise to a large group of colored progeny cells, so the colored patch will be large and will sometimes encompass the entire kernel. On the other hand, if the mutation takes place late in seed development (figure 12.2*b*), when there are many cells and few cell divisions left to go, the colored spot will be tiny. It is obvious that these mutations must have occurred during the lifetime of this corn plant and in the endosperm tissue, which is made up of somatic cells. Therefore, they must be somatic mutations; we will see later in this chapter that these particular corn mutations are caused by mobile pieces of DNA called *transposable genetic elements.*

It is also likely that somatic mutations play a major role in cancer. Many cancers are derived from a single cell that has changed its behavior from normal to malignant.

FIGURE 12.3
An albino mouse (top) with a wild-type littermate.
R. L. Brinster, from the cover of Cell, *vol. 27, November 1981. Reprinted with permission from Cell Press.*

It appears that somatic mutations frequently contribute to such behavior changes. We will explore this topic further later in this chapter.

G erm-line mutations affect the gametes, so the mutation can be transmitted to the next generation. Somatic mutations affect the nonsex, or somatic cells. Their effects are felt in the organism that experiences the mutation, but are passed no further.

■

MORPHOLOGICAL MUTATIONS

Complex multicellular organisms carrying **morphological mutations,** or **visible mutations,** can usually be distinguished readily from wild-type organisms because of their altered appearance. For example, *albino* mammals, including humans, have a mutation in a gene that is responsible for dark coat (or skin) pigment. This is ordinarily the gene for *tyrosinase,* the key enzyme in the pathway that leads to *melanin,* the black pigment in hair, eyes, and skin. A mutated tyrosinase gene may produce no active enzyme, so no melanin can be made. As a result, albino humans have very fair skin and hair, and light blue eyes; albino mice have white fur and pink eyes (figure 12.3).

Many morphological mutants of the fruit fly *Drosophila* have been observed. Among these are *white* flies, with white eyes instead of the usual red (figure 12.4); *miniature,* with tiny wings; *Bar eye,* with bar-shaped eyes instead of round; and *curly,* with curly wings instead of straight.

Microorganisms and even viruses also show a few morphological mutations. The mutant bacterium *Streptococcus pneumoniae,* used by Avery and his colleagues to demonstrate that DNA is the genetic material (chapter 6), had a different morphology (rough) than the wild-type (smooth). A mutant yeast is called *petite* because it forms

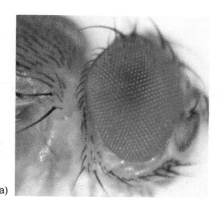

(a)

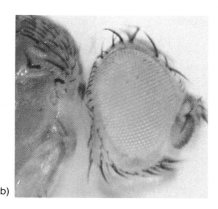

(b)

FIGURE 12.4
Morphological mutation in *Drosophila*. (*a*) A wild-type fly with red eyes is contrasted with (*b*) a white-eyed mutant fly.
Gerald M. Rubin, from the cover of Cell, *vol. 40, April 1985. Reprinted with permission from Cell Press.*

much smaller than normal colonies. The bacterial virus T4 can be mutated to the *rapid lysis,* or *r* form, which clears out a bigger and clearer hole, or plaque, in a lawn of infected bacteria than does the wild-type phage (chapter 13). Thus, the altered morphology we observe in microorganisms is usually not in a single organism, but in colonies of organisms—or, in the case of a virus, in plaques.

M orphological mutations cause a visible change in the affected organism.

■

NUTRITIONAL MUTATIONS

Although some morphological mutations can be distinguished in microorganisms, these are necessarily few in number. That is because these simple organisms have a limited number of observable morphological features. Thus, we must usually rely on more subtle mutant phenotypes. The most popular of these is the inability to grow on simple growth media. Certain wild-type organisms can grow on "minimal medium," which contains just salts and an energy source such as glucose. These organisms, called **prototrophs,** can make all the amino acids, nucleotides, vitamins, and other substances they need to sustain life. Nutritional mutants, also called **auxotrophs,** need an extra substance in order to survive, so they die on minimal medium.

For example, a *bio* mutant of *E. coli* will die unless it is supplied with the vitamin biotin, while a *leu* mutant will die without the amino acid leucine. Similarly, as we saw in chapter 7, a pantothenateless mutant of *Neurospora* will die without the vitamin pantothenate. The reason for these deficiencies, of course, is that genes for enzymes in the biotin, leucine, and pantothenate synthesis pathways, respectively, are defective. Since these genes do not make active enzymes, no biotin, leucine, or pantothenate can be made and the cells die. Such mutations are also sometimes called biochemical mutations, but that is not a very useful term, since all mutations, once we understand them, are seen to have biochemical effects.

> **N** utritional mutations render an organism dependent on added nutrients that a wild-type organism does not require, since it can make these substances itself.

∎

LETHAL MUTATIONS

Some mutations are so severe that an organism carrying them cannot survive at all; these mutations are called **lethals** (chapter 3). An example would be an inactivating mutation in the gene for one of the subunits of RNA polymerase. If the mutation rendered the RNA polymerase inactive, the mutant organism could make no RNA and therefore could not live. Haploid organisms with lethal mutations die immediately because they have no wild-type gene to compensate. But diploid organisms can carry a lethal mutation, masked by a wild-type allele, for generations with no ill effects. It is only when two heterozygotes mate that we may discover the lethal consequences of the mutation. In chapter 3 we saw one such example in yellow mice. The heterozygotes are viable, but homozygous yellow mice are never seen—they die in the mother's uterus. Thus, most lethal mutations in diploid organisms are recessive. If one parent contributes a defective gene for an essential protein and the other contributes a wild-type gene, the latter will usually allow the cell to make enough protein to compensate. It is only when two defective genes are brought together in a homozygote that lethality ordinarily results.

Huntington's disease in humans is an example of a dominant lethal. You may be wondering how a dominant lethal mutation allows the victim to live long enough to have a disease, let alone pass the mutant gene along to his or her children. We do not know how the disease operates, but we do know that heterozygotes typically live normal lives well past the childbearing years, then begin to show signs of the degenerative disease that ultimately kills them. This is the fate that befell folk singer Woody Guthrie.

> **L** ethal mutations occur in indispensable genes. Haploid organisms with lethal mutations die immediately, but lethal mutations are usually recessive, so diploid organisms can tolerate them in heterozygotes.

∎

CONDITIONAL MUTATIONS

As we have seen, lethal mutations in haploid organisms usually cause death immediately. Hence, they are of little use to geneticists who are interested in studying the altered gene product. With no organism, there can be no gene product. One way around this problem is to look for **conditional lethals.** These are mutations that are lethal under certain circumstances, but not under others.

Consider, for example, a **temperature-sensitive (ts) mutation.** This allows growth at low temperature (the **permissive temperature),** but not at normal growth temperature (the **restrictive,** or **nonpermissive temperature),** so lethality in this case is conditional on temperature. It is important to realize that the gene itself is not temperature-sensitive; its protein product is. The mutation creates an altered protein that is more easily unfolded, or **denatured,** than the wild-type protein. Since heat is a powerful protein denaturant, the mutant protein will therefore be heat-sensitive. One has only to consider what happens to a boiled egg to appreciate heat denaturation of protein: The transparent solution of protein (mostly ovalbumin) in the egg white becomes a rubbery, opaque mass upon heating. In effect, this is what happens to the protein product of a temperature-sensitive gene at the nonpermissive temperature. It can happen to the wild-type protein too, but at a considerably higher temperature.

Another example of a conditional mutation is the amber mutation introduced in chapter 11. This mutation creates a premature translation stop signal—a "nonsense" mutation—in the middle of a gene, an event that would be lethal in a vital gene of an ordinary, nonpermissive strain. It is only conditionally lethal because the gene will usually still work in an amber suppressor strain (a permissive strain). Recall from chapter 11 that such a strain contains a tRNA with the ability to recognize an amber codon and insert an amino acid, thus preventing termination. When a phage bearing an amber mutation infects an amber suppressor strain, it will therefore usually survive.

Mutations can be conditional without being conditional lethal. An example is the Siamese cat (figure 12.5). These animals are genotypically albinos, which accounts for their blue eyes. But Siamese cats are not pure

FIGURE 12.5
A Siamese cat.
© Carl W. May/Biological Photo Service.

white like albino mice. Instead, they have dark patches on their feet, noses, and ears—areas of the body where the temperature is somewhat lower than the rest. This indicates that the mutation in the color gene is temperature-sensitive. Most of the cat's coat is warm enough to inactivate the color-producing enzyme, so it is white, but the extremities are cool enough that the enzyme operates, producing the dark color.

C onditional mutations only affect the mutant organism under certain conditions. Temperature-sensitive mutations, for example, cause a protein to be inactive at high temperature but active at lower temperature. Amber mutations do not produce active product under ordinary circumstances but do so in organisms with an amber suppressor tRNA. If a conditional mutation causes death under nonpermissive conditions, it is a conditional lethal.

■

EFFECTS OF MUTATIONS ON THE GENETIC MATERIAL

A second way of classifying mutations is to consider their effects on DNA. The mutations we are considering in this chapter involve the alteration, insertion, or deletion of one or a few bases at a time, and so are called **point mutations.** Many of these are **missense mutations.** In a missense mutation, a base change alters the sense of a codon from one amino acid to another. This causes an improper amino acid to be inserted into the protein product of the mutated gene. For example, a missense mutation might change the proline codon CCG to the arginine codon CGG.

Sometimes, as we saw in chapter 11, a codon can be converted to a nonsense codon, which causes termination of translation. For example, the tryptophan codon UGG can be converted by a one-base change to either UGA or UAG, both of which are stop (or nonsense) codons. Of course, the mutation occurs at the DNA level, so TGG would be converted to TGA or TAG in this example. We present the RNA codons here because they are the familiar way of writing codons. In chapter 11 we also learned that single-base insertions or deletions can alter the translation reading frame in the middle of a gene; these are called frameshift mutations.

We can also consider the chemistry of a single base change. Viewed in the simplest terms, such changes fall into two classes: (1) **transitions,** in which a pyrimidine replaces a pyrimidine (C→T or T→C) or a purine replaces a purine (A→G or G→A), or (2) **transversions,** a more drastic kind of change, in which a purine replaces a pyrimidine or vice versa.

Why do we say that a transversion is "more drastic" than a transition? This makes sense chemically, because one purine resembles another purine (or one pyrimidine another pyrimidine) far more than a purine resembles a pyrimidine. It also makes sense genetically, because of the degeneracy of the genetic code. Recall from chapter 11 that two related codons ending in a pyrimidine, such as UUU and UUC, are more likely to code for the same amino acid (phenylalanine, in this case) than are two related codons, one of which ends in a pyrimidine and the other of which ends in a purine (e.g., UUU and UUG, which code for phenylalanine and leucine, respectively). Similarly, two related codons ending in a purine are likely to code for the same amino acid (e.g., GAA and GAG both code for glutamic acid). Thus, changing one pyrimidine to another or one purine to another, at least in the third position, is less likely to change the sense of a codon than is changing a purine to a pyrimidine, or vice versa.

SPONTANEOUS MUTATIONS

All mutations have a cause, of course, but sometimes mutations occur without our having to apply a **mutagen,** a mutation-causing agent. Such spontaneous mutations have several causes.

Mutations Caused by the DNA Replication Machinery

We saw in chapter 8 that replication is very faithful, but it is not perfect, so a certain number of mutations will occur simply due to fallible DNA synthesis.

We can see the effects of unfaithful replication especially clearly in mutant strains of *E. coli* called **mutators.** These bacteria make more than the usual number of mistakes during DNA replication, so their mutation rates are higher than normal. Mutator mutations map in several different genes, some of which have been identified. For example, *mutD* encodes the epsilon (ε) subunit of the DNA polymerase III holoenzyme. This is the polypeptide that, together with the α subunit, gives the holoenzyme its 3′→5′ exonuclease activity. Without this activity,

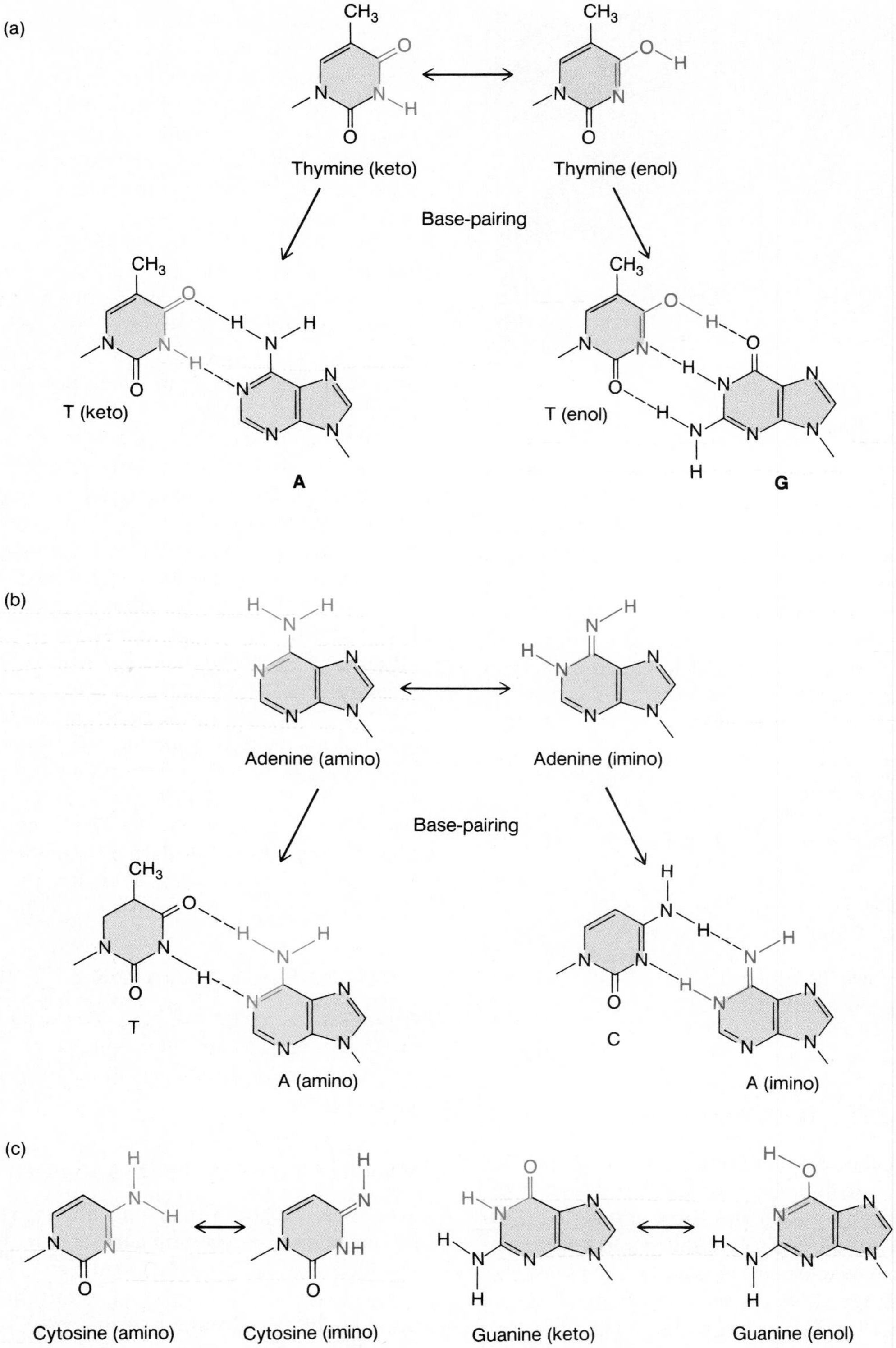

FIGURE 12.6

Spontaneous mutation induced by tautomerization. (*a*) The normal (keto) tautomer of thymine (left) base-pairs with adenine. After a rare conversion to the enol tautomer (right), the thymine base-pairs with guanine. The dashed lines represent hydrogen bonds between bases. (*b*) The normal (amino) tautomer of adenine (left) base-pairs with thymine. After a rare conversion to the imino tautomer (right), the adenine base-pairs with cytosine. (*c*) Left: The normal (amino) tautomer of cytosine can sometimes convert to the rare imino tautomer, which pairs with adenine (not shown). Right: The normal (keto) tautomer of guanine sometimes converts to the rare enol tautomer, which pairs with thymine (not shown). The properties of these altered base pairs are the same as those shown in (*b*) and (*a*), respectively.

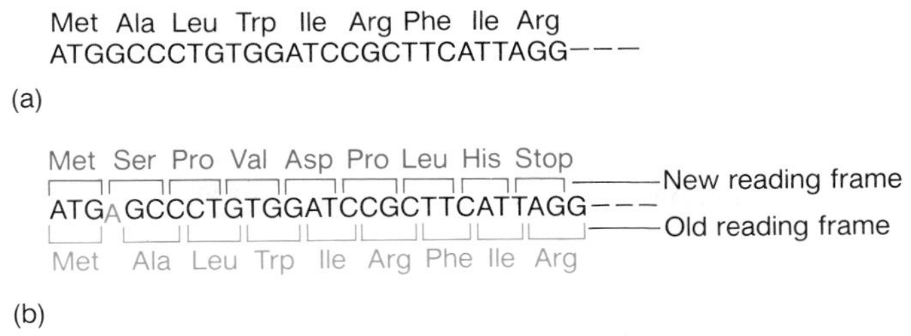

Met Ala Leu Trp Ile Arg Phe Ile Arg
ATGGCCCTGTGGATCCGCTTCATTAGG———

(a)

Met Ser Pro Val Asp Pro Leu His Stop
ATGₐGCCCTGTGGATCCGCTTCATTAGG——— New reading frame
——— Old reading frame
Met Ala Leu Trp Ile Arg Phe Ile Arg

(b)

FIGURE 12.7

Frameshift mutation. (*a*) Normal reading frame of part of a gene. (*b*) The adenine nucleotide inserted in the fourth position of the coding region (red) shifts the reading frame one base to the left. The new reading frame with the corresponding codon meanings is shown by the brackets above the base sequence (green); the old reading frame, which corresponds to the unmutated base sequence, is denoted by brackets below the sequence (blue). A deletion of a base, instead of an insertion, would have shifted the reading frame one base to the right.

proofreading cannot occur, so the newly replicated DNA is left with an excess of mutations (chapter 8). Mutations in *mutH, mutL,* and *mutS* are also mutator mutations. They impair mismatch repair, a mechanism for repairing mismatches that the proofreading system missed. This mechanism will be discussed later in the chapter.

As unlikely as it may seem, there are also mutations that have the opposite effect—they make DNA replication even more faithful than normal. This raises a question: If a more faithful DNA replication system is available, why have *E. coli* evolved with a less effective one? The answer is probably that mutants with these extra-faithful systems have too slow a rate of evolution; they are not flexible enough to compete with more plastic organisms in adapting to a changing environment.

Mistakes in Replication Caused by DNA Bases

The bases in DNA ordinarily exist in one of two possible forms, or **tautomers.** For example, the base thymine usually assumes the **keto** form, as illustrated in figure 12.6*a*. This is the structure presented in chapter 6 that base-pairs naturally with adenine. Occasionally, the keto form of thymine switches to the **enol** form, also shown in figure 12.6. Note that both tautomers have the same atoms but are arranged slightly differently. The enol form pairs naturally with guanine instead of adenine, so if a thymine happens to be in the enol form at the moment it takes a partner during replication, a guanine will be inserted in place of an adenine. If this error goes uncorrected, DNA replication will perpetuate it and a mutation will result.

Figure 12.6*b* shows the two tautomeric forms of adenine. The **amino** form is by far the more common one; as we have seen, it base-pairs with thymine. Occasionally, adenine can assume the alternate, **imino** form, which base-pairs with cytosine. If this happens to an adenine in DNA just as DNA polymerase is supplying it with a partner, the polymerase will insert a C instead of a T. Again, if this mistake persists until the next round of replication, it will cause a mutation that will last: a GC pair instead of an AT pair. Figure 12.6*c* shows the **tautomerization** of cytosine to its rare imino form and of gaunine to its rare enol form. By inspecting these rare tautomers you can satisfy yourself that they form abnormal CA and GT base pairs, respectively.

So far, we have only considered what would happen if the bases in the replicating DNA strand form the rare tautomer during base pairing. It is also possible that a newly inserted base would exist in the rare tautomeric form just at the moment of base pairing. If so, it would result in the same kind of faulty base pairing we have just observed.

Spontaneous Frameshift Mutations During Replication

Sometimes DNA replication causes the insertion or deletion of one or more bases in the middle of a coding region, which changes the translational reading frame from that point on (chapter 11). These **frameshift mutations** are very severe because they change every codon from the point of the mutation to the end of the mRNA. Figure 12.7 shows how this works: (*a*) is a hypothetical gene fragment, showing the translation of each codon. (*b*) is the same gene fragment, showing how insertion of a single A in the fourth position of the gene shifts the reading frame one base to the left. All the codons after that point are different. In fact, one of these new codons, at the far right, is a stop codon, so protein synthesis will stop prematurely.

If small insertions and deletions cause frameshift mutations, the next question is: How do these insertions and deletions occur? We do not understand exactly how this works, but figure 12.8 presents one hypothesis that calls for the DNA replication machinery to "slip a cog" every now and then. The idea is that a base in one strand sometimes fails to pair with its partner in the complementary strand. This *looping out* of a base seems especially likely in stretches of DNA where one base is repeated over and over. Note that if a base in the template strand loops out,

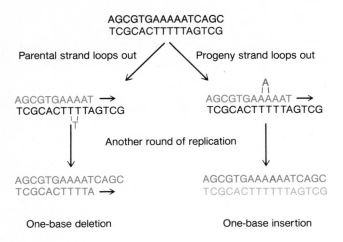

```
            AGCGTGAAAAATCAGC
            TCGCACTTTTTAGTCG
```

Parental strand loops out ↙   ↘ Progeny strand loops out

```
                                          A
                                         ⌐⌐
AGCGTGAAAAT →                    AGCGTGAAAAAT →
TCGCACTTTTTAGTCG                TCGCACTTTTTAGTCG
       ⌐⌐
        T
```

Another round of replication

```
AGCGTGAAAATCAGC                 AGCGTGAAAAAATCAGC
TCGCACTTTTA →                   TCGCACTTTTTTAGTCG
```

One-base deletion   One-base insertion

FIGURE 12.8

Hypothetical mechanism for frameshift mutation. A parental DNA duplex (top) contains a string of five A–T pairs. During replication of this DNA, transient looping out of bases within this A–T stretch can occur. On the left, one of the T's (green) in the parental strand loops out. This leaves only four T's to pair with A's in the growing progeny strand (red). After one more round of replication, these four A's serve as the template for the incorporation of four T's in the new progeny strand (blue). Now we have four A–T pairs instead of five; a one-base deletion has occurred. On the right, one of the newly incorporated A's (gray) in the progeny strand loops out, allowing six A's to be inserted instead of five. When this DNA replicates again, the strand with six A's will dictate the incorporation of six T's in the new progeny strand (light blue). A one-base insertion results.

one too few bases will be incorporated into the progeny strand, resulting in a one-base deletion. On the other hand, if a base in the progeny strand loops out, one too many bases will be incorporated, producing a one-base insertion.

Spontaneous Mutations Caused by Deamination

Spontaneous mutations can occur by mechanisms other than mistakes in DNA replication. For example, bases, especially cytosine, have a slight tendency to lose their amino groups in a process called **deamination.** When cytosine is deaminated, it receives a carbonyl oxygen in place of its amino group; this converts it to uracil, which base-pairs with adenine instead of guanine. Adenine can also be deaminated, yielding the base hypoxanthine, which base-pairs with cytosine instead of thymine. In both of these cases, deamination can potentially cause a mutation because a new base, with new pairing properties, is created. Figure 12.9 shows the changes and new base pairing caused by deamination of cytosine and adenine.

The most common kind of deamination, the conversion of cytosine to uracil, does not usually lead to a mutation, because cells have a mechanism for removing uracils that find their way into DNA by mistake. It involves an enzyme called uracil-DNA-glycosidase (figure 12.10), which cuts the bond between the uracil and its deoxyribose, thus removing the uracil and leaving behind

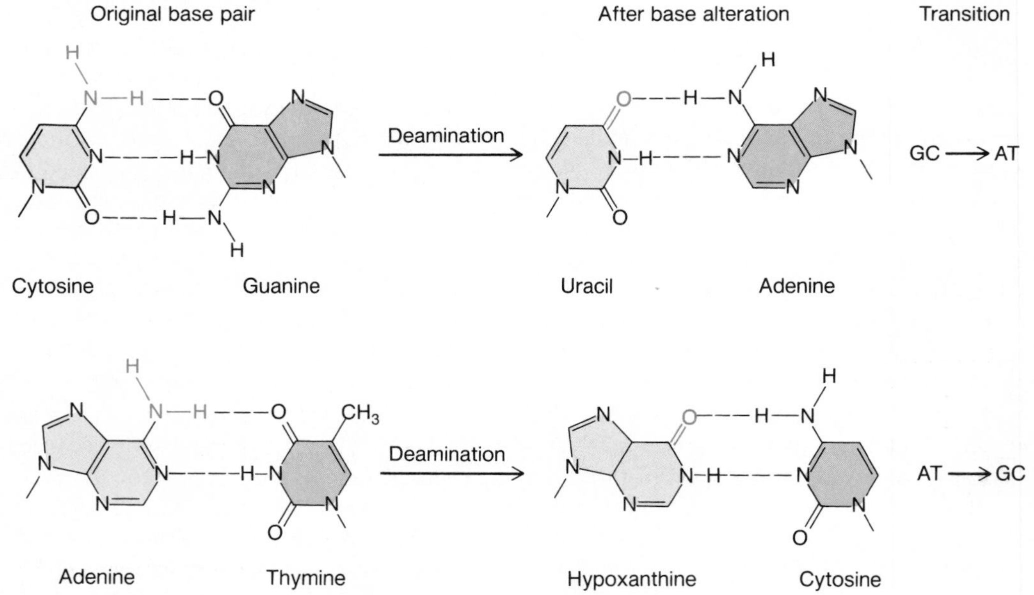

FIGURE 12.9

Deamination of cytosine and adenine. After deamination, the bases uracil and hypoxanthine are formed. These pair with adenine and cytosine, respectively, so transitions occur in each case.

a DNA strand containing one sugar without a base. Another enzyme soon supplies a cytosine to pair with the guanine on the opposite strand.

The DNA of some organisms contains a small number of modified bases in addition to the usual four. The most common of these is 5-methylcytosine (figure 12.11a), which base-pairs with guanine exactly as ordinary cytosine does. However, 5-methylcytosine does not behave in every respect like cytosine. In particular, the sites in DNA that contain 5-methylcytosine can become *hot spots* for spontaneous mutation via deamination.

If deamination of cytosine is usually not mutagenic, why is deamination of 5-methylcytosine so likely to cause mutations? The answer is simple: Although the deaminated product of cytosine is uracil, which is easily recognized and removed, the deaminated product of 5-methylcytosine is thymine, one of the ordinary DNA bases and therefore not recognizable as foreign. This means that once 5-methlycytosine deaminates, a GT mismatch appears. After this DNA replicates, one duplex will have the wild-type GC pair, but the other will have a mutant AT pair (figure 12.11b).

The pattern of methylation of cytosines in bacteria and in higher eukaryotes is specific. That is, a certain few cytosines are selected for methylation, while all the others remain unmodified. In mammals, for example, cytosines in certain CpG sequences are targets for methylation. This specificity has predictable consequences. The sequence

CpG has become much rarer than any other dinucleotide sequence, presumably because deamination has led to conversion of most of the CpGs to TpGs during evolution. In bacteria, too, methylated C's become hot spots for mutation.

Measuring Spontaneous Mutation Rates

How frequently do spontaneous mutations occur? This is a difficult determination to make. To begin with, many mutations are undetectable by ordinary means, so they are not counted. We will consider such silent mutations later. Furthermore, mutations are rare, so we must examine many organisms to find a significant number of mutants. Once we have detected a given number of mutants, we must determine the rate at which they are produced—that is, the number of mutations per unit of time. But geneticists measure time in different ways than physicists do. Instead of mutations per minute, day, or year, we speak of mutations per generation or per cell division. This is the **mutation rate.** In higher eukaryotes, including humans,

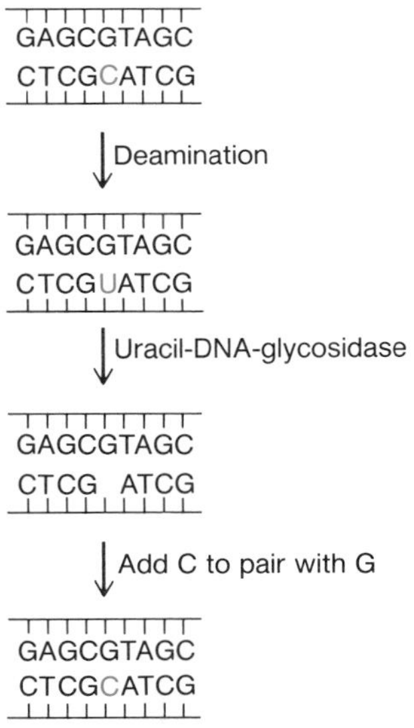

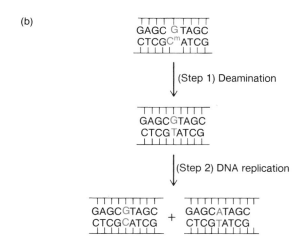

(a) 5-methylcytosine → Deamination → Thymine

(b)

GAGC G TAGC
CTCGC^mATCG

(Step 1) Deamination

GAGCGTAGC
CTCGTATCG

(Step 2) DNA replication

GAGCGTAGC   GAGCATAGC
CTCGCATCG +   CTCGTATCG

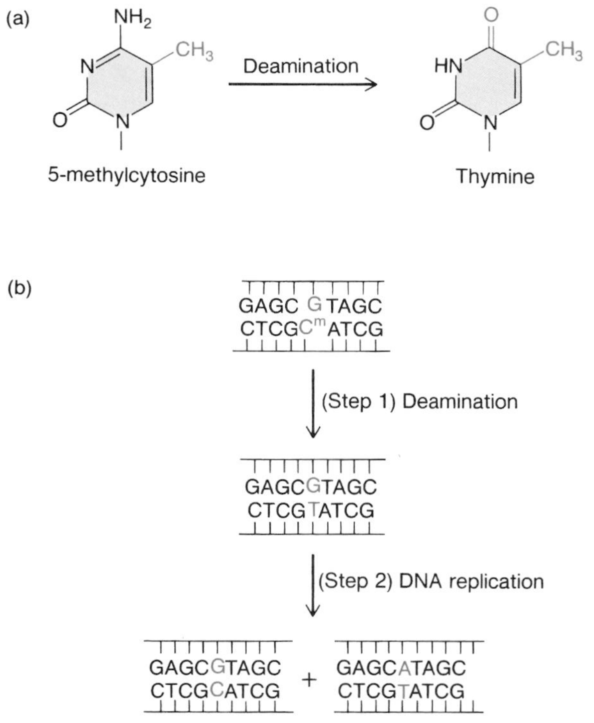

FIGURE 12.11

Mutation by deamination of 5-methylcytosine. (*a*) The deamination step converts 5-methylcytosine to thymine. (*b*) Making the change permanent. In step 1, deamination converts the 5-methylcytosine (red) to thymine (blue). Since thymine cannot be recognized as foreign, it is not removed. In step 2, DNA replication yields one duplex with a wild-type GC pair (red) and one with a mutant AT pair (blue).

GAGCGTAGC
CTCGCATCG

↓ Deamination

GAGCGTAGC
CTCGUATCG

↓ Uracil-DNA-glycosidase

GAGCGTAGC
CTCG ATCG

↓ Add C to pair with G

GAGCGTAGC
CTCGCATCG

FIGURE 12.10

Repair of deaminated cytosine. A deamination event (top) has converted a C (red) to a U (blue). The enzyme uracil-DNA-glycosidase recognizes the U as foreign and removes it (middle), leaving a sugar without a base in the bottom DNA strand. In the second step of the repair process (bottom), a C (red) is added to pair with the G in the top strand.

TABLE 12.1 A Sample of Spontaneous Mutation Rates in Different Organisms

Organism	Character	Rate	Units
Bacteriophage: *T2*	Lysis inhibition, $r \rightarrow r^+$	1×10^{-8}	Per gene per replication
	Host range, $h^+ \rightarrow h$	3×10^{-9}	
Bacteria: *Escherichia coli*	Lactose fermentation, $lac^- \rightarrow lac^+$	2×10^{-7}	
	Phage $T1$ sensitivity, $T1\text{-}s \rightarrow T1\text{-}r$	2×10^{-8}	
	Histidine requirement, $his^- \rightarrow his^+$	4×10^{-8}	
	$his^+ \rightarrow his^-$	2×10^{-6}	Per cell per division
	Streptomycin sensitivity, $str\text{-}s \rightarrow str\text{-}d$	1×10^{-9}	
	$str\text{-}d \rightarrow str\text{-}s$	1×10^{-8}	
Algae: *Chlamydomonas reinhardi*	Streptomycin sensitivity, $str\text{-}s \rightarrow str\text{-}r$	1×10^{-6}	
Fungi: *Neurospora crassa*	Inositol requirement, $inos^- \rightarrow inos^+$	8×10^{-8}	Mutant frequency among asexual spores
	Adenine requirement, $ade^- \rightarrow ade^+$	4×10^{-8}	
Corn: *Zea mays*	Shrunken seeds, $Sh \rightarrow sh$	1×10^{-5}	
	Purple, $P \rightarrow p$	1×10^{-6}	
Fruit fly: *Drosophila melanogaster*	Yellow body, $Y \rightarrow y$, in males	1×10^{-4}	Mutant frequency per gamete per sexual generation
	$Y \rightarrow y$, in females	1×10^{-5}	
	White eye, $W \rightarrow w$	4×10^{-5}	
	Brown eye, $Bw \rightarrow bw$	3×10^{-5}	
Mouse: *Mus musculus*	Piebald coat color, $S \rightarrow s$	3×10^{-5}	
	Dilute coat color, $D \rightarrow d$	3×10^{-5}	
Man: *Homo sapiens*	Normal $\rightarrow$ hemophilic	3×10^{-5}	
	Normal $\rightarrow$ albino	3×10^{-5}	
Human bone marrow cells in tissue culture	Normal $\rightarrow$ 8-azoguanine resistant	7×10^{-4}	Per cell per division
	Normal $\rightarrow$ 8-azoguanosine resistant	1×10^{-6}	

From R. Sager and F. J. Ryan, *Cell Heredity.* Copyright © 1961 John Wiley & Sons, Inc., New York, NY.

the number of generations is usually taken as the number of gametes in the population under study. For example, if we are considering a population of 50,000 people, the number of gametes that produced this population is 100,000.

Table 12.1 presents mutation rates for several organisms. We do not fully understand why these numbers vary so much from one organism to another, but one contributing factor is likely to be the efficiency of proofreading during DNA synthesis (chapter 8). Another consideration may be the efficiency of mismatch repair.

As an example of how these data were collected, consider human achondroplasia, the dominant form of dwarfism mentioned in chapter 2. The records of 94,075 children at Lying-In Hospital in Copenhagen show that ten were afflicted with achondroplasia. Two of the ten came from families with a history of such dwarfism, so

only eight represent new mutations. Since the total number of gametes involved is twice the total number of individuals, the mutation rate is $8/(2)(94,075) = 4.2 \times 10^{-5}$.

*S*pontaneous mutation rates—the number of mutations arising per gamete each generation—vary considerably from one organism to another. Spontaneous mutations can occur in several ways: (1) The DNA replicating machinery can simply make mistakes that go uncorrected. (2) The bases in the DNA template strand or in the newly inserted nucleotide can shift to an alternate tautomeric form that base-pairs incorrectly. (3) Too many or too few bases can be inserted, causing frameshift mutations. (4) Certain bases can deaminate, altering their base-pairing properties. ∎

Chapter 12

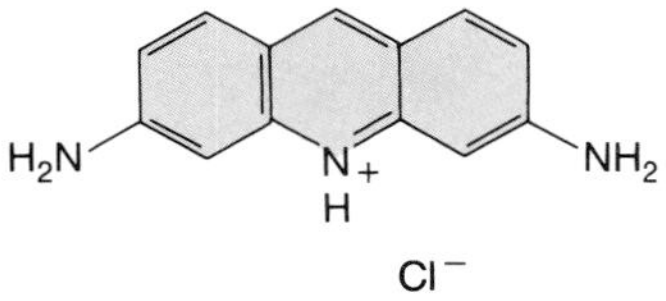

FIGURE 12.12
Proflavin (chloride), one of the acridine dyes.

CHEMICAL MUTAGENESIS

We have just seen that mutations can occur spontaneously, but they occur much more frequently in response to environmental agents: chemicals and radiation. We will consider chemical agents first.

Some chemicals act by accelerating the rate of mutations that occur spontaneously. We have already discussed several mechanisms of spontaneous mutation, including deamination, frameshifting, and tautomerization. Deamination can be greatly enhanced by agents such as nitrous acid or bisulfite, or even by heat. Frameshift mutations can be accelerated by planar molecules like the acridine dyes (figure 12.12), which *intercalate,* or insert themselves between the flat base pairs of DNA. This disruption may stabilize the looping out process that seems to be important in frameshift mutations (see figure 12.8).

Mutations Caused by Base Analogues

Some synthetic compounds can enhance the frequency of tautomerization and thereby induce mutations. The classic example of such a mutagen is **5-bromodeoxyuridine (BrdU),** which resembles thymidine except for the substitution of a bromine atom for a methyl group in the 5-position (figure 12.13). However, once BrdU is incorporated in place of thymidine, it can cause trouble. The trouble derives from the enhanced tendency of BrdU to switch occasionally to the enol tautomer that base-pairs as C instead of T. As we have seen, natural thymidine does this upon rare occasion; BrdU still switches to the enol tautomer only rarely, but it does so more frequently than thymidine.

Mutations Caused by Alkylation of Bases

Some substances in our environment, both natural and man-made, are **electrophilic.** This name means electron (or negative charge)-loving; thus, electrophiles seek centers of negative charge in other molecules and bind to them. Many other environmental substances are metabolized in the body to electrophilic compounds. One of the most obvious centers of negative charge in biology is the DNA molecule. Every nucleotide contains one full

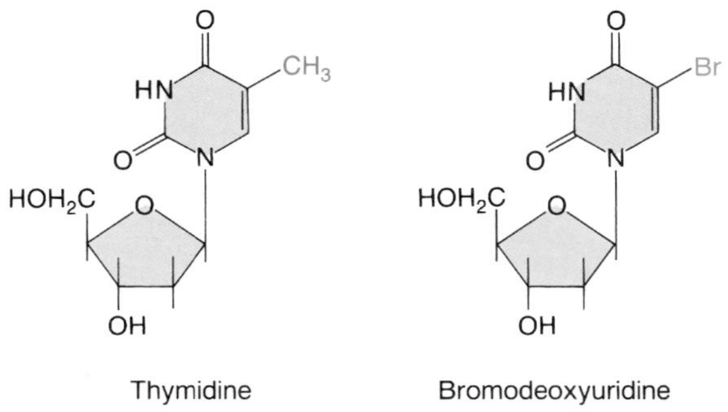

FIGURE 12.13
Comparison of thymidine and bromodeoxyuridine.

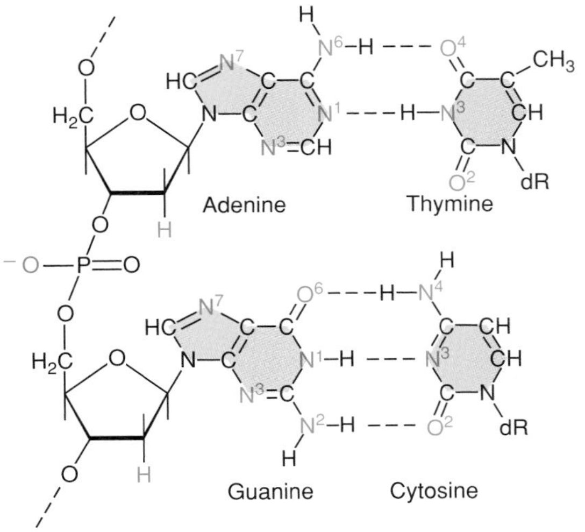

FIGURE 12.14
Electron-rich centers in DNA. The targets most commonly attacked by electrophiles are the phosphate groups, N^7 of guanine and N^3 of adenine (red); other targets are in blue.

negative charge on the phosphate and partial negative charges on the bases. When **electrophiles** encounter these negative centers, they attack them, usually adding carbon-containing groups called *alkyl groups.* Thus, we refer to this process as **alkylation.**

Aside from the phosphodiester bonds, the favorite sites of attack by alkylating agents are the N^7 of guanine and the N^3 of adenine (figure 12.14). Since neither of these positions is involved in base pairing, such alkylations do not lead immediately to mispairing. However, they do make the bond between sugar and base more labile, or more apt to break. When this break occurs, it leaves an apurinic site, a sugar without its purine. This obviously cannot be replicated properly unless it is first repaired, and this repair itself can sometimes be mutagenic. Alkylation can also enhance the tendency of a base to form the rare

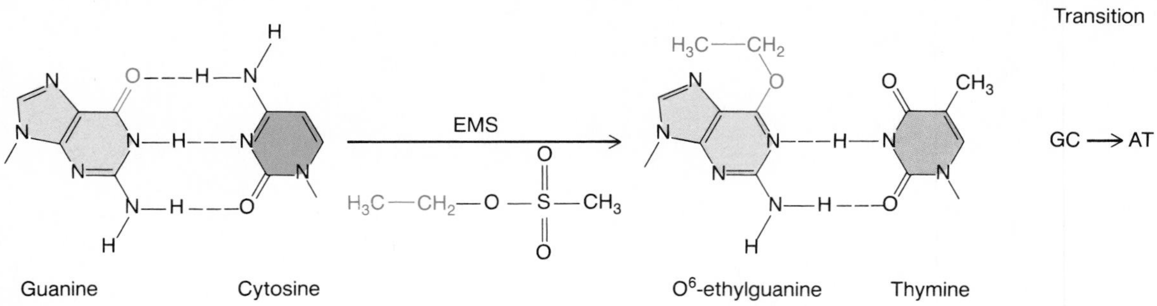

Transition

GC → AT

FIGURE 12.15
Alkylation of guanine by EMS. At the left is a normal guanine-cytosine base pair. Note the free O⁶ oxygen (red) on the guanine. Ethylmethane sulfonate (EMS) donates an ethyl group (blue) to the O⁶ oxygen, creating O⁶-ethylguanine (right), which base-pairs with thymine instead of cytosine.

(wrong) tautomer. As we have seen, this can lead to mutations. Moreover, all of the nitrogen and oxygen atoms involved in base pairing are also subject to alkylation, which can directly disrupt base pairing and lead to mutation (figure 12.15).

Most environmental *carcinogens,* or cancer-causing agents, are electrophiles that seem to act by attacking DNA and alkylating it as we will see later in this chapter. Many of the favorite mutagens used in the laboratory for the express purpose of creating mutations are also alkylating agents. One example is *ethylmethane sulfonate (EMS)*, which transfers ethyl (CH_3-CH_2) groups to DNA (figure 12.15).

D ifferent chemicals induce different kinds of DNA damage. Nitrous acid and bisulfite cause deamination of bases, especially cytosine, causing them to base-pair differently. Alkylating agents like ethylmethane sulfonate add bulky alkyl groups to bases, either disrupting base pairing directly or causing loss of bases, either of which can lead to faulty DNA replication or repair. Base analogues such as BrdU are incorporated into DNA and then can lead to abnormal base pairing. Certain planar molecules, such as the acridine dyes, cause addition of extra bases or too few bases during DNA replication, which can alter the reading frame of a gene.

■

RADIATION-INDUCED MUTATIONS

Ultraviolet, gamma-, and X-radiation are the common types of mutagenic radiation found in nature and used in mutagenesis experiments. These kinds of radiation differ greatly in energy; therefore, they differ greatly in the kinds of DNA damage they cause. We will examine each in turn.

Ultraviolet Radiation

Ultraviolet radiation is relatively weak, so the damage it causes is relatively modest; it cross-links adjacent pyrimidines on the same DNA strand, forming dimers, usually **thymine dimers.** Figure 12.16 shows the structure of a thymine dimer and illustrates how it interrupts base pairing between the two DNA strands. This blocks DNA replication because the replication machinery cannot tell which bases to insert opposite the dimer. As we will see, replication sometimes proceeds anyway, and bases are inserted at random. If these are the wrong bases (and they usually are), a mutation results.

The kind of ultraviolet radiation that is most damaging to DNA has a wavelength of about 260 nm, which is not surprising, since this is the wavelength of radiation that is absorbed most strongly by DNA. Such radiation has great biological significance; it is abundant in sunlight, so most forms of life are exposed to it to some extent. The mutagenicity of ultraviolet light explains why sunlight causes skin cancer: Its ultraviolet component damages the DNA in skin cells, which sometimes causes those cells to lose control over their division.

Given the dangers of ultraviolet radiation, we are fortunate to have a shield—the ozone layer—in the earth's upper atmosphere to absorb the bulk of such radiation. However, scientists have recently noticed alarming holes in this protective shield—the most prominent one located over Antarctica. The causes of this ozone depletion are still somewhat controversial, but they probably include compounds used in air conditioners and in plastics. Unless we can arrest the destruction of the ozone layer, we are destined to suffer more of the effects of ultraviolet radiation, including skin cancer.

Chapter 12

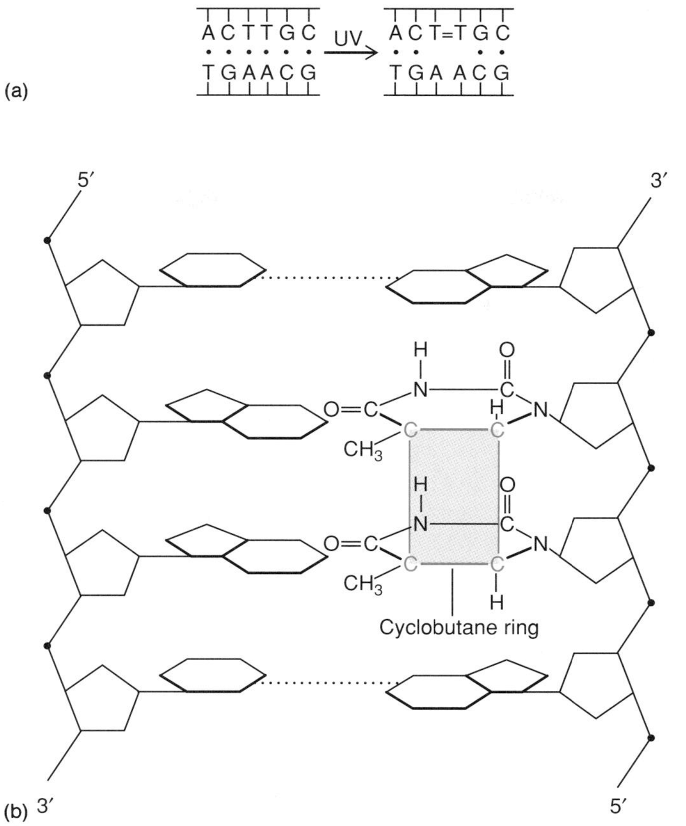

FIGURE 12.16
Thymine dimers. (*a*) Ultraviolet light cross-links the two thymine bases on the top strand. This distorts the DNA so that these two bases no longer pair with their adenine partners. (*b*) The two bonds joining the two thymines form a four-membered cyclobutane ring (red).

Gamma and X Rays

The much more energetic **gamma rays** and **X rays,** like ultraviolet rays, can interact directly with the DNA molecule. However, they cause most of their damage by ionizing the molecules surrounding the DNA, especially water. This forms *free radicals,* chemical substances with an unpaired electron. These free radicals, especially those containing oxygen, are extremely reactive, and they immediately attack neighboring molecules. When such a free radical attacks a DNA molecule, it can change a base, but it frequently causes a single- or double-stranded break. Single-stranded breaks are ordinarily not serious because they are easily repaired, but double-stranded breaks are very difficult to repair properly, so they frequently cause a lasting mutation. Because ionizing radiation can break chromosomes, it is referred to not only as a **mutagen,** or mutation-causing substance, but also as a **clastogen,** which means "breaker."

*D*ifferent kinds of radiation cause different kinds of mutations. Ultraviolet rays have comparatively low energy, and they cause a moderate type of damage: thymine dimers. Gamma and X rays are much more energetic. They ionize the molecules around DNA and form highly reactive free radicals that can attack DNA, altering bases or breaking strands.

SILENT MUTATIONS

Most mutations are harmful, leading to the production of faulty proteins or RNAs. Still, life is not perfect; there is always room for improvement. Accordingly, mutations can sometimes actually improve genes—change their products so that they work better or allow their hosts to survive better. These mutations allow evolution to occur.

Many mutations are not detectable by ordinary genetic means. We call them **silent mutations.** For example, consider the change of a codon from UCA to UCG. This is a base change, so it is a real mutation. But both triplets code for the same amino acid (serine), so no change occurs in the protein product of the gene. The mutation is silent. The only way we would ever know for sure that it had happened would be to clone the gene and determine its exact base sequence. Such mutations in the third base of a codon (the *wobble position*) are more likely to be silent since they frequently do not change the sense of the codon. Therefore, these mutations are tolerated and we find them much more frequently than mutations in the first two positions of a codon. Other types of silent mutations occur within introns, where they usually have no effect on the gene's function, or entirely outside of genes—in *intergenic regions*—where they are also usually without phenotypic effect.

REVERSION

Point mutations, especially frameshift mutations, may have drastic effects, but they are reversible. **Reversion,** or **back-mutation,** to the original phenotype can occur in two ways: A *true reversion* is the alteration of the mutated base back to its original identity; alternatively, a change can occur elsewhere in the same gene to compensate for the original mutation. For example, consider a gene that codes for a protein having a positively charged amino acid (arginine) that interacts with a negatively charged amino acid (glutamic acid), as in figure 12.17. This interaction helps hold part of the protein in the proper shape—a fold that is vital to the protein's function.

Suppose that we cause a mutation (a **forward mutation**—away from wild-type) in the codon for the arginine. This will prevent proper folding of the protein, so the protein will not function. Now suppose we cause another mutation that creates a new positively charged amino acid (lysine) near the position of the lost arginine. If this lysine can substitute for the missing arginine in holding the protein in the right shape, it will restore its function. The mutation that created the new lysine codon will therefore be a reversion. It is not a true reversion, because the reverted gene is different from the original. Instead, we call it a *second-site reversion.*

Second-site reversion can also be considered *suppression,* in that it is a compensation by one mutation for the effects of another. In particular, it is an *intragenic suppression,* since both mutations occur in the same gene. *Intergenic suppression*—suppression of a mutation in one gene by a mutation in another—also happens. Because two genes are involved, intergenic suppression is not the same as reversion. A familiar example of intergenic suppression is suppression of nonsense codons, such as amber codons, by mutant tRNAs (e.g., amber suppressors) that recognize these codons (chapter 11).

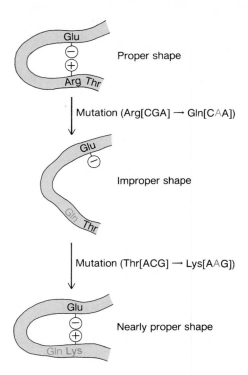

FIGURE 12.17
Second-site reversion. The original mutation (blue) changed an arginine, which bears a positive charge, to a glutamine (blue), which is uncharged. This destroyed an important interaction with a negatively charged glutamic acid and resulted in an inactive protein. The reversion (red) changed a neighboring threonine to a positively charged lysine (red), restoring the vital interaction with the glutamic acid and therefore restoring the protein's function.

Frameshift mutations can revert in a number of ways (figure 12.18): adding a base to compensate for a deleted base; losing a base to balance an extra one that has been inserted; losing two more bases if one has been lost, bringing the total lost to three and eliminating the frameshift; or gaining extra bases to bring the total number gained to three. However, such gains and losses are not likely to happen naturally, so further mutagenesis with frameshifting agents is generally required to obtain reversion of frameshift mutations.

DNA REPAIR

We have discussed many different ways that DNA can be damaged. Since much of this damage is serious enough to threaten life, it is not surprising that living things have evolved mechanisms for dealing with it. One way to cope with a mutation is to repair it, or restore it to its original, undamaged state. There are two basic ways to do this, and they can be illustrated by comparing a mutation in a strand of DNA to a knot in a piece of wire: (1) Directly undo the damage (untie the knot in the wire), or (2) remove the damaged section of DNA and fill it in with new, undamaged DNA (cut out the knot in the wire and solder in a fresh piece of straight wire).

Chapter 12

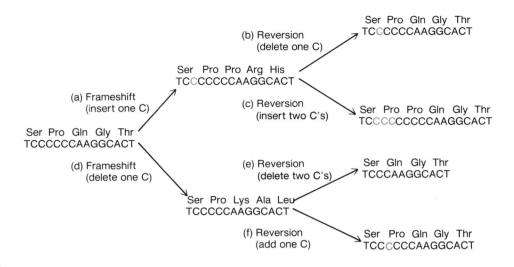

FIGURE 12.18

Reversion of frameshift mutation. The original coding strand is presented at left. It contains a string of six C's. (*a*) A frameshift occurs, involving the insertion of an extra C (blue). This shifts the reading frame at the point of the insertion and alters the coding, starting with the third codon. (*b*) This frameshift can revert simply by deleting one of the C's, returning the reading frame and coding to their exact original state. (*c*) The frameshift can also revert by inserting two more C's (red). Now the total number of inserted bases is three, so one whole new codon (CCC) has been created and the original reading frame has been restored. This results in a slightly altered protein with one extra proline. (*d*) A frameshift occurs, involving the deletion of one C. This shifts the reading frame at the point of the deletion and alters the coding, starting with the third codon. (*e*) This frameshift can revert by deleting two more C's, restoring the original reading frame but resulting in the loss of one codon (CCC). The corresponding protein will lack one proline. (*f*) The frameshift can also revert simply by inserting one C (green) to compensate for the one that was lost. This restores the exact original reading frame and coding.

DIRECTLY UNDOING DNA DAMAGE

In the late 1940s, Albert Kelner was trying to measure the effect of temperature on repair of ultraviolet damage to DNA in the mold *Streptomyces*. However, he noticed that damage was repaired much faster in some mold spores than in others kept at the same temperature. Obviously, some factor other than temperature was operating. Finally, Kelner noticed that the spores whose damage was repaired fastest were the ones kept most directly exposed to light from a laboratory window. When he performed control experiments with spores kept in the dark, he could detect no repair at all. Renato Dulbecco soon observed the same effect in bacteria infected with ultraviolet-damaged phages, and it now appears that most, if not all, forms of life share this important mechanism of repair, which is termed **photoreactivation,** or **light repair.**

It was discovered in the late 1950s that photoreactivation is catalyzed by an enzyme called *photoreactivating enzyme* or *DNA photolyase*. This enzyme operates by the mechanism sketched in figure 12.19. First, the enzyme detects and binds to the ultraviolet-damaged DNA site (a pyrimidine dimer). Then the enzyme absorbs visible light, which activates it so it can break the bonds holding the pyrimidine dimer together. This restores the pyrimidines to their original independent state. Finally, the enzyme dissociates from the DNA, the damage repaired.

Organisms ranging from *E. coli* to human beings exhibit another mechanism for the direct reversal of a mutation, this time an alkylation of the O⁶ of guanine. After

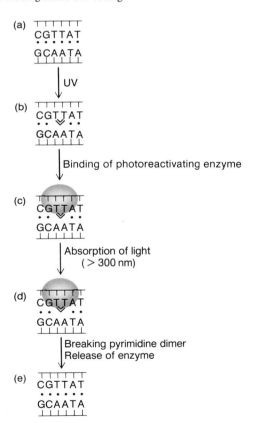

FIGURE 12.19

Model for photoreactivation. (*a*) and (*b*) Ultraviolet radiation causes a thymine dimer to form. (*c*) The photoreactivating enzyme (red) binds to this region of the DNA. (*d*) The enzyme absorbs visible light, breaks the dimer, and (*e*) finally dissociates.

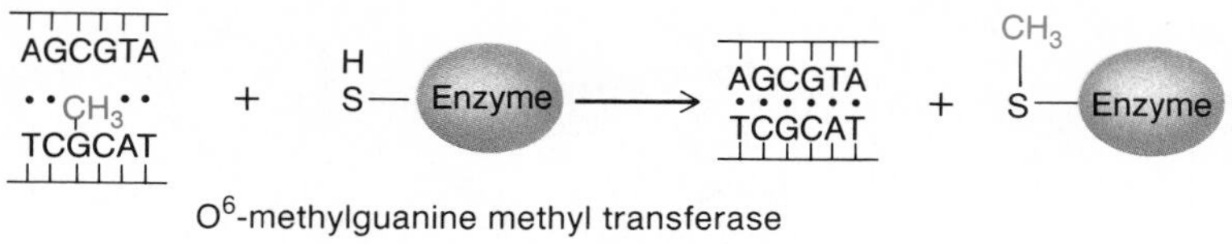

AGCGTA
··CH₃··   +   H—S—Enzyme → AGCGTA + CH₃—S—Enzyme
TCGCAT   TCGCAT

O⁶-methylguanine methyl transferase

FIGURE 12.20

Mechanism of O⁶-methylguanine methyl transferase. A sulfhydryl group of the enzyme accepts the methyl group (green) from a guanine on the DNA, thus inactivating the enzyme.

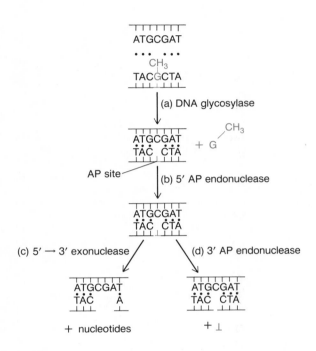

FIGURE 12.21

Removal of methylated base with a DNA glycosylase. (*a*) DNA glycosylase removes the methylguanine base, leaving an apurinic (AP) site. (*b*) A 5′ AP endonuclease cuts the DNA strand just upstream from the AP site. From this point on, two different pathways are possible: (*c*) A 5′ → 3′ exonuclease removes nucleotides, including the AP site, or (*d*) A 3′ AP endonuclease cuts the DNA on the other side of the AP site, removing the apurinic sugar. Both procedures leave a gap that must be filled.

DNA is methylated or ethylated, an enzyme called *O⁶-methylguanine methyl transferase* comes on the scene to repair the damage. It does this by accepting the methyl or ethyl group itself, as outlined in figure 12.20. The acceptor site on the enzyme for the alkyl group is the sulfur atom of an amino acid called cysteine. Strictly speaking, this means that the methyl transferase does not fulfill one part of the definition of an enzyme—that it be regenerated unchanged after the reaction. Instead, this protein seems to be irreversibly inactivated, so we call it a *suicide enzyme* to denote the fact that it "dies" in performing its function. The repair process is therefore expensive; each repair event costs one protein molecule.

One more property of the O⁶-methylguanine methyl transferase is worth noting. The enzyme, at least in *E. coli,* is induced by DNA alkylation. This means bacterial cells that have already been exposed to alkylating agents are much more resistant to DNA damage than cells that have just been exposed to such mutagens for the first time.

U ltraviolet damage to DNA (pyrimidine dimers) can be directly repaired by a photoreactivating enzyme that uses energy from visible light to break the bonds holding the two pyrimidines together. O⁶-alkylations on guanine residues can be directly reversed by the suicide enzyme O⁶-methylguanine methyl transferase, which accepts the alkyl group onto one of its amino acids, a cysteine.

EXCISION REPAIR

The percentage of mutations that can be handled by direct reversal is necessarily small. Most mutations involve neither pyrimidine dimers nor O⁶-alkylguanine, so they must be handled by a different mechanism. Most are removed by a process called **excision repair.** Again, excision repair is like cutting out a knotted wire segment and replacing it with fresh wire. The damaged DNA is first removed, then replaced with fresh DNA. This occurs by one of several mechanisms; some notable examples will be described here.

Using DNA Glycosylase

Certain mutations are recognized by an enzyme called *DNA glycosylase,* which breaks the **glycosidic bond** between the damaged base and its sugar (figure 12.21). This leaves an **apurinic** or **apyrimidinic site (AP site),** which is a sugar without its purine or pyrimidine base. Once the AP site is created, it is recognized by a *5′ AP endonuclease* that cuts, or **nicks,** the DNA strand on the 5′-side of the AP site. (The "endo" in endonuclease means the enzyme cuts *inside* a DNA strand, not at a free end; Greek: *endo* = within.) This step, known as **incision,** creates free ends within the DNA strand that signal other enzymes to complete the excision process. One possibility is for a **5′ → 3′ exonuclease** to start at the nick and move left to right (5′ to 3′), removing nucleotides, including the damaged one. (The "exo" in exonuclease means the enzyme must start at a free end and degrade the DNA one nucleotide at a time, working toward the other end; Greek: *exo* = outside.) Another possibility is for a *3′ AP endonuclease* simply to cut just after the AP site, releasing the AP deoxyribose phosphate. Both of these mechanisms leave a gap in the DNA that must be repaired, using the opposite strand as the template.

Pyrimidine dimers can also be removed by a DNA glycosylase in a variation on the theme described above. This time, a DNA glycosylase specific for pyrimidine dimers cuts the glycosidic bond between the first pyrimidine and its sugar, leaving an AP site (figure 12.22). However, the enzyme does not release a free base because the two pyrimidines are covalently bonded together. In the incision step, a 3′ AP endonuclease cuts the phosphodiester bond just after the AP site. Then either a 5′ AP endonuclease or a 3′ → 5′ exonuclease excises the AP sugar. Finally, a 5′ → 3′ exonuclease removes a few nucleotides, including the one that contains the pyrimidine dimer, leaving a gap in the DNA just like the one in figure 12.21.

Repairing Damage Without Creating AP Sites

Bulky base damage, such as thymine dimers, can also be removed directly, without help from a DNA glycosylase. In principle, there are two ways to do this; both are illustrated in figure 12.23. In the one-nick pathway (*a*), a damage-specific DNA incising enzyme recognizes the bulky damage and cuts the DNA strand upstream from

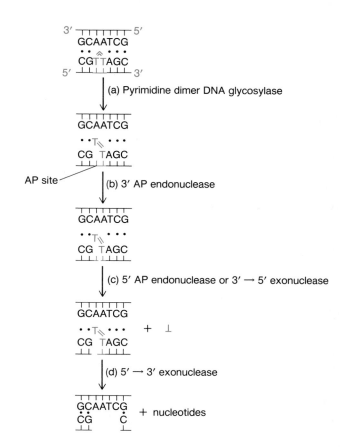

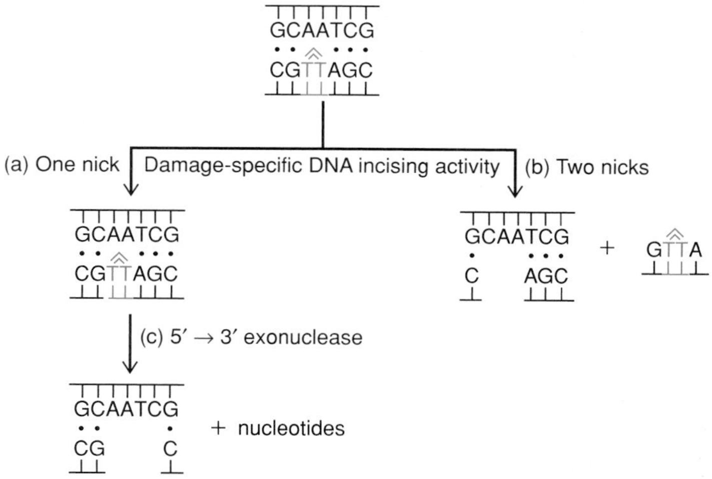

FIGURE 12.22

Removal of pyrimidine dimer with a DNA glycosylase. (*a*) The pyrimidine dimer DNA glycosylase cuts the glycosidic bond between one thymine and its sugar, creating an AP site and leaving the dimer still attached to the DNA. (*b*) A 3′ AP endonuclease cuts just downstream from the AP site. (*c*) A 5′ AP endonuclease or a 3′ → 5′ exonuclease removes the AP sugar. (*d*) A 5′ → 3′ exonuclease removes several nucleotides, including the thymine dimer. As in figure 12.21, a gap remains that must be filled.

FIGURE 12.23

DNA incision without AP sites. A bulky, damage-dependent DNA incising enzyme attacks the damaged DNA strand (*a*) at one site or (*b*) at two sites. In the latter case, the bulky damage is removed as part of an oligonucleotide (twelve to thirteen bases long in *E. coli*). (*c*) In the former case, a 5′ → 3′ exonuclease is needed to remove the damaged DNA. Both procedures leave a gap that must be filled.

the damage. Then a $5' \rightarrow 3'$ exonuclease removes the damaged DNA as described previously. In the two-nick pathway (*b*), the incising enzyme makes cuts on either side of the mutation, removing an oligonucleotide with the damage. *E. coli* cells seem to use the latter mechanism primarily, at least for repair of pyrimidine dimers and other bulky damage. The key enzyme in this process is called the *uvrABC endonuclease* because it contains three polypeptides, the products of the *uvrA, uvrB,* and *uvrC* genes. This enzyme generates an oligonucleotide that is twelve to thirteen bases long, rather than the four shown for convenience here. Other repair systems may use the one-nick pathway.

In all cases just described, we wind up with a gapped DNA strand across from an intact strand. Recall from chapter 8 that this is exactly the same situation the DNA replication apparatus faces after an RNA primer is removed. And the same mechanism for filling in the gap is probably used here: DNA polymerase I synthesizes new DNA, using the intact strand as the template, and DNA ligase forms the last phosphodiester bond to complete the job (figure 12.24). In fact, DNA polymerase I contains both a $5' \rightarrow 3'$ exonuclease and a DNA polymerase activity. Therefore, in principle at least, it can remove damage and fill in the gap at the same time. In that case, the long gaps shown in figure 12.24 would not occur. In fact, fairly long gaps do appear in *E. coli* DNA undergoing repair, so simultaneous DNA degradation and resynthesis apparently do not always take place.

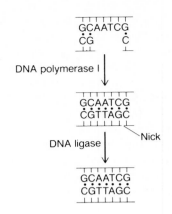

FIGURE 12.24
Filling in the gaps left by damage excision. First, DNA polymerase I links together nucleotides to fill the gap, leaving a nick (an unformed phosphodiester bond). Finally, DNA ligase seals the nick by forming the last phosphodiester bond.

M	ost DNA damage is corrected by excision repair, in which the mutated DNA is removed and replaced with normal DNA. This can occur by a number of different mechanisms. (1) The damaged base can be clipped out by a DNA glycosylase, leaving an apurinic or apyrimidinic site that attracts the DNA cutting enzymes that remove the damaged region. (2) A pyrimidine dimer can also attract a specific DNA glycosylase to initiate the same kind of repair pathway. (3) The damaged DNA can be clipped out directly either by cutting on one side with an endonuclease and then removing the mutated DNA with an exonuclease, or by cutting on both sides with an endonuclease to remove the damaged DNA as part of an oligonucleotide.

∎

MISMATCH REPAIR

So far, we have been dealing with repair of DNA damaged by external agents. What about DNA that simply has a mismatch due to incorporation of the wrong base and failure of the proofreading system (chapter 8)? At first, it would seem tricky to repair such a mistake, because of the apparent difficulty in determining which strand is the newly synthesized one that has the mistake and which is the parental one that should be left alone.

At least in *E. coli* this is not a problem, because the parental strand has identification tags that distinguish it from the progeny strand. These tags are methylated adenines, created by a methylating enzyme that recognizes the sequence GATC and places a methyl group on the A. Since this four-base sequence occurs approximately every 250 base pairs, there is usually one not far from a newly created mismatch.

Moreover, GATC is a palindrome, so the opposite strand also reads GATC in the $5' \rightarrow 3'$ direction. This means that a newly synthesized strand across from a methylated GATC is also destined to become methylated, but a little time elapses before this can happen. The **mismatch repair** system takes advantage of this delay; it uses the methylation on the parental strand as a signal to leave that strand alone and correct the nearby mismatch in the unmethylated progeny strand. This process must occur fairly soon after the mismatch is created, or both strands will be methylated and no distinction between them will be possible. It is not clear how eukaryotes carry out mismatch repair; some of them, such as yeast and *Drosophila,* do not seem to methylate their DNA.

COPING WITH DNA DAMAGE WITHOUT REPAIRING IT

The direct reversal and excision repair mechanisms described so far are all true repair processes. They eliminate the defective DNA entirely. However, cells have other means of coping with mutations that do not remove the damage but simply skirt around it. These are usually called repair mechanisms, even though they really are not.

Recombination Repair

So-called **recombination repair** is the most important of these mechanisms. Because it requires DNA replication

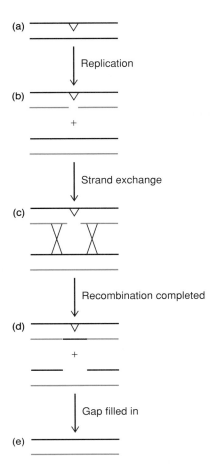

(a) ▽
↓ Replication
(b) ▽
+
↓ Strand exchange
(c) ▽
↓ Recombination completed
(d) ▽
+
↓ Gap filled in
(e)

FIGURE 12.25
Recombination repair. (*a*) We begin with DNA with a pyrimidine dimer, represented by a V shape. (*b*) During replication, the replication machinery skips over the region with the dimer, leaving a gap; the complementary strand is replicated normally. The two newly synthesized strands are shown in red. (*c*) Strand exchange between homologous strands occurs. (*d*) Recombination is completed, filling in the gap opposite the pyrimidine dimer, but leaving a gap in the other daughter duplex. (*e*) This last gap is easily filled, using the normal complementary strand as template.

before it can operate, it is also sometimes called **post-replication repair.**

Figure 12.25 shows how recombination repair works. First, the DNA must be replicated. This creates a problem for DNA with pyrimidine dimers because the dimers stop the replication machinery. Nevertheless, after a pause, replication continues, leaving a gap across from the dimer. (A new primer is presumably required to restart DNA synthesis.) Next, recombination occurs between the gapped strand and its homologue on the other daughter DNA duplex. This recombination depends on the *recA* gene product, which exchanges the homologous DNA strands (chapter 8). The net effect of this recombination is to fill in the gap across from the pyrimidine dimer but create a new gap in the other DNA duplex. However, since the other duplex has no dimer, the gap can easily be filled

in by DNA polymerase and ligase. Note that the DNA damage still exists, but the cell has at least managed to replicate its DNA. Sooner or later, true DNA repair could presumably occur.

Error-prone Repair

So-called **error-prone repair** is another way of dealing with damage without really repairing it. This pathway is induced by DNA damage, including ultraviolet damage, and depends on the product of the *recA* gene. We have encountered *recA* before in our discussion of the induction of a λ prophage during the SOS response (chapter 9). Indeed, error-prone repair is also part of the SOS response.

The chain of events seems to be as follows (figure 12.26): Ultraviolet light or another mutagenic treatment somehow activates the RecA protease activity. This protease has several targets. One we have studied already is the λ repressor; another is the product of the *lexA* gene. This product, **LexA**, is a repressor for many genes, including repair genes; when it is cleaved by RecA protease, all these genes are induced.

Two of the newly induced genes are *umuC* and *umuD,* which make up a single operon (*umuDC*). Somehow the products of these genes promote error-prone repair. We know this is so because mutations in either *umuC* or *umuD* prevent error-prone repair, but we do not know how these genes work. One thing seems clear: Error-prone repair requires DNA replication. When a gene for the replicating enzyme DNA polymerase III is mutated, no error-prone repair occurs. Our best guess at this time is that error-prone repair involves replication of DNA across from the pyrimidine dimer even though correct "reading" of the defective strand is impossible. This avoids leaving a gap, but it usually puts the wrong bases into the new DNA strand. When the DNA replicates again, these errors will be perpetuated.

> *C*ells can employ nonrepair methods to circumvent DNA damage. One of these is recombination repair, in which the gapped DNA strand across from a damaged strand recombines with a normal strand in the other daughter DNA duplex after replication. This solves the gap problem but leaves the original mutation unrepaired. Another mechanism to deal with DNA damage is to induce the SOS response. The effect of this on DNA damage seems to be to replicate the DNA, even though the damaged region cannot be read correctly. This results in errors in the newly made DNA, so the process is called error-prone repair.
>
> ∎

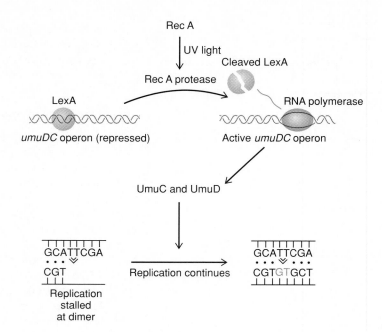

FIGURE 12.26
Error-prone (SOS) repair. Ultraviolet light activates the RecA protease, which cleaves the LexA protein (blue), releasing it from the *umuDC* operon. This results in synthesis of UmuC and UmuD proteins, which somehow allow DNA synthesis across from a thymine dimer, even though mistakes (green) will usually be made.

SEVERE CONSEQUENCES OF DEFECTS IN DNA REPAIR MECHANISMS

Wild-type *E. coli* cells can tolerate as many as fifty pyrimidine dimers in their genome without ill effect because of their active repair mechanisms. Bacteria lacking one of the *uvr* genes cannot carry out excision repair, so their susceptibility to ultraviolet damage is greater. However, they are still somewhat resistant to DNA damage. On the other hand, double mutants in *uvr* and *recA* can perform neither excision repair nor recombination repair, and they are very sensitive to ultraviolet damage. Only one to two pyrimidine dimers per genome is a lethal dose.

A similar situation is found in humans. Congenital defects in DNA repair cause a group of human diseases, including xeroderma pigmentosum, ataxia telangiectasia, Fanconi's anemia, and Bloom's syndrome. The latter three conditions are characterized by increased incidence of chromosomal abnormalities, especially gaps and breaks. This is understandable in the case of Bloom's syndrome, because that condition involves a deficiency in DNA ligase, the enzyme that repairs DNA breaks. Whether this defect explains the failure to repair damaged chromosomes is still unclear, but the result of this failure is that patients with any of these diseases are more susceptible to cancer than is the general population.

Xeroderma pigmentosum is better understood. Patients with this disease are thousands of times more likely to develop skin cancer than normal people if they are exposed to the sun. In fact, their skin can become literally freckled with skin cancers. This reflects the fact that xeroderma pigmentosum cells are defective in excision repair and therefore cannot repair pyrimidine dimers effectively. In addition, patients with xeroderma pigmentosum have a somewhat higher than average incidence of

internal cancers. This probably means that these people are also defective in repairing DNA damage caused by chemical mutagens. Notice the underlying assumption here that unrepaired genetic damage can lead to cancer. We will expand on this theme later in this chapter.

What repair steps are defective in xeroderma pigmentosum cells? There is no simple answer to this question because the defect varies from one patient to the next. The problem has been investigated by fusing cells from different patients to see if the fused cells still show the defect. Frequently they do not; instead, the genes from two different patients complement each other. Of course, this probably means that a different gene was defective in each patient. So far, eight different complementation groups have been identified this way, suggesting that the defect can lie in any of eight different genes. Most often, the first step in excision repair (incision) seems to be defective. There is even a variant of the disease in which excision repair occurs normally; in these patients, about 10% of the total, some other repair process must be blocked.

F ailure to correct genetic damage is very harmful to an organism. *E. coli* cells that are blocked in both excision repair and recombination repair can tolerate only a small fraction of the ultraviolet radiation that would be innocuous to wild-type bacteria. Humans defective in DNA repair suffer from a variety of conditions, the best understood of which is xeroderma pigmentosum. Most patients with this disease cannot carry out excision repair. As a result, ultraviolet damage persists in their skin cells and leads to multiple skin cancers.

■

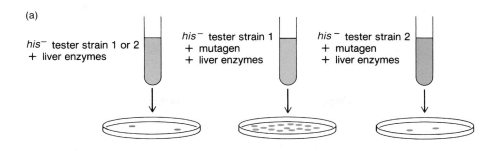

(a)

his⁻ tester strain 1 or 2
+ liver enzymes

his⁻ tester strain 1
+ mutagen
+ liver enzymes

his⁻ tester strain 2
+ mutagen
+ liver enzymes

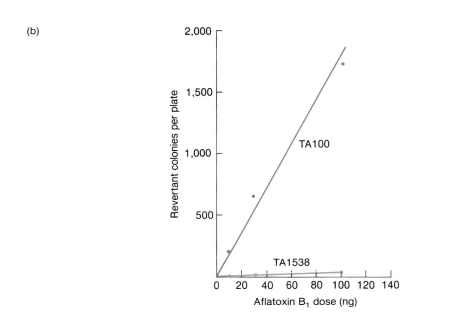

(b)

FIGURE 12.27

The Ames test. (*a*) Outline of the procedure. In this example, two *his⁻* tester strains are used. Strain 1 contains a missense mutation in the *his* operon and so will detect mutagens that can cause missense back-mutations, or reversions. Strain 2 contains a frameshift mutation in the *his* operon and so is sensitive to frameshift reversions. These strains are mixed with liver enzymes, which can metabolize compounds to their mutagenic forms. In the control on the left, no mutagen is added, thus only a few spontaneous reversions occur. In the middle, the missense mutation tester strain has experienced multiple reversions, showing that the compound under study can cause missense mutations. On the right, the frameshift mutation tester strain shows no significant difference from the control, so this compound does not cause frameshift mutations. (*b*) Data from a test of the mutagen aflatoxin B₁. The tester strain TA100, like the hypothetical strain 1 above, tests for missense mutations. Clearly, aflatoxin is a powerful mutagen in this strain. By contrast, the frameshift tester strain, TA1538, detects little if any tendency of aflatoxin to cause frameshift mutations.

DETECTING MUTAGENS

We have seen that many chemicals are now recognized as mutagens, and that most mutations are deleterious. In fact, as we will see later in this chapter, environmental mutagens are thought to be responsible for the majority of human cancers. It is important, therefore, to identify mutagenic chemicals so that we can limit our exposure to them. But because there are so many potential mutagens to test and because massive numbers of organisms must be examined to find significant numbers of mutants, using test animals such as mice is extremely expensive and terribly time-consuming.

Fortunately, excellent mutagen testing systems have been developed using bacteria. As a test organism, bacteria have tremendous advantages: They are cheap and easy to culture, and billions of cells can be tested in a day or two. The best known of the bacterial assays was created by Bruce Ames and is known as the **Ames test.** Ames started with auxotrophic histidine mutants—strains of *Salmonella typhimurium* with mutations in the *his* operon that rendered them incapable of growing in the absence of the amino acid histidine. The basic idea of the assay is to observe the frequency of reversions to wild-type in the *his* operon (figure 12.27). This is very simple to do. One just looks for colonies that can grow on plates lacking histidine. Without any mutagen, these reversions occur only rarely, so few colonies can grow. On the other hand, a powerful mutagen can greatly accelerate the mutation rate—including back-mutations, or reversions—and many colonies can be observed.

This concept sometimes gives students trouble because the mutation we are looking for is a beneficial back-mutation to wild-type, rather than a deleterious mutation. Nevertheless, it is a mutation, and it is much easier to observe than the forward mutation. To understand this, consider the ease of counting the colonies produced by a few prototrophs that can grow, among a vast excess of auxotrophs that cannot; then compare this with the hopelessness of finding a few auxotrophs that cannot grow, among countless prototrophs that can.

Ames had to make some additions and modifications to his bacteria in order to make them work well in detecting mutagens. First, he took note of the fact that most chemical mutagens must be metabolized by the body before they become mutagenic. This metabolism, carried out by liver enzymes, is part of the normal method by which we excrete toxic materials. Since bacteria do not perform such metabolism, Ames added a rat liver extract to his bacteria to mimic the metabolism a potential mutagen would experience in the body. In addition, Ames mutated his bacteria to make them leaky to exogenous chemicals and deficient in excision repair. He also added a plasmid bearing an aberrant *umuDC* operon, which produces enzymes that are extra efficient in error-prone SOS repair. This converts most kinds of DNA damage to mutations.

How valid are the results of this kind of mutagenicity testing? Are these strains too sensitive to mutation? Are compounds that cause mutations in bacteria also mutagens in humans? It is true that we have much better defenses against mutagens than these severely weakened bacteria possess, but the Ames test does give a good idea about the relative mutagenicities of different chemical agents. And since bacteria and humans share the same kind of genetic material, agents that mutate their genes are likely to do the same to ours.

T he Ames test uses a set of bacterial tester strains to screen for chemical mutagens. These *Salmonella* cells are histidine auxotrophs and are exquisitely sensitive to mutation by chemicals. To perform the Ames test, the researcher adds a suspected mutagen to these cells (plus liver enzymes to metabolize the chemical to its mutagenic form) and observes the appearance of histidine prototrophic colonies. The number of revertant colonies is a measure of the potency of the mutagen.

■

BACTERIAL TRANSPOSONS

We have already learned that an organism's genome is not absolutely fixed from the beginning to the end of its life. In addition to the point mutations already discussed, rearrangements of genetic material take place. This can be useful, as in the rearrangement of immunoglobulin gene segments to form a functioning gene, or in the activation of a trypanosome surface protein gene by moving it to an expression site (chapter 14). But pieces of DNA with no apparent function besides their own replication also move around the genomes of both prokaryotes and eukaryotes. These **transposable elements,** or **transposons,** are mobile in the sense that they can insert copies of themselves throughout the genome. They can also cause rearrangements of host DNA, giving rise to chromosome mutations. We will discuss the characteristics of bacterial and eukaryotic transposons in turn, showing how they can cause mutations, including chromosomal mutations.

Bacterial transposons were originally noticed by James Shapiro and others in the late 1960s as mutations that did not behave normally. For example, they did not revert readily the way point mutations do, and the mutant genes contained long stretches of extra DNA. Shapiro demonstrated this by taking advantage of the fact that a λ phage will sometimes pick up a piece of host DNA during lytic infection of *Escherichia coli* cells, incorporating this "passenger" DNA into its own genome. He allowed λ phages to pick up either a wild-type *E. coli* galactose utilization gene (gal⁺) or its mutant counterpart (gal⁻), then measured the sizes of the **recombinant DNAs,** which contained λ DNA plus host DNA (figure 12.28). He measured the DNA sizes by measuring the densities of the two types of phage using cesium chloride gradient centrifugation. Since the phage coat is made of protein and always has the same volume, and since DNA is much denser than protein, the more DNA the phage contains, the denser it will be. It turned out that the phages harboring the gal⁻ gene were denser than the phages with the wild-type gene, and therefore held more DNA. The simplest explanation, which turned out to be correct, is that foreign DNA had inserted into the gal gene and thereby mutated it. In fact, later experiments revealed 800–1,400-bp inserts in the mutant gal gene, which were not found in the wild-type gene. In the rare cases when such mutants did revert, they lost the extra DNA. These extra DNAs that could inactivate a gene by inserting into it were the first transposons discovered in bacteria. They are called **insertion sequences (ISs).**

INSERTION SEQUENCES: THE SIMPLEST TRANSPOSONS

Bacterial insertion sequences contain only the elements necessary for transposition. The first of these elements is a set of special sequences at a transposon's ends, one of

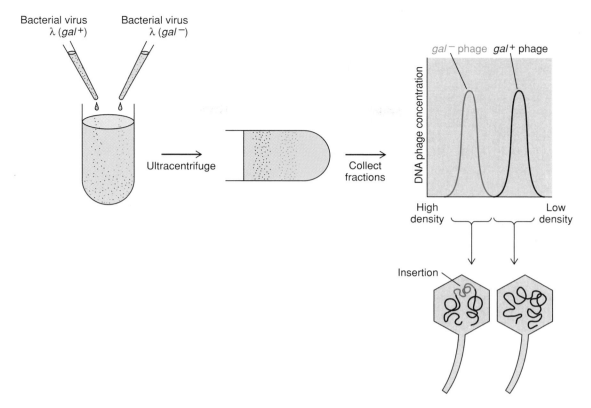

FIGURE 12.28

Demonstration of mutation by insertion. λ phage bearing a wild-type (*gal*⁺) or mutant (*gal*⁻) gene from *E. coli* was subjected to CsCl gradient centrifugation. Two bands of phage separated clearly, the denser (red) bearing the mutant phage. Since the denser phage must contain more DNA, this means that the mutation in the *gal* gene was caused by an insertion of extra DNA.

From "Transposable Genetic Elements" by Stanley N. Cohn and James A. Shapiro. Copyright © 1980 by Scientific American, Inc. All rights reserved.

which is the inverted repeat of the other. The second element is the set of genes that code for the enzymes that catalyze transposition.

Since the ends of an insertion sequence are inverted repeats, then if one end of an insertion sequence is 5′-ACCGTAG, the other end of that strand will be the reverse complement: CTACGGT-3′. This is akin to the palindromes, or inverted repeats, discussed in chapter 9, but the two parts of the palindrome are interrupted by hundreds of base pairs, corresponding to the body of the insertion sequence. The seven-nucleotide inverted repeats given here are hypothetical and are presented to illustrate the point. Typical insertion sequences have somewhat longer inverted repeats, from fifteen to twenty-five base pairs long. IS1, for example, has inverted repeats twenty-three base pairs long. Larger transposons can have inverted repeats hundreds of base pairs long.

The main body of an insertion sequence codes for at least two proteins that catalyze transposition. These proteins are collectively known as **transposase;** their mechanism of action is still being worked out. We do know that these proteins are necessary for transposition, because mutations in the body of an insertion sequence can render that transposon immobile.

One other feature of an insertion sequence, shared with more complex transposons, is found just outside the transposon itself. This is a pair of short direct repeats in the DNA immediately surrounding the transposon. These repeats did not exist before the transposon inserted; they result from the insertion process itself and tell us that the transposase cuts the target DNA in a staggered fashion rather than with two cuts right across from one another. Figure 12.29 shows how staggered cuts in the two strands of the target DNA at the site of insertion lead automatically to direct repeats. The length of these direct repeats depends on the distance between the two cuts in the target DNA strands. This distance depends in turn on the nature of the insertion sequence. The transposase of IS1 makes cuts nine base pairs apart and therefore generates direct repeats that are nine base pairs long.

I nsertion sequences are the simplest of the transposons. They contain only the elements necessary for their own transposition: short inverted repeats at their ends and at least two genes coding for an enzyme called transposase that carries out transposition. Transposition involves duplication of a short sequence in the target DNA; one copy of this short sequence flanks the insertion sequence on each side after transposition.

FIGURE 12.29

Generation of direct repeats in host DNA flanking a transposon. (*a*) The arrows indicate where the two strands of host DNA will be cut in a staggered fashion, nine base pairs apart. (*b*) After cutting. (*c*) The transposon (orange) has been ligated to one strand of host DNA at each end, leaving two nine-base gaps. (*d*) After the gaps are filled in, there are nine base-pair repeats of host DNA (green boxes) at each end of the transposon.

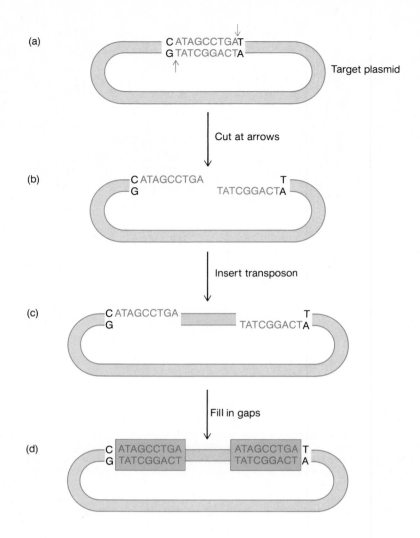

MORE COMPLEX TRANSPOSONS

Insertion sequences are sometimes called "selfish DNA." You should not interpret this anthropomorphic term to mean that ISs think about what they do; of course they cannot. Rather, it means that an IS replicates at the expense of its bacterial host and apparently provides nothing useful in return. Other transposons do carry genes that are valuable to their hosts, the most familiar being genes for antibiotic resistance. Not only is this a clear benefit to the bacterial host, it is also valuable to geneticists, because it makes the transposon much easier to track. No longer is the transposon detectable only by the genes it inactivates; it now carries a conspicuous marker of its own.

For example, consider the situation in figure 12.30, in which we start with a donor plasmid containing a gene for kanamycin resistance (Km) and harboring a transposon (**Tn3**) with a gene for ampicillin resistance (Ap); in addition, we have a target plasmid with a gene for tetracycline resistance (Tc). After transposition, Tn3 has replicated, and a copy has moved to the target plasmid. Now the target plasmid confers both tetracycline and ampicillin resistance, properties that we can easily monitor

by transforming antibiotic-sensitive bacteria with the target plasmid and growing these host bacteria in medium containing both antibiotics. If the bacteria survive, they must have taken up both antibiotic resistance genes; therefore, Tn3 must have transposed to the target plasmid.

MECHANISMS OF TRANSPOSITION

One sometimes hears transposons referred to as "jumping genes," because of their ability to move from one place to another. However, the term is a little misleading, since it implies that the DNA always leaves one place and jumps to the other. This mode of transposition does occur and is called **conservative transposition,** because both strands of the original DNA are conserved as they move together from one place to the other. However, transposition frequently involves DNA replication, so one copy of the transposon remains at its original site as another copy inserts at the new site. This is called **replicative transposition** because when a transposon moves by this route, it also replicates itself.

Chapter 12

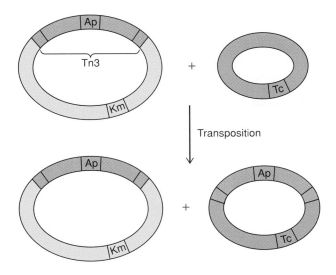

FIGURE 12.30
Tracking transposition with antibiotic resistance markers. We begin with two plasmids: The larger (yellow) encodes kanamycin resistance (Km) and bears the transposon Tn3 (blue), which codes for ampicillin resistance (Ap); the smaller (green) encodes tetracycline resistance (Tc). After transposition, the smaller plasmid bears both the Tc and Ap markers.

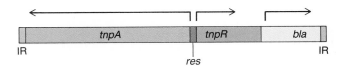

FIGURE 12.31
Structure of Tn3. The *tnpA* and *tnpR* genes are necessary for transposition; *res* is the site of the recombination that occurs during the resolution step in transposition; the *bla* gene encodes β-lactamase, which protects bacteria against the antibiotic ampicillin. This gene is also called Ap and Amp^r. Inverted repeats (IR) are found on each end. The arrows indicate the direction of transcription of each gene.

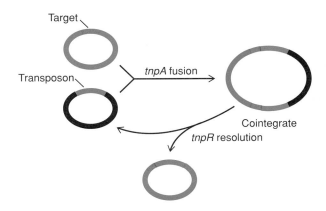

FIGURE 12.32
Simplified scheme of the two-step Tn3 transposition. In the first step, catalyzed by the *tnpA* gene product, the plasmid (black) bearing the transposon (blue) fuses with the target plasmid (green, target in red) to form a cointegrate. During cointegrate formation, the transposon replicates. In the second step, catalyzed by the *tnpR* gene product, the cointegrate resolves into the target plasmid, with the transposon inserted, plus the original transposon-bearing plasmid.

Tn3, whose structure is shown in figure 12.31, illustrates the replicative mechanism of transposition. In addition to the *bla* gene, which encodes ampicillin-inactivating β-**lactamase,** Tn3 contains two genes that are instrumental in transposition. Tn3 transposes by a two-step process, each step of which requires one of the Tn3 gene products. Figure 12.32 shows this sequence of events. We begin with two plasmids: the donor, which harbors Tn3, and the target. In the first step, the two plasmids fuse, with Tn3 replication, to form a **cointegrate** in which they are coupled through a pair of Tn3 copies. This step requires recombination between the two plasmids, which is catalyzed by the product of the Tn3 transposase gene *tnpA.*

Most of the recombinations we have encountered so far required homology between the two recombining DNAs, but the recombination catalyzed by the Tn3 transposase requires little if any homology between donor and target DNAs. This means that transposition is relatively independent of DNA sequence, so Tn3 can transpose to a wide variety of target loci. This kind of **illegitimate recombination,** which shows little dependence on DNA sequence, is a common feature of transposons. Figure 12.32 illustrates transposition between two plasmids, but the donor and target DNAs can be other DNAs, including phage DNAs or the bacterial chromosome itself.

The second step in Tn3 transposition is a **resolution** of the cointegrate, in which the cointegrate breaks down into two independent plasmids, each bearing one copy of Tn3. This step, catalyzed by the product of the **resolvase** gene *tnpR,* is a recombination between homologous sites on Tn3 itself, called *res* sites. Several lines of evidence show

that Tn3 transposition is a two-step process. First, mutants in the *tnpR* gene cannot resolve cointegrates, so they cause formation of cointegrates as the final product of transposition. This demonstrates that the cointegrate is normally an intermediate in the reaction. Second, even if the *tnpR* gene is defective, cointegrates can be resolved if a functional *tnpR* gene is provided *in trans* by another DNA molecule—the host chromosome or another plasmid, for example.

*M*any transposons contain genes aside from the ones necessary for transposition. These are commonly antibiotic resistance genes. For example, Tn3 contains a gene that confers ampicillin resistance. Tn3 and its relatives transpose by a two-step process (replicative transposition). First the transposon replicates and the donor DNA fuses to the target DNA, forming a cointegrate. In the second step, the cointegrate is resolved into two DNA circles, each of which bears a copy of the transposon. An alternate pathway, not used by Tn3, is conservative transposition, in which no replication of the transposon occurs.

Transposition Impacts on Public Health

E ven more complex than Tn3 and its relatives are **composite transposons.** These consist of two parts: a central region and two arms. The central region comprises the genes for transposition plus one or more antibiotic resistance genes. The arms are insertion sequences or IS-like elements. **Tn9** is an example of a composite transposon. Its central region carries a chloramphenicol resistance gene and is flanked by two IS1 elements. The ISs can cooperate to transpose the whole Tn9 element, or they can transpose on their own.

Figure B12.1 shows an antibiotic resistance plasmid that has been the target for a variety of different transposons, including the composite transposon **Tn10.** Such a plasmid, with multiple antibiotic resistance genes, is obviously a great advantage to the bacteria that harbor it, as long as antibiotic use is widespread. In fact, these plasmids may help explain how some pathogenic (disease-causing) bacteria are becoming resistant to the antibiotics we use to kill them. Once these plasmids are assembled, they can be shared with other bacteria, instantly conferring multiple antibiotic resistance.

As an example of the potential severity of the antibiotic resistance problem, consider the case of a huge, modern dairy in Illinois that became contaminated with the bacterium *Salmonella typhimurium* in 1985. As a result, over 14,000 people in six states suffered salmonellosis, or salmonella poisoning. This thoroughly unpleasant condition, one of the diseases commonly known as "food poisoning," is characterized by nausea, abdominal cramps, vomiting, diarrhea, fever, and even death. In this outbreak, five people died. Fortunately, the bacteria involved in this epidemic were not antibiotic-resistant. If they had been, the death toll would probably have been much higher.

Another outbreak of salmonellosis had occurred two years earlier in the Midwest, mostly in Minnesota. It was much more limited in scope—only eighteen patients—but was potentially much more dangerous, because the bacteria, *Salmonella newport* in this case, were antibiotic-resistant. Where did these bacteria come from? Scott A. Holmberg of the Centers for Disease Control (CDC) in Atlanta used a nice bit of detective work to identify the culprit.

Holmberg's only clues at the outset were that *Salmonella newport* is unusual in the north, and that most of the victims had taken antibiotics shortly before they fell ill. This pointed to the likelihood of a contaminated antibiotic, but Holmberg eliminated that possibility by discovering that different antibiotics were involved, and that some of the patients had not even taken antibiotics. The big break came when Holmberg learned that four other cases of drug-resistant salmonellosis had occurred somewhat earlier in South Dakota. Three of the patients were related, and one of them was a farmer who raised beef cattle.

The farmer told Holmberg that his herd had been infected by an epidemic of diarrhea several months before, and that several calves had died. He mentioned in passing that he thought *Salmonella* was involved. Was it the same strain of *Salmonella* implicated in the human cases? To find out, the plasmids in the *Salmonella* from the dead calf and those from the human patients in Minnesota were compared by electrophoresis. The pattern from the calf matched those of ten of the eleven patients, and was clearly different from ninety-one other samples of *Salmonella newport* from unrelated cases.

This suggested strongly that there was a relationship between the South Dakota cattle herd and all the cases of drug-resistant salmonellosis. Moreover, the beef from this

TRANSPOSONS AS MUTAGENIC AGENTS

A bacteriophage called **Mu** that infects *E. coli* also behaves as a transposon. In fact, its replication depends on transposition. Phage Mu particles contain a linear DNA genome 37 kb long that can replicate either lytically or lysogenically (figure 12.33). (Recall the lytic and lysogenic replication of phage λ described in chapter 9.) In the lysogenic phase, the Mu genome inserts into the host chromosome, and a phage repressor prevents most phage

gene expression, just as in the λ lysogenic phase. However, the lytic phase of Mu is very different from that of λ. Instead of remaining free, the phage DNA integrates into the host chromosome. More strikingly, the phage DNA remains integrated throughout the lytic phase and replicates by transposing.

Austin Taylor discovered phage Mu because of the mutations it caused when it transposed. In fact, the name Mu stands for mutator. Obviously, the phage can mutate

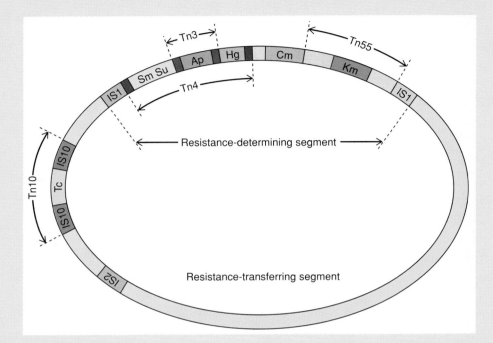

FIGURE B12.1
Hypothetical antibiotic resistance plasmid. Several different transposons, each bearing its own resistance marker, have clustered in a single plasmid. Tn4, which includes Tn3, confers resistance to streptomycin (Sm), sulfonamide (Su), ampicillin (Ap), and mercury (Hg). Tn55 confers resistance to kanamycin (Km), and Tn10 confers resistance to tetracycline (Tc). All but the last transposon comprise the resistance-determining segment, which is bounded by two IS1 elements and also contains a gene for chloramphenicol resistance (Cm). The two IS1 elements allow the resistance-determining segment to transpose independently, carrying all its valuable resistance genes with it.
From "Transposable Genetic Elements" by Stanley N. Cohn and James A. Shapiro. Copyright © 1980 by SCIENTIFIC AMERICAN, Inc. All rights reserved.

herd could be traced directly to a meat processor in Nebraska who sold boxed meat to two meat brokers in Minnesota and Iowa. These brokers in turn sold the meat to supermarkets in Minneapolis and St. Paul at which all the Minnesota victims had bought hamburger.

Why did antibiotics contribute to most of the cases? By taking antibiotics, the victims eliminated most of the natural bacteria in their bodies, allowing the drug-resistant bacteria to proliferate without competition. More importantly, how did this strain of *Salmonella* come to be antibiotic-resistant? Here the scenario becomes somewhat more controversial, but most

epidemiologists now believe that the use of low levels of antibiotics in livestock feed selects for antibiotic-resistant bacteria. It is probably significant that the South Dakota farmer, following a widespread practice dating back more than thirty years, had added the antibiotic tetracycline to his cattle feed to promote growth and prevent disease. The result seems to be that his cattle bred a strain of *Salmonella* resistant not only to tetracycline, but to penicillin and its derivative, amoxycillin. They probably obtained this multiple drug resistance by acquiring a resistance plasmid such as the one in figure B12.1. ∎

a gene simply by inserting into it and interrupting it, but Taylor found that Mu could cause many other types of mutations beyond simple insertions. It could cause inversions, deletions, translocations, and fusions of independent DNAs. We have already seen that other transposons can direct the formation of cointegrates (DNA fusion) and transposition (translocation). Figure 12.34 illustrates how Mu and other transposons can provide sites for ordinary homologous recombination (general recombination) to operate. Depending on the orientation of the two transposons, this can result in deletions or inversions.

B acteriophage Mu can behave as a transposable element. By moving from one site to another and simply providing sites of homology for recombination, Mu can cause rearrangements of host DNA, including replicon fusion, translocation, deletion, and inversion. These mutations in host DNA can also be directed by other transposons.

∎

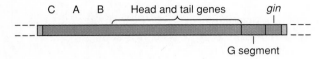

(a) Free phage DNA

C A B Head and tail genes *gin*

Host DNA G segment Host DNA

(b) Prophage DNA

C A B Head and tail genes *gin*

G segment

FIGURE 12.33

Map of phage Mu. (*a*) The DNA found in phage particles is linear and contains unequal lengths of host DNA (pink) attached to each end of the phage DNA (purple). (*b*) The prophage DNA has the same linear map and includes a 5-base-pair direct repeat of host DNA at each end (pink). The host DNA present in the free phage DNA has been lost upon integration.

EUKARYOTIC TRANSPOSABLE ELEMENTS

It would be surprising if prokaryotes were the only organisms to harbor transposable elements, especially since these elements have powerful selective forces on their side. First, many transposons carry genes that are an advantage to their hosts. Therefore, their hosts can multiply at the expense of competing organisms, and can multiply the transposons along with the rest of their DNA. Second, even if transposons are not advantageous to their hosts, they can replicate themselves within their hosts in a "selfish" way.

THE FIRST EXAMPLES OF TRANSPOSABLE ELEMENTS

Barbara McClintock discovered the first transposable elements in a study of maize (corn) in the late 1940s. It had been known for some time that the variegation in color observed in the kernels of so-called Indian corn was caused by an unstable mutation. For example, in figure 12.35*a,* we see a kernel that is colored. This color is due to a factor encoded by the maize *C* locus. Figure 12.35*b* shows what happens when the *C* gene is mutated; no purple pigment is made, and the kernel appears almost white. The spotted kernel in figure 12.35*c* shows the results of reversion in some of the kernel's cells. Wherever the mutation has reverted, the revertant cell and its progeny will be able to make pigment, giving rise to a dark spot on the kernel. It is striking that there are so many spots in this kernel. That means the mutation is very unstable. It reverts at a rate much higher than we would expect of an ordinary mutation.

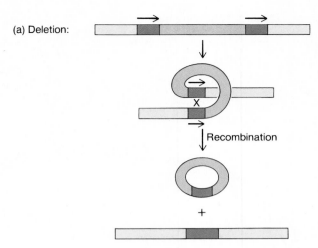

(a) Deletion:

X

Recombination

+

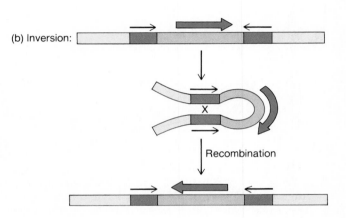

(b) Inversion:

X

Recombination

FIGURE 12.34

Deletion and inversion promoted by transposons. (*a*) When two identical transposons (purple) are in the same orientation (arrows) on a chromosome, they can pair as shown in the middle figure. Recombination as indicated by the X releases the DNA between the two transposons (pink) plus one of the transposons (purple). This leaves the original DNA with one transposon and a deletion. (*b*) When two identical transposons are in opposite orientations, they can pair as shown in the middle figure. Recombination as shown by the X inverts the DNA between the transposons (large green arrow).

In this case, McClintock discovered that the original mutation resulted from an insertion of a transposable element, called ***Ds*** for "dissociation," into the *C* gene (figure 12.36*a* and *b*). Another transposable element, ***Ac*** for "activator," could cause *Ds* to transpose out of *C,* causing reversion (figure 12.36*c*). In other words, *Ds* can transpose, but only with the help of *Ac. Ac,* on the other hand, is an autonomous transposon. It can transpose itself and therefore inactivate other genes without help from other elements.

Now that molecular genetics tools are available, we can isolate and characterize these genetic elements decades after McClintock found them. Nina Fedoroff and her collaborators have obtained the structure of *Ac* and three different forms of *Ds*. *Ac* resembles the bacterial transposons we have already studied (figure 12.37). It is about 4,500 base pairs long and is bounded by short, imperfect inverted repeats. It contains a transposase gene and another gene of unknown function. The various forms of *Ds* are derived from *Ac* by deletion. *Ds-a* is very similar to *Ac*, except that a piece of the transposase gene has been deleted. This explains why *Ds* is unable to transpose itself. *Ds-b* is more severely shortened, retaining only a small fragment of the transposase gene, and *Ds-c* retains only the inverted repeats in common with *Ac*. These inverted repeats are all that *Ds-c* needs to be a target for transposition directed by *Ac*.

*T*he variegation in the color of Indian corn kernels is caused by multiple reversions of an unstable mutation in the *C* locus, which is responsible for the kernel's coat color. The mutation and its reversion result from a *Ds* (dissociator) element, which transposes into the *C* gene, mutating it, and then transposes out again, causing it to revert to wild-type. *Ds* cannot transpose on its own; it must have help from an autonomous transposon called *Ac* (for activator), which supplies the transposase. *Ds* is an *Ac* element with more or less of its middle removed. All *Ds* needs in order to be transposed is a pair of inverted terminal repeats that the *Ac* transposase can recognize.

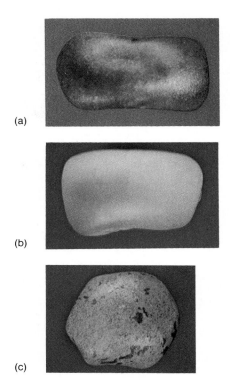

(a)

(b)

(c)

FIGURE 12.35

Effects of mutations and reversions on maize kernel color. (*a*) Wild-type kernel has an active *C* locus that causes synthesis of purple pigment. (*b*) The *C* locus has mutated, preventing pigment synthesis, so the kernel is colorless. (*c*) The spots correspond to patches of cells in which the mutation in *C* has reverted, again allowing pigment synthesis.

From "Transposable Genetic Elements in Maize" by Nina V. Fedoroff.

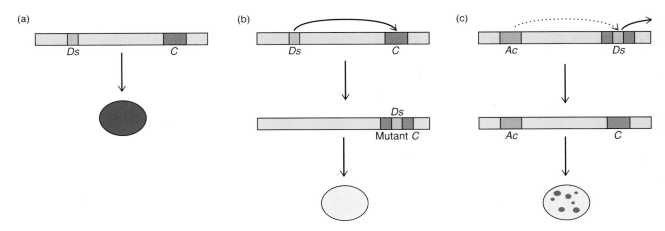

FIGURE 12.36

Transposable elements cause mutations and reversions in maize. (*a*) A wild-type maize kernel has an uninterrrupted, active *C* locus (purple) that causes synthesis of purple pigment. (*b*) A *Ds* element (pink) has inserted into *C*, inactivating it and preventing pigment synthesis. The kernel is therefore colorless. (*c*) *Ac* (green) is present, as well as *Ds*.

This allows *Ds* to transpose out of *C* in many cells, giving rise to groups of cells that make pigment. Such groups of pigmented cells account for the purple spots on the kernel. Of course, *Ds* must have transposed into *C* before it became defective, or else it had help from an *Ac* element.

Gene Mutation, Transposable Elements, and Cancer

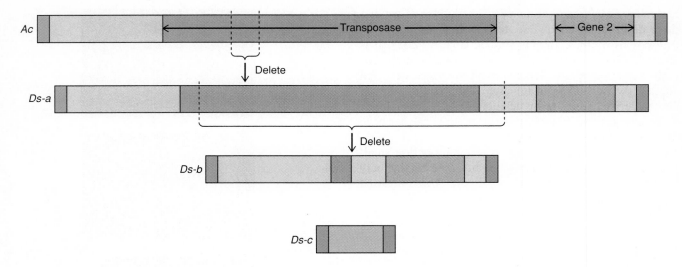

FIGURE 12.37

Structures of *Ac* and *Ds*. *Ac* contains two genes (purple and orange) and two imperfect inverted terminal repeats (blue). *Ds-a* is missing a 194-base-pair region from the transposase gene (dotted lines); otherwise, it is almost identical to *Ac*. *Ds-b* is missing a much larger segment of *Ac*. *Ds-c* has no similarity to *Ac* except for the inverted terminal repeats.

YEAST AND *DROSOPHILA* TRANSPOSONS

In spite of the early discovery of the transposable elements in maize, the best-studied eukaryotic transposons are still those of yeast (*Saccharomyces cerevisiae*) and the fruit fly (*Drosophila melanogaster*). The prototype *Drosophila* transposon is called *copia,* because it is present in copious quantity. In fact, *copia* and related transposons called **copia-like elements** account for about 1% of the total fruit fly genome. Similar transposable elements in yeast are called **Ty** for "transposon yeast."

The physical structures of yeast and fruit fly transposons are somewhat related to those of bacterial transposons. They contain genes that are necessary for their own transposition and they cause duplication of host DNA at either end. For example, Ty is a 6,300-base-pair element flanked by 5-base-pair repeats of host DNA. However, there are major differences: Ty contains 330-base-pair direct terminal repeats called **delta repeats.** More importantly, Ty transposes by a pathway very different from the one taken by bacterial transposons. This pathway involves an RNA intermediate and is very similar to the replication scheme of eukaryotic tumor viruses called **retroviruses.**

RETROVIRUSES

In order to understand the mode of transposition of Ty, it is first necessary to introduce the retroviruses, since transposition of Ty resembles so closely the replication of this class of viruses. In fact, retroviruses can be considered transposons in that their genomes (or more precisely, DNA copies of their genomes) are found inserted randomly into host DNA. Retroviruses have a profound impact on public health, since they can cause both cancer and AIDS.

The most salient feature of a retrovirus, indeed the feature that gives this class of viruses its name, is its ability to make a DNA copy of its RNA genome. This reaction, RNA → DNA, is the reverse of the transcription reaction, so it is commonly called **reverse transcription.** In 1970, Howard Temin and, simultaneously, David Baltimore convinced a skeptical scientific community that this reaction takes place. They did so by finding that the virus particles contain an enzyme that catalyzes the reverse transcription reaction. Inevitably, this enzyme has been dubbed **reverse transcriptase.** A more proper name is **RNA-dependent DNA polymerase.**

Figure 12.38 illustrates the retrovirus replication cycle. We start with a virus infecting a cell. The virus contains two copies of its RNA genome, linked together by base pairing at their 5'-ends. When the virus enters a cell, its reverse transcriptase makes a double-stranded DNA copy of the viral RNA. This DNA cyclizes through its **long terminal repeats,** or **LTRs,** and fuses with the host genome. This integrated form of the viral genome is called the **provirus.** To complete virus replication, host RNA polymerase II makes viral mRNAs, which are then translated to viral proteins. Polymerase II also makes full-length RNA copies of the provirus, which of course are new viral genomes. These RNAs are packaged into virus particles (figure 12.39) that bud out of the infected cell and go on to infect other cells.

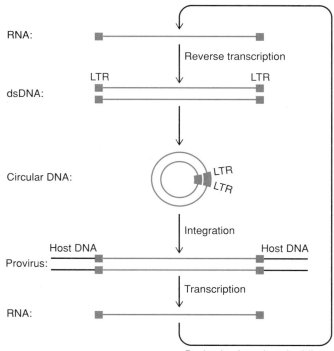

FIGURE 12.38

Retrovirus replication cycle. The viral genome is an RNA, with parts of long terminal repeats (LTRs, red) at each end. Reverse transcriptase makes a linear, double-stranded DNA copy of the RNA, which then cyclizes and integrates into the host DNA (black), creating the provirus form. The host RNA polymerase II transcribes the provirus, forming genomic RNA. The viral RNA is packaged into a virus particle, which buds out of the cell and infects another cell, starting the cycle over again.

Ty Transposition by a Retrovirus-like Mechanism

Several lines of evidence indicate that Ty transposition resembles the replication of a retrovirus:

1. Ty encodes a reverse transcriptase. This seems likely because the *tyb* gene in Ty codes for a protein with an amino acid sequence closely resembling that of the reverse transcriptases encoded in the *pol* genes of retroviruses. If the Ty element really codes for a reverse transcriptase, this enzyme should appear when Ty is induced to transpose; moreover, mutations in *tyb* should block the appearance of reverse transcriptase. Gerald Fink and his colleagues have performed experiments that bear out both of these predictions.

2. Full-length Ty RNA and reverse transcriptase activity are both associated with particles that closely resemble retrovirus particles (figure 12.40). These particles only appear in yeast cells that are induced for Ty transposition.

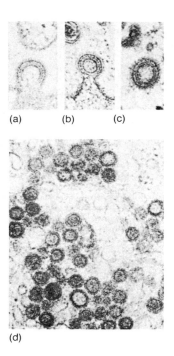

FIGURE 12.39

Retroviruses. (*a*) A B-type retrovirus budding from a mouse mammary tumor cell. (*b*) Later in the budding process. (*c*) Free, but still immature B-type retrovirus. (*d*) Intracellular, A-type particles in a mouse mammary tumor.

Maugh, T. H. "RNA Virus: The Age of Innocence Ends." Science, March 22, 1974, Vol. 183, pp. 1181–85, fig. 1. Copyright 1974 by AAAS.

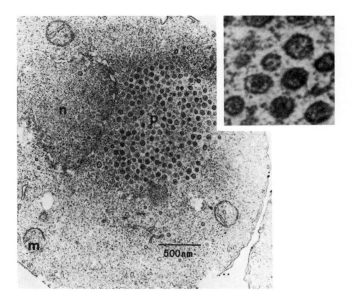

FIGURE 12.40

Retrovirus-like particles in a yeast cell induced for Ty transposition. The particles in this electron micrograph are labeled *p*. The inset shows the particles at higher magnification. The nucleus and a mitochondrion are labeled *n* and *m*, respectively.

Garfinkle and Boeke. "Retrovirus-like Particles in a Yeast Cell Induced for Ty Transposition." Cell 42:513, 1985. Reprinted with permission from Cell Press.

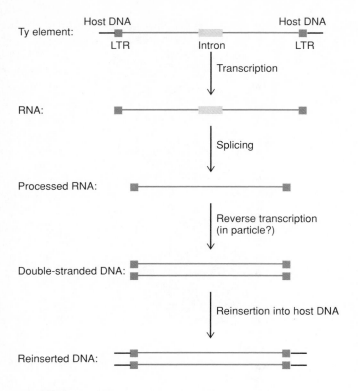

Ty element: Host DNA | Host DNA
LTR Intron LTR

Transcription

RNA:

Splicing

Processed RNA:

Reverse transcription
(in particle?)

Double-stranded DNA:

Reinsertion into host DNA

Reinserted DNA:

FIGURE 12.41

Model for transposition of Ty. The Ty element has been experimentally supplied with an intron (yellow). The Ty element is transcribed to yield an RNA copy containing the intron. This transcript is spliced, and then the processed RNA is reverse transcribed, possibly in a viruslike particle. The resulting double-stranded DNA then reinserts into the yeast genome. (LTR = long terminal repeat.)

3. Finally, in a very clever experiment, Fink inserted an intron into a Ty element and then analyzed the element again after transposition. The intron was gone! This finding is incompatible with the kind of transposition prokaryotes employ, in which the transposed DNA looks just like its parent. But it is consistent with the following mechanism (figure 12.41): The Ty element is first transcribed, intron and all; then the RNA is spliced to remove the intron; and finally, the spliced RNA is reverse transcribed within a viruslike particle, and the resulting DNA is inserted back into the yeast genome at a new location.

The similarity between this proposed mechanism of Ty transposition and retrovirus replication leads to the term **retrotransposon** (or even **retroposon**), which describes this and similar transposable elements. *Copia* and its relatives share many of the characteristics we have described for Ty; it is very likely that these are also retrotransposons. Vertebrates too have agents that probably qualify as retrotransposons. Unlike Ty and *copia,* these are not elements that were first discovered because of their transposition. Rather, they were originally noticed as particles that looked and behaved like retroviruses, except that they were not infectious. These are called **A-type particles.** Their proviruses constitute as much as 0.3% of vertebrate genomes, and they sometimes transpose to interrupt vertebrate genes. The similarity between retrotransposons and retroviruses suggests that these genetic elements share a common evolutionary origin. Does it also mean that the viruslike retrotransposon particles are infectious? Apparently not; all attempts to make them infect other cells have failed.

A retrovirus replicates by reverse transcribing its RNA genome to double-stranded DNA, inserting this DNA into its host's genome, then using the host RNA polymerase II to make new RNA genomes that are packaged into virus particles. Several eukaryotic transposons, including Ty of yeast, *copia* of *Drosophila,* and probably the proviruses of A-type particles of vertebrates, apparently transpose in a similar way. They start with DNA in the host genome, make an RNA copy, then reverse transcribe it—probably within a viruslike particle—to DNA that can insert in a new location.

■

TRANSPOSONS AND THE MUTABILITY OF CHROMOSOMES

In our discussion of bacteriophage Mu, we pointed out that this and other prokaryotic transposons could cause rearrangements of their host genomes. Such mutations can come as a by-product of the act of transposition itself, or simply because transposons provide multiple homologous sites on a genome that can serve as targets for general recombination. This mutation enhancement is also a property of eukaryotic transposons, of course, and we have already seen one example: the unstable mutations caused by McClintock's maize transposons.

The phenomenon called **hybrid dysgenesis** illustrates one obvious kind of mutation enhancement caused by a eukaryotic transposon. In hybrid dysgenesis, one strain of *Drosophila* mates with another to produce hybrid offspring that suffer so much chromosomal damage that they are dysgenic, or sterile. Hybrid dysgenesis requires a contribution from both parents; for example, in the **P-M system** the father must be from strain **P** (**paternal contributing**) and the mother must be from strain **M** (**maternal contributing**). The reverse cross, with an M father and a P mother, produces normal offspring, as do crosses within a strain (P × P or M × M).

What makes us suspect that a transposon is involved in this phenomenon? First, any P male chromosome can cause dysgenesis in a cross with an M female. Moreover, recombinant male chromosomes derived in part from P males and in part from M males usually can cause dysgenesis, showing that the P trait is carried on multiple sites on the chromosomes. This is characteristic of transposable elements. Margaret Kidwell and her colleagues have investigated the **P elements** that inserted into the *white* locus of dysgenic flies. They found that these elements had great similarities in base sequence but differed considerably in size (from about 500 to about 2,900 base pairs). Furthermore, the P elements had direct terminal repeats and were flanked by short direct repeats of host DNA, both signatures of transposons. Finally, the *white* mutations reverted at a high rate by losing the entire P element—again a property of a transposon.

If P elements act like transposons, why do they transpose and cause dysgenesis only in hybrids? The mechanism is not clear, but the M female cytoplasm apparently contributes something (the **M cytotype**) that is necessary for P element transposition. Thus P × P crosses produce chromosomes with inactive P elements that cannot transpose, and M × M crosses produce offspring with the ability to promote P transposition, but no P elements to transpose. Only when these two factors, the P element from the male and the M cytotype from the female, are brought together can transposition and therefore dysgenesis occur. It is as if the P elements are the gunpowder and the M cytotype is the spark; only when they combine does an explosion take place. Geneticists now introduce P elements intentionally into fruit flies to cause mutations or to carry extra genes into the flies.

Haig Kazazian has now identified transposition as the cause of at least one human genetic disease: hemophilia. In two of 240 hemophiliacs screened, Kazazian found a transposable element in place of the gene for a blood clotting protein, factor VIII. This destroyed the factor VIII gene, which in turn impaired the clotting process in the patients' blood. Now that transposons have been observed in humans, it seems certain that they will be linked to other mutations and therefore to other human genetic diseases.

GENE MUTATIONS AND CANCER

We have examined some of the causes and effects of gene mutations, but it is becoming clear that these mutations can have another serious consequence: They can cause cancer. What changes turn a well-behaved normal cell into a renegade cancer cell? We certainly do not have the full answer to that question, but examinations of normal and tumor cells in culture show that malignant cells differ in three important respects from their normal counterparts.

First, malignant cells are immortal; they will go on dividing indefinitely. Consider, for example, the HeLa cells cultured from a human cervical carcinoma in 1952 (see chapter 11). These malignant cells have already far outlived their donor, and we have every reason to believe they will continue to prosper as long as people are interested in

growing them. This immortality stands in stark contrast to the behavior of normal vertebrate cells. Just as complex multicellular organisms have finite lifetimes, so their cells can divide only a limited number of times in culture before they too "grow old" and die.

In addition to immortality, cancer cells exhibit a suite of characteristics that contrast with those of normal cells, which we lump together under the name **transformed.** Normal cells need hard surfaces on which to grow and serum to supply growth factors. Also, when they sense that they are crowded, they cease dividing. But transformed cells typically do not show these limitations. They can usually grow in soft agar, as well as on hard surfaces; they usually do not need growth factors; and they continue to divide when they are crowded, even piling up on top of one another until their medium is exhausted.

Besides immortality and the transformed phenotype, cancer cells usually exhibit a progressive loss of the characteristics of the normal cells from which they derived. Thus, liver tumor cells may at first produce many of the myriad enzymes that normal liver cells do, but they gradually lose those capabilities. Also, cancer cells are usually aneuploid, and this aneuploidy becomes more and more pronounced with time. Thus, tumor cells come to resemble their tissue of origin less and less, both biochemically and genetically.

It is fairly clear that the aneuploidy of cancer cells is an effect, rather than a cause, of malignancy. However, it is also obvious that genes are intimately involved in the malignant transformation of cells. Sometimes the genes themselves change; other times only gene expression seems to change. In this section we will explore the relationship between genes and **carcinogenesis** (the creation of a cancer cell).

---

*C*ancer cells differ from normal cells in several respects: They are immortal; they are transformed—capable of growing in physical and chemical environments where normal cells could not survive; and they progressively lose their biochemical and genetic resemblance to their normal ancestors.

■

CARCINOGENS AS MUTAGENS

One strong indication of the relationship between genes and cancer is the fact that many agents that cause cancer (**carcinogens**) are also mutagens. Since the late eighteenth century, we have suspected that chemicals can cause cancer. It was then that Sir Percival Pott noticed a high incidence of otherwise rare scrotal cancer in chimney sweeps. Pott concluded that these boys' constant exposure to soot had something to do with the disease. In fact, soot

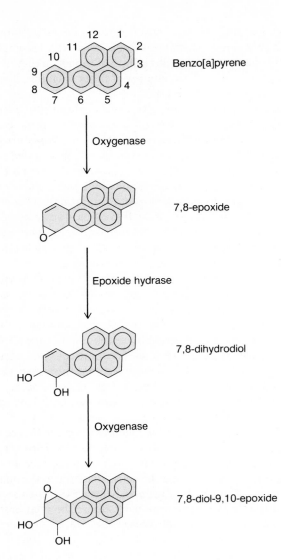

FIGURE 12.42
Conversion of benzo[a]pyrene to the ultimate carcinogenic form.

Benzo[a]pyrene

Oxygenase

7,8-epoxide

Epoxide hydrase

7,8-dihydrodiol

Oxygenase

7,8-diol-9,10-epoxide

Guanine

Sugar

Diol epoxide

Adduct

Sugar

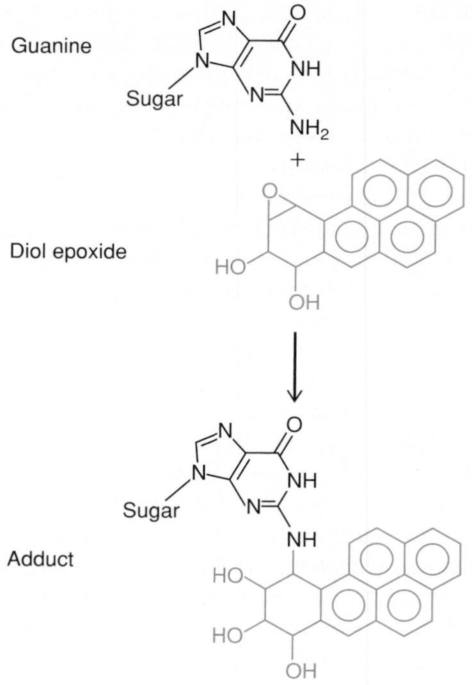

FIGURE 12.43
Interaction between a benzo[a]pyrene diol epoxide (red) and the DNA base guanine.

has the same kinds of mutagenic combustion by-products found in cigarette tar, and these very likely caused the trouble.

Earlier in this chapter we examined the mechanism behind most chemical carcinogens. They are electrophilic compounds that attack centers of negative charge in DNA, causing mutations. These mutations, in turn, apparently start the chain of events that leads to cancer. For example, consider benzo[a]pyrene, the most likely carcinogenic culprit in cigarette tar. Benzo[a]pyrene (figure 12.42) belongs to a class of organic compounds called hydrocarbons, which are composed of hydrogen and carbon. Gasoline is primarily a mixture of hydrocarbons, although simpler ones than benzo[a]pyrene. Chemically, hydrocarbons are not very reactive, and they certainly are not electrophilic, so it is at first a little difficult to understand how benzo[a]pyrene could be carcinogenic. But hydrocarbons are also very insoluble in water, thus the body

needs to change them chemically in order to excrete them in the urine. This chemical change is what converts the unreactive hydrocarbon to a dangerous carcinogen.

Figure 12.42 shows the enzymatic pathway whereby benzo[a]pyrene is converted to an electrophilic carcinogenic product called the **ultimate carcinogen.** (This is not the most carcinogenic compound in the world, just the most carcinogenic compound to which benzo[a]pyrene is metabolized.) The first step in the pathway is oxidation of the benzo[a]pyrene to an epoxide by an oxygenase. The epoxide then adds the elements of water under the direction of an epoxide hydrase. The product, a dihydrodiol, is oxidized again by the oxygenase to the ultimate carcinogen, a diol epoxide. Figure 12.43 demonstrates how this diol epoxide can attack an electron-rich center on a DNA base, altering the DNA.

Another environmental agent known to cause cancer is radiation. We have seen that radiation is mutagenic; several lines of evidence also show that it is carcinogenic. For example, consider the fact that people with light complexions, who have relatively little defense against ultraviolet radiation from the sun, are much more likely than dark-skinned people to develop skin cancer. Also, as mentioned earlier in this chapter, some people, born with xeroderma pigmentosum, have defective DNA repair machinery. This suggests that the unrepaired DNA lesions caused by the ultraviolet radiation are responsible for their skin cancers. However, the situation may be a bit more complex than that.

Recall that bacteria have an error-prone DNA repair system. If this system is blocked, the bacteria have a much lower mutation rate. In other words, the improper repair of DNA damage, rather than the damage itself, seems to cause many permanent mutations. If eukaryotes also behave this way, the real mutagenic (and carcinogenic) event in xeroderma pigmentosum patients may not be unrepaired damage, but misrepaired damage. The same may be true of the DNA damaged by chemical carcinogens. It may be the cell's inaccurate attempts to repair bases damaged by these chemical agents that cause the permanent mutations leading to cancer.

A related argument may apply to the damage caused by more energetic radiation such as X rays. Remember that X rays cause double-stranded DNA breaks (chromosome breaks). This is obviously a mutagenic event, but it may become a carcinogenic event when the cell attempts to repair the damage by sticking together the severed ends of chromosomes. If more than one chromosome break has occurred, the wrong chromosome segments may be rejoined, resulting in translocations. Such translocations could activate or inactivate genes inappropriately, giving rise to a cancer cell.

M utagenic chemicals and radiation can lead to cancer, probably because of the DNA damage they cause. Chemicals can sometimes act directly, but usually they need to be metabolized to an electrophilic ultimate carcinogen that can attack DNA. Direct damage due to chemicals or radiation appears to be carcinogenic, but the inaccurate repair of such damage may actually be the root cause.

■

ONCOGENES

Viruses, especially retroviruses, represent a third kind of environmental agent capable of causing cancer. By studying **oncogenic** (tumor-causing) viruses, molecular virologists have revealed the identities of many of the viral genes (**oncogenes**) that cause tumors. More surprisingly, they have also discovered cellular counterparts of these viral oncogenes known as cellular oncogenes, or more accurately, cellular **proto-oncogenes.** We now have abundant evidence that these harmless cellular proto-oncogenes can convert to true oncogenes, which at least contribute to **oncogenesis,** the formation of tumors. We even know some of the conditions under which this conversion takes place.

Conversion Caused by Mutation

The most straightforward mechanism for converting a cellular proto-oncogene to an oncogene is mutation. Since viral oncogenes resemble their cellular cousins so closely,

it is not difficult to imagine a few well-chosen mutations causing the conversion. In fact, in at least one case—the **Ha-*ras*** oncogene—a single point mutation is sufficient.

The c-Ha-*ras* proto-oncogene is the cellular version of the Harvey rat sarcoma virus oncogene (v-Ha-*ras*). Here is the story of how this gene was implicated in human cancer: First, Michael Wigler and his colleagues demonstrated that human bladder carcinoma cells contain an oncogene capable of transforming NIH 3T3 cells. Specifically, this team of scientists introduced DNA from a line of human bladder cancer cells into 3T3 cells. The 3T3 cells are mouse cells that are immortal and apparently poised on the brink of transformation. Something in the bladder cancer cell DNA pushed them over into the transformed state (figure 12.44). DNA from normal human cells could not have transformed them.

Which gene has the transforming activity? Robert Weinberg and his coworkers used radioactive oncogenes to probe DNA fragments from 3T3 cells transformed by human bladder carcinoma DNA. These experiments revealed that the transformed 3T3 cells have a new DNA fragment not found in ordinary 3T3 cells, which hybridizes with Ha-*ras* DNA (figure 12.44). Therefore, the human cancer DNA has introduced a Ha-*ras* oncogene into the 3T3 cells. Furthermore, we know that this new gene is active in the transformed cells because RNAs made in transformed cells, but not in untransformed 3T3 cells, include an RNA that hybridizes with the Ha-*ras* probe.

Since human bladder cancer cells, but not ordinary bladder cells, contain a transforming agent, and the agent seems to be the Ha-*ras* oncogene, we would like to know how this oncogene differs from the harmless Ha-*ras* proto-oncogene in normal bladder cells. To answer this question, Mariano Barbacid and his colleagues took fragments of the oncogene and proto-oncogene, recombined them in vitro, and used the recombinant genes to try to transform 3T3 cells (figure 12.45). This process narrowed down the important region to a 350-base-pair segment of the gene. Whenever this region came from the cancer cell oncogene, the recombinant gene had transforming activity. When it came from normal cell DNA, it did not transform. A single base change (a G → T transversion) is the only difference between this region in the oncogene and the same region in the proto-oncogene. Therefore, it appears that this change is what converts the proto-oncogene to the oncogene.

The product of the Ha-*ras* oncogene is a protein with a molecular weight of 21,000, called **p21.** The G → T transversion that converts the proto-oncogene to the oncogene in the human bladder carcinoma changes a glycine (the twelfth amino acid in p21) to a valine. When does the change occur? It seems to be a somatic mutation, since it appears in human lung cancer cells, but not in normal trachea cells from the same patient. Is this the only cause

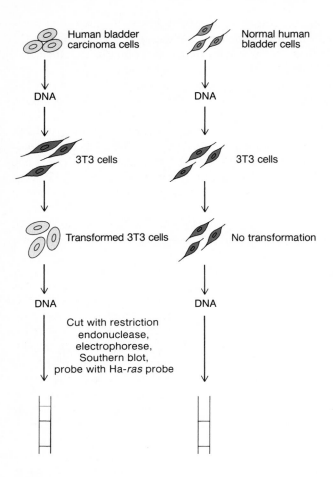

FIGURE 12.44
Human bladder carcinoma cells have DNA with transforming
potential. Starting at the top of this figure, DNA is extracted from
either human bladder carcinoma cells (left) or normal human bladder
cells (right) and introduced into NIH 3T3 cells. The cells that receive
the carcinoma DNA become transformed, but those that receive
normal cell DNA remain normal. Then, DNA is extracted from both
sets of cells that have received exogenous DNA, cut with a restriction
endonuclease, electrophoresed, Southern blotted, and hybridized to a
radioactive Ha-*ras* DNA probe. Two common bands "light up" in
both Southern blots, corresponding to an endogenous mouse *ras* gene,
but a third band (blue) is found only in the transformed cells. This
band corresponds to the human Ha-*ras* gene that transformed these
cells.

of the lung cancer? That seems very doubtful, because oncogenesis appears to be a multistep process that takes years
to complete. The mutation in c-Ha-*ras* is probably just
one step in that process.

Since the pioneering work of Wigler, Weinberg, and
others, mutated *ras* genes have been found in a variety of
human tumors, including 40% of colon cancers. These
mutations are not confined to codon 12, but almost all of
them are clustered in three regions of the *ras* gene: codons
12–16; codons 59–63; and codons 116–119.

Certain human tumor cells contain an oncogene
called Ha-*ras* that can transform mouse 3T3 cells
in culture. Normal human DNA lacks this transforming
activity. The cellular Ha-*ras* proto-oncogene in the first
tumor cells studied had suffered a one-base mutation, a
G → T transversion in the twelfth codon, which changed a
glycine to a valine in the protein product. This was enough
to convert the proto-oncogene to an oncogene. Almost all
the *ras* mutations found in human tumor cells are clustered
in three small regions of the gene.

■

THE c-Ha-*ras* GENE PRODUCT:
A G PROTEIN-LIKE SUBSTANCE

Why are these subtle changes so important? Clues come
from the base sequence of the Ha-*ras* gene, which predicts
the amino acid sequence of the protein product p21. This
amino acid sequence is very similar to those of proteins
belonging to a class called **G proteins.**

The G proteins all bind GTP, enabling them to carry
out their functions. They also have GTPase activity, so they
can cleave the bound GTP to GDP, whereupon they
become inactive until they bind a new molecule of GTP.
The functions of the G proteins are quite varied. Some
reside in cell membranes and respond to hormones and
other substances. When activated by these signalling
agents plus GTP, the G proteins stimulate the cell to appropriate activity. Another G protein responds to light and
converts the light signal into a chemical signal inside the
cell. Still another substance with G protein-like characteristics is EF-Tu, which participates in protein synthesis.
This protein binds GTP, which enables it to bind aminoacyl-tRNA and carry it to the ribosome where the amino
acid can be joined to a growing protein chain.

The three-dimensional structure of normal human p21
has been solved by X-ray crystallography. Figure 12.46
shows this structure, along with a molecule of GDP bound
to p21. Since the normal p21 has GTPase activity and the
mutant p21 has much less, we would predict that the center
of GTPase activity (the GTP or GDP binding site) would
lie close to the amino acids that are altered in the mutant
proteins. This expectation is borne out beautifully. Recall,
for example, that many of the altered amino acids in the
mutant p21 proteins are clustered between amino acids 12
and 16. This region is highlighted in pink in figure 12.46,
and we can see that it forms a loop that wraps around the
phosphates of GDP (or GTP, if it were there instead). It
is not too surprising, therefore, that alterations in these
amino acids affect the reaction with GTP. In fact, these
mutations apparently block GTPase activity. Since GTP
stimulates the activity of G proteins and p21, the inability
of the protein to remove the GTP locks it in the activated
state. In order to understand why this predisposes a cell
to become malignant, we need to know one more key piece
of the puzzle: the exact function of p21. Unfortunately,
this information is still unavailable, but we can speculate
that it has something to do with controlling cell division.

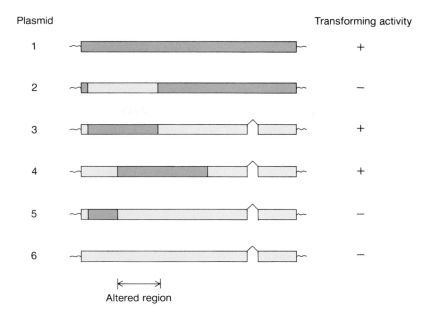

Plasmid   Transforming activity

1   +

2   −

3   +

4   +

5   −

6   −

Altered region

FIGURE 12.45

Pinpointing the difference between the Ha-*ras* oncogene and proto-oncogene. Cloned plasmids containing the oncogene (red) or proto-oncogene (yellow) were cut and rejoined in vitro in different combinations, as shown. In addition, deletions (∧) were introduced into four of the plasmids. These recombinant plasmids were then tested for transformation activity, and the results are reported in the far right column. Obviously, the deletion did not affect transformation activity, since plasmids 3 and 4, with the deletion, retain full activity.

Furthermore, since plasmids 3 and 4 have activity, the alteration in the oncogene must lie in the region of overlap between the oncogene parts of these two plasmids ("altered region" shown at bottom). This conclusion is reinforced by plasmids 2 and 5, which have proto-oncogene sequences in this region and no transformation activity. Further work revealed only one base difference between oncogene and proto-oncogene in this region.

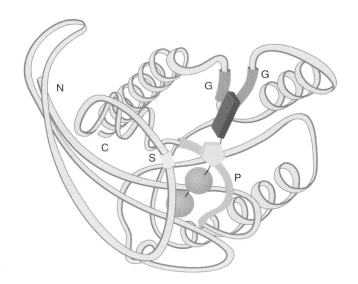

FIGURE 12.46

Interaction between the *ras* proto-oncogene product and GDP. The protein is shown as a blue ribbon, interacting with GDP (dark purple, yellow, and red). The dark purple rectangle represents the guanosine; the dark yellow pentagon, the ribose; and the dark red circles, the phosphates of GDP. The domains of the protein interacting with each of these parts of GDP are represented by sleeves on the protein ribbon, color coded to match the corresponding part of GDP. Thus,

the light purple sleeves (G) interact with the guanosine, the light yellow sleeve (S) interacts with the ribose, and the light red sleeve (P) interacts with the phosphates of GDP. The P domain of the protein contains glycine-12, so this amino acid is in a critical position next to the phosphates of the nucleotide. N and C represent the N- and C-termini of the protein.

From L. Tong, et al., Science *245:244, 1989. Copyright © 1989 by the American Association for the Advancement of Science, Washington, D.C. Reprinted by permission of the author and publisher.*

*T*he c-Ha-*ras* gene codes for a protein called p21, which binds GTP, carries out some unknown function, and then hydrolyzes GTP, turning off the protein's function. Conversion of the proto-oncogene to the oncogene causes an amino acid substitution in a key position near the presumed GTP binding site of p21. This probably accounts for the reduction in the GTPase activity of the protein, which in turn means the protein spends a higher proportion of its time than normal in the activated state. This may predispose the cell to lose its control over growth.

■

Summary

*M*utations are heritable changes in the genetic material. They can be categorized according to their effects on the mutant organism's phenotype. Germ-line mutations affect the gametes, so the mutation can be transmitted to the next generation. Somatic mutations affect the nonsex, somatic cells, thus are not passed on. Morphological mutations cause a visible change in the affected organism. Nutritional mutations render an organism dependent on added nutrients that a wild-type organism does not require. Lethal mutations occur in indispensable genes. Haploid organisms with lethal mutations die immediately, but lethal mutations are usually recessive, so diploid organisms can tolerate them in heterozygotes. Conditional mutations only affect the mutant organism under certain conditions, such as high temperature. If a conditional mutation causes death under nonpermissive conditions, it is a conditional lethal.

A number of different kinds of agents can cause genetic mutations. Radiation can either damage DNA directly or produce free radicals that do the damage. Ultraviolet light causes pyrimidine dimers, while the more energetic X rays and gamma rays usually cause breaks in DNA. Chemicals can injure DNA in a variety of ways, including deamination and alkylation of bases or by forcing the incorporation of the wrong base, or too many or too few bases, into a replicating DNA. Chemicals usually cause small lesions called point mutations that can revert readily to wild-type.

All living things have means of reversing genetic damage. Some mutations can be directly undone. For example, the photoreactivating enzyme can break pyrimidine dimers, and the suicide enzyme O^6-methylguanine methyl transferase can remove methyl and ethyl groups from guanine. However, most DNA damage is corrected by excision repair, in which the mutated DNA is removed and replaced with normal DNA. Cells can also employ nonrepair methods to replicate their DNA even though it is damaged. Two examples are recombination repair and error-prone (SOS) repair.

Transposable elements, or transposons, are pieces of DNA that can move from one site to another. Some transposable elements replicate, leaving one copy at the original location and placing one copy at a new site; others transpose conservatively, leaving the original location altogether. Bacterial transposons include the following types: (1) insertion sequences such as IS1 that contain only the genes necessary for transposition, flanked by inverted terminal repeats; (2) transposons such as Tn3 that are like insertion sequences but contain at least one extra gene, usually a gene that confers antibiotic resistance; (3) composite transposons such as Tn9 that have the added complexity of insertion sequences at each end; (4) bacteriophage Mu, which can transpose in its prophage form.

Some examples of eukaryotic transposons are: (1) *Ds* and *Ac* elements of maize; (2) Ty of yeast; (3) *copia* of *Drosophila;* and (4) retroviruses. The latter three apparently transpose by a mechanism very different from the one used by prokaryotic transposons. RNA polymerase II from the host cell transcribes the transposon; this RNA is then reverse transcribed back to double-stranded DNA, probably within a viruslike particle, and finally inserted into the host genome at a new location.

Transposons enhance the mutability of both prokaryotic and eukaryotic genomes. They can directly disrupt genes by inserting into them during transposition; they can promote chromosome breakage, translocation, and inversion during transposition; and they can provide sites of homology for generalized recombination, even without transposing.

Mutagenic chemicals and radiation can lead to cancer, probably because of the DNA damage they cause. In addition, viruses—both DNA and RNA types—can cause tumors. All known RNA tumor viruses are retroviruses. Their RNA genomes usually contain an oncogene that can transform host cells, making them behave as tumor cells rather than as normal cells. The viral oncogenes appear to have descended from closely related cellular proto-oncogenes.

Various processes can convert a cellular proto-oncogene to a tumor-inducing oncogene. For example, the cellular Ha-*ras* proto-oncogene in certain human tumor cells has suffered a one-base mutation that is enough to convert the proto-oncogene to an oncogene. Other proto-oncogenes suffer translocation, which converts them to oncogenes. In Burkitt's lymphoma, the *myc* oncogene is translocated next to a cluster of antibody H-chain genes.

The Ha-*ras* gene codes for a G protein called p21, which binds GTP, carries out some unknown function, then hydrolyzes GTP, turning off the protein's function. Conversion of the proto-oncogene to the oncogene causes an amino acid substitution in a key position near the presumed GTP binding site of p21, leading to a reduction in the GTPase activity of the protein, which in turn forces the protein to spend a higher proportion of its time than normal in the activated state. This seems to predispose the cell to lose its control over growth.

■

Problems and Questions

1. You notice a rose that is all red except for a small patch of white. Is this likely to be due to a somatic or a germ-line mutation? Why?

2. Two mice, heterozygous for the yellow trait, mate. The progeny are ⅔ yellow and ⅓ wild-type color. The yellow mice are heterozygotes; the homozygous mutants died in utero. Is the yellow color dominant or recessive? Is the lethality of the yellow gene dominant or recessive?

3. You are mapping a DNA polymerase III gene in a new strain of bacteria. Can you use nonconditional mutations that inactivate the gene? Why, or why not?

4. What kind of mutations would be useful in the mapping studies in question 3?

5. Is a temperature-sensitive gene more easily melted (denatured) than a wild-type gene?

6. A Siamese cat was wounded and had to have a patch of white fur shaved. The fur grew in dark instead of white. With time, this dark fur was gradually replaced with white. Present a plausible explanation for these observations.

7. Does tautomerization of thymine cause transitions or transversions?

8. (a) What kind of DNA damage does ultraviolet light cause? (b) Why is this damage harmful to a cell?

9. (a) What kind of DNA damage do X rays cause? (b) Why is this damage harmful to a cell?

10. A hypothetical mRNA, AUGCGCCUAAAGAGG, codes for fMet-Arg-Leu-Lys-Arg. What happens to the coding if you delete the first C?

11. What would be the simplest way to cause a second-site reversion of the mutation in question 10?

12. Which two enzymes catalyze direct reversal of DNA damage?

13. What is the difference between an exonuclease and an endonuclease?

14. In excision repair in *E. coli*, two mechanisms could be used for removing damaged DNA and filling in the gap: (1) The old DNA could be removed and replaced in a single pass of the DNA polymerase I, or (2) the old DNA could be removed and then, in a separate step, replaced. What DNA feature would occur in the second case but not in the first? Which mechanism does the evidence favor?

15. Why are recombination repair and error-prone repair not real repair processes?

16. (a) What elements, at a minimum, must a transposon contain? (b) What are the minimum elements that a transposable element must have in order to be transposed in a cell that harbors a separate, complete transposon?

17. Mutations in the transposase genes of IS1 render the transposon immobile (unable to transpose). What can we conclude from this finding?

18. Why is "jumping gene" a misleading nickname for some transposons?

19. (a) A certain transposon's transposase creates staggered cuts in the host DNA five base pairs apart. What consequence does this have for the host DNA surrounding the inserted transposon? (b) Draw a diagram to explain how the staggered cuts affect the host DNA.

20. What is the meaning of the terms "selfish" or "parasitic" DNA?

21. You are interested in measuring the rate of transfer of transposon Tn3 from one plasmid, carrying no antibiotic resistance genes, to another, which carries a gene for chloramphenicol resistance. (Tn3 itself confers resistance to ampicillin.) Describe an experiment you would perform to assay for this transposition.

22. Would a transposon that requires homologous recombination with its target DNA be able to transpose to more or fewer sites than one that uses illegitimate recombination? Why?

23. (a) What are the two steps in Tn3 transposition? (b) What is the intermediate called?

24. Identify the end product of abortive transposition carried out by Tn3 transposons with mutations in the following genes: (a) transposase; (b) resolvase.

25. What is the relationship between bacterial transposons and public health?

26. Transposon TnT in plasmid A transposes to plasmid B. How many copies of TnT are in the cointegrate? Where are they with respect to the two plasmids?

27. If the transposable element *Ds* of maize transposed by the same mechanism as Tn3, would we see the speckled kernels with the same high frequency? Why, or why not?

28. Assume you have two cell-free transposition systems that have all the enzymes necessary for transposition of Tn3 and Ty, respectively. What effect would the following inhibitors have on these two systems, and why? (a) Inhibitors of double-stranded DNA replication; (b) inhibitors of transcription; (c) inhibitors of reverse transcription; (d) inhibitors of translation.

29. Assume that Ty transposes by a mechanism analogous to that used by Tn3. (Of course, it really does not.) Assume further that you have inserted an intron into a Ty element. What would the element look like after transposing?

30. Suspecting that Ty transposes by a mechanism analogous to the replication of a retrovirus, you have introduced an intron into a Ty element. What would be the fate of this intron upon transposition?

31. How do we know that unrepaired DNA damage can lead to cancer?

32. List three classes of environmental agents that are carcinogenic.

33. The term "transformation" has two meanings in biology. What is its meaning in the context of tumor biology?

34. Oncogenesis is a multistep process, yet the Ha-*ras* oncogene from human tumor cells is capable of transforming NIH 3T3 cells in a single step. Why is this statement not contradictory? *Hint:* Consider the nature of 3T3 cells.

35. Is the mutation in Ha-*ras* that contributes to human cancer a germ-line or a somatic mutation? How do we know?

Answers appear at end of book.

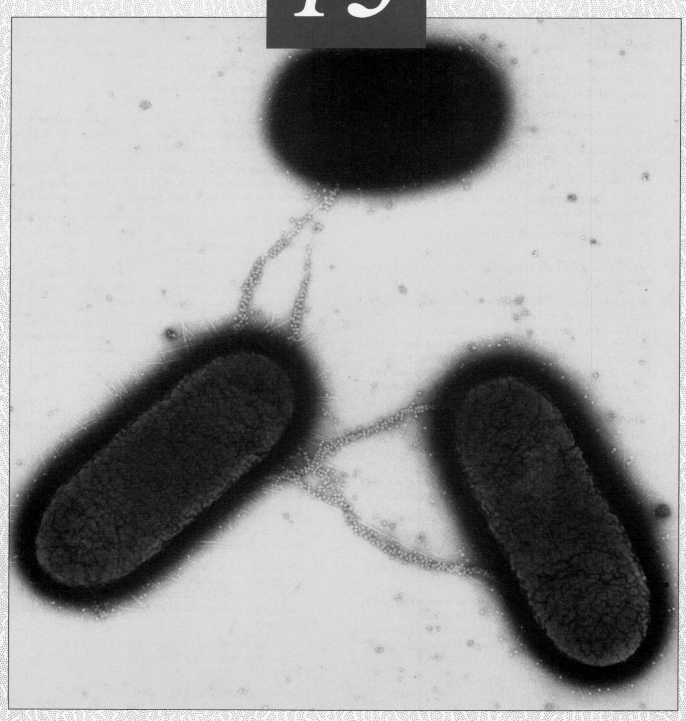

Genetics of Bacteria and Phages

N ature is to be found in her entirety nowhere more than in her smallest creatures.

■

_____

Pliny
Roman writer

Learning Objectives

In this chapter you will learn:

1. How bacterial cells exchange genes by conjugation.

2. How geneticists can take advantage of bacterial conjugation to map the exchanged genes.

3. How phages can transfer genes between bacteria, thus shedding light on the genetics of their hosts.

4. How genetic mapping experiments can be performed in phages.

False-color transmission electron micrograph of a male *Escherichia coli* bacterium (bottom left) conjugating with two females. Magnification: ×5,000.
© *Dr. L. Card/Science Photo Library/Photo Researchers, Inc.*

he study of the genetics of bacteria and their viruses, the phages, began as recently as the 1940s. Nevertheless, the story is an amazingly rich one, and includes the main body of what we call molecular genetics. Our repeated references to *Escherichia coli,* the favorite model of bacterial geneticists, make this importance obvious. Why should this simple intestinal bacterium loom so large as a tool for understanding genetics? One reason is its very simplicity. It contains only about three thousand genes, compared to as many as one hundred thousand in humans. Another advantage is the fact that it exists for its whole life cycle in the haploid state, so mutations that might be masked by the other allele in a diploid cell show up clearly in bacteria. But the most important qualification of bacteria for genetic studies is their extremely rapid rate of growth.

Given ideal growth conditions, *E. coli* cells divide every twenty minutes. Consider what that means: One ounce (30 grams) of bacteria growing constantly at that rate would surpass the mass of the earth in less than a day and a half! On a more practical level, a single *E. coli* cell will grow overnight into a visible colony containing millions of cells, even under relatively poor growth conditions. Thus, genetic experiments on *E. coli* usually last one day, whereas experiments on corn, for example, take months. It is no wonder that we know so much more about the genetics of *E. coli* than about the genetics of corn, even though we have been studying corn much longer.

BACTERIAL CONJUGATION

In spite of the inherent advantages of bacteria, they are worthless as genetic tools unless we can perform genetic experiments on them. In other words, we must be able to cross them somehow. In 1946, Joshua Lederberg and E. L. Tatum figured out a way to do this. Lederberg reasoned that if bacteria exchange DNA, he might be able to detect this by mixing two different mutant strains and looking for wild-type recombinants.

BACTERIAL GENETIC NOTATION

In order to understand the experiments Lederberg and others performed, we should begin by defining some terms: Wild-type bacteria can grow on **minimal medium,** which contains only salts and the sugar glucose. These bacteria can make all their own amino acids and vitamins; they are called **prototrophs** (Greek: *proto* = first + *trophe* = nourishment). Mutants that lack the ability to make one or more nutrients are called **auxotrophs** (Latin: *auxilium* = help).

In the following discussion, you will also encounter the system of notation used by bacterial geneticists. According to this system, the names of genes or operons are rendered in italics and begin with lowercase letters (e.g.,

the *ara* operon). When we refer to the genotype of a bacterium, we again use this convention, with a + to denote the wild-type or a − to denote the mutant genotype. For example, *leu*− bacterium has a mutation in its *leu* operon and an *ara*+ bacterium has a wild-type *ara* operon. When we write the phenotype of a bacterium, we capitalize and use standard letters. For example, Leu− bacteria cannot make the amino acid leucine, while Ara+ bacteria can break down the sugar arabinose.

BACTERIAL EXCHANGE OF DNA

Lederberg foresaw difficulty with an experiment using auxotrophs having only a single mutated gene. These mutants reverted to wild-type with a high enough frequency (about one in a million) that the revertants might have been mistaken for recombinants, which occur at about the same rate. For instance, if Lederberg started with two strains (A+ B− and A− B+) and looked for A+ B+ recombinants, revertants of A− B+ to A+ B+ would appear the same as recombinants and would therefore confuse the analysis. Accordingly, Lederberg decided to use double mutants in his crosses. The chance of both mutations reverting simultaneously would be negligible. The mutations used in these and other genetic experiments are sometimes called **markers** because they "mark" an organism's genome with a change that shows up clearly in the phenotype.

The decision to use two markers instead of one led Lederberg to a collaboration with Tatum, because the latter already had a variety of double mutants of *E. coli* available. Fortunately, Tatum's bacteria were *E. coli* strain *K12,* which just happened to be a strain that does exchange DNA. Almost all the other *E. coli* strains in common use at that time would not have worked.

Figure 13.1 shows the plan of the first experiment. One auxotrophic strain (A) required the amino acid methionine and the vitamin biotin; the other (B) required the amino acids threonine and leucine. Both strains grew on rich medium containing amino acids and vitamins, but not on minimal medium. However, if one strain could transfer DNA to the other, recombinants could presumably form that would bring together the wild-type genes for threonine and leucine synthesis (*thr*+ and *leu*+) from strain A and the genes for methionine and biotin synthesis (*met*+ and *bio*+) from strain B. These prototrophic recombinants would survive on minimal medium. In fact, that is just what Lederberg and Tatum found. One in ten million cells in the mixture of the two auxotrophs was a prototroph. On the other hand, if the two auxotrophs were grown separately, they produced no prototrophs.

This experiment showed that the DNAs from the two auxotrophs mixed and recombined, but it did not demonstrate conclusively that DNA transfers directly from one cell to the other. Instead, one auxotroph could have burst

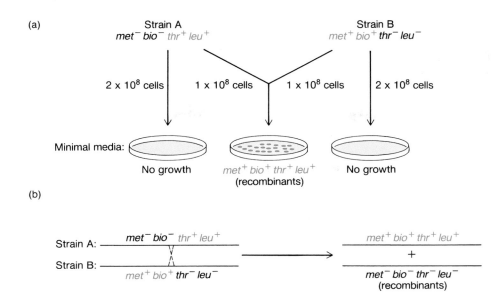

(a)

Strain A
met⁻ bio⁻ thr⁺ leu⁺

Strain B
met⁺ bio⁺ thr⁻ leu⁻

2 x 10⁸ cells   1 x 10⁸ cells   1 x 10⁸ cells   2 x 10⁸ cells

Minimal media:

No growth

met⁺ bio⁺ thr⁺ leu⁺
(recombinants)

No growth

(b)

Strain A: —————— *met⁻ bio⁻ thr⁺ leu⁺* ——————   —————— *met⁺ bio⁺ thr⁺ leu⁺* ——————
                                                                                                    +
Strain B: —————— *met⁺ bio⁺ thr⁻ leu⁻* ——————   —————— *met⁻ bio⁻ thr⁻ leu⁻* ——————
                                                                                            (recombinants)

FIGURE 13.1

The classic bacterial conjugation experiment. (*a*) Two auxotrophic strains, A and B, were mixed. Strain A had the genotype *met⁻ bio⁻ thr⁺ leu⁺*; strain B had the genotype *met⁺ bio⁺ thr⁻ leu⁻*. Neither of these auxotrophic strains could grow by itself on minimal medium; however, when the two strains were mixed, prototrophic recombinants (*met⁺ bio⁺ thr⁺ leu⁺*) arose, which could grow on minimal medium. These gave rise to the colonies growing on the petri dish at center. (*b*) The recombination event that produced the prototrophs. A crossover occurred between the *bio* and *thr* genes, bringing together all the wild-type alleles on one chromosome and the mutant alleles on the other.

open and its DNA could have transformed the other, as in Griffith's and Avery's experiments with *Streptococcus pneumoniae* (chapter 6). To rule this out, Lederberg and Tatum lysed one strain and filtered it to remove whole cells. Then they used the cell-free filtrate, which contained DNA, to try to transform the other strain. It did not. Therefore, they concluded that the two strains of bacteria in the original experiment had shared DNA by direct contact.

Direct contact between bacterial cells is called **conjugation;** it is made possible by a tubular structure, called a sex pilus, on the surface of one of the strains (figure 13.2). The sex pilus is now usually called the **F pilus,** where F stands for "fertility." Strain A in the original experiment had the F pilus and, presumably, some fertility factor in its genome, so it was designated **F⁺**. Strain B had no F pilus, lacked the fertility factor, and was designated **F⁻**. Figure 13.3 shows two conjugating bacteria, joined by the F pilus. The pilus is hollow and could, in principle, allow DNA to pass through, but it is retracted during conjugation. Thus, it appears that this structure serves not as a direct conduit for passing DNA, but as a contractile element that draws the two cells close enough together to allow DNA to pass through their membranes by an alternate route.

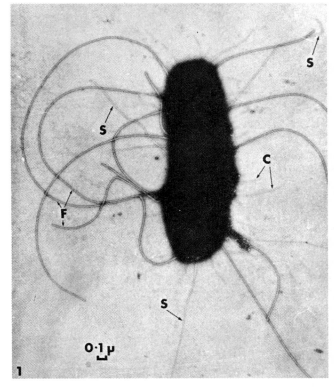

FIGURE 13.2

E. coli cell with sex pili. Three kinds of filamentous structures project from the surface of the cell: common pili (C), flagella (F), and sex pili (S).

Elinor Meynell, G. G. Meynell, and Naomi Datta, Bacteriological Reviews *32:58, 1968.*

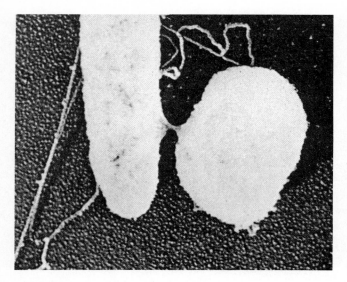

FIGURE 13.3
Conjugating *E. coli*. The two cells, joined by a connecting bridge, are an elongated donor (left) and a round recipient (right).
E. L. Wollman, F. Jacob, and W. Hayes. Symposia on Quantitative Biology *(Cold Spring Harbor), vol. 21, p. 153, 1956.*

> *T*wo mutant (auxotrophic) bacterial strains can share DNA by direct contact, producing wild-type (prototrophic) recombinants. This process is called conjugation.

■

One-way DNA Transfer During Conjugation

Lederberg and Tatum interpreted their experiments in terms of classical crosses between haploid gametes, and therefore assumed that the recombinant prototrophs were true diploids. However, William Hayes and others produced evidence that this could not be the case. For one reason, the F⁺ partner could be killed with the antibiotic streptomycin and still function in conjugation, but if the F⁻ partner was treated this way, no conjugation occurred. This suggested that the F⁺ partner served as the donor and could continue to do so after streptomycin treatment, even though it was no longer viable. By contrast, a dead F⁻ recipient would hardly provide fertile ground for the donated genes. Moreover, if the genotype of an F⁺ donor differed in several markers from that of the F⁻ recipient, the recombinant usually wound up with mostly recipient markers. This indicated that the donor usually transferred only a portion of its genome to the recipient. Therefore, in spite of all the sexual nomenclature associated with fertility in bacteria, it appears that these organisms rarely share their whole genomes to form true diploid zygotes. Instead, they usually form partial diploids called **merodiploids** or **merozygotes** (Greek: *meros* = part).

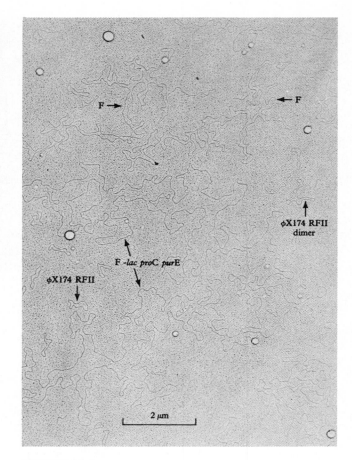

FIGURE 13.4
The F plasmid. This electron micrograph shows two circular molecules of F plasmid DNA among other circular DNAs of different sizes.
© *Dr. Richard Deonier.*

The F Factor: A Plasmid

Hayes also found that an F⁺ cell could convert an F⁻ to F⁺ very efficiently, with a frequency of about one in ten, compared to one in ten million for transferring other genes. This means the F factor can be passed easily, probably by a different mechanism than that used with other genes. How do we know the F factor consists of genes? In one experiment, *E. Coli* F⁺ cells were mixed with another species of bacteria (*Serratia marcescens*) containing a DNA of different density. After conjugation, a new species of DNA appeared in the recipient cells, and its density matched that of *E. coli*, not that of the recipient bacterium. This new DNA must be the F factor, since it is the component that is transferred readily during conjugation. As we will see, this is not true conjugation because only a small amount of DNA, the F factor, is transferred from the *E. coli* donor to the *S. marcescens* recipient. Subsequent work has shown that the F factor is a **plasmid**, a circular extra-chromosomal DNA (figure 13.4) that replicates independently of the host chromosome. It is about 2% as large as the host chromosome, or 94,500 base pairs (94.5 kb). Now that we know what it is, we refer to the F factor as the **F plasmid,** and transfer of fertility can be seen as simply transferring the F plasmid from an F⁺ cell to an F⁻ cell.

> T he fertility of F⁺ *E. coli* is due to an F plasmid of 94.5 kb that can transfer at high frequency into F⁻ bacteria.

■

Hfr STRAINS

In the 1950s, several investigators discovered a new type of *E. coli* in addition to the F⁺ and F⁻ already under study. The new type was called **Hfr,** which stands for "high-frequency recombinant." Hfr bacteria are mutant F⁺ strains, but they have lost the ability to transfer fertility at high frequency. Instead, they transfer host chromosomal genes at high frequency.

For example, consider an Hfr strain isolated by Hayes in 1953. When Hayes conjugated this Hfr with auxotrophic recipient cells, it generated prototrophic recombinants a thousand times more efficiently than did its F⁺ parent. Hayes also found that this transfer of host genes closely resembled the transfer by F⁺ cells in two respects: It was not inhibited by treating the Hfr donor with streptomycin, and the recombinant genotype came mostly from the recipient, not from the donor. We will see that an F⁺ cell converts to an Hfr when its F plasmid inserts into the host chromosome.

Hfr Cell Transfer of Host Genes

The fact that F⁺ strains contain a few Hfr cells raises the question: Does the F plasmid in every F⁺ cell have a small, but significant potential to transfer host genes, or does this ability belong only to the few Hfr cells contaminating an F⁺ population? By the mid 1950s, Elie Wollman and François Jacob had resolved this question using the **fluctuation test** devised by Max Delbrück and Salvador Luria. Here is how this test works: The investigators took a very dilute F⁺ culture (only 250 cells per milliliter), split it in half, and then split the first half into fifty equal parts. The second half was not divided. After the bacteria in the cultures had multiplied to a million times their original concentration, each was tested for the ability to transfer host genes, thereby forming recombinant prototrophs.

The premise of the experiment is that if each F⁺ cell has an equal probability of transferring host genes, each of the fifty subcultures should show an equal ability to do this. On the other hand, if the occasional Hfr cells in the F⁺ population are the only ones that can transfer genes, the fifty cultures should vary widely in this ability. This is because the original dilute cultures contained so few cells that most of them included no Hfr cells at all. Of course, as the cells multiplied, there was a good chance of Hfr cells appearing in the cultures by mutation, and once an Hfr appears, it divides to yield Hfr progeny. Thus, a culture in which an F⁺ → Hfr conversion occurred early in its growth would wind up with the most cells competent to transfer host genes. This is represented on a small scale in figure 13.5*b,* where culture 2 experiences an F⁺ → Hfr conversion in generation 2 and therefore accumulates eight Hfr cells after four generations, while culture 4, with no F⁺ → Hfr conversions, finishes with no Hfr cells.

When the experiment was performed, the small cultures did indeed show great variation—from one prototrophic recombinant in one culture to 116 in another. As a control, the half of the culture that was not split before the cells multiplied was finally divided into fifty equal parts and each of these was tested in the same way. They varied much less: from a low of ten prototrophic recombinants to a high of twenty-three. Taken together, these results demonstrate that a genetic change in a small proportion of the total F⁺ population renders them capable of transferring genes to auxotrophic F⁻ cells to convert them to prototrophs. This small subpopulation is composed of Hfr cells.

F Plasmid Insertion into the Host Chromosome

How can Hfr cells transfer host genes so efficiently? A clue comes from the similarity between transfer of the F plasmid and transfer of the host genes. It is the F plasmid itself that causes both. An F⁺ cell becomes an Hfr cell when the F plasmid ceases its solitary existence and inserts into the host chromosome. When this happens, the F plasmid loses the ability to transfer itself directly into a recipient cell, but confers this ability on the host chromosome. This leads to the hypothesis of F⁺ → Hfr conversion shown in figure 13.6. The F plasmid recombines with the host chromosome at any of a number of different locations. This mobilizes the chromosome, which is presumably nicked within the F plasmid part, initiating replication by the rolling circle mechanism (see chapter 8). During replication, part of the integrated F plasmid acts as a locomotive and leads the newly synthesized chromosomal DNA into the F⁻ cell.

The bridge between the two cells established by the F pilus is fragile and usually breaks before DNA transfer is complete. The part of the Hfr chromosome that has entered the recipient cell then recombines with the F⁻ chromosome, changing its genotype. Occasionally, the whole Hfr chromosome makes it across before the bridge breaks. When this happens, the recipient cell is converted to Hfr,

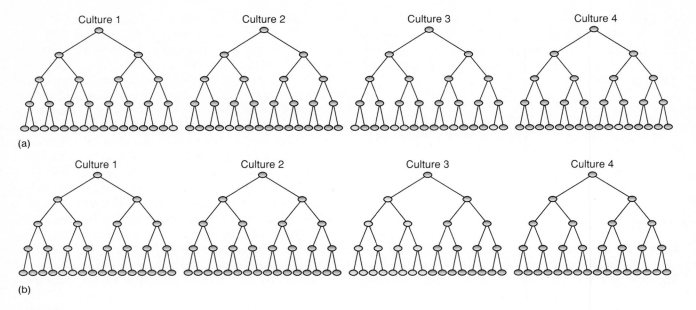

(a)

(b)

FIGURE 13.5

The fluctuation test. Two possibilities are illustrated: (*a*) Each culture has the same number of F⁺ cells, and each cell has a small, but equal chance of transferring host genes during conjugation (represented by the yellow cells). Therefore, each culture should give roughly the same amount of host gene transfer. (*b*) Only Hfr cells (mutant F⁺ cells, yellow) are competent to transfer host genes. Therefore, cultures in which an F⁺→ Hfr conversion occurs early in their growth (culture 2) will wind up with many competent cells, whereas other cultures can easily wind up with few (if the F⁺ → Hfr conversion occurs late in the growth phase; culture 1) or none (if no conversion occurs; culture 4). In other words, the competence of the cells in host gene transfer will fluctuate widely. That is what Wollman and Jacob demonstrated.

because the entire integrated F plasmid has been transferred. Part of it led the chromosome across the bridge; the rest of it came across last, like a caboose.

> *O* ccasionally, the F plasmid recombines with its host's chromosome, converting the cell to an Hfr. This mobilizes the Hfr chromosome to transfer to recipient cells, presumably by the same mechanism the independent F plasmid would normally use. The favorite hypothesis for this mechanism involves a rolling circle-style replication of the Hfr chromosome, with one of the daughter DNA molecules entering the recipient cell and recombining with it to change its genotype.

■

MAPPING BY INTERRUPTED CONJUGATION

The natural interruption of conjugation comes from random movements of the conjugating partners that can rupture the fragile link connecting them. Elie Wollman and François Jacob learned that they could stop conjugation at a defined time simply by placing the bacteria in a blender. This vigorous treatment guaranteed that all the conjugating partners would be separated. Furthermore, Wollman and Jacob took advantage of this procedure to build genetic maps.

Here is the basic strategy used by Wollman and Jacob: They conjugated two auxotrophs for a defined length of time, interrupted their conjugation by placing them in a Waring blender, then checked to see whether a given gene had been transferred. For example, consider an F⁻ Leu⁻ Lac⁻ Gal⁻ Str^r auxotroph (an F⁻ auxotroph incapable of growing on media lacking the amino acid leucine or on media containing the sugars lactose or galactose, but resistant to streptomycin). If it is conjugated with an Hfr Leu⁺ Lac⁺ Gal⁺ Str^s (streptomycin-sensitive) auxotroph for two minutes, no Str^r recombinants that are Leu⁺ will be formed. This tells us that in two minutes, transfer of the *leu* genes from the streptomycin-sensitive Hfr cell to the streptomycin-resistant F⁻ cell has not yet occurred. On the other hand, if conjugation is allowed for four minutes, a significant number of these recombinants will arise. Therefore, transfer of the *leu* genes begins about four minutes after conjugation begins.

(a) F-plasmid transfer:

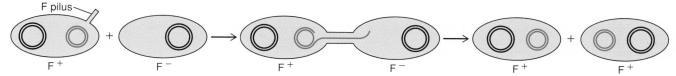

(b) F $^+$ → Hfr conversion:

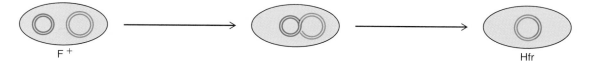

(c) Hfr transfer of host genes:

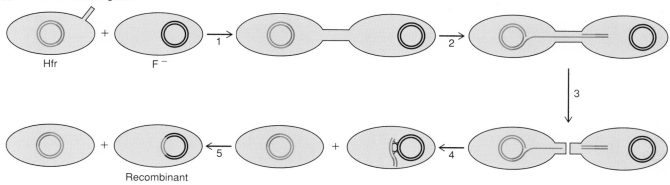

Recombinant

FIGURE 13.6

Model for F$^+$ → Hfr conversion. (*a*) The F plasmid (blue) normally passes from cell to cell, leaving a copy in the donor cell. This presumably involves rolling circle-style replication of the plasmid, as pictured. The F plasmid and host chromosome are not drawn to scale; the host chromosome is actually about fifty times larger than the F plasmid. (*b*) In F$^+$ → Hfr conversion, the F plasmid (blue) recombines with the host chromosome (red), mobilizing the whole chromosome. (*c*) Hfr transfer of host genes: 1. The F pilus establishes a bridge between the conjugating cells. 2. The Hfr chromosome begins replicating, with one daughter DNA molecule passing across the bridge to the recipient cell. 3. The bridge breaks during conjugation, leaving a piece of donor DNA in the recipient cell. 4. The donor DNA recombines with the recipient chromosome. 5. After the recipient cell replicates, it contains a recombinant DNA, with part (red) derived from the donor and part (black) from the recipient.

Taking the procedure a step further, if conjugation proceeds for ten minutes, some Leu$^+$ Lac$^+$ Str^r recombinants appear. This means that the Hfr cell begins to transfer its *lac* genes about ten minutes after the onset of conjugation. Furthermore, if conjugation proceeds for nineteen minutes, some Leu$^+$ Lac$^+$ Gal$^+$ Str^r recombinants appear. Therefore, the Hfr cell has begun to transfer its *gal* genes by this time. Figure 13.7 presents an idealized picture of the results of such an experiment. Why does the recipient cell remain streptomycin-resistant? There are two reasons: (1) The *str^s* locus lies beyond gal on the donor chromosome, and (2) the *str^r* allele is dominant.

These time values tell us something very important about the *E. coli* genetic map. If we assume that the rate of transfer is constant along the whole length of the bacterial chromosome, then the difference in the times it takes to transfer a given pair of genes is proportional to their separation on the genetic map. For example, the difference in the transfer times of the *leu* and *lac* genes is ten minutes minus four minutes, or six minutes; we therefore say that the *leu* and *lac* loci lie about "6 **minutes**" apart on the genetic map. Similarly, the differences in times of transfer of the *gal* and *lac* genes is 19 − 10, or 9 minutes. Thus these two loci lie 9 minutes apart, and the *leu* and *gal* loci are 9 + 6, or 15 minutes apart. It takes just one hundred minutes for the entire *E. coli* chromosome to be transferred, including its F "caboose." Therefore, the *E. coli* map is composed of 100 minutes.

In his Nobel address, François Jacob made the following tongue-in-cheek observation about this long conjugation time: "Marvellous organism, in which conjugal bliss can last for nearly three times the life span of the

individual." Actually, this is an underestimate; *E. coli* cells can divide five times in the time it takes for conjugation to occur. However, Jacob forgot to note that this "conjugal bliss" is almost always interrupted.

The Circular *E. coli* Genetic Map

Not all Hfr strains transfer the *leu* genes first; each strain has its own characteristic starting point. Moreover, some strains transfer genes in an order opposite to what we have just described. Thus, *gal* would be transferred before *lac,* and *leu* would be transferred last of all. Table 13.1 lists eleven different Hfr strains and the order in which each transfers host genes. Even though the orders vary, you can see that they can all be made the same by considering them not as open-ended lines, but as closed circles. For example, *thr* is the first and *thi* is the last gene transferred by Hfr Hayes, whereas these two genes are transferred one after the other (*thr,* then *thi*) in Hfr 1. All we must do to make these two orders identical is to form a circle of each one. Then the only difference will be the location of *O,* the origin of replication, which determines which gene will be transferred first. Now look at the order of Hfr 5. You will see that the order of transfer of these two genes is reversed: *thi,* then *thr.* Again, the order of gene transfer by these two strains can be reconciled by forming a circle of each one and turning one of them over as you would flip a coin.

What does all this mean in physical terms? First, it tells us that the *E. coli* chromosome is circular and that the F plasmid inserts in different places in different Hfr strains. This accounts for the fact that the *order* of transfer is the same in different strains, but that the starting points (defined by the site of F plasmid insertion) vary. Second, the F plasmid can insert in either orientation, thus inverting the order of gene transfer. These phenomena are

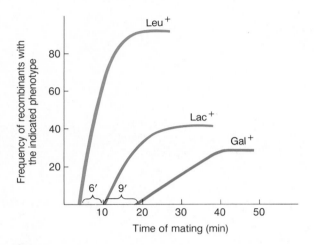

FIGURE 13.7

Timing of gene transfer in a bacterial conjugation experiment. An Hfr Leu+ Lac+ Gal+ was mated with an F- Leu- Lac- Gal- recipient and the conjugation was interrupted at various times by blending. The phenotypes of the recipients were tested by growing in medium lacking leucine or containing lactose or galactose, instead of glucose, as the sole carbon source. Some Leu+ recombinants (red line) began to appear at four minutes after the onset of conjugation, whereas Lac+ and Gal+ recombinants began to appear at ten and nineteen minutes, respectively (blue and green lines). This means that the time between transfer of the *leu+* and *lac+* markers was six minutes, while that between *lac+* and *gal+* was nine minutes. This tells us the relative positions and spacing of these three markers.

T ABLE 13.1 *Order of Gene Transfer During Conjugation with a Variety of Hfr Strains*

Hfr Strain	Order of Gene Transfer
Hayes	*O-thr-leu-azi-ton-pro-lac-pur-gal-trp-his-gly-str-mal-xyl-mtl-ile-met-thi*
Hfr 1	*O-ieu-thr-thi-met-ile-mtl-xyl-mal-str-gly-his-trp-gal-pur-lac-pro-ton-azi*
Hfr 2	*O-pro-ton-azi-leu-thr-thi-met-ile-mtl-xyl-mal-str-gly-his-trp-gal-pur-lac*
Hfr 3	*O-pur-lac-pro-ton-azi-leu-thr-thi-met-ile-mtl-xyl-mal-str-gly-his-trp-gal*
Hfr 4	*O-thi-met-ile-mtl-xyl-mal-str-gly-his-trp-gal-pur-lac-pro-ton-azi-leu-thr*
Hfr 5	*O-met-thi-thr-leu-azi-ton-pro-lac-pur-gal-trp-his-gly-str-mal-xyl-mtl-ile*
Hfr 6	*O-ile-met-thi-thr-leu-azi-ton-pro-lac-pur-gal-trp-his-gly-str-mal-xyl-mtl*
Hfr 7	*O-ton-azi-leu-thr-thi-met-ile-mtl-xyl-mal-str-gly-his-trp-gal-pur-lac-pro*
AB311	*O-his-trp-gal-pur-lac-pro-ton-azi-leu-thr-thi-met-ile-mtl-xyl-mal-str-gly*
AB312	*O-str-mal-xyl-mtl-ile-met-thi-thr-leu-azi-ton-pro-lac-pur-gal-trp-his-gly*
AB313	*O-mtl-xyl-mal-str-gly-his-trp-gal-pur-lac-pro-ton-azi-leu-thr-thi-met-ile*

From F. Jacob and E. L. Wollman, *Sexuality and the Genetics of Bacteria.* Copyright © 1961 Academic Press, Inc., Orlando, Florida. Reprinted by permission of the publisher and author.

illustrated in figure 13.8. In the 1960s, John Cairns used electron micrographs (chapter 8) to show that *E. coli* DNA is circular, so we accept the proposition of a circular map naturally. Nevertheless, in 1957, when Jacob and Wollman proposed the idea, it was a radical one. A partial circular map of the *E. coli* chromosome is presented in figure 13.9.

B y interrupting conjugation between an Hfr donor and an F⁻ recipient, we can determine the length of time it takes to transfer a given gene. The differences in the transfer times of different genes tell us their relative positions on the circular genetic map of *E. coli*.

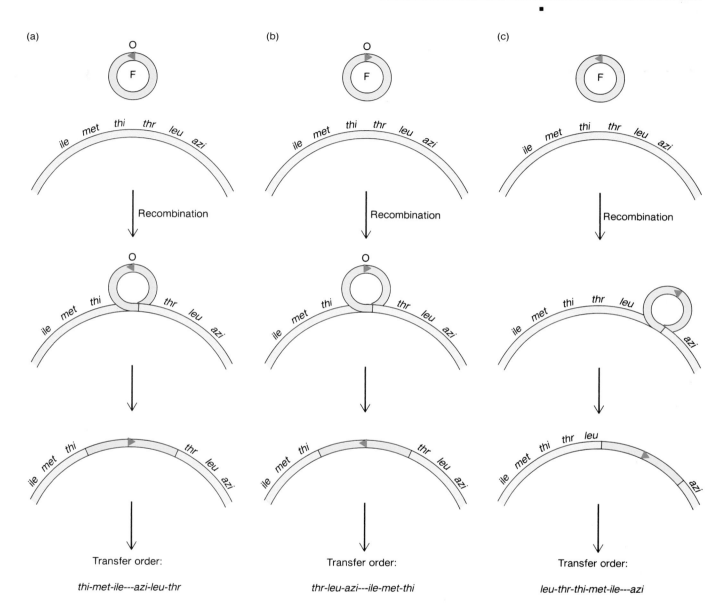

(a)

Recombination

O

ile met thi thr leu azi

Transfer order:

thi-met-ile---azi-leu-thr

(b)

Recombination

O

ile met thi thr leu azi

Transfer order:

thr-leu-azi---ile-met-thi

(c)

Recombination

ile met thi thr leu azi

Transfer order:

leu-thr-thi-met-ile---azi

FIGURE 13.8

The F plasmid portion of an Hfr chromosome determines the direction and order of host gene transfer. (*a*) The F plasmid (yellow circle with red arrowhead showing orientation) inserts between *thi* and *thr* in a rightward orientation; thus, the host genes will be transferred in the order *thi-met-ile---azi-leu-thr*, where the three dashes in the middle represent the entire remainder of the Hfr chromosome (not shown in this picture). The last three genes are therefore transferred only in those rare conjugations that last throughout the entire one hundred minutes. (*b*) The F plasmid inserts between *thi* and *thr* in a leftward orientation; thus, the host genes will be transferred in the order *thr-leu-azi---ile-met-thi*. (*c*) The F plasmid inserts between *leu* and *azi* in a rightward orientation; thus, the order of transfer is *leu-thr-thi-met-ile---azi*.

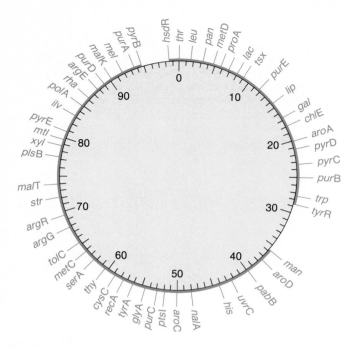

FIGURE 13.9
Circular map of the *E. coli* genome. This map was prepared in 1976 and contains only a fraction of the genes that have now been mapped.

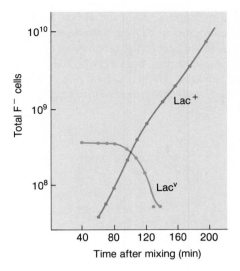

FIGURE 13.10
Recombination is the final step in conjugation. The Lac^V phenotype, indicative of unstable association between donor and recipient chromosomes, decays, beginning about fifty minutes after the onset of conjugation. At the same time, the Lac^+ phenotype, indicative of a stable association that results from recombination between the two chromosomes, begins a steady rise.

COMPLETION OF CONJUGATION

Just because a given gene enters an F^- cell during conjugation, it does not mean that the gene will be expressed indefinitely in the recipient. In order to become a stable part of the recipient cell's genetic makeup, the newly arrived gene must recombine with the recipient cell's chromosome. If it does not, it has no way of replicating and will be lost. J. Tomizawa demonstrated this phenomenon clearly in a conjugation experiment with an Hfr Lac^+ donor and an F^- Lac^- recipient. After various times, he interrupted conjugation and tested the merozygotes for the Lac^+ phenotype by growing them on indicator plates containing EMB-lactose agar. Lac^+ cells metabolize the lactose, causing chemical changes that turn the indicators in the medium red; Lac^- cells cannot metabolize the lactose and so remain white.

At the earliest times that Lac^+ cells appear, they produce colonies that are not pure red, but red and white *variegated,* a phenotype called **Lac^V**. This is because the *lac* genes in the merozygotes are not yet stably integrated into their new hosts' chromosomes. Thus, they can be lost, giving rise to Lac^- cells that will be white. On the other hand, when Tomizawa tested the merozygotes for longer periods of time, he observed a drop in the Lac^V phenotype, beginning about fifty minutes after the onset of conjugation, and a corresponding rise in the number of pure Lac^+ colonies (figure 13.10). This tells us that integration of the *lac* genes from the Hfr cell into the F^- chromosome begins

about fifty minutes after conjugation begins, or about forty minutes after the *lac* genes entered the recipient cell. This integration of the *lac* genes into the *lac^-* genome stabilizes the Lac^+ phenotype. The name of the integration process, of course, is recombination.

> *T*he completion of conjugation in *E. coli* requires recombination between the donated piece of Hfr chromosome and the F^- chromosome. Otherwise, the donated DNA cannot replicate and will be lost. This recombination requires extensive sequence similarity between the recombining DNAs and therefore follows the generalized recombination pathway.

GENETIC MAPPING WITH *E. COLI*

We have seen that conjugation occurs in *E. coli* and that this sharing of DNA allows us to construct genetic maps by following the progressive entry of one gene after another into the recipient F^- cell. This technique works fairly well when the distance separating the markers is great, but it is too crude a method for mapping closely spaced genes. For such fine mapping, more delicate genetic tools have been developed. We will examine the most important of these.

FIGURE 13.11
Formation of F-*lac*. The integrated F plasmid (yellow) in an Hfr chromosome is clipped out along with some host DNA, including the nearby *lac* operon (green). This produces a recombinant DNA called F-*lac*, part F plasmid and part host DNA, including *lac*.

F' PLASMIDS

In 1959, E. H. Adelberg discovered that one of his Hfr strains was contaminated with an F⁺ revertant. At first this seemed like an impediment, but it turned out that the F⁺ revertant had very interesting properties. In fact, this was one of those occasions when an annoyance turned out to be a boon.

Adelberg's F⁺ revertant did not behave normally. Whereas an F plasmid usually inserts randomly into the host chromosome to generate a new Hfr, the F plasmid in this revertant always went back to the same locus at which it had resided in the original Hfr. And it generated Hfrs with a much higher frequency than normal. Adelberg recognized that the F factor in his F⁺ revertant had suffered a genetic change, so he gave it a new name: **F'.** He also guessed the reason for the unusual behavior of the F' plasmid: It had picked up a piece of the host chromosome during its sojourn there in the Hfr. This host DNA gave the F' plasmid a "memory" of its former location, since it provided homologous sequences on the plasmid and the host chromosome that could serve as sites for generalized recombination. This ensured that recombination between F' plasmid and host chromosome would be rapid and would always occur at the same locus on the chromosome (figure 13.11).

Apart from its theoretical interest, the F' plasmid also provides a handy genetic tool. We can use it to introduce a defined small segment of the *E. coli* genome into an F⁻ cell very efficiently. This process, sometimes called **F-duction,** or even **sex-duction,** is so efficient because the segment of host DNA is now part of the F plasmid and transfers with the same high frequency as the F plasmid itself. Conjugation between an F' cell and an F⁻ therefore produces a merodiploid for genetic analysis.

For example, consider an F' called **F-*lac*** that carries part of the *lac* operon. When an F⁺ cell bearing F-*lac* is conjugated with an F⁻ Lac⁻ cell, it converts the recipient to F⁺ Lac⁺ (figure 13.12). Thus, even though the recipient retains a mutant *lac* gene in its chromosome, the wild-type gene on the F-*lac* plasmid provides the missing

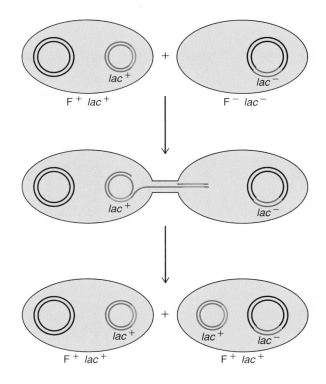

FIGURE 13.12
Transfer of F-*lac* to an F⁻ *lac*⁻ cell. The F-*lac* plasmid (blue, with *lac*⁺ genes in red) crosses the connecting bridge, just as an F⁺ normally would. This converts the F⁻ *lac*⁻ recipient to F⁺ *lac*⁺. The host chromosomes are black, with *lac*⁻ genes in green. Again, the host chromosome and F-*lac* plasmid are not drawn to scale.

enzyme. Note that no recombination is necessary; the F-*lac* plasmid reproduces along with the host cell. This cross tells us that the wild-type *lac* gene is dominant over the *lac*⁻ gene in the F⁻ chromosome. Of course, merodiploids make possible many other, more sophisticated genetic experiments, which we will also investigate. These experiments have been aided considerably by a collection of F' plasmids that represent the entire *E. coli* genome (figure 13.13).

FIGURE 13.13

F′ plasmids cover the whole *E. coli* map.

Modified from Barbara J. Bachman and K. B. Low, in Microbiological Review *44:1–56, 1980. Copyright © 1980 American Society for Microbiology, Washington, D.C. Used with permission.*

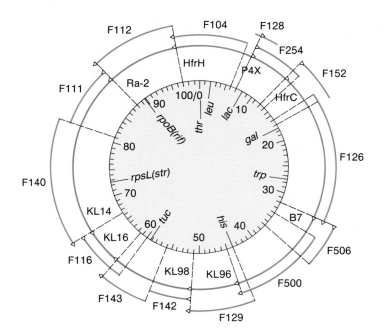

THE *CIS-TRANS* COMPLEMENTATION TEST

S ometimes a recombination event within an Hfr chromosome clips out the F plasmid along with some of the host chromosome. This produces an F′ plasmid that includes the clipped out piece of host DNA. The host genes thus become part of a highly mobile element that can deliver them to another cell. We can take advantage of this efficient sharing of limited pieces of DNA to perform genetic experiments.

Suppose we have isolated two strains of *E. coli* that cannot grow on the sugar lactose. Two genes (*lacZ* and *lacY*) code for the two enzymes a cell needs to metabolize this sugar. Therefore, an important question about these two Lac⁻ mutants is: Do both strains have mutations in the same gene, or are two different genes affected? We can answer this by conjugating cells to form a merodiploid, thus placing both mutant *lac* operons together in the same cell. If the two mutations are in different genes, they will complement each other in the merodiploid to give a Lac⁺ cell. In other words, the donor provides the gene product that the recipient is missing, and vice versa. In the case illustrated in figure 13.14*a,* the donor (F-*lac*) cell contained a wild-type *lacZ* gene, but a mutant *lacY* gene. Thus, the wild-type *lacZ* gene product from the donor cooperated with the wild-type *lacY* gene product from the recipient to allow the merodiploid to use lactose.

On the other hand, if the mutations are in the same gene in the two cells, they will usually not complement each other, so the merodiploid will be Lac⁻. Figure 13.14*b* illustrates what happens when both mutations are in the

same gene—the *lacY* gene, in this case. Both the donor plasmid and the recipient chromosome provide wild-type *lacZ* products, but mutant *lacY* products. They cannot complement because neither can provide a functional *lacY* product. Notice the emphasis here on gene *products.* In order for complementation to work, the gene products have to be **diffusible.** That is, they must be capable of moving within the cell so that they can cooperate.

As an aid to understanding the complementation concept, consider the following analogy: Two factories in the same town make the same kind of automobile. Each has two assembly lines, one for the engines and one for the bodies. One day the engine line of one factory and the body line of the other factory break down. Nevertheless, the two factories are able to continue producing cars by cooperating—workers put the engines from the second factory together with the bodies from the first. This is complementation, and you can see how it depends on the portable nature of the product. Of course, if the engine assembly lines of both factories break down at the same time, the two defective factories cannot complement each other and there will be no way to continue making cars.

Several qualifications must be made about the complementation test. First, it must be performed in a situation where recombination does not occur to a significant extent. If recombination is common, it can take place between two homologous mutant genes, creating one gene with no mutations and one with two (figure 13.14*c*). Thus, the two mutations would seem to complement each other and would be scored as existing in different genes even though they actually started out in homologous genes in the two chromosomes. One way to eliminate this problem is to use a RecA⁻ recipient that cannot carry out generalized recombination.

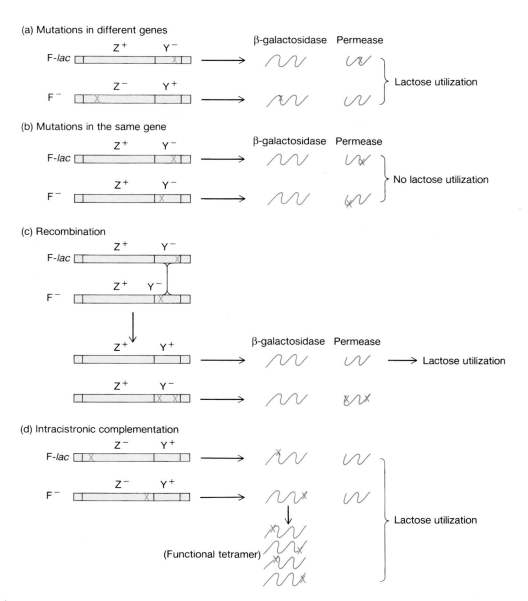

FIGURE 13.14

The *cis-trans* complementation test. (*a*) Mutations, represented by small red x's, are in different genes—*lacZ* and *lacY*, respectively. Therefore, their products can complement each other, and the cell will be *lac+*. (*b*) The mutations are in the same gene (*lacY*). Now both chromosomes will produce active *lacZ* product, but inactive *lacY* product, so the cell will be *lac−*. (*c*) and (*d*) are complications that must be considered when running the *cis-trans* test. (*c*) Sometimes recombination can occur between markers in the same gene on homologous chromosomes. This produces a wild-type gene and a double-mutant gene. The cell is therefore *lac+*, which makes it look as if the two mutations were in different genes. (*d*) Sometimes the products of two homologous genes with different mutations (two different *lacZ* mutants in our example) can get together to make an active, multi-subunit protein, in this case an active β-galactosidase. This makes it look as if the two mutations were in different genes.

Another kind of problem can arise if one of the mutations affects the expression of several other genes. Such a **pleiotropic** mutation behaves as if it were mutations in several genes instead of just one, which complicates the complementation test. A final problem is that mutations in homologous genes in two different chromosomes can sometimes complement one another. This occurs when the gene product is a polypeptide that aggregates with one or more identical polypeptides to form an **oligomeric protein.**

β-galactosidase, a **tetramer** (composed of four identical polypeptides, or monomers), is a good example. Each *lacZ* gene in figure 13.14*d* is nonfunctional; each gives rise to an inactive tetramer of four identical defective β-galactosidase polypeptides. However, when the two different *lacZ* genes are present in the same cell, they produce two different defective polypeptides that somehow cooperate to form an active tetramer. Therefore, these two *lacZ* mutations complement each other; if we did not know better, we would say they were on different genes. Instead, they are on the same gene, or **cistron,** and they therefore engage in **intracistronic complementation.**

We can use the complementation examples just given to illustrate some important genetic terms. The mutant genes in these cases were supplied *in trans*—on separate chromosomes (Latin: *trans* = across). Alternatively, we could supply the mutant genes *in cis*—on the same chromosome (Latin: *cis* = here). These two possible configurations of the mutant genes—*cis* and *trans*—have given the complementation test another name: the **cis-trans test.** That in turn has provided the term *cistron,* which Seymour Benzer coined to describe a genetic unit that could be distinguished by the *cis-trans* test from another genetic unit. Of course, this is a unit of genetic function—a piece of chromosome that gives rise to a protein (or RNA) product. As such, it is synonymous with our current understanding of the word "gene."

We also use the terms *cis* and *trans* to describe the way two genetic loci interact. Promoters, the binding sites for RNA polymerase (chapters 7, 9, and 10), are **cis-acting** elements: They must be on the same DNA molecule as the gene in question in order to influence that gene's activity. In our factory analogy, they are like the main power switch. If it breaks down, the factory cannot operate and the defect cannot be complemented by a product of the other factory. By contrast, proteins are *trans*-acting. Each is a product of one gene and can interact with other genes or their products. The mobility of proteins allows them to act at a distance.

> *T*wo different mutations that affect the same function can be subjected to the *cis-trans* test, or complementation test, to determine whether they occur in the same or in different genes. In this test, the two mutant genes are supplied to the same cell *in trans* (on separate chromosomes). If the mutations complement each other to give wild-type function, they are probably in separate genes. If the mutations fail to complement, they affect the same gene.

■

TRANSDUCTION

The term **transduction** refers to the use of phages to carry bacterial genes into new host cells. Two different transduction mechanisms are widely used: **specialized transduction,** in which λ or a similar phage DNA carries the bacterial genes, and **generalized transduction,** in which **P1** or other phage heads carry the bacterial genes instead of phage genes.

Phages as Carriers of Host Genes

Lambda (λ) is a temperate phage that infects *E. coli* cells. Temperate phages, unlike T2, T4, and other virulent phages, do not necessarily kill their hosts. Instead, as we saw in chapter 9, λ DNA can incorporate itself into the host genome. This gives rise to a *lysogenic* infection in which the phage DNA becomes a passenger and replicates along with the host DNA. On the other hand, the λ phage can initiate a *lytic* infection in which the phage DNA remains independent of the host chromosome. Under these circumstances, host cells fill up with progeny phage particles and finally burst, or lyse, releasing the new phages to infect other cells. The spreading infection started by a single phage replicating lytically on a lawn of bacteria can produce a hole, or *plaque,* devoid of living cells.

During lysogenic infection, λ DNA inserts into a site called **att B,** which lies between the *gal* and *bio* genes in the *E. coli* DNA. We also saw that mutagens can induce the λ lytic cycle, causing the phage DNA to be excised from the host chromosome. Usually, the λ DNA excises perfectly, but sometimes it comes out carrying some excess baggage: a piece of host DNA.

Since the λ DNA usually inserts between the host *gal* and *bio* genes, these are the genes that λ DNA picks up when it excises improperly (figure 13.15). And since the room available inside a λ phage head is limited, the amount of extra DNA that can be accommodated is only about 10% of the size of the λ DNA itself, or about 5 kb. In practice, more host DNA than this is usually included, which means that some phage DNA must be excluded. If a large person sits down on one end of a crowded bench, someone on the other end gets pushed off. By the same token, if the λ DNA picks up the *E. coli bio* locus at one end, it loses some of its regulatory region from the other. The result is a **transducing phage,** a particle that contains a recombinant DNA: part *E. coli bio* and part λ. The λ part is missing a regulatory region required for lysogeny but not for the λ lytic cycle, so the resulting transducing phage can replicate lytically to form plaques. It is therefore called λ*pbio,* where the *p* stands for "plaque-forming."

On the other hand, if the λ DNA picks up the *gal* locus on one end, it loses genes on the other end that are essential for the lytic cycle, and the resulting transducing phage (λ*dgal*) is defective. The *d* stands for "defective," of course. Such a defective transducing phage needs a nondefective **helper phage** to supply the functions it requires to replicate.

What can we do with specialized transduction? We can use it instead of conjugation to form merodiploids for complementation tests. It is somewhat more limited than conjugation in that the *gal* and *bio* loci are the only ones that usually participate. However, λ DNA sometimes goes

326   Chapter 13

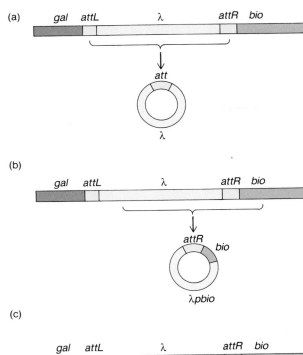

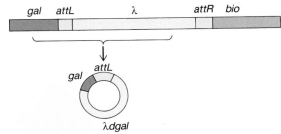

FIGURE 13.15
Creation of λ transducing phages. (*a*) Normal λ excision: The λ DNA is correctly cut out of the host genome (colored purple on the left and green on the right) by clipping in the *att* sites (yellow) on either side of the λ DNA. (*b*) The cutting occurs too far to the right, thus including the host *bio* operon (green), but missing the left end of the λ genome. The resulting phage will still be able to replicate and can transduce the *bio* genes; it is called λ*pbio*. (*c*) The cutting occurs too far to the left, thus including the host *gal* genes (purple), but missing the right end of the λ genome, which is essential for replication. The resulting transducing phage is called λ*dgal*.

to alternate sites in the host genome, so it can pick up and transfer a variety of genes. This makes specialized transduction more versatile than it seems at first glance.

There is a significant difference between the physical relationship of the two genes in merodiploids resulting from F-duction and specialized transduction. After F-duction, the two complementing genes are on different DNA molecules: one on the host chromosome and the other on the F' plasmid. After specialized transduction, both complementing genes are part of the host chromosome, because

the λ DNA, with its piece of host DNA attached, inserts into the recipient genome. Nevertheless, both genes are presumably active in each case, so complementation can occur equally well in both.

> W hen λ DNA escapes from the host chromosome after lysogenic infection, it sometimes carries a piece of host DNA along with it. A phage carrying such a recombinant DNA is called a transducing phage because it can be used to "transduce" a piece of host DNA into a new host cell to form a merodiploid. Geneticists use such merodiploids for complementation tests. λ transduction is called specialized transduction because it usually carries only the *gal* or *bio* genes that flank the λ integration site.

■

P1 Phage Mediation

In specialized transduction, a temperate phage such as λ carries host genes as passengers on its genome. In generalized transduction, little if any phage DNA is involved. Instead, the phage head forms around a piece of host DNA, producing a transducing particle that can deliver this DNA to a new host. Of course, the resulting infection cannot produce new phages in the absence of phage genes, but the donated host DNA can recombine with the recipient's chromosome and change its genotype. In fact, recombination is necessary for successful generalized transduction, just as it is for mating between Hfr and F⁻ strains. In both cases, linear pieces of DNA enter new host cells and must recombine with the host chromosome or be lost.

The favorite phage for generalized transduction is P1, whose genome normally contains 91.5 kb of DNA. This means that a P1 phage head can hold up to 91.5 kb of donor DNA if it does not have to accommodate any phage DNA. This is about 2.4% of the whole *E. coli* genome, enough room for about seventy-five average-size genes—many times more DNA than can be transferred by specialized transduction.

About 0.3% of the phages from a given P1 infection are transducing phages; the rest are normal. Since a given transducing phage contains at most 2.4% of the host genome, picked up more or less at random, the chance of finding a given host locus in a given phage particle is at most 0.024×0.003, or 7.2×10^{-5}. This is less than one in ten thousand; in practice, it is only about one in one hundred thousand, so the experimenter must perform an efficient selection to find this "needle in a haystack."

Consider an experiment in which we want to transduce a *leu*⁺ gene into a *leu*⁻ recipient (figure 13.16). This is relatively straightforward. We simply infect the *leu*⁺

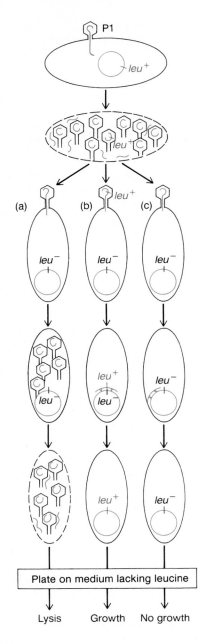

donor with P1, collect the progeny phage, including the transducing phage, then infect the *leu*⁻ recipient and transfer the cells to medium lacking the amino acid leucine. Cells infected by normal P1 phage will lyse; cells infected by transducing phages carrying genes other than *leu*⁺, and cells not infected at all, will fail to grow in the absence of leucine. Only those rare cells infected by transducing phages carrying the *leu*⁺ gene will survive. These will be easy to spot as opaque colonies on petri dishes, surrounded by transparent regions of nongrowing cells. This type of procedure is called a **selection.** It removes all undesirable cells—in this case, by preventing them from reproducing. Only the desired cells (*leu*⁺ in this example) can grow.

Now suppose we want to transduce a *leu*⁻ gene into a *leu*⁻ recipient. (We will see the usefulness of such an experiment shortly.) This poses a problem. We cannot select for the Leu⁻ phenotype by leaving out leucine, because the Leu⁻ cells fail to grow under those conditions. Instead, we can pick a selectable marker near the nonselectable one. Figure 13.17 shows how this works in the *leu*⁻ case we are trying to solve. Since the *leu* locus is very close to the *ara* locus, which codes for enzymes that metabolize the sugar arabinose, chances are that a transducing phage containing *leu* will also contain *ara*. Therefore, we can infect an Ara⁺ Leu⁻ donor strain with P1, collect the transducing phage, infect an Ara⁻ Leu⁻ recipient strain with them, and select for Ara⁺ transductants (cells made Ara⁺ by transduction) by plating on medium containing arabinose, but no glucose. Some of the Ara⁺ transductants will also be carrying the *leu*⁻ gene from the donor cell, since *leu* and *ara* are closely linked. This process, called **cotransduction,** narrows our search considerably.

FIGURE 13.16

Generalized transduction. Infection of a *leu*⁺ bacterium by the phage P1 generates three kinds of particles: (*a*) a normal P1 particle containing only phage DNA (green); (*b*) a transducing particle containing host DNA (red) that includes the *leu* operon (blue); (*c*) a transducing particle containing non-*leu* host genes only (red). These three kinds of particles can be tested by using them to infect *leu*⁻ hosts, then plating on medium lacking leucine. The normal P1 particles (left) will infect and destroy these cells. The transducing particles with the *leu* genes (middle) will render the host cells *leu*⁺, so they will grow on the leucine-free medium. The transducing particles lacking the *leu* genes (right) cannot confer the *leu*⁺ genotype, so the infected cells will die on the leucine-free medium.

> **P**1 phage can mediate generalized transduction by incorporating host DNA instead of phage DNA. Such particles can deliver up to 91.5 kb of host DNA to a new cell, allowing recombination between the donor DNA and the recipient chromosome. Since only a small fraction of the phages from a P1 infection will carry a given gene, a selection must be performed to locate the transductant of interest. This is easiest if transduction creates a prototrophic recombinant.

∎

Mapping Mutations

Why would we want to transduce a DNA from a *leu*⁻ cell into yet another *leu*⁻ cell? One good reason would be to see if the two *leu*⁻ mutations are at identical nucleotide sites, and if not, how far apart they are; in other words, to map the mutations. This can be done by trying to grow many of our cotransductants that were *ara*⁺ on plates

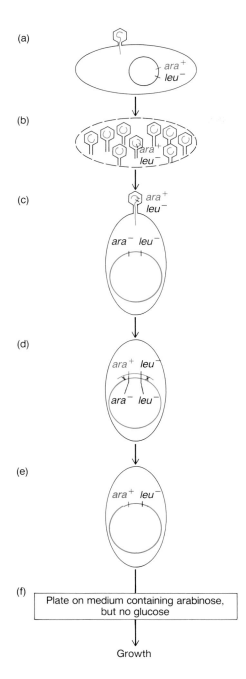

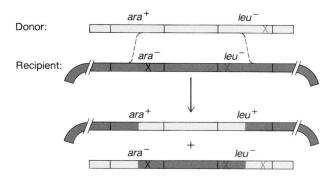

FIGURE 13.18

Crossover between two mutations. The donor, transducing DNA, contains the *ara*⁺ and *leu*⁻ alleles; the chromosome of the recipient cell (purple) contains the *ara*⁻ and *leu*⁻ alleles. Recombination occurs between these two DNAs (dashed lines), resulting in an Ara⁺ Leu⁺ cell. The cell becomes Leu⁺ because the recombination crossover occurred between the two mutations (red) in the two *leu* genes, placing both of them on the short, linear piece of DNA (bottom), which is lost. This leaves the chromosomal DNA free of mutations in both *ara* and *leu*.

FIGURE 13.17

Cotransduction screen. The goal is to transduce a *leu*⁻ locus, which is nonselectable. Therefore, we select for the nearby *ara* locus, which confers the ability to grow on the sugar arabinose. (*a*) Cells with the phenotype Ara⁺ Leu⁻ are infected with phage P1. (*b*) Some of the particles resulting from this infection are transducing phages, containing host DNA (red), including the *ara*⁺ (blue) and *leu*⁻ loci. (*c*) These particles are used to infect Ara⁻ Leu⁻ hosts; most particles will be normal P1 phage or transducing particles with irrelevant genes, but a few, like the one pictured, will contain the host *ara* and *leu* genes. (*d*) Recombination occurs between the transduced DNA and the new host chromosome. (*e*) The result is an Ara⁺ Leu⁻ cell that can grow on medium containing only arabinose as energy source (*f*). Those cells that received the *ara* genes from the transducing particle are likely to have received the *leu* genes as well, since these loci are closely linked.

lacking leucine. If all of these are still *leu*⁻, chances are the *leu*⁻ mutations in both the donor and recipient cell are in precisely the same place. But this is unlikely; it is much more probable that some of the *ara*⁺ transductants will also be *leu*⁺. This would mean that a recombination occurred between the two *leu* mutations, as shown in figure 13.18.

We learned in chapter 5 that the farther apart two genes are, the better the chance that a recombination event will occur between them. This applies equally well to the two *leu* mutations in figure 13.18. They are very far apart in the figure, so the chance of a crossover between them is relatively high. As the figure shows, the crossover results in a *leu*⁺ recombinant. Generalized transduction makes this type of fine mapping within a gene relatively convenient. In fact, Charles Yanofsky used this technique to map the mutations in the *trpA* gene in his classic experiment that demonstrated the colinearity of gene and protein. The linkage between two or more different genes can also be examined by this same kind of recombination analysis. Again, the farther apart the genes, the greater the frequency of recombination between them.

A Direct Mapping Scheme Based on Cotransduction

We have seen that cotransduction is a useful way to "piggyback" a gene we want to study with another that is selectable. Cotransduction also gives us another mapping method. First, if two genes cotransduce, we immediately know that they lie within 91.5 kb of each other and are therefore fairly closely linked. Furthermore, the closer the linkage between two genes, the more likely they are to cotransduce. Notice that this is just the opposite of the recombination test. There, the closer the linkage between

(a)

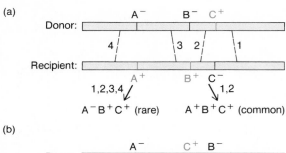

(b)

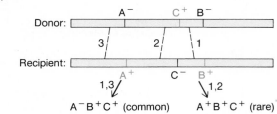

FIGURE 13.19

Three-factor cross to determine the order of genes A, B, and C. We know that genes B and C are closer together than either is to gene A. We want to know if the order is A B C as shown in (a), or A C B as shown in (b). We perform a transduction with an A⁻ B⁻ C⁺ donor and an A⁺ B⁺ C⁻ recipient, and screen for B⁺ C⁺ transductants. Then we check to see whether among these transductants we get more A⁺ or A⁻. In (a) we know that crossovers 1 and 2 have occurred to bring together the B⁺ and C⁺ alleles on the recipient chromosome. If those are the only crossovers, the transductant will also be A⁺; if crossovers 3 and 4 also occur, the transductant will be A⁻. But two crossovers will occur much more frequently than four, so A⁺ will be much more common if (a) is correct. In (b) we again know that crossovers 1 and 2 or 1 and 3 have occurred to bring together B⁺ and C⁺. But 1 and 3 are more likely than 1 and 2, because there is much more space for 3 to the left (not fully shown) than for 2, which must occur between A and C. And 1 and 3 give A⁻, while 1 and 2 give A⁺. Therefore, A⁻ will be more common if (b) is correct.

two genes, the less likely they are to recombine. T. T. Wu has proposed a formula for calculating the frequency (x) of cotransduction of two markers:

$$x = (1 - d/L)^3$$

where d is the distance between the two markers, and L is the length of the DNA that can be accommodated by the transducing particle. Using this expression, we find that markers separated by 60 kb (kilobase pairs), 30 kb, and 10 kb are expected to cotransduce with frequencies of about 4%, 30%, and 70%, respectively. An average bacterial gene is only about 1 kb long, which means that two markers within a gene cotransduce with a frequency higher than 96%.

We can also use transduction to deduce the order of genes. For example, consider three genes: A, B, and C. We know from previous experiments that genes B and C are close together and A is farther away. The question is this: Is the order of these genes A B C, as shown in figure 13.19a, or A C B, as shown in figure 13.19b? We can answer this by doing a three-factor cross as illustrated in figure 13.19.

We start with a recipient cell with the phenotype A⁺ B⁺ C⁻ and transduce DNA from a cell with the phenotype A⁻ B⁻ C⁺. Then we screen for transductants that are B⁺ C⁺. In order to produce this phenotype, crossovers 1 and 2, at least, must occur. Among these B⁺ C⁺ transductants, we then test for the A alleles. We ask: How many transductants are A⁺ B⁺ C⁺, and how many are A⁻ B⁺ C⁺? Figure 13.19a shows us what to expect if the gene order is A B C: Crossovers 1 and 2 alone will produce A⁺ B⁺ C⁺, but two additional crossovers (3 and 4) must occur to bring the A⁻, B⁺, and C⁺ alleles together. Clearly, two crossovers are much more likely to occur than four, so if the gene order is A B C, we would expect to find many more A⁺ B⁺ C⁺ transductants than A⁻ B⁺ C⁺.

On the other hand, if the gene order is A C B, figure 13.19b shows us what to expect. Here, crossovers 1 and 2 or 1 and 3 will produce the B⁺ C⁺ phenotype we first screen for. The difference is that crossovers 1 and 2 will produce A⁺ B⁺ C⁺, whereas 1 and 3 will give A⁻ B⁺ C⁺. And crossover 3 is much more likely to occur than 2 because there is much more room at the left end of both DNAs (not fully shown in the figure) than between genes A and C. Therefore, if the gene order is A C B, we would expect to find more A⁻ B⁺ C⁺ transductants than A⁺ B⁺ C⁺. This is the opposite of what we expect for the A B C gene order.

*T*ransduction allows us to do several types of genetic mapping. We can transduce two genes, or two markers within a gene, and observe the frequency of recombination between them. The closer together they are, the lower their frequency of recombination. Alternatively, we can observe the frequency of cotransduction of two markers. The closer together the markers, the higher their frequency of cotransduction. We can also deduce the order of the markers on the genome by performing three-factor crosses in which we transduce three markers.

■

GENETICS OF T-EVEN PHAGES

Scientists began to work on bacterial viruses as early as 1915, when F. W. Twort reported that very small, invisible agents could destroy bacteria. He proposed that these agents were viruses that parasitized their bacterial hosts. Twort's work, like other genetic studies we have encountered, escaped recognition, allowing F. d'Herelle to rediscover the same phenomenon two years later. D'Herelle popularized the notion of a bacterial virus and gave this agent the name **bacteriophage** (Greek: *phagein* = to devour; hence, bacteriophage = bacteria-devourer). Delbrück and his colleagues initiated the modern era of phage investigation in 1938 when they began to focus on the genetics of a small group of phages called **T1-T7**, where the T stands for "type." This focusing of attention was important, especially since it was directed mostly on the

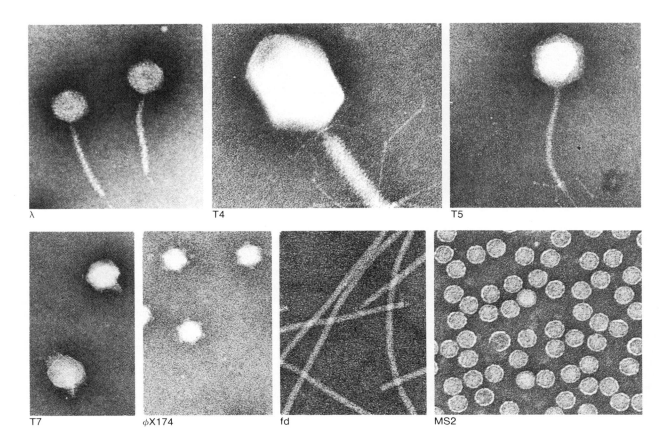

FIGURE 13.20
Structures of some phages of *E. coli*. All are shown at approximately equal magnification.

From Robley C. Williams and Harold W. Fisher. An Electron Micrographic Atlas of Viruses, *pp. 26, 70, 104, 112, 116, 120; 1974. Courtesy of Charles C. Thomas, Publisher, Springfield, Illinois.*

closely related **T-even** phages (T2, T4, and T6). Thus, new findings could be compared with one another and organized into a coherent picture, rather than remaining isolated fragments of information about a bewildering variety of very different phages. Figure 13.20 illustrates some of the various forms of *E. coli* phages.

STRUCTURE AND GROWTH OF T-EVEN PHAGES

Figure 13.21 is a schematic diagram of the structure of T4, one of the T-even phages. The most obvious features are a polyhedral *head*, which contains the phage genome— 175 kb of double-stranded DNA tightly coiled inside; a *tail*, which the phage uses to inject its DNA into the host bacterium *E. coli;* and *tail fibers* and a *baseplate,* by which the phage attaches to the surface of the host cell. Figure 13.22 dramatically illustrates the tight coiling of phage DNA inside the head by showing that when the head bursts upon osmotic shock, the DNA spews out in all directions. Indeed, it is difficult to believe all that DNA could fit into the phage head until you remember that the DNA is heavily shadowed with metal to make it show up; it is really far thinner than it appears here.

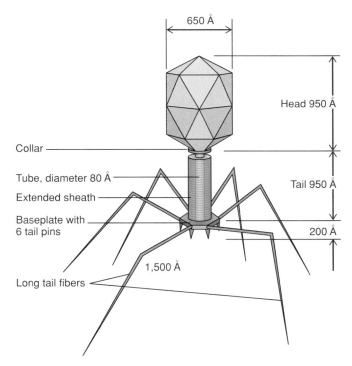

FIGURE 13.21
Schematic diagram of a T-even phage particle.

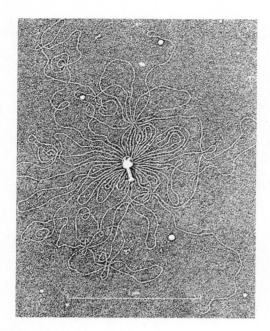

FIGURE 13.22
Lysed T4 phage. The phage ghost is at center, with escaped phage DNA all around.

A. K. Kleinschmidt et al. "Darstellung und Längenmessungen des gesamten Desoxyribosenucleinsäure-Inhaltes von T2-Bacteriophagen," Biochemica et Biophysica Acta *61(1962):857–64. Reproduced by permission.*

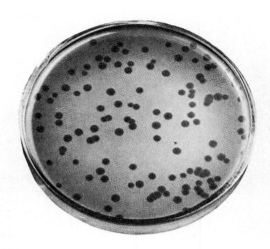

FIGURE 13.23
Plaque assay for T2 phage. The phages were mixed with *E. coli* cells and plated on nutrient agar in petri dishes. Each hole, or plaque, in the lawn of bacteria started from a single phage infecting a single cell. The infection spread to surrounding cells to form the plaque.

From Molecular Biology of Bacterial Viruses *by Gunther S. Stent. Copyright © 1963 by W. H. Freeman and Company. Reprinted with permission.*

Even though phages are invisible to all but the electron microscopist, they can be easily counted by observing their effects on their hosts. In this procedure, called a **plaque assay,** an experimenter mixes phage particles with bacterial cells, suspends the mixture in molten agar containing nutrients for the bacteria, then pours the mixture into a petri dish. The agar quickly solidifies, and the bacteria grow overnight to form a uniform, turbid "lawn." However, as figure 13.23 shows, phages make their presence known by creating clear holes, or **plaques,** in the bacterial lawn.

Here is how phages create plaques: Each phage that infects a bacterial cell in the original mixture takes over the host machinery for DNA, RNA, and protein synthesis, forcing the cell to make phage products instead of its own. Ultimately, the phage DNAs and proteins assemble to form complete progeny phages, and about twenty-five minutes after the start of infection, the cell lyses, releasing hundreds of progeny phages. By that time, however, the agar has hardened, so the progeny phages cannot float away; instead, they are restricted to infecting adjacent cells. This process repeats itself over and over, until the phages, by lysing one host cell after another, succeed in clearing out a visible plaque in the bacterial lawn.

Each infectious phage in the original suspension will, in principle, cause one plaque to form. Therefore, by counting plaques, we get an idea of the number of infectious phages, or **plaque-forming units (pfu),** in a given volume. For example, suppose we start with a phage suspension, dilute it 1:100,000 (a 10^5-fold dilution), and spread ¼ ml on a plate of bacteria. If we obtain one hundred plaques, that means we had $100 \times 10^5 \times 4 = 4 \times 10^7$ pfu, or infectious phages, per milliliter of the original suspension. Why did we do the dilution? If we had plated the concentrated suspension, there would have been so many plaques that they all would have blended into one, and there would have been no way to count them. In practice, we perform several different dilutions so that some of the plates in every experiment have a reasonable number of plaques.

Our description of phage replication contains a flavor of discontinuity. One phage infects a cell, and after a period in which no new phages appear, gives rise to a burst of hundreds of progeny phages. This idea comes from the "single-step growth experiment" performed by Delbrück and E. Ellis at the beginning of their investigation into phage genetics. In this study, Delbrück and Ellis infected *E. coli* cells with T4 phages and then measured the number of phages (pfu) per milliliter at various times thereafter. Figure 13.24 shows that there was no increase in the phage concentration for almost twenty-five minutes. This is the **latent period** (or **lag period**). Then, very rapidly, the concentration rose from about 10^5 per milliliter to more than 10^7, an increase of more than 100-fold. This is the **rise period.** The difference between the final and initial phage concentrations (over 100) is the **burst size.**

Inadvertent Sharing of a Phage

Many stories circulate in the scientific world, some apocryphal, some true. Here is one concerning phages that shows how cleverness can triumph over selfishness. It is probably true, but no names will be used.

It is traditional for biologists to share materials with each other; when one scientist reads about a new mutant, plasmid, or virus that another has isolated or made, all he or she usually has to do is call or write the first scientist requesting a sample, and it will be delivered.

This system has obvious advantages: It greatly accelerates the pace of research because it puts new materials into many hands, instead of bottling them up in one lab. However, nowadays these biological materials, especially clones of bacteria harboring useful eukaryotic genes, can be exploited commercially. This makes them potentially very valuable, and regrettably, leads to less sharing than there used to be.

Our story predates the era of genetic engineering, so no immediate profit motive was involved. Instead, it simply reflects the quite understandable desire to avoid the intense competition that characterizes much genetic research.

It seems that a certain phage geneticist (let us call him A) discovered a novel phage, and a second scientist (B) soon got wind of the discovery. B wrote to A and asked for a sample of the new phage. A wrote back and declined to cooperate, claiming the right to exploit the phage himself, without competition from B.

Nevertheless, A had already unwittingly cooperated, simply by writing back, because B had a clever idea. He cut the letter into pieces and dropped it into a culture of *E. coli*. Soon, the bacteria were infected and producing the new phage. How can a letter be infectious? Since A was already actively growing the phage in his lab, the air in the lab contained significant numbers of phage particles. Some of these settled on the letter A sent to B, so without intending to, A actually did send B the phage!

■

The actual production of infectious particles within the host cell may be somewhat more gradual than suggested by the rise period, but each infected cell will be counted as only one pfu in the plaque assay until it lyses. In other words, the rise period really reflects **lysis,** or rupture (Greek: *lysis* = a loosening), of infected cells, which allows each progeny phage to produce a separate infection. Before lysis, all the infectious particles are contained in one cell, thus that cell and all the infectious particles within still behave as only one infectious unit.

T-even phages have polyhedral heads, containing the double-stranded DNA genome of about 175kb, and tails through which they inject the DNA into their host, *E. coli*. The phages take over their hosts' DNA, RNA, and protein synthesizing machinery, reproduce themselves, lyse their hosts, and go on to infect other cells. Since T-even phages lyse their host, they can be conveniently counted by observing the number of plaques, or areas devoid of living cells, they produce on a lawn of bacteria.

■

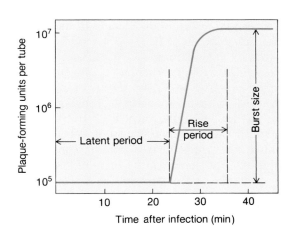

FIGURE 13.24

The one-step growth experiment. Plaque-forming units per tube were recorded as a function of time after infection. No change occurred for almost twenty-five minutes (the latent period). At the end of this time, a rapid rise in phage numbers began (the rise period). The difference in phage number at the beginning and end of the rise period is the burst size.

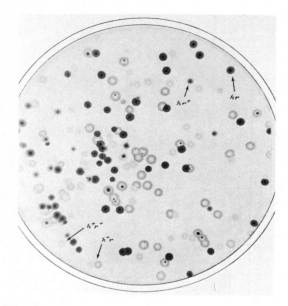

FIGURE 13.25
Distinguishing four phenotypes of phage T2. The phages were plated on a mixture of Tto^s and Tto^r *E. coli*. Four different kinds of plaques are visible, corresponding to $h^+ r^+$, $h^+ r$, $h r^+$, and $h r$ phages.
From Molecular Biology of Bacterial Viruses *by Gunther S. Stent. Copyright*
© 1963 by W. H. Freeman and Company. Reprinted with permission.

CROSSING MUTANT T-EVEN PHAGES

Since phages are detected by the plaques they form, it is perhaps not surprising that some of the first T-even phage mutants discovered were those that formed abnormal plaques. Among these was the *r* mutant, which, for reasons we still do not fully understand, forms extra-large plaques. A. D. Hershey (the same Hershey who later participated in the famous Hershey-Chase experiment, chapter 6) discovered this mutant in 1946 and called it *r*, for "rapid lysis," because it seemed that rapid lysis of the host cells was what allowed the plaques to grow so large. The wild-type allele of the *r* gene is r^+. Hershey also demonstrated that the *r*-mutant phenotype was indeed due to a mutated gene, since phages from an *r*-mutant plaque gave rise only to *r* plaques, not to wild-type plaques. In other words, the *r* phenotype was heritable.

Another early T-even mutant, also discovered in 1946, was T2*h,* where the *h* stands for **host-range.** This is a T2 phage that can infect *E. coli* cells (Tto^r) that cannot be infected by wild-type T2 phage. Tto is simply a genetic way of writing T2, so Tto^r is T2-resistant. (Similarly, Ton^r means T1-resistant.) Since wild-type T2 phages (h^+) cannot infect Tto^r cells, they will not produce plaques on this strain of cells. Nevertheless, we can get them to form plaques by plating them on a mixture of Tto^r and Tto^s cells. The T2 phages will infect the Tto^s cells in the lawn and therefore form turbid plaques that still contain un-lysed Tto^r cells.

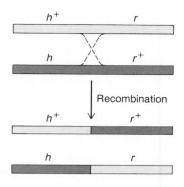

FIGURE 13.26
Recombination between *h* and *r* genes of phage T2. Bacteria were co-infected with phages of genotypes $h^+ r$ and $h r^+$. Recombination between these two markers produced progeny phage of genotype $h^+ r^+$ and the double-mutant *h r.*

The existence of two different T2 mutants allowed phage geneticists to do crosses and look for recombinants. Here is how the experiments were performed: A mixture of Tto^r and Tto^s cells was infected with a high concentration of two different phages, T2*h* r^+ and T2$h^+ r$. The high concentration of phages gave a high ratio of phages to bacteria (a high **multiplicity of infection,** or **m.o.i.**) and ensured that most bacteria would be infected by more than one phage. This allowed the two different phage DNAs to recombine to produce the nonparental types, T2$h^+ r^+$ and T2*h r,* in the progeny.

The mixed lawn of Tto^r and Tto^s cells offers a neat way to distinguish all four possible combinations of the two pairs of genotypes in the progeny phages (figure 13.25): T2$h^+ r^+$ phages will produce small, turbid plaques (small because of the r^+ gene, turbid because of the h^+ gene); T2$h^+ r$ phages will produce large, turbid plaques; T2*h* r^+ phages will produce small, clear plaques; and T2*h r* phages will produce large, clear plaques. In fact, all four combinations were observed. Does this prove that recombination has occurred? Since the progeny phages went on producing nonparental plaques, this new plaque morphology bred true. Thus, recombination, not just complementation, has indeed occurred (figure 13.26).

Recombination between two different strains of a T-even phage makes it possible to map phage genes by methods analogous to those used for bacterial genes. For example, Hershey and his colleagues isolated several different *r* mutants of phage T2, numbered *r*1, *r*2, etc., and studied the recombination frequency of each one with an *h* mutant. Let us consider two *r* mutants whose recombination frequencies with *h* differ significantly: *r*7 and *r*13. In fact, recombinants between *r*7 and a particular *h* mutation are found in 12.3% of the progeny phages, but recombinants between *r*13 and this same *h* mutation are found in only 1.7% of the progeny. This means that the *r*13 and *h* mutations lie much closer on the T2 genetic map than do the *r*7 and *h* mutations. This can be explained by two possible arrangements of *r*7, *r*13, and *h,* as illustrated in figure 13.27.

To distinguish between these two possibilities, we need to examine the recombination frequency between *r*13 and *r*7 and see if it is greater or less than that between *h* and *r*7. If it is greater, *r*13 and *r*7 are farther apart than *h* and *r*7, and figure 13.27*b* is correct. If it is less, *r*13 and *r*7 are closer than *h* and *r*7, and figure 13.27*a* is correct. In fact, it is less, so diagram (*a*) is the right one. You may be thinking that this was a foregone conclusion, since the two *r* mutations surely lie in the same gene and that gene is not likely to be interrupted by another gene, *h*. However, it turns out that *r*7 and *r*13 lie in *different* genes, but mutations in either gene lead to the same mutant plaques. So the experiment was not so trivial after all.

> T-even phages can be crossed by coinfecting host cells with phages of two different genotypes that differ at two distinct loci. Recombination occurs, producing nonparental genotypes. Recombination frequencies in these experiments can be measured to determine the order and linkage of phage genes.

■

FINE STRUCTURE MAPPING

We have already examined some mapping of T-even phage *r* mutants. Now we are going to focus on a class of T4 *r* mutants that Hershey mapped to a region he named the **rII** locus. These *rII* mutants were especially interesting because they exhibited a phenotype in addition to the *r*-mutant plaque morphology. They could grow on *E. coli* strain B, but not on strain K(λ), which harbors a λ phage DNA in its chromosome. Thus, *rII* mutants are a kind of conditional lethal—the lethality in this case depends on the host strain—and Benzer exploited this trait to perform an exhaustive study of the *rII* locus, mapping a large number of very closely linked *rII* mutants.

Benzer's *rII* Mapping

This kind of detailed mapping, or **fine structure mapping,** required that Benzer isolate many mutants in a single locus and map them relative to one another. Obviously, if all these mutations are in one gene, they will all be relatively close together. In fact, some will be only one or a few nucleotides apart. Mutations that close together will recombine with a very low frequency. Thus, Benzer had to hunt through thousands of unrecombined, parental (T4*rII*) phages to find one recombinant phage. This is where the conditional lethality of *rII* mutants came in handy. The

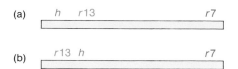

FIGURE 13.27
Two possible arrangements of the markers: *h*, *r*13, and *r*7.

unrecombined parental phages could not grow on *E.coli* strain K(λ); only the rare, recombinant, wild-type (T4*r⁺*) phages could do so. This simplified the search tremendously. It was much easier to see a wild-type plaque against a smooth background than to pick one out of a sea of mutant plaques.

Benzer's conditional lethal screening method was so powerful that he could have detected a recombination frequency of one in a million. In fact, the lowest real frequency of r^+ recombinants he observed was one in ten thousand, or 0.01%. This seems to represent the closest that two mutations can be and still recombine.

How close are two mutations that give r^+ recombinants with a frequency of 0.01%? We can make an estimate as follows: First, we must realize that if r^+ recombinants appeared with a frequency of 0.01%, another 0.01% of double-mutant recombinants went undetected. So the real rate of recombination was 0.02%. We know from other mapping data that the circular phage map consists of about 1,500 **map units,** where a map unit is the distance separating two loci that recombine with a frequency of 1% (chapter 5). Thus, two loci that recombine with a frequency of 0.02% are separated by a distance equal to 0.02/1,500, or 1.3×10^{-5} of the total phage genome, assuming that the recombination frequency is constant throughout the genome. Since the T4 phage genome contains about 1.75×10^5 base pairs, the distance between recombinable sites is one or two base pairs.

The mechanisms we envision for recombination (chapter 8) predict that it should be able to occur between mutations on adjacent nucleotides, and Benzer's experiments showed that in all likelihood, it can.

The finding that recombination could occur *within* a gene was an important one. It destroyed the classical notion of the gene as an indivisible unit. Instead of the gene itself being the unit that recombines, a subgenic unit that Benzer called the *recon* was recognized as the real indivisible unit of recombination. When it became obvious in later years that a recon is really just a base pair, Benzer's term fell into disuse.

Figure 13.28 plots the locations of hundreds of Benzer's spontaneous mutations in the *rII* region. One fact is immediately obvious. The mutations are not evenly distributed throughout the region. Instead, some sites are not mutated at all, whereas others are mutated repeatedly. One site was affected in more than five hundred of the 1,612

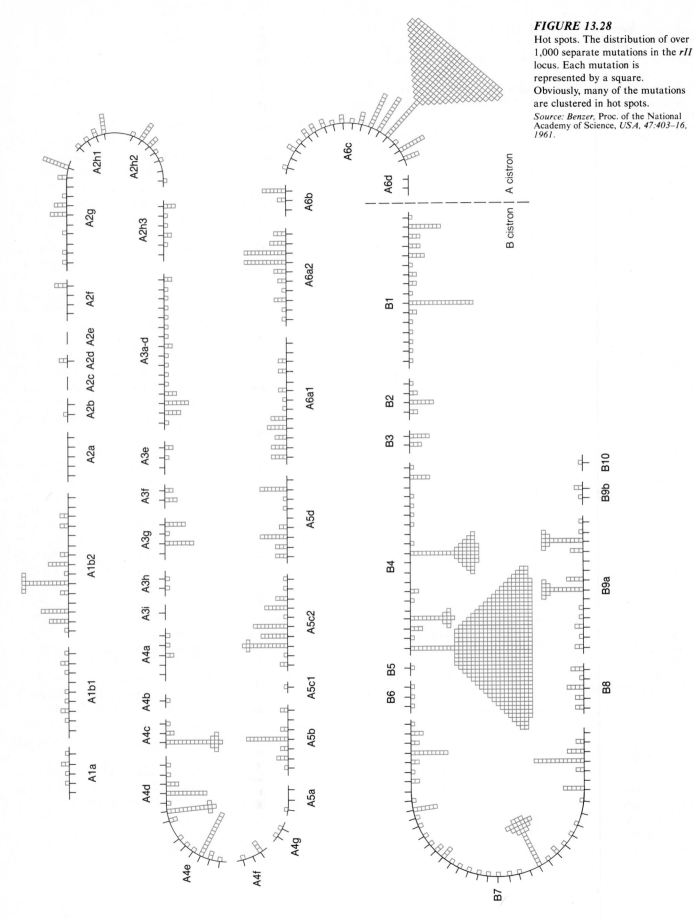

FIGURE 13.28
Hot spots. The distribution of over 1,000 separate mutations in the *rII* locus. Each mutation is represented by a square. Obviously, many of the mutations are clustered in hot spots.

Source: Benzer, Proc. of the National Academy of Science, *USA, 47:403–16, 1961.*

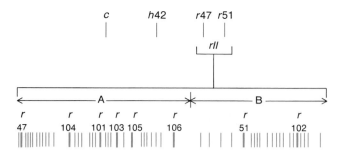

FIGURE 13.29
The two cistrons of the *rII* locus. The *rII*A and B genes and a few of their mutations are depicted.
Source: Benzer, Proc. of the National Academy of Science, USA, 41:344, 1955.

mutants screened. Benzer called these highly mutable sites **hot spots.** The same phenomenon has been observed with mutagen-induced mutations.

> **S**eymour Benzer collected thousands of mutants in the *rII* region of phage T4 and mapped their mutations exhaustively. This fine structure mapping revealed that recombination can occur between mutations as close as two (and probably only one) nucleotide(s) apart. Benzer also showed that his mutations were not evenly distributed throughout the *rII* region. Instead, many were clustered in hot spots.
>
> ■

APPLYING THE *CIS-TRANS* TEST TO THE *rII* REGION

We have already observed, in the context of bacterial genetics, the usefulness of the *cis-trans* complementation test in determining whether two mutations affect the same or different genes. The idea is to bring the two mutant genomes together in one organism and see whether a wild-type phenotype results. If so, the two mutant genomes have complemented one another, demonstrating that their mutations lie in different genes. On the other hand, if they cannot complement each other, the mutations must be in the same gene. You may already have recognized that this is really the same thing that geneticists do when they mate two mutant eukaryotic organisms. They place two mutant genomes in the same offspring and observe the phenotype of that offspring.

In 1955, Benzer applied this approach to phage genetics—in particular to the *rII* region of phage T4. We have already seen that recombination between some *rII* mutants is quite frequent, which tells us that the mutated loci are relatively far apart, and therefore suggests that more than one gene is involved in the *rII* phenotype. Benzer proved this point as follows: He coinfected bacteria with pairs of *rII* mutants at a high ratio of phages to bacteria

to ensure that each host cell would receive both mutant phage genomes. (Notice that this is the equivalent of producing a diploid by crossing bacteria or higher organisms.) The question was: Would the two *rII* mutants complement each other? The answer was that many of them would. Thus, more than one gene was indeed involved. How many genes? Benzer found that all of his *rII* mutants fell into only two **complementation groups,** or cistrons. That is, members of one group would complement members of the other, but not members of their own group. These two cistrons represent the two genes of the *rII* region, which Benzer called *rII*A and *rII*B.

Figure 13.29 shows the arrangement of these two genes and some of the mutations that lie in each one. At the time Benzer carried out these studies his term cistron became popular as a substitute for the word gene, because the latter term was tied to the outdated notion of an indivisible unit. However, *gene* has proven to be a very durable word and is now widely used by geneticists of all stripes.

> **B**enzer applied the *cis-trans* test to the *rII* region of phage T4 and identified two complementation groups, or cistrons, that are responsible for the *rII*-mutant phenotype. Thus, the *rII* region consists of two genes.
>
> ■

Summary

Two mutant (auxotrophic) bacterial strains can share DNA by direct contact, producing wild-type (prototrophic) recombinants. This process, called conjugation, is mediated in F⁺ *E. coli* by an F plasmid of 94.5 kb that can transfer at high frequency into F⁻ bacteria. Occasionally, the F plasmid recombines with its host's chromosome, converting the cell to an Hfr and mobilizing the Hfr chromosome to transfer to recipient cells. By interrupting mating between an Hfr donor and an F⁻ recipient, we can determine the length of time it takes to transfer a given gene. The differences in the transfer times of different genes tell us their relative positions on the circular genetic map of *E. coli*. The completion of mating in *E. coli* requires recombination between the donated piece of Hfr chromosome and the F⁻ chromosome.

Sometimes a recombination event brings together a host DNA and an F plasmid, creating an F′ plasmid. The host genes thus become part of a highly mobile element that can deliver them to another cell. We can take advantage of this efficient sharing of limited pieces of DNA to perform genetic experiments. For example, two different bacterial mutations that affect the same function can be subjected to the *cis-trans* test, or complementation test. If they complement one another to restore wild-type function, the mutations are usually in separate genes.

Phages can also be used to carry, or transduce, DNA from one bacterium to another. Transduction by phage λ is called specialized transduction because it usually carries only the *gal* or *bio* genes that link the λ integration site. P1 phage heads can mediate generalized transduction by incorporating up to 91.5 kb of host DNA instead of phage DNA. Transduction allows us to do several types of genetic mapping. We can transduce two genes, or two markers within a gene, and observe the frequency of recombination between them. The closer together they are, the lower their frequency of recombination. Alternatively, we can observe the frequency of cotransduction of two markers. The closer together the markers, the higher their frequency of cotransduction. We can also perform three-factor crosses, transducing three markers and then deducing the order of the markers on the genome.

T-even phages can be crossed by coinfecting host cells with phages of two different genotypes differing at two different loci. Recombination occurs, producing nonparental genotypes. Recombination frequencies in these experiments can be measured to determine the order and linkage of phage genes. Conditional lethal mutations are used in essential genes to avoid destroying the phage's viability. The two common kinds of conditional lethal mutations are temperature-sensitive and amber.

Benzer's fine structure mapping of the *rII* region of phage T4 revealed that recombination can occur between mutations as close as two (and probably only one) nucleotide(s) apart, and that the mutations were clustered in hot spot regions. The *cis-trans* test, applied to the *rII* region of phage T4, identified two complementation groups, or cistrons, that are responsible for the *rII* phenotype. Thus, the *rII* region consists of two genes. ■

Problems and Questions

1. Write the genotypes and phenotypes of bacteria having the following characteristics:
   (*a*) Unable to metabolize the sugar lactose (lac).
   (*b*) Able to make the amino acid leucine (leu).
   (*c*) Unable to make the vitamin thiamine (thi).
   (*d*) Resistant to the antibiotic streptomycin (str).
   (*e*) Sensitive to the antibiotic puromycin (pur).
   (*f*) Able to make the vitamin biotin (bio).

2. Which of the strains in problem 1 are prototrophs, and which are auxotrophs, assuming that they are wild-type in all genes but the given one?

3. Given a reversion rate of one in a million and a recombination rate of one in ten million in Lederberg and Tatum's first experiment, was Lederberg justified in his use of double mutants? Why, or why not?

4. Why is a prototrophic recombinant in a bacterial conjugation experiment rarely a true zygote? What do we call it instead?

5. You have discovered that certain *E. coli* cells are resistant to infection by phage T1 (*ton^r*). You would like to know whether this characteristic is caused by a mutation in the bacterial genome or by an adaptation of the cells to the presence of the phage. To answer this question, you perform a fluctuation test and find that twenty individual cultures vary quite a bit in their resistance to T1—from a low of zero to a high of 107 *ton^r* colonies per culture. Is the characteristic caused by mutation or adaptation? Why?

6. What event converts an F$^+$ cell to an Hfr? How does this work?

7. Why does an Hfr transfer its whole genome only rarely?

8. The giant sandworms of the planet Arrakis harbor a gut bacterium *(Xenobacterium gigantii)* that undergoes conjugation. However, like everything in these worms, the circular bacterial chromosome is large, and conjugation takes 200 minutes. You perform a conjugation experiment with this organism and find that the following markers are transferred at the respective times: *ple,* 25 minutes; *bar,* 75 minutes; *gap,* 100 minutes; *zap,* 150 minutes; *chr,* 200 minutes. Draw a genetic map of this exotic bacterium, assuming a linear rate of transfer of DNA during conjugation, and ignoring any lag at the beginning of conjugation.

9. In the experiment in problem 8, you find that the efficiency of transfer of the *ple* marker is very much greater than that of the *chr* marker. What is the probable explanation for this phenomenon?

10. In the experiment in problem 8, if you used a different Hfr that begins its transfer just after *zap*, what difference in transfer times would you expect between *chr* and *ple*?

11. In another Hfr of *X. gigantii*, the fertility plasmid inserts between *gap* and *zap* in the opposite orientation to that in problem 8. What would be the order of transfer of markers by this Hfr?

12. One strain of *X. gigantii* harbors an F′ plasmid, F-*zap*, that contains the *zap* operon and is composed of two genes: *zapA* and *zapB*. You have two Zap⁻ mutants and you want to determine whether their mutations lie in the same or different genes. Assuming that you can readily place one of these mutant genes on the F-*zap* plasmid, outline an experiment to answer your question.

13. You perform the experiment described in the answer to problem 12 (see answer section at end of book) and obtain only a few Zap⁺ colonies. What are two possibilities for the cause of these Zap⁺ cells?

14. Why should we not expect to find both *bio* and *gal* markers in a λ transducing phage?

15. In *E. coli*, the markers *thr* and *leu* are separated by about 2% of the total bacterial genome. Would you expect these genes to cotransduce in P1 phage? Why, or why not? If so, at what frequency?

16. The genes *zip* and *zap* lie less than 10 kb apart on the *X. gigantii* genome. A generalized transducing phage called Q2 infects this organism. You have isolated two different *zip* mutants and want to map the locations of the corresponding mutations within the *zip* gene. Outline the transduction experiment you would perform.

17. You know that a gene called *bop* is tightly linked to both *zip* and *zap* in the *X. gigantii* genome and readily cotransduces with these other markers in Q2 phage. You want to know whether the gene order is *bop zip zap* or *zip zap bop*. Therefore, you perform a three-factor cross by transduction as follows: You transduce DNA from a Zip⁺ Zap⁻ Bop⁺ strain into a Zip⁻ Zap⁺ Bop⁻ recipient and screen for Bop⁺ transductants. The lowest frequency transductant is Zip⁺ Zap⁺ Bop⁺. What is the order of these three genes? What results would you expect if the other gene order were correct? (Drawing diagrams of the two possibilities should help you figure out the answer.)

18. You perform a three-factor cross experiment in which you use Q2 phage to transduce DNA with these three markers: Awk⁺ Kat⁺ and Nrd⁺ into an Awk⁻ Kat⁻ Nrd⁻ recipient. You select for Awk⁺ transductants and observe the following:

    Awk⁺ Kat⁻ Nrd⁻: 638
    Awk⁺ Kat⁻ Nrd⁺: 309
    Awk⁺ Kat⁺ Nrd⁺: 115
    Awk⁺ Kat⁺ Nrd⁻:   1

    What is the order of these markers? What are the cotransduction frequencies between *awk* and *kat* and between *awk* and *nrd*?

19. You have a suspension of phages and you want to know its titer (plaque-forming units per milliliter). You dilute the suspension by a factor of 10^6 and plate 0.05 ml. You observe 250 plaques. What was the titer of the original suspension?

20. You perform a phage growth experiment with phage T2. Instead of waiting for the phage to lyse the host cells, you break the cells open at various times after infection and assay for plaque-forming units. How will your growth curve differ from the one-step growth curve? Why?

21. You infect *X. gigantii* cells with different strains of Q2 phage and observe the following frequencies of recombination among the three markers K, L, and M: between K and L, 18%; between K and M, 2.4%; between L and M, 15%. What is the order of these three markers?

Answers appear at end of book.

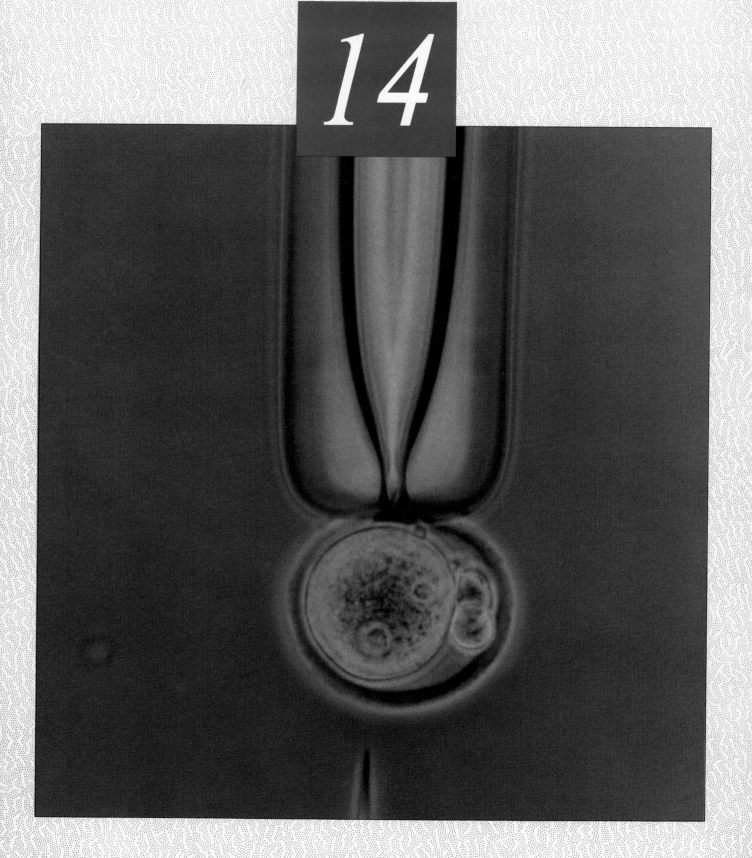

Developmental Genetics

he trouble with a kitten is

THAT

Eventually it becomes a

CAT.

■

Ogden Nash
American poet and humorist

Learning Objectives

In this chapter you will learn:

1. How gene expression and cell differentiation are related.

2. The difference between cell differentiation and determination.

3. Theories to explain determination.

4. How a hierarchical group of genes cooperates to control development in one organism: the fruit fly.

5. How genes themselves can sometimes be altered during development.

*O*ne of the greatest mysteries, if not the central mystery, of modern biology is this: How does a fertilized egg develop into a human being? The question is phrased in human terms because we tend to be most curious about ourselves, but it is just as mystifying when we ask about the development of any other complex organism. Through the ages, scientists have offered explanations of increasing sophistication, but we are still a long way from solving the mystery.

In the seventeenth century, biologists believed that a tiny, preformed person, or *homunculus* (figure 14.1), resided in each sperm cell. They assumed that the homunculus simply grew in the fertilized egg until it reached the proportions of a newborn baby. Of course there were serious theoretical problems with this idea—one being that the homunculus itself would also have to contain miniature sperm cells enclosing even tinier homunculi, and so on, to include all possible future generations.

Scientists studying the embryology of chickens found no support for the homunculus theory. In fact, as they looked farther and farther back in the development of the chick, they noticed that the embryo not only became smaller, it also became simpler. This gave rise to the idea of *epigenesis,* which holds that tissues do not exist in the earliest embryo, but arise from simpler structures by developmental change. Taken to its logical conclusion, this means that a single fertilized egg cell develops into an adult human composed of trillions of cells that are organized into dozens of different organs, each of which performs a different, highly specialized function.

THE GENETIC BASIS OF CELL DIFFERENTIATION

We have already seen that specialized cellular functions depend on proteins: structural proteins to give cells their shapes; enzymes to catalyze specific chemical reactions; hormones to carry signals from cell to cell; and so forth. For example, consider two different specialized cells in the human body: a red blood cell and a β cell from the pancreas (figure 14.2). The blood cell is highly specialized to carry oxygen through the bloodstream. Its smooth disk shape allows it to glide easily through tiny capillaries, and it contains a large concentration of a single protein, hemoglobin. As blood flows through the lungs, this protein binds oxygen and turns bright red. Later, it releases the oxygen to peripheral tissues, sustaining their lives.

By contrast, β cells do not move throughout the body; they are stuck in one place—in small conglomerations of cells called *islets of Langerhans* in the pancreas. The main job of the β cells is to produce a hormone called *insulin.* The cells secrete this small protein into the bloodstream, after which it travels to remote areas of the body, signaling cells to alter their metabolism. For example, insulin responds to a high level of sugar (glucose) in the blood

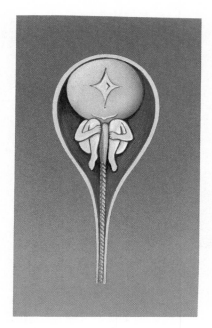

FIGURE 14.1
Homunculus.

by promoting the uptake of glucose by cells and depressing the breakdown of fats in adipose (fat) tissue that would normally produce more sugar. The result of these and other actions of insulin is a lowering of the blood glucose level.

Obviously, the functions of red blood cells and β cells are vastly different. These different functions are clearly reflected in the proteins the two types of cells make. Immature red blood cells make huge amounts of hemoglobin and very little else; in particular, they do not make insulin. On the other hand, β cells make lots of insulin, but no hemoglobin. Yet both of these cell types descended directly from the same fertilized egg cell. The early embryo contained many cells, all of which looked more or less alike, and none of which made either hemoglobin or insulin. However, at some point during development, the cells **differentiated** so that they began to manufacture these specialized products. Some became blood cells and made hemoglobin; others became β cells and made insulin.

All cells, even though they begin to synthesize different specialized products, continue to make the key proteins essential for life. For example, they all make RNA polymerase and ribosomal proteins. Because of their maintenance function, the genes for these fundamental proteins are sometimes called **housekeeping genes.** By contrast, genes that code for specialized cell products like hemoglobin or insulin are sometimes called **luxury genes.** They are a "luxury" only in that their products are not immediately vital to the survival of the cells that make them; they are certainly not a luxury to the whole organism.

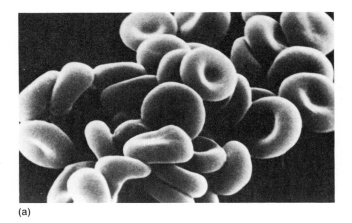

(a)

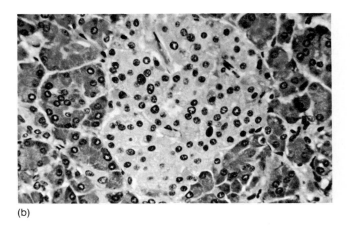

(b)

FIGURE 14.2
Contrasting cell morphologies. (*a*) Mature red blood cells; (*b*) an islet of Langerhans (light area at center) composed of insulin-producing β cells.

(a) © *MEOLA/ARS-USDA/Science Source/Photo Researchers, Inc.*
(b) © *Edwin A. Reschke.*

In short, different specialized cells make different specialized proteins, and those proteins determine what the cells can do. Moreover, as we have already learned, proteins are products of genes. This means that the genes for *globins,* the proteins in hemoglobin, are turned on in immature red blood cells, and the gene for insulin is turned on in β cells of the pancreas. Cell differentiation is therefore a reflection of differential gene activity, and those who seek to understand differentiation and development must ultimately ask: What allows expression of certain genes in one kind of cell and other genes in a second kind of cell? We will examine two theories that attempt to answer this question—mosaic development and regulative development.

MOSAIC DEVELOPMENT VERSUS REGULATIVE DEVELOPMENT

One simple explanation of gene differentiation would be that a differentiating cell loses all its genes, except the ones it needs for housekeeping and for its specialized function. This would be a true genetic change, since the genetic material itself would be altered or lost. A version of this theory, presented by Wilhelm Roux, held that embryonic cells begin losing genetic potential from the very beginning, so that each cell division during development involves the unequal partitioning of genetic material between the daughter cells. This would produce an embryo whose genetically dissimilar cells resemble a mosaic of many different pieces. Hence the term **mosaic embryo.**

Mosaic Development

The mosaic development theory found support in the fact that several invertebrate embryos, even at a very early stage, are composed of cells that behave quite differently;

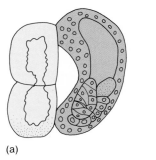

(a)

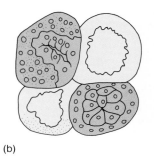

(b)

FIGURE 14.3
Demonstration of mosaic development. *Styela* embryos were forced through a pipette at the four-cell stage to kill some of the blastomeres, then allowed to develop. (*a*) Both cells on the far left (yellow) were killed, but the two remaining cells (red) continued to divide and differentiate. (*b*) The lower left and upper right cells (yellow) were killed, but again the remaining cells (red) developed into partial embryos.

in fact, they seem to be irreversibly programmed to develop along preset lines. A favorite example of such a mosaic embryo is the ascidian *Styela.* Even at the two-cell stage, the two embryonic cells (**blastomeres**) in this organism are different. These differences continue at the four-cell stage, as shown by an experiment performed by E. G. Conklin in 1905, which is depicted in figure 14.3. Conklin forced four-cell *Styela* embryos through a pipette to destroy some of the blastomeres, then looked to see what would happen to the remaining live cells.

In figure 14.3*a,* the two cells on the left (anterior cells) were killed, but the two on the right (posterior cells) went on to differentiate as they would have in a complete embryo, giving rise to chorda cells and muscle cells. In figure 14.3*b,* the bottom left (anterior-vegetal) and top right (posterior-animal) cells were killed, but again the two remaining cells carried out their own developmental programs. The top left cell made neural plate cells, and the bottom right cell again made muscle cells. In principle, this could be caused by the embryo's dividing up its genes with each cell division, so that the top left cell received the genes for making neural plate cells and the bottom right cell got the genes for muscle. No evidence exists for such a partitioning of genes among embryonic cells, and it is increasingly unlikely that any will be found.

Regulative Development

Even if we someday find an unusual organism that does divide its genes among embryonic cells, this cannot be a general phenomenon. Embryos from higher eukaryotes, such as amphibians, can be separated into individual cells as late as the eight-cell stage, and all of the resulting cells are **totipotent,** which means they can go on to produce whole, normal organisms. Clearly, these individual embryonic cells have not lost any of their genetic potential. Even in humans, we know that embryonic cells retain all of their genes at least through the first three cell divisions, because identical quintuplets have occurred. Humans and frogs are examples of organisms that have **regulative** embryos. The cells in these embryos look and behave identically much longer than those of mosaic embryos. Of course, other organisms, such as sea urchins, fall in between these extremes, and sometimes it is difficult to label an embryo either mosaic or regulative.

We have established the fact that early regulative embryos contain totipotent cells. But what about differentiated cells? Are they still totipotent, or have they lost some of their genes? A number of experiments have shown that they are sometimes totipotent. Figure 14.4 illustrates John Gurdon's classic nuclear transplantation experiment, which worked as follows: First, Gurdon used ultraviolet light to kill the nuclei in several eggs from the South African clawed frog (*Xenopus laevis*), producing **enucleate eggs** (eggs without nuclei). Then he extracted nuclei from the intestinal epithelium of a *Xenopus* tadpole and transplanted these nuclei into the enucleate eggs by **microinjection** with a very fine glass pipette.

The transplanted nuclei were diploid, just like **zygote** (fertilized egg) nuclei, and many of the eggs began dividing, just like zygotes, to produce early "hollow ball" embryos called **blastulas.** Gurdon took nuclei from these blastulas and transplanted them to new enucleate eggs. Some of the resulting eggs actually developed into tadpoles, and a few made it all the way to the adult frog stage. This demonstrated that even a differentiated intestinal cell nucleus retained the genetic power to produce a whole adult frog. In other words, the nucleus was still totipotent. It could not have lost all the superfluous genes it did not

need for housekeeping and intestinal function. In fact, since it gave rise to every cell type in a whole frog, it seems not to have lost any genes at all.

A potential objection to this experiment could be that the ultraviolet light failed to kill all of the original egg nuclei, and these nuclei provided the genes that caused **embryogenesis** (embryonic development). To counter this argument, Gurdon used transplanted nuclei that were genetically different from the original nuclei in that they had only one nucleolus, whereas the original nuclei had two. The resulting tadpoles and frogs consisted of cells with only one nucleolus, proving that the transplanted nuclei really had directed their development.

The nuclear transplantation experiment has been remarkably difficult to duplicate in other animal species. Even before Gurdon's experiment on *Xenopus,* Robert Briggs and Thomas King showed that another frog species, *Rana pipiens,* does not contain totipotent nuclei after gastrulation, which is a relatively early phase of embryogenesis. However, totipotent adult cells are abundant in plants. F. C. Steward provided the first such example in the early 1950s by regenerating a whole carrot plant from a single carrot root cell. Since then, several different kinds of plants, including tobacco, petunias, and peas, have been grown from single cells. Notice that this furnishes us with a way to *clone* (produce many identical copies of) a plant. In fact, the word clone comes from the greek *klon,* meaning "twig," and refers to a centuries-old method of propagating plants by taking cuttings.

We are left with the somewhat messy conclusion that, whereas some animals and plants do contain totipotent differentiated cells, many seem not to. At least, we can state that loss of genetic material is not necessary for cell differentiation. Specialized cells that are not totipotent may have suffered genetic changes, and we will see some examples of these later in this chapter, but there is certainly no evidence for wholesale loss of genes. Instead, most cell differentiation seems to be caused by **epigenetic** changes. These are changes in gene expression, rather than changes in the genes themselves.

*M*ulticellular organisms with specialized tissues develop from a single cell, or zygote, by a process called embryogenesis. During embryogenesis, cells differentiate to produce tissues with specialized functions. Differentiation seems to require little if any change in a cell's genetic makeup, since differentiated cells or nuclei from some plants and animals are still totipotent. Instead, the pattern of expression of a common set of genes varies from one kind of differentiated cell to another.

■

Chapter 14

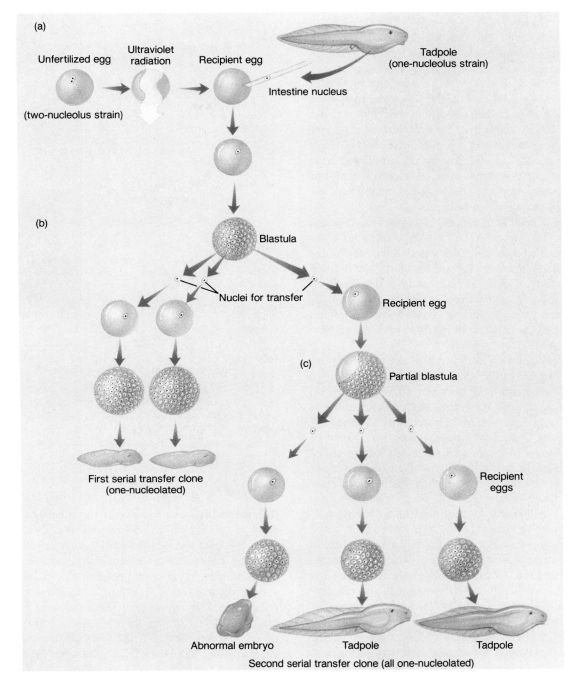

FIGURE 14.4

Demonstration of regulative development. (*a*) An unfertilized frog egg from a two-nucleolus strain was enucleated by ultraviolet radiation, then given a replacement nucleus from the intestine of a one-nucleolus strain. The artificially fertilized egg subsequently developed into a blastula. (*b*) Nuclei from this blastula were transplanted to new enucleate eggs from the two-nucleolus strain. These developed into partial embryos or tadpoles. (*c*) Nuclei from a partial embryo (a blastula) were transplanted to new enucleate eggs, some of which developed into normal, one-nucleolated tadpoles. The first transfer rarely produces normal tadpoles because the differentiated nuclei cannot divide fast enough to support normal development. After the first transfer, the nuclei seem to be geared up to divide more rapidly.

DETERMINANTS

Even before a cell becomes differentiated and starts expressing its specialized functions, it undergoes a process called **determination.** This is the stage at which the cell passes the point of no return in its developmental program. It may not yet look or behave any differently than its neighbors, because it has not differentiated, but it has become irreversibly committed to its fate. Indeed, determination is a prerequisite for differentiation. We will look in detail at a possible mechanism for determination later in the chapter; for now, let us consider some examples of determination that imply the existence of cytoplasmic effector substances called **determinants.**

The very existence of mosaic embryos, in which the position of a blastomere in the embryo determines absolutely the fate of that cell, suggests that determinants are present in the egg and that they are unequally distributed. The simplest form of unequal distribution of determinants would be a gradient from the top (**animal pole**) to the bottom (**vegetal pole**) of the egg. But a second gradient, from one side to the other, could coexist with the vertical one (figure 14.5).

Styela provides a good example of unequal distribution of determinants. The top left blastomere in figure 14.3b must have contained the determinants for neural plate development, while the bottom right blastomere must have held muscle determinants. This point was made even more powerfully by J. R. Whittaker, who pushed some of the cytoplasm from the posterior-vegetal (bottom right) blastomere into the posterior-animal (top right) blastomere by squeezing the embryo during the third cleavage. He then blocked further cell division with the drug cytochalasin B and looked for the appearance of the muscle protein acetylcholinesterase (AChE). It appeared in a new site, the posterior-animal part of the embryo, in addition to the posterior-vegetal part, where it is normally made. Only the cytoplasm moved in this experiment, not the nuclei, yet this drastically altered the fate of the cells containing those nuclei. This means that the determinants at work here are cytoplasmic, not nuclear.

What is the chemical nature of determinants? Some are no doubt **maternal messages**—mRNAs made and stored in large quantity in the developing egg. However, embryonic development can be blocked at an early stage by inhibitors of RNA synthesis, proving that there are not nearly enough maternal messages to direct all of the embryogenic processes. For instance, synthesis of AChE in *Styela* embryos depends on mRNA synthesis. In short, maternal messages may account for some of the determinants, but others must be something else, such as transcription factors. Many of these are probably not found in the egg, but are made by the embryonic cells themselves.

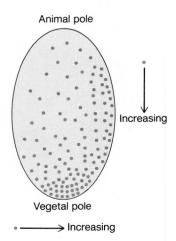

FIGURE 14.5
Hypothetical 2-fold gradient of determinants in an egg. One type of molecule (red dots) is arranged in a gradient from top to bottom, where the highest concentration is at the bottom. The other type of molecule (blue dots) is arranged in a horizontal gradient, where the highest concentration is on the right.

D eterminants are substances, some of which are present and unequally distributed in the egg. Determinants cause cells to become determined, or irreversibly committed, to a given fate.

INDUCERS

Substances that one cell makes to influence the development of another cell are called **inducers.** This property of signaling from one cell to another sets the inducers apart from the determinants we have just discussed. An inducer can producer rather startling effects if its signal goes to the wrong address. For example, consider the development of feathers in an embryonic chicken. Feathers are products of one of the three classical embryonic tissue layers, the **ectoderm.** But the underlying **mesoderm** plays an inductive role; that is, it produces substances (inducers) that dictate what kind of feathers the ectoderm will make. We know this because we can transplant embryonic thigh mesoderm to the wing of a developing chick and produce thigh feathers on that wing. The reverse experiment yields wing feathers on the thigh.

Another example of induction sheds additional light on determination. We can transplant embryonic cells of the amphibian *Triturus* to new locations in the embryo and follow their fates. Up to the gastrula stage, the transplanted cells are still totipotent; they will differentiate just like their neighbors in their new location, not like the cells in their old location. Clearly, the transplanted cells are responding to inducers they would not have experienced in their original home. After gastrulation, something quite different happens. The transplanted cells have become determined in their original locations, so they are no longer responsive to external influences; they follow their predetermined program in spite of the inducers they encounter after transplantation.

---

*E*mbryonic inducers are substances that diffuse from one embryonic site to another and cause differentiation of the target cells.

■

CONTROL OF GENE EXPRESSION

Given that cell differentation occurs via the turn-on of selected genes, differentiation is largely a question of how gene expression is controlled. Chapters 9, 10, and 11 demonstrated that structural gene expression involves three main levels: transcription, translation, and, especially in eukaryotes, an intervening, **posttranscriptional** phase during which transcripts are processed (by splicing, clipping, and modification) into mature form. In addition, many proteins experience posttranslational modifications of various sorts.

Prokaryotes undergo radical changes in gene expression and cellular activity that resemble cell differentiation in higher organisms. A prime example, discussed in chapter 9, is sporulation in bacteria. In that case, changes in RNA polymerase σ factors allow recognition of new promoters, shifting transcription from vegetative genes to sporulation genes. Other examples of massive gene control are found in the replication cycles of bacteriophages, which effect complex, time-dependent shifts in gene expression. Again, these changes in gene expression take place at the transcription level.

Developmental Activation of Genes

In chapter 10 we learned that eukaryotic genes are controlled at many levels, but primarily at the transcription level. As a model of transcriptional control, we examined the chicken ovalbumin gene and saw that estrogen can control this gene by binding to a protein receptor and then, together with the receptor, binding to a DNA sequence in the 5′-flanking region of the gene.

Does this hormone model teach us anything about the activation of genes during embryonic development? As we will see, a multitude of factors activate the genes in a de-

veloping organism. Some of these arise and do their work in the same cell; others are signals one cell uses to influence its neighbors' behavior. In this latter respect, at least, they resemble hormones. Another echo of the hormone story is that 5′-flanking regions also seem to play a crucial role in the activation of genes during embryonic development.

Gary Felsenfeld and his colleagues have provided us with some insight into this development process with their studies on the expression of the adult β-globin gene (β^A) in the chick embryo. In *erythrocytes* (red blood cells) from a five-day embryo, this gene is not yet turned on, but by nine days, it has become active. Felsenfeld's group found that the β^A gene in erythrocyte chromatin from nine-day embryos is DNase-hypersensitive, whereas the same gene in five-day embryos is not. (DNase hypersensitivity correlates well with gene activity.) The hypersensitive site in this case, as in most such cases, is in the 5′-flanking region of the gene; it extends approximately 200 base pairs, covering the region between 60 and 260 base pairs upstream from the start of transcription. Furthermore, the DNase-hypersensitive region is devoid of nucleosomes (figure 14.6a).

What causes the DNase hypersensitivity in this gene? Whatever it is, we strongly suspect that it also plays a major role in turning the gene on. If the hormone model holds true, we would expect the factor to be a protein that binds to the 5′-flanking region of the gene. This is just what Felsenfeld found: a factor in fourteen-day embryos, in which the gene is active, that confers DNase hypersensitivity on the globin gene.

The assay for the factor works as follows (figure 14.6b): We mix a cloned chicken β^A globin gene with a frog oocyte extract to assemble nucleosomes on the bare, cloned DNA. If no other substances are added, nucleosomes form and the globin gene does not become DNase-hypersensitive. If we add an extract from cells that do not express the globin gene (oviduct cells or five-day embryonic erythrocytes), we see the same result—no hypersensitivity. but if we add an extract from fourteen-day embryonic erythrocytes, along with the frog oocyte extract, the hypersensitive region appears just in the right spot in the gene's 5′-flank. The same extract has no effect if it is added after the nucleosomes have already had a chance to form. This tells us that the factor acts by preventing nucleosomes from forming, thereby preserving a small, nucleosome-free, DNase-hypersensitive region in the gene. We assume that this helps activate the gene and that these experiments have therefore discovered a new transcription factor, or factors, but these assumptions have not been proven yet. The factor behaves like a protein: Its activity is abolished by proteinase, an enzyme that destroys protein.

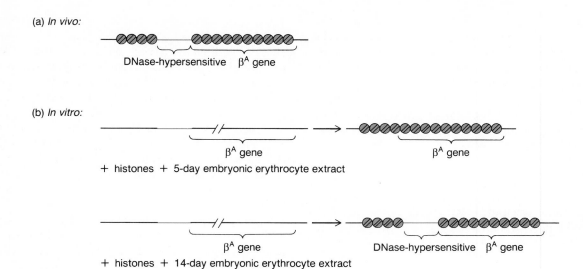

(a) *In vivo:*

DNase-hypersensitive β^A gene

(b) *In vitro:*

β^A gene

β^A gene

+ histones + 5-day embryonic erythrocyte extract

β^A gene

DNase-hypersensitive β^A gene

+ histones + 14-day embryonic erythrocyte extract

FIGURE 14.6

Activation of the chicken adult β-globin gene. (*a*) When the gene is active in vivo, it contains a nucleosome-free, DNase-hypersensitive region (blue) upstream from the coding sequence of the gene. (*b*) When the gene is mixed with histones and an extract from early (five-day) embryo erythrocytes in vitro, nucleosomes (red) cover the entire locus and no DNase-hypersensitive site results. However, when the extract is from fourteen-day embryo erythrocytes, where the adult

β-globin gene is active, the same DNase-hypersensitive site as seen in vivo appears. Factors in late embryos, not found in earlier embryos, can generate the DNase hypersensitivity. The discontinuities in the DNA in the β^A genes on the right in (*b*) denote DNA not drawn. This was done to avoid having to draw the gene six times longer on the right. (Remember that nucleosomes cause a 6-fold compacting of DNA.)

T he adult β-globin gene (β^A) is inactive in red blood cells from the five-day chick embryo; neither is it DNase-hypersensitive. The same gene is both active and DNase-hypersensitive in red blood cells from nine-day and older embryos. The DNase-hypersensitive region is found in a nucleosome-free zone in the 5'-flanking region of the gene. The hypersensitivity, and presumably the gene activity, are caused by a factor that is found only in cells where the adult β-globin gene is active. The factor seems to be a protein, and it apparently acts by preventing nucleosomes from forming just upstream from the transcription start site.

■

Light-induction of Plant Genes

When plant seedlings grow in the dark, they do not make the products needed for photosynthesis. This is economical, because photosynthesis is impossible in the dark and it would be a waste of energy to make the photosynthetic apparatus. Since one of the key components of this apparatus is the green pigment *chlorophyll,* an obvious consequence of this behavior is that the seedlings remain pale yellow. When the experimenter turns the light on, the seedling rapidly turns green, reflecting the activation of genes. Chlorophyll is not the direct product of a gene, since it is not a protein. It is a small molecule whose synthesis depends on enzymes that are, of course, gene products.

These genes are obviously among those activated by light. Another light-activated gene is one that codes for the small polypeptide subunit of an enzyme called **ribulose bisphosphate carboxylase,** or **rubisco,** a vital element in photosynthesis.

How does the plant control this rubisco gene in response to light? We are not sure of all the details, but at least we know that the 5'-flanking region of the gene is important to the light response. Nam-Hai Chua and his colleagues cloned the rubisco gene from the pea and introduced it into petunia cells. They found that, although the transplanted gene was expressed only at a low level in its new surroundings, it was still light-inducible. Next, they pared away various amounts of the 5'-flanking region of the gene to see what would affect light inducibility (figure 14.7*a*). As long as at least 35 base pairs of DNA were preserved upstream from the start of transcription, the gene remained light-inducible. However, as soon as the experimenters encroached on this 35-base-pair region, which contains the gene's TATA box as well as a sequence that is conserved in other light-inducible genes, they destroyed light inducibility. This indicates that a factor that carries light's message to the rubisco gene must interact with this small DNA region.

In further, more sophisticated experiments (figure 14.7*b*), Chua introduced the rubisco gene into a whole tobacco plant, forming what we call a **transgenic** plant. In this case, the transplanted gene was expressed just as actively as in the pea, its natural home, and was responsive

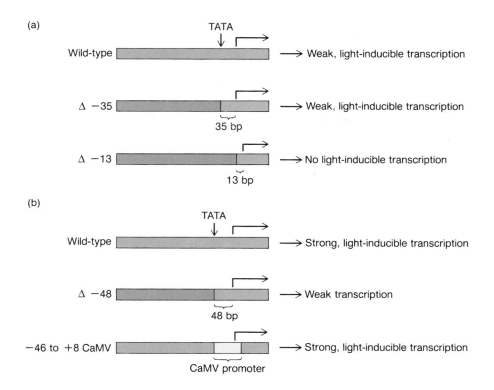

(a)

Wild-type → Weak, light-inducible transcription

Δ −35   35 bp → Weak, light-inducible transcription

Δ −13   13 bp → No light-inducible transcription

(b)

Wild-type → Strong, light-inducible transcription

Δ −48   48 bp → Weak transcription

−46 to +8 CaMV   CaMV promoter → Strong, light-inducible transcription

FIGURE 14.7

Light activation of a plant gene. (*a*) The rubisco gene from the pea was cloned and introduced into petunia cells. The wild-type gene (top) was still light-inducible in its new host, though activity was weak. When all but the thirty-five base pairs upstream from the transcription start site were deleted (middle), the gene was still light-inducible. However, when only thirteen base pairs remained upstream from the transcription start site (bottom), light inducibility was lost. The orange color denotes deleted DNA. (*b*) The cloned rubisco gene was introduced into a whole tobacco plant (see chapter 15 for details).

The wild-type gene (top) was strongly active and light-inducible. When all but the forty-eight base pairs upstream from the transcription start site were deleted (middle), the gene was only weakly active. When the region between −48 and +8 was replaced with a non-light-responsive cauliflower mosaic virus (CaMV) promoter (yellow, bottom), strong, light-inducible transcription was restored. Thus, the upstream sequence contains a light-responsive enhancer.

to light. Moreover, the expression of the gene in tobacco was tissue-specific, being most active in leaves, weakly active in stems, and essentially inactive in roots. This is the same pattern of gene expression observed in the pea.

In contrast to the previous experiment, deletions of the region upstream from the −35 position in the transplanted gene greatly reduced the gene's expression in transgenic plants. When this 5'-flanking region was added back, the gene's activity was restored and it was light-dependent. Further probing revealed that the region from −327 to −48 could be inverted or moved and still exert its effect. Therefore, it appears that this region contains an enhancer that stimulates rubisco transcription.

Is this enhancer only responsible for stimulation of gene transcription, or is it also responsive to light? To answer this question, Chua removed the region from position −46 to +8, previously shown to be light-responsive, and substituted a promoter from cauliflower mosaic virus, which is not light-responsive. He knew that if the gene still responded to light, the region upstream from position −46 must be responsible—and indeed it did respond to light. Apparently this enhancer also helps mediate the stimulation of the rubisco gene by light.

Thus in plants, as well as in animals, the 5'-flanking regions of genes contain sequences that modulate expression of the nearby genes. These regulatory sequences are distinct from promoters and function by interacting with regulatory proteins.

> *T*he gene for the small subunit of the rubisco enzyme is responsive to light. Two regions in the 5'-flank of this gene mediate this light stimulation: (1) a short region immediately upstream from the cap site and (2) an enhancer located further upstream.

THE NATURE OF DETERMINATION

Determination, as we previously defined it, is the process by which a cell makes the irreversible "decision" to follow a certain developmental path. Donald Brown and Robert Roeder and their colleagues have performed detailed studies on the control of transcription of the 5S rRNA genes in the frog *Xenopus laevis*. Brown believes this simple system gives us important insights into the more general phenomenon of determination. First, let us examine the details of the control of the 5S rRNA genes, then we will analyze Brown's hypothesis of determination.

CONTROL OF THE 5S rRNA GENES IN *XENOPUS LAEVIS*

Three proteins cooperate to activate the 5S rRNA genes so they can be transcribed by RNA polymerase III. Roeder named one of these proteins TFIIIA. The haploid number of 5S rRNA genes is quite large, about 20,000, and these genes comprise two different families. The first family includes approximately 98% of the total; since the transcription of these genes occurs only in oocytes, they are called the **oocyte 5S rRNA genes.** The smaller family contains only about four hundred genes, which are transcribed in both oocytes and somatic cells and are therefore called the **somatic 5S rRNA genes.** These observations raise a crucial question: What causes the oocyte genes to be active in oocytes but inactive in somatic cells?

If this system operates the same way as the chicken globin model, we would predict that nucleosomes play a role in gene inactivation. In particular, we would expect the oocyte gene, or at least its internal control region, to be complexed with nucleosomes in somatic cells and therefore repressed, but to be relatively free of nucleosomes in oocytes and therefore active. In fact, Brown's group has found this to be true.

Evidence for this assertion comes from the following in vitro transcription experiments. First, Brown transcribed purified DNA from oocyte and somatic 5S genes and found that RNA polymerase III plus the three transcription factors could transcribe both genes very well. Next, he gently extracted chromatin from oocytes and from somatic cells, and tried to transcribe both in vitro. Here he obtained a much different result: The oocyte 5S gene in oocyte chromatin was active in vitro, but the same gene in somatic cell chromatin was not. Something in the somatic cell chromatin was repressing the oocyte gene.

As we have just suggested, that "something" involves histones and, therefore, nucleosomes. When Brown mixed purified 5S genes with histones and then added transcription factors and RNA polymerase, no transcription occurred. But when he added the factors before the histones, RNA polymerase III could transcribe the gene very well. This suggested a sort of race between factors and histones (figure 14.8). If the factors get to the gene first, they form a stable complex that prevents nucleosomes from forming and repressing the gene. Thus the gene stays on. But if histones win the race, they form nucleosomes that bind stably to the gene's control region and therefore prevent the factors from binding to activate the gene. In that case, the gene stays off.

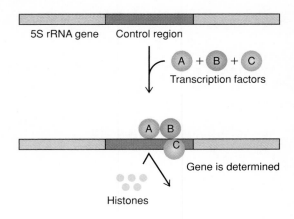

(a) Transcription factors win:

5S rRNA gene   Control region

A + B + C
Transcription factors

Gene is determined

Histones

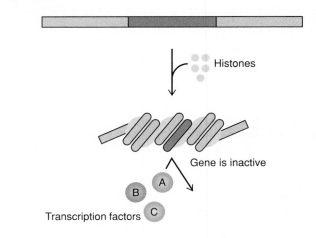

(b) Histones win:

Histones

Gene is inactive

Transcription factors

FIGURE 14.8
The race between transcription factors and histones for the 5S rRNA control region. (*a*) Transcription factors (blue, red, and green) win the race, associate with the control region (purple), and prevent histones from binding; the gene is determined. (*b*) Histones (yellow) win the race, associate with the control region, and tie it up in a nucleosome; the gene is unavailable to transcription factors and remains inactive.

Brown also identified histone H1 as a key player in this mechanism. Any procedure that removed histone H1 from oocyte chromatin activated the repressed gene. How might histone H1 exert such profound influence? We have already discussed (chapter 10) how histone H1 does not itself form part of the nucleosomal core, but cross-links nucleosomes in chromatin. This cross-linking fixes the nucleosomes in place, and nucleosomes lacking histone H1 are therefore presumably free to slide along the DNA. This means the 5S gene's control region will sometimes be uncovered as the histone H1-deficient nucleosomes slide away from it; at those moments, transcription factors can bind, turning the gene on.

The real developmental question is: How do the somatic 5S rRNA genes escape the repression that befalls the oocyte 5S rRNA genes in the same somatic cells? Clues to this mechanism come from a comparison of the activity of oocyte 5S genes in embryonic somatic cells and the concentration of TFIIIA in those same cells. We find that the amount of factor available to each 5S rRNA gene drops dramatically during embryogenesis. When it falls below a threshold level, the oocyte 5S genes become stably repressed, while the somatic 5S genes remain active.

More specifically, we see that immature oocytes contain an excess of factors and make tremendous quantities of oocyte 5S rRNA, which is saved for use during embryogenesis. Later in the oocyte's development, all 5S rRNA synthesis ceases, then switches back on again at a time when the developing embryo, a blastula, contains about four thousand cells. In the meantime, the total concentration of TFIIIA has remained constant, but the number of 5S genes this factor must service has increased 4,000-fold. Still, there is an approximate 10-fold excess of factor to 5S genes, and the oocyte genes remain somewhat active. After three more cell divisions, however, only about one molecule of TFIIIA exists per 5S gene in the embryo, now a gastrula, and the oocyte genes have become stably repressed.

The simplest explanation for this behavior is that there is no longer enough factor to go around and the somatic 5S genes compete successfully with the oocyte genes for the limited supply. Without transcription factors, the oocyte genes are vulnerable to histones, which form nucleosomes, repressing the genes.

Brown believes that the important difference between the somatic and oocyte 5S rRNA genes lies in their relative ability to form stable complexes with transcription factors. The somatic genes form more stable complexes and are therefore able to resist being repressed. By contrast, the oocyte genes form relatively unstable complexes that allow histones to tie up and repress the genes.

We can see that this behavior of the 5S rRNA genes can be generalized to give us a hypothesis of determination: Every time a cell divides, the DNA replication machinery strips all proteins—histones and transcription factors alike—off the DNA. This leaves a clean slate, and the histones and factors then compete for re-binding. If the transcription factors are in low concentration, or if they form a weak complex with the DNA, the histones will tie up the gene in nucleosomes and it will not be determined. On the other hand, if the transcription factors are plentiful, or if they bind tightly to the DNA, they prevent nucleosomes from tying up the gene's control region and the gene is determined. A cell can therefore become determined by increasing its transcription factor concentration and replicating. This hypothesis might explain most determination, but it now appears that repressed genes can sometimes be activated without DNA replication. We do not understand how this happens.

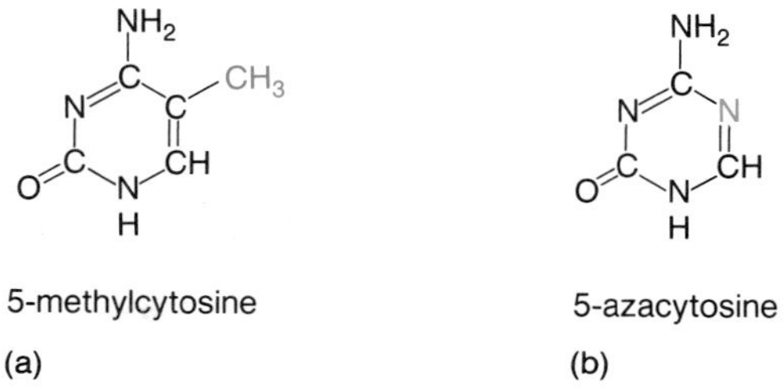

FIGURE 14.9
Structures of (a) 5-methylcytosine and (b) 5-azacytosine.

The two families of 5S rRNA genes in the frog *Xenopus laevis* are oocyte genes and somatic genes. The former are expressed only in oocytes; the latter are turned on in both oocytes and somatic cells. The apparent reason for this difference is that the somatic genes form more stable complexes with transcription factors, including TFIIIA. The transcription factors seem to keep the somatic genes active by preventing nucleosomes from forming a stable complex with the internal control region. This stable complex requires the participation of histone H1. Once it forms, transcription factors are excluded and the gene is repressed. Genes can therefore become determined by binding tightly to transcription factors, and one or more determined genes can determine a cell.

■

THE ROLE OF DNA METHYLATION

Rollin Hotchkiss first discovered methylated DNA in 1948. He found that DNA from certain sources contained, in addition to the standard four bases, a fifth: 5-methylcytosine (figure 14.9a). Watson and Crick ignored this base when they built their DNA model; fortunately, the model is still valid, because the methyl group does not change the base-pairing characteristics of cytosine. It took almost thirty years after Hotchkiss discovered DNA methylation for molecular biologists to begin to find a role for it. That role is in development.

In the mid-1970s, Harold Weintraub and his colleagues noticed that active genes were low in methyl groups, or **undermethylated.** Therefore, a relationship between undermethylation and gene activity seemed likely, as if methylation helped repress genes. This would be a valuable means of keeping genes inactive if methylation is passed on from parent to daughter cells during cell division. And, in fact, it turned out that methylation is inherited from cell to cell: Each parental strand retains its methyl groups, which serve as signals to the methylating apparatus to place methyl groups on the newly made progeny strands (figure 14.10). Notice that the methyl groups are on the cytosines of 5'-CpG-3' pairs—that is, cytosines followed by guanines. This is the general rule for methylation in higher eukaryotes.

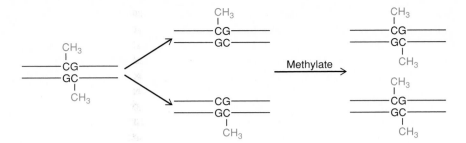

FIGURE 14.10
Methylation persists after DNA replication. After replication, only one methyl group remains on each daughter molecule. This attracts the methylation machinery, which methylates the opposite strand.

Thus, methylation has two of the requirements for a mechanism of determination: (1) It represses gene activity, and (2) it is permanent. Once a gene is turned off by methylation, it remains that way. This is an epigenetic mechanism. Strictly speaking, the DNA is altered, since methyl groups are attached, but because methyl cytosine behaves the same as ordinary cytosine, the genetic coding remains the same. Thus, methylation cannot qualify as a true genetic change. A striking example of such a role for methylation is seen in the inactivation of the X chromosome in female mammals. One of the two X chromosomes in each cell of a female placental mammal is turned off early in development. The inactive X chromosome becomes heterochromatic and appears as a dark fleck under the microscope; this chromosome is said to be **Lyonized,** in honor of Mary Lyon, who first postulated the effect in mice. The dark fleck is called a **Barr body,** for Murray Barr, its discoverer. The Lyonized X chromosome can be either the maternal or paternal one; the selection process is completely random. One consequence of this random inactivation of X chromosomes is the patchy coat pattern in calico cats.

Of course, we can explain the heterochromatization of the X chromosome by the cross-linking of chromatin by histone H1 discussed in chapter 10. But the repression of the genes in this chromosome goes beyond merely covering up with protein; something permanent has happened to the DNA itself. This conclusion comes from the following type of experiment: DNA is taken from the inactive X chromosome and stripped of all protein, then introduced into another cell and checked for activity. It is one hundred times less active than DNA taken from active X chromosomes. Again, this difference cannot be caused by protein, since all protein has been removed. It has to be due to the DNA itself.

An obvious explanation is that the DNA in the Lyonized X chromosome is methylated, whereas the DNA in the active X chromosome is not. To check this hypothesis, Peter Jones and Lawrence Shapiro and their colleagues grew cells in the presence of the drug 5-azacytosine (figure

14.9*b*), which prevents DNA methylation. This reactivated the Lyonized X chromosome. Furthermore, Shapiro showed that these reactivated chromosomes could be transferred to other cells and still remain active. It appears that methylation can be a potent and lasting inhibitor of gene activity, and this has led some to speculate about methylation as a mechanism of determination.

As attractive as this idea may be, it cannot be a general explanation. While 80% of DNA in higher vertebrates is methylated, only 20% of that in lower vertebrates is methylated. The situation in lower eukaryotes is even more depressing for methylation enthusiasts. *Drosophila* seems to lack methylated DNA altogether. We are left with the conclusion that methylation can be a mode of inactivating genes, at least in mammals, but that it is probably a mechanism of recent origin in evolutionary history. As such, its role is most likely supplementary rather than fundamental.

M ethylation of cytosines in CpG sequences in mammals is associated with inactive genes and can indeed turn genes off. The inactivation of methylated genes is perpetuated by methylation of progeny DNA strands across from a methylated parental strand after every round of DNA replication. Methylation can therefore play a role in determination of mammalian cells during embryogenesis; however, it is unlikely that his mechanism is a general one, since lower eukaryotes have little methylated DNA and sometimes none at all.

■

DETERMINATION IN THE FRUIT FLY *DROSOPHILA*

We have discussed some genes that become determined during embryogenesis, but a more important, and more difficult question is this: What are the genes that cause determination? The best clues to this mystery so far have come from one of the standard subjects of genetics research, the fruit fly *Drosophila*. In fact, the genes that are

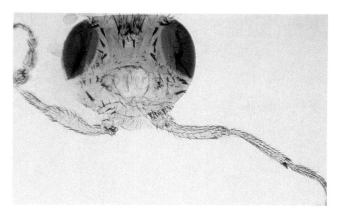

FIGURE 14.11
The *Antennapedia* phenotype. Legs appear on the head where antennae would normally be.
Courtesy of Walter J. Gehring, University of Basel.

causing so much excitement today—called **homoeotic** or, more recently, **homeotic** genes—came to light in mutants in the late nineteenth century. William Bateson coined the term **homoeosis** in 1894 to describe these mutants. He defined the term as a transformation in which "something has been changed into the likeness of something else." That is what seems to happen in a homeotic mutant. A body part appears in a location normally occupied by another part. For example, in *Antennapedia (Antp)* mutants, legs appear on the head in place of antennae (figure 14.11). It seems that antennae have been changed into the likeness of legs, in Bateson's words. Actually, cells that would normally have developed into antennae have been programmed to produce legs instead.

In other *Drosophila* mutants, different transformations occur. In some, wings replace eyes; in others, legs appear where mouth parts ought to be. The drastic transformations seen in homeotic mutants must involve the activity of many genes, yet only one gene is mutated in each case. This suggests that the mutated genes normally control many other genes, which act during embryogenesis to build body parts. The homeotic genes are therefore "master genes" that are involved in determination. Mutating these genes diverts determination to a new path.

Since these homeotic genes play such a central role in development, we would like to know what kinds of products they encode and how they function. The powerful tools of molecular genetics give us a means of attacking this problem, and a start has already been made. Several laboratory groups have cloned homeotic genes and have found that some of them fall into two very large clusters: One of these, the *Antennapedia* complex (ANT-C), spans more than 100 kb of DNA and includes the *Antp* gene, among at least four others. The other is the *Bithorax* complex (BX-C), which controls the segmentation pattern in the fly's thorax and abdomen. Other homeotic genes exist outside these two major complexes.

The Homeo Box

DNA sequencing studies on these homeotic genes have uncovered a short (about 180-base-pair) sequence that most of them have in common. Several other genes that are, strictly speaking, not homeotic also contain these sequences, called **homeo boxes.** Even though these other genes do not share the homeotic genes' dramatic effect of transforming one body part into another, all of them are just as important as homeotic genes in the physical organization of the developing embryo. An example is the *engrailed (en)* gene, which participates in the division of the embryo's segments into anterior and posterior compartments.

Another gene with a homeo box is the *ftz* gene, which maps to the ANT-C region. The ftz stands for *fushi tarazu,* meaning "not enough segments" in Japanese; homozygous *ftz* mutations are lethal, but the mutant embryos live long enough to reveal that they have only half the normal number of segments. Thus, the *ftz* gene apparently helps establish the correct segmentation pattern in the embryo. Although this function is not characteristic of a homeotic gene, *ftz* could be considered homeotic because mutants in *ftz* can also have a homeotic phenotype.

Several features of the homeo box stand out. First, it is very well conserved. The homeo boxes of *ftz, Antp,* and *Ubx* (one of the genes in BX-C) are 75% to 79% conserved, which means that their base sequences correspond in almost eighty out of one hundred positions. The conservation of amino acids is even higher, approaching 90%. This indicates all the homeo box genes share a common function that depends on the homeo box itself. That function seems to be to control other genes so as to orchestrate development. In this discussion, bear in mind that a homeo box makes up only part of a gene, so it encodes only a part, or domain, of the protein encoded by the whole gene.

Another striking feature of the homeo boxes is that they appear to code for DNA-binding domains. Several lines of evidence lead to this conclusion. First, the homeo boxes code for very basic domains (about 30% of the amino acids are basic). Such a basic (positively charged) protein region would have a natural affinity for acidic (negatively charged) DNA. Second, the products of homeo box genes (genes with homeo boxes) migrate to the cell nucleus where the DNA is located; they have been found there with antibodies designed to react with the protein products of homeo boxes. Furthermore, computer modeling of the protein products of the homeo boxes shows a probable secondary structure (the helix-turn-helix motif) very similar to that of known prokaryotic DNA-binding proteins such as CAP, Cro, and λ repressor (chapter 9). Of course, a DNA-binding protein would be strategically placed to influence the activities of other genes, which is precisely the function postulated for the homeo box genes.

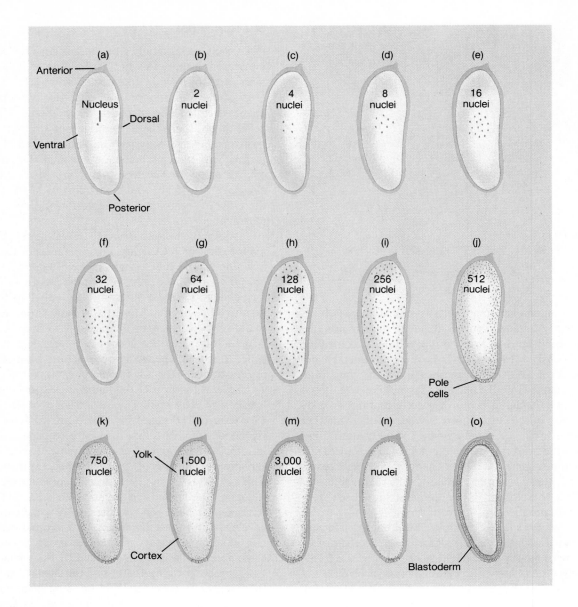

FIGURE 14.12

Early embryogenesis of *Drosophila*. (*a–i*) Nuclei have divided eight times without cell division, producing 256 free nuclei. After the seventh division, the nuclei begin to migrate toward the periphery of the embryo (the cortex, brown). (*j*) A group of nuclei are surrounded by cell membranes and isolated in the posterior end of the embryo. These pole cells will become the germ cells. (*k–n*) The nuclei continue to divide in the cortex. (*o*) When there are about 6,000 nuclei in the cortex, cell membranes form, producing blastoderm cells.

A final line of evidence linking homeo boxes with a DNA-binding function comes from the similarity in sequence between *Drosophila* homeo boxes and regions of the MAT genes of yeast. These MAT genes control the differentiation of yeast cells into one of the two possible mating types (hence the name) or into spores. The MAT gene products effect this differentiation by binding to the 5'-flanking regions of the genes they control. Thus, the MAT genes are another example of "master genes" that control development; they have a homeo box and they code for a DNA-binding protein.

> *H* omeotic genes of *Drosophila* appear to control other genes that determine the nature of segments in the animal. Mutations in these genes cause dramatic transformations of one body part to another. These, and related genes that influence the insect's body plan, frequently contain regions of about 180 base pairs called homeo boxes. The homeo boxes in different *Drosophila* genes are very well conserved and code for protein domains that probably bind to DNA.

■

Expression of Homeo Box Genes During Embryogenesis

We do not yet know the precise functions of the homeo box genes. However, the simplest explanation is that they are master control genes that switch on batteries of other genes whose products specify certain body parts. To understand this, it will help to look at an outline of early *Drosophila* development (figure 14.12). One characteristic of this process worth noting is that the nuclei of the early embryo divide several times without cell division. This continues until the 512-nucleus stage, when membranes form around some of the nuclei in the posterior end of the embryo, giving rise to *pole cells* that will eventually differentiate into the animal's germ cells. The majority of the nuclei are still free at this stage, but they have begun to migrate toward the periphery, or *cortex,* of the embryo. The nuclei go on dividing in the cortex until their number reaches about six thousand. At this point, cell membranes finally form to segregate the nuclei into individual **blastoderm** cells.

Later in embryogenesis (figure 14.13), the embryo extends, and the posterior region, which will develop into the abdomen, folds back on itself. Later still, the embryo shortens again, and overt segmentation occurs. After one day of development, the embryo hatches into a larva that feeds, molts twice, then pupates. The pupa metamorphoses into an adult fly. The external adult structures (legs, wings, eyes, antennae, and the like) derive from groups of larval cells called **imaginal disks** (Latin: *imago* = copy or likeness). One disk gives rise to a leg, another to a wing,

and so forth. The disks are further divided into regions, each responsible for part of an adult structure, such as a leg segment. The imaginal disks find expression as adult structures only when their cells differentiate under the influence of the appropriate hormone. It is important to note that imaginal disks and body segments are not the same structures. A given body segment may contain several imaginal disks.

The genes that contribute to the fruit fly's body plan can be grouped into three classes. First, and most fundamental, are the **maternal effect genes,** which are active in the mother during oogenesis. Thus, only the mother contributes the products of this first class of genes to the egg. These gene products establish the egg's overall spatial characteristics, such as the anterior-posterior axis. The second group is comprised of the **segmentation genes,** which establish the segmentation pattern of the animal. The homeotic genes, responsible for the identity of each segment, comprise the third class of genes.

> *H* omeo box genes are apparently master control genes. They switch on the groups of genes that ultimately determine the insect's body plan. These genes can be divided into three hierarchical groups: the maternal effect genes, the segmentation genes, and the homeotic genes.

■

Maternal Effect Genes

If the maternal effect genes establish polarity in the egg, we might expect to find a polarized gradient of products of these genes: high concentration in one part of the egg changing continuously to low concentration in another. This is precisely what Walter Gehring and his colleagues observed with the maternal effect **bicoid (bcd)** gene. *Bicoid* mutant mothers produce mutant embryos that have only abdomens—no heads or thoraxes. However, by injecting such a mutant with cytoplasm from a wild-type embryo, we can induce an almost complete head and thorax at the site of injection. Cytoplasm from the anterior pole of the donor embryo works best; the activity drops off rapidly as one moves toward the posterior pole. This suggests that the product of the *bcd* gene is responsible for making anterior body parts and is polarized from the anterior to the posterior end of the wild-type embryo.

Markus Noll and his colleagues demonstrated that *bcd* gene products are indeed polarized in *Drosophila* oocytes and early embryos. These researchers constructed a radioactive piece of the *bcd* gene that could be used to hybridize to *bcd* RNA *in situ* (meaning in a whole embryo, its natural situation, rather than in a test tube; Latin: *situs*

FIGURE 14.13

Late embryogenesis of *Drosophila*. (*a*) "Fate map" of the embryo, in which vertical dashed lines delineate the regions of the embryo that will develop into segments. (*b*) The embryo has extended and folded back on itself, with most of the abdominal segments (A4–A8) lying on the dorsal surface. Md, Mx, and Lb will develop into the head, T1–T3 will become thoracic segments, and A1–A8 will form abdominal segments. (*c*) The embryo has shortened again and segments have become obvious. (*d*) Adult fly, showing the ultimate positions of the segments.

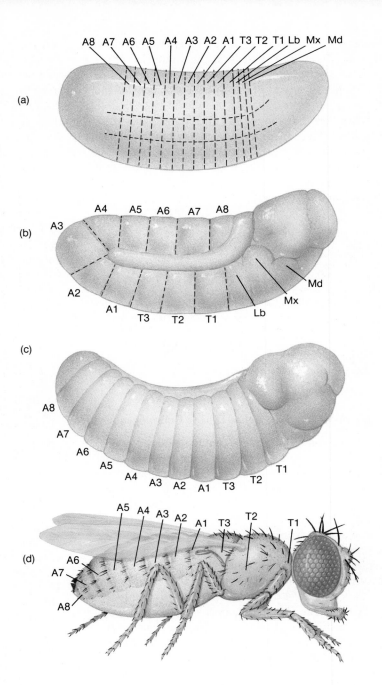

= position or situation). After the radioactive *bcd* probe hybridized to the mRNA in a thin, longitudinal section of the embryo, autoradiography of the section revealed where the radioactivity had gone. Figure 14.14 shows the results: The radioactivity, and therefore the *bcd* mRNA, was found in highest concentration at the anterior end of the oocyte or embryo. We do not yet know how the gradient is established, but it appears that an "anchor" in the anterior part of the egg attracts the maternal *bcd* messages and holds them there.

M aternal effect genes are responsible for the general spatial organization of the fly. For example, *bicoid* is responsible for anterior structures, and its product is distributed in a gradient, with the highest concentration at the anterior end.

■

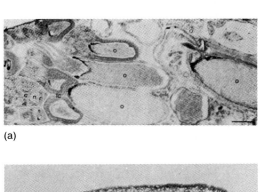
(a)

(b)

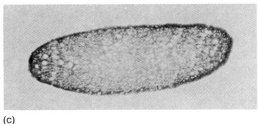
(c)

(d)

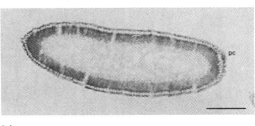
(e)

(f)

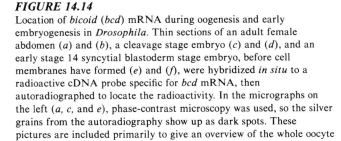

FIGURE 14.14

Location of *bicoid* (*bcd*) mRNA during oogenesis and early embryogenesis in *Drosophila*. Thin sections of an adult female abdomen (*a*) and (*b*), a cleavage stage embryo (*c*) and (*d*), and an early stage 14 syncytial blastoderm stage embryo, before cell membranes have formed (*e*) and (*f*), were hybridized *in situ* to a radioactive cDNA probe specific for *bcd* mRNA, then autoradiographed to locate the radioactivity. In the micrographs on the left (*a, c,* and *e*), phase-contrast microscopy was used, so the silver grains from the autoradiography show up as dark spots. These pictures are included primarily to give an overview of the whole oocyte or embryo. Dark-field illumination was used in the micrographs on the right (*b, d,* and *f*), so the silver grains appear as bright spots on a dark background and are much easier to see. Embryos are oriented in all cases with their anterior ends to the left and their dorsal sides up. In the oocytes, designated "o" in micrograph (*a*), *bcd* mRNA is deposited primarily at the anterior end, as can be seen in (*b*). This anterior localization of *bcd* mRNA persists in the cleavage embryo (*d*) and in the stage 14 embryo (*f*).

Gabriella Frigero, Maya Burri, Daniel Bopp, Stefan Baumgartner, and Markus Noll, Cell 47:740, 1986. Reprinted with permission from Cell Press.

Segmentation Genes

The segmentation genes, as their name implies, direct the organization of the fly into discrete segments. *Fushi tarazu* is a good example. As mentioned previously, *ftz* mutants have only half the normal number of segments; this is because every other segment is missing. Since the studies performed so far on homeo box gene expression have revealed a very satisfying correspondence between the site of gene expression and the phenotype of mutants in that gene, we could predict that the expression of *ftz* in wild-type embryos would occur in every other segment. Gehring performed *in situ* hybridization experiments with a radioactive *ftz* probe to detect *ftz* mRNA; figure 14.15 shows that the radioactivity, and therefore the expression of the *ftz* gene, was confined to seven neat bands in the early wild-type embryo. Moreover, these bands correspond precisely to the parts of the embryo that are missing in *ftz* mutants, which is just what we would expect if expression of this gene is required for those parts to form. (*Fushi tarazu* is just one of a class of segmentation genes called **pair-rule** genes. Pair-rule mutants lack body parts that are roughly segment-sized, and these deletions occur at two-segment intervals. *Engrailed* is another pair-rule gene.)

It is also interesting that the *ftz* gene is only expressed early in embryogenesis, from the early blastoderm stage to gastrulation. In fact, expression begins even before cell

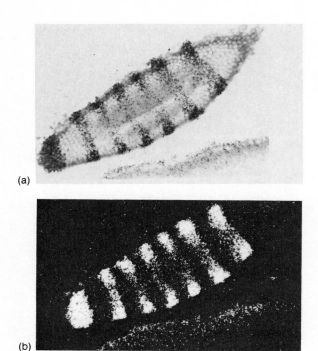

(a)

(b)

FIGURE 14.15
Expression of *fushi tarazu* in *Drosophila*. A radioactive *ftz* cDNA
was hybridized to a thin section of an early embryo, then
autoradiographed to detect *ftz* transcripts. (*a*) Autoradiograph
showing seven bands of dark silver grains that indicate seven zones of
transcription of the *ftz* gene. (*b*) Dark-field micrograph showing the
bands more clearly. The bands do not extend across the embryo
because by this stage the nuclei are all located in the cortex of the
embryo.
From Dr. Walter J. Gehring, University of Basel, Cover of Cell *37:838, July
1984. Reprinted by permission of Cell Press.*

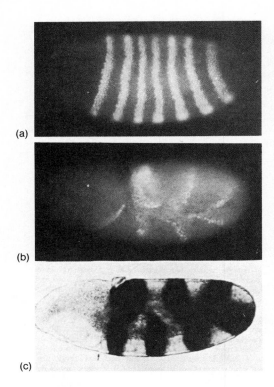

(a)

(b)

(c)

FIGURE 14.16
Activity of the *ftz* promoter in early *Drosophila* embryos. (*a*) A
fluorescent antibody was used to detect the *ftz* gene product in the
typical seven-banded pattern. (*b*) At a later stage, the embryo has
extended, and the posterior abdominal segments are doubled back on
top of the embryo, as shown in figure 14.13*b*. We can still see the
seven bands of *ftz* gene product. (*c*) The *ftz* promoter was fused to the
β-galactosidase coding region and introduced into a fly embryo. Such
a transformed embryo at the same stage as in (*b*) was stained for
β-galactosidase activity. The usual seven-banded pattern was observed,
demonstrating that these seven-band regions contain transcription
factors capable of activating the *ftz* promoter.
Courtesy of Dr. Walter J. Gehring. Figures 5, 6, and 7, Science, *Vol. 236,
pp. 1245–52, 5 June 1987, "Homeo Boxes in the Study of Development."
Copyright 1987 by the AAAS.*

membranes have formed between the embryonic nuclei.
In spite of this, *ftz* expression at this early stage has already assumed a characteristic banded pattern. This
means the *ftz* genes are responding to determinants in the
early embryo that must also be present in a banded pattern. If a nucleus finds itself in a band of high determinant
concentration, it will express the *ftz* gene; but a short distance away is a band of low determinant concentration
where little *ftz* expression will occur.

What are these determinants? Gehring and his coworkers performed an experiment that suggests they are
transcription factors that act on the 5'-flanking region of
the *ftz* gene. Here is how the experiment worked: Gehring
coupled the 5'-flanking region of the *ftz* gene to a "reporter" gene—the coding region of the bacterial *lacZ* gene,
which codes for β-galactosidase. The product of this gene
can be measured easily, since β-galactosidase cleaves a
colorless dye into two parts, one of which is bright blue.
Next, Gehring injected the fused gene into an early *Drosophila* embryo. When the indicator dye was added, blue
bands appeared in the same locations that *ftz* mRNAs

would normally occupy (figure 14.16*c*). This showed that
transcription factors were already arranged in these bands,
poised to activate the *ftz* gene. In this case, however, by
binding to the *ftz* 5'-flanking region in the fused gene, the
factors activated the *lacZ* reporter gene.

> *O*ne of the first segmentation genes to be expressed
> in *Drosophila* is *fushi tarazu*. This gene responds
> to determinants, probably transcription factors, arranged in
> a banded pattern in the embryo even before cell
> membranes have formed around the nuclei. The expression
> of *fushi tarazu* therefore assumes the same banded pattern
> and probably participates in establishing the segmentation
> of the embryo.

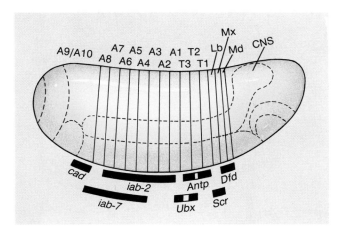

FIGURE 14.17
Patterns of expression of various homeo box genes. The black bars represent the sites on the fate map where each gene is expressed.

Homeotic Genes

The true homeotic gene, *Ubx,* is also expressed only in the segments affected by *Ubx* mutations, although this situation is more complex than those just described. *Ubx* mutations cause obvious transformations in four contiguous compartments of the fruit fly: the posterior compartment of the second thoracic segment (T2p); the anterior and posterior compartments of the third thoracic segment (T3a and T3p); and the anterior compartment of the first abdominal segment (A1a). In addition, these mutations cause more subtle changes of the larval cuticle, or skin, in the anterior compartments of abdominal segments 2–7. We would therefore predict that *Ubx* would be expressed most strongly in T2p–A1a, and weakly in the posterior regions of A2–A7. Using fluorescent antibodies, David Hogness and his coworkers have confirmed this hypothesis.

Similar experiments with other homeotic genes have shown a good correlation between the locus of expression of a gene and the sites affected by mutants in the same gene. Figure 14.17 summarizes these data on a **fate map** of a *Drosophila* embryo at the blastoderm stage. A fate map shows the parts of the embryo that are "fated" to differentiate into each segment of the adult. This particular fate map also shows the parts of the embryo that will express each homeotic gene. We can see that the sites of expression of *Antp* and *Ubx* overlap somewhat, but that *Antp* will generally be expressed more anteriorly than *Ubx.* This is satisfying, because recessive loss-of-function mutants in *Antp* affect more anterior segments than do similar mutations in *Ubx.*

Furthermore, inactivation of the *Ubx* gene leads to the turn-on of *Antp* in segments where *Ubx,* not *Antp,* would normally be expressed. This suggests that one of the functions of *Ubx* is to keep *Antp* turned off. Other studies suggest that *Ubx* is repressed by *iab-2,* the homeotic gene expressed just posterior to *Ubx.* These findings lead to the hypothesis that there is a sequential pattern of control among the homeotic genes, with the posteriorly expressed genes repressing the more anteriorly expressed ones.

*T*he true homeotic genes, such as *Antennapedia,* switch on after the segmentation genes and determine the nature of each segment. Although we do not understand the details of these genes' operations, one of their functions seems to be to repress the homeotic gene that operates in the neighboring anterior section of the embryo.

∎

A Model for Genetic Control of *Drosophila* Development

The phenomenon of control of one gene by another probably extends to all the homeo box genes, not just the homeotic genes. In fact, we can envision the following model: The maternal effect genes, such as *bcd,* are expressed very early, beginning during oogenesis, and they give products that become nonuniformly distributed in the embryo. This distribution reflects the function of these genes: to establish the general structure of the embryo. The segmentation genes, such as *ftz,* are expressed later, during early embryogenesis, and help establish the segments of the embryo. These genes may well respond to signals put in place by the earlier expressed maternal effect genes. The homeotic genes, such as *Ubx* and *Antp,* are expressed later still and determine the identities of the segments they control. These genes could in turn be controlled by products of the segmentation genes. Thus, we see a hierarchy of homeo box genes: Those that are expressed first accomplish the most fundamental developmental steps in embryogenesis, while the more specialized genes are turned on later and control finer steps.

VERTEBRATE HOMEO BOXES

Homeo box genes clearly play a crucial role in *Drosophila* development, but what relevance do they have for higher organisms? We have no definitive answer to this question yet, but it is likely that similar genes play similar roles in vertebrates, including humans. The strongest evidence for this hypothesis is the very great similarity in sequence between homeo boxes in *Drosophila* and in vertebrates. For example, there is 95% similarity between the amino acid sequences encoded in homeo boxes in the mouse and in the *Antp* gene of the fruit fly. The similarity extends for

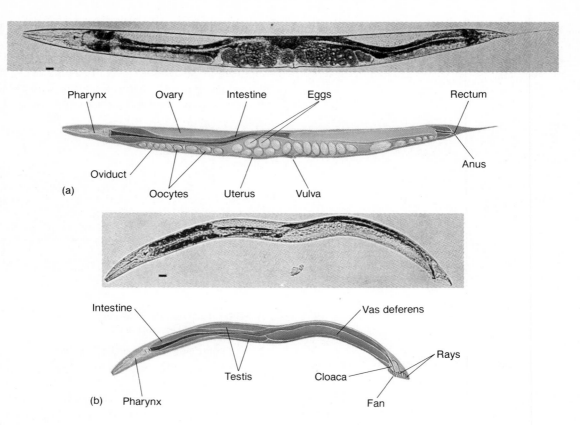

Pharynx   Ovary   Intestine   Eggs   Rectum

Oviduct

Oocytes   Uterus   Vulva   Anus

(a)

Intestine   Vas deferens

Rays

Testis   Cloaca

(b)   Pharynx   Fan

FIGURE 14.18

Caenorhabditis elegans. (*a*) Self-fertilizing hermaphrodite; (*b*) male.

From J. E. Sulston and H. R. Horvitz, Developmental Biology *56:111, 1977.
Copyright © 1977 Academic Press, Orlando, FL. Reprinted by permission of
the author and the publisher.*

a considerable distance outside the homeo box itself, especially toward the 3′-end of the gene. This is a startling likeness between genes in two such different species, and it suggests an important function for these genes. This function is apparently so vital that changes in the genes are lethal and are therefore not perpetuated.

What can this vital function be? It is tempting to guess that it is the same as the apparent function of the homeo box genes in fruit flies: determination, especially determination of segments or compartments of the animal. If this were so, we would expect to see expression of mammalian homeo box genes at the time of determination of a tissue, before overt differentiation begins. But early results with cells that differentiate in culture suggest that these cells' homeo box genes are expressed during and after differentiation. Perhaps homeo box genes play a different role in mammalian development, but it is hardly plausible that they are not related to mammalian development at all.

DETERMINATION IN THE ROUNDWORM

The roundworm **Caenorhabditis elegans** is an especially attractive model for studying development for several reasons. First, although the adult worm has well-defined organ systems, it contains a total of only 959 cells, a third of which are nerve cells. This is approximately the square root of the number of cells in *Drosophila,* which is in turn roughly the square root of the number of cells in a human (several trillion). Obviously, tracing the development of fewer than one thousand cells is incalculably simpler than doing the same with trillions of cells. The job is simplified further by the fact that *C. elegans* is transparent (figure 14.18), so the cells can be seen in the intact animal. Moreover, *C. elegans* is hermaphroditic, or self-fertilizing. This greatly simplifies genetic experiments because a mutant worm can be "selfed," or mated with itself, like a pea plant. Nevertheless, mapping all those cells was an overwhelming task, requiring years of attention by several research groups. The result is that we now know the fate of every cell at every stage in this animal's development. Conversely, we can trace the development of every one of the adult's 959 cells back to the zygote that started it all. We also know about hundreds of *C. elegans* mutations, some of which affect the animal's development.

One would have expected this kind of detailed knowledge about an organism's development to shed a great deal of light on the genetic processes underlying that development, especially if a genetic program of development really exists. But in fact, although there have been some interesting insights, no really incisive simplifications have resulted. Sydney Brenner, one of the pioneers in *C. elegans* research, believes this is because development is not written in a simple genetic program the way the sequence of a protein is written in a gene. In other words, even though we are beginning to understand how genes themselves are controlled, we may be no closer to understanding how these genes in turn control development. This is because genes do not operate in a neat, linear sequence during development, but in concert; thus the contribution of each individual gene cannot be understood except in the context of all the other genes' contributions.

Brenner illustrates this point with a simple example: "The icosahedral head of a bacteriophage is a precise geometric object. This geometry is inherited. So you say, there must be a definition of an icosahedron in the genome somewhere: Where is it? You won't find a gene that says, 'Make an icosahedron.' The icosahedron is encoded in a distributed form throughout the entire genome. That's a pretty good picture of what we have to say when we ask, what does it take to make a hand, make a foot, make a liver. The specifications for these structures are scattered throughout the genome." Brenner concludes that we may not understand the development of even a simple animal like *C. elegans* until we know everything about it. That is a rather depressing prospect, but Brenner concludes on an optimistic note: "Molecular biology is the art of the inevitable. If you do it, it's inevitable you will find out how it works—in the end" (in R. Lewin, *Science* 224: 1327 & 1329, 1984).

One of the interesting findings to come out of the *C. elegans* work so far is the identification of homeotic genes. Certain mutations in one of these, called *lin-12*, cause failure of vulva development in the hermaphroditic (self-fertilizing) form of the worm. Robert Horvitz and his colleagues have traced the reason for this failure to the lack of a gonadal anchor cell that would normally induce the development of the vulva. By consulting the developmental fate map of *C. elegans*, we can see what went wrong. One cell (Z1.ppp) would normally develop into the anchor cell, and another (Z1.aaa) would become a uterine

precursor cell. But the *lin-12* mutants we are discussing develop either two anchor cells or two uterine precursor cells, not one of each.

Mutations that cause hyperactivity of the *lin-12* gene give rise to two uterine precursor cells; mutations that cause depression of *lin-12* activity result in two anchor cells. This suggests that the *lin-12* gene is like a two-way switch. If it is on, a cell goes one way; if it is off, the cell follows the other path. We would therefore predict that *lin-12* will be expressed strongly in uterine precursor cells but weakly in anchor cells at the time of determination. Furthermore, *lin-12* affects the development of other pairs of cells that follow quite different fates from those we have just considered. Thus, the switching activity of *lin-12* is not confined to just one branch point, but is more general.

> *T*he entire fate map of the roundworm *C. elegans* has been worked out so that the lineage of every cell in the adult is known. This has revealed the existence of homeotic genes whose malfunction can change the fate of one cell to that of another. One of these genes is *lin-12*. Unfortunately, this elegant system has so far given us few underlying, simplifying principles.

■

GENE REARRANGEMENT DURING DEVELOPMENT

We have been assuming that an organism's genome remains constant during development as gene expression changes. This is generally true, but there are conspicuous exceptions.

REARRANGEMENT OF IMMUNOGLOBULIN GENES

One example of rearrangement involves the mammalian genes that produce **antibodies,** or **immunoglobulins,** the proteins that help us fight infections. As mentioned in chapter 7, an antibody is composed of four polypeptides: two heavy chains and two light chains. Figure 14.19 illustrates an antibody schematically and shows the sites that combine with an invading antigen. These sites, called **variable regions,** vary from one antibody to the next and give these proteins their specificities; the rest of the molecule (the **constant region**) does not vary from one antibody to the next within an antibody class, though there is some variation between the few classes of antibodies. Any given immune cell can only make antibody with one kind of specificity. Remarkably enough, humans have immune

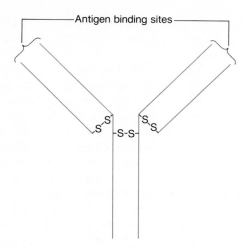

FIGURE 14.19
The antibody is composed of two light chains (blue) bound through disulfide bridges to two heavy chains (red), which are themselves held together by a disulfide bridge. The antigen binding sites are at the ends of the protein chains, where the variable regions lie.

cells capable of producing antibody to react with virtually any foreign substance we would ever encounter. That means we can make literally millions, perhaps even billions, of different antibodies.

Does this imply that we have millions or billions of different antibody genes? That is an untenable hypothesis; it would place an impossible burden on our genomes to carry all the necessary genes. So how do we solve the antibody diversity problem? As unlikely as it may seem, a cell that is destined to make antibody rearranges its genome to bring together separate parts of its antibody genes. The machinery that puts together the gene selects these parts at random from heterogeneous groups of parts, rather like ordering from a Chinese menu ("Choose one from column A and one from column B"). This arrangement greatly increases the variability of the genes. For instance, if there are 200 possibilities in "column A" and 5 in "column B," the total number of combinations of A + B is 200 × 5, or 1,000. Thus, from 205 gene fragments, we can assemble 1,000 genes. And this is just for one of the antibody polypeptides. If a similar situation exists for the other, the total number of antibodies will be 1,000 × 1,000, or one million. This description, though correct in principle, is actually an oversimplification of the situation in the antibody genes; as we will see, they have somewhat more complex mechanisms for introducing diversity, which lead to an even greater number of possible antibody products.

Antibody genes have been studied most thoroughly in the mouse because of very handy mouse tumors called myelomas. A myeloma is a clone of malignant immune cells that overproduce one kind of antibody. Studies on the antibodies made by these myelomas have revealed two

families of mouse antibody light chains called kappa (κ) and lambda (λ). Figure 14.20a illustrates the arrangement of the gene parts for a κ light chain. "Column A" of this Chinese menu contains 90–300 variable region parts (V); "Column B" contains 5 **joining region** parts (J). The J segments actually encode the last thirteen amino acids of the variable region, but they are located far away from the rest of the V region and close to a single constant region part. This is the situation in the animal's germ cells, before the antibody-producing cells differentiate and before rearrangement brings the two unlinked regions together. The rearrangement and expression events have been worked out by several groups, including those led by Susumo Tonegawa, Philip Leder, and Leroy Hood; these events are depicted in figure 14.20b.

First, a recombination event brings one of the V regions together with one of the J regions. In this case, V_3 and J_3 fuse together, but it could just as easily have been V_1 and J_4; the selection is random. After the two parts of the gene assemble, transcription occurs, starting at the beginning of V_2 and continuing until the end of C. Next, the splicing machinery joins the J_3 region of the transcript to C, removing the extra J regions and the intervening sequence between the J regions and C. It is important to remember that the rearrangement step took place at the DNA level, but this splicing step occurs at the RNA level by mechanisms we studied in chapter 10. The messenger RNA thus assembled moves into the cytoplasm to be translated into an antibody light chain with a variable region (encoded in both V and J) and a constant region (encoded in C).

Why does transcription begin at the beginning of V_3 and not further upstream? The answer seems to be that an enhancer in the intron between the J regions and the C region activates the promoter closest to it: the V_3 promoter in this case. This also provides a convenient way of activating the gene after it rearranges; only then is the enhancer close enough to turn on the promoter.

The rearrangement of the heavy chain gene is even more complex, because there is an extra set of gene parts in between the V's and J's. These gene fragments are called D, for "diversity," and they represent a third column on our Chinese menu. Figure 14.21 shows that the heavy chain is assembled from 100 to 200 V regions, approximately 20 D regions, and 4 J regions. On this basis alone, we could put together up to 200 × 20 × 4, or 16,000 different heavy chain genes. Furthermore, 16,000 heavy chains combined with 1,000 light chains yield more than ten million different antibodies.

But there are even more sources of diversity. The first derives from the fact that the mechanism joining V, D, and J segments is not precise. It can delete a few bases on either side of the joining site, or even add bases that are not found in the original segments. This leads to extra differences in antibodies' amino acid sequences.

(a)

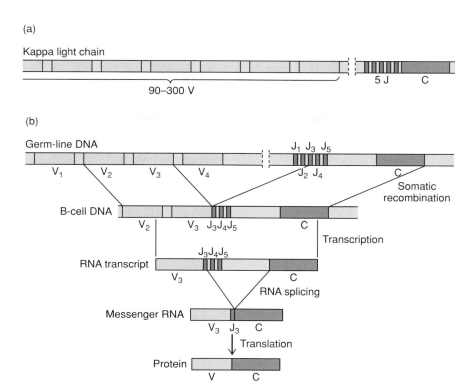

Kappa light chain

90–300 V   5 J   C

(b)

Germ-line DNA

V₁   V₂   V₃   V₄   J₁ J₃ J₅   C
                        J₂ J₄

Somatic recombination

B-cell DNA

V₂   V₃ J₃J₄J₅   C

Transcription

RNA transcript

J₃J₄J₅
V₃   C

RNA splicing

Messenger RNA

V₃ J₃ C

Translation

Protein

V   C

FIGURE 14.20

Rearrangement of an antibody light chain gene. (*a*) The kappa antibody light chain is encoded in 90–300 variable gene segments (V; light green), five joining segments (J; orange), and one constant segment (C; blue). (*b*) During maturation of an antibody-producing cell, a DNA segment is deleted, bringing a V segment (V₃, in this case) together with a J segment (J₃, in this case). The gene can now be transcribed to produce the mRNA precursor shown here, with extra J segments and intervening sequences. The material between J₃ and C is then spliced out, yielding the mature mRNA, which is translated to the antibody protein shown at the bottom. The J segment of the mRNA is translated into part of the variable region of the antibody.

Heavy chain

100–200 V   20 D   4 J   C

FIGURE 14.21

Structure of an antibody heavy chain gene. The heavy chain is encoded in 100–200 variable segments (V; yellow), 20 diversity segments (D; green), four joining segments (J; red), and one constant segment (C; blue).

Another source of antibody diversity is *somatic mutation,* or mutation in an organism's somatic (nonsex) cells. In this case, the mutations occur in antibody genes, probably at the time that a clone of antibody-producing cells proliferates to meet the challenge of an invader. This offers an even wider variety of antibodies and, therefore, a better response to the challenge. Assuming that the extra diversity introduced by imprecise joining of antibody gene segments and somatic mutation is a factor of about 100, this brings the total diversity to one hundred times ten million, or about one billion, surely enough different antibodies to match any attacker.

*T*he immune systems of vertebrates can produce millions, if not billions, of different antibodies to react with virtually any foreign substance. These immune systems generate such enormous diversity by three basic mechanisms: (1) assembling genes for antibody light chains and heavy chains from two or three component parts, respectively, each part selected from heterogeneous pools of parts; (2) joining the gene parts by an imprecise mechanism that can delete bases or even add extra bases, thus changing the gene; and (3) causing a high rate of somatic mutations, probably during proliferation of a clone of immune cells, thus creating slightly different genes. ∎

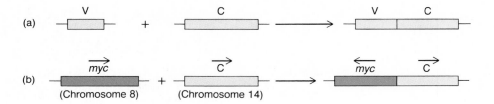

FIGURE 14.22

The 8;14 translocation in Burkitt's lymphoma as a corruption of a normal antibody gene rearrangement. (*a*) Normal antibody gene rearrangement. The variable and constant regions (V and C, respectively) start out as separate entities; recombination brings them together as a functioning antibody heavy chain gene. Introns have been omitted throughout this figure for simplicity. (*b*) Abnormal rearrangement involving *myc* from chromosome 8 and an antibody heavy chain constant region from chromosome 14. The *myc* gene, minus its first exon, becomes attached to part of the antibody constant region gene, with the two gene fragments in opposite orientations.

REARRANGEMENT OF AN ONCOGENE

As we have just seen, antibody genes normally undergo a translocation in which a variable region is juxtaposed with a constant region. In certain cells, this process seems to be corrupted, resulting in cancer. For example, in Burkitt's lymphoma cells, an 8;14 translocation occurs (figure 14.22). However, instead of a variable region, it is the *myc* oncogene that winds up next to an H-chain constant region. In the process, two things happen to the *myc* gene that may disturb its normal function. First, the gene loses the first of its three exons. This has no effect on the protein product of the gene, since the first exon does not encode any amino acids, but it may affect the regulation of the gene. Second, the translocated gene has moved close to the very active antibody H-chain genes. This could have an activating effect on *myc*, which is indeed frequently overactive in Burkitt's lymphoma cells.

Thus, the *myc* story provides an example, not only of altering a proto-oncogene by translocation, but of activating an oncogene. And that is not all. The translocated *myc* gene may also escape from the mechanisms that normally keep it under control. Which of these factors is important in the transformation process? We do not know the answer yet; it may be that all are important.

*S*ome proto-oncogenes suffer translocations that convert them to oncogenes. In Burkitt's lymphoma, the *myc* oncogene is translocated next to a cluster of antibody H-chain genes. In the process, it loses its first (noncoding) exon, becomes activated, and loses its normal control pattern.

■

REARRANGEMENT OF TRYPANOSOME COAT GENES

Do the mechanisms for introducing diversity during development of the immune system apply to development in general? Probably not. In most cases, a gene in the germ cells is the same as the corresponding one in somatic cells in the adult organism. However, changes during an animal's lifetime are not unique to antibody genes. Another well-known example is the rearrangement of surface protein genes during the life of the **trypanosomes,** the protozoa that cause African sleeping sickness. These parasites have protein coats that are easily recognized as foreign, yet they manage to evade human immune defenses by changing the nature of their coat proteins just as our bodies are mounting an effective response to the previous coat. The change in coat comes from switching on new coat protein genes, but in an unorthodox way: The genome rearranges periodically to bring new coat protein genes into an **expression site** at the end of a chromosome where they can be transcribed efficiently. This is rather like taking a light bulb out of a dead socket and plugging it into a live one.

SEQUENTIAL ACTIVATION OF MEMBERS OF A GENE FAMILY

Most eukaryotic genes, even those we call unique, or single-copy genes, are actually represented by a family of closely related DNA sequences. Such a family is usually modest in size. For example, the globin genes in vertebrates exist in two families, one for the α-like globins and one for the β-like globins. The members of each of these families are arranged together in **gene clusters.** In the human, the α-globin gene cluster is located on chromosome 16 and contains five members. The β-globin cluster, with six members, lies on chromosome 11. Both of these clusters

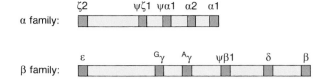

α family: ζ2   ψζ1 ψα1 α2 α1

β family: ε   Gγ Aγ ψβ1 δ β

FIGURE 14.23

The human globin gene families. The α gene family, arranged on the genome in order of expression, consists of a ζ gene, two pseudogenes, and two α genes. The pseudogenes are not expressed. The β family, also arranged in order of expression, consists of an ε gene, two γ genes, a β pseudogene, a δ gene, and a β gene. Again, the pseudogene is not expressed.

probably arose by repeated duplication of a single ancestral globin gene. After duplication, the extra gene copies were free to mutate and take on somewhat different functions. At present, we are interested in the way the cell activates these genes in turn during development.

Figure 14.23 shows the structures of the α- and β-globin gene clusters. All the genes are transcribed from left to right, as drawn here. On the far left of the α cluster is a zeta gene (ζ), which is active only very early in embryogenesis. This is followed by two **pseudogenes,** which have collected mutations that render them nonfunctional. Finally, on the right, lie two α genes that code for exactly the same protein. Such genes are obviously not alleles, since they exist on the same chromosome; they are called *nonallelic copies.* One or both of these genes are active from embryonic through adult life.

The β-globin genes are subject to even finer control. On the left is an epsilon gene (ε) that turns on only during embryonic life. Next come two gamma genes (γ) whose products differ from each other in only one amino acid. These genes are active from embryonic through fetal life. A pseudogene comes next, followed by the delta (δ)- and β-globin genes, which are active only during adult life (β being about fifty times more active than δ).

What is the developmental advantage of this gene switching? Remember that the purpose of hemoglobin is to carry oxygen through the bloodstream to the various tissues of the body. If an embryo or fetus's hemoglobin had the same affinity for oxygen as its mother's hemoglobin, it would not capture enough oxygen to support life. Fortunately, embryonic and fetal hemoglobins have a higher affinity for oxygen than does adult hemoglobin, so they can accept oxygen from the mother's blood in the placenta.

T he globin genes exist in two clusters on separate chromosomes. During embryonic life, the embryonic and adult α-like genes (ζ and α), as well as the embryonic and fetal β-like genes, are active; during the fetal stage, the fetal β-like genes (γ) and the adult α gene are active; finally, just before birth, expression switches over almost exclusively to the adult genes (α and β plus δ).

■

Summary

M ulticellular organisms with specialized tissues develop from a single cell, or zygote, by a process called embryogenesis. During embryogenesis, cells differentiate to produce tissues with specialized functions. Differentiation seems to require little if any change in a cell's genetic makeup, since differentiated cells or nuclei from some plants and animals are still totipotent. Instead, the pattern of expression of a common set of genes varies from one kind of differentiated cell to another. Even before they differentiate, cells become determined, or irreversibly committed to a given fate.

Cell determination in turn depends on the determination of one or more genes. Brown's model of gene determination invokes tissue-specific transcription factors that associate with the control region of a gene and keep it free of nucleosomes. This allows RNA polymerase and any other effector molecules, such as hormones and their receptors, to locate and transcribe the gene, completing the differentiation process. Methylation of cytosines in CpG sequences in mammals is associated with inactive genes and represents another potential mechanism for controlling determination, at least in mammals.

Homeotic genes of *Drosophila* appear to control other genes that determine the nature of segments in the animal. Mutations in these genes cause dramatic transformations of one body part to another. These, and related genes that influence the body plan of the insect, frequently contain regions of about 180 base pairs called homeo boxes. The homeo boxes in *Drosophila* genes are very well conserved and code for protein domains that probably bind to DNA. Thus, these homeo box proteins are well equipped to bind to other genes and to control their activity.

The immune systems of vertebrates can produce millions, if not billions, of different antibodies to react with virtually any foreign substance. These immune systems generate this enormous diversity by three basic mechanisms: (1) assembling genes for antibody light chains and heavy chains from two to three component parts, respectively, each part selected from heterogeneous pools of parts; (2) joining the gene parts by an imprecise mechanism that can delete bases or even add extra bases, thus changing the gene; (3) causing a high rate of somatic mutations, probably during proliferation of a clone of immune cells, thus creating slightly different genes.

■

1. In what sense is the vertebrate hemoglobin gene a "luxury" to the red blood cell? Why is it not a luxury to the whole organism?

2. Is the chicken ovalbumin gene determined, differentiated, or both in the following instances? (*a*) When a young chick's oviduct cells are competent to produce ovalbumin in response to estrogen, but the chick does not yet make estrogen. (*b*) After the chicken matures and begins producing estrogen and ovalbumin.

3. Why does the existence of identical human quintuplets prove that human embryonic cells remain totipotent at least through three cell divisions?

4. Based on the results of experiments with the 5S rRNA gene, what would happen to activity and DNase hypersensitivity if the chicken β^A gene were mixed first with histones, then with extract from fourteen-day embryos?

5. The experiment cited in figure 14.7 examined the location of the control region of the rubisco gene that is light-sensitive. How does having a cloned gene greatly simplify this kind of experiment?

6. In the experiment cited in figure 14.7, why was it important to move the mutated rubisco gene to a new species, petunia, rather than to reinsert it into a pea plant?

7. If the cloned *Xenopus* 5S rRNA gene is mixed with histones H2A, H2B, H3, and H4, and finally with transcription factors, would you predict that the gene would be active or inactive. Why?

8. Why do you think it is important for an early embryo to have a good supply of 5S rRNA?

9. (*a*) Which genes are determined in the liver of an immature chick that has not begun producing estrogen—the ovalbumin gene and/or the vitellogenin gene? (*b*) Which genes are differentiated?

10. In what kinds of *Drosophila* genes are homeo boxes found? How do you think this is related to the fact that a homeo box encodes a DNA-binding domain of a protein?

11. How does an *in situ* hybridization experiment show us where a particular gene is expressed in a *Drosophila* embryo?

12. How did the *ftz* promoter-β-galactosidase construct help Gehring demonstrate the presence of bands of transcription factors in the *Drosophila* embryo?

13. How do mutations in homeotic genes and other homeo box genes show that these genes are important in development?

Answers appear at end of book.

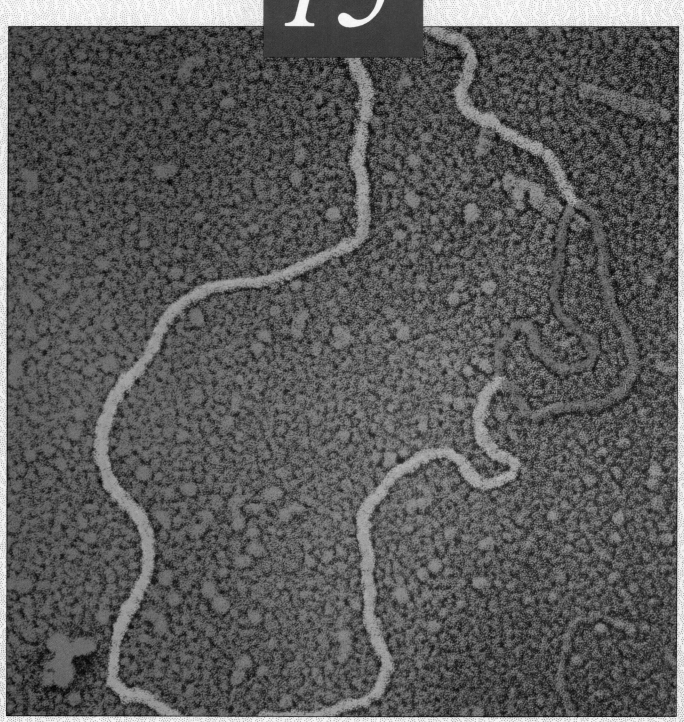

Gene Cloning and Manipulation

*T*he heart of the most recent revolution in biology has been the development of the technology of genetics. Its achievements have simply changed what biologists do and, perhaps even more important, the way they think.

■

Philip Leder
American geneticist

Learning Objectives

In this chapter, you will learn:

1. How we can produce pure genes in virtually limitless quantities by cloning them in bacteria or higher organisms.

2. How we can obtain the products of cloned genes.

3. How we can use cloned genes as probes to investigate the structure and function of genes in their natural environment.

4. How we can obtain the sequence of bases in a cloned gene.

5. How cloned genes provide powerful tools for improving medicine and agriculture.

ntil the mid-1970s, molecular genetics was practiced only on bacteria. The reason: Only bacteria yielded purified genes in quantities large enough to study. In fact, the bacteria themselves would not have been nearly as useful without the phages that infected them. By linking bacterial genes to phage genomes, geneticists could produce significant quantities of these genes for detailed study. No such systems were available for eukaryotes.

Why was this a handicap? Imagine that you are a geneticist in the year 1972. You want to investigate the function of eukaryotic genes at the molecular level. In particular, you are curious about the molecular structure and function of the human globin genes: What are the base sequences of these genes? What do their promoters look like? How does RNA polymerase interact with these genes? What changes occur in these genes to cause diseases like thalassemia, where one of the globin polypeptides is not produced at all?

These questions cannot be answered unless you can purify enough of your gene to study—probably about a milligram's worth. A milligram does not sound like much, but it is an overwhelming amount when you imagine purifying it from whole human DNA. Consider that the DNA involved in one globin gene is less than one part per million in the human genome. This means you will need kilograms of human DNA to start with, which is a very depressing prospect. And even if you could round up that much material somehow, you would not know how to separate the one gene you are interested in from all the rest of the DNA.

Gene cloning neatly solves these problems. By linking eukaryotic genes to small bacterial or phage DNAs and inserting these recombinant molecules into bacterial hosts, we can put the molecular genetics of eukaryotes on an equal footing with that of bacteria. In this chapter we will see how to clone genes in bacteria and in eukaryotes. We will also learn the basic manipulations molecular geneticists use to discover their cloned genes' secrets.

GENE CLONING

The purpose of any cloning experiment is to produce a **clone,** a group of identical organisms. In chapter 14 we saw that some plants can be cloned simply by taking cuttings (Greek: *klon* = twig), and that others can be cloned by growing whole plants from single cells collected from one plant. Even vertebrates can be cloned. John Gurdon produced clones of identical frogs by transplanting many nuclei from one frog embryo to many enucleate eggs (chapter 14). Identical twins also constitute a clone.

The procedure we usually follow in a gene cloning experiment is to place a foreign gene into a bacterial cell, then to grow a clone of these modified bacteria, with each cell in the clone containing the foreign gene. Thus, as long as we ensure that the foreign gene can replicate, we can clone the gene by cloning its bacterial host. Stanley Cohen, Herbert Boyer, and their colleagues performed the first cloning experiment in 1973.

THE ROLE OF RESTRICTION ENDONUCLEASES

Cohen and Boyer's elegant plan depended on invaluable enzymes called **restriction endonucleases.** Stewart Linn and Werner Arber discovered restriction endonucleases in *E. coli* in the late 1960s. These enzymes get their name from the fact that they "restrict" invasion by foreign DNA, such as viral DNA, by cutting it up. Furthermore, they cut at sites within the foreign DNA, rather than chewing it away from the ends, so we call them endonucleases (Greek: *endo* = within), rather than exonucleases (Greek: *exo* = outside). Linn and Arber hoped that their enzymes would cut DNA at specific sites, giving them finely honed molecular knives with which to slice DNA. Unfortunately, these particular enzymes did not fulfill that hope.

However, an enzyme from *Haemophilus influenzae,* discovered by Hamilton Smith, did show specificity. This enzyme is called HindII (pronounced Hin-dee-2) because it comes from *H. influenzae* strain R_d. All restriction enzymes derive the first three letters of their names from the Latin name of the microorganism that produces them. The first letter is the first letter of the genus and the next two letters are the first two letters of the species (hence: *Haemophilus influenzae* yields Hin). In addition, the strain designation is sometimes included; in this case, the "d" from R_d is used. Finally, if the strain of microorganism produces just one restriction enzyme, the name ends with the Roman numeral I. If more than one enzyme is produced, they are numbered I, II, III, etc.

HindII recognizes this sequence:

and cuts in the middle as shown by the arrows. Py stands for either of the pyrimidines (T or C), and Pu stands for either purine (A or G). Wherever this sequence occurs, and *only* when this sequence occurs, HindII will make a cut. Happily for molecular geneticists, HindII turned out

TABLE 15.1	*Recognition Sequences and Cutting Sites of Selected Restriction Enzymes*
Enzyme	**Recognition Sequence**
AluI	A G ↓C T
BamHI	G↓G A T C C
BglII	A↓G A T C T
ClaI	A T↓C G A T
EcoRI	G↓A A T T C
HaeIII	G G↓C C
HindII	G T Py↓Pu A C
HindIII	A↓A G C T T
HpaII	C↓C G G
KpnI	G G T A C↓C
MboI	↓G A T C
PstI	C T G C A↓G
PvuI	C G A T↓C G
SalI	G↓T C G A C
SmaI	C C C↓G G G
XmaI	C↓C C G G G

to be only one of hundreds of restriction enzymes, each with its own specific recognition site. Table 15.1 lists the sources and cutting sites for several popular restriction enzymes. Note that some of these enzymes recognize four-base sequences instead of the more common six-base sequences. As a result, they cut much more frequently. This is because a given sequence of four bases will occur about once in every $4^4 = 256$ bases, while a sequence of six bases will occur only about once in every $4^6 = 4,096$ bases. Notice also the three possible positions of the cutting sites, indicated by the arrows on the table: in the middle of the sequence, therefore leaving blunt ends (AluI, SmaI); at the 5′-end, leaving protruding 5′-ends (BamHI, HpaII); or at the 3′-end, leaving protruding 3′-ends (PstI, PvuI). Finally, notice that the recognition sequences for SmaI and XmaI are identical, although the cutting sites are different. We call such enzymes that recognize identical sequences **isoschizomers** (Greek: *iso* = equal; *schizo* = split).

The main advantage of restriction enzymes is their ability to cut a DNA reproducibly in the same places. This property is the basis of many techniques used to analyze genes and their expression. We have seen examples of some of these in previous chapters; in this chapter we will see several more. But this is not the only advantage. Many restriction enzymes make staggered cuts in the two DNA strands (table 15.1), leaving single-stranded overhangs, or

sticky ends, that can base-pair together briefly. Note, for example, the complementarity between the ends created by EcoRI (pronounced Eeko R-1 or Echo R-1):

$$\begin{array}{l}5'\text{---GAATTC---}3' \\ 3'\text{--CTTAAG---}5'\end{array} \mapsto \begin{array}{l}\text{---G} \\ \text{---CTTAA}\end{array} + \begin{array}{l}\text{AATTC---} \\ \text{G---}\end{array}$$

Restriction enzymes can make staggered cuts because the sites they recognize are usually symmetrical. That is, they read the same forwards and backwards. Thus, EcoRI cuts between the G and the A in the top strand (on the left), and between the G and the A in the bottom strand (on the right).

These symmetrical sites are also called **palindromes.** In ordinary language, palindromes are sentences that read the same forwards and backwards. Examples are Napoleon's lament: "Able was I ere I saw Elba," or a wart remedy: "Straw? No, too stupid a fad; I put soot on warts." DNA palindromes also read the same forwards and backwards, but you have to be careful to read in the same sense ($5' \rightarrow 3'$) in both directions. This means that you read the top strand left to right and the bottom strand right to left.

Cohen and Boyer took advantage of the sticky ends created by a restriction enzyme in their cloning experiment (figure 15.1). They cut two different DNAs with the same restriction enzyme, EcoRI. Both DNAs were plasmids. The first, called pSC101, carried a gene that conferred resistance to the antibiotic tetracycline; the other, RSF1010, conferred resistance to both streptomycin and sulfonamide. Both plasmids had just one EcoRI **restriction site,** or cutting site for EcoRI. Therefore, when EcoRI cut these circular DNAs, it converted them to linear molecules and left them with the same sticky ends. These sticky ends then base-paired with one another, at least briefly. Finally, DNA ligase completed the task of joining the two DNAs.

The result was a **recombinant DNA,** two previously separate pieces of DNA linked together. This new, recombinant plasmid was easy to detect. When introduced into bacterial cells, it conferred resistance to both tetracycline, a property of pSC101, and to streptomycin, a property of RSF1010. We have seen several examples of recombinant DNA in this text, but this one differs from the others in that it was not created naturally in a cell. Instead, geneticists put it together in a test tube.

R estriction endonucleases recognize specific sequences in DNA molecules and make cuts in both strands. This allows very specific cutting of DNAs. Also, because the cuts in the two strands are frequently staggered, restriction enzymes create sticky ends that help link together two DNAs to form a recombinant DNA in vitro.

■

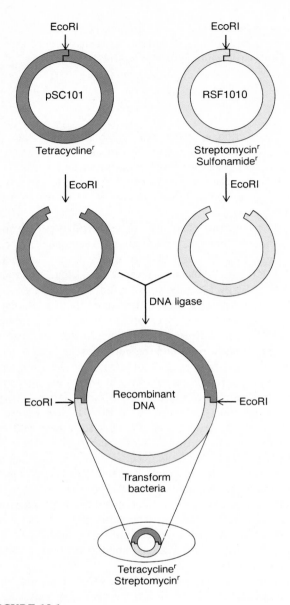

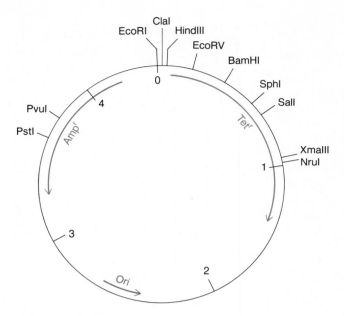

FIGURE 15.2
The plasmid pBR322, showing the locations of eleven unique restriction sites that can be used to insert foreign DNA. The locations of the two antibiotic resistance genes (Amp^r = ampicillin resistance; Tet^r = tetracycline resistance) and the origin of replication (Ori) are also shown. Numbers refer to kilobase pairs (kb) from the EcoRI site.

FIGURE 15.1
The first cloning experiment involving a recombinant DNA assembled in vitro. Boyer and Cohen cut two plasmids, pSC101 and RSF1010, with the same restriction endonuclease, EcoRI. This gave the two linear DNAs the same sticky ends, which were then linked in vitro using DNA ligase. The investigators reintroduced the recombinant DNA into *E. coli* cells by transformation, and screened for clones that were resistant to both tetracycline and streptomycin. These clones were therefore harboring the recombinant plasmid.

VECTORS

Both plasmids in the Cohen and Boyer experiment are capable of replicating in *E. coli*. Thus, both can serve as carriers to allow replication of recombinant DNAs. All gene cloning experiments require such carriers, which we call **vectors,** but a typical experiment involves only one vector, plus a piece of foreign DNA that depends on the vector

for its replication. Since the mid-1970s, many vectors have been developed; these fall into two classes: plasmids and phages.

Plasmids as Vectors

In the early years of the cloning era, Boyer and his colleagues developed a set of very popular vectors known as the pBR plasmid series. One of these plasmids, **pBR322** (figure 15.2), contains genes that confer resistance to two antibiotics, ampicillin and tetracycline. Between these two genes lies the origin of replication. The plasmid has been engineered to contain only one cutting site for several of the common restriction enzymes, including EcoRI, BamHI, PstI, HindIII, and SalI. This is convenient because it allows each enzyme to be used to create a site for inserting foreign DNA without losing any of the plasmid DNA.

For example, let us consider cloning a foreign DNA fragment into the PstI site of pBR322 (figure 15.3). First, we would cut the vector with PstI to generate the sticky ends characteristic of that enzyme. In this example, we also cut the foreign DNA with PstI, thus giving it PstI sticky ends. Next, we combine the cut vector with the foreign DNA and incubate them with DNA ligase. As the sticky ends on the vector and on the foreign DNA base-pair momentarily, DNA ligase seals the nicks, attaching the two DNAs together covalently. Once this is done, they cannot come apart again unless they are recut with PstI.

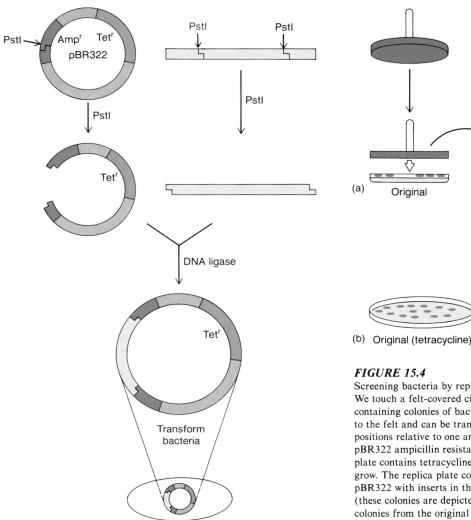

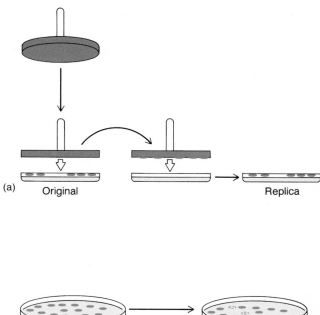

(a) Original   Replica

(b) Original (tetracycline)   Replica (ampicillin)

FIGURE 15.4
Screening bacteria by replica plating. (*a*) The replica plating process. We touch a felt-covered circular tool to the surface of the first dish containing colonies of bacteria. Cells from each of these colonies stick to the felt and can be transferred to the replica plate in the same positions relative to one another. (*b*) Screening for inserts in the pBR322 ampicillin resistance gene by replica plating. The original plate contains tetracycline, so all colonies containing pBR322 will grow. The replica plate contains ampicillin, so colonies bearing pBR322 with inserts in the ampicillin resistance gene will not grow (these colonies are depicted by dotted circles). The corresponding colonies from the original plate can then be picked.

FIGURE 15.3
Cloning foreign DNA using the PstI site of pBR322. We cut both the plasmid and the insert (yellow) with PstI, then join them through these sticky ends with DNA ligase. Next, we transform bacteria with the recombinant DNA and screen for tetracycline-resistant, ampicillin-sensitive cells. The plasmid no longer confers ampicillin resistance because the foreign DNA interrupts that resistance gene (blue).

In the following step, we transform *E. coli* with our DNA mixture. It would be nice if all the cut DNA had been ligated to plasmids to form recombinant DNAs, but that never happens. Instead, we get a mixture of re-ligated plasmids and re-ligated inserts, along with the recombinants. How do we sort these out? This is where the antibiotic resistance genes of the vector come into play. First we grow cells in the presence of tetracycline, which selects for cells that have taken up either the vector or the vector with inserted DNA. Cells that received no DNA, or that received insert DNA only, will not be tetracycline-resistant and will fail to grow.

Next, we want to find the clones that have received recombinant DNAs. To do this, we screen for clones that are both tetracycline-resistant and ampicillin-sensitive. Figure 15.3 shows that the PstI site, where we are inserting DNA in this experiment, lies within the ampicillin resistance gene. Therefore, inserting foreign DNA into the PstI site inactivates this gene and leaves the host cell vulnerable to ampicillin.

How do we do the screening? One way is to transfer copies of the clones from the original tetracycline plate to an ampicillin plate. This can be accomplished with a felt transfer tool as illustrated in figure 15.4. We touch the tool lightly to the surface of the tetracycline plate to pick up cells from each clone, then touch the tool to a fresh ampicillin plate. This deposits cells from each original clone in the same relative positions as on the original plate. We look for colonies that do *not* grow on the ampicillin plate and then find the corresponding growing clone on the tetracycline plate. Using that most sophisticated of scientific tools—a sterile toothpick—we transfer cells from this positive clone to fresh medium for storage or immediate use. Notice that we did not call this procedure a selection, because it does not remove unwanted clones automatically. Instead, we had to examine each clone individually. We call this more laborious process a **screen.**

Nowadays we can choose from many plasmid cloning vectors besides the pBR plasmids. One useful class of plasmids is the **pUC** series (figure 15.5). These plasmids are based on pBR322, from which almost half the DNA, including the tetracycline resistance gene, has been deleted. Furthermore, the pUC vectors have their cloning sites clustered into one small area called a **multiple cloning site (MCS).** The pUC vectors contain the pBR322 ampicillin resistance gene to allow selection for bacteria that have received a copy of the vector. Moreover, to compensate for the loss of the other antibiotic resistance gene, they have genetic elements that permit a very convenient screen for clones with inserts in their multiple cloning sites.

Figure 15.5*b* shows the multiple cloning sites of pUC18 and pUC19. Notice that they lie within a DNA sequence coding for the amino portion of β-galactosidase (denoted *lacZ'*). The host bacteria used with the pUC vectors carry a gene fragment that encodes the carboxyl portion of β-galactosidase. By themselves, the β-galactosidase fragments made by these partial genes have no activity. But they can complement each other by intracistronic complementation, which we encountered in chapter 13. In other words, the two partial gene products can cooperate to form an active enzyme. Thus, when pUC18 by itself transforms a bacterial cell carrying the partial β-galactosidase gene, active β-galactosidase is produced. If we plate these clones on medium containing a β-galactosidase indicator, colonies with the pUC plasmid will turn color. The indicator X-gal, for instance, is a synthetic, colorless galactoside; when β-galactosidase cleaves X-gal, it releases galactose plus a dye that stains the bacterial colony blue.

On the other hand, if we have interrupted the plasmid's partial β-galactosidase gene by placing an insert into the multiple cloning site, the gene is usually inactivated. It can no longer make a product that complements the host cell's β-galactosidase fragment, so the X-gal remains colorless. Thus, it is a simple matter to pick the clones with inserts. They are the white ones; all the rest are blue. Notice that, in contrast to selection with pBR322, this is a one-step process. We look simultaneously for a clone that (1) grows on ampicillin and (2) is white in the presence of X-gal. The multiple cloning sites have been carefully constructed to preserve the reading frame of β-galactosidase. Thus, even though the gene is interrupted

by eighteen codons (placed in parentheses, with amino acid names in lowercase), a functional protein still results. Further interruption by large inserts, especially those that shift the reading frame, is usually enough to destroy the gene's function.

The multiple cloning site also allows us to cut it with two different restriction enzymes (say, EcoRI and BamHI) and then to clone a piece of DNA with one EcoRI end and one BamHI end. This is called **directional cloning,** or **forced cloning,** because we force the insert DNA into the vector in only one orientation. (Obviously, the EcoRI and BamHI ends of the insert have to match their counterparts in the vector.) Knowing the orientation of an insert has certain benefits, which we will explore later in this chapter. Forced cloning also has the advantage of preventing the vector's simply re-ligating by itself, since its two restriction sites are incompatible.

Notice that the multiple cloning site of pUC18 is just the reverse of that of pUC19. That is, the restriction sites are in opposite order. This means we can clone our fragment in either orientation simply by shifting from one pUC plasmid to the other.

A mong the commonly used plasmid cloning vectors are pBR322 and the pUC plasmids. The former has two antibiotic resistance genes and a variety of unique restriction sites into which we can introduce foreign DNA. Most of these sites interrupt one of the antibiotic resistance genes, making selection straightforward. Selection is even easier with the pUC plasmids. These have an ampicillin resistance gene and a multiple cloning site that interrupts a β-galactosidase gene. We select for ampicillin-resistant clones that do not make β-galactosidase and therefore do not turn the indicator, X-gal, blue. The multiple cloning site also makes it convenient to carry out forced cloning into two different restriction sites.

▪

Phages as Vectors

We have already seen how phages serve as natural vectors in transducing bacterial DNA from one cell to another. It was only natural, then, to engineer phages to do the same thing for *all* kinds of DNA.

λ *phage vectors* Fred Blattner and his colleagues constructed the first phage vectors by modifying the well-known λ phage. They took out the region in the middle of the phage DNA, which codes for proteins needed for lysogeny, but retained the genes needed for lytic infection. These phages are no longer capable of lysogenic infection, but their missing genes can be replaced with foreign DNA. Blattner named these vectors **Charon phages** after Charon, the boatman on the river Styx in classical mythology. Just as Charon carried souls to the underworld, the Charon

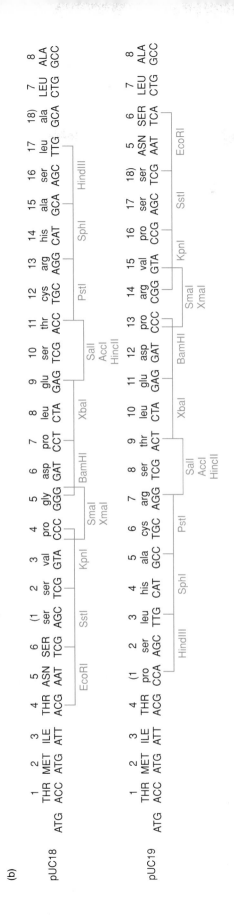

FIGURE 15.5

Architecture of a pUC plasmid. (*a*) The pUC plasmids retain the ampicillin resistance gene and the origin of replication of pBR322. In addition, they include a multiple cloning site (MCS) inserted into a gene encoding the amino part of β-galactosidase (*lacZ'*). (*b*) The MCS's of pUC18 and pUC19. In pUC18, the MCS, containing thirteen restriction sites, is inserted after the sixth codon of the *lacZ'* gene. In pUC19, it comes after the fourth codon and is reversed. In addition to containing the restriction sites, the MCS's preserve the reading frame of the *lacZ'* gene, so plasmids with no inserts still support active expression of this gene. Therefore, clones harboring the vector alone will turn blue in the presence of the synthetic β-galactosidase substrate X-gal, while clones harboring the vector plus an insert will remain white.

Reprinted with permission of Life Technologies, Inc., Gaithersburg, MD.

Gene Cloning and Manipulation

375

FIGURE 15.6

Cloning in Charon 4. (*a*) Forming the recombinant DNA. We cut the vector (yellow) with EcoRI to remove the stuffer fragments, and save the arms. Next, we ligate partially digested insert DNA (blue) to the arms. (*b*) Packaging and cloning the recombinant DNA. We mix the recombinant DNA from (*a*) with an in vitro packaging extract that contains λ phage head and tail components and all other factors needed to package the recombinant DNA into functional phage particles. Finally, we plate these particles on *E. coli* and collect the plaques that form.

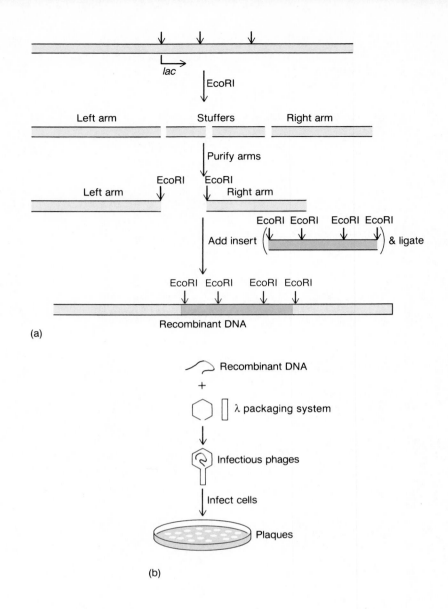

(a)

(b)

phages carry foreign DNA into bacterial cells. Charon the boatman is pronounced "Karen," but Charon the phage is usually pronounced "Sharon."

One clear advantage of the Charon phages over plasmid vectors is that they can accommodate much more foreign DNA. For example, Charon 4 can accept up to about 20 kb of DNA, a limit imposed by the capacity of the λ phage head. When would we need such high capacity? The most common use for Charon phage vectors is in constructing **genomic libraries.** Suppose we wanted to clone the entire human genome. This would obviously require a great many clones, but the larger the insert in each clone, the fewer total clones would be needed. In fact, such genomic libraries have been constructed for humans and for a variety of other organisms, and the Charon phages have been popular vectors for this purpose.

Aside from their high capacity, some of the Charon phages have the advantage of a minimum size requirement for their inserts. Figure 15.6 illustrates the reason for this requirement: To get the Charon 4 vector ready to accept an insert, we cut it with EcoRI. This cuts at two sites near the middle of the phage DNA, yielding two "arms" and two "stuffer" fragments. Next, we purify the arms by ultracentrifugation and throw away the stuffers. The final step is to ligate the arms to our insert, which then takes the place of the discarded stuffers.

At first glance, it may appear that the two arms could simply ligate together without accepting an insert. Indeed, this may happen, but it will not produce a clone, because the two arms constitute too little DNA and will not be packaged into a phage head. The packaging is done in vitro; we simply mix the ligated arms plus inserts with all the components needed to put together a phage particle.

This extract has rather stringent requirements as to the size of DNA it will package. It must have at least 12 kb of DNA in addition to Charon 4 arms, but no more than 20 kb, or the phage head will overflow.

Since we can be sure that each clone in our genomic library has at least 12 kb, we know we are not wasting space in our library with clones that contain insignificant amounts of DNA. This is an important consideration since, even at 12–20 kb per clone, we need about half a million clones to be sure of having each human gene represented at least once. It would be much more difficult to make a human genomic library in pBR322, since plasmids tend to ligate selectively to small inserts, and bacteria selectively reproduce small plasmids. Therefore, most of the clones would contain inserts of a few thousand, or even just a few hundred base pairs. Such a library would have to contain many millions of clones to be complete.

Since EcoRI produces fragments whose average size is about 4 kb and yet the vector will not accept any inserts smaller than 12 kb, it is obvious that we cannot cut our DNA completely with EcoRI, or most of the fragments will be too small to clone. Furthermore, EcoRI, or any other restriction enzyme, cuts in the middle of most eukaryotic genes one or more times, so a complete digest would contain only fragments of most genes. We can avoid these problems by performing an incomplete digestion with EcoRI; if the enzyme only cuts about every fourth or fifth site, the average length of the resulting fragments will be about 16–20 kb, just the size the vector will accept and big enough to include the entirety of most eukaryotic genes, introns and all.

Another vector designed especially for cloning large DNA fragments is called a **cosmid**. Cosmids resist classification because they behave both as plasmids and as phages. They contain the cos sites, or cohesive ends, of λ phage DNA, which allow the DNA to be packaged into λ phage heads (hence the "cos" part of the name cosmid). They also contain a plasmid origin of replication, so they can replicate as plasmids in bacteria (hence the "mid" part of the name).

Because almost the entire λ genome, except for the cos sites, has been removed from the cosmids, they have room for very large inserts (40–50 kb). Once these inserts are in place, the recombinant cosmids are packaged into phage particles. These particles cannot replicate as phages because they have almost no phage DNA, but they are infectious, so they carry their recombinant DNA into bacterial cells. Once inside, the DNA replicates as a plasmid, using its plasmid origin of replication.

A genomic library is very handy. Once it is established, we can search it for any gene we want. The only problem is that there is no card catalog for such a library, so we need some kind of probe to tell us which clone contains the gene of interest. An ideal probe would be a labeled nucleic acid whose sequence matches that of the gene we are trying to find. We would then carry out a **plaque**

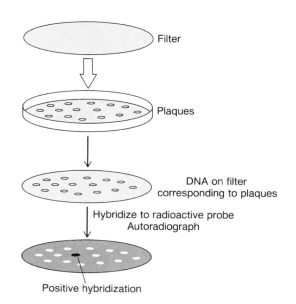

FIGURE 15.7

Selection of positive genomic clones by plaque hybridization. First, we touch a nitrocellulose or similar filter to the surface of the dish containing the Charon 4 plaques from figure 15.6. Phage DNA released naturally from each plaque will stick to the filter. Next, we denature the DNA with alkali and hybridize the filter to a radioactive probe for the gene we are studying, then autoradiograph to reveal the position of any radioactivity. Cloned DNA from one plaque near the center of the filter has hybridized, as shown by the dark spot on the autoradiograph.

hybridization procedure in which the DNA from each of the thousands of λ phages from our library is hybridized to the labeled probe. The DNA that forms a labeled hybrid is the right one.

We have encountered hybridization before in several chapters, and we will discuss it again later in this chapter. Figure 15.7 shows how plaque hybridization works. We grow thousands of plaques on each of several petri dishes (only a few plaques are shown here for simplicity). Next, we touch a filter made of a DNA-binding material such as **nitrocellulose** to the surface of the petri dish. This transfers phage DNA from each plaque to the filter. The DNA is then denatured with alkali and hybridized to the radioactive probe. When the probe encounters complementary DNA, which should only be the DNA from the clone of interest, it will hybridize, making that DNA spot radioactive. This radioactive spot is then detected by autoradiography. The black spot on the autoradiograph shows us where to look on the original petri dish for the plaque containing our gene. In practice, the original plate may have been so crowded with plaques that it is impossible to pick out the right one, so we pick several plaques from that area, replate at a much lower phage density, and rehybridize to find the positive clone.

M13 phage vectors Another phage frequently used as a cloning vector is the filamentous phage M13. Joachim Messing and his coworkers endowed the phage DNA with the same β-galactosidase gene fragment and multiple cloning sites found in the pUC family of vectors. In fact, the M13 vectors were engineered first; then the useful cloning sites were simply transferred to the pUC plasmids.

What is the advantage of the M13 vectors? The main factor is that the genome of this phage is a single-stranded DNA, so DNA fragments cloned into this vector can be recovered in single-stranded form. As we will see later in this chapter, single-stranded DNA is an invaluable aid to site-directed mutagenesis, by which we can introduce specific, premeditated alterations into a gene.

Figure 15.8 illustrates how we can clone a double-stranded piece of DNA into M13 and harvest a single-stranded DNA product. The DNA in the phage particle itself is single-stranded, but after infecting an *E. coli* cell, it is converted to a double-stranded replicative form (RF). This double-stranded replicative form of the phage DNA is what we use for cloning. After it is cut by one or two restriction enzymes at its multiple cloning site, foreign DNA with compatible ends can be inserted. This recombinant DNA is then used to transform host cells, giving rise to progeny phages that bear single-stranded recombinant DNA. The phage DNA, along with phage particles, is secreted from the transformed cells and can be collected from the growth medium.

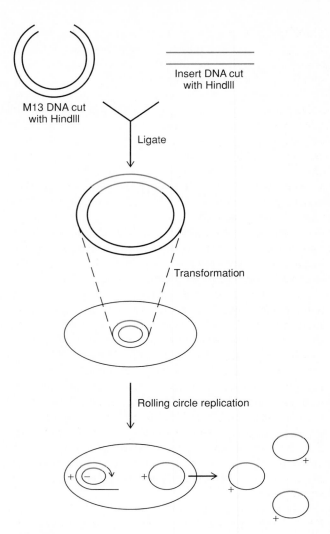

FIGURE 15.8
Obtaining single-stranded DNA by cloning in M13 phage. Foreign DNA (red), cut with HindIII, is inserted into the HindIII site of the double-stranded phage DNA. The resulting recombinant DNA is used to transform *E. coli* cells, whereupon the DNA replicates by a rolling circle mechanism, producing many single-stranded product DNAs. These product DNAs are called positive (+) strands, by convention. The template DNA is therefore the negative (−) strand.

T wo kinds of phages have been especially popular as cloning vectors. The first of these is λ, from which certain nonessential genes have been removed to make room for inserts. In some of these engineered phages, called Charon phages, inserts up to 20 kb can be accommodated. This makes the Charon phages useful for building genomic libraries, in which it is important to have large pieces of genomic DNA in each clone. Cosmids can accept even larger inserts—up to 50 kb—making them a favorite choice for genomic libraries. The second major class of phage vectors is composed of the M13 phages. These vectors have the convenience of a multiple cloning site and the further advantage of producing single-stranded recombinant DNA, which can be used for DNA sequencing and for site-directed mutagenesis.

■

IDENTIFYING A SPECIFIC CLONE WITH A SPECIFIC PROBE

We have already mentioned the need for a probe to identify the clone we want among the thousands we do not want. What sort of probe could we employ? Two different kinds are widely used: polynucleotides and antibodies. Both are molecules able to bind very specifically to other molecules. We will discuss oligonucleotide probes here and antibody probes later in this chapter.

Polynucleotide Probes

If the gene we are trying to locate is a very active one, appropriate cells may make its corresponding mRNA in large enough quantities to purify. In principle, we could use this mRNA itself as a probe, but it is more practical to make a cDNA copy by reverse transcription (chapter 10). Once we have made the cDNA, we can clone it or use it directly as a probe.

More often than not, the gene we want is not active enough to allow this direct approach. In that case, we might use the homologous gene from another organism if someone has already managed to clone it. For example, if we were after the human insulin gene and another research group had already cloned the rat insulin gene, we could ask them for their clone to use as a probe. We would hope the two genes have enough similarity in sequence that the rat probe could hybridize to the human gene. This hope is usually fulfilled. However, we generally have to lower the **stringency** of the hybridization conditions so that the hybridization reaction can tolerate some mismatches in base sequence between the probe and the cloned gene.

Researchers use several means to control stringency. High temperature, high organic solvent concentration, and low salt concentration all tend to promote the separation of a DNA double helix. We can therefore adjust these conditions until only perfectly matched DNA strands will form a duplex; this is high stringency. By relaxing these conditions (lowering the temperature, for example), we lower the stringency until DNA strands with a few mismatches can hybridize.

Without cDNA or homologous DNA from another organism, what could we use? There is still a way out if we know at least part of the sequence of the protein product of the gene. We faced a problem just like this in our lab when we cloned the gene for a plant toxin known as ricin. Fortunately, the entire amino acid sequences of both polypeptides of ricin were known. That meant we could examine the amino acid sequence and, using the genetic code, deduce the nucleotide sequence that would code for these amino acids. Then we could construct the nucleotide sequence chemically and use this synthetic probe to find the ricin gene. This sounds easy, but there is a hitch. The genetic code is degenerate, so for most amino acids we would have to consider several different nucleotide sequences.

Fortunately, we were spared much inconvenience because one of the polypeptides of ricin includes this amino acid sequence: Trp-Met-Phe-Lys-Asn-Glu. The first two amino acids in this sequence have only one codon each, and the next three only two each. The sixth gives us two free bases because the degeneracy occurs only in the third base. Thus, we only had to make eight 17-base oligonucleotides (*17-mers*) to be sure of getting the exact coding sequence for this string of amino acids. This degenerate sequence can be expressed as follows:

		U	G	U	
UGG	AUG	UUC	AAA	AAC	GA
Trp	Met	Phe	Lys	Asn	Glu

Using this mixture of eight 17-mers (UGGAUGU-UCAAAAACGA, UGGAUGUUUAAAAACGA, etc.), we quickly identified several ricin-specific clones.

*S*pecific clones can be identified using polynucleotide probes that bind to the gene itself. Knowing the amino acid sequence of a gene product, one can design an oligonucleotide—a short polynucleotide that encodes part of this amino acid sequence. This can be one of the quickest and most accurate means of identifying a particular clone.

■

cDNA CLONING

Molecular geneticists use a variety of techniques to clone cDNAs; here we will consider a fairly simple, yet effective strategy, which is illustrated in figure 15.9. The central part of any cDNA cloning procedure is synthesis of the cDNA from an mRNA template using **reverse transcriptase.** This reverse transcriptase is like any other DNA-synthesizing enzyme in that it cannot initiate DNA synthesis without a primer; give it only mRNA and you get nothing in return. To get around this problem, we take advantage of the poly(A) tail at the end of most mRNAs and use oligo(dT) as the primer. The oligo(dT) is complementary to poly(A), so it binds to the poly(A) at the 3′-end of the mRNA and primes DNA synthesis, using the mRNA as the template.

After the mRNA has been copied, yielding a single-stranded DNA (the "first strand"), we remove the mRNA with alkali or **ribonuclease H** (RNase H). This enzyme degrades the RNA part of an RNA/DNA hybrid—just what we need to remove the RNA from our first strand cDNA. Next, we must make a second DNA strand, using the first as the template. Again, we need a primer, and this time we do not have a convenient poly(A) to which to hybridize the primer. Instead, we build an oligo(dC) tail at the 3′-end of the first strand, using the enzyme **terminal transferase** and the substrate dCTP. The enzyme adds dCs, one at a time, to the 3′-end of the first strand. To this tail, we hybridize a short oligo(dG), which primes second strand synthesis. We can use reverse transcriptase again to make the second strand, but DNA polymerase also works. Actually, the most successful enzyme is a fragment of DNA polymerase called the **Klenow fragment.** This piece of enzyme is generated by cleaving *E. coli* DNA polymerase I with a proteolytic enzyme. The Klenow fragment contains the DNA polymerase activity and the 3′ → 5′ exonuclease activity, but it lacks the 5′ → 3′ exonuclease activity normally associated with DNA polymerase I. The latter activity is undesirable because it degrades DNA from the 5′-end, which is damage that DNA polymerase cannot repair.

Once we have a double-stranded cDNA, we must ligate it to a vector. This was easy with our pieces of genomic DNA, since they had sticky ends, but the cDNA

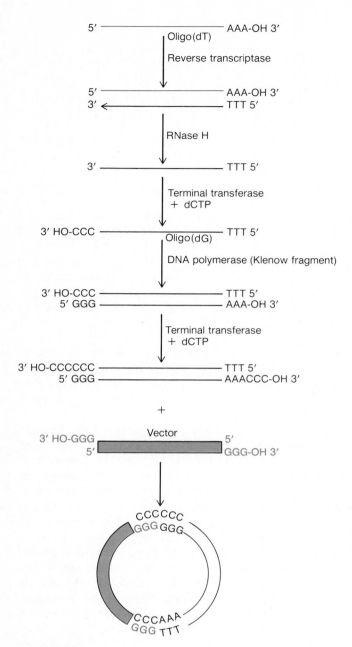

FIGURE 15.9
cDNA cloning. We start with a eukaryotic messenger RNA (green) having poly(A) at its 3′-end. Oligo(dT) hybridizes to the poly(A) and primes reverse transcription, forming the first cDNA strand. We remove the mRNA template, using RNase H, then add an oligo(dC) tail to the 3′-end of the cDNA, using terminal transferase. Oligo(dG) hybridizes to the oligo(dC) and primes second strand cDNA synthesis by reverse transcriptase or by the Klenow fragment of DNA polymerase I. In order to give the double-stranded cDNA sticky ends, we add oligo(dC) with terminal transferase, then anneal these ends to complementary oligo(dG) ends of a suitable vector (red). The recombinant DNA can then be used to transform bacterial cells.

has no sticky ends. This problem is easily solved. We simply tack sticky ends (oligo[dC]) onto the cDNA, again using terminal transferase and dCTP. In the same way, we attach oligo(dG) ends to our vector and allow the oligo(dC)s to anneal to the oligo(dG)s. This brings the vector and cDNA together in a recombinant DNA that can be used directly for transformation. The base pairing between the oligonucleotide tails is strong enough that no ligation is required before transformation. The DNA ligase inside the transformed cells finally performs this task.

What kind of vector should we use? Several choices are available, depending on the way we wish to detect positive clones (those that bear the cDNA we want). We can use a simple vector such as one of the pBR or pUC plasmids; if we do, we usually identify positive clones by **colony hybridization** with a radioactive DNA probe. This procedure is analogous to the plaque hybridization described previously. Or we can use a λ phage, such as λgt11, as a vector. This vector places the cloned cDNA under the control of a β-galactosidase promoter, so that transcription and translation of the cloned gene can occur. We can then use an antibody to screen directly for the protein product of the correct gene. We will describe this procedure in more detail later in this chapter.

> *A* cDNA can be synthesized one strand at a time, using mRNA as template for the first strand and this first strand as template for the second. The double-stranded cDNA can be endowed with oligonucleotide tails that base-pair with complementary tails on a cloning vector. This recombinant DNA can then be used to transform bacteria. Positive clones can be detected by colony hybridization with radioactive DNA probes, or with antibodies if an expression vector such as λ gt11 is used.

METHODS OF EXPRESSING CLONED GENES

Why would we want to clone a gene? An obvious reason, suggested at the beginning of this chapter, is that cloning allows us to produce large quantities of pure eukaryotic (or prokaryotic) genes so that we can study them in detail. Several specific studies using cloned genes have been described in previous chapters. For example: Cloned genes gave us labeled probes to localize the expression of homeo box genes in fruit flies (chapter 14). Cloned genes were manipulated to discover the important control regions of the chicken ovalbumin gene, the pea rubisco gene, and the frog 5S rRNA genes (chapters 10 and 14). Thus, the gene itself can be a valuable product of gene cloning. Another goal of gene cloning is to make a large quantity of the gene's product, either for investigative purposes or for profit.

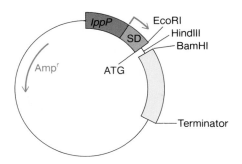

FIGURE 15.11

The pIN-I-A1 expression vector. The coding region of a gene to be expressed is inserted into one of the three unique restriction sites (EcoRI, HindIII, or BamHI). Transcription starts at the arrow and is directed by the strong lipoprotein promoter (*lppP*) and terminated by the transcription terminator. Translation, abetted by the Shine-Dalgarno ribosome binding site (SD), begins at the ATG just upstream from the inserted DNA, so the ultimate product will be a fusion protein containing just a few extra amino acids at the amino end. The Amp[r] gene confers resistance to ampicillin for selection purposes.

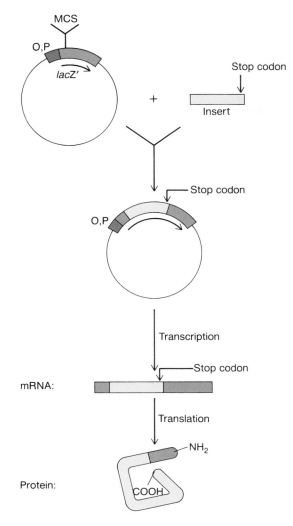

FIGURE 15.10

Formation of a fusion protein by cloning in a pUC plasmid. We insert foreign DNA (yellow) into the multiple cloning site (MCS); transcription from the *lac* promoter (purple) gives a hybrid mRNA beginning with a few *lacZ* codons, changing to insert sequence, then back to *lacZ* (red). This mRNA will be translated to a fusion protein containing a few β-galactosidase amino acids at the beginning (amino end), followed by the insert amino acids for the remainder of the protein. Since the insert contains a translation stop codon, the remaining *lacZ* codons will not be translated.

EXPRESSION VECTORS

The vectors we have examined so far are meant to be used primarily in the first stage of cloning—when we first put a foreign DNA into a bacterium and get it to replicate. By and large, they work well for that purpose, growing readily in *E. coli* and producing high yields of recombinant DNA. Some of them even work as **expression vectors** that can yield the protein products of the cloned genes. For example, the pUC vectors place inserted DNA under the control of the *lac* promoter, which lies upstream from the multiple cloning site. If an inserted DNA happens to be in the same reading frame as the *lac* gene it

interrupts, a **fusion protein** will result. It will have a partial β-galactosidase protein sequence at its amino end and another protein sequence, encoded in the inserted DNA, at its carboxyl end (figure 15.10).

However, if we are interested in high expression of our cloned gene, specialized expression vectors usually work better. These typically have two elements that are required for active gene expression: a strong promoter and a ribosome binding site that includes a Shine-Dalgarno sequence near an initiating ATG codon.

Expression Vectors with Strong Promoters

The main function of an expression vector is to yield the product of a gene—usually, the more product the better. Therefore, expression vectors are ordinarily equipped with very strong promoters; the rationale is that the more mRNA is produced, the more protein product will be made.

One of the strongest of all *E. coli* promoters is the *lpp* promoter, which directs transcription of the lipoprotein gene. A consequence of the strength of this promoter is that the product of the gene, the major membrane lipoprotein, is one of the most abundant proteins in the bacterial cell. Thus, this promoter is a natural for an expression vector, and Masayori Inouye and his colleagues have constructed a set of such vectors, called the *pIN* plasmids. Figure 15.11 shows the architecture of a typical pIN vector. Just behind the *lpp* promoter lie three cloning sites—restriction sites for the enzymes EcoRI, BamHI, and HindIII. If we ligate a gene's coding region into any of these three sites, being careful to preserve the reading frame of the *lpp* coding region, we can get a protein with just a few extra amino acids at the amino end.

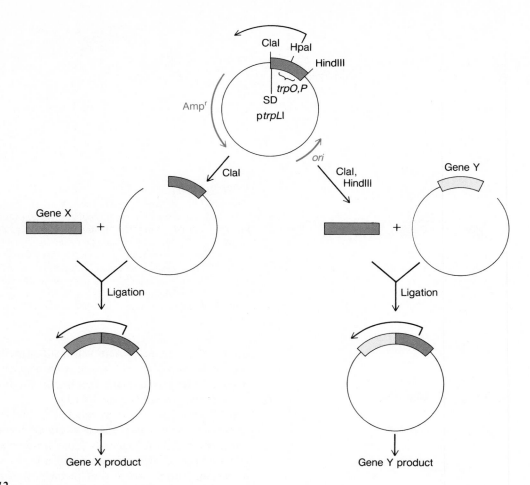

FIGURE 15.12

Two uses of the *ptrpL*1 expression vector. The vector contains a ClaI cloning site, preceded by a Shine-Dalgarno ribosome binding site (SD) and the *trp* operator/promoter region (*trpO,P*). Transcription occurs in a counterclockwise direction as shown by the arrows (top). The vector can be used as a traditional expression vector (left) simply by inserting a foreign coding region (X, red) into the unique ClaI site. Alternatively (right), the *trp* control region (purple) can be cut out with ClaI and HindIII and inserted into another plasmid bearing the coding region (Y, yellow) to be expressed.

Another strong promoter, though not as strong as the one from the *lpp* gene, is the *trp* (tryptophan operon) promoter. It forms the basis for several expression vectors, including *ptrpL1*. Figure 15.12 shows how this vector works. It has the *trp* promoter/operator region, followed by a Shine-Dalgarno sequence, and can be used directly as an expression vector by inserting a foreign gene into the ClaI site. Alternatively, the *trp* control region can be made "portable" by cutting it out with ClaI and HindIII and inserting it in front of a gene to be expressed in another vector.

Inducible Expression Vectors

It is usually advantageous to keep a cloned gene turned off until we are ready to express it. One reason is that eukaryotic proteins produced in large quantities in bacteria can be toxic. Even if these proteins are not actually toxic, they can build up to such great levels that they interfere with bacterial growth. In either case, if the cloned gene were allowed to remain turned on constantly, the bacteria bearing the gene would never grow to a great enough concentration to produce meaningful quantities of protein product. The solution is to keep the cloned gene turned off by placing it behind an inducible promoter.

The *lac* promoter is inducible to a certain extent, presumably remaining off until stimulated by the inducer allolactose or by its synthetic analog IPTG. However, the repression wrought by the *lac* repressor is incomplete, and some expression of the cloned gene will be observed even in the absence of inducer. One way around this problem is to express our gene in a host cell that has an overactive *lac* repressor gene, I^q. The excess repressor produced by such a cell keeps our cloned gene turned off until we are ready to induce it.

Another strategy is to use a very tightly controlled promoter such as the λ phage promoter P_L. Expression vectors with this promoter/operator system are cloned into

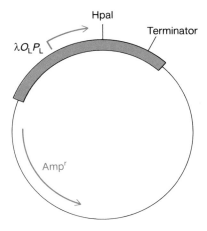

FIGURE 15.13
The inducible expression vector pKC30. A gene to be expressed is inserted into the unique HpaI site, downstream from the λ $O_L P_L$ operator-promoter region. The host cell is a λ lysogen bearing a temperature-sensitive λ repressor gene (*cI*857). To induce expression of the cloned gene, the temperature is raised from 32° to 42° C, which inactivates the temperature-sensitive λ repressor, removing it from O_L and allowing transcription to occur.

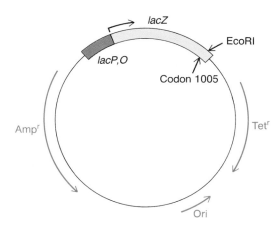

FIGURE 15.14
The pβ-gal13C expression vector. The unique EcoRI cloning site lies near the end of the *lacZ* coding region, just after codon number 1005. Therefore, products of genes cloned into this site will contain a 1,005-amino acid tag of β-galactosidase at their amino ends.

host cells bearing a temperature-sensitive λ repressor gene (*cI*857). As long as the temperature of these cells is kept relatively low (32° C), the repressor functions, and no expression takes place. However, when we raise the temperature to the nonpermissive level (42° C), the temperature-sensitive repressor can no longer function and the cloned gene is induced. Figure 15.13 illustrates this mechanism for the expression vector pKC30.

> **E**xpression vectors are designed to yield the protein product of a cloned gene, usually in the greatest amount possible. To optimize expression, these vectors provide strong bacterial promoters and bacterial ribosome binding sites that would be missing on cloned eukaryotic genes. Most cloning vectors are inducible, to avoid premature overproduction of a foreign product that could poison the bacterial host cells.

■

Expression Vectors That Produce Fusion Proteins

When most expression vectors operate, they produce fusion proteins. This might at first seem a disadvantage because the natural product of the inserted gene is not made. However, the extra amino acids at the amino terminus of the fusion protein can be useful. Consider the expression vector pβ-gal13C. It has the *lac* operator and promoter, and the EcoRI cloning site lies near the end of the *lacZ* gene (figure 15.14). This means the product will be a fusion protein with a large piece of β-galactosidase at its amino end.

How can fusion proteins be useful? One reason is that they are sometimes more stable in bacterial cells than normal eukaryotic proteins are. Furthermore, if we do not have a good way of purifying the protein product of our cloned gene, the long β-galactosidase addition can be a big help. For example, at a molecular weight of 116,000, the β-galactosidase monomer is already among the largest polypeptides in an *E. coli* cell and is correspondingly easy to purify by techniques that separate molecules by size. A fusion protein, with another polypeptide tacked onto the carboxyl terminus, will be even larger, and it is usually a simple matter to identify and purify such a protein.

λ phages have also served as the basis for expression vectors. Even Charon 4 has a promoter in front of the cloning site, so inserted DNA could, in principle, be expressed. On the other hand, this vector only accepts large DNA fragments, thus it is not suitable for cloning most gene-sized pieces of cDNA. A λ phage that has been designed specifically as an expression vector is **λgt11.** This phage (figure 15.15) contains the *lac* control region followed by the *lacZ* gene. The cloning sites are located within the *lacZ* gene, so products of a gene inserted into this vector will be fusion proteins with a leader of β-galactosidase.

The expression vector λgt11 has become a popular vehicle for making and screening cDNA libraries. In the examples of screening presented earlier, we looked for the proper DNA sequence, either directly, by probing with a labeled oligonucleotide or polynucleotide, or indirectly, by discovering which cloned DNA would hybridize to the appropriate mRNA. By contrast, λgt11 allows us to directly screen a group of clones for the expression of the right protein. The main ingredients required for this procedure are a cDNA library in λgt11 and an antiserum directed against the protein of interest.

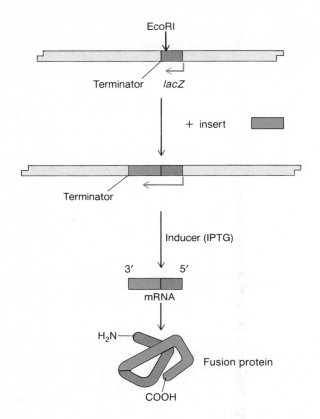

EcoRI

Terminator *lacZ*

+ insert

Terminator

Inducer (IPTG)

3'   5'

mRNA

H₂N

Fusion protein

COOH

FIGURE 15.15
Forming a fusion protein in λgt11. The gene to be expressed (blue) is inserted into the EcoRI site near the end of the *lacZ* coding region (red) just before the transcription terminator. Thus, upon induction of the *lacZ* gene by IPTG, a fused mRNA results, containing the inserted coding region just downstream from that of β-galactosidase. This mRNA is translated by the host cell to a fusion protein.

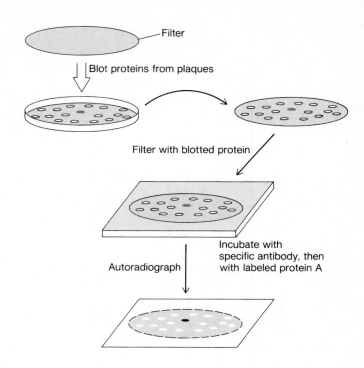

Filter

Blot proteins from plaques

Filter with blotted protein

Autoradiograph

Incubate with specific antibody, then with labeled protein A

FIGURE 15.16
Detecting positive λgt11 clones by measuring expression. A filter is used to blot proteins from phage plaques on a petri dish. One of the clones (red) has produced a plaque containing a fusion protein including β-galactosidase and a part of the protein we are interested in. The filter with its blotted proteins is incubated with an antibody directed against our protein of interest, then with radioactive *Staphylococcus* protein A, which binds specifically to antibodies. It will therefore bind only to the antibody-antigen complexes at the spot corresponding to our positive clone. A dark spot on the autoradiograph of the filter reveals the location of our positive clone.

Figure 15.16 shows how this works. We plate our λ phages with various cDNA inserts and blot the proteins released by each clone onto a support such as nitrocellulose. The nice thing about a λ clone in this regard is that the host cells lyse, forming plaques. In so doing, they release their products, making it easy to transfer proteins from thousands of clones simultaneously, simply by touching a nitrocellulose filter to the surface of a petri dish containing the plaques.

Once we have transferred the proteins from each plaque to nitrocellulose, we probe with our antiserum. Next, we probe for antibody bound to protein from a particular plaque, using radioactive protein A from *Staphylococcus aureus*. This protein binds tightly to antibody and makes the corresponding spot on the nitrocellulose radioactive. We detect this radioactivity by autoradiography, then go back to our master plate and pick the corresponding plaque. Note that we are detecting a fusion protein, not the cloned protein by itself. Furthermore, it does not matter if we have cloned a whole cDNA or not. Our antiserum is a mixture of antibodies that will react with several different parts of our protein, so even a partial gene will do, as long as its coding region is cloned in the same orientation and reading frame as the leading β-galactosidase coding region.

E xpression vectors frequently produce fusion proteins, with the first part of the protein coming from coding sequences in the vector and the latter part from sequences in the cloned gene itself. Fusion proteins have certain advantages: They may be more stable in prokaryotic cells than unmodified eukaryotic proteins, and they may be much simpler to isolate.

Eukaryotic Expression Systems

Eukaryotic genes are not really "at home" in prokaryotic cells, even when they are expressed under the control of their prokaryotic vectors. One reason is that *E. coli* cells frequently recognize the protein products of cloned eukaryotic genes as outsiders and destroy them. Another is that prokaryotes do not carry out the same kinds of post-translational modifications as do eukaryotes. For example, a protein that would ordinarily be coupled to sugars in a eukaryotic cell will be expressed as a bare protein when cloned in bacteria. This can affect a protein's activity or stability, or at least its response to antibodies. A more serious problem is that the interior of a bacterial cell is not as conducive to proper protein folding as the interior of a eukaryotic cell. Frequently, the result is improperly folded, inactive products of cloned genes. Perhaps related to this phenomenon is the fact that we can sometimes express a cloned gene at a stupendously high level in bacteria, but the product forms highly insoluble, inactive granules that are of no use unless we can somehow get the protein to dissolve and regain its activity. Finally, it is obviously hopeless to try to express genes with introns in prokaryotes, since only eukaryotes have the machinery to splice introns out.

In order to avoid the incompatibility between a cloned gene and its host, we can express our gene in a eukaryotic cell. In such cases, we usually do the initial cloning in *E. coli,* using a **shuttle vector** that can replicate in both bacterial and eukaryotic cells. We then transfer the recombinant DNA to the eukaryote of choice by transformation. The traditional eukaryote for this purpose is yeast. It shares the advantages of rapid growth and ease of culture with bacteria, yet it is a eukaryote, and thus it carries out the protein folding and glycosylation (adding sugars) expected of a eukaryote. In addition, by splicing our cloned gene to the coding region for a yeast export signal peptide, we can usually ensure that the gene product will be secreted to the growth medium. This is a great advantage in purifying the protein. We simply remove the yeast cells in a centrifuge, leaving relatively pure secreted gene product behind in the medium.

The yeast vectors are based on a plasmid, called the *2-micron plasmid,* that normally inhabits yeast cells. It provides the origin of replication needed by any vector that must replicate in yeast. Yeast-bacterial shuttle vectors also contain the pBR322 origin of replication, so they can also replicate in *E. coli*. In addition, of course, a yeast expression vector must contain a strong yeast promoter.

Another eukaryotic vector that has been remarkably successful is derived from the nuclear polyhedrosis virus (NPV) that infects the caterpillar known as the alfalfa looper. Viruses in this class have a rather large circular DNA genome, approximately 130 kb in length. The major viral structural protein, polyhedrin, is made in copious quantities in infected cells. In fact, it has been estimated that when a caterpillar dies of an NPV infection, up to 10% of the dry mass of the dead insect is this one protein. This indicates that the polyhedrin gene must be very active, and indeed it is—apparently due to its powerful promoter. When Max Summers and his colleagues placed a human interferon gene under the control of the polyhedrin promoter in an NPV, they harvested five

hundred times as much interferon as had previously been obtained from the same gene expressed in bacterial cells.

MANIPULATING CLONED GENES

Besides simply obtaining the products of cloned genes, we can put them to many uses, as shown in the following three examples: (1) We do not have to be satisfied with the natural product of a cloned gene; once the gene is cloned, we can change it any way we want and collect the correspondingly changed gene product. (2) We can make cloned genes radioactive and use them as probes for many purposes. For example, we can find out how many similar genes exist in an organism's genome or how actively the gene is expressed in living cells. (3) We can determine the exact base sequence of a cloned gene.

PROTEIN ENGINEERING WITH CLONED GENES

Traditionally, protein biochemists have relied on chemical methods to alter certain amino acids in the proteins they study, so that they can then observe the effects of these changes on protein activities. But chemicals are rather crude tools for manipulating proteins; it is difficult to be sure that only one amino acid, or even one kind of amino acid, has been altered. Cloned genes make this sort of investigation much more precise, allowing us to perform microsurgery on a protein. By replacing specific bases in a gene, we also replace amino acids at selected spots in the protein product and observe the effects of those changes on that protein's function.

How do we perform such **site-directed mutagenesis?** First, we need a cloned gene whose base sequence is known. (We will see later in this chapter how DNA sequencing is done; for the moment, let us assume that we have already determined the sequence of the gene we want to manipulate.) Then, we need to obtain our gene in single-stranded form. This can be done by cloning it into M13 phage and collecting the single-stranded progeny phage DNA (plus strands) as described in figure 15.8.

Our next task is to change a single codon in this gene. Let us suppose the gene contains the sequence of bases given in figure 15.17, which codes for a sequence of amino acids that includes a tyrosine. The amino acid tyrosine contains a phenolic group:

To investigate the importance of this phenolic group, we can change the tyrosine codon to a phenylalanine codon. Phenylalanine is just like tyrosine except that it lacks the phenolic group; instead, it has a simple phenyl group:

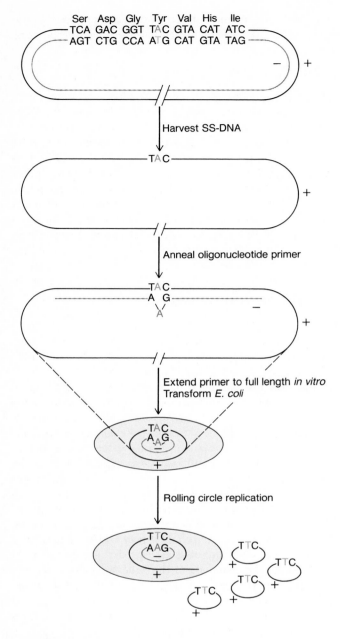

Ser Asp Gly Tyr Val His Ile
TCA GAC GGT TAC GTA CAT ATC
AGT CTG CCA ATG CAT GTA TAG
— +

Harvest SS-DNA

TAC
+

Anneal oligonucleotide primer

TAC
A G
A
— +

Extend primer to full length *in vitro*
Transform *E. coli*

TAC
A A G
—
+

Rolling circle replication

TTC
AAG
—
+

FIGURE 15.17
Site-directed mutagenesis using an oligonucleotide. We begin with the
gene we want to mutagenize, cloned into double-stranded phage M13
DNA. Our goal is to change the tyrosine codon TAC (blue) to the
phenylalanine codon TTC. We transform *E. coli* cells with the
recombinant phage DNA and harvest single-stranded (plus strand)
progeny phage DNA. To this plus strand, we anneal a synthetic
primer that is complementary to the region of interest, except for a
single base change, an A (red) for a T. This creates an A–A mismatch
(blue and red). We extend the primer to form a full-length minus
strand in vitro and use this double-stranded DNA to transform *E. coli*
cells. The minus strand with the single base change then serves as the
template for making many copies of the plus strand, each bearing a
TTC phenylalanine codon (pink) in place of the original TAC tyrosine
codon.

If the tyrosine phenolic group is important to a protein's
activity, replacing it with phenylalanine's phenyl group
should diminish that activity.

We want to change the DNA codon TAC (Tyr) to TTC
(Phe). The simplest way is to use an automated DNA syn-
thesizer to make an oligonucleotide (a "21-mer") with the
following sequence:

3'-AGTCTGCCAAAGCATGTATAG-5'

This has the same sequence as a piece of the original minus
strand, except that the central triplet has been changed
from ATG to AAG. Thus, this oligonucleotide will hybridize to the plus strand we harvested from the M13 phage,
except for the one base we changed, which will cause an
A–A mismatch. The oligonucleotide in this hybrid can
serve as a primer for DNA polymerase to complete the
negative strand in vitro; then we can transform *E. coli* with
the resulting double-stranded replicative form (RF), which
still contains the one-base mismatch.

Once inside the host bacterium, the RF replicates by
a rolling circle mechanism, producing many copies of the
positive strand and using the mutated negative strand as
template. The first positive strand released will be the
original, with its TAC tyrosine codon. However, all subsequent positive strands will be made using the altered
negative strand as template, so they will contain a TTC
phenylalanine codon instead. These mutated positive
strands will be packaged into phage particles (not shown
in figure 15.17), which can be used to infect more bacterial cells. In the final step of the process, we can collect
the mutant RF molecules produced by these cells, cut out
the mutant gene with restriction enzymes, and clone it into
an appropriate expression vector. The protein product will
be identical to the wild-type product except for the single
change of a tyrosine to a phenylalanine.

Note the precision of this technique compared to traditional chemical mutagenesis. With chemical mutagenesis, we assault organisms with an often fatal dose of
mutagen, then examine progeny organisms for the desired
mutant characteristics. Before DNA sequencing was possible, we could not be certain of the molecular nature of
such a mutation without laborious protein sequencing.
Even with modern DNA sequencing techniques, we would
have to determine the sequence of an entire mutant gene
before we could be sure what specific damage a chemical
had caused.

By contrast, site-directed mutagenesis enables us to
decide in advance what parts of a protein we want to
change and to tailor our mutants accordingly. Nevertheless, in spite of its lack of precision, traditional mutagenesis still plays an important role; one such experiment can
quickly and easily create a rich variety of mutants, some
of which involve changes we might never think of—or
never get around to making by site-directed mutagenesis.

Chapter 15

sing cloned genes, we can introduce changes at will, thus altering the amino acid sequences of the protein products. This is most conveniently done with single-stranded cloned DNA from a bacteriophage vector and synthetic oligonucleotide primers containing the desired base change.

■

USING CLONED GENES AS PROBES

The phenomenon of hybridization—the ability of one single-stranded nucleic acid to form a double helix with another single strand of complementary base sequence—is one of the backbones of modern molecular genetics. So pervasive are techniques using hybridization that it would be difficult to overestimate its importance. We have already encountered plaque and colony hybridization earlier in this chapter, we will illustrate here two further examples of hybridization techniques.

Using DNA Probes to Count Genes

In chapter 14 we learned that many eukaryotic genes are not found just once per genome, but are parts of families of closely related genes. Suppose we want to know how many genes of a certain type exist in a given organism. If we have cloned a member of that gene family—even a partial cDNA—we can estimate this number.

We begin by using a restriction enzyme to cut genomic DNA that we have isolated from the organism (figure 15.18). It is best to use a restriction enzyme such as EcoRI or HindIII that recognizes a 6-base-pair cutting site. These enzymes will produce thousands of fragments of genomic DNA, with an average size of about 4,000 base pairs. Next, we electrophorese these fragments on an agarose gel. The result, if we visualize the bands by staining, will be a blurred streak of thousands of bands, none distinguishable from the others. Eventually, we will want to hybridize a radioactive probe to these bands to see how many of them contain coding sequences for the gene of interest. First, however, we must transfer the bands to a medium on which we can conveniently perform our hybridization.

E. Southern was the pioneer of this technique; he transferred, or blotted, DNA fragments from an agarose gel to nitrocellulose by diffusion as depicted in figure 15.18. This process has been called **Southern blotting** ever since. Nowadays, blotting is frequently done by electrophoresing the DNA bands out of the gel and onto the blot. Figure 15.18 also illustrates this process. Before blotting, the DNA fragments are denatured with base so that the resulting single-stranded DNA can bind to the nitrocellulose, forming the Southern blot. Media superior to nitrocellulose are now available; some of them use nylon supports and have clever trade names like Gene Screen. Next, we apply DNA polymerase to our cloned gene in the presence of radioactive DNA precursors to make it radioactive. Then we hybridize this labeled probe to the Southern blot. Wherever the probe encounters a DNA fragment bearing the same sequence (actually the complementary sequence), it hybridizes, forming a band of radioactivity corresponding to the band of DNA containing our gene of interest. Finally, we visualize these bands by autoradiography.

If we find only one band, the interpretation is relatively easy; there is probably only one gene corresponding to our cDNA. If we find multiple bands, there are probably multiple genes, but it is difficult to tell exactly how many. One gene can give more than one band if it contains one or more cutting sites for the restriction enzyme we used. We can minimize this problem by using a short probe, such as a 100–200-base-pair restriction fragment of the cDNA, for example. Chances are, a restriction enzyme that only cuts on average every 4,000 base pairs will not cut within the 100–200-base-pair region of the genes that hybridize to such a probe (unless these regions contain introns, of course).

Using DNA Probes to "Fingerprint" DNA

In 1985, Alec Jeffreys and his colleagues were investigating a DNA fragment from the gene for a human blood protein, α-globin, when they discovered that this fragment contained a sequence of bases repeated several times. For reasons that are not important here, this repeated DNA is called a *minisatellite*. More interestingly, this same minisatellite sequence was found in other places in the human genome, again repeated several times. This simple finding turned out to have far-reaching consequences, because individuals differ in the pattern of repeats of the basic sequence. In fact, they differ enough that two individuals have only a remote chance of having exactly the same pattern. That means that these patterns are like fingerprints; indeed, they are called DNA fingerprints.

How do we make a DNA fingerprint? We first cut the DNA under study with a restriction enzyme such as HaeIII. Jeffreys chose this enzyme because the repeated sequence he had found did not contain a HaeIII recognition site. That means that HaeIII will cut on either side of the minisatellite regions, but not inside, as shown in figure 15.19a. In this case, the DNA has three sets of repeated regions, containing four, three, and two repeats, respectively. Thus, we will get three different-sized fragments bearing these repeated regions.

Next, we electrophorese the fragments we have produced. This separates all the fragments into bands according to their sizes, including the three bearing the minisatellites. Then we separate the two strands of the DNA fragments and blot the single-stranded DNA as just described. At this point, if we have used human DNA, we have millions of fragments, but none of them is visible on the blot; all we can see is a smear of DNA.

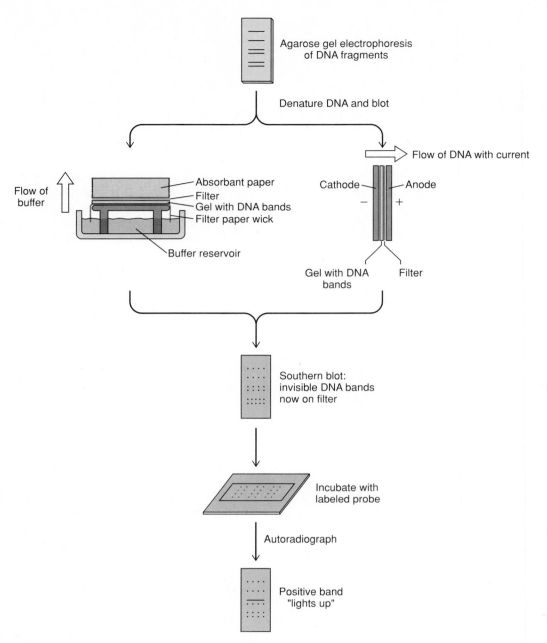

Agarose gel electrophoresis
of DNA fragments

Denature DNA and blot

Flow of DNA with current

Flow of buffer

Absorbant paper
Filter
Gel with DNA bands
Filter paper wick

Buffer reservoir

Cathode   Anode

−   +

Gel with DNA
bands   Filter

Southern blot:
invisible DNA bands
now on filter

Incubate with
labeled probe

Autoradiograph

Positive band
"lights up"

FIGURE 15.18
Southern blotting. First, we electrophorese DNA fragments on an
agarose gel. Next, we denature the DNA with base, and transfer the
single-stranded DNA fragments from the gel (yellow) to a sheet of
nitrocellulose (red) or similar material. This can be done in two ways:
by diffusion, in which buffer passes through the gel, carrying the
DNA with it (left), or by electrophoresis (right). The blot can then be
hybridized to a radioactive probe and autoradiographed to visualize
bands corresponding to the probe.

We can find the bands with the minisatellites, among
millions of irrelevant bands, by hybridizing the blot to a
radioactive single-stranded DNA containing the minisatellite sequence. This labeled probe DNA will form a radioactive double helix with any single-stranded fragments
on the blot containing the complementary minisatellite sequence, making them radioactive. Finally, we can detect
the radioactive bands by autoradiography. In this case,
there are three radioactive bands, so we will see three dark
bands on the film (figure 15.19*d*).

Of course, real animals have a much more complex
genome than the simple piece of DNA in this example, so
they will have many more than three fragments that react
with the probe. Figure 15.20 shows an example of the
DNA fingerprints of several unrelated people and a set of
monozygotic twins. As we have already mentioned, this is
such a complex pattern of fragments that the patterns for
two individuals are extremely unlikely to be identical,
unless they come from monozygotic twins. This complexity makes DNA fingerprinting a very powerful identification technique (see box 15.1).

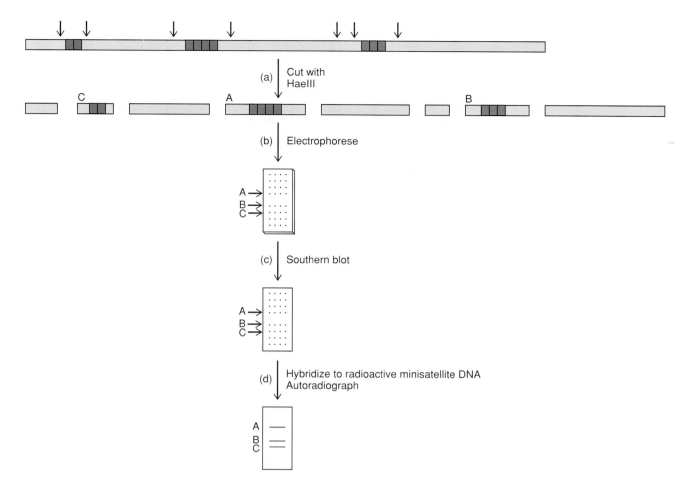

FIGURE 15.19

DNA fingerprinting. (*a*) First, we cut DNA with a restriction enzyme. In this case, the enzyme HaeIII cuts the DNA in seven places (arrows), generating eight fragments. Only three of these fragments (labeled A, B, and C according to size) contain the minisatellites, represented by blue boxes. The other fragments (yellow) contain unrelated DNA sequences. (*b*) We electrophorese the fragments from part (*a*), which separates them according to their sizes. All eight fragments are present in the electrophoresis gel, but we cannot see any of them. The positions of all fragments, including the three (A, B, and C) with minisatellites are indicated by dotted lines. (*c*) We denature the DNA fragments and Southern blot them. (*d*) We hybridize the DNA fragments on the Southern blot to a radioactive DNA with several copies of the minisatellite. This probe will bind to the three fragments containing the minisatellites, but with no others. Finally, we autoradiograph the blot, which reveals the three positive, radioactive bands.

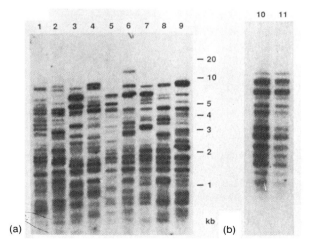

FIGURE 15.20

A DNA fingerprint. (*a*) The nine parallel lanes contain DNA from nine unrelated Caucasians. Note that no two patterns are identical, especially at the upper end. (*b*) The two lanes contain DNA from monozygotic twins, so the patterns are identical.

From G. Vassart, et al. "A Sequence in M13 Phage Detects Hypervariable Minisatellites in Human and Animal DNA." Science, February 6, 1987, p. 683, fig. 1A,B. Copyright by the AAAS.

Uses of DNA Fingerprinting

A valuable feature of DNA fingerprinting is the fact that, although almost all individuals have different patterns, parts of the pattern (sets of bands) are inherited in a Mendelian fashion. Thus, fingerprints can be used to establish parentage. A recent immigration case in England illustrates the power of this technique. A Ghanaian boy born in England had moved to Ghana to live with his father. When he wanted to return to England to be with his mother, British authorities questioned whether he was a son or a nephew of the woman. Information from blood group genes was equivocal, but DNA fingerprinting of the boy demonstrated that he was indeed her son.

Scientists studying family relationships in other vertebrate species are finding DNA fingerprinting a very valuable tool. Alec Jeffrey's human probe works surprisingly well with a wide variety of vertebrate species, and probes for other repeated sequences have also been developed.

Aside from testing parentage, DNA fingerprinting has already become a valuable tool for identifying criminals. One dramatic case involved a man who murdered another man and a woman as they slept in a pickup truck, then about forty minutes later went back and raped the woman. This act not only compounded the crime, it also provided forensic scientists with the means to convict the perpetrator. They obtained DNA from the sperm he had left behind, fingerprinted it, and showed that the fingerprint matched that of the suspect's DNA. The chance that another person's DNA fingerprint would be identical was claimed to be about one in 9.34 billion, more than the entire population of the earth. Thus, this procedure can in principle eliminate all other suspects and therefore offers something very close to absolute identification. In this case, it helped convince the jury to convict the defendant and recommend severe punishment. Jeffreys has now developed probes that hybridize to single sites in minisatellite regions, rather than to a whole repeat. As a result, these probes react with fewer bands and give simpler patterns. They are not as powerful in identification by themselves, but a panel of probes, used in separate experiments, can be definitive. In a related DNA-identifying procedure, the probe hybridizes to specific *variable number tandem repeat* (VNTR) sequences, thus this method is called VNTR typing. These sequences resemble minisatellites in that they are composed of repeats of a core sequence and different individuals have different numbers of repeats. However,

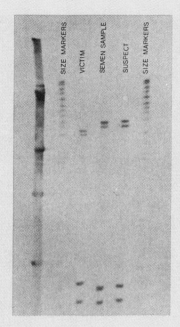

FIGURE B15.1
A VNTR analysis used to identify a rape suspect. Note that the banding pattern of the suspect's DNA is identical to that of the DNA in the rapist's semen, but clearly different from the pattern of the victim's DNA. This very simple pattern was produced by only one probe; to get a more precise identification, three or four probes would be used.
© *Lifecodes Corporation.*

unlike minisatellites, there is usually only one VNTR of a given type in an individual. Thus, VNTR analysis gives patterns that are simpler and easier to analyze than DNA fingerprinting, but several different probes must be used to be convincing. Figure B15.1 presents an example of DNA typing that was used to identify a rape suspect. The pattern from the suspect clearly matches that from the sperm DNA.

One advantage of DNA typing of the kind described here is its extreme sensitivity. Only a few drops of blood or semen are sufficient to perform a test. However, sometimes forensic scientists have even less to go on—a hair pulled out by the victim, for example. While this is not enough for fingerprinting, it still contains a considerable amount of information, especially if accompanied by hair follicle cells. Selected segments of DNA from these cells can be amplified by a technique called *polymerase chain reaction* (PCR) invented by a team of scientists at Cetus Corporation. As figure B15.2 explains, PCR uses the enzyme DNA polymerase to make a copy of a defined region of the DNA. You can select the part of the DNA you want to amplify by putting in short pieces of DNA called primers that hybridize to DNA sequences on either side of a given DNA region and cause initiation (priming) of DNA synthesis through it.

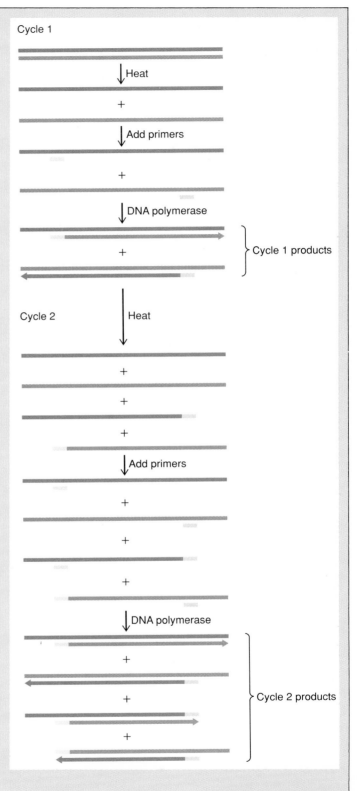

FIGURE B15.2

Amplifying DNA by polymerase chain reaction (PCR). Cycle 1: We start with a DNA duplex (top), and heat it to separate its two strands (red and blue). We then add short, single-stranded DNA primers (pink and blue) complementary to sequences on either side of the region we want to amplify. The primers hybridize to the appropriate sites on the separated DNA strands; now DNA polymerase can use these primers to start synthesis of complementary DNA strands. The arrows represent newly made DNA. At the end of cycle 1, we have two DNA duplexes including the region we want to amplify, while we started with only one. Cycle 2: We repeat the process, heating to separate DNA strands, adding primers (or simply relying on the primers we already added), and adding DNA polymerase (or relying on a heat-stable DNA polymerase added at the beginning). Now each of the four DNA strands, including the two newly made ones, can serve as templates for complementary DNA synthesis. The result is four DNA duplexes that have the region we want to amplify. Notice that each cycle doubles the number of molecules of DNA, because the products of each cycle join the parental molecules in serving as templates for the next cycle. This exponential increase leads to very high numbers in only a short time.

The copies of both strands of the given DNA region, as well as the original DNA strands, then serve as templates for the next round of amplification. In this way, the amount of the selected DNA region doubles over and over—up to at least a million times the starting amount—until there is enough to analyze. Of course, it helps to choose a variable DNA region, such as an HLA region, to amplify, since this increases the information we can obtain. Using a variety of probes on such amplified DNA may in the future allow us to say that a given pattern would only be found once in 10,000 to 100,000 people. This is not as good as the one in ten billion possible with DNA fingerprinting, but coupled with other evidence it could still be convincing.

In spite of its potential accuracy, DNA fingerprinting has recently been effectively challenged in court. The problem is that the technique is tricky and must be performed very carefully to give reliable results. In two celebrated cases, commercial forensic laboratories did not exercise enough care, and the DNA fingerprinting evidence had to be thrown out. The legal profession will have to establish standards for acceptable DNA fingerprint evidence. ■

Using DNA Probes to Measure Gene Activity

Suppose we have cloned a cDNA and want to know how actively the corresponding gene (gene X) is transcribed in a number of different tissues of organism Y. We could answer that question in several ways, but the method we describe here will also tell us the size of the mRNA the gene produces.

We begin by collecting RNA from several tissues of the organism in question. Then we electrophorese these RNAs in an agarose gel and blot them to a suitable support. Sometimes it is necessary to use activated paper that can bind RNA covalently; at other times, the conditions of the transfer ensure that the RNAs will bind tightly, though perhaps not covalently, to nitrocellulose or nylon membranes. Since a similar blot of DNA is called a Southern blot, it was natural to name a blot of RNA a **Northern blot.** (By analogy, a blot of protein is called a Western blot.)

Next, we hybridize the Northern blot to a radioactive cDNA probe. Wherever an mRNA complementary to the probe exists on the blot, hybridization will occur, resulting in a radioactive band that we can detect by autoradiography. If we run marker RNAs of known size next to the unknown RNAs, we can tell the sizes of the RNA bands that "light up" when hybridized to the probe.

Furthermore, the autoradiograph tells us how actively gene X is transcribed. The more RNA the band contains, the more probe it will bind and the darker the band will be on the autoradiograph. We can quantify this darkness by measuring the amount of light it absorbs in an instrument called a densitometer. Figure 15.21 shows a hypothetical Northern blot of RNA from the liver, spleen, brain, muscle, and kidney of a vertebrate, hybridized to a probe for gene X. Clearly, gene X is most actively expressed in the liver.

C loned DNAs make excellent probes because of their ability to hybridize to DNAs or RNAs with the same, or very similar, sequences. These cloned DNAs can be made radioactive and hybridized to DNAs on a Southern blot to reveal the sizes of DNA fragments containing similar sequences. Or they can be hybridized to Northern blots to tell us the sizes and amounts of RNAs containing similar sequences.

∎

DETERMINING THE BASE SEQUENCE OF A GENE

In 1975, Alan Maxam and Walter Gilbert, working as a team, and Frederick Sanger and his colleagues developed two methods for determining the exact base sequence of

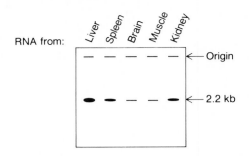

FIGURE 15.21

A Northern blot. mRNAs from vertebrate liver, spleen, brain, muscle, and kidney were electrophoresed and then blotted to nitrocellulose. The blot was hybridized to a radioactive probe for gene X, then autoradiographed. A 2.2-kb mRNA is observed in all tissues, but it is most abundant in the liver.

a defined piece of DNA. These were spectacular breakthroughs. They allowed us to dissect the fine structure of a gene much faster than Benzer could using his traditional genetic tools (chapter 13). Moreover, although Benzer's fine structure mapping placed many mutants close together on a genetic map, it did not reveal anything about base sequence. As a result, Benzer's elegant experiments are unlikely to be applied to any other genetic systems; they are already obsolete.

The key to both the Maxam-Gilbert and Sanger methods is gene cloning. Only cloned genes yield enough of a well-defined piece of DNA to sequence. Another indispensable ingredient of the Maxam-Gilbert sequencing procedure is the ability to cut DNA with restriction enzymes. Only these precise molecular knives allow us to produce DNA molecules all having the exact same ends. Without these identical ends, the DNA sequences obtained would be unintelligible babble.

Let us consider the modern Sanger method that has evolved from the original. We focus on this method because it is easier to understand than the Maxam-Gilbert procedure, although in practice, both are widely used. To sequence a piece of DNA by the Sanger method (figure 15.22), we begin the same way we did for site-directed mutagenesis. We clone the DNA into a vector, such as M13 phage, that can give us the cloned DNA in single-stranded form. To the single-stranded DNA we hybridize an oligonucleotide primer about twenty bases long. This primer is designed to hybridize right beside the multiple cloning region of the vector and is oriented with its 3'-end pointing toward the insert in the multiple cloning region.

If we extend the primer using the Klenow fragment of the DNA polymerase, we will produce DNA complementary to the insert. The trick to Sanger's method is to carry out such DNA synthesis reactions in four separate tubes, but to include in each tube a different chain ter-

Chapter 15

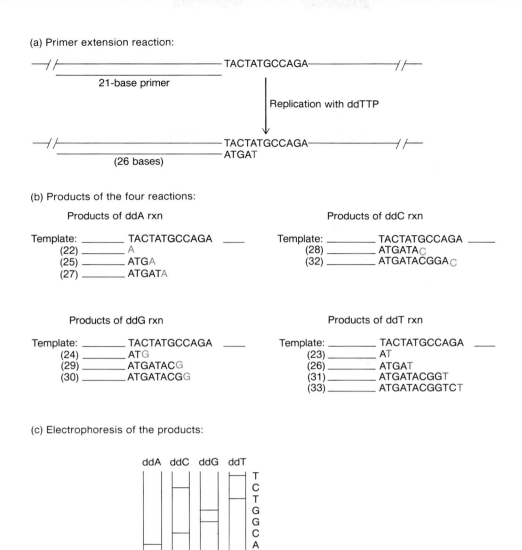

FIGURE 15.22

The Sanger dideoxy method of DNA sequencing. (*a*) The primer extension (replication) reaction. A primer, 21 bases long in this case, is hybridized to the single-stranded DNA to be sequenced, then placed with DNA polymerase Klenow fragment and dNTPs to allow replication. One dideoxy NTP is included to terminate replication after certain bases; in this case, ddTTP is used, and it has caused termination at the second position where dTTP was called for. (*b*) Products of the four reactions (rxn's). In each case, the template strand is shown at the top, with the various products underneath. Each product will begin with the 21-base primer and will have one or more nucleotides added to the 3′-end. The last nucleotide is always a dideoxy nucleotide (color) that terminated the chain. The total length of each product is given in parentheses at the left end of the fragment. Thus, we wind up with fragments ranging from 22 to 33 nucleotides long. (*c*) Electrophoresis of the products. The products of the four reactions are loaded into parallel lanes of a high-resolution electrophoresis gel and electrophoresed to separate them according to size. By starting at the bottom and finding the shortest fragment (22 bases in the A lane), then the next shortest (23 bases in the T lane), and so forth, we can read the sequence of the product DNA. Of course, this is the complement of the template strand.

minator. The chain terminator is a **dideoxy ribonucleotide** such as **dideoxy ATP (ddATP)**. Not only is this terminator 2′-deoxy, like a normal DNA precursor, it is 3′-deoxy as well. Thus, it cannot form a phosphodiester bond, since it lacks the necessary 3′-hydroxyl group. That is why we call it a chain terminator; whenever a dideoxy nucleotide is incorporated into a growing DNA chain, that growth stops.

If we add only dideoxy nucleotides, the chains will not grow at all, so we must put in normal deoxy nucleotides mostly, with just enough dideoxy nucleotide to stop chain growth once in a while. In each tube we place a different

FIGURE 15.23

A typical sequencing film. The sequence begins CAAAAAACGG. How much more of the DNA sequence can you read?

Courtesy of Bethesda Research Laboratories, Life Technologies, Inc. Gaithersburg, MD.

dideoxy nucleotide; ddATP in tube 1, so chain termination will occur with A's; ddCTP in tube 2, so chain termination will occur with C's; and so forth. We also include radioactive dATP in all the tubes so the DNA products will be radioactive.

The result is a series of fragments of different lengths in each tube. In tube 1, all the fragments end in A; in tube 2, in end in C; in tube 3, all end in G; and in tube 4, all end in T. Next, we electrophorese all four reaction mixtures in parallel lanes in a high-resolution polyacrylamide gel. Finally, we perform autoradiography to visualize the DNA fragments, which appear as horizontal bands on an X-ray film.

Figure 15.23 shows a typical sequencing film. To begin reading the sequence, we start at the bottom and find the first band. In this case, it is in the C lane, so we know that this short fragment ends in C. Now we move to the next longer fragment, one step up on the film; the gel electrophoresis used to separate these fragments has such good resolution that it can separate fragments differing by only one base in length. And the next fragment, one base longer than the first, is found in the A lane, so it must end in A. Thus, so far, we have found the sequence C–A. We simply continue reading the sequence in this way as we work up

the film. At first we will just be reading the sequence of part of the multiple cloning region of the vector. However, before very long, the DNA chains will extend into the insert—and unknown territory. An experienced sequencer can continue to read sequence from one film for hundreds of bases.

C loned genes give us a homogeneous population of DNA molecules whose base sequence can be determined by either of two methods. The Sanger method uses dideoxy nucleotides to terminate DNA synthesis, yielding a series of DNA fragments whose sizes can be measured by electrophoresis. The last base in each of these fragments is known, since we know which dideoxy nucleotide was used to terminate each reaction. Therefore, ordering these fragments by size—each fragment one (known) base longer than the next—tells us the base sequence of the DNA.

■

DNA SEQUENCING: THE ULTIMATE IN GENETIC MAPPING

We have already studied the traditional methods for mapping genes. However, the ultimate in fine structure genetic mapping is to obtain the exact base sequence of a gene, a goal that has already been realized for hundreds of genes. By the same token, the ultimate in genetic mapping of a phage would be to learn the base sequence of its whole genome. This has been accomplished for only a handful of simple phages.

The first phage to have its entire base sequence revealed was ϕX174. Frederick Sanger applied his DNA sequencing technique to this single-stranded, circular phage DNA and determined the total sequence of its 5,375 nucleotides.

What kind of information can we glean from this sequence? First, we can locate exactly the coding regions for all the genes. This tells us the spatial relationships among genes and the distances between them (to the exact nucleotide) without any guesswork about recombination frequency. How do we recognize a coding region? It contains an **open reading frame,** a sequence of bases that, if translated in one frame, contains no stop codons for a relatively long distance—long enough to code for one of the phage proteins. Furthermore, the open reading frame must start with an ATG (or occasionally a GTG) triplet, corresponding to an AUG (or GUG) translation initiation codon. In other words, an open reading frame is the same as a gene's coding region.

The base sequence of the phage DNA also tells us the amino acid sequence of all the phage proteins, in case we did not have this information before. All we have to do is use the genetic code to translate the DNA base sequence of each open reading frame into the corresponding amino

FIGURE 15.24
The genetic map of phage φX174. Each letter stands for a phage gene.

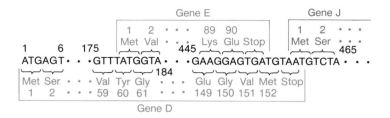

FIGURE 15.25
Overlapping reading frames of φX174. Gene D (red) begins with the base numbered 1 in this diagram and continues through base number 459. This corresponds to amino acids 1–152 plus the stop signal TAA. Dots represent bases or amino acids not shown. Only the coding strand is shown. Gene E (blue) begins at base number 179 and continues through base number 454, corresponding to amino acids 1–90 plus the stop signal TGA. This gene uses the reading frame one space to the right, relative to the reading frame of Gene D. Gene J (green) begins at base number 459 and uses the reading frame one space to the left, relative to Gene D.

acid sequence. This sounds like a laborious process, but we now have computer programs that allow a microcomputer to perform this task in a split second.

Sanger's analysis of the open reading frames of the φX174 DNA revealed something unexpected and fascinating: Some of the phage genes overlap. Figure 15.24 shows that the coding region for gene B lies within gene A and the coding region for gene E lies within gene D. Furthermore, genes D and J overlap by one base pair. How can two genes occupy the same space and code for different proteins? The answer is that the two genes are translated in different reading frames (figure 15.25). Since entirely different sets of codons will be encountered in these two frames, the two protein products will also be quite different.

This was certainly an interesting finding, and it raised the question of how common this phenomenon would be. So far, major overlaps seem to be confined to viruses, which is not surprising since these simple genetic systems have small genomes and there is a premium on efficient use of the genetic material. Moreover, viruses have prodigious power to replicate, so enormous numbers of generations have passed during which evolution has honed the viral genomes.

> *T*he base sequence of the simple phage φX174 reveals all its genes' coding regions, or open reading frames, and shows that two of the phage genes lie within other genes. These genes-within-genes still code for different products because they are translated in different frames.

■

PRACTICAL APPLICATIONS OF GENE CLONING

Our discussion of expression vectors demonstrated that it is possible to engineer bacteria or other cells to produce the products of cloned genes. Obviously, some of these protein products may be very useful, or very scarce, or both, and are therefore potentially valuable. In this section we will see how cloned gene products, and the genes themselves, have the potential to reshape medicine and agriculture.

PROTEINS: THE NEW PHARMACEUTICAL FRONTIER

What sorts of valuable proteins would we want to produce? The first cloned pharmaceutical product to be sold was human insulin, which is valuable for two reasons: First, some diabetics become allergic to the pig and bovine insulin traditionally used. Since pig insulin is a little different from the human variety, our bodies can detect that difference and mount an immune response, or even an allergic response, to the foreign protein. In principle, human insulin should not cause a problem, because the body does not recognize it as foreign (although a few patients have suffered unexplained reactions). Furthermore, the cloned gene assures an unlimited supply of insulin, so we do not have to depend on the vagaries of the livestock market.

Another example of a useful protein is **human growth hormone (hGH)**. Some children inherit a condition called **hypopituitary dwarfism.** They do not grow normally because their bodies make too little growth hormone. They are destined to be dwarves unless they are treated with hGH. Growth hormone from other animals will not help. In the past, these children were treated with growth hormone extracted from the pituitary glands of human cadavers. But this was very expensive; hormone from more than seventy cadavers was required in order to treat one child for one year. Moreover, early preparations of hGH were apparently contaminated with the agent that causes a lethal brain condition called Creutzfeld-Jacob disease. The solution was to obtain hGH by cloning its gene. This has been accomplished, and the product is effective.

IMPROVED VACCINES

Novel, safer vaccines may be another valuable product of cloned genes. Usually, vaccines are made from whole viruses that have been either killed or weakened so that they can no longer cause disease but can still arouse the body's immune defenses. The problem is that sometimes these viruses are not killed or are not weakened enough, thus they end up causing the disease, and sometimes even death, after all. For example, smallpox vaccine is now considered more dangerous than the almost nonexistent chance of catching smallpox, so it is no longer given routinely in the United States.

One of the most fearful viral diseases in the United States is AIDS (acquired immune deficiency syndrome). The most promising long-term solution to the AIDS problem is vaccination, but consider the potential dangers of an AIDS vaccine containing live virus. No pharmaceutical company would risk selling such a product to the general public.

Gene cloning may provide a solution to this problem. Several research groups have already cloned the genes for the surface proteins of the AIDS virus (**human immunodeficiency virus, or HIV**). They are now harvesting the proteins from the bacteria harboring those genes and using them to trigger the body's immune system in much the same way a whole virus would. In other words, by immunizing people against a viral surface protein, we ought to be able to immunize them against the virus and therefore against AIDS. Notice that these proteins do not come directly from the virus but from relatively harmless bacteria. Thus, the vaccine should be safe.

A related strategy would be to insert a cloned AIDS surface protein gene into a harmless virus such as vaccinia. This virus, otherwise known as the cowpox virus, has been used since the end of the eighteenth century to immunize people against smallpox. With an added gene for HIV surface protein, it could become the basis for an effective AIDS vaccine. However, some observers have expressed concern that the HIV surface protein might direct the recombinant vaccinia virus to the very helper T-cells that HIV normally attacks. This could disrupt those cells' functions and possibly cause an AIDS-like condition.

In addition to this reservation, there are other reasons to doubt that a preventive for AIDS is just around the corner. Vaccines are tricky to develop, even in the best of circumstances, and researchers are finding many different forms of AIDS virus, each of which would require a different vaccine. Furthermore, the virus seems capable of rapid genetic change, so it may change its coat to frustrate vaccines, the way flu virus does.

*T*he most obvious use of cloned genes is to express them and harvest their products. This has already yielded proteins that have pharmaceutical value, including human insulin and human growth hormone. In the future, it may provide purified viral proteins that can serve as safe, effective vaccines against deadly diseases. We may even be able to engineer harmless viruses that can immunize us against unrelated dangerous viruses.

INTERVENING IN HUMAN GENETIC DISEASE

Ultimately, we may be able to use cloned genes to treat patients suffering from genetic diseases. The idea would be to implant a normal gene to correct the defect.

The classic example of a genetic disease is sickle-cell disease (chapter 7). Recall that sickle-cell patients are homozygous for a defective β-globin gene; as a result, they make abnormal hemoglobin, which causes the trouble. This suggests that a genetic engineering solution is possible. Since blood cells are made in the bone marrow, we should be able to remove all of a patient's abnormal marrow and replace it with marrow that has a normal gene for hemoglobin. It would be best if we could use the patient's own marrow, to avoid a rejection problem. But how can we get a normal gene into abnormal marrow cells? Again, cloning is the key. We have already cloned human globin genes, and these can be added to cells in much the same way that we add recombinant DNAs to bacterial cells.

An experimental procedure might work like this: (1) Take out some of the patient's bone marrow cells. (2) Add a cloned normal hemoglobin gene to the extracted cells. (3) Grow these engineered cells in culture. (4) Irradiate the patient with a high enough dose of X rays to kill all the remaining marrow cells. (5) Rescue the patient, who is destined to die without marrow cells, by implanting the engineered cells; these will grow and repopulate the bone marrow. (6) Presumably, the normal gene that was added will function well enough to prevent sickling.

Do we know this procedure will work? Actually, we don't, but we do know that it is possible to cure sickle-cell disease with a marrow transplant. An eight-year-old girl in Boston suffered from both leukemia and sickle-cell disease. Doctors treated the leukemia by destroying the patient's cancerous bone marrow and replacing it with normal marrow from another person. Not only did this seem to cure the girl's leukemia, but it had the side effect of curing her sickle-cell disease as well, because of course the donor cells had come from a person who didn't suffer from that disease. So we know the transplant part of the procedure can work.

The current roadblock is getting a cloned gene to function normally in the patient's cells. This has been a problem, not only for globin genes, but for all genes transplanted into animals. For some reason, they are not usually expressed and regulated as they would be in their natural state. This might be especially serious for the globin genes, which are normally regulated so that their α- and β-globin products are produced in roughly equal amounts. For this reason, sickle-cell disease will probably not be the first genetic disorder attacked with cloned genes.

More attractive targets for cloned genes will be diseases caused by failure of a gene to function at all, where a donor gene producing any amount of product will help and overproduction will not hurt. An obvious example is Lesch-Nyhan syndrome, which is characterized by mental retardation and bizarre self-mutilation such that the patient may literally bite off his own fingers and lips. This genetic disease results from a defective gene for the enzyme hypoxanthine-guanine phosphoribosyl transferase (HPRT). Nobody knows why lack of this enzyme causes such dramatic effects, but placement of a normal HPRT gene in transplanted bone marrow might yield at least some of the enzyme and therefore partially alleviate the symptoms.

I n the future, it may be possible to treat genetic defects by transplanting cloned, normal genes into patients whose own genes are damaged. A logical target tissue for transplanted genes is the bone marrow, which can be removed, manipulated, and replaced.

∎

USING CLONED GENES IN AGRICULTURE

In 1983, Richard Palmiter and Ralph Brinster placed the gene for human growth hormone in an early mouse embryo. The result, dubbed "supermouse" by the popular press, was a giant mouse—twice normal size. We call such animals **transgenic** because they contain a transplanted gene from another organism. It is important to realize that a true transgenic animal receives the foreign gene (the *transgene*) at a very early stage in development so that the transgene becomes incorporated into the host animal's chromosomes and can be passed on to progeny as any normal gene would be. This is different from adding a gene

to one tissue in an adult animal, as would be done in the sickle-cell disease case we just discussed. In that situation, the transplanted gene would not enter the germ cells, and so would not be passed on to the next generation.

As soon as people heard about "supermouse," they began wondering, "Why not 'supercow,' or 'superpig?'" (Never mind "superman.") Palmiter, Brinster, and their colleagues have now attempted to create a "superpig," and so far their experiment is only a qualified success. These workers have created transgenic pigs by injecting genes for human growth hormone (hGH) and bovine growth hormone (bGH) into pig eggs. The bGH worked best, but the transgenic pigs containing bGH probably do not deserve the name "superpig." They did not grow to supernormal size, but they did show a greater than normal feed efficiency—that is, weight gain per unit of feed. This is important, since feed accounts for up to 70% of the day-to-day cost of raising pigs. Furthermore, the transgenic pigs had a much smaller amount of subcutaneous fat, the layer of fat just beneath the skin. This is a benefit during the current era of concern about the adverse effects of animal fat in our diet.

Offsetting these benefits, however, are some serious health problems in the transgenic pigs. They have an abnormally high incidence of stomach ulcers, arthritis, enlarged heart, dermatitis, and kidney disease. These conditions depress appetite and lead to higher mortality, already a significant problem in domestic pigs. Clearly, more work is required to produce a real superpig.

PROTEIN PRODUCTS FROM TRANSGENIC ORGANISMS

Geneticists have recently begun using transgenic farm animals for a more mundane purpose: producing valuable proteins from cloned genes. Consider, for example, the gene for a human blood clotting protein called factor IX. People homozygous for a defect in this gene suffer the blood clotting disorder hemophilia B. They have traditionally been treated with human plasma or partially purified factor IX, but these human blood products can be contaminated with hepatitis virus, or even HIV, the AIDS virus, making that method of treatment somewhat risky. Thus, a cloned source of the protein would be a big advantage.

However, factor IX is produced by human cells as a precursor that requires extensive modification to make it active. The usual cells in which we express cloned genes, bacteria or yeast, do not perform these key alterations. Therefore, expression in mammalian cells is called for, but cultured mammalian cells are expensive to maintain. An alternative is to make a transgenic animal do the job, and A. J. Clark and his colleagues have done just that. They linked the human factor IX gene to the control sequences of a sheep milk protein gene called β-lactoglobulin (BLG). Then they injected this fused gene into sheep eggs and implanted the eggs in ewes. The rationale for this, of course, is that the factor IX gene will be expressed in the mammary gland, where BLG is normally produced, and the factor IX can therefore be conveniently recovered in the milk of transgenic ewes. Two transgenic ewes were born; after they were mated, they did indeed produce human factor IX in their milk. Admittedly, the amounts of factor IX produced were very modest, but the idea is such a good one that people will try hard to make it successful. Indeed, the idea is not confined to transgenic animals. Recent experiments have shown that transgenic tobacco plants can also be used as protein factories.

GENETICALLY ENGINEERED CROPS

The ability to place cloned genes into plants raises the possibility of engineering new, improved strains of crops. Of course, humankind has been improving crops by selective breeding for millenia, but it has been a time-consuming, hit-and-miss process in which many undesirable genes are transferred along with the advantageous ones. Now we can identify useful genes, isolate them by cloning, and insert them directly into a new plant host. If these genes behave normally in their new environment (a big "if," indeed), they may be able to effect dramatic improvements in the recipient plants.

One of the first useful genes to be transferred this way was one that confers herbicide resistance. This trait is useful because an herbicide-resistant plant can survive treatment while weeds all around it are dying. The herbicide in question was **glyphosate,** the active ingredient in Monsanto's weed killer "Roundup." Glyphosate kills by inhibiting an enzyme called EPSP synthase, which is necessary for making the essential amino acids phenylalanine, tyrosine, and tryptophan. Without these amino acids, plants cannot live, making glyphosate a very effective herbicide. But one would hardly want to spray it on fields planted with ordinary crops, since it kills all kinds of plants—weeds and crops alike. The aim, therefore, was to make crops resistant to glyphosate.

David Stalker and his coworkers managed to engineer glyphosate-resistant tobacco plants. First, they grew *Salmonella* bacteria in the presence of glyphosate and isolated a resistant mutant that could still grow. The EPSP synthase of the mutant has suffered a single amino acid change, which accounts for its glyphosate resistance. Next, these workers cloned the mutant gene and transferred it to a tobacco plant.

The transfer is not as easy as it sounds. In order to make the foreign gene function in a plant, genetic engineers had to clone it in a vector that could function effectively in plant cells. The common bacterial vectors, such as pBR322, would not serve this purpose, because plant cells cannot recognize their prokaryotic promoters and replication origins. Instead, the research team placed the glyphosate resistance gene into a plasmid containing so-called **T-DNA.** This is a piece of DNA from a plasmid known as **Ti** (tumor-inducing).

The Ti plasmid inhabits the bacterium *Agrobacterium tumefaciens,* which causes tumors called **crown galls** (figure 15.26) in dicotyledonous plants. When this bacterium infects a plant, it transfers its Ti plasmid to the host cells, whereupon the T-DNA integrates into the plant DNA, causing the abnormal proliferation of plant cells that gives rise to a crown gall. This is advantageous for the invading bacterium, because the T-DNA has genes directing the synthesis of unusual organic acids called **opines;** these opines are worthless to the plant, but the bacterium has enzymes that can break down opines so that they can serve as an exclusive energy source for the bacterium.

The T-DNA genes coding for the enzymes that make opines (e.g., mannopine synthase) have strong promoters. Plant molecular geneticists take advantage of them by putting T-DNA into small plasmids, then placing foreign genes under the control of one of these promoters. Stalker and his colleagues used the mannopine synthase promoter. After they had formed their recombinant DNA containing the herbicide resistance gene, they placed it back into *Agrobacterium,* which they then used to infect plant cells and thereby transfer the cloned gene.

Figure 15.27 outlines the process these investigators used to transfer the herbicide resistance gene to a tobacco plant. They punched out a 2-mm diameter disk from a tobacco leaf and placed it in a dish with nutrient medium. Under these conditions, tobacco tissue will grow around the edge of the disk. Next, they added *Agrobacterium* containing the cloned glyphosate resistance gene; these bacteria infected the growing tobacco cells and introduced the cloned gene.

When the tobacco tissue grew roots around the edge, Stalker and his coworkers transplanted those roots to medium that encouraged shoots to form. These plantlets gave rise to full-sized tobacco plants that, if they contained a functioning glyphosate resistance gene, should be resistant to the herbicide. The next step was to spray the plants with herbicide to find out if they really were resistant. Figure 15.28 shows the results of the test. The wild-type tobacco plants at the top were obviously sensitive, even to intermediate levels of the herbicide, while the engineered plants at the bottom were quite resistant to intermediate levels and somewhat resistant to high levels.

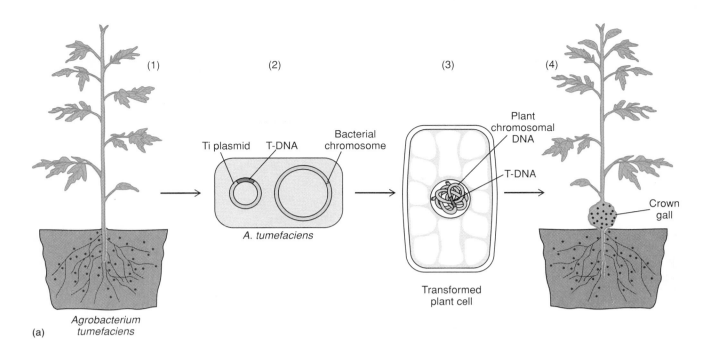

(a)

(b)

FIGURE 15.26

(*a*) Formation of a crown gall. 1. *Agrobacterium* cells enter a wound in the plant, usually at the crown, or the junction of root and stem. 2. The *Agrobacterium* contains a Ti plasmid in addition to the much larger bacterial chromosome. The Ti plasmid has a segment (the T-DNA, red) that promotes tumor formation in infected plants. 3. The bacterium contributes its Ti plasmid to the plant cell, and the T-DNA from the Ti plasmid integrates into the plant's chromosomal DNA. 4. The genes in the T-DNA direct the formation of a crown gall, which nourishes the invading bacteria. (*b*) Photograph of a crown gall tumor generated by cutting off the top of a tobacco plant and inoculating with *Agrobacterium*. This crown gall tumor is a teratoma, which generates normal as well as tumorous tissue. Note the disorganized, more or less normal plant tissues springing from the tumor.

(a) From "A Vector for Introducing New Genes into Plants" by Mary-Dell Chilton. Copyright © 1983 by SCIENTIFIC AMERICAN, Inc. All rights reserved. Photo (b) courtesy of Dr. Robert Turgeon and Dr. B. Gillian Turgeon, Cornell University.

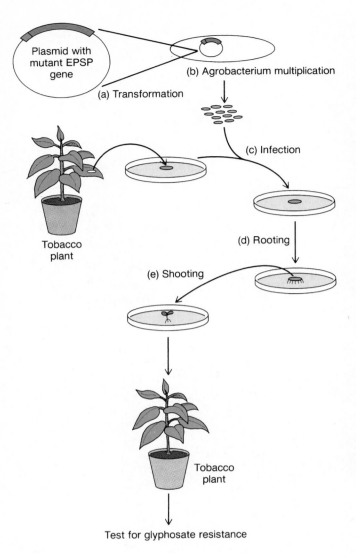

FIGURE 15.27
Using a T-DNA plasmid to introduce a gene into tobacco plants.
(*a*) A plasmid is formed with the mutant EPSP gene (red) under the
control of the mannopine synthetase promoter (blue). This plasmid is
used to transform *Agrobacterium* cells. (*b*) The transformed bacterial
cells divide repeatedly. (*c*) A disk of tobacco leaf tissue is removed and
incubated in nutrient medium, along with the transformed
Agrobacterium cells. These cells infect the tobacco tissue, transferring
the plasmid bearing the mutant EPSP gene. (*d*) The disk of tobacco
tissue sends out roots into the surrounding medium. (*e*) One of these
roots is transplanted to another kind of medium, where it forms a
shoot. This plantlet grows into a tobacco plant that can be tested for
glyphosate resistance.

FIGURE 15.28
Testing transformed tobacco plants for herbicide resistance. The top
nine plants are wild-type; the bottom nine have been transformed with
a glyphosate-resistant EPSP gene. The left vertical row received no
glyphosate. The middle vertical row received the equivalent of 0.6 kg
glyphosate per hectare. The right vertical row received the equivalent
of 1.0 kg glyphosate per hectare. Note the resistance of the
transformed plants, especially to the lower concentration of
glyphosate.

*From L. Comai, D. Facciotti, W. R. Hiatt, G. Thompson, R. E. Rose, and D. M.
Stalker. Reprinted by permission from* Nature *317:744. Copyright © 1985
Macmillan Magazines Limited.*

This is only one example of plant genetic engineering.
Other plant geneticists have made additional strides, including the following: (1) conferring glyphosate resistance on petunias by inserting extra copies of a normal
EPSP gene; (2) conferring virus resistance on tobacco
plants by inserting a gene for the viral coat protein;
(3) conferring pest resistance on tobacco plants by inserting a bacterial gene for an insecticidal toxin; and
(4) inserting the gene for firefly luciferase into tobacco
plants. The latter experiment has no practical value, but
it does have the arresting effect of making the plant glow
in the dark (figure 15.29).

Why all this emphasis on tobacco? Would it not be
more productive to work on an important food crop such
as wheat, corn, or rice? It certainly would, but tobacco
happens to be a very convenient plant to manipulate, for
two important reasons: First, it can be infected with *Agrobacterium,* providing an easy way to insert foreign DNA.
(By contrast, all cereal plants are *monocots* that, unfortunately, cannot be infected with *Agrobacterium.*) Furthermore, in the experiments just described, it was
necessary to regrow a whole plant from individual plant

Chapter 15

FIGURE 15.29
A tobacco plant that has been transformed with the firefly luciferase gene. The green light emitted by the plant indicates the activity of luciferase. Note the concentration of the enzyme in the roots and stem.

D. W. Ow, Keith Wood, Marlene DeLuca, Jeffrey R. DeWet, Donald R. Helinski, Steven Howell. "Transient and Stable Expression of the Firefly Luciferase Gene in Plant Cells and Transgenic Plants." Science 234:856–59, fig. 5, November 1986. Copyright 1986 by AAAS.

cells that had been transformed with the cloned gene. This is easy with tobacco and its relative, the petunia, but could not be done with cereals until the mid-1980s.

I t is possible to transfer cloned genes to plants, thereby changing the properties of the plants. In dicotyledonous plants, this is generally done using a gene under control of a promoter from a T-DNA plasmid derived from *Agrobacterium.* The recombinant DNA is then placed back in the bacterium, which is used to infect plant cells. Finally, the infected cells, with their transplanted genes, are grown into whole plants and tested for new traits, such as herbicide or disease resistance.

■

MAPPING GENETIC DEFECTS IN HUMANS

Certain debilitating diseases are heritable, indicating that they are caused by genetic defects. In chapter 7, for example, we saw that a defective gene for the β-globin blood protein can cause sickle-cell disease. There is great interest in locating the defective genes responsible for such human genetic diseases because knowing the nature of the defect may lead us to the cure. But this is more difficult than it sounds. Finding the defective gene was relatively simple in the case of sickle-cell disease, since we knew the identity of the defective protein. But in most cases, we only know the symptoms of the disease, not the cause.

Consider *Huntington's disease,* the degenerative nerve disorder that killed folksinger Woody Guthrie; *Duchenne muscular dystrophy,* a muscle-wasting disease that almost always attacks boys rather than girls; or *cystic fibrosis,* a common disease in which thick mucous secretions build up in the lungs of young patients. In none of these disorders did we know at the outset what gene products were defective, so we could not rely on that kind of information to lead us to the mutant genes. But in each case, imagination, hard work, and a measure of luck have brought us close to the answer after all.

HUNTINGTON'S DISEASE

As mentioned, Huntington's disease is a progressive nerve disorder. It begins almost imperceptibly with small tics and clumsiness. Over a period of years, these symptoms intensify and are accompanied by emotional disturbances. Nancy Wexler describes the advanced disease as follows: "The entire body is encompassed by adventitious movements. The trunk is writhing and the face is twisting. The full-fledged Huntington's patient is very dramatic to look at." Finally, after ten to twenty years, the patient dies.

Huntington's disease is almost certainly controlled by a single dominant gene (or conceivably, by two or more very tightly linked genes). A child of a Huntington's patient has a 50:50 chance of being afflicted with the disease. People who have the disease could avoid passing it on by not having children, except that the first indications of trouble usually do not appear until after the childbearing years.

RESTRICTION FRAGMENT LENGTH POLYMORPHISMS

Because we do not know the nature of the product of the Huntington's disease gene, we cannot look for the gene directly. The next best approach is to look for a gene that is tightly linked to the Huntington's disease gene. Michael Conneally and his colleagues spent more than a decade trying to find such a linked gene, but with no success.

Another approach, which does not depend on finding linkage with a known gene, is to establish linkage with an "anonymous" stretch of DNA that may not even contain any genes. We recognize such a piece of DNA by its pattern of cleavage by DNA-cutters called restriction enzymes, which recognize very specific sequences in DNA and cut at or near those sequences.

Ethical Issues

D iscussion of gene cloning raises ethical questions about genetic engineering, especially as it applies to changing human genes. These questions have been with us ever since the first gene cloning experiments. Actually, the early debate about cloning revolved around the safety issue. Some scientists were so alarmed about the potential dangers of introducing new genes into bacteria that they convinced the molecular biology community to stop this kind of research until it could be studied more fully. The moratorium lasted for eighteen months, at which time it began to be clear that dangerous germs have evolved along with their hosts for thousands of years and that dozens of genes are probably needed to make a bacterium pathogenic. It is extremely improbable that we could create a dangerous new germ simply by adding one new gene to an innocuous microbe. Solid evidence now indicates that those early fears were unfounded; thousands of scientists have been cloning genes during the last decade, with no untoward effects.

But as the safety debate has subsided, a new one has heated up regarding the ethics of altering human genes. It is not too hard to identify sickle-cell disease as a genetic defect, but where do you draw the line? Is a below-average IQ a genetic defect? Is left-handedness? Who would make these decisions? Fortunately, we are probably at least a century away from knowing the genetic basis of intelligence, let alone being able to change it genetically, so worries of this sort are probably a bit premature.

What about creating transgenic humans—changing not only the genes of patients, but the genes in their germ cells, so that their descendants' genes will be changed as well? This is tantamount to interfering in human evolution, and some would argue that it is too much like playing God. Others point out that we have a duty to minimize human suffering, and thus, that eliminating a defective allele from the population could only be good. Whatever your feelings on these issues, it is clear that they raise questions for which there may never be easy answers, let alone correct ones. ■

Since one person differs genetically from another, the sequences of their DNAs will differ somewhat, as will the pattern of cutting by restriction enzymes. Consider the restriction enzyme HindIII, which recognizes the sequence AAGCTT. One individual may have three such sites separated by 4 and 2 kilobases (kb), respectively, in a given region of a chromosome (figure 15.30). Another individual may lack the middle site but have the other two 6 kb apart. This means that if we cut the first person's DNA with HindIII, we will produce two fragments 2 kb and 4 kb long. The second person's DNA will yield a 6-kb fragment instead. In other words, we are dealing with a restriction fragment length polymorphism (see chapter 5). This very clumsy term simply means that cutting the DNA from any two individuals with a restriction enzyme may yield fragments of different lengths. The abbreviated term, RFLP, is usually pronounced "rifflip."

Because geneticists failed to find a standard genetic marker, or gene, linked to the Huntington's disease gene, they looked for a RFLP that is linked. But how do we go about looking for a RFLP? Clearly, we cannot analyze the whole human genome at once. It contains hundreds of thousands of cut sites for a typical restriction enzyme, so each time we cut the whole genome with such an enzyme, we release hundreds of thousands of fragments. No one would relish sorting through that morass for subtle differences between individuals.

Fortunately, there is an easier way. We can cut the DNA with a restriction enzyme, electrophorese the hundreds of thousands of fragments to separate them by size, then transfer them to a piece of special DNA-binding paper by Southern blotting. During the blotting process, we denature the fragments (separate their strands). Now, how can we identify the one or two bands we are interested in among thousands of irrelevant bands? Here we take advantage of the ability of one strand of DNA to form a double helix with a complementary strand but not with an unrelated strand. We isolate a piece of DNA that spans at least part of our region of interest (about 4 kb in figure 15.30), make it radioactive, and denature it. The separated strands of this radioactive DNA **probe** will now **hybridize,** or form a new double helix, with the separated

Chapter 15

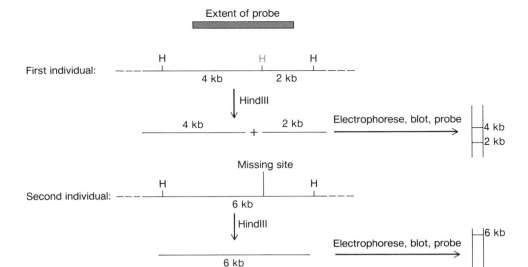

FIGURE 15.30

Detecting a RFLP. Two individuals are polymorphic with respect to a HindIII restriction site (blue). The first individual contains the site, so cutting the DNA with HindIII yields two fragments, 2 and 4 kb long, that can hybridize with the probe, whose extent is shown at top. The second individual lacks this site, so cutting that DNA with HindIII yields only one fragment, 6 kb long, which can hybridize with the probe. The results from electrophoresis of these fragments, followed by blotting, hybridization to the radioactive probe, and autoradiography, are shown at right. The fragments at either end, represented by dotted lines, do not show up because they cannot hybridize to the probe.

DNA strands on the Southern blot. But this hybridization will be very specific; the probe will only hybridize to a complementary DNA strand. That means the 4-kb and 2-kb bands, but no others, from the first individual should hybridize to the probe. Similarly, only the 6-kb band from the second individual should hybridize. Since the probe is radioactive, we can easily detect it by placing a piece of X-ray film over the Southern blot and allowing the radioactive emissions to expose the film by autoradiography (chapter 8).

Using RFLPs to Map the Huntington's Disease Gene

The approach just described was used by a team of scientists, including Wexler, Conneally, and James Gusella, to find a RFLP linked to the Huntington's disease gene. Of course, there is a catch. Each radioactive probe only hybridizes to a small fraction of the total human DNA, so hundreds of probes may have to be screened before the right one is found. In this case, the investigators were lucky; they found a promising probe (called G8) among the first dozen they tried.

They were also lucky to have a very large family to study. Living around Lake Maracaibo in Venezuela is a family whose members have suffered from Huntington's disease since the early nineteenth century. The first member of the family to be so afflicted was a woman whose father, presumably a European, carried the defective gene. So the pedigree of this family can be traced through seven generations, and the number of individuals is unusually large, a typical nuclear family having fifteen to eighteen children.

The Gusella et al. group used their probe to examine RFLPs in the members of this Venezuelan family, then checked to see if the RFLPs correlated with the disease. Indeed, the probe located a RFLP that is very tightly linked to the Huntington's disease gene. Figure 15.31 shows the locations of HindIII sites in the stretch of DNA that hybridizes to the probe. We can see seven sites in all, but only five of these are found in all family members. The other two, marked with asterisks and numbered 1 and 2, may or may not be present. These latter two sites are therefore **polymorphic**, or variable.

Let us see how the presence or absence of these two restriction sites gives rise to a RFLP. If site 1 is absent, a single fragment 17.5 kb long will be produced. However, if site 1 is present, the 17.5-kb fragment will be cut into two pieces having lengths of 15 kb and 2.5 kb. Only the 15-kb band will show up on the autoradiograph, since the 2.5-kb fragment lies outside the region that hybridizes to the G8 probe. If site 2 is absent, a 4.9-kb fragment will be produced. On the other hand, if site 2 is present, the 4.9-kb fragment will be subdivided into a 3.7-kb fragment and a 1.2-kb fragment.

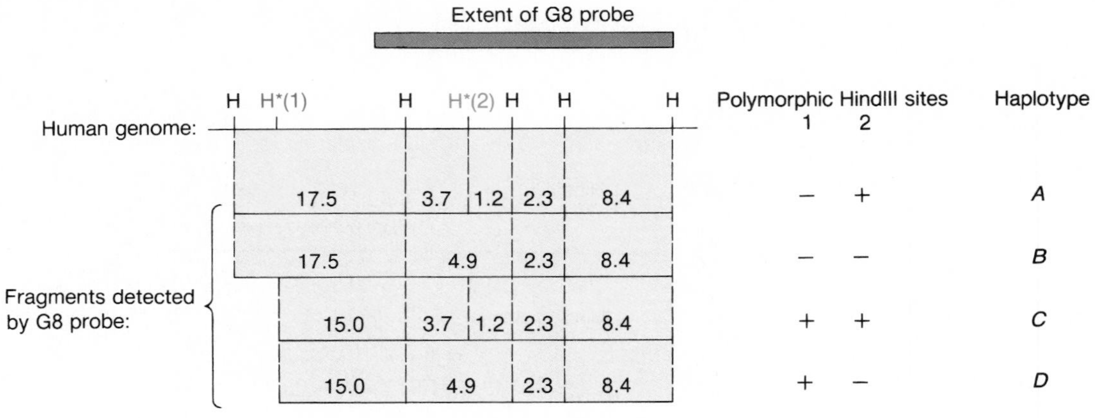

Extent of G8 probe

Human genome:	H	H*(1)		H	H*(2)	H	H	H	Polymorphic HindIII sites 1	2	Haplotype
		17.5		3.7	1.2	2.3	8.4		−	+	A
Fragments detected by G8 probe:		17.5		4.9		2.3	8.4		−	−	B
		15.0		3.7	1.2	2.3	8.4		+	+	C
		15.0		4.9		2.3	8.4		+	−	D

FIGURE 15.31

The RFLP associated with the Huntington's disease gene. The HindIII sites in the region covered by the G8 probe are shown. The families studied show polymorphisms in two of these sites, marked with an asterisk and numbered 1 (blue) and 2 (red). Presence of site 1 results in a 15-kb fragment plus a 2.5-kb fragment that is not detected because it lies outside the region that hybridizes to the G8 probe. Absence of this site results in a 17.5-kb fragment. Presence of site 2 results in two fragments of 3.7 and 1.2 kb. Absence of this site results in a 4.9-kb fragment. Four haplotypes (A–D) result from the four combinations of presence or absence of these two sites. These are listed at right, beside a list of polymorphic HindIII sites and a diagram of the HindIII restriction fragments detected by the G8 probe for each haplotype. For example, haplotype A lacks site 1 but has site 2. As a result, HindIII fragments of 17.5, 3.7, and 1.2 are produced. The 2.3- and 8.4-kb fragments are also detected by the probe, but we ignore them because they are common to all four haplotypes.

There are four possible haplotypes (sets of linked genetic loci) with respect to these two polymorphic HindIII sites, and they have been labeled A–D:

Haplotype	Site 1	Site 2	Fragments observed
A	Absent	Present	17.5; 3.7; 1.2
B	Absent	Absent	17.5; 4.9
C	Present	Present	15.0; 3.7; 1.2
D	Present	Absent	15.0; 4.9

Of course, each member of the family will inherit two haplotypes, one from each parent. This means that a person with the AD genotype will have the same RFLP pattern as one with the BC genotype, since all five fragments will be present in both cases. However, the true genotype can be deduced by examining the parents' genotypes. Figure 15.32 shows autoradiographs of Southern blots of two families, using the radioactive G8 probe. The 17.5- and 15-kb fragments migrate very close together, so they are difficult to distinguish when both are present, as in the AC genotype; nevertheless, the AA genotype with only the 17.5-kb fragment is relatively easy to distinguish from the CC genotype with only the 15-kb fragment. The B haplotype in the first family is obvious because of the presence of the 4.9-kb fragment.

Now we would like to know which haplotype is associated with the disease in the Venezuelan family; figure 15.33 demonstrates that it is C. Nearly all individuals with this haplotype have the disease. Those who do not will almost certainly develop it later. Equally telling is the fact that no individual lacking the C haplotype has the disease. Thus, this is a very accurate way of predicting whether a member of this family is carrying the Huntington's disease gene. A similar study of an American family showed that the A haplotype was linked with the disease. Therefore, each family varies in the haplotype associated with the disease, but within a family, the linkage between the RFLP site and the Huntington's disease gene is so close that recombination between these sites is very rare.

Aside from its use as a genetic counseling tool, what does this analysis reveal? For one thing, it has allowed us to locate the Huntington's disease gene on chromosome 4. Gusella et al. did this by making mouse-human hybrid cell lines, each containing only a few human chromosomes, as described in chapter 5. They then prepared DNA from each of these lines and hybridized it to the radioactive G8 probe. Only the cell lines having chromosome 4 hybridized; the presence or absence of all other chromosomes was irrelevant. Therefore, human chromosome 4 carries the Huntington's disease gene.

So far, we know the chromosomal location of this gene, and we know from the tightness of its linkage to the RFLP that it is within a few million base pairs of the RFLP. Now we want to identify the gene itself. That will be a painstaking process. First, we must find new RFLPs closer to the gene and flanking it on either side. Then, once we have narrowed the search to a piece of DNA containing 500 kb or less, we can begin a more detailed scrutiny.

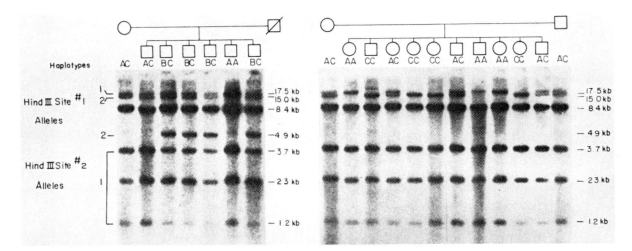

FIGURE 15.32

Southern blots of HindIII fragments from members of two families, hybridized to the G8 probe. The bands in the autoradiograms represent DNA fragments whose sizes are listed at the right. The genotypes of all the children and three of the parents are shown at the top. The fourth parent was deceased, so his genotype could not be determined.

From James F. Gusella, et al. Reprinted by permission from Nature *306:236. Copyright © 1983 Macmillan Magazines Limited.*

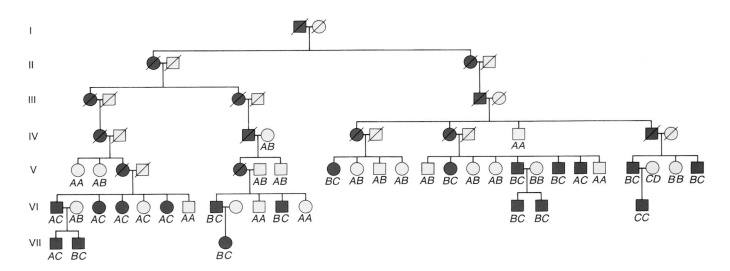

FIGURE 15.33

Pedigree of the large Venezuelan Huntington's disease family. Family members with confirmed disease are represented by purple symbols. Notice that most of the individuals with the *C* haplotype already have the disease, and that no sufferers of the disease lack the *C* haplotype. Thus, the *C* haplotype is strongly associated with the disease, and the corresponding RFLP is tightly linked to the Huntington's disease gene.

Pinpointing the Huntington's Disease Gene

A detailed study will likely follow two strategies. In the first approach, the DNA sequence is determined and examined for regions that obviously encode proteins. Telltale signs of such regions are **open reading frames (ORFs),** sequences of DNA beginning with an initiation codon (ATG) followed by many codons in the same reading frame and uninterrupted by a stop codon until the very end (unless, of course, an intron intervenes). However, the Huntington's disease gene, if it is indeed a single gene, will contain only one of perhaps dozens of open reading frames in our 500-kb piece of DNA. The hope is that one of these will be similar to a coding region for a known nerve cell protein. That would make it a prime candidate. In the second approach, the DNA from the Huntington's disease patient is compared to that from a normal individual in hopes of finding a deletion or insertion—a red flag marking the gene.

Problems in Genetic Screening

N ow that we have identified RFLPs and other genetic markers closely associated with genes for Huntington's disease, cystic fibrosis, and other serious maladies, how will we use this information? We are in a position to identify carriers of genetic diseases or people who are at risk for developing these debilitating conditions. This sounds like a good thing; in general, it undoubtedly will be, but there are some ethical problems.

Consider, for example, the actual case of an adoption agency that was trying to place a baby girl whose mother suffered from Huntington's disease. Understandably, potential adoptive parents wanted to know if the baby had inherited the gene. However, Michael Conneally, the geneticist consulted in the case, decided not to test the child for the following reason: If the results were positive and were later shared with the girl, she would know that she was doomed to get the disease. Many people at risk of developing this terrible disease feel that they cannot cope with knowing for sure that they have the gene, so they elect not to take the test. In this case, the child was too young to choose for herself.

Another dilemma would occur in the following situation: A woman is at risk of developing Huntington's disease because her father has it. She has decided she cannot bear to know that she has inherited the gene, so she will not submit to a test. Then she gets pregnant. She and her husband want the child, but not if it carries the defective gene. They could check the fetus's genotype by amniocentesis. A negative result would end their worries about the child and allow the mother to cling to hope for her own future, but a positive result would confirm her worst fears, not only for the baby but for herself.

In the case of the autosomal recessive condition cystic fibrosis, a different set of problems arises. Cystic fibrosis is the most common lethal genetic disease afflicting American children. One in twenty-five Caucasians is a CF carrier, and therefore, one in 2,500 children will be homozygous for CF and suffer the disease. Now that the CF gene has been identified, one would predict that millions of people would want to be tested to help them with their family planning. Nevertheless, in 1989 the American Society for Human Genetics called for a moratorium on widespread screening, and companies that perform CF tests for a fee in the $100–$200 range are not pushing it.

The immediate reason for this reluctance is that the mutation found by Tsui, Riordan, Collins, and colleagues is present in only 70% of the CF carriers. The other 30% have a variety of other, unknown mutations and will therefore test negative. This means that only about 50% (70% $\times$ 70% = 49%) of couples at risk will be identified, and that is deemed too low to warrant mass screening.

However, the equation will probably change rapidly, since dozens of labs are working together to identify the remaining mutations. At what point will it be appropriate to begin widespread testing? This is a difficult enough question, but even if the test were perfect, problems would remain. The major difficulty arises from the sheer

Mapping the Alzheimer's Disease and Cystic Fibrosis Genes

A similar method was used to locate the gene for Alzheimer's disease, a fatal senile dementia, on chromosome 21. Curiously, this is the same chromosome implicated in Down syndrome, and Down patients, if they live long enough, often develop Alzheimer's disease.

Investigators have used RFLPs and a variety of other methods to identify the cystic fibrosis (CF) gene. One of the techniques used in this fine mapping takes advantage of the fact that active human genes tend to be associated with unmethylated CG sequences, whereas the CG's in inactive regions are almost always methylated (chapter 14). Furthermore, the restriction enzyme HpaII cuts at the sequence CCGG, but only if the second C is unmethylated. In other words, it will cut active genes that have unmethylated CG's within CCGG sites, releasing small DNA fragments, but it will leave inactive genes (with methylated CCGG's) alone. Thus, Robert Williamson and colleagues scanned a relatively large region of DNA linked to the CF gene for "islands" of sites that could be cut with HpaII in a "sea" of other DNA sequences that could not be cut. They found such a DNA region extremely tightly linked to the CF gene. (This region is called an *HTF island* because it yields HpaII tiny fragments.)

Once the gene had been located within a 1.5 megabase (million base pairs) region of the q31 band on the long arm of chromosome 7, a large team of researchers, headed by Lap-Chee Tsui, Jack Riordan, and Francis Collins, obtained the base sequence of more than 250 kb of this DNA. Within this sequence they found the CF gene. Its relationship with the disease is quite clear. First, it codes for a membrane-spanning protein that presumably mediates the flow of chloride ions across the membrane. (Ab-

magnitude of the problem. In the United States alone, there are 100–200 million people in the reproductive age group. Even assuming we could test that many people, someone has to explain to them what the results mean, and there are not nearly enough professional genetic counselors to do that.

Our experience with screening blacks for sickle-cell disease teaches us what can go wrong with a genetic screening. In the early 1970s, many states passed laws requiring testing of blacks for sickle-cell disease, but did not provide enough counseling to make the program successful. This had several unfortunate outcomes. For one, many carriers of the sickle-cell gene came to the mistaken and distressing conclusion that they actually had the disease. Also, the suggestion to black couples who were both carriers that they forego having children, coming as it frequently did from whites, provoked charges of racism. Moreover, in some cases confidentiality was not respected. This led to the denial of health insurance to some carriers. The result of all this was a loss of confidence in the screening process among those it was designed to help. Thus, few people took part, and even fewer used the information they received to help in their family planning.

These cases exemplify the tough decisions that people must make about their own genetic testing and the psychological impact such tests may have on them. Quite a different kind of question arises concerning the mandatory testing of others. Would it be ethical, for example, for insurance companies to demand genetic tests? Those who test positive for such conditions as Huntington's disease or Alzheimer's disease would either be denied health insurance or forced to pay exorbitant premiums. We have already seen that this happened to carriers of the sickle-cell gene. If it is not fair to burden people because of their genes, over which they have no control, would it be fair to require them to submit to tests before obtaining life insurance? Keep in mind that life insurance companies have traditionally required proof of good health before issuing insurance and that genetic screening could simply be interpreted as an extension of this practice.

Let us take the problem one step further and assume that a life insurance company has paid for genetic testing of an individual and found him likely to develop a debilitating disease such as Alzheimer's. Would it be ethical for the insurance company to share such information with others—the person's employer or prospective employer, for instance? This information would certainly give an indication of the person's long-term fitness for employment, but it would just as certainly place an enormous burden on that person and maybe even render him unemployable. The problem becomes more difficult if the insurer is also the employer, as is the case with some large corporations.

These thorny questions have no easy answers, but the development of more and more tests for genetic diseases will force us to deal with them. ∎

normal chloride ion transport in lung and sweat gland tissue is a hallmark of cystic fibrosis.) Second, the normal allele of this gene is active in sweat glands. Finally, CF patients show a mutation in this gene. In fact, 68% of them show exactly the same mutation: the loss of one codon out of a total of 1,480 in the whole gene. This results in the loss of one amino acid from the protein, but it is in a key position—a part of the protein that binds ATP. Apparently this disrupts ATP binding, preventing the protein from obtaining energy from ATP to drive chloride ion transport in the lung and other tissues.

Identifying the Duchenne Muscular Dystrophy Gene

Geneticists have also found the exact gene responsible for Duchenne muscular dystrophy (DMD)—without using RFLPs. Several methods were employed, including clas-

sical mapping, but a key piece of the puzzle came from studying the rare cases of girls afflicted with the disease. Ronald Worton and coworkers examined the X chromosomes of twelve such patients and found that all had X chromosomes broken in about the same place—band Xp21 near the middle of the chromosome's short arm. This strongly suggests that the break in every case occurred in the DMD gene, inactivating it and thereby causing the disease.

L. M. Kunkel and his colleagues have cloned the DMD gene and have found that it is enormous. It contains at least sixty exons and spans 1–2 megabases of DNA. The mature mRNA produced from this gene comprises more than 14 kilobases and encodes a gigantic polypeptide having a molecular weight of about 400,000. This protein, called *dystrophin,* is about ten times the size of an average polypeptide.

Dystrophin has been located in human muscle cells using antibodies directed against the product of the very similar mouse DMD gene. It is satisfying that these antibodies react with a protein in normal muscle cells but not in muscle cells from DMD patients. It is also gratifying that this protein has a relative molecular mass of about 400,000. The geneticists studying DMD have thus completed an important part of their job. The next step will be for cell biologists to determine the role dystrophin plays in normal muscle cells. This, in turn, may give clues about how to intervene in the disease. It may even be possible someday to perform gene therapy by providing DMD patients with a functional dystrophin gene.

W e can map genes for human diseases even if we do not know the genes' functions. A common way to do this is by identifying a pattern of cutting by a restriction enzyme (a restriction fragment length polymorphism, or RFLP) that is found in people afflicted by the disease but not in other members of the same family. Once linkage between the disease gene and such a RFLP has been established, the gene can be localized to a chromosome, and more detailed mapping can begin.

■

―――――――――――― **Summary** ――――――――――――

T o clone a gene, we must insert it into a vector that can carry the gene into a host cell and ensure that it will replicate there. The insertion is usually carried out by cutting the vector and the DNA to be inserted with the same restriction nucleases to endow them with the same "sticky ends." Vectors for cloning in bacteria come in two types: plasmids and phages.

Among the plasmid cloning vectors are pBR322 and the pUC plasmids. pBR322 has two antibiotic resistance genes and a variety of unique restriction sites into which we can introduce foreign DNA. Most of these sites interrupt one of the antibiotic resistance genes, making selection straightforward. Selection is even easier with the pUC plasmids. These have an ampicillin resistance gene and a multiple cloning site that interrupts a β-galactosidase gene whose product is easily detected with a color test. We select for ampicillin-resistant clones that do not make β-galactosidase.

Two kinds of phage have been especially popular as cloning vectors. The first is λ, which has had certain nonessential genes removed to make room for inserts. In some of these engineered phages, called Charon phages, inserts up to 20 kb in length can be accommodated. Cosmids, a cross between phage and plasmid vectors, can accept inserts as large as 50 kb. This makes these vectors very useful for building genomic libraries. The second major class of phage vectors is the M13 phages. These vectors have the convenience of a multiple cloning region and the further advantage of producing single-stranded recombinant DNA, which can be used for DNA sequencing and for site-directed mutagenesis.

Expression vectors are designed to yield the protein product of a cloned gene, usually in the greatest amount possible. To optimize expression, these vectors provide strong bacterial promoters and bacterial ribosome binding sites that would be missing on cloned eukaryotic genes. Most cloning vectors are inducible, to avoid premature overproduction of a foreign product that could poison the bacterial host cells. Expression vectors frequently produce fusion proteins, which often have the advantages of stability and ease of isolation. We can introduce changes into cloned genes at will, thus changing the amino acid sequences of the protein products.

Cloned DNAs have many uses. They make excellent probes because of their ability to hybridize to DNAs or RNAs with the same, or very similar, sequences. Because of their homogeneity, they also provide ideal material for determining a DNA's base sequence. Some products of cloned genes (human insulin, human growth hormone) are already finding uses as pharmaceuticals. In the near future, they may also provide safe, effective vaccines. In the more distant future, it may be possible to treat genetic defects by transplanting cloned, normal genes into patients whose own genes are damaged. Genes are already being transferred to plants, and there is great potential for using this technique to improve crops.

Genes for human diseases can be mapped even if we do not know their functions. A common tool for performing such mapping is a restriction fragment length polymorphism (RFLP). A particular pattern of cutting by a given restriction endonuclease can be treated as a genetic marker; if it is tightly linked to the disease gene, it can be used as a first step in locating the gene.

■

―――――――――――― **Problems and Questions** ――――――――――――

1. The hypothetical bacterium *Xenobacterium giganticus* produces one restriction endonuclease. According to the convention for naming these enzymes, what would you call it?

2. Here is the sequence of a small piece of DNA that contains a 6-base palindromic recognition site for a restriction endonuclease: GACGATATCAACT. What is the sequence of this recognition site?

3. Some restriction endonucleases recognize 5-base sites in target DNA. Approximately how frequent are these 5-base sites?

4. Here are the recognition sites of four restriction enzymes:

| BamHI | G↓GATCC |
| | CCTAG↑G |

| EcoRI | G↓AATTC |
| | CTTAA↑G |

| HindIII | A↓AGCTT |
| | TTCGA↑A |

| BglII | A↓GATCT |
| | TCTAG↑A |

Chapter 15

Each cuts between the first two bases on the top strand and between the last two bases on the bottom strand, as indicated by the arrows. Which of these enzymes will create compatible (complementary) sticky ends? What is the sequence of this compatible sticky end?

5. Suppose you cut two different DNAs, one with BamHI and one with BglII, then ligate them together through their compatible sticky ends. Once joined, could you separate these two DNAs again with either restriction enzyme? Why, or why not?

6. Suppose you transform bacteria with a recombinant DNA, and by mistake, plate the transformed cells on medium lacking any antibiotic. What result would you observe? Why?

7. You want to insert a gene into the BamHI site of pBR322. Which antibiotic would you use to distinguish between cells that have taken up plasmid without insert and those that have taken up plasmid with insert? Consult the map of pBR322 (figure 15.2).

8. Is it possible to use antibiotics to determine whether pBR322 has received an insert in its EcoRI site? Why, or why not?

9. Would a plasmid like pUC8 be a good choice as a vector with which to construct a genomic library of a complex organism? Why, or why not?

10. Would a plasmid like pUC8 be a good choice as a vector with which to construct a genomic library of a relatively simple viral genome? Why, or why not?

11. You have constructed a cDNA library in λgt11 using mRNAs from a pea plant. You want to identify a rubisco clone among the 50,000 clones in the library. You have a rubisco cDNA from tobacco to use as a probe. When hybridizing the probe to the plaques in the library, would you use higher, lower, or the same temperature as for hybridizing the tobacco DNA to a tobacco rubisco clone? Why?

12. Here is the amino acid sequence of part of a hypothetical protein whose gene you want to clone:

Pro-Arg-Tyr-Met-Cys-Trp-Ile-Leu-Met-Ser

(a) What sequence of five amino acids would give a 14-mer probe with the least degeneracy? (b) How many different 14-mers would you have to make in order to be sure that your probe matches the corresponding sequence in your cloned gene perfectly? (c) If you started your probe one amino acid to the left of the one you chose in (a), how many different 14-mers would you have to make? Use the genetic code to determine degeneracy.

13. When and why can you ignore the ligation step in forming a recombinant DNA for transforming host cells?

14. What is the advantage of inducibility in an expression vector?

15. When we use λgt11 as a cDNA cloning vector, we can screen for expression of our inserted cDNA as a fusion protein, using antibody as described in figure 15.16. In order for this to work, the cDNA must be in the right orientation and in the same reading frame as the λ gene to which it is fused. From a population of λgt11 cDNA clones containing fragments of our gene of interest, what is the theoretical fraction that would satisfy these two conditions? Why?

16. Here is the base and amino acid sequence of a portion of a gene and its protein product:

```
Thr   Asp   Ala   Val   Gly
ACC   GAT   GCA   GTG   GGC
```

You are performing a site-directed mutagenesis experiment to change the gene so that it encodes Ser instead of Ala in this region. What 15-mer would you construct to serve as primer for making this change? Use the genetic code to find the Ser codons.

17. A certain animal virus contains a circular DNA that has five recognition sites for the restriction enzyme EcoRI. Infected cells make a large quantity of a viral surface protein. A cDNA corresponding to the mRNA for this surface protein has been cloned. How would you use a Southern blot to determine which EcoRI fragment(s) contain the gene for the viral surface protein?

18. You have cloned a gene from the petunia and want to know which tissue (leaves, stems, roots, flowers, or seeds) transcribes this gene to the greatest extent. How would you use a Northern blot to answer this question?

19. What will the Northern blot in question 18 tell you besides the extent of transcription of the petunia gene in each tissue?

20. You find a family that is polymorphic for the second HindIII site from the right in figure 15.31. What would be the size of the new restriction fragment observed in family members lacking this site?

21. If the family in problem 21 is polymorphic for the two other HindIII sites, as well as the new one, how many haplotypes could there be with respect to this region of the genome?

22. If the rightmost HindIII site lay just to the right of the edge of the G8 probe, would polymorphisms with respect to this site be detectable by this probe? Why, or why not?

23. Imagine the situation in problem 23 and still another HindIII site even further to the right. Would polymorphisms in this site be detectable by the G8 probe? Why, or why not?

24. Find the longest open reading frame in this sequence of bases:

CCAGATGCCTAAATGAGTTGGCCAGCAGAGCGAGCATGGATGTAATCAG

Answers appear at end of book.

Population Genetics

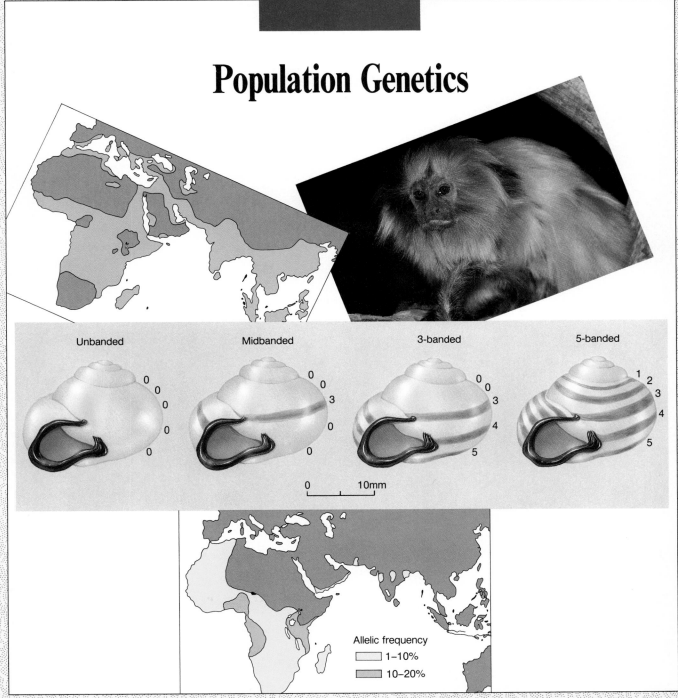

Unbanded   Midbanded   3-banded   5-banded

0 10mm

Allelic frequency
1–10%
10–20%

C·H·A·P·T·E·R

16

An Introduction to Population Genetics

T hings cannot be other than they are. . . .
Everything is made for the best purpose. Our noses
were made to carry spectacles, so we have spectacles.
Legs were clearly intended for breeches, and we wear
them.

■

François Voltaire
French author and philosopher

Learning Objectives

In this chapter you will learn:

1. That there is extensive genetic variation in most natural populations.

2. That genetic variation can be estimated by calculating the proportion of polymorphic loci or the level of heterozygosity in a population.

3. That the Hardy-Weinberg principle shows the relationship between allelic and genotypic frequencies.

4. That inbreeding increases the frequency of homozygotes.

5. That mutation is the original source of genetic variation.

6. That genetic drift may result in loss of genetic variation.

7. That gene flow can introduce new alleles into a population.

8. That natural selection can either reduce or maintain genetic variation in a population.

Galápagos tortoises in a pond on Isla Isabela at dawn.
© 1984 Frans Lanting/Minden Pictures.

*U*ntil now, we have focused on genetic variants that occur rarely in a given group of individuals—for example, mutations that result in inborn errors of metabolism, such as albinism in humans. However, natural populations of most species have alleles in substantial frequencies that affect such traits as body color, protein characteristics, or pesticide resistance. Why do these different genetic types exist within a population, and what factors determine their frequencies? Answering these questions necessitates an understanding of **population genetics,** the field that focuses on the extent and pattern of genetic variation in natural populations.

Natural selection has often been considered the major force determining genetic variation, an idea tracing back to Charles Darwin's *On the Origin of Species by Means of Natural Selection,* first published in 1859. However, other factors also appear to be important in given situations or in certain species. In this chapter we will introduce those factors—inbreeding, mutation, genetic drift, and gene flow. In chapter 17 we will combine the effects of several of these factors in order to understand specific topics in population genetics, namely, the incidence of genetic disease in humans, the evolution of molecules, conservation genetics, agricultural genetics, and pesticide resistance.

We will find Mendelian genetic principles essential to understanding the genetic composition of populations. For example, the principle of segregation (that two alleles in a heterozygous individual are equally represented in the gametes produced by that individual), when applied to all individuals in a population, will allow us to understand the overall genetic variation in that population. The genetic constitution of a population is sometimes referred to as its **gene pool.** Let us first describe some examples of genetic variation before discussing the principles of population genetics that explain the basis of this variation and how it may change.

TYPES OF GENETIC VARIATION

Genetic variants can directly affect the morphology or color of an organism in a natural or wild population. A classic example of genetic variation that has fascinated scientists in England and France for decades is the color and banding pattern of *Cepaea nemoralis,* a European land snail common in gardens and uncultivated fields. The morphological variation is determined by the alleles at a group of closely linked genes, so that the snails vary strikingly in color and pattern from yellow, unbanded types to brown, banded ones (see figure 16.1). The frequencies of the alleles that determine yellow color and bandedness vary on a large scale from one part of Europe to another and on a small scale from one field to another.

In one detailed study, the frequency of yellow, unbanded types over a 200-meter stretch on an earthen dam in the Netherlands varied from 60% at one end to 40% at the other. Furthermore, these populations remained at the same frequencies over a twelve-year period, suggesting that evolutionary forces were maintaining the differences. For example, it appears from other studies that the different forms confer protective camouflage from the predation of birds, depending upon the habitat in which they are found. When the snails are resting on grass blades, the bands are vertical, so the banded snails blend in with the grass. But when the snails are in the ground litter, the unbanded brown snails are the most difficult to detect.

In a number of species, variation in chromosome structure has been observed among different individuals. Recently, new techniques developed to examine morphological details of chromosomes (see chapter 4) have uncovered many new chromosomal variants, and it is likely that numerous other small variations in chromosome structure, such as deletions, duplications, and inversions, are still to be discovered. One of the most thoroughly studied examples of chromosomal variation was done by Theodosius Dobzhansky, a Russian geneticist who moved

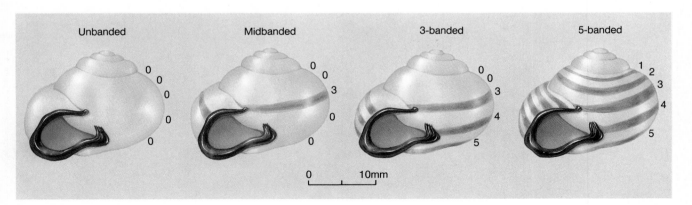

FIGURE 16.1
Several shell banding patterns observed in the snail *Cepaea nemoralis.*
Reproduced, with permission, from the Annual Review of Ecology and Systematics, *vol. 8. © 1977 by Annual Reviews, Inc.*

to the United States in the 1920s. He and his students analyzed the inversion variants on the third chromosome of the fruit fly *Drosophila pseudoobscura*, a widespread species in the mountains of western North America (see figure 4.25).

In the 1930s, Dobzhansky began a long-term study of variation in this inversion. Some of these data, showing the frequencies of four inversion types over thirty years of sampling at a site in New Mexico, are given in figure 16.2. Initially, the inversion type AR (Arrowhead) was most common, but in 1965, type PP (Pikes Peak) increased in frequency, only to decline in later years. Two other inversion types, CH (Chiricahua) and ST (Standard), were always present, but at low frequencies in the samples from this population.

These inversions, although inherited as single units like alleles (see chapter 4), generally contain hundreds of genes. Because different inversions may have different alleles at these genes, they can substantially affect the survival of individual flies. In fact, some of the changes observed over time may have been caused by differential survival of various inversions to DDT, an insecticide that was first applied in the 1940s.

Most alleles that cause genetic diseases in humans occur in very low frequencies. However, some disease alleles have higher frequencies in certain human populations than in others. For example, the frequency of the sickle-cell allele at the β-globin locus is quite high in many African populations and in populations with African ancestry, such as American blacks. The sickle-cell allele also occurs in some Mediterranean and Asian populations (see figure 16.3b).

The presence of the sickle-cell allele can be ascertained by biochemical examination of the β-globin molecule (see a photo of normal and sickle cells, figure 7.20). Variant forms of a protein can be detected by **gel electrophoresis** (see also chapter 8), a technique that allows the

separation of different proteins extracted from blood, tissues, or whole organisms. The process is carried out by imposing an electric field across a gelatinous supporting medium (either a polyacrylamide or a starch gel) into which the protein has been placed. The proteins are allowed to migrate for a specific amount of time and are then stained with various protein-specific chemicals, resulting in bands on the gel that permit the relative mobility of a specific protein to be determined (see figure 16.4). The relative mobility is generally a function of the charge, size, and shape of the molecule. If two proteins that are the products of different alleles have different amino acid sequences, they often have different mobilities as well, because the differences in sequence result in a change in charge, size, or shape of the molecule.

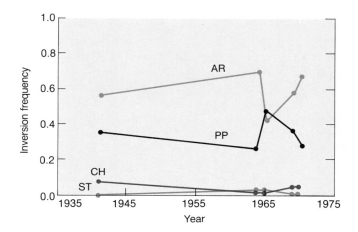

FIGURE 16.2
Frequencies of inversions on the third chromosome in *Drosophila pseudoobscura* that have taken place over a period of three decades in the Capitan area of New Mexico. (AR = Arrowhead; PP = Pikes Peak; CH = Chiricahua; ST = Standard.)

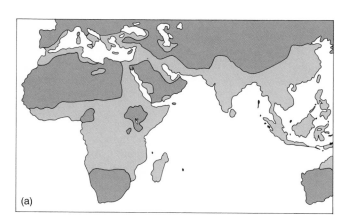

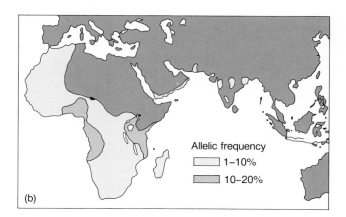

FIGURE 16.3
(*a*) The geographic distribution of malaria (green) and (*b*) the geographic distribution of the sickle-cell allele, indicating the regions of low frequency (yellow) and higher frequency (orange).

Reprinted from Macmillan Publishing Company from Genetics, *3d ed., by Monroe W. Strickberger. Copyright © 1985 by Monroe W. Strickberger.*

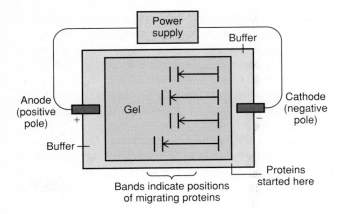

FIGURE 16.4
Diagram of a gel electrophoresis apparatus. The buffers are used to conduct electricity and ensure a given pH.

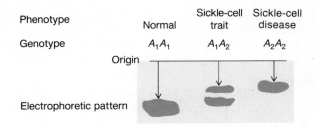

FIGURE 16.5
The phenotypes, genotypes, and electrophoretic patterns of the β-globin gene.

The three β-globin genotypes (homozygotes for the normal and sickle-cell allele plus a heterozygote) have different banding patterns when they are run on an electrophoretic gel as shown in figure 16.5. The normal A_1A_1 genotype has only one band that migrates relatively fast as in the first column, while individuals with sickle-cell disease, genotype A_2A_2 in column 3, also have only one band, but it is slower migrating. The heterozygous individuals, A_1A_2, who have the sickle-cell trait (they are carriers) have two bands (see column 2), each band representing the protein synthesized from the respective allele. Sometimes these genotypes at the β-globin locus are represented as *AA, AS,* and *SS,* instead of A_1A_1, A_1A_2, A_2A_2.

The reason for the high frequency of this disease allele is that heterozygotes (those who carry this allele and the normal allele) have a relatively high resistance to malaria. The relative immunity of sickle-cell carriers to malaria occurs because the red blood cells of heterozygotes are sickle-shaped, a state that inhibits infection by the malarial parasite. When we compare the frequency of the sickle-cell allele and the past incidence of malaria on a geographic scale (before the extensive eradication of the mosquito that carries the malarial parasite), we find an extremely high concordance (see figure 16.3).

> *E* xtensive genetic variation occurs in most populations at several different levels, including variants affecting color, chromosomal structure, and protein characteristics.

∎

MEASURING GENETIC VARIATION

How much genetic variation exists in a given population? How can we quantify this variation in a standardized manner? To properly answer these questions, we need to introduce some quantitative measures and the symbols and statistics that accompany them. However, let us first define a **population** as a group of interbreeding individuals of the same species that exist together in time and space. For example, from a genetic viewpoint, all the breeding largemouth bass in a small lake in a given year constitute a population.

To determine the frequencies of the genotypes or alleles at a given locus in a population, we must first count the number of individuals having different genotypes. For example, at the β-globin gene, let N_{11}, N_{12}, and N_{22} indicate the number of individuals counted with genotypes A_1A_1, A_1A_2, and A_2A_2, respectively. If there are N total individuals in the sample, the estimated frequencies of these three genotypes in the population are:

$$P = \frac{N_{11}}{N}$$

$$H = \frac{N_{12}}{N}$$

$$Q = \frac{N_{22}}{N}$$

Because there are only three genotypes, the sum of the frequencies of the three genotypes must be one; that is, $P + H + Q = 1$.

Notice that all the alleles in the homozygote A_1A_1 are A_1 and that all the alleles in the homozygote A_2A_2 are A_2. In addition, half of the alleles in the heterozygotes are A_1

and half are A_2. Therefore, we can calculate the frequency of allele A_2 (designated by q) as the sum of half the frequency of the heterozygote plus the frequency of the A_2A_2 homozygote. In other words:

$$q = \frac{1}{2}H + Q$$

If we substitute the numbers of different genotypes given in the first formulas, then:

$$q = \frac{\frac{1}{2}N_{12}}{N} + \frac{N_{22}}{N}$$
$$= \frac{\frac{1}{2}N_{12} + N_{22}}{N}$$

If we designate the frequency of the other allele, A_1, as p, then using the same logic:

$$p = P + \frac{1}{2}H$$

and

$$p = \frac{N_{11} + \frac{1}{2}N_{12}}{N}$$

Because there are only two alleles, the sum of their frequencies must be one, or $p + q = 1$. Furthermore, if we subtract q from both sides of this equation, we get the useful relationship $p = 1 - q$.

Let us illustrate the use of these formulas by a numerical example. In parts of Africa, the frequency of the sickle-cell allele is quite high, considering that it causes a severe disease when homozygous. In one sample of 400 individuals from western Africa, 320 were normal (genotype A_1A_1), 72 were carriers of the sickle-cell trait (A_1A_2), and 8 actually had sickle-cell disease (A_2A_2). In other words, N_{11}, N_{12}, and N_{22} were 320, 72, and 8, respectively. The frequencies of the three genotypes in this population would be expressed as:

$$P = \frac{320}{400} = 0.80$$

$$H = \frac{72}{400} = 0.18$$

$$Q = \frac{8}{400} = 0.02$$

This means that 2% of the individuals have sickle-cell disease and 18% are carriers of the sickle-cell allele. Using the formula given for the frequency of allele A_2, the frequency of the sickle-cell allele is:

$$q = \frac{1}{2}(0.18) + 0.02$$
$$= 0.11$$

The frequency of the normal allele, A_1, is $p = 1 - q = 0.89$.

Now that we have introduced these formulas, we can discuss two useful ways to quantify the extent of genetic variation in a population. Using the allelic frequencies calculated for a given gene, we can define a gene as **polymorphic** (having many forms; that is, having two or more alleles in substantial frequency) or **monomorphic** (having only one allele in high frequency). The simplest measure of genetic variation is to categorize loci as polymorphic or monomorphic and then to calculate the proportion of all polymorphic loci. By general (although somewhat arbitrary) definition, if the frequency of the most common allele is 0.99 or greater (or if all other alleles combined have a frequency of 0.01 or less), the gene is considered monomorphic; otherwise, it is polymorphic. For example, the sickle-cell gene in the example just given is polymorphic for the two alleles A_1 and A_2 because the frequency of A_1 is less than 0.99.

A second approach for measuring genetic variation in a population is to calculate the average heterozygosity over all loci. The measure of heterozygosity is useful because it takes into account all levels of genetic variation rather than just classifying loci into two categories as does the first method. The average heterozygosity for a sample of n different loci is:

$$H = \frac{1}{n}(H_A + H_B + H_C \ldots)$$

where H_A is the heterozygosity for gene A, H_B for gene B, etc. As a simple example, if the heterozygosities at five loci are $H_A = 0.5$, $H_B = 0.0$, $H_C = 0.1$, $H_D = 0.0$, and $H_E = 0.2$, then $n = 5$, and the average heterozygosity is:

$$H = (\frac{1}{5})(0.5 + 0.0 + 0.1 + 0.0 + 0.2)$$
$$= 0.16$$

Note that in this case three loci, A, C, and E, are polymorphic, making the proportion of polymorphic loci $\frac{3}{5} = 0.6$.

The first attempts to quantify the overall extent of genetic variation in a given species were made in the mid-1960s by Richard Lewontin in *D. pseudoobscura* and by Harry Harris in human beings. Both surveys used protein electrophoresis and demonstrated substantial genetic variation from a sample of genes. As an illustration, some of Harris's data from a Caucasian population in England is given in table 16.1 for seventy-one different genes. Twenty of these loci are polymorphic; that is, they have the most common allele with a frequency of 0.99 or less. Therefore, the proportion of polymorphic loci in this sample is $\frac{20}{71} = 0.282$. Notice that the level of heterozygosity ranges from 0.00 for the monomorphic loci to around 0.5 for five polymorphic loci, most of which have multiple alleles. The average heterozygosity over all seventy-one loci is 0.067. Another way to interpret this value is that on the average, about 7% of the loci in a given individual are heterozygous for such biochemical variants.

A relatively recent discovery in population genetics (and one that caused a radical reexamination of its foundations) is the high degree of polymorphism and heterozygosity that was found in nearly all natural populations once modern biochemical techniques were applied. At first thought, this level of genetic variation is surprising, because little variation appears in the wild-type morphology of most populations. However, the success of plant and animal breeding programs in changing morphological traits such as size (see chapters 3 and 17) suggests that there is extensive background genetic variation.

Surveys of the genetic variation in various species demonstrate that up to 20% of the genes examined by electrophoresis are polymorphic and that 10% of these genes are heterozygous in a given individual. Table 16.2 lists a number of species with different levels of polymorphism. Generally, the highest levels of genetic variation occur in invertebrates and in some plants, while vertebrates generally have lower levels of variation. Several organisms are of particular interest. The horseshoe crab, an animal that has not changed morphologically in millions of years according to fossil records, has a high level of genetic variation. Thus, **stasis** (no change) in morphology over geological time does not necessarily indicate lack of genetic variation on a biochemical level. On the other hand, the Northern elephant seal, which was nearly hunted to extinction in the nineteenth century, and the cheetah, also an endangered species, show no genetic variation in these electrophoretic surveys. In both cases, small population numbers appear to have caused a loss of genetic variation via genetic drift, a phenomenon that will be discussed later in this chapter and in chapter 17.

TABLE 16.1   *Heterozygosity for Seventy-one Human Loci*

Locus	Heterozygosity (H)
51 monomorphic loci	0.00
Peptidase C	0.02
Peptidase D	0.02
Glutamate-oxaloacetate transaminase	0.03
Leukocyte hexokinase	0.05
6-phosphogluconate dehydrogenase	0.05
Alcohol dehydrogenase-2	0.07
Adenylate kinase	0.09
Pancreatic amylase	0.09
Adenosine deaminase	0.11
Galactose-1-phosphate uridyl transferase	0.11
Acetylcholinesterase	0.23
Mitochondrial malic enzyme	0.30
Phosphoglucomutase-1	0.36
Peptidase A	0.37
Phosphoglucomutase-3	0.38
Pepsinogen	0.47
Alcohol dehydrogenase-3	0.48
Glutamate-pyruvate transaminase	0.50
RBC acid phosphatase	0.52
Placental alkaline phosphatase	0.53

From Harris and Hopkinson, in *Annals of Human Genetics* 35:9–20, 1972. Copyright © 1972 Cambridge University Press, New York, NY. Reprinted by permission.

By counting the numbers of different genotypes in a population, we can calculate genotypic frequencies, allelic frequencies, and heterozygosity. Two measures of genetic variation, the proportion of polymorphic loci and the level of heterozygosity, are used to quantify the extent of genetic variation. In many species, the amount of genetic variation observed for genes coding for proteins is substantial.

■

TABLE 16.2   *Average Proportions of Polymorphic Loci and Average Heterozygosity for Various Organisms*

Organism	Number of Genes Examined	Proportion Polymorphic	Heterozygosity
Humans	71	0.28	0.067
Northern elephant seal	24	0.0	0.0
Horseshoe crab	25	0.25	0.057
Elephant	32	0.29	0.089
Drosophila pseudoobscura	24	0.42	0.12
Barley	28	0.30	0.003
Tree frog	27	0.41	0.074

THE HARDY-WEINBERG PRINCIPLE

The description of the genetic variation in a population is often greatly simplified when allelic frequencies, rather than genotypic frequencies, can be used. In 1908, shortly after the rediscovery of Mendel's work, an English mathematician, Godfrey Hardy, and a German physician, Wilhelm Weinberg, showed that a simple relationship often exists between allelic frequencies and genotypic frequencies. (Actually, William Castle, an American geneticist, had published a special case of this relationship in 1903.) This relationship, generally known today as the **Hardy-Weinberg principle,** states that the genotypic frequencies for a gene with two different alleles are a binomial function of the allelic frequencies.

To understand the Hardy-Weinberg principle, assume that a population is segregating for two alleles, A_1 and A_2, at gene A, with frequencies of p and q (remember p + q = 1), respectively. We will assume that the male and female gametes unite at random to form zygotes; that is, the product rule of probability discussed in chapter 2 is applicable. In addition, we should assume that mutation, genetic drift, gene flow, and selection are not affecting genetic variation.

The random union of gametes can be illustrated graphically, as in figure 16.6. The unit length across the top of the square in this figure is divided into proportions p and q, representing the frequencies of the female gametes, A_1 and A_2, respectively. Likewise, the side of the square is divided into proportions representing the frequencies of the male gametes. (The frequencies of the alleles in female and male gametes are assumed to be identical.) To envision random union of gametes, one must imagine a pool composed of both female and male gametes, p with A_1 alleles and q with A_2 alleles—and then realize that zygote formation occurs by chance collision in this gametic pool.

The areas within the unit square represent the probabilities (or proportions) of different progeny zygotes. For example, the upper left square has an area of p^2, the frequency of genotype A_1A_1, and the lower right square has an area of q^2, the frequency of A_2A_2. The other two gametic combinations both have frequencies of pq and result in heterozygotes, making the total proportion of heterozygotes 2pq. Overall, therefore, the three progeny zygotes (A_1A_1, A_1A_2, A_2A_2) are formed in proportions (p^2, 2pq, q^2). If we square the sum of the allelic frequencies (that is, $[p + q]^2 = p^2 + 2pq + q^2$), we can see that the Hardy-Weinberg proportions are really binomial proportions.

Of course, different populations may have different allelic frequencies. To illustrate the effect of different allelic frequencies on genotypic frequencies, figure 16.7 gives the frequencies of the three different genotypes, assuming Hardy-Weinberg proportions for the total range of allelic frequencies. Notice that the heterozygote is the most common genotype for intermediate allelic frequencies, while one of the homozygotes is the most common genotype for nonintermediate frequencies. The maximum frequency of the heterozygote occurs when q = 0.5 (p = 0.5). In this case, half of the individuals are heterozygotes. When the locus is monomorphic (p > 0.99 or q > 0.99), the amount of heterozygosity is very low.

In most organisms, random union of gametes does not occur. Instead, the parental genotypes pair and mate; then these mated pairs produce gametes that unite independently of the alleles they contain. Let us illustrate this situation when **random mating** occurs between different female and male genotypes. We will consider a diploid organism and assume that the three possible genotypes, A_1A_1, A_1A_2, and A_2A_2, are present in the population in frequencies P, H, and Q, respectively (remember that P + H + Q = 1). Given the three genotypes in each sex, nine combinations of matings between the male and female genotypes are possible, as shown in table 16.3. The frequency of a particular mating type, given random mating and using the product rule of probability again, is equal to the product of the frequencies of the genotypes that constitute a particular mating type. For example, the frequency of mating type $A_1A_1 \times A_1A_1$ is P^2. Only six mating types will be distinguished here, because reciprocal matings—for example, $A_1A_1 \times A_1A_2$ and $A_1A_2 \times A_1A_1$, where the first genotype is the female—have the same genetic consequences.

Female gametes (frequency)

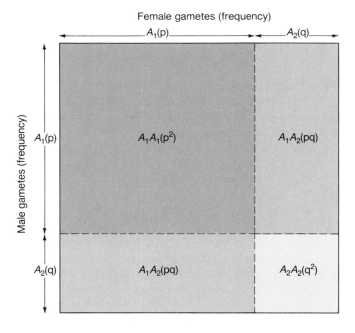

FIGURE 16.6
Hardy-Weinberg proportions as generated from the random union of gametes, using a unit square.

An Introduction to Population Genetics

419

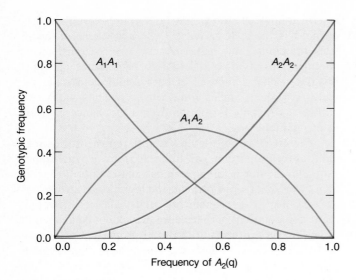

FIGURE 16.7
The relationship between the allelic frequency and the three genotypic frequencies for a population in Hardy-Weinberg proportions.
Reprinted with permission of Jones and Bartlett Publishers, Inc., Boston, MA, from Population Biology, *1984, written by Philip Hedrick.*

Female Genotypes (Frequencies)	Male Genotypes (Frequencies)		
	$A_1A_1(P)$	$A_1A_2(H)$	$A_2A_2(Q)$
$A_1A_1(P)$	P^2	PH	PQ
$A_1A_2(H)$	PH	H^2	HQ
$A_2A_2(Q)$	PQ	HQ	Q^2

TABLE 16.3 *Frequencies of Different Mating Types for Two Alleles, with Random Mating*

The mating types, their frequencies, and the expected frequencies of their offspring genotypes, using the principle of segregation, are given in table 16.4. For example, the mating $A_1A_1 \times A_1A_1$ produces only A_1A_1 progeny; the mating $A_1A_1 \times A_1A_2$ produces ½ A_1A_1 and ½ A_1A_2; and so on. If the frequency of A_1A_1 progeny contributed by each mating type is summed (adding down the first progeny column), we find that $P^2 + PH + ¼H^2 = (P + ½H)^2$ of the progeny are A_1A_1. From the expression for the frequency of A_1 given earlier, we know that $p = P + ½H$ and, therefore, that p^2 of the progeny are genotype A_1A_1. Similarly, the frequencies of A_1A_2 and A_2A_2 progeny are $2(P + ½H)(Q + ½H) = 2pq$ and $(Q + ½H)^2 = q^2$, respectively. In other words, the Hardy-Weinberg principle is also true for random mating in the population, an assumption that is entirely reasonable for most genes.

The crucial point of the Hardy-Weinberg principle is that, given any set of initial genotypic frequencies (P, H, Q) after one generation of random mating, the genotypic frequencies in the progeny are in the proportions (p^2, 2pq, q^2). For example, given the initial genotypic frequencies (0.2, 0.4, 0.4) in which p = 0.4 and q = 0.6, after one generation, the genotypic frequencies become $[(0.4)^2, 2(0.4)(0.6), (0.6)^2 = 0.16, 0.48, 0.36]$. Furthermore, the genotypic frequencies will stay in these exact proportions generation after generation, given continued random mating and the absence of factors that change allelic or genotypic frequencies. (After all this discussion of p's and q's, one might think that it is the source of the phrase "Mind your p's and q's." A more probable origin, however, is the tradition of British bartenders who tell their customers, "Mind your pints and quarts"—p's and q's for short.)

In the allelic frequency estimates given earlier in this chapter, all alleles were assumed to be expressed (observable) in heterozygotes. However, in many cases—for example, metabolic disorders in humans and color polymorphisms in various other organisms—the heterozygote is generally indistinguishable from one of the homozygotes because one of the alleles is completely dominant. In these instances, the population is usually assumed to be in Hardy-Weinberg proportions, to allow the estimation of allelic frequencies. For example, with two alleles and complete dominance, it is assumed that the proportion of A_2A_2 genotypes in the population is:

$$q^2 = \frac{N_{22}}{N}$$

By taking the square root of both sides of this equation, an estimate of the frequency of A_2 is:

$$q = \left(\frac{N_{22}}{N}\right)^{1/2}$$

(The superscript ½ means square root.)

This approach is applied in many human metabolic diseases where it is often important to know the proportion of heterozygotes (carriers) in a population—individuals who usually cannot be distinguished from homozygotes for the normal allele. An estimate of the proportion of carriers in a population is:

$$H = 2pq$$

where q is the estimate of allelic frequency and p = 1 − q. For example, assuming that only 1 in 40,000 individuals has a particular rare recessive disease such as albinism (or is an A_2A_2 homozygote), then from the equation, $q = (1/40,000)^{1/2} = 0.005$. Therefore, p = 0.995, and the frequency of carriers is H = 2(0.005)(0.995) = 0.00995. In other words, about 1% of such a population carries the disease allele. People who are not geneticists are often surprised at how many individuals are carriers of recessive disease alleles when the actual frequency of the disease itself is so low.

	Parents		Progeny	
Mating Type	Frequency	A_1A_1	A_1A_2	A_2A_2
$A_1A_1 \times A_1A_1$	P^2	P^2	—	—
$A_1A_1 \times A_1A_2$	$2PH$	PH	PH	—
$A_1A_1 \times A_2A_2$	$2PQ$	—	$2PQ$	—
$A_1A_2 \times A_1A_2$	H^2	$\frac{1}{4}H^2$	$\frac{1}{2}H^2$	$\frac{1}{4}H^2$
$A_1A_2 \times A_2A_2$	$2HQ$	—	HQ	HQ
$A_2A_2 \times A_2A_2$	Q^2	—	—	Q^2
Total	1	$(P + \frac{1}{2}H)^2 = p^2$	$2(P + \frac{1}{2}H)(Q + \frac{1}{2}H) = 2pq$	$(Q + \frac{1}{2}H)^2 = q^2$

THE CHI-SQUARE TEST

The observed genotypic numbers are often quite close to the expected genotypic numbers calculated from the Hardy-Weinberg principle. But in some organisms, such as self-fertilizing plants, the observed and expected genotypic frequencies in a population may differ greatly. Therefore, a statistical test is necessary to determine whether the fit is sufficiently close to expected Hardy-Weinberg proportions. The common test employed for this purpose is the chi-square (χ^2) test used in chapter 2 to examine Mendel's principles. To determine whether the observed numbers are consistent with Hardy-Weinberg predictions, a chi-square value can be calculated as:

$$\chi^2 = \Sigma \frac{(O - E)^2}{E}$$

O and E are the observed and expected *numbers* of a particular genotype (remember to use numbers and not proportions), and Σ indicates that it is summed over all genotypic classes. In other words, this chi-square value is the squared difference of the observed and expected genotypic numbers, divided by the expected genotypic numbers, and summed over all genotypic classes (k). Knowing the calculated value of chi-square and the degrees of freedom, the probability that the observed numbers will deviate from the expected numbers by chance can be obtained from a chi-square table (see table 2.5).

Using the Hardy-Weinberg expected frequencies for the two-allele case, where all genotypes are distinguishable, the chi-square expression becomes:

$$\chi^2 = \frac{(N_{11} - p^2N)^2}{p^2N} + \frac{(N_{12} - 2pqN)^2}{2pqN} + \frac{(N_{22} - q^2N)^2}{q^2N}$$

The three terms represent the squared, standardized deviations from the expected numbers for genotypes A_1A_1,

A_1A_2, and A_2A_2, respectively. For example, N_{11} is the observed number of genotype A_1A_1, and p^2N is the expected number of A_1A_1.

As an example of this procedure, let us use the hemoglobin data given previously. The observed numbers of the three genotypes were: $N_{11} = 320$, $N_{12} = 72$, and $N_{22} = 8$, giving p = 0.89 and q = 0.11. Therefore, the expected numbers of the three genotypes are $p^2N = (.89)^2400 = 316.8$, $2pqN = 2(.89)(.11)400 = 78.3$, and $q^2N = (.11)^2400 = 4.8$. Using the chi-square formula:

$$\chi^2 = \frac{(320 - 316.8)^2}{316.8} + \frac{(72 - 78.3)^2}{78.3}$$
$$+ \frac{(8 - 4.8)^2}{4.8}$$
$$= 0.03 + 0.51 + 1.28$$
$$= 1.82$$

To determine whether this value is statistically significant, we first need to know the degrees of freedom. As in chapter 2, we start by counting the number of categories (or genotypes)—three in this case. Next, we subtract 1 from this total because the numbers of all genotypes sum to N. Furthermore, in this application of the chi-square, we subtract an additional 1 because we had to estimate p in order to calculate the expected numbers of the genotypes. Overall then, the remaining degrees of freedom are 3 − 1 − 1 = 1.

Now we can look at table 2.5 to determine whether a chi-square value of 1.82 with 1 degree of freedom is significant. Because 1.82 is less than 3.84 (the lowest level of statistical significance for 1 degree of freedom), we can conclude that the observed numbers in the sample are consistent with those expected under the Hardy-Weinberg principle.

TABLE 16.5	The Hardy-Weinberg Principle Applied to the ABO Blood Group System		
Female Gametes (Frequencies)	**Male Gametes (Frequencies)**		
	I^A(p)	I^B(q)	I^O(r)
I^A(p)	I^AI^A(p^2)	I^AI^B(pq)	I^AI^O(pr)
	A	AB	A
I^B(q)	I^AI^B(pq)	I^BI^B(q^2)	I^BI^O(qr)
	AB	B	B
I^O(r)	I^AI^O(pr)	I^BI^O(qr)	I^OI^O(r^2)
	A	B	O

TABLE 16.6	Frequencies of Gametes and Genotypes for an X-linked Gene		
Female Gametes (Frequencies)	**Male Gametes (Frequencies)**		
	Female Progeny		*Male Progeny*
	A_1(p)	A_2(q)	Y(l)
A_1(p)	A_1A_1(p^2)	A_1A_2(pq)	A_1Y(p)
A_2(q)	A_1A_2(pq)	A_2A_2(q^2)	A_2Y(q)

MORE THAN TWO ALLELES

Many genes have more than two alleles, as we discussed in chapter 3. For example, the β-globin gene containing the sickle-cell allele actually has more than fifty variants, most of which are rare. Also, many genes coding for proteins and examined by electrophoresis have more than two alleles.

In the case of multiple alleles, the frequency of an allele can be calculated as the sum of the frequency of its homozygote plus half the frequency of each heterozygote. To illustrate, assume that a sample of N individuals is categorized for the genotypes from three alleles. As discussed in chapter 3, six genotypes are possible: three homozygotes (A_1A_1, A_2A_2, and A_3A_3) and three heterozygotes (A_1A_2, A_1A_3, and A_2A_3). The frequency of allele A_1 is calculated in the same manner as for two alleles, except that A_1 alleles are in two different heterozygotes, A_1A_2 and A_1A_3. Therefore, the frequency of A_1 is:

$$p = \frac{N_{11} + \frac{1}{2}N_{12} + \frac{1}{2}N_{13}}{N}$$

where N_{13} is the number of A_1A_3 heterozygotes.

The Hardy-Weinberg principle can also be extended to more than two alleles. Let us illustrate this with the human ABO blood group system, which has three alleles: I^A, I^B, and I^O. Table 16.5 gives the frequencies of these alleles in gametes p, q, and r, respectively, and the frequencies of the resulting genotypes and phenotypes. As for two alleles (see figure 16.6), the genotypic frequencies can be calculated as the product of particular male and female gametic frequencies. For example, the frequency of genotype I^AI^A is (p)(p) = p^2. Overall, the genotypic frequencies are equal to the square of the following trinomial (because there are three alleles):

$$(p + q + r)^2 = p^2 + 2pq + q^2 + 2pr + 2qr + r^2$$

Check table 16.5 to see that you understand the genotypes associated with these frequencies.

Note that because of the recessivity of allele I^O, two genotypes have the A phenotype (I^AI^A and I^AI^O) and two have the B phenotype (I^BI^B and I^BI^O). The frequency of the O phenotype, using the Hardy-Weinberg principle, is the square of the frequency of the recessive allele I^O, that is, r^2; the frequencies of the A and B phenotypes are p^2 + 2pr and q^2 + 2qr, respectively; and the frequency of the AB phenotype is 2pq. In the United States white population, the frequencies of alleles I^A, I^B, and I^O are approximately 0.28, 0.06, and 0.66, resulting in phenotypic frequencies of 0.45, 0.08, 0.03, and 0.44 for ABO blood group types A, B, AB, and O, respectively. Luckily, the rarest phenotype, type AB, is also the universal recipient, so people with this phenotype can receive transfusions from individuals of any ABO blood group type.

An important consequence of having many alleles at a locus is that in general, most individuals in a population are heterozygotes. For example, at the *HLA-A* and *HLA-B* loci, over 90% of the individuals are heterozygotes and less than 10% are homozygotes (see chapter 3). To illustrate this calculation, the Hardy-Weinberg heterozygosity for the ABO system is:

$$\begin{aligned} H &= 2pq + 2pr + 2qr \\ &= 2(.28)(.06) + 2(.28)(.66) + 2(.06)(.66) \\ &= 0.482 \end{aligned}$$

When there are many alleles (which means many more heterozygotes than homozygotes), the simplest approach is to calculate the Hardy-Weinberg homozygosity and subtract it from unity. The remainder is the Hardy-Weinberg heterozygosity.

X-LINKED GENES

A number of human diseases, such as hemophilia and muscular dystrophy, are determined by recessive genes on the X chromosome. These diseases are much more common in males than in females, a fact that can be explained by the Hardy-Weinberg principle. For example, table 16.6 gives the expected frequencies of different genotypes, assuming random union of gametes for an X-linked trait. The frequency of affected males, A_2Y, is q, while the frequency of affected females, A_2A_2, is q^2. If q is low, as it is for most disease alleles (say, 0.005), the frequencies of affected males and females are 0.005 and 0.000025, respectively. In other words, the ratio of affected males to females is $q:q^2$ or $1:q$. Thus, if $q = 0.005$, the ratio is 200:1. Because of the high number of males affected by X-linked traits and the relatively high frequency of the hemophilia allele for a disease allele, a high proportion of individuals with a recessive genetic disease (nearly one individual in four in some populations) is a male hemophiliac. On the other hand, very few females have hemophilia.

T he Hardy-Weinberg principle states that genotypic frequencies are a binomial function of allelic frequencies after one generation of random mating. Genotypic numbers can be examined statistically, using the chi-square test to determine whether they are consistent with Hardy-Weinberg expectations. The Hardy-Weinberg principle and estimation of allelic frequencies can also be applied to multiple alleles and X-linked genes.

■

INBREEDING

So far in this chapter, we have considered populations in which mating is random, but there are obvious exceptions to this assumption in some species, such as self-fertilizing plants or invertebrates. **Inbreeding,** nonrandom mating that results from mating of close relatives, is of particular importance in humans, because many individuals who have recessive diseases are the products of matings between relatives. Furthermore, inbreeding is used in both plant and animal breeding to develop varieties or lines that have certain desired characteristics. In chapter 17 we will discuss the joint effects of inbreeding and selection that result in the phenomenon of inbreeding depression.

Nonrandom mating with respect to genotype occurs in populations where the mating individuals are either more closely or less closely related than mates drawn by chance (at random) from the population. The results of these two types of matings are called inbreeding and **outbreeding,** respectively. As we will see, neither inbreeding nor outbreeding can by itself cause a change in allelic frequency, but both do cause a change in genotypic frequencies. In an inbred population, the frequency of homozygotes is increased and the frequency of heterozygotes is reduced relative to random mating (or Hardy-Weinberg) proportions. In outbreeding, the opposite occurs, with the frequency of heterozygotes increased and the frequency of homozygotes reduced relative to random mating proportions. These genotypic changes affect all loci in the genome.

Nonrandom mating based on phenotypes rather than on genotypes may also occur. If the mated pairs in a population are composed of individuals with the same phenotype more often than would be expected by chance,

TABLE 16.7 Positive Assortative Mating in Studies of Five Human Traits

Trait	Number of Studies Showing Correlation Coefficients of:			
	<0.1	0.1–0.2	0.2–0.3	>0.3
Height	7	8	7	5
Weight	1	2	3	1
Cephalic index	14	5	3	—
Hair color	—	2	2	1
Eye color	2	1	1	1

positive assortative mating has occurred. By the same token, if mated pairs share the same phenotype less often than expected, we say that **negative assortative mating** has occurred. For example, positive assortative mating occurs in humans for traits such as height and eye color (see table 16.7). Assortative mating generally affects the genotypic frequencies of only those loci involved in determining the phenotypes used in mate selection. Because positive assortative mating has effects similar to those of inbreeding, except that it generally affects only one or a few loci rather than all loci, we will consider only the effects of inbreeding in the following discussion.

SELF-FERTILIZATION

In general, the most extreme type of inbreeding is self-fertilization, or "selfing," where both pollen and egg (or sperm and egg) are produced by the same individual. With complete self-fertilization, a population is divided into a series of inbred lines that quickly become highly homozygous. For example, the lines of pea plants used by Mendel were homozygous for different alleles that affected morphology or color because they were allowed to self-fertilize.

To illustrate how such high homozygosity occurs, assume that the frequencies of the three genotypes in the parents are P, H, and Q for A_1A_1, A_1A_2, and A_2A_2, respectively (table 16.8). Notice that A_1A_1 and A_2A_2 are true-breeding genotypes; that is, all selfed progeny produced by them are A_1A_1 and A_2A_2, respectively. However, only half the progeny of A_1A_2 are A_1A_2, with the other half split equally between homozygotes A_1A_1 and A_2A_2 due to Mendelian segregation.

If we sum the columns of table 16.8 for the different progeny types, we find that the frequency of heterozygotes is decreased by one-half compared to the parents, while the frequencies of the two homozygotes are each increased. For example, if the initial frequencies were 0.25, 0.5, and 0.25 for genotypes A_1A_1, A_1A_2, and A_2A_2, these frequencies would be 0.375, 0.25, and 0.375 after one generation of complete selfing.

TABLE 16.8 *Frequency of Parents and Progeny with Complete Self-fertilization*

Parent		Progeny		
	Frequency	A_1A_1	A_1A_2	A_2A_2
A_1A_1	P	P	—	—
A_1A_2	H	¼H	½H	¼H
A_2A_2	Q	—	—	Q
	1.0	P + ¼H	½H	Q + ¼H

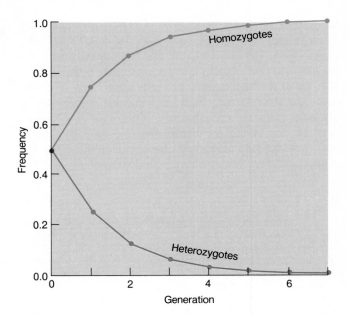

FIGURE 16.8
Change in heterozygosity and homozygosity over time with self-fertilization.

To understand how inbreeding changes the genotypic proportions, let us examine the frequency of heterozygotes when complete self-fertilization occurs. The relationship for heterozygosity in two succeeding generations is a general one; thus, after one generation, the heterozygosity is halved and after two generations, it is halved again, so that the result is one-fourth the initial value. After t generations of complete selfing, heterozygosity is halved t times, so that it is $(½)^t$ of its initial value. For example, if H was initially 0.5, after five generations, the expected heterozygosity would be $(½)^5 0.5 = 0.016$. This is a reduction of heterozygosity from 50% to less than 2% in five generations, showing that the reduction in heterozygosity through selfing is very rapid. The change in heterozygosity and homozygosity $(1 - H)$ is given in figure 16.8.

One of the most important aspects of inbreeding is that, even though genotypic frequencies may be greatly altered, allelic frequencies remain unchanged overall. To illustrate this, let us indicate frequencies of the alleles and genotypes in the progeny generation by primes ('). Therefore, the frequency of allele A_1 in the next generation, p', is:

$$p' = P' + ½H'$$

where P' and H' are the frequencies of genotypes A_1A_1 and A_1A_2 in the next generation. By substituting the values of P' and H' from table 16.8, then:

$$p' = P + ¼H + ¼H$$
$$= p$$

In other words, as stated previously, no change in allelic frequency occurs from generation to generation due to self-fertilization.

THE INBREEDING COEFFICIENT

In order to describe the effect of inbreeding on genotypic frequencies in general, we will use a measure called the **inbreeding coefficient** (f). This value is defined as the probability that the two alleles at a gene in an individual are identical by descent. (Alleles are identical by descent

when the two alleles in a diploid individual are derived from one particular allele in their ancestry, a situation we will illustrate later using pedigrees.) The largest value f can take is 1, when all genotypes in a population are homozygotes containing alleles identical by descent, as in the pure-breeding lines of Mendel's peas. The lowest value f can take is 0, when no alleles are identical by descent. In large, random mating populations, it is generally assumed that f is close to 0, because any inbreeding that may occur is between very distant relatives and thus will have little effect on the inbreeding coefficient.

A general formulation of the proportion of the three genotypes A_1A_1, A_1A_2, and A_2A_2, using the inbreeding coefficient and the allelic frequencies, can be derived as shown in table 16.9. Here it is assumed that the population can be divided into an inbred proportion (f) and a random mated proportion $(1 - f)$. In the inbred proportion, the frequencies of A_1A_1, A_1A_2, and A_2A_2 are p, 0, and q, respectively. These frequencies are the proportions of lines expected for each of the genotypes if, for example, complete selfing continues. When the inbred and random mated proportions are added together (using the relationship $q = 1 - p$), the genotypic frequencies become:

$$P = p^2 + fpq$$

$$H = 2pq - 2fpq$$

$$Q = q^2 + fpq$$

The first term in each equation is the Hardy-Weinberg proportion of the genotypic frequencies, and the second term is the deviation from that value. Note that an

	Inbred (f)	Random mated (1 − f)	Total	
A_1A_1	fp	$(1 − f)p^2$	$fp + (1 − f)p^2$	$= p^2 + fpq$
A_1A_2	—	$(1 − f)2pq$	$(1 − f)2pq$	$= 2pq − 2fpq$
A_2A_2	fq	$(1 − f)q^2$	$fq + (1 − f)q^2$	$= q^2 + fpq$
	f	1 − f	1	1

individual may be homozygous either because the two alleles are identical by descent (the fpq terms) or because the two alleles are identical in kind through random mating (the p^2 and q^2 terms in the equations for P and Q, respectively). For example, A_2A_2 homozygotes that occur unrelated to inbreeding contain alleles identical in kind, while A_2A_2 homozygotes resulting from the same ancestral allele are termed identical by descent. The size of the coefficient of inbreeding reflects the genotypes' deviation from Hardy-Weinberg proportions such that, when f is 0, the zygotes are in Hardy-Weinberg proportions, and when f is positive, as from inbreeding, a deficiency of heterozygotes and an excess of homozygotes occur.

CALCULATING THE INBREEDING COEFFICIENT

We will discuss two ways the inbreeding coefficient can be estimated—from genotypic frequencies or from pedigrees. By the first method, we estimate the inbreeding coefficient in a natural population using the expression for the frequency of heterozygotes given previously. Thus, we can solve the expression for f as:

$$H = 2pq − 2fpq$$
$$1 − f = \frac{H}{2pq}$$
$$f = 1 − \frac{H}{2pq}$$

From the bottom equation, it is clear that the inbreeding coefficient (f) is a function of the ratio of the observed heterozygosity (H) over the Hardy-Weinberg heterozygosity (2pq). With inbreeding, H is less than 2pq, so f is greater than 0. As expected, if no heterozygotes are observed (H = 0), the inbreeding coefficient is 1.

Many species of plants have mating systems that include both self-fertilization and random mating (often called outcrossing) with other individuals. If the proportion of self-fertilization is high, nearly all individuals in the population should be homozygotes. An example is *Avena fatua*, a highly self-fertilizing species of wild oats, which is widespread in many areas around the Mediterranean Sea and was introduced into California where it is responsible for the "golden" hills in summer. A sample from the plant revealed that the frequencies of the three genotypes were P = 0.67, H = 0.06, and Q = 0.27. In order to estimate the inbreeding coefficient, we must first calculate the frequencies of the alleles. As before, the frequency of A_2 (q) is:

$$q = ½H + Q$$
$$= (½)(0.06) + 0.27$$
$$= 0.3$$

The value of p is therefore 1 − q = 0.7. Using the expression given previously:

$$f = 1 − \frac{0.06}{2(0.3)(0.7)}$$
$$= 0.86$$

This is an extremely high inbreeding coefficient, not far from 1, the maximum value possible. This high value suggests that most of the population reproduces by self-fertilization and that very little mating is random.

As stated earlier, the inbreeding coefficient is known as the probability of identity by descent, which implies that homozygosity is the result of the two alleles in an individual having descended from the same ancestral allele. The probability of identity by descent varies, depending upon the relationship of the parents of the individual being examined. For example, if the parents are unrelated (or not closely related), there is no possibility (or a very low one) that the individual will be homozygous by descent. The other extreme (given one generation of sexual reproduction) occurs when the same individual produces both gametes; that is, when self-fertilization occurs. In this case, the probability of an offspring having identical alleles by descent is 0.5. To illustrate, assume that the parent has the genotype A_1A_2. Progeny are produced in the proportions ¼ A_1A_1, ½ A_1A_2, and ¼ A_2A_2; thus, half the progeny, the A_1A_1 and the A_2A_2, will have alleles identical by descent. We assumed the parent was heterozygous so that we could identify the alleles in the progeny. However, even if the parent were homozygous for alleles identical in kind, we could arbitrarily designate them as different in order to calculate the probability of identity by descent.

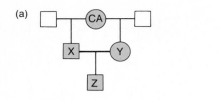 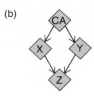

FIGURE 16.9
A pedigree illustrating a mating between two half sibs, X and Y. (a) With all individuals; (b) without the father. (CA = common ancestor.)

A second way to obtain the inbreeding coefficient for a progeny is from a pedigree in which a **consanguineous mating** (a mating between relatives) has occurred. We use the pedigree to calculate the probability of identity by descent in the progeny. As an example, let us figure the inbreeding coefficient for an offspring of two half sibs, individuals who share one common parent. Figure 16.9a gives the pedigree for this type of mating, where X and Y are two half sibs having the same mother but different fathers. The two fathers are indicated as open squares, even though they do not contribute to the inbreeding coefficient. The mother of X and Y is indicated as the common ancestor (CA). In figure 16.9b, the same pedigree appears in a different form, with the fathers omitted and diamonds symbolizing all individuals, because sex is not important in determining the inbreeding coefficient here.

Let us assume that the mother (CA) has the genotype A_1A_2. To calculate the inbreeding coefficient, we need to know the probability that her grandchild (Z) is either A_1A_1 or A_2A_2—that is, identical by descent for either of her alleles. The first alternative, that Z is A_1A_1, can only occur if X and Y each contribute to Z a gamete containing A_1. The probability of an A_1 allele in X is the probability that an A_1 allele came from CA, or 0.5. Because the probability of transmission of A_1 to Z from X is also 0.5, the joint probability of both events (that is, A_1 from CA to X and then to Z) is $(0.5)(0.5) = 0.25$, using the product rule. Likewise, the probability of A_1 to Z through Y is 0.25. Therefore, the probability of an A_1A_1 progeny, who has received an A_1 from X and Y each, is $(0.25)(0.25) = 0.0625$. Using the same approach, the probability of an A_2A_2 progeny is 0.0625. The overall probability of identity by descent in Z is then $0.0625 + 0.0625 = 0.125$.

A more straightforward way to calculate the inbreeding coefficient from a pedigree is the **chain-counting technique.** A chain for a given common ancestor starts with one parent of the inbred individual, goes up the pedigree to the common ancestor, and comes back down to the other parent. For the example in figure 16.9, the chain is quite simple and is expressed X-CA-Y. The number of individuals in the chain (N) can be used in the following formula to calculate the inbreeding coefficient as:

$$f = (½)^N$$

For example, for the pedigree in figure 16.9, the inbreeding coefficient would be $(½)^3 = 0.125$, the same result we obtained by the longer approach, using the definition of the inbreeding coefficient.

> **I**nbreeding increases the frequency of homozygotes and reduces the frequency of heterozygotes, without changing allelic frequencies. The inbreeding coefficient can be measured either from genotypic frequencies or from pedigrees.

MUTATION

Mutation is a particularly important process in population genetics and evolution because it is the major source of genetic variation in a species or population. Mutation can take many forms (see chapter 12). For example, it may involve only one DNA base, several bases, the major part of a chromosome, whole chromosomes, or sets of chromosomes. The immediate cause of a particular mutation might be a mistake in DNA replication, the physical breakage of the chromosome by a mutagenic agent, the insertion of a transposable element, or a failure in disjunction of meiosis. Our discussion will center on **spontaneous mutations,** which appear without apparent explanation, rather than on **induced mutations,** which are caused by some mutagenic agent.

In order to examine the effect of mutation on genetic variation in a population, let us assume that the rate of mutation from wild-type alleles (A_1) to deleterious (or detrimental) alleles (A_2) for a given gamete per generation is u. This is known as the **forward mutation rate.** We will call the rate of mutation from deleterious alleles to wild-type alleles v, or the **backward mutation rate.** Because only A_1 alleles can mutate to A_2 and the A_1 allele has a frequency of p, the increase in the frequency of A_2 from forward mutation is (u)(p). Using the same logic, the frequency of A_2 alleles can be decreased by backward mutation by the amount (v)(q). Overall then, the change in the frequency of A_2 (Δq) due to mutation is:

$$\Delta q = up - vq$$

The maximum positive value for this change is u, when $p = 1$ and $q = 0$ (all alleles are wild-type). The maximum negative value is v, when $p = 0$ and $q = 1$. However, because the mutation rates u and v are generally small (see chapter 12), the expected change due to mutation is also quite small. For example, if $u = 10^{-5}$, $v = 10^{-6}$, and $q = 0.0$, then:

$$\Delta q = (0.00001)(1.0) - (0.000001)(0.0)$$
$$= 0.00001$$

Chapter 16

This is obviously quite a small change in allelic frequency.

Even so, mutation is of fundamental importance in determining the incidence of rare genetic diseases (see chapter 17). In fact, the balance between mutation (increasing the frequency of the disease allele) and selection (reducing the frequency of the disease allele) can explain the observed incidence of diseases such as albinism. In addition, mutation in concert with genetic drift provides a reasonable explanation for the large amounts of molecular variation recently observed in many species (see chapter 17).

M utation is the original source of genetic variation. It can cause only a small change in allelic frequency per generation.

■

GENETIC DRIFT

We have assumed until now that the population under study is large. But in fact, many populations are small, such as the number of individuals in an endangered species or in a species colonizing a new habitat. As a result of small population numbers, chance effects, called **genetic drift,** which are introduced by sampling gametes from generation to generation in a small population, can change allelic frequencies. In a large population, usually only a small chance change in the allelic frequency will be due to genetic drift. But in a small population, allelic frequency can undergo large fluctuations in different generations in a seemingly unpredictable pattern. Genetic drift is fundamental to efforts toward genetic conservation and understanding molecular evolution (see chapter 17).

An example will illustrate the type of allelic frequency change that can be expected in a small population. Assume that a diploid population has five individuals (ten copies of the gene, or ten gametes) and that initially, half of these are A_1 and half are A_2. The ten gametes for the next generation are randomly chosen from the gametic pool, a process analogous to flipping a coin ten times and counting the number of heads and tails. Sometimes the proportion of A_1 alleles is greater than the proportion of A_2, sometimes the proportions are equal, and sometimes the proportion of A_2 is greater than the proportion of A_1. If many small populations, each containing ten gametes, are randomly chosen from a population with equal numbers of A_1 and A_2, the proportions of populations with different frequencies of A_2 will be as shown in figure 16.10. For example, a frequency of 0.4 indicates that four of the ten gametes were A_2. Notice that in most of the populations, the frequencies are close to the initial frequency of 0.5, but that significant variation occurs around that value.

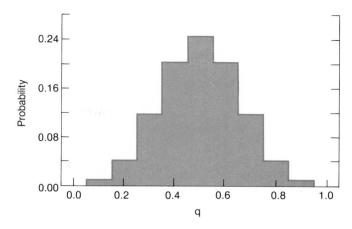

FIGURE 16.10
The probability that a population with ten gametes will have a given allelic frequency after one generation of genetic drift.

To illustrate the potential for cumulative genetic change over time due to genetic drift, let us follow four particular populations of size twenty (forty gametes) over a number of generations (figure 16.11). These values were obtained using computer simulation in which random numbers represented the chance segregation that took place from generation to generation. The solid lines indicate the four replicates, and the broken line is the mean frequency of allele A_2 over the four replicates. All of the replicates were initiated with the frequency of A_2 equal to 0.5. After one generation of genetic drift, the frequencies were 0.625, 0.55, 0.55, and 0.475. By generation 15, they were even more widely dispersed—0.8, 0.5, 0.475, and 0.3. In one of the replicates, the frequency of A_2 went to 1.0 in generation 19, meaning that all of the alleles in the population were A_2. In another replicate, the A_2 allele frequency went to 0.0 in generation 28. The other two replicates were still segregating for both alleles at the end of thirty generations. The mean of the four replicates varied from 0.625 in generation 19 to 0.475 in generation 30, but was generally near the initial frequency of 0.5. (In cases where there are enough replicate populations, no change occurs in the mean allelic frequency over all populations, because increases in allelic frequency in some populations are cancelled by reductions in allelic frequency in others.)

From this example, it is obvious that genetic drift can cause large and erratic changes in the allelic frequencies of individual populations in a rather short time and can lead to differentiation among populations. Within a small population, there is a high probability that a gene will become fixed for one allele or the other over a period of time. But in a large population over the same time span, there would be little change in allelic frequency due to genetic drift, and consequently, little cumulative differentiation among populations.

FIGURE 16.11
The allelic frequencies over thirty generations for four replicate populations of size twenty. The mean is represented by the broken line.

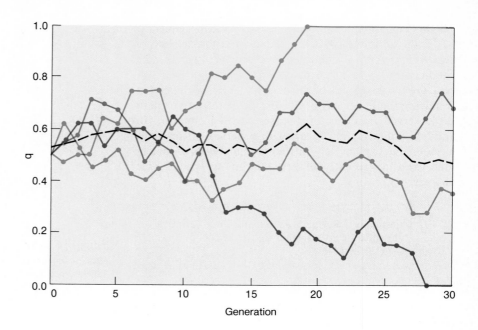

FIGURE 16.12
The northern elephant seal.
© *Richard Humbert/Biological Photo Service.*

One prediction of genetic drift—that a small population will become highly homozygous over time—is supported by the findings in several endangered species. For example, the Northern elephant seal was hunted to near-extinction at the turn of the century, until fewer than one hundred individuals survived (see figure 16.12). Although the population size has rebounded to over 70,000 individuals in recent years, it was low for a number of generations. Apparently due to this "bottleneck," genetic variation was lost, resulting in an estimated zero heterozygosity for the twenty-four genes mentioned in table 16.2.

Another example of a seemingly genetically depauperate species is the Pere David's deer. This species was discovered in the hunting compound of a Chinese emperor in 1856, presumably having been sequestered there for three hundred years. Approximately four hundred of the deer are living today, all descended from eleven animals taken from China. A survey of this species has also shown no heterozygosity for enzyme genes.

We can identify the effects of genetic drift in several ways. First, populations eventually go either to fixation for $A_2(q = 1)$ or to loss of $A_2(q = 0.0)$. The probability that a population will eventually become fixed for A_2 is equal to the initial frequency of A_2. For example, if q is initially 0.2, the probability that the population will be fixed for A_2 is 0.2, and the probability that it will lose the A_2 allele is 0.8. We can understand this point intuitively by realizing that, for A_2 to become fixed, it must change by chance from 0.2 to 1.0, a much bigger change than for the loss of A_2 (from 0.2 to 0.0).

Second, because the mean frequency over replicate populations does not change and the distribution of the allelic frequencies over replicate populations does, the overall effect of drift can be understood by examining the variation of the allelic frequency over replicate populations. One statistic used to measure the extent of variation is the variance (see chapter 3). The expected variance in allelic frequency over replicate populations, due to one generation of genetic drift, is:

$$V = \frac{pq}{2N}$$

In a large population (large N), V is small, indicating only a minor effect from genetic drift. The impact of genetic drift can also be understood by examining the rate of loss of heterozygosity. In chapter 17 we will use the decay of heterozygosity to measure the effectiveness of genetic conservation in endangered species.

To establish a general yardstick for measuring the effect of genetic drift on allelic frequency relative to other factors, we note that the square root of the variance (the standard deviation, see chapter 3) is approximately equal to the average absolute value of the allelic frequency change, or:

$$|\Delta q| = V^{1/2}$$

For example, if $p = q = 0.5$ and $N = 50$, then:

$$V = \frac{(0.5)(0.5)}{2(50)} = 0.0025$$

Taking the positive square root of V, then:

$$|\Delta q| = 0.05$$

Obviously, if N is relatively small, the effect of genetic drift on allelic frequency may be substantial, much larger than that of mutation and comparable to that of selection or gene flow, as we will see later.

When N is large, as in most species, the expected per-generation change in allelic frequency from genetic drift is small. However, the cumulative change over many generations may be quite important, as we will find out when we discuss molecular evolution in chapter 17.

G enetic drift in small populations can result in large chance changes in allelic frequency. Genetic drift may be the cause of the low heterozygosity observed in some endangered species.

■

GENE FLOW

Previously, we have considered only a single large population in which random mating occurred throughout. Many species, however, have smaller populations that are divided because of ecological or behavioral factors. For example, the populations of fish in tide pools, deciduous trees in farmlands, and insects on host plants are separated from each other because the suitable habitat for the species is not continuous. With this type of population structure, differing amounts of genetic connectedness can exist among the separated populations. The extent of this genetic connection depends primarily on the amount of **gene flow**—that is, genetically effective migration—among the populations. As a result of gene flow, genetic variants can be introduced from one population to another.

For example, the genetic variants that confer resistance to insecticides like DDT generally are initially present in only a limited number of populations (see

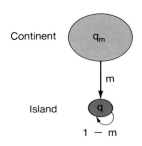

FIGURE 16.13
A model of the gene flow from a continental population to an island population, where m is the rate of gene flow and q_m is the allelic frequency in the migrants.

chapter 17). However, the movement of individuals between populations introduces resistant alleles to populations that did not have them previously. As a result, DDT resistance can spread rapidly throughout mobile species such as mosquitoes or houseflies.

The simplest type of gene flow takes place when an allele enters a single population from an outside source. For example, such unidirectional gene flow occurs in an island population that receives migrants from a continental population. This description basically applies to species having populations on islands and nearby large landmasses, to aquatic species that live in ponds and have lakes as a source of migration, and to peripheral populations that are constantly replenished by individuals from the main part of the species range.

To examine the effect of this type of population structure, let us assume that an island population receives migrants from a large source (continent) population, as shown in figure 16.13. Although reciprocal gene flow may occur, we will assume it is so low that it has a negligible effect on the allelic frequency in the source population. Let the proportion of migrants moving into the island population each generation be m, so that the proportion of nonmigrants (or residents) in the island population is $1 - m$. If the frequency of A_2 in the migrants is q_m and the frequency of A_2 on the island before migration is q, the allelic frequency of A_2 after migration is:

$$\begin{aligned} q' &= (1 - m)q + mq_m \\ &= q - m(q - q_m) \end{aligned}$$

The change in allelic frequency after one generation of gene flow is the difference between the frequencies after gene flow and before gene flow, or:

$$\Delta q = q' - q$$

If we substitute q' from above into this expression, then:

$$\Delta q = -m(q - q_m)$$

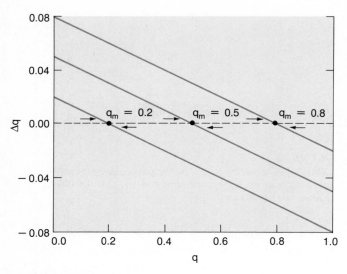

FIGURE 16.14
The expected change in allelic frequency in one generation due to gene flow when m = 0.1 and q_m = 0.2, 0.5, or 0.8. The arrows indicate the direction of changes in allelic frequency.

From these expressions, it is obvious that no change in allelic frequency will occur; that is, $\Delta q = 0$ if m = 0 or if $q = q_m$. Only the second alternative is of interest here, because a value of m = 0 indicates no gene flow. Note that q_m and m are assumed to be constant over time and that they have values between zero and unity. If q is smaller than q_m, the frequency of A_2 increases on the island because of the gene flow; that is, Δq is positive. If q is greater than q_m, the frequency of A_2 will decrease, and Δq is negative. As a result of these two effects, there is a **stable equilibrium frequency** of A_2 on the island at $q_m = q$. (A stable equilibrium occurs where the population frequency returns to a given value after being perturbed either above or below that value.)

The effect of different allelic frequencies in the migrants on Δq when m = 0.1 can be seen in figure 16.14. Note that the change in allelic frequency increases linearly as the frequency moves away from the equilibrium value and that it reaches an absolute maximum at either zero or unity, depending upon the value of q_m. From the values in this figure, it is obvious that the change in allelic frequency due to gene flow can be important. It may equal the size of changes caused by genetic drift in small populations and may be much larger than changes caused by mutation.

Many populations are composed of individuals who have descended from different populations. If we assume that only two ancestral populations have contributed to the gene pool in a given population (for clarity, we will call this population the **hybrid population**), we can mea-

sure the amount of **admixture,** the proportion of the gene pool descended from one of the ancestral populations. We achieve this estimate by rearranging the first equation so that it reads:

$$q' = (1 - m)q + mq_m$$

Then, we solve for m so that:

$$m = \frac{q - q'}{q - q_m}$$

Such admixture estimates have been made for human populations that have descended from different racial groups. For example, the American black population is primarily of African origin, mixed with some more recent Caucasian ancestry. How can we estimate the proportion of Caucasian ancestry in American black populations? One approach is to use a gene that has quite different allelic frequencies in the two ancestral populations, such as an allele at the Duffy blood group locus, another gene that codes for red blood cell antigens. Table 16.10 gives estimates of the ancestral frequencies of this allele and estimates of admixture (m) in two United States cities. For example, to calculate m in the Oakland sample:

$$m = \frac{0.0 - 0.094}{0.0 - 0.429}$$
$$= 0.220$$

Notice that the estimate of admixture is much higher (0.22) in the northern locality (Oakland) and only 0.037 in Charleston, indicating that certain southern populations may have little Caucasian admixture.

> **G**ene flow between populations can introduce new alleles and, in some cases, can rapidly change allelic frequencies. The extent of past gene flow between populations can be estimated.

∎

TABLE 16.10 *Estimates of Admixtures (m) in American Blacks*

Locality	Blacks (q')	African (q)	Caucasian (q_m)	m
Oakland, CA	0.094	0.0	0.429	0.220
Charleston, SC	0.016	0.0	0.429	0.037

NATURAL SELECTION

Charles Darwin is often considered the chief architect of the theory of natural selection. He provided strong evidence for it from his collection of plants and animals gathered on an around-the-world voyage on the *H.M.S. Beagle*. Darwin's theory held that variation among individuals of a species was the raw material through which selection could eventually produce individuals better adapted to their environment. Darwin was primarily concerned with the evolution of quantitative traits (discussed in chapter 3), but the same general principles operate regarding traits determined by single genes.

Selection can occur in several different ways (affecting fertility or mating success, for example), but here we will assume for simplicity that it results from differential survival of genotypes. To illustrate, assume 50% of the haploid genotype A_1 survive to adulthood, while only 40% of A_2 survive to adulthood, making the survival rate of A_1 higher than that of A_2. Intuitively, one would expect the frequency of A_1 to increase and the frequency of A_2 to decrease over time in a population initially composed of these two genotypes.

Selection has two basic effects on the amount of genetic variation. First, selection favoring a particular allele may lead to the reduction of genetic variation and, consequently, to homozygosity for the favored allele, as for DDT resistance alleles in mosquitoes exposed to DDT for many generations. Second, selection may result in the maintenance of two or more alleles in a population, such as for the banding polymorphism in snails or for the sickle-cell allele in humans.

In order to understand the effects of selection on genetic variation, we must consider the relative fitnesses of the different genotypes. **Relative fitness** can be defined as the relative ability of different genotypes to pass on their alleles to future generations. If, as in the example given previously, the relative fitnesses of two genotypes are based solely on their viabilities, these values can be used to calculate relative fitness values. Generally, the highest relative fitness value is assumed to be 1.0, and the others are standardized against this value. (As we will see, this is done to simplify the algebra.) Therefore, in the previous example, the relative fitness of genotype A_1 is $0.5/0.5 = 1.0$ and that of genotype A_2 is $0.4/0.5 = 0.8$.

SELECTION AGAINST A HOMOZYGOTE

We will begin by considering the situation in which selection operates against a homozygote, with the other homozygote and the heterozygote having equal relative fitnesses. This is sometimes called purifying selection because it purifies the gene pool of detrimental alleles. Let

Natural selection at work

"The Far Side" cartoon by Gary Larson is reprinted by permission of Chronicle Features, San Francisco, CA.

the relative fitnesses of the three possible genotypes, A_1A_1, A_1A_2, and A_2A_2, be designated as 1, 1, and $1 - s$, respectively, where the fitness of genotype A_2A_2 is smaller by an amount s, called the **selective disadvantage** (or **selection coefficient**) of the homozygote. If $s = 1$, then A_2 is a recessive lethal like a number of disease alleles in humans. If the relative fitness of A_2A_2 is 0.8, the selective disadvantage is $s = 1 - 0.8 = 0.2$.

Because we are considering only selection resulting from different viabilities of the genotypes, the relative fitness of a genotype is its relative probability of surviving from the newly formed zygote stage until reproduction. We can give the genotypic frequencies before selection in terms of allelic frequencies, assuming Hardy-Weinberg proportions. The weighted contribution of the three genotypes to the next generation is the product of the frequency of the genotypes before selection and their relative fitnesses. The **mean fitness** of the population (w) is the sum of the weighted contributions of the different genotypes, or:

$$w = p^2(1) + 2pq(1) + q^2(1 - s)$$
$$= p^2 + 2pq + q^2 - sq^2$$

	Genotype			**Total**
	A_1A_1	A_1A_2	A_2A_2	
Relative fitness	1	1	$1 - s$	—
Frequency before selection	p^2	$2pq$	q^2	1.0
Weighted contribution	p^2	$2pq$	$q^2(1 - s)$	$1 - sq^2$
Frequency after selection	$\dfrac{p^2}{1 - sq^2}$	$\dfrac{2pq}{1 - sq^2}$	$\dfrac{q^2(1 - s)}{1 - sq^2}$	1.0

TABLE 16.11 *Frequency of Genotypes Before and After Selection, Assuming Hardy-Weinberg Proportions Before Selection*

Because $p^2 + 2pq + q^2 = 1$ (that is, the sum of the frequencies of all the genotypes is 1), the mean fitness is:

$$w = 1 - sq^2$$

If the relative contributions are standardized by the mean fitness, the frequencies of the genotypes after selection can be obtained as shown in the bottom line of table 16.11. The frequency of A_2 after selection (q') is equal to half the frequency of the heterozygote A_1A_2 (because half the genes in A_1A_2 are A_2 alleles) plus the frequency of the homozygote A_2A_2, or:

$$q' = (\tfrac{1}{2})\left(\frac{2pq}{1 - sq^2}\right) + \frac{q^2(1 - s)}{1 - sq^2}$$

$$= \frac{pq + q^2(1 - s)}{1 - sq^2}$$

$$= \frac{q(p + q - sq)}{1 - sq^2}$$

$$= \frac{q(1 - sq)}{1 - sq^2}$$

Therefore, the frequency of A_2 after selection is a function of the frequency before selection and of the selection coefficient.

As an example, let us assume that the homozygote A_2A_2 is inviable; in other words, that A_2 is a recessive lethal. In order to make the fitness of A_2A_2 zero, then $1 - s = 0$, or $s = 1.0$. Let us now calculate the change in the frequency of A_2 over several generations. If the initial frequency of A_2 is 0.5, then from the formula:

$$q' = \frac{0.5[1 - (1.0)(0.5)]}{1 - (1.0)(0.5)^2}$$

$$= 0.333$$

The same formula can be used to calculate the frequency in the next generation, assuming that $q = 0.333$ before selection, so that:

$$q' = \frac{0.333[1 - (1.0)(0.333)]}{1 - (1.0)(0.333)^2}$$

$$= 0.250$$

This process can be repeated over subsequent generations to calculate the allelic frequency. After twenty generations of selection against A_2A_2, the frequency of A_2 has been reduced from 0.5 to approximately 0.045.

Many of the principles of population genetics have been investigated using *Drosophila* laboratory populations as a model. For example, the *Glued* mutant is a lethal homozygote that also reduces eye size and affects eye appearance in heterozygotes. Michael Clegg and several co-workers followed the decline in the frequency of the *Glued* mutant in artificial populations initially composed of all heterozygotes, that is, those having an initial frequency of 0.5. Figure 16.15 gives the expected change (broken line) just as we calculated previously and the observed change in two replicates, A and B. Notice that the general trend observed is close to that expected for the first few generations but that there is a difference in the later generations. Further investigation also showed selection against the heterozygotes, explaining why the observed frequency declined faster than expected.

The amount of allelic frequency change in one generation due to selection is defined as:

$$\Delta q = q' - q$$

By substituting q' given previously, then:

$$\Delta q = \frac{q(1 - sq)}{1 - sq^2} - q$$

$$= \frac{q(1 - sq) - q(1 - sq^2)}{1 - sq^2}$$

$$= \frac{q(1 - sq - 1 + sq^2)}{1 - sq^2}$$

$$= \frac{-spq^2}{1 - sq^2}$$

Chapter 16

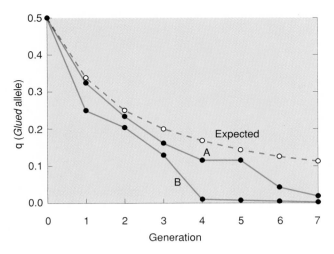

FIGURE 16.15

The expected change in the frequency of the lethal allele *Glued* (blue line) and the observed change in the replicate populations (red lines). *Data from Clegg, et al., Genetics 83:793–810, 1976.*

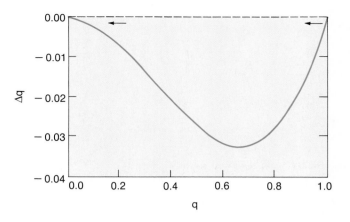

FIGURE 16.16

The change in the frequency of A_2 in one generation for different initial frequencies (q) when there is selection against the recessive genotype A_2A_2. The arrows at top indicate the direction of allele frequency change.

Notice that the change in allelic frequency is negative, indicating that selection in this case is always reducing the frequency of A_2.

To illustrate the connection between this formula and the one given for q′, let us assume again that q = 0.6 and that s = 0.2. Putting these values in the formula for change in allelic frequency:

$$\Delta q = \frac{-(0.2)(0.4)(0.6)^2}{1 - (0.2)(0.6)^2}$$

$$= -0.031$$

In other words, the allelic frequency has been reduced from 0.6 to 0.569, as we found earlier.

The change in allelic frequency is a function of allelic frequency and selective disadvantage. Figure 16.16 illustrates how Δq changes for different initial frequencies when the selective disadvantage is 0.2. The change in frequency is 0.0 when either p = 0 or q = 0, because the population is monomorphic for either the A_2 or the A_1 alleles. The change in frequency is greatest for an intermediate allelic frequency and becomes quite low when q approaches 0.0. This occurs because most of the A_2 alleles are in heterozygotes and are not subject to selection.

HETEROZYGOUS ADVANTAGE

The mode of selection just described, that is, selection against homozygotes, leads to the eventual fixation of one allele and, as a result, reduces the genetic variation in the population. However, when the heterozygote has a higher fitness than both homozygotes, two alleles can be maintained in the population. The classic example of this type

of selection is sickle-cell disease, mentioned earlier; another example is resistance to the rodenticide warfarin in rats, which will be discussed in chapter 17.

To investigate the heterozygous advantage model, we will use a fitness array where the heterozygote, A_1A_2, has the maximum relative fitness (1) and where the homozygotes are $1 - s_1$ and $1 - s_2$ (s_1 and s_2 being the selective disadvantages, or selection coefficients, of the homozygotes A_1A_1 and A_2A_2, respectively). If we follow the same procedure used before, the change in allelic frequency due to selection is:

$$\Delta q = \frac{pq(s_1p - s_2q)}{w}$$

In order to maintain both alleles in the population, Δq must be zero for some value of q between 0 and 1. The point where this occurs is called the **equilibrium frequency** of allele A_2—that is, the allelic frequency for which the change is zero.

The value of q, when this expression equals zero and both alleles are present, occurs when the term inside the parentheses in the equation is zero, or:

$$s_1p - s_2q = 0$$

By solving this equation for q, one can show that an equilibrium occurs when:

$$q_e = \frac{s_1}{s_1 + s_2}$$

where q_e signifies the equilibrium frequency of allele A_2.

Industrial Melanism

T he peppered moth, *Biston betularia,* is generally a mottled light color, but in a number of areas of England, a dark, melanic form is common (see figure B16.1). The melanics were quite rare a century ago, but now they constitute nearly 100% of some urban populations. The melanic moths have increased because their dark color serves as protective camouflage from bird predation in the polluted areas they inhabit. This phenomenon, which has occurred in a number of other species, is called **industrial melanism.** A further confirmation of this hypothesis is the increasing frequency of the nonmelanic forms in areas where the air has become cleaner due to pollution control. In other words, there are both spatial and temporal associations between air pollution and the frequency of melanic peppered moths.

To measure the extent of selection between the two moth phenotypes, British biologists Cedric Clarke and Phillip Sheppard pinned dead moths of the two types on dark and light backgrounds and measured their survival from bird predation. On the dark background, fifty-eight of seventy melanic moths survived, while only thirty-nine of seventy typical moths survived (see table 16.12a). As a result, the relative fitness of the typical moths was only 0.67, compared to 1.0 for the melanics in this environment. On the other hand, on the light background (table 16.12b), the relative fitness of typicals was highest (or 1) and that of melanics was only 0.75.

Let us use the relative fitnesses estimated on the dark background to calculate the expected change in the typical allele. In this case, the selective disadvantage

(a)

(b)

FIGURE B16.1
Melanic and typical forms of the peppered moth *Biston betularia* (*a*) on a soot-covered trunk and (*b*) on a lichen-coated tree.
Both photos © Michael Tweedie/Photo Researchers.

In other words, for this allelic frequency, there is no change, and both alleles are maintained in the population. Notice that the equilibrium is a function only of the selection coefficients for the two homozygotes.

The properties of the equilibrium can be seen in another way by rewriting Δq using the equilibrium frequency so that:

$$\Delta q = \frac{-pq(s_1 + s_2)(q - q_e)}{w}$$

From this formulation, one can see that when q is greater than q_e, Δq is negative; thus, the allelic frequency will decrease toward the equilibrium. Likewise, when q is less than q_e, Δq is positive and the allelic frequency will increase toward the equilibrium. In other words, as in the gene flow example discussed earlier, the equilibrium is stable because after a perturbation away from it, the allelic frequency will change back toward the equilibrium frequency again.

against typical homozygotes is s = 0.33, because 1 − s = 0.67. If we assume that the frequency of the typical allele is 0.2, then using the formula:

$$q' = \frac{0.2[1 - (0.33)(0.2)]}{1 - (0.33)(0.2)^2}$$

$$= 0.189$$

In other words, the allelic frequency will decline from 0.2 to 0.189, a significant but relatively small change. ∎

TABLE 16.12 *Survival of Melanic and Typical Moths on Dark and Light Backgrounds*

(a) Dark Background

	Melanic		Typical	
	Exposed	Survived	Exposed	Survived
Number	70	58	70	39
Survival	$\frac{58}{70} = 0.83$		$\frac{39}{70} = 0.56$	
Relative fitness	$\frac{0.83}{0.83} = 1$		$\frac{0.56}{0.83} = 0.67$	

(b) Light Background

	Melanic		Typical	
	Exposed	Survived	Exposed	Survived
Number	40	24	40	32
Survival	$\frac{24}{40} = 0.6$		$\frac{32}{40} = 0.8$	
Relative fitness	$\frac{0.6}{0.8} = 0.75$		$\frac{0.8}{0.8} = 1$	

The change in allelic frequency is a function of the allelic frequency and the selection coefficients. An example is given in figure 16.17, where the fitnesses of the three genotypes are 0.8, 1.0, and 0.8, so that $s_1 = s_2 = 0.2$. As a result of these fitness values, the equilibrium frequency is $q_e = {}^{0.2}/_{(0.2 + 0.2)} = 0.5$. If q is smaller than 0.5, the change in allelic frequency is positive, with a maximum value just above q = 0.2. When the value of q is greater than 0.5, the change in allelic frequency is negative, with a maximum value just below 0.8.

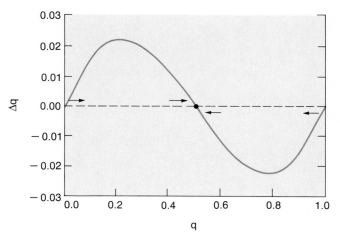

FIGURE 16.17
The change in the frequency of A_2 in one generation for different initial frequencies when there is a symmetric heterozygote advantage ($S_1 = S_2$). The arrows indicate the direction of allele frequency change.

We mentioned earlier that heterozygote advantage is the basis of the maintenance of the sickle-cell allele at the β-globin gene in humans. A number of studies have shown that heterozygotes have a higher viability than normal homozygotes in areas where malaria is prevalent. Heterozygotes are apparently resistant to malaria because their abnormal red blood cells, sickle-shaped rather than donut-shaped, are hard for the parasite to penetrate. If we let the relative fitness of heterozygotes be 1, that of normal homozygotes should be about 0.8, making $s_1 = 0.2$. In malarial areas, sickle-cell disease is generally fatal, so that the fitness of genotype A_2A_2 is 0, making $s_2 = 1$. Putting these selection coefficients into the formula given previously, the equilibrium is $q_e = {}^{0.2}/_{(0.2 + 1.0)} = 0.17$. In fact, the frequency of the sickle-cell allele is between 0.15 and 0.2 in many regions of Africa where malaria is endemic, making these calculations consistent with observations in human populations (see figure 16.3).

Transferrin is an iron-binding protein that is polymorphic in most vertebrate species. In pigeons there are two common alleles, called Tf^A and Tf^B, which vary in frequency from about 0.38 to 0.59 in different breeds. Jeffrey Frelinger found that the eggs from female pigeons heterozygous for these two alleles were better able to inhibit microbial growth than were the eggs from homozygous females. His data, given in table 16.13, show that 67% of the eggs from heterozygous females survived infection while only 46% and 52% of the two homozygotes survived. If we standardize these values to calculate relative fitnesses, then $s_1 = 0.31$ and $s_2 = 0.22$, and the predicted equilibrium is 0.584, a value that is at the upper end of the range observed in different pigeon breeds.

TABLE 16.13 Number of Eggs Laid and Hatched by Pigeons of Three Different Transferrin Genotypes

	Female Genotype		
	$Tf^A Tf^A$	$Tf^A Tf^B$	$Tf^B Tf^B$
Number of eggs laid	128	267	144
Number of eggs hatched	59	180	75
Proportion hatched	0.46	0.67	0.52
Relative fitness	0.69	1.0	0.78

N atural selection can result in the elimination of detrimental alleles by selection against homozygotes or in the maintenance of genetic variation by selection for heterozygotes.

▪

Summary

E xtensive genetic variation occurs in most populations at several different levels, including variants affecting color, chromosomal structure, and protein characteristics. By counting the numbers of different genotypes in a population, genotypic frequencies, allelic frequencies, and heterozygosity can be calculated. Two measures of genetic variation, the proportion of polymorphic loci and the level of heterozygosity, can be used to quantify the extent of genetic variation. In many species, the amount of genetic variation observed for genes coding for proteins is substantial. The Hardy-Weinberg principle states that genotypic frequencies are a binomial function of allelic frequencies after one generation of random mating. Genotypic numbers can be examined statistically using the chi-square test to determine whether they are consistent with Hardy-Weinberg expectations. The Hardy-Weinberg principle and the methods of estimating allelic frequencies also apply to multiple alleles and X-linked genes.

Inbreeding increases the frequency of homozygotes and reduces the frequency of heterozygotes, without changing allelic frequencies. The inbreeding coefficient can be measured either from genotypic frequencies or from pedigrees. Mutation is the original source of genetic variation, but it can cause only a small change in allelic frequency per generation. Genetic drift in small populations can result in larger chance changes in allelic frequency. Genetic drift may be the cause of the low heterozygosity observed in some endangered species. Gene flow between populations introduces new alleles, and in some cases, rapidly changes allelic frequencies. The extent of past gene flow between populations can be estimated. Natural selection eliminates detrimental alleles by selection against homozygotes, or maintains genetic variation by selection for heterozygotes.

▪

Problems and Questions

1. Calculate the frequency of alleles A_1 and A_2 for the following populations:

Population	N_{11}	N_{12}	N_{22}
a	20	40	40
b	30	8	12
c	72	122	91

What are the Hardy-Weinberg frequencies and expected genotypic numbers for the three populations? Calculate the chi-square values for these populations. Are these populations statistically different from Hardy-Weinberg proportions?

2. Assume that the frequencies of genotypes A_1A_1, A_1A_2, and A_2A_2 are 0.1, 0.4, and 0.5, respectively. Use these values and table 16.3 to illustrate that random mating will produce Hardy-Weinberg proportions in one generation. Using a chi-square test and assuming that there are 200 individuals, is this population statistically different from Hardy-Weinberg proportions?

3. If q = 0.3 and there are Hardy-Weinberg proportions, what is the most common genotype and what is its frequency? What is the least frequent genotype and its frequency?

4. If A_2 is recessive and 4 out of 400 individuals are A_2A_2 (the rest are the dominant phenotype), what is the estimated frequency of A_2? What proportion of the population is expected to be heterozygous?

5. Calculate the frequency of alleles A_1, A_2, and A_3 in the following populations:

Population	N_{11}	N_{12}	N_{13}	N_{22}	N_{23}	N_{33}
a	10	20	30	10	20	10
b	40	—	—	40	—	120

What are the expected genotypic frequencies for these two populations? Calculate the chi-square values for these populations. Are these populations statistically different from Hardy-Weinberg proportions? (There are 3 degrees of freedom in this case.)

6. In some populations, approximately 8% of the males have a particular type of color blindness (an X-linked recessive trait). What proportion of females would you expect to have this trait? What proportion of females would you expect to be carriers for this allele?

7. If the initial genotypic frequencies are 0.04, 0.32, and 0.64 for A_1A_1, A_1A_2, and A_2A_2, what are the frequencies after one generation of complete self-fertilization? What is the frequency of heterozygotes after five generations of complete self-fertilization?

8. Calculate the frequency of the three genotypes, A_1A_1, A_1A_2, and A_2A_2, given $f = 0.2$ and $q = 0.4$. Assuming that the frequencies of the three genotypes are 0.35, 0.1, and 0.55, what is f?

9. (a) Draw the pedigree for a mating between two half first cousins who share only one grandparent. What is the expected inbreeding coefficient? (b) Draw a pedigree for a mating between full sibs. What is the expected inbreeding coefficient? (There are two chains in this case because there are two common ancestors. To calculate the overall inbreeding coefficient, you need to add the two f values together.)

10. If $u = 10^{-6}$ and $v = 10^{-7}$, what is Δq for $q = 0.0$, 0.25, 0.5, 0.75, and 1.0? When is Δq largest? Explain why.

11. If the initial frequency in a small population is 0.3 for A_2, what is the probability that the population will become fixed for A_2? For A_1? Explain your answers.

12. In a population of 100 individuals with $q = 0.4$, what are V_q and the approximate value of $|\Delta q|$ after one generation? How does this value of Δq compare to that for mutation? (For example, see problem 10.)

13. Assume that $m = 0.05$ and $q_m = 0.1$. (a) Draw a Δq curve (as in figure 16.14) with these values for different levels of q. (b) What is the expected equilibrium frequency?

14. In two ancestral populations, the frequency of an allele is 0.0 and 0.6. In a hybrid population, the frequency of the allele is 0.08. What is the estimate of admixture? What kind of assumptions need to be made for such estimates of gene flow?

15. (a) For a recessive with $s = 0.4$, what is Δq when $q = 0.0$, 0.25, 0.5, 0.75, and 1.0? (b) Graph these results on a Δq curve. (c) For a recessive with $s = 0.1$, what is the allelic frequency after one and two generations when the initial frequency is 0.2?

16. In another experiment like that described in table 16.12a, 120 melanic moths were exposed and 96 survived, while only 74 of 120 typicals survived. Estimate the relative fitness of the two morphs. What is the expected frequency of the melanic allele after one generation, given that its initial frequency is 0.4?

17. If the relative fitnesses of genotypes A_1A_1, A_1A_2, and A_2A_2, are 0.8, 1.0, and 0.2, respectively, what is the expected equilibrium frequency of A_2? Calculate Δq for $q = 0.1$ and $q = 0.8$. What does this tell you about the expected change in frequency?

18. Compare in general the expected change in allelic frequencies that result from mutation, gene flow, genetic drift, and selection. Which factors will have the largest effect on Δq, and when? In what ways is the effect on allelic frequency from genetic drift different from that due to mutation, selection, and gene flow?

19. Inbreeding affects genetic variation in a different way than do other factors. Describe its influence, and tell how you could detect whether high inbreeding is present in a population. How might inbreeding be used in plant or animal breeding?

20. In the study of snails mentioned at the beginning of this chapter, it was suggested that differential selection via protective camouflage is important in determining the frequencies of different snail types. Design an experiment that would illustrate this directly. Some people suggest that genetic drift is important in determining snail types. How would you experimentally measure the importance of genetic drift? (In this experiment, you can remove snails, add snails, look at different sites, etc.)

Answers appear at end of book.

Extensions and Applications of Population Genetics

*I*t is the triumph of scientific men . . . to desire tests by which the value of beliefs may be ascertained, and to feel sufficiently masters of themselves to discard contemptuously whatever may be found untrue.

■

Sir Francis Galton
British scientist

Rattus norvegicus. A Norway rat washing its tail with its tongue.
© Tom McHugh/Photo Researchers, Inc.

Learning Objectives

In this chapter you will learn:

1. That the incidence of many human genetic diseases can be explained by a mutation-selection balance.

2. That selection, inbreeding, and genetic drift can also influence the incidence of genetic diseases.

3. That the theory of neutrality is consistent with many findings in molecular evolution.

4. How the molecular clock and phylogenetic trees are used to understand the relationships between species.

5. That conservation genetics is concerned with the preservation of genetic variation in agricultural plants and animals as well as in rare and endangered species.

6. That the primary goals of conservation genetics are to avoid inbreeding depression and loss of genetic variation.

7. That pesticide resistance occurs in many pest species.

8. That pesticide resistance can provide excellent case studies of evolutionary change.

I n the introduction to population genetics in chapter 16, we discussed each of the important population genetic factors separately. However, in reality, more than one factor is often at work in a given situation. In this chapter we will illustrate the combined effects of two or more population genetic factors as they operate in several important areas of biological research.

The principles of population genetics can provide insight into a number of areas of biology, medicine, and agriculture, including human genetic disease, molecular evolution, genetic conservation, animal and plant breeding, and pesticide resistance. Here we will use these areas to illustrate the applications of population genetics and to show why it is important to consider several population genetic factors simultaneously. For example, understanding the incidence of human genetic disease necessitates considering both mutation and selection, while studying molecular evolution involves discussing both mutation and genetic drift. Furthermore, a broad comprehension of conservation genetics, animal or plant breeding, or pesticide resistance requires understanding and integrating nearly all the population genetic factors.

FREQUENCY OF HUMAN GENETIC DISEASE

Generally, human genetic diseases are quite rare, with some occurring only in a few families, such as Mal de Meleda and Ellis-van Crevald syndrome, which will be discussed later in this chapter. Such diseases have sometimes been referred to as **orphan diseases.** Because of their rarity, they do not have advocates for research, as do more common diseases such as cancer or AIDS. However, other inherited diseases, such as sickle-cell disease and cystic fibrosis, reach relatively high incidences in certain populations (see table 17.1). Here we will consider some of the factors that determine the incidence of both the rare and the more common genetic diseases. Overall, the combined effect of genetic disease is substantial—nearly 2% of newborns in the United States have a single-gene defect. We will begin this discussion by examining the joint effects of mutation and selection in determining the frequency of recessive diseases. In addition, we will mention the influences of inbreeding, genetic drift, and limited gene flow on the incidence of genetic disease in various populations.

MUTATION-SELECTION BALANCE

Selection is the major force that keeps detrimental alleles from increasing in frequency in a population. For example, individuals with cystic fibrosis, an autosomal recessive condition, usually do not survive to adulthood and very seldom reproduce (although recent advances in treatment are delaying their mortality). As a result of this strong selection, the frequency of the detrimental alleles is kept quite low. On the other hand, new mutations are

TABLE 17.1 *Frequency of Alleles Causing Diseases in Different Human Populations*

Disease	Population	Allelic Frequency
Tay-Sachs	Israel	0.014
	U.S. (Jewish)	0.013
	U.S. (non-Jewish)	0.0013
Cystic fibrosis	United Kingdom	0.022
	Ohio	0.013
	Africa	0.0036
	Hawaii (Asian)	0.0033
Sickle-cell disease	West Africa	0.10–0.20
	U.S. (black)	0.08
	U.S. (white)	<0.001
Phenylketonuria	Ireland	0.0141
	U.S. (white)	0.0079
	Israel	0.0063
	U.S. (black)	0.0023

constantly occurring, so that the cystic fibrosis allele is never completely eliminated. The counterbalancing effects of selection and mutation provide a reasonable explanation for the frequency of many genetic diseases.

To understand this situation, let us assume that there are two alleles and that one of them, the recessive mutant A_2, causes a reduction in viability. If the detrimental allele is recessive, the reduction in allelic frequency due to selection is:

$$\Delta q_s = \frac{-spq^2}{1 - sq^2}$$

where s is the selective disadvantage of homozygote A_2A_2 (see chapter 16). The increase in allelic frequency due to mutation is approximately:

$$\Delta q_{mu} = up$$

if we assume that v, the backward mutation rate (A_2 to A_1), is small compared with u, the forward mutation rate (A_1 to A_2) (see chapter 16). Because these two forces have opposite effects on allelic frequency, they balance each other, so that at some point, the increase in allelic frequency from mutation is the same magnitude as the decrease from selection, or:

$$\Delta q_{mu} + \Delta q_s = 0$$

If we substitute the expressions just given into this equation, then:

$$up - \frac{spq^2}{1 - sq^2} = 0$$

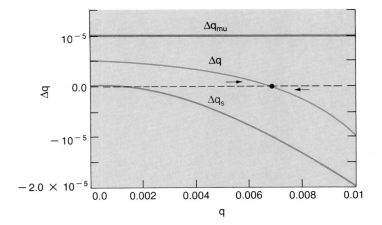

FIGURE 17.1

The change in allelic frequencies where both mutation to and selection against recessives occur. The line labeled Δq is the summation of the other two curves.

Reprinted with permission of Jones and Bartlett Publishers, Inc., Boston, MA, from Population Biology, *1984, written by Philip Hedrick.*

If we assume q^2 is small, because the frequency of the detrimental allele is low (if $q = 0.01$, then $q^2 = 0.0001$), the denominator of the expression giving the change due to selection will be approximately unity ($sq^2 = 0$). Thus:

$$up - spq^2 = 0$$

This expression can then be solved for the frequency of the recessive genotype as:

$$q^2_e = \frac{u}{s}$$

The equilibrium allelic frequency is then:

$$q_e = \left(\frac{u}{s}\right)^{1/2}$$

The allelic frequency is obviously increased with a higher mutation rate and decreased with a higher selective disadvantage. Of course, if the allele is a recessive lethal, then $s = 1.0$, and the genotypic and allelic frequencies are the mutation rates, u and $(u)^{1/2}$, respectively. Because the mutation rate is generally quite low (see the estimates in table 12.1), we would expect the frequency of lethal disease alleles to be quite low also.

The balance (or equilibrium) between selection and mutation can be understood by examining Δq_{mu} and Δq_s at low frequencies of A_2. Figure 17.1 gives these values when $u = 10^{-5}$ and $s = 0.2$. Notice that Δq_{mu} is about 10^{-5} throughout this low range in allelic frequencies, while Δq_s is 0.0 when $q = 0.0$ and becomes increasingly negative as q increases. The equilibrium for this example is:

$$q_e = \left(\frac{10^{-5}}{0.2}\right)^{1/2}$$

$$= 0.00707$$

In other words, this is the point at which $\Delta q_{mu} + \Delta q_s = 0$. The two curves are summarized in the resultant Δq curve, and the solid circle indicates the equilibrium.

As we mentioned in chapter 2, albinism is a recessive condition caused by a biochemical error that results in a lack of the pigment melanin. Individuals with albinism often have poor eyesight and are sensitive to sunlight, but their relative fitness is probably not greatly reduced. One estimate (based on relative viabilities) is that the fitness of albinos is about 0.9 that of unaffected individuals, making $s = 0.1$. Estimation of the mutation rate per gene is generally between 10^{-5} and 10^{-6} (see chapter 12). If we use 5×10^{-6} as the mutation rate, the expected frequency of albino individuals from mutation-selection balance, using the equation just given, is:

$$q^2_e = \frac{5 \times 10^{-6}}{0.1}$$

$$= 0.00005$$

or about 1 in 20,000 individuals. In most populations, the incidence of albinism ranges between 1 in 10,000 and 1 in 40,000, and is thus consistent with our mutation-selection balance explanation.

We can also use the mutation-selection balance to predict the future incidence of genetic disease. For example, the frequency of a genetic disease may rise either because of improved medical care that enables affected individuals to reproduce or because of an increased mutation rate that results in the production of more mutant alleles. When the selective disadvantage against the disease decreases due to improved medical care so that s' is the new selective disadvantage, the new equilibrium becomes:

$$q^2_e = \frac{u}{s'}$$

For example, if a disease was formerly lethal before the age of reproduction (s = 1.0) and now, with better medical care, is only slightly disadvantageous (say, s' = 0.1), its equilibrium frequency should increase 10-fold. Phenylketonuria is a disease for which the fitness has increased from almost zero when untreated to nearly that of unaffected individuals after diet therapy. As a result, we might expect the frequency of individuals with PKU to increase over time.

Likewise, an increase in the mutation rate should also increase the equilibrium value. Assume that because of environmental pollutants, food additives, or some other mutagenic factor, there is an elevated mutation rate, u'. The new equilibrium is then:

$$q^2_e = \frac{u'}{s}$$

If u' = 10u, a 10-fold increase in mutation rate, the equilibrium is also increased 10-fold. However, an increase in mutation rate may influence a number of loci simultaneously. We should caution that higher incidence of disease as a result of both improved medical care and increased mutation rate would generally occur slowly over many, many generations.

OTHER FACTORS AFFECTING DISEASE FREQUENCY

Although mutation-selection balance appears to be a good general explanation for the frequency of many genetic diseases, other evolutionary factors can also be important. For example, the frequency of the sickle-cell allele is increased by the advantage of heterozygotes (the homozygotes are still at a disadvantage) in areas with endemic malaria (see chapter 16). However, the sickle-cell allele is affected by different evolutionary factors in contemporary United States populations. Although malaria was once endemic in parts of the southern United States, the present rare incidence of the disease has resulted in a different set of fitness values here than in Africa; that is, there is no heterozygous advantage, and the homozygote is often no longer lethal.

Genetic counseling may lead to lower reproduction of sickle-cell carriers and selective abortion of sickle-cell homozygotes. In addition, past and present gene flow into American black populations from Caucasians, in which the frequency of the sickle-cell allele is near zero, further dilutes its frequency. Overall then, one would predict that the frequency of the sickle-cell allele will eventually decline to a very low level among blacks in the United States.

Examining the DNA of sickle-cell alleles suggests that these favorable mutants have been incorporated into populations only a few times. Support for this conclusion comes from the fact that in most individuals the β-globin gene is contained in a 7.6-kb restriction fragment when the

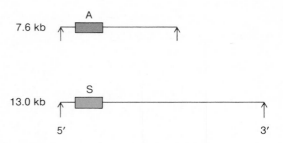

FIGURE 17.2
A diagram of the restriction fragments for the normal β-globin allele and the sickle-cell allele. The arrows indicate the cleavage sites caused by restriction enzyme HpaI.

enzyme HpaI is used (see figure 17.2). However, in American black populations, a longer type of restriction fragment is almost always associated with the sickle-cell allele. In fact, the S allele is really associated with absence of a restriction site that is present on the 3' end of the fragment with the normal allele.

The association of particular alleles at different loci, as in the example of allele S lacking a restriction site at a linked region, is called **linkage disequilibrium.** Linkage disequilibrium between disease alleles and linked restriction sites is used to locate disease genes on particular DNA fragments. Furthermore, examining other sickle-cell alleles in Africa and checking linkage disequilibria with various DNA markers suggests that nearly all of the African sickle-cell alleles descend from only three independent mutants. In other words, mutation (or lack of mutation) may actually limit the evolutionary response in this case.

Another factor influencing the incidence of genetic disease is inbreeding. In isolated human populations, consanguineous matings often occur because of the limited number of available mates. Such inbreeding will increase the incidence of recessive diseases in these populations (see further discussion later in this chapter). For example, in the Kel Kummer Taureg tribe of the remote South Sahara, which is composed of only three hundred people, approximately 30% of marriages are between first cousins. Excellent documentation is also available in Italy, where the Catholic church requires special permission for consanguineous marriages. Records show that consanguineous marriages are three to four times more frequent in the remote, less densely populated mountain villages than in the more densely populated areas.

To further illustrate inbreeding, figure 17.3 gives a pedigree of an isolated island population off the coast of Yugoslavia. Circles and squares indicate females and males, respectively, and purple symbols indicate affected individuals. The disease Mal de Meleda is a rare, autosomal recessive disorder that results in thickened skin on the hands and feet. In this pedigree, two first-cousin matings are indicated by heavy lines connecting individuals 1 and 2 in generation III and individuals 1 and 2 in generation IV. Individuals III-1 and III-2 must both have been heterozygotes, because two of their seven offspring were affected. The other first-cousin mating, between heterozygote IV-1 and diseased homozygote IV-2, resulted in two out of three progeny being affected. (Although this disease is rare elsewhere, its commonness in this population results in many heterozygotes.) The high proportion of consanguinity in this population has resulted in a much higher incidence of Mal de Meleda than would occur in a random mating population (or on the mainland).

Some isolated religious groups have relatively high incidences of particular rare diseases. In fact, a number of diseases were originally described in populations such as the Amish in the United States. Often, groups like this were originally established by a small number of founders and have subsequently kept their community closed to immigrants. As a result, genetic drift, or **founder effect,** is a primary factor accounting for the abnormally high incidence of certain genetic diseases in their populations. Of course, in the case of the Amish, the high incidence is relatively temporary, and selection will reduce the frequency over a number of generations.

The Amish population of Lancaster County, Pennsylvania, has a high incidence of a recessive disorder known as six-fingered dwarfism, or Ellis-van Crevald (EvC) syndrome (see figure 17.4). In this population of about thirteen thousand, eighty-two affected individuals in forty affected sibships have been diagnosed. Pedigrees of the eighty parents in these sibships showed that all trace their ancestry to Samuel King and his wife, early members of the community. Thus, it appears quite certain that the high incidence of EvC syndrome is due primarily to the founder effect; that is, either Samuel King or his wife carried the recessive allele, and because many individuals in the population are their descendants, the incidence of the disease is high.

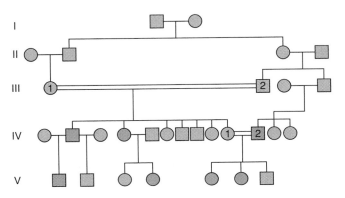

FIGURE 17.3
A pedigree in which two first-cousin matings (double lines) have taken place, III-1 with III-2 and IV-1 with IV-2. Notice the high incidence of a rare recessive disease (purple).

Reprinted with permission of Jones and Bartlett Publishers, Inc., Boston, MA, from Population Biology, *1984, written by Philip Hedrick.*

*T*he incidence of human genetic disease can be generally explained by a mutation-selection balance in which mutation increases the frequency of the mutant and selection reduces it. In addition to mutation and selection, the frequency of disease alleles is sometimes influenced by genetic drift and gene flow. The incidence of recessive genetic diseases is also affected by inbreeding.

∎

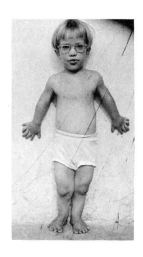
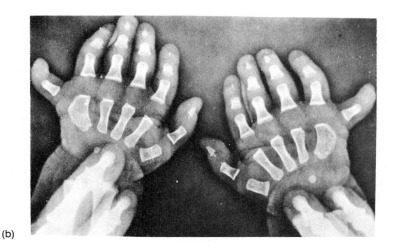

(a)   (b)

FIGURE 17.4
(*a*) A child with Ellis-van Crevald syndrome, a form of dwarfism found in the Amish of Pennsylvania. Affected individuals have short limbs, malformed hearts, and six fingers on each hand (see X ray in *b*).
Both photos © Victor McKusick.

MOLECULAR EVOLUTION

One of the most exciting and potentially useful applications of population genetics is in understanding the process of **molecular evolution**—the factors that have caused changes in the nucleotide sequence or amount of DNA and in the amino acid sequence of proteins over evolutionary time. In addition, the differences between these macromolecules can explain the relationship between species. In the ensuing discussion, we will assume that most differences in DNA sequences and many changes in protein sequences are not under strong selective constraints. As a result, other evolutionary factors that affect genetic variation, such as genetic drift and mutation, appear to have predominant roles in influencing the variation and evolution of these molecular sequences.

NEUTRALITY

Perhaps the best perspective for understanding molecular evolution is the theory based on mutation and genetic drift that was developed by Japanese geneticist Motoo Kimura and his colleagues. Although genetic drift has only a small effect in a large population each generation, its cumulative effect may be substantial if there are many generations. Likewise, the mutation rate in any one generation is small, but the likelihood of a mutation becomes substantial over a number of generations. Thus, the joint effects of genetic drift and mutation may be important over long periods of time and may be an important influence on molecular evolution.

In order to examine mutation in a finite population, let us assume selective differences are not important—that different alleles (or sequences) have no (or a very small) selective advantage or disadvantage. This concept is called **neutrality.** Generally, neutrality refers to a situation where the effects of genetic drift on allelic frequency are larger than the effects of selection. For example, alleles A_1 and A_2 are neutral with respect to each other when the relative fitnesses of A_1A_1, A_1A_2, and A_2A_2 are all equal. Another approach is to say that the alleles are all equally important.

Let us assume that a new mutant, A_2, occurs in a population of size N that otherwise consists only of A_1 alleles. The initial frequency of the mutant is:

$$q = \frac{1}{2N}$$

where 2N is the number of alleles in a diploid population. Assuming the two alleles are neutral with respect to each other, the probability of fixation of the new mutant is equal to its initial frequency, or $1/(2N)$, as discussed in chapter 16. The probability that the new mutant will be lost (equal to the fixation of the original allele) is then $1 - 1/(2N)$. In

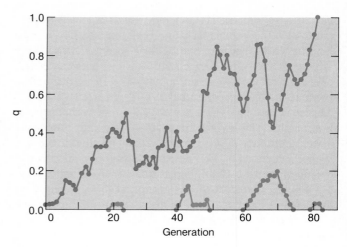

FIGURE 17.5
The frequency of five new mutants over time in a finite population. In this example, only the first mutant eventually became fixed. The rest were lost due to genetic drift.
Reprinted with permission of Jones and Bartlett Publishers, Inc., Boston, MA, from Population Biology, *1984, written by Philip Hedrick.*

other words, unless the population is quite small, a new neutral mutant will nearly always be lost from the population. If N = 50, the probability of fixation of a new mutant is only 0.01.

The two alternatives, fixation or loss of the new mutant, take quite different amounts of time to occur. Because the change of frequency necessary when loss occurs is small, from $1/(2N)$ to zero, the average amount of time for loss to occur is short. On the other hand, fixation of the new mutant requires a substantial change in allelic frequency, from $1/(2N)$ to unity, and the time required for fixation is much longer. Figure 17.5 illustrates the time necessary for fixation and loss to occur for five separate mutants. In this example, the first mutant eventually becomes fixed, while the others are lost from the population due to genetic drift.

If a large number of potential alleles exist at a locus, mutation will increase the number of alleles, and genetic drift will reduce the number of alleles. In fact, the heterozygosity resulting from the balance of these two factors may be rather large if the population is big. As we discussed in chapter 16, extensive variation occurs for electrophoretically determined alleles in many species. The neutrality theory suggests that much molecular variation is primarily due to the joint effects of genetic drift and mutation.

However, some electrophoretic variation observed in natural populations is probably maintained by selection. The actual proportion of loci affected by selection or by the forces of neutrality has been a controversial topic. One of the most thoroughly studied biochemical polymorphisms is the variation at the alcohol dehydrogenase (*Adh*)

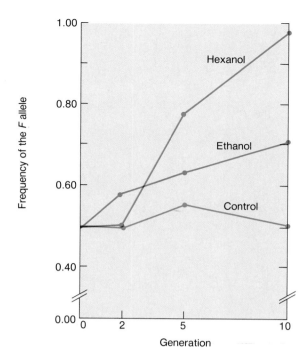

FIGURE 17.6
The change in frequency of the Adh^F allele in *D. melanogaster* when the flies are exposed to different alcohols.

TABLE 17.2 *Hypothetical Array of Amino Acids from Two Species*[1]

	Position				
Species	*1*	*2*	*3*	*4*	*5*
a	lys	glu	ala	asp	arg
b	lys	glu	lys	asp	iso

[1]Abbreviations indicate different amino acids.
Reprinted with permission of Jones and Bartlett Publishers, Inc., Boston, MA, from *Population Biology*, 1984, written by Philip Hedrick.

locus in *Drosophila melanogaster*. The two common alleles, *F* and *S,* differ only in one out of 255 amino acids; that is, in position 192, the *F* allele has a threonine and the *S* allele has a lysine. Because alcohols are an important component of the rotting fruit and vegetables in which *Drosophila* live, breakdown of alcohols by alcohol dehydrogenase is probably necessary for fly survival. A number of researchers have found that the frequency of the two alleles is affected by alcohols. Furthermore, the rate of change appears to be affected by the type of alcohol present, as illustrated in figure 17.6 for the change in allele *F* for different alcohols. For nearly all the alcohols tested, the *F* allele increased in frequency. In other words, strong selection seems to favor the *F* allele under these conditions, making neutrality an unlikely explanation for the maintenance of this variation.

MOLECULAR CLOCK

The amino acid sequences for a number of proteins from different organisms are known. Paleontological information dating the splits between the ancestors of various organisms indicates that substitutions accumulate at a regular rate over evolutionary time for a given protein. Of course, this apparent regular replacement is an average of a number of events over a long period of time. Assuming that such a regular replacement occurs, the differences in amino acid sequence between two species with an unknown common ancestor may serve as a **molecular clock,** indicating the time since the two species diverged. In addition, by comparing the amino acid sequences of homologous proteins (or nucleotide sequences) from different species, we can estimate the rate of substitution of molecular variants, given the approximate time since the species diverged.

As an example, let us consider the hypothetical array of homologous amino acids from two species in table 17.2. The five amino acids are identical at positions 1, 2, and 4 and different at positions 3 and 5. Therefore, the proportion of sites that are different is $2/5 = 0.4$. A neutrality explanation of this difference would state that a mutation occurred in one of the species at these positions and has subsequently become fixed by genetic drift. If the two species were very closely related, they would have the same amino acid at nearly every site, whereas if they were distantly related, they would be different at many different sites.

The molecular clock hypothesis is consistent with the neutrality theory. In other words, if the replacement of molecular variants over time is a function of genetic drift and the mutation rate, relatively regular turnover (replacement) of molecular variants will take place over a long period of time. To illustrate, assume that genetic drift and mutation are the only factors that change the frequencies of molecular variants. Let the mutation rate to a new molecular variant be u, so that in a diploid population of size N, there are 2Nu new mutants per generation. Remember that the probability of fixation of a new neutral mutant is equal to its initial frequency, $1/(2N)$. Assuming an equilibrium situation between mutation and genetic drift, the **rate of allelic substitution** per generation (k) is the product of the number of mutants generated per generation and their probability of fixation, or:

$$k = 2Nu \frac{1}{2N}$$

$$= u$$

In other words, the average rate of allelic substitution in this simple model is equal to the mutation rate at the locus.

If we assume that the mutation rate per gene is about 10^{-6} and that about one hundred amino acids exist in a typical protein, the mutation rate per amino acid should be about 10^{-8}. When there are no selective constraints anywhere in the amino acid sequence, this should be an approximate estimate of the rate of amino acid substitution. However, the rate of amino acid substitution appears to differ for different protein molecules. For example, the amount of divergence for three molecules that have been sequenced in a number of organisms is given in figure 17.7. From these comparisons, it appears that the rate of amino acid substitution is very fast for fibrinopeptides, intermediate for hemoglobin, and very slow for cytochrome c. The estimated rates of amino acid substitutions for fibrinopeptides, hemoglobin, and cytochrome c are 9.0×10^{-9}, 1.4×10^{-9}, and 0.3×10^{-9}, respectively.

These differences may result from the relative importance of the three-dimensional structure of these molecules and the proportion of amino acids involved in active enzymatic sites. For example, histone IV, which has a substitution rate of 0.006×10^{-9}, fits in tightly with the DNA molecule, thus nearly the whole amino acid sequence is critical for the three-dimensional structure. At the other extreme, the fibrinopeptide molecule appears to be hardly more than a spacer molecule, with no enzymatic sites. One explanation is that, where constraints are greatest, the rate of amino acid substitution is reduced. In other words, the rate for fibrinopeptides is close to that expected from neutrality for the whole molecule, while the rate for histones is much lower, presumably because of the functional constraints on a large part of the histone molecule.

PHYLOGENETIC TREES

The differences among related species may be measured for morphological, biochemical, immunological, or other traits, with more closely related species presumably being more similar. Thus, one important use of genetic or phenotypic measures of differences is in organizing related populations or species in a biologically meaningful way—that is, arranging these groups to illustrate ancestral relationships. Such relationships are usually shown on a **phylogenetic tree,** in which the root or ancestral type is at the top of the tree and the branches, or evolutionarily derived types, are at the bottom. The similarities and differences in molecular structure among different organisms can help us understand evolutionary relationships. Presumably, organisms having similar molecules are closely related, and those having quite different molecules are only distantly related.

To show how to construct a simple phylogenetic tree, let us examine a number of groups that are either different populations or species. Assume that the genetic differences between these groups have been measured, using

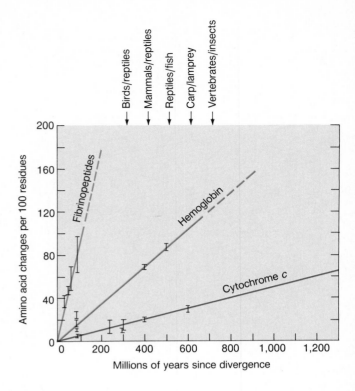

FIGURE 17.7
The number of amino acid changes for three different molecules in organisms that diverged at different times.

Reprinted with permission of Jones and Bartlett Publishers, Inc., Boston, MA, from Population Biology, *1984, written by Philip Hedrick.*

such genetic data as allelic frequencies, amino acid differences, or restriction map information, so that a matrix of values is available.

These differences can be organized in a way that reflects biological relationships. For example, proteins are still intact in specimens of some extinct animals. Comparing the immunological properties of these proteins to those of related extant animals gives a measure of the molecular differences among these organisms. In one case, the remains of a 40,000-year-old baby mammoth were discovered in the Soviet Union in 1977. Comparison of the immunological properties of the protein albumin from the mammoth, two species of elephants, and a manatee yielded the phylogenetic tree shown in figure 17.8. Here the differences in the properties of albumin between the mammoth, the Indian elephant, and the African elephant are small, corresponding to the current belief that the three species are closely related. The molecular similarity of the three species suggests that they diverged from each other at about the same time.

In figure 17.8, we cannot really tell whether the mammoth is more closely related to the Indian or the African elephant. In fact, we would need more detailed data to determine which of the three species first branched off from

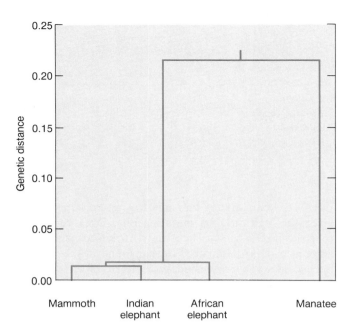

FIGURE 17.8
A phylogenetic tree constructed from immunological genetic distance data, indicating a close relationship between the mammoth, the Indian elephant, and the African elephant.

Reprinted with permission of Jones and Bartlett Publishers, Inc., Boston, MA, from Population Biology, *1984, written by Philip Hedrick.*

T ABLE 17.3	*Proportion of Difference for Each Species Pair at the 7,100-Base Sequence*		
	Chimpanzee	**Gorilla**	**Orangutan**
Human	0.016	0.017	0.034
Chimpanzee	—	0.021	0.038
Gorilla	—	—	0.037

the other two. However, the exact sequence of branching is a major controversy regarding the relationship of humans, chimpanzees, and gorillas. As we saw in chapter 4, the detailed chromosomal patterns of humans, chimpanzees, and gorillas show high similarity. Recently, the DNA sequences of these three species and of the orangutan were compared for a 7,100-base-pair sequence in the β-globin region. The proportion of divergence between all of them is small, but the orangutan has the greatest difference from the others, at over 3% (see table 17.3). Notice that the least different pair, the human-chimpanzee, is slightly more similar than the human-gorilla pair.

These data can be organized into a phylogeny to illustrate the ancestral relationships between the species (see figure 17.9). The numbers in this most parsimonious phylogeny (there are other arrangements that do not explain the data quite as well) indicate the number of DNA changes along each branch of the tree. For example,

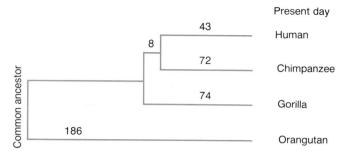

FIGURE 17.9
The phylogeny between humans, chimpanzees, gorillas, and orangutans, based on their base sequences in the β-globin region. The numbers indicate the number of changes that have occurred along each lineage.

humans have accumulated forty-three changes and chimpanzees seventy-two since they diverged. Interestingly, humans and chimpanzees share eight changes that are not present in gorillas, supporting the hypothesis that humans are most closely related to chimpanzees. In the near future, other comparative nucleotide sequences from unlinked regions should be available to help further resolve the ancestry of these species.

USE OF mtDNA IN STUDYING GENETIC RELATIONSHIPS

Because mitochondrial DNA can be separated from nuclear DNA, most of the recent studies of DNA variation have used mtDNA. Remember that mtDNA is maternally inherited and that there appears to be little genetic exchange between mitochondria, so one can assume that individuals sharing the same mtDNA type have a common female ancestor. In addition, the rate of substitution for mtDNA in mammals is approximately 5- to 10-fold higher than that of nuclear DNA, possibly due to inefficient DNA repair in mitochondria that results in a high mutation rate.

Because of the maternal inheritance of mitochondria, mtDNA and nuclear DNA may show quite different patterns of variation. For example, two species of house mice form a hybrid zone about 50 kilometers wide in Denmark, with a marked anatomical difference between the species. A number of nuclear genes identified by protein electrophoresis have different frequencies above and below this zone; that is, they are concordant with the known anatomical differences. However, mtDNA shows a different pattern; mtDNA frequencies above and below the hybrid zone are quite similar.

The power of examining mtDNA is illustrated in the recent study of green sea turtles by John Avise and his colleagues. Each year female turtles migrate 2,000 kilometers from their feeding grounds near Brazil to Ascension Island in the mid-Atlantic. It has been hypothesized

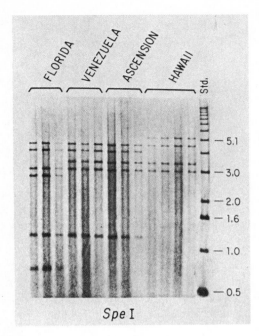

FIGURE 17.10
mtDNA digested by restriction endonuclease SpeI from green turtles at four different rookeries. The turtles from Venezuela and Ascension are identical, while the mtDNA from the turtles in Florida and Hawaii is different.

From Brian W. Bowen, Anne B. Meylan, and John C. Avise. AN ODYSSEY OF THE GREEN SEA TURTLE: Ascension Island Revisited. Proceedings of the National Academy of Science 86:573-76, January 1989, Evolution fig. 2A.

that this migration has endured over forty million years since a time when South America and the mid-Atlantic were near each other (they have since moved apart by continental drift). The female turtles are believed to migrate to these beaches to lay eggs because of *natal homing,* a behavior in which adults return to their birthplace. If in fact the Ascension Island population has been isolated from other Atlantic populations for millions of years, it should be genetically quite distinct.

When the mtDNA of turtles from Hawaii, Florida, Venezuela (as a sample of South American turtles), and Ascension Island was examined, it was clear that the Ascension Island population mtDNA was identical to the most common type found in Venezuela. For most restriction sites (ninety-five were assayed), the mtDNA from all populations was identical. Figure 17.10 gives an example showing that individuals from Venezuela and Ascension Island were identical, while individuals from Florida and Hawaii were different. Therefore, it appears that the Ascension Island population has not been isolated for long and is probably descended from other Atlantic-breeding populations.

Recent studies using phylogenetic techniques in human mtDNA have suggested that the current mtDNA descended from a female some 200,000 years ago, prompting speculation about a human Eve. This suggestion is based on the observation that less variation in

TABLE 17.4 *Frequencies of Single and Triplicated Copies of the α-globin Gene*

Population	Number of α-globin Genes		
	1	2	3
American Black	0.16	0.8364	0.0036
Sardinian	0.18	>0.816	<0.004
Greek Cypriot	0.07	0.88	0.05

mtDNA occurs among humans than would be expected, given the high mtDNA mutation rate. First, remember that the mtDNA consists of only 16 kilobases, a small amount of DNA compared to that found in the nucleus. Second, as we just discussed, the patterns of variation observed for mtDNA and nuclear DNA may be quite different. Finally, remember that genetic variation is reduced by genetic drift. A small population size 50,000 years ago could have the same effect as the "bottleneck" of a single woman 200,000 years ago. When similar information is available about nuclear DNA, we will have a better picture of recent human evolution.

MULTIGENE FAMILIES

A number of genes, such as the immunoglobins, the histones, rRNA genes, and tRNA genes, are now known to exist as multigene families—multiple copies that have apparently arisen as a series of tandem duplications (see chapter 4) on a chromosome. The number of copies in such a multigene family may vary in different individuals, and unequal crossing over may lead to either the contraction or expansion of the gene number in a family.

In humans, two copies of the α-globin gene ordinarily occur 3.7 kilobases apart on chromosome 16. If one or both of these loci are deleted, the individual has thalassemia, a blood disorder that causes anemia. The loss of the α-globin genes from unequal crossing over can also produce gametes with three copies of the α-globin gene. A survey using endonuclease mapping on three populations with high incidences of thalassemia for triplicated α-globins found seven individuals heterozygous for a triplicated chromosome and a normal chromosome (table 17.4). None of the individuals had clinical symptoms related to the triplication.

Understanding the number and types of genes in a multigene family requires knowing about the relevant genetic processes—unequal crossing over, gene conversion, and transposition. In addition, the size and makeup of a multigene family appear to be affected by genetic drift, mutation, and selection. The overall scenario, therefore, becomes quite complex, but the combination of genetic and population genetic factors should explain the evolution of these important sequences.

Chapter 17

■

CONSERVATION GENETICS

In recent decades, habitat destruction, hunting, and other factors have resulted in the extinction of a number of species and the near-extinction of many others. In fact, some animals, such as the condor and the black-footed ferret, now only exist in small numbers in captivity. In other parts of the world, such as the Amazonian river basin, whole ecosystems containing many species of plants and animals are in danger. Furthermore, the number of animal breeds and crop varieties used in agriculture has precipitously declined in recent years. All of these factors have stimulated attempts to maintain genetic variation in many different species, a discipline known as **conservation genetics.**

CROP SPECIES

In the past, a number of different varieties of each crop plant were grown, many of which were adapted to local environmental conditions. Now, fewer and fewer varieties are being planted, resulting in a restriction of the gene pool of these crops. Part of the reason for this change is that in many areas of the world, newly developed high-yield strains have replaced the local plant varieties, or **land races.** For example, in some areas, a "green revolution" has caused new strains of rice and wheat to be substituted for the land races. Such high-yield varieties often require large-scale planting and mechanized spreading of fertilizer and pesticides. As a result, local varieties that were selected to yield well in a labor-intensive situation are no longer used. However, these local varieties often have alleles that are well adapted to local conditions, such as particular soil types, diseases, and pests, which could be valuable in developing future varieties. Unfortunately, many land races may already have been lost, and their wild relatives may now be the only source of advantageous alleles.

Two related phenomena can result from such developments. First, a crop species can become genetically vulnerable to a particular disease, because in a large planting, the varieties have the same alleles at the loci for pathogen or insect resistance. For example, in 1970, genetic uniformity in planted corn led to a loss of approximately 15%

FIGURE 17.11
Corn suffering from southern corn leaf blight, a disease that severely damaged crops until a blight-resistant strain of corn was developed. *Courtesy of U.S. Department of Agriculture.*

of the United States crop due to southern corn leaf blight (see figure 17.11). Recovery from this disaster only occurred when a blight-resistant strain of corn was developed.

The second possible result of such developments is that genetic variation in the total species may be reduced in land races, germ plasm (or seed) collections, and wild progenitors of the crop. Several different approaches have been advocated to ensure the maintenance of genetic variation in crop species. One system would conserve genetic variation by growing the land races *in situ*—that is, in the area where they developed. Obviously, this is preferred (if possible), so that further adaptive change may occur while present genetic types are maintained.

A second approach to genetic conservation is the accumulation of germ plasm collections of many cultivated varieties and their wild relatives. The general strategy of collection should consider three factors: the number of seeds to be collected per local population, the number of populations sampled, and the distribution of these populations over the species range. For example, in a species with high gene flow, local populations may be quite similar, while in species with high amounts of self-fertilization and low gene flow, local populations may differ greatly. Where knowledge of the species is inadequate, it is usually advisable to sample as many populations as possible, basing the distribution of the samples upon both the structure of the species and the heterogeneity of the environment.

FIGURE 17.12
The variation in tomatoes developed for various purposes.
Cover of Science of Food and Agriculture, *Vol. 4, No. 1, January 1986. Photo by Dr. R. Clark, Plant Introduction Station, Iowa State University.*

When measuring genetic diversity in either land races or the wild relatives of crop plants, three major types of traits are considered: economic or other metric traits, disease or pest resistance genes, and Mendelian morphological markers. All of these traits, plus the adaptation of different varieties to a range of environments, would be important considerations in a sampling regime. In addition, electrophoretic variants may be useful, since the distribution of alleles over populations can indicate the population structure of the species and help identify local populations. Electrophoretic variants may also be important in identifying genotypes common in only one or a few populations.

One crop that exemplifies the importance of genetic variation in wild relatives is the tomato (see figure 17.12). The tomato was presumably domesticated from a wild species present in the Andean region of South America. It is actually rather new as a crop, having been avoided even into the early twentieth century because of its alleged toxicity (the plant is a member of the night-shade family). However, today the tomato yields more tons of produce and has a greater value than any other vegetable grown for human consumption in the United States.

A number of important phenotypic qualities have been incorporated into tomato cultivars in recent decades. These include an allele that causes uniform ripening of the fruit (no green area near the stem) and another that prevents the plant from growing in an indeterminate fashion. Certain tomato varieties also have traits that make them easily harvestable by machine; for instance, they are uniformly short and have tough-skinned fruit that ripens in a concentrated period of time.

The extensive phenotypic variation among the wild species of tomatoes also has potential breeding importance. One type of tomato lives only 5 meters from the high tide line in the Galápagos Islands. These plants can survive in seawater, suggesting that it might be feasible to develop a tomato variety capable of using water containing a high proportion of salt. A second species exists in very dry habitats in Peru, where it obtains most of its water from fog or mist. These plants appear to have extreme resistance to water loss, a trait of potential importance for irrigated tomatoes.

INBREEDING DEPRESSION

As we discussed in chapter 16, inbreeding increases the frequency of homozygotes and decreases the frequency of heterozygotes. Thus, if recessive genotypes have lower fitness, the incidence of recessive diseases or traits may increase, and consequently, the mean fitness of a population may be reduced.

To understand the joint effect of inbreeding and selection, let us first examine the incidence of recessive genetic diseases among individuals from consanguineous matings. Inbreeding increases the frequency of homozygous recessives, which in this case also increases the incidence of the recessive disease. The genetic basis for this effect becomes clear when the proportion of recessive homozygotes in a noninbred population ($Q = q^2$) is compared to that for a given inbreeding coefficient (Q_f). The ratio of these two quantities is:

$$\frac{Q_f}{Q} = \frac{q^2 + fpq}{q^2}$$

$$= \left(\frac{q + fp}{q}\right)$$

which is greater than unity for any allelic frequency but very large for low allelic frequencies. For example, when $f = \frac{1}{16}$ (the inbreeding coefficient that results from a first-cousin mating) and $q = 0.01$, as it does for albinism, then:

$$\frac{Q_f}{Q} = \frac{0.01 + (0.0625)(0.99)}{0.01}$$

$$= 7.2$$

In other words, there are over seven times as many affected individuals in the inbred group as in the random mating group. If $q = 0.001$, as it does for many rarer genetic diseases, this ratio becomes 63, illustrating why many individuals with very rare recessive diseases are the progeny of consanguineous matings.

Because individuals are more likely to be homozygous for recessive deleterious alleles in an inbred population, the mean fitness of the total population may decline. **Inbreeding depression**—the decline of population fitness due to inbreeding—seems to be a nearly universal phenomenon. The extent of inbreeding depression varies among different organisms. For example, three generations of mating full sibs in Japanese quail result in complete loss of reproductive fitness, while many plant species with partial self-fertilization show little decline in fitness when inbred.

Let us examine how inbreeding can affect the mean fitness of the population. Assume, as in chapter 16, that selection causes the relative fitness values of genotypes A_1A_1, A_1A_2, and A_2A_2 to be 1, 1, and $1 - s$, respectively. If we use the genotypic frequencies given for inbreeding in chapter 16, the mean fitness with inbreeding is:

$$w = (1)(p^2 + fpq) + (1)(2pq - 2fpq)$$
$$+ (1 - s)(q^2 + fpq)$$
$$= 1 - sq^2 - sfpq$$

using the fact that $p^2 + 2pq + q^2 = 1$. Remember from chapter 16 that the mean fitness for a recessive gene when there is random mating is $1 - sq^2$. In other words, the mean fitness is lowered from inbreeding by the amount $sfpq$. Obviously, this formula predicts that the fitness is lowered (inbreeding depression increased) linearly as the inbreeding coefficient is increased. In fact, results of extensive studies in *Drosophila,* humans, and other species support this prediction.

Until recent years, some individuals breeding endangered species in captivity were not aware of, or thought unimportant, the extent to which fitness is reduced by inbreeding. One reason for this misconception was that, until recently, few records (stud books) were kept to show the relationships between breeding animals in zoos. Using pedigree information from a number of captive species, Kathy Ralls and her coworkers at the National Zoo documented that the overall mortality of inbred offspring is nearly double that of noninbred offspring. As an example, table 17.5 gives the infant mortality for eight mammalian species with relatively large numbers of both noninbred and inbred progeny. The species are ordered by their juvenile mortality in noninbreds, which ranges from 0.106 in the degu to 0.501 in the golden lion tamarin (see figure 17.13). For all the species, the mortality in inbreds is higher, and in six of the eight, the difference is significant, using the chi-square test.

The maintenance of endangered species may result in a number of detrimental evolutionary changes. In addition to the general reduction in fitness caused by inbreeding, inadvertent selection may produce animals more adapted to captive life than to natural conditions. That is, selection may result in tame and docile animals or individuals that are not resistant to the diseases or parasites encountered in nature. These animals might not be able to survive if reintroduced to nature or to nature preserves. Furthermore, some captive animals appear to have hereditary diseases, due either to their small founder number or to inbreeding. For example, the captive golden lion tamarin population, derived from fifty-one founder individuals, has a high frequency of hereditary diaphragmatic hernia.

FIGURE 17.13
The golden lion tamarin. This species shows an increased mortality rate when inbred.
© *Michael Dick/Animals Animals.*

T ABLE 17.5	Proportion of Juvenile Mortality in Noninbred and Inbred Progeny	
	Juvenile Mortality (Sample Size)	
	Noninbred	*Inbred*
Degu	0.106	0.235*
Elephant shrew	0.153	0.233
Pygmy hippopotamus	0.245	0.549*
Cheetah	0.263	0.442*
Dorcas gazelle	0.280	0.595*
Japanese serow	0.288	0.565*
Greater galago	0.308	0.386
Golden lion tamarin	0.501	0.634*

*Indicates $p < 0.05$.

LOSS OF GENETIC VARIATION

Because the population size of a species in captivity is by necessity small (about one hundred Bengal tigers can be maintained by the zoos in the United States), future genetic drift may cause the loss of genetic variation or the possible fixation of detrimental alleles. The best way to understand the loss of genetic variation from genetic drift is to examine the change in heterozygosity from one generation to the next as the result of finite population size. A formula that predicts this change is:

$$H_1 = \left(1 - \frac{1}{2N}\right)H_0$$

where H_0 is the initial heterozygosity. H_1 is the heterozygosity after one generation, and N is the number of individuals. For example, if $N = 50$, then:

$$H_1 = 0.99H_0$$

TABLE 17.6	Proportion of Initial Heterozygosity Remaining After 100 Years			
Population Size	**Generation Length (in Years)**			
N	1	2	5	10
10	0.01	0.08	0.36	0.60
50	0.37	0.61	0.82	0.90
250	0.82	0.90	0.96	0.98

Thus, the heterozygosity is reduced 1% in one generation because of genetic drift.

However, for captive animals, we are generally concerned about the reduction in genetic variation over a longer period of time, possibly several generations. In this case, we can use the formula:

$$H_t = \left(1 - \frac{1}{2N}\right)^t H_0$$

where t is the number of generations. For example, if the average generation length of an endangered antelope is two years, there would be about fifty generations in one hundred years. If we assume again that the breeding population is composed of fifty animals, then after fifty generations (one hundred years):

$$H_{50} = \left(1 - \frac{1}{100}\right)^{50} H_0$$

$$= 0.61 \, H_0$$

In other words, the heterozygosity would be only 61% of the initial value.

Table 17.6 gives the proportions of heterozygosity retained for combinations of four population sizes and four generation lengths. To preserve 90% of the original heterozygosity after one hundred years for an organism with a two-year generation length, the population size would have to be 250 individuals. Obviously, the population size can be smaller for species with a longer generation length. However, these are optimistic predictions that don't include problems, such as disease, which may drastically reduce the population size.

In theory, the loss of genetic variation in a captive population can be counteracted by periodically introducing captured wild animals. However, often it is not possible to obtain wild animals or to induce them to breed in captivity. Another strategy for reducing the effect of genetic drift is to exchange individuals or their gametes (semen or eggs) between zoos. This artificial gene flow may delay the reduction of genetic variation, but it is not a long-term solution. Obviously, genetic conservation is based on a thorough understanding of population genetics fundamentals, including genetic drift, selection, and inbreeding.

C onservation genetics is concerned with maintaining genetic variation in agricultural plants and animals and in rare and endangered species. The term inbreeding depression refers to a decline in fitness resulting from inbreeding. Loss of genetic variation occurs in captive populations when genetic drift has a large effect.

■

ANIMAL AND PLANT BREEDING

Population genetic principles are widely used to develop new breeds of livestock and new varieties of crop plants. They are also useful in understanding and predicting changes in the traits important in a breeding program. The phenotypic mean in a population is most frequently changed by animal and plant breeding, using selection for particular types. Some of the characteristics that are important in animal and plant breeding, such as resistance to diseases, are often controlled by only a few genes, while other traits, such as crop yield, appear to be influenced by literally hundreds of genes. In addition to selection, inbreeding is used to develop lines that have particular traits, and gene flow is used to bring in new alleles from other breeds or varieties. Genetic drift is often a problem in breeding programs because favorable alleles may be lost by chance and unfavorable alleles may be accidentally increased in frequency. In the following section, we will discuss artificial selection, inbred lines, and heterosis to illustrate the applications of population genetic principles to animal and plant breeding.

ARTIFICIAL SELECTION

Generally, the increase of a favorable trait in a population is accomplished by **artificial selection**—that is, selecting as parents for the next generation those individuals that are closest to the ideal phenotype. As we discussed in chapter 3, the rate of progress from artificial selection is a result of the heritability of a trait and the selection differential, the difference between the mean of the selected parents and the mean of all the individuals in the parental population. If the heritability is high (most phenotypic differences are genetically determined) and the selection differential is large, progress can be made quickly.

The impact of directional selection is to increase the frequency of alleles that cause a change in the phenotypes being selected. To illustrate, let us assume that gene A is one of the genes influencing a given economically important trait, say milk production in dairy cattle. With two alleles, A_1 and A_2, the effect of this gene on the phenotypic value of milk production can be visualized as in figure 17.14. In this case, genotype A_1A_1 has the highest phenotypic value, so artificial selection should eventually increase the frequency of A_1 to fixation.

After many generations of selection, say for high milk yield, there often appears to be a **selection limit,** a point beyond which selection appears unable to change the phenotypic value. For example, the two-way (up and down) selection experiment for body weight in six-week-old mice conducted by Richard Roberts (figure 17.15) reaches a selection limit in the down line between 10 and

15 grams. On the other hand, the up line reached a plateau at slightly above 30 grams for many generations and then unexpectedly responded again to selection in generations 40–45, reaching a new level at 35 grams.

There are a number of reasons why a selection limit occurs during artificial selection. First, there may be no more genetic variation in the population because of fixation for favorable alleles at all the loci that were originally variable in the population. Second, fertility or viability may be reduced in the extreme phenotypic individuals so that, for example, very large animals have fewer progeny.

The problem of reaching a selection limit may be overcome if new genetic variation is introduced into the population. For instance, mutation to favorable variants or recombination between favorable and unfavorable genes may allow selection to proceed further. In the selection experiment for mouse body weight in figure 17.15, the delayed response from 30 grams to 35 grams probably resulted either from a new mutation or from recombination.

Another approach is to cross new genes from other breeds into the selected populations. This artificial gene flow generally brings in genes that initially reduce the phenotypic value as well as genes that increase it. As a result, there may be an initial loss before the new and old favorable alleles together produce superior plants or animals. Molecular genetic techniques, such as those discussed in chapter 15, promise a way to introduce only the favorable alleles.

Often more than one trait is of economic importance in a selection program. For example, in pig breeding programs, both growth rate per day and back fat thickness

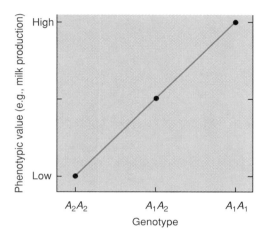

FIGURE 17.14
Effect of a gene on the phenotypic value of milk production. In this case, the A_1A_1 genotype has the highest milk production.

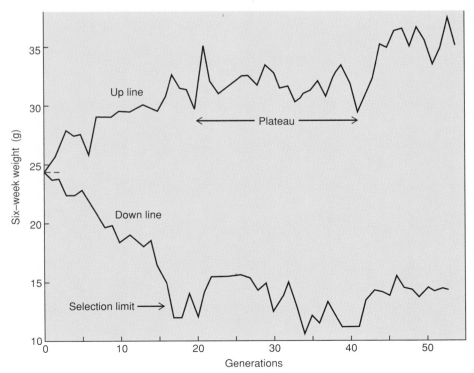

FIGURE 17.15
Two-way selection experiment for body weight in mice. The down line reached a limit of 13 g after seventeen generations, while the up line reached a temporary plateau of about 32 g after thirteen generations.

From F. W. Nicholas, Veterinary Genetics. *Copyright © Oxford University Press, Oxford, England. Reprinted by permission.*

FIGURE 17.16
The growth rate and back fat thickness for
one hundred pigs. The twenty pigs with the
combination of highest growth rate and
lowest back fat thickness are indicated by
X, while the rest are indicated by •.
From F. W. Nicholas, Veterinary Genetics.
Copyright © Oxford University Press, Oxford,
England. Reprinted by permission.

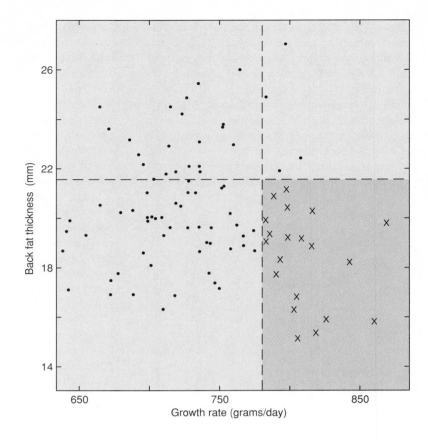

(an indicator of body fat) are important. Figure 17.16 gives
the values for these two traits in one hundred pigs. The
selection scheme is to reduce back fat and increase growth
rate, so the "best" pig is in lower right with a back fat
value of under 16 millimeters and a growth rate of 856
grams per day. To continue the breeding program, twenty
pigs are needed out of the one hundred. One way to pick
the appropriate number of animals is by **independent
culling levels,** in which animals must exceed a given value
for each trait or be culled (not allowed to breed). In our
example, 710 grams per day is the culling level for growth
rate while 21.4 millimeters is that for back fat. The twenty
selected pigs are indicated by X's in the shaded part of
figure 17.16.

When a particular combination of a number of traits
is desired, selection response may be quite slow. An example of such a long-term selection experiment for multiple traits was supervised by Gordon Dickerson for eleven
characteristics in leghorn chickens that together constitute *performance*. This flock was selected for twenty years
for viability, egg production, egg weight, freedom from
blood spots in the eggs, and seven other traits. Heritability
estimates for many of these traits were above 0.2, suggesting that response to selection would be substantial over
a period of time. However, the response to selection was
much less than expected; for example, egg production actually declined during the experiment, and egg weight
showed response to selection much smaller than expected
(figure 17.17).

INBRED LINES

In plant breeding, medical research, and to some extent
animal breeding, inbred lines are useful because they are
genetically uniform for known traits. Inbred lines of mice
and rats (lines that have been brother-sister mated for over
a hundred generations) are used to test new medical treat-
ments because they provide a genetically uniform test
animal.

What happens genetically when a plant is self-
fertilized repeatedly or when an animal is brother-sister
mated generation after generation? As we discussed in
chapter 16, continuous inbreeding results in homozygosity
for a given locus. As an example, let us assume that an
individual genotype $A_1A_2B_3B_3C_2C_4D_2D_2E_1E_3$ . . . is self-
fertilized and that we take three progeny to start inbred
lines (figure 17.18). These progeny after one generation
are already more homozygous than the parent, and after

Chapter 17

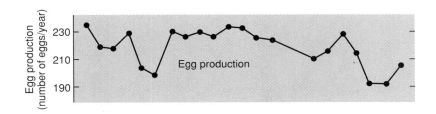

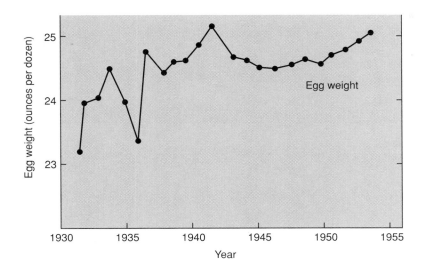

FIGURE 17.17
Changes in egg production and egg weight in long-term selection experiments for eleven traits in chickens.

Reprinted with permission of Jones and Bartlett Publishers, Inc., Boston, MA, from Population Biology, *1984, written by Philip Hedrick.*

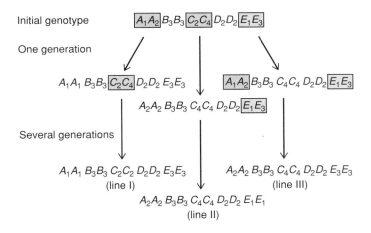

Initial genotype A_1A_2 B_3B_3 C_2C_4 D_2D_2 E_1E_3

One generation

A_1A_1 B_3B_3 C_2C_4 D_2D_2 E_3E_3 A_1A_2 B_3B_3 C_4C_4 D_2D_2 E_1E_3

A_2A_2 B_3B_3 C_4C_4 D_2D_2 E_1E_3

Several generations

A_1A_1 B_3B_3 C_2C_2 D_2D_2 E_3E_3 A_2A_2 B_3B_3 C_4C_4 D_2D_2 E_3E_3
(line I)   (line III)

A_2A_2 B_3B_3 C_4C_4 D_2D_2 E_1E_1
(line II)

FIGURE 17.18
The differentiation of three inbred lines produced by self-fertilization from an initial genotype. Heterozygous loci are indicated in colored boxes.

several generations, they are completely homozygous for the genes we have shown. If we want to obtain a group of genetically identical individuals, all we have to do is raise individuals from a given inbred line, say line I, all of which would be $A_1A_1B_3B_3C_2C_2D_2D_2E_3E_3$. . . .

We showed that all three lines survived inbreeding and could be maintained. In most cases, many of the lines die out from inbreeding because of low fertility or viability resulting from fixation of lethal or deleterious alleles. For example, if the genotype C_2C_2 were lethal, then inbred

line I would have died out. Douglas Falconer illustrated this phenomenon in an experiment in which he started with twenty lines of mice that he maintained by brother-sister mating. After five generations, only half of the lines survived (figure 17.19), and after twelve generations, only one survived. This one line survived twenty generations (until the experiment was stopped) and still had a litter size of nearly eight, similar to the average before inbreeding began.

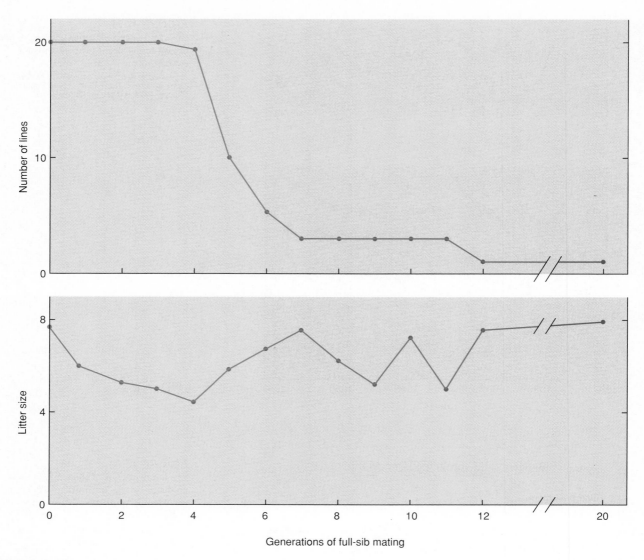

FIGURE 17.19
Results of an inbreeding experiment with twenty lines of mice,
showing the number of lines surviving and their average litter size as a
result of continuous full-sib mating.

HETEROSIS

Long before Mendelism was understood, many researchers noticed the increase in vigor and productivity associated with crossing plant varieties. Even Darwin experimented briefly with corn and noted, "The first and most important conclusion . . . is that cross-fertilization is generally beneficial and self-fertilization injurious." Two early maize geneticists, George Shull and Edward East, documented the detrimental effects of inbreeding and the positive effects of crossbreeding on corn yield. The increase in yield of an F_1 cross over its parental inbred lines is termed **heterosis.**

The extent of heterosis, a term nearly synonymous with **hybrid vigor,** can be measured by the difference in the yield (or other traits) for an F_1 and the mean of the parents, or:

$$H = F_1 - \frac{1}{2}(P_1 + P_2)$$

For example, the yields in bushels per acre in two parental corn strains, P_1 and P_2, were 20, while the F_1 cross between them was 101. The heterotic effect is then 81 bushels per acre. Today, nearly all of the United States corn crop is produced from hybrid corn to capitalize upon this heterotic effect.

Often in recent years, the market animal is also crossbred rather than purebred. Although different breeds are not really inbred lines, they may be somewhat homozygous because of breed selection. If we examine the increase in calf weight weaned per cow for purebred Hereford, Angus, and Shorthorn cattle, and the crosses between them, the effect of heterosis is 8.5% (see figure 17.20), a small but economically important amount. In addition, if we have a three-breed cross in which the mother is an F_1 and is crossed to a bull of the third breed, say (Hereford × Angus) ♀ × Shorthorn ♂, an additional 14.8% increase is observed. The maternal heterotic effect is often substantial because large F_1 females produce large offspring.

What is the genetic basis of heterosis? This topic has been controversial for many decades, and only in recent years does there seem to be some consensus. Initially, many researchers thought that heterosis was caused by heterozygous advantage at some loci in the F_1. However, it is now generally accepted that inbreeding (or selection) leads to divergence between lines or breeds and to chance fixation of different somewhat deleterious alleles. Because of the general dominance of the favorable or wild-type allele assumed to be present in the other line, this effect is covered up in the progeny when two lines are crossed, resulting in an increased yield.

PESTICIDE RESISTANCE

The rapid development of pesticide resistance in many organisms, including the yellow fever mosquito, the Colorado potato beetle, the Norway rat, and pigweed, is a lesson in population genetics in itself. The genetic variation that confers resistance must initially arise by mutation, then spread to other populations by selection and gene flow. Obviously, to control species with pesticides, one must understand basic population genetics—or expect eventual failure if resistance arises. (Perhaps we should learn to use other control techniques for pest species and thereby save our environment from toxic chemicals as well.)

Agricultural crops have always been beset by insects and other pests, but modern agricultural techniques have changed the environment in a number of ways that have made diligent pest control even more necessary. For example, planting large expanses of land with the same crop (a monoculture) allows a pest or pathogen easier dispersal from plant to plant or from field to field. In addition, using only a few varieties of a particular plant reduces the natural genetic variation that confers resistance to pest infestation. In a similar vein, human infectious diseases passed by insect and animal vectors have also become more of a problem now that greater concentrations of people live in cities.

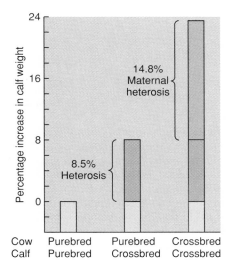

FIGURE 17.20

The percentage increase in calf weight due to heterosis in the calf and in the cow that produced the calf.

From F. W. Nicholas, Veterinary Genetics. *Copyright © Oxford University Press, Oxford, England. Reprinted by permission.*

Following World War II, synthetic organic pesticides were introduced to control both agricultural and public health pests. These pesticides greatly increased crop production and reduced the incidence of many diseases. In addition, they were inexpensive, thus they quickly replaced other suppressive measures. By now, however, the use of pesticides has greatly altered the biological environment, reducing or eliminating predators or parasites that originally helped control the pest species.

Studies of evolutionary response to pesticides have shown that many pest organisms are capable of quickly evolving resistance to pesticides. Many pests, such as the mosquitoes that transmit malaria, are vectors of human diseases, but the problems of control are similar to those for agricultural pests. For example, in 1969, fifteen species of anopheline mosquitoes, the malarial vector, were resistant to DDT, and in 1976, forty-three species were resistant. A rather discouraging observation is that malaria is rapidly increasing in many areas where it had nearly been eradicated; for example, India reported twenty-seven million cases in 1977 compared with only a half million in 1969. This trend is apparently due to the increased use of pesticides associated with new agricultural practices and to the consequent increased pesticide resistance. The housefly seems to be the organism most broadly resistant to pesticides geographically, but *Tribolium,* a genus of flour beetle, and seven species of rodents, including two rats, are also widely resistant to various pesticides. Until the late 1970s, no weeds were resistant to herbicides, but by 1985, forty-eight species had evolved resistance to at least one.

Lysenko—Soviet Anti-Geneticist

P robably no person has had so great an effect on a country's science as Trofim Lysenko (1898–1976) in the Soviet Union and other Communist countries. Lysenko, born in a peasant family, had a slow educational start but finally obtained a degree by correspondence from Kiev Agricultural Institute. This background fit perfectly with Stalin's intent to replace the old specialists of the tsar with newly trained persons of worker or peasant origin.

Lysenko's scientific training was minimal, and many of his experimental results showed an obvious lack of rigor. For example, his first experiment was to plant peas for pasture in the winter before a cotton crop. The year he tried this, 1925–26, was a mild winter and produced a good yield of peas to use for pasture. However, he did not replicate the experiment in other more typical years, and today this approach has proven impractical.

One of Lysenko's "discoveries" was vernalization, a practice of treating seeds before germination to cause them to mature more quickly. For example, he suggested soaking the seeds of wheat and other crops for several days before sowing in order to protect the plants from later drought. However, many technical problems and extra work resulted from this procedure, and in fact vernalization often actually lowered the yield.

Stalin created a personality cult and sought to destroy all those who did not agree with him. In keeping with this, Lysenko brought criticism of his controversial ideas into the polemics of Communism. For example, in a 1935 speech, he stated, "Tell me, comrades, was there not a class struggle on the vernalization front?" At the end of this speech, Stalin exclaimed, "Bravo, Comrad Lysenko, bravo!"

Lysenko went on to claim that the inheritance of characteristics in plants could be altered by changing their environment. Of course, this notion was completely at odds with modern genetics and with the opinions of the many outstanding Soviet geneticists in the 1930s. However, the idea was compelling to Stalin because it fit with how he wanted to change Soviet society and offered a solution to the perennial agricultural failure of the collective farms. As a result, Lysenko moved up the Soviet scientific hierarchy quickly, denouncing and persecuting his critics as adherents of the old Mendel-Morgan genetic "heresies." As director of the Genetic Institute of the Academy of Sciences, he even ordered all the fruit flies to be destroyed by boiling!

Lysenko's impact was widespread until 1964 when Khrushchev fell. He had managed not only to extinguish Soviet genetics for nearly three decades but also to inhibit progress in many other scientific areas. In addition, other Communist countries, such as China, suffered similar suppression of scientific thought. The Soviet Union, China, and other Communist countries today have experienced a resurgence of activity in genetics, but recovery from the Lysenko era has been slow and difficult. ∎

Trofim Lysenko acquainting collective farmers with a new product called branched wheat.
© SOVFOTO/EASTFOTO AGENCY.

FACTORS IMPORTANT IN PESTICIDE RESISTANCE

A number of factors can affect the evolution of pesticide resistance. One basic requirement is that an organism have some physiological, anatomical, or other mechanism that enables it to resist a pesticide. In numerous examples, pesticide resistance takes different forms in different strains of the same species. If the potential for resistance is present, a number of other factors help determine whether a population actually becomes resistant.

First, the new genetically advantageous variant or type must already be present in the population, be generated by mutation, or migrate in from other populations. Once the resistant form is established in the population, its increase is dependent upon such factors as level of dominance, selective advantage, and initial frequency. For example, a variant that is dominant (expressed in heterozygotes) and has a large selective advantage should increase quickly. Also, many pest species have short generation lengths, multiple generations per year, and high fecundity, all of which are advantageous for the evolution of resistance. In addition, it is possible that chemical resistance genes sometimes reside in the DNA of chloroplasts, mitochondria, or plasmids (small nonnuclear pieces of DNA). Bacterial resistance to antibiotics is often present in plasmids, and transmission of plasmids may occur horizontally (at times other than reproduction) both between members of the same species and of different species; thus, antibiotic-resistant bacteria have become a major medical problem.

To illustrate the potential for developing resistance, consider the use of juvenile hormone, once thought to be a panacea pesticide for insects. Because it is a necessary chemical for insects' normal development, researchers thought insects could not develop a mechanism of resistance to it. However, in the past few years, several pest insects have developed resistance, making juvenile hormone pesticides ineffective.

In addition, the spread of antibiotic resistance may have increased in many bacteria due to the presence of preadapted genotypes. Specimens of bacteria from the preantibiotic era, which have been stored since the 1930s, do have some resistant strains. Furthermore, collections from remote parts of the world where antibiotics have never been used also show resistant strains. These strains may occur because they confer resistance to certain factors in the environment.

EXAMPLES OF PESTICIDE RESISTANCE

To better understand the genetic basis of pesticide resistance, let us examine several case studies. The Norway or brown rat, *Rattus norvegicus,* a widespread agricultural and domestic pest, was controlled for about a decade by

TABLE 17.7	Relative Fitness at the R_w Locus in the Norway Rat		
	SS	*RS*	*RR*
Warfarin	Susceptible	Resistant	Resistant
Vitamin K dependence	No	Intermediate	Yes
Relative fitness	0.68	1.00	0.37

Reprinted with permission of Jones and Bartlett Publishers, Inc., Boston, MA, from *Population Biology,* 1984, written by Philip Hedrick.

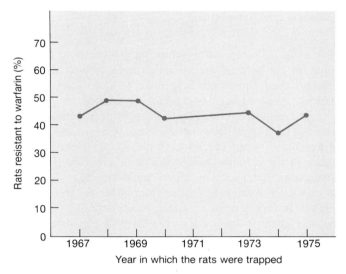

FIGURE 17.21
The frequency of rats resistant to warfarin over an eight-year period along the England-Wales border.

the anticoagulant warfarin. However, resistance to this chemical evolved quickly, and resistant rats have now been found in the United States and in many European countries. The mechanism of resistance has been examined in some detail and appears to be the result of a dominant resistant allele, *R,* at an autosomal locus. Most of the resistant animals are heterozygotes for this allele and for the wild-type or susceptible allele, *S,* because homozygotes for the resistant allele have quite low viability (see table 17.7). The lowered viability in *RR* individuals results from a 20-fold increase in vitamin K requirement.

As a result, there appears to be a stable polymorphism when warfarin is applied, owing to the overall net survival advantage of the heterozygote. Further support for this conclusion comes from a survey of a population where the frequency of the resistant allele appeared to be stable over a number of years (see figure 17.21) and showed a large excess of heterozygotes compared to Hardy-Weinberg expectations. In this population, where warfarin was regularly applied, the overall relative fitnesses of the three genotypes were estimated to be 0.68,

Resistant	CAA	TAT	GCT	GGT	TTC	AAC
	Gln	Tyr	Ala	Gly	Phe	Asn
Susceptible	CAA	TAT	GCT	AGT	TTC	AAC
	Gln	Tyr	Ala	Ser	Phe	Asn
Amino acid position	225	226	227	228	229	230

FIGURE 17.22
The DNA and amino acid sequence of pbsA-resistant and
pbsA-susceptible strains of pigweed. The boxed nucleotide and amino
acid indicate the difference in the resistant strain. The amino acid
position is shown at the bottom.

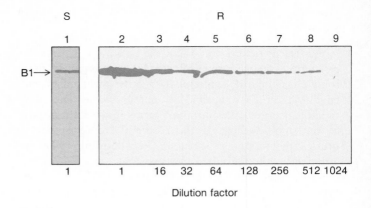

FIGURE 17.23
The amounts of esterase B1 produced in susceptible and resistant
mosquitoes are shown in lanes 1 and 2, respectively. Lanes 3 through 9
give the amount when the resistant mosquito extract was subjected to
serial dilution.

1.0, and 0.37, respectively, showing a substantial advantage for the heterozygote. Furthermore, in another population, the frequency of the resistant allele declined very
quickly when warfarin was no longer applied.

Let us mention two other instances of pesticide resistance in which detailed genetic information is known. First,
a number of weed species have quickly developed resistance to the commonly used herbicide atrazine. The resistance is maternally inherited, indicating that the trait
is coded for by a gene in the chloroplast (or mitochondrion). In fact, a protein called pbsA, encoded by a chloroplast gene, is involved in the part of photosynthesis
affected by atrazine. The amino acid and nucleotide sequences of both resistant and susceptible *Amaranthus*
plants (pigweed) were compared. Three differences in nucleotide sequence were found, two of which were silent
and resulted in no amino acid change. However, the third,
the replacement of an adenine by a guanine at position
682, caused the substitution of a glycine for a serine at
amino acid position 228 (see figure 17.22). This simple
change apparently resulted in a pbsA protein that has lost
affinity for the herbicide atrazine. Further evidence that
this is the critical change comes from two other species in
which resistant and susceptible strains also differ at amino
acid position 228.

Instead of undergoing a qualitative change, some insects apparently become resistant to organophosphates by
producing large amounts of detoxifying esterases. For example, the amount of esterase in susceptible and resistant
mosquitoes was evaluated using immunological techniques (see figure 17.23). The amount of B1 esterase in
the electrophoretic gel is given in the first column, while
the amount of esterase under different dilutions is given
in columns 2–9 for the resistant strain. It appears that the
resistant strain has approximately a 500-fold increase in
the level of the esterase; that is, a dilution of 1 to 512 gives
about the same amount of esterase in the susceptible strain
as in the resistant strain. It is not known whether the overproduction of esterase results from a change in regulation
or from an increased number of genes.

CONTROL OF PESTICIDE RESISTANCE

To control pesticide resistance, all of the evolutionary factors just discussed must be considered. For example, to
localize warfarin resistance in rats, efforts were made to
eradicate every rat within a 5-kilometer diameter area;
however, the gene flow of the rat is such that resistant rats
escaped beyond this border. Atrazine resistance is hard to
control because independent resistant types have arisen;
this is due either to a large number of independent mutations (some individual plants can produce 100,000 seeds,
making this a realistic possibility) or to some variation in
populations prior to exposure to atrazine.

One way to control insect pests on crops without using
insecticides is to plant crop varieties that are themselves
toxic or avoided by the pest. However, even with this
strategy, pests have evolved virulence to resistant crop varieties. For example, the Hessian fly, a pest of wheat, has
caused severe damage to susceptible varieties. Certain alleles in wheat can inhibit larvae development, but the flies
may have corresponding virulence alleles. However, only
when there is a virulence allele for each resistance allele
can the flies develop normally.

The first definitive studies on the genetics of plant resistance to a rust were done in flax. Two varieties of flax,
Ottawa and Bombay, were tested for their susceptibility
(S) or resistance (R) to two rust strains, 22 and 24. As
shown in table 17.8, Ottawa is resistant to rust strain 24
and Bombay is resistant to rust strain 22. When the flax
varieties are crossed, the F_1 is resistant to both rust strains,
demonstrating that the alleles conferring resistance are
dominant over the alleles showing susceptibility.

Obviously, reducing applications of particular pesticides will slow the development of resistance. In fact,
integrated pest management strategies, such as using pest-
resistant strains of crops, introducing predators or parasites to control pests, managing the agroecosystem, etc.,

	Flax		
TABLE 17.8	**Resistance (R) or Susceptibility (S) of Two Flax Varieties to Flax Rust**		
Rust Strain	**Flax**		
	Ottawa	*Bombay*	*F₁ Progeny*
22	S	R	R
24	R	S	R

Reprinted with permission of Jones and Bartlett Publishers, Inc., Boston, MA, from *Population Biology*, 1984, written by Philip Hedrick.

should reduce reliance on chemical control. Pesticides could then be introduced effectively at critical times or places. In addition, resistant alleles often have associated negative pleiotropic effects (lowered survival), such as for the warfarin resistance allele, which should result in a natural decrease of resistance in the absence of the pesticide. In other words, prudent use may actually make a given pesticide more useful over a longer period of time.

P esticide resistance is common in many agricultural and medical pests. Pesticide resistance spreads quickly in species with preadaptation, large population numbers, and high gene flow. Warfarin resistance in rats, atrazine resistance in pigweed, and organophosphate resistance in mosquitoes are case studies of evolutionary change as well as examples of pesticide resistance. Pesticide resistance can be controlled by using pest-resistant crops or other integrated pest management strategies.

Summary

T he incidence of human genetic disease can be generally explained by a mutation-selection balance in which mutation increases the frequency of the mutant and selection reduces it. In addition to mutation and selection, the frequency of disease alleles is sometimes influenced by genetic drift, gene flow, and inbreeding.

The evolution of molecules is generally consistent with the neutrality model, a theory based on the combined effects of genetic drift and mutation. Changes in DNA and proteins generally occur at a regular rate, resulting in a molecular clock. Molecular differences in different species can be used to construct phylogenetic trees that indicate the evolutionary relationships between the species. Mitochondrial DNA has been employed to determine phylogenetic relationships.

Conservation genetics is concerned with maintaining genetic variation in agricultural plants and animals as well as in rare and endangered species. Inbreeding depression refers to the decline in fitness that results from inbreeding. Loss of genetic variation occurs in captive populations when genetic drift has a large effect. Pesticide resistance is common in many agricultural and medical pests. Pesticide resistance spreads quickly in species with preadaptation, large population numbers, and high gene flow. Warfarin resistance in rats, atrazine resistance in pigweed, and organophosphate resistance in mosquitoes are case studies of evolutionary change as well as examples of pesticide resistance. Pesticide resistance can be controlled by using pest-resistant crops or other integrated pest management strategies.

Problems and Questions

1. If u $= 10^{-5}$, v $= 0.0$, and the selection coefficient against a recessive (s) $= 0.6$, what are the expected equilibrium frequencies of genotype A_2A_2 and allele A_2? If u is doubled, how much are the equilibrium genotypic and allelic frequencies changed?

2. Assume that u $= 10^{-6}$ for a lethal recessive gene. What is the equilibrium genotypic frequency? Assume that with better medical care, homozygotes have only a slight disadvantage, so that s $= 0.05$. What is the new equilibrium genotypic frequency?

3. Discuss the factors that affected the frequency of the sickle-cell allele in African blacks in the past and those that are influencing its frequency in American blacks today.

4. Cystic fibrosis is a recessive disease that has a high frequency in Caucasians (about 1 in 2,500 newborns). Until recently, it was lethal in childhood. Given this information, calculate the expected mutation rate. Compare this value to known mutation rates, and give a better explanation for the frequency of this disease.

5. Compare the predictions of the neutrality theory and the selection theory in influencing the amount of genetic variation. Suggest an experiment that might differentiate between them as explanations for variation at a given polymorphic locus.

6. What is the molecular clock? What is the rate of amino acid substitution? Would you expect the rate of nucleotide substitution to be faster or slower than the rate of amino acid substitution? Why?

7. If two species of rodents differ at 40 out of 90 amino acid positions for one protein and at 20 out of 80 amino acid positions for another, what proportion of sites differ for the two proteins? Why might these values differ for the two proteins?

8. Assume that four closely related species of birds (a, b, c, and d) were analyzed and that the genetic distances (a measure of the biochemical differences) between them were:

	b	c	d
a	0.2	0.12	0.55
b		0.3	0.7
c			0.65

From these data, which pair of species is most closely related? Which species is least related to the other species?

9. Table 17.3 suggests that molecular data may be able to resolve the deviation of closely related species. How could this information be used to understand evolutionary changes, including differences in morphology or physiology between species?

10. How do you think the population genetics of mtDNA differs from that of nuclear DNA? (In answering, remember that mtDNA is maternally inherited and haploid.)

11. The amount of the inbreeding depression is equal to $sfpq$. What is the expected inbreeding depression when $f = 0.25$, $s = 0.5$, and $q = 0.01$? If $q = 0.5$? Why is one of these values larger than the other?

12. In a population of 100 individuals where $q = 0.4$ and there are initial Hardy-Weinberg proportions, what are H_1 and H_2? If $N = 10$, what would H_1 and H_2 be? Discuss your answers.

13. How would you identify different land races or wild varieties of rice having genetic variation that might be useful in future plant breeding? How could you maintain these samples so that none of the genetic variation would be lost?

14. The progeny from a mating of half-first cousins have an inbreeding coefficient of $1/32$. For a recessive disease like PKU, with an allelic frequency of about 0.008, what is the expected frequency of PKU cases in the progeny of half-first cousin matings? Compare this to the frequency expected in the progeny of unrelated individuals.

15. Some animals, such as the black-footed ferret and the North American condor, now exist only in captivity. What kind of plan would you propose to conserve these species, considering such factors as inbreeding depression, loss of genetic variation, and reintroduction to the wild?

16. Calculate the expected equilibrium frequency of the allele that confers resistance to warfarin in rats, using the relative fitnesses in table 17.7. Some populations do not have this allele. Give two possible reasons why.

17. From table 17.8, it appears that resistance to two rust strains is dominant in flax. If the frequencies of the two alleles for resistance to rust strains 22 and 24 were 0.1 and 0.2, what proportions of the population would be susceptible to rust strain 22? To both rust strains?

18. In some species, pesticide resistance has spread very quickly. For example, the malaria mosquito is resistant to DDT nearly everywhere. Discuss the population genetic factors that may be important in causing such a speedy spread of resistance.

19. It is tempting to imagine using genetic engineering to produce new crops. However, in light of our discussion on genetic conservation and pesticide resistance, what problems may result from this approach?

Answers appear at end of book.

Chapter 17

CHAPTER 2

Problem

A number of researchers repeated Mendel's crosses in order to confirm his genetic principles. For example, Correns and Tschermak both crossed yellow and green varieties of peas and selfed the subsequent F_1 progeny to get F_2 progeny arrays. Correns observed 1,394 yellow and 453 green, while Tschermak observed 3,580 yellow and 1,190 green. Are these values consistent with the principle of segregation? (Calculate χ^2 values to determine consistency.)

Solution

Set up a table with the observed and expected numbers for the two crosses, then calculate the χ^2 values:

		Observed Numbers	Expected Numbers	$\left(\dfrac{(0-E)}{E}\right)^2$
Correns	Yellow	1,394	$1,847 \times (0.75) = 1,385.25$	0.055
	Green	453	$1,847 \times (0.25) = 461.75$	0.166
		1,847	1,847	$\chi^2 = 0.221$
Tschermak	Yellow	3,580	$4,770 \times (0.75) = 3,577.5$	0.002
	Green	1,190	$4,770 \times (0.25) = 1,192.5$	0.032
		4,770	4,770	$\chi^2 = 0.074$

In this case, there is 1 degree of freedom (2 categories minus 1 = 1 degree of freedom). As a result, the χ^2 values are not significant because both are less than 3.84 (see table 2.5), making the observations consistent with the principle of segregation.

CHAPTER 3

Problem

A woman with blood type A and a man with blood type B had children that are blood types A, B, AB, and O. What are the genotypes of the parents?

Solution

Without any knowledge about the offspring, the mother could be either $I^A I^A$ or $I^A I^B$, and the father could be either $I^B I^B$ or $I^A I^O$. There are then four possible mating types, with the following possible progeny phenotypes:

Mother	Father	Offspring
$I^A I^A$	$I^B I^B$	All AB
$I^A I^A$	$I^B I^O$	½AB, ½A
$I^A I^O$	$I^B I^B$	½AB, ½B
$I^A I^O$	$I^B I^O$	¼AB, ¼A, ¼B, ¼O

In other words, only if the mother is $I^A I^O$ and the father $I^B I^O$, can they have progeny of all four types.

CHAPTER 4

Problem

A wild-type strain of flies was crossed to another strain homozygous for the recessive X-linked mutants m, r, and v. The F_1 females were backcrossed to recessive strain males, and the following numbers of progeny were observed:

+	+	+	331
m	+	+	2
+	r	+	73
+	+	v	14
m	r	+	10
m	+	v	81
+	r	v	0
m	r	v	309
			820

What is the gene order and the map distance between these mutants?

Solution

There are three possible orders of the genes: *m–r–v, m–v–r, and r–m–v*. The two smallest progeny classes must be progeny produced by double recombination; that is, classes *m+ +*, and *+rv*. The only way that these progeny can be produced from *+m +r +v* females by double recombination is if gene *m* is in the center, the gene order *v–m–r*. (See figure 5.6 as an example for genes *y, w*, and *m*.)

To calculate the map distance, rewrite the gene order for these progeny with *m* in the middle, count up all the progeny with recombination in a given region, and divide by the total number of progeny:

Genotype			Numbers	Region of recombination
+	+	+	331	None
+	+	+	2	*v–m, m–r*
+	+	*r*	73	*m–r*
v	+	+	14	*v–m*
+	*m*	*r*	10	*v–m*
v	*m*	*r*	81	*m–r*
v	*m*	+	0	*v–m, m–r*
v	*m*	*r*	309	None
			820	

For the region between *v* and *m*, the rate of recombination is:

$$v - m = (2 + 14 + 10 + 0)/820 = 0.032$$

For the region between *m* and *r*, the rate of recombination is:

$$m - r = (2 + 73 + 81 + 0)/820 = 0.190$$

In other words, the numbers of map units between *v* and *m* is 3.2; between *m* and *r*, 19; and between *v* and *r*, 22.2.

CHAPTER 6

Problem

A certain phage has a double-stranded DNA with 45,000 base pairs (45 kb). (*a*) Approximately how many double-helical turns does this DNA contain? (*b*) How long is the DNA in microns (1 micron = 10 angstroms)? (*c*) What is the molecular weight of this DNA? (*d*) How many phosphorous atoms does the DNA contain?

Solution

(*a*) One double-helical turn encompasses about 10 base pairs, so all we have to do is divide the total number of base pairs by 10:

$$45,000 \text{ bp} \times 1 \text{ helical turn}/10 \text{ bp} = 4,500 \text{ helical turns}$$

(*b*) One base pair is about 3.4 angstroms, or 3.4×10^{-4} microns long, so we just need to multiply the total number of base pairs by 3.4×10^{-4}:

$$45,000 \text{ bp} \times 3.4 \times 10^{-4} \text{ microns/bp} = 15.3 \text{ microns}$$

(*c*) One base pair has a molecular weight of about 660, so we need to multiply the total number of base pairs by 660:

$$45,000 \text{ bp} \times 660/\text{bp} = 2.97 \times 10^7$$

(*d*) Each base pair has two phosphorous atoms, one on each strand. Therefore, we simply multiply the total number of base pairs by 2:

$$45,000 \text{ bp} \times 2 \text{ phosphorous atoms/bp} = 90,000 \text{ phosphorous atoms}$$

CHAPTER 7

Problem 1

Table 6.2 shows that the DNAs of bacteriophages T2 and T4 contain 2×10^5 base pairs. How many genes of average size (encoding proteins of about 40,000 molecular weight) can these phages contain?

Solution

Since the average molecular weight of an amino acid is about 100, a protein with a molecular weight of 40,000 contains about 400 amino acids. It takes three base pairs to encode one amino acid, so each protein will be encoded in about 1,200 base pairs ($3 \times 400 = 1,200$). Thus, the number of genes of average size in a DNA having 2×10^5 base pairs will be $2 \times 10^5/1.2 \times 10^3 = 1.67 \times 10^2$, or 167.

Problem 2

Here is the sequence of the anticoding strand of a DNA fragment:

5'ATGCGTGACTAATTCG3'

(*a*) Write the sequence of this DNA in double-stranded form.

(*b*) Assuming that transcription of this DNA begins with the first nucleotide and ends with the last, write the sequence of the transcript of this DNA in conventional form (5' to 3' left to right).

Solution

(*a*) Following the Watson-Crick base-pairing rules, and remembering that the two DNA strands are antiparallel, the double-stranded DNA would look like this:

5'ATGCGTGACTAATTCG3'
3'TACGCACTGATTAAGC5'

Note that the anticoding strand is still on top and that we have matched every base with its complement in the coding strand on the bottom.

(b) Since the sequence of an RNA is conventionally written 5' to 3' left to right, and the coding strand of the DNA, which looks like the RNA transcript, is written here in the opposite orientation, it would help to begin by inverting the DNA left to right as follows:

3'GCTTAATCAGTGCGTA5'
5'CGAATTAGTCACGCAT3'

Now the coding strand is still on the bottom. Since we know that the transcript of a gene has the same sequence and orientation as the coding strand (with U's in place of T's), we simply need to write the sequence of the coding strand, replacing all T's with U's as follows:

5'CGAAUUAGUCACGCAU3'

CHAPTER 9

Problem 1

Here is a DNA sequence from *E. coli:*

5'CGATTAGCATCGTA3'
3'GCTAATCGTAGCAT5'

The coding strand is on top. The transcription of this DNA begins with the first purine and ends with the last pyrimidine. What is the sequence of the RNA product of this DNA?

Solution

The top strand is the coding strand, so it has the same polarity and sense as the RNA product. The first purine is the first G, and the last pyrimidine is the last T. So, we simply write the sequence of the top strand, beginning with the first G and ending with the last T, but remembering to replace T's with U's. Thus, the sequence of the RNA product is:

5'GAUUAGCAUCGU3'

Problem 2

Here is a hypothetical promoter from *E. coli*. The starting point of transcription is marked in boldface. The coding strand is on top.

5'ACTTGACGACTGGCGCTTAATGCAGCATATAGTGCCGTCGCTAG3'
3'TGAACTGCTGACCGCGAATTACGTCGTATATCACGGCAGCGATC5'

(a) What is the sequence of the −10 box (coding strand)?
(b) What is the sequence of the −35 box (coding strand)?

Solution

(a) The middle of a typical −10 box lies about 10 bases upstream (to the 5' side) of the start of transcription, and has a sequence resembling the consensus sequence TATAAT. Thus, we start at the transcription start site (G) on the coding strand and count back ten bases. This brings us to the middle of the sequence TATAGT, which differs in only one base from the consensus sequence. Therefore, this is very likely to be this gene's −10 box.

(b) The middle of a typical −35 box lies about 35 bases upstream of the start of transcription, and has a sequence resembling the consensus sequence TTGACA. Counting backwards 35 bases from the start site of the coding strand of this gene brings us to the sequence TTGACG. This is probably this gene's −35 box.

CHAPTER 10

Problem

Here is the sequence of the 5'-end of the coding strand of a hypothetical mammalian gene recognized by RNA polymerase II:

5'CGATGCCGTATATAACGATCAGTACGTAGCATCCTCACTGA

Identify the most likely transcription start site (cap site).

Solution

First, find the TATA box. There is a sequence (TATATAA) near the 5'-end of the sequence that matches closely the TATA box consensus sequence (TATAAAA). Next, since the cap site of a gene transcribed by mammalian RNA polymerase II usually lies about 25 bases downstream (to the 3'-side) of the TATA box, count 25 bases to the 3' side of the middle of this sequence (the second A). This brings us to a region (CCTCACTGA) in which the boldfaced A is exactly 25 bases from the middle of the TATA box. Since transcription usually begins with a purine, this is the most likely start site, although the G and A just downstream are also candidates, and could be alternate start sites in a real gene.

CHAPTER 11

Problem 1

(a) Use the genetic code to predict the amino acid sequence of a peptide encoded in this mini-message, assuming that only physiologically significant initiation signals are used.

5'AUGAGAUACCAUGGGCUAAUGUGAAAA3'

What amino acid sequences would result if the following changes were made?
(b) The first C is changed to a G.
(c) The first U is changed to a G.
(d) The first C is changed to a U.
(e) The second G is changed to an A.
(f) The first C is deleted.
(g) An extra G is added after the first G.

Solution

(a) Since physiologically significant initiation signals (AUGs) are the only ones used, translation of this message will begin with the first codon (AUG). In order to simplify the decoding of the message, it helps to write the codons in order, separated by spaces, as follows:

AUG AGA UAC CAU GGG CUA AUG UGA AAA

Next, we consult the genetic code for the meaning of each of these codons:

fMet Arg Tyr His Gly Gln Met Stop Lys
AUG AGA UAC CAU GGG CUA AUG UGA AAA

The next-to-last codon (UGA) is a stop signal, so translation will stop after the methionine (Met) codon (AUG). Thus, the last codon (AAA, lysine) will remain untranslated, and the amino acid sequence of the peptide encoded in this little message is: fMet-Arg-Tyr-His-Gly-Gln-Met.

(b) The first C is changed to a G. The new message will read AUG AGA UAG, etc., where the boldface indicates the altered base. Now UAG is a stop codon, so only the first two codons will be translated, and the product will be fMet-Arg.

(c) The first C is changed to a U. The new message will read AUG AGA UAU, etc., where the boldface indicates the altered base. UAU codes for tyrosine (Tyr), just as UAC does, so there will be no change in the amino acid sequence.

(d) The second G is changed to an A. The new message will read AUG AAA, etc., where the boldface indicates the altered base. AAA codes for lysine (Lys), so the product will have Lys in the second position, and the rest will remain the same: fMet-Lys-Tyr-His-Gly-Gln-Met.

(e) The first G is changed to a U. This changes the first codon from AUG to AUU, which is no longer an initiation codon. Thus, initiation will occur at the next AUG to the right. Note that this does not have to be in the original reading frame, and in fact it is not in this case. The first AUG in this new message begins at the 11th base: AUUAGAUACC AUG GGC UAA UGU GAA AA. The coding of this new reading frame is:

fMet Gly Stop
AUG GGC UAA UGU GAA AA

Thus, this message codes for fMet-Gly.

(f) The first C is deleted. The new message will read:

AUG AGA UAC AUG GGC UAA UGU GAA AA

where the boldfaced bases lie on either side of the deleted base. The coding is as follows:

fMet Arg Tyr Met Gly Stop Cys Glu
AUG AGA UAC AUG GGC UAA UGU GAA AA

Thus, this message codes for fMet-Arg-Tyr-Met-Gly.

(g) An extra G is added after the first G. The new message will read:

AUG GAG AUA CCA UGG GCU AAU GUG AAA A

where the boldface represents the inserted base. This message will have the following coding properties:

fMet Glu Ile Pro Trp Ala Asn Val Lys
AUG GAG AUA CCA UGG GCU AAU GUG AAA A

Thus, this message codes for fMet-Glu-Ile-Pro-Trp-Ala-Asn-Val-Lys.

Problem 2

A certain amber suppressor inserts tryptophan (Trp) in response to the amber codon. What change in the anticodon of tRNA^{Trp} probably created this suppressor strain?

Solution

The tryptophan codon is 5'UGG3', which is recognized by the anticodon 3'ACC5' of the wild-type tRNA^{Trp}. The amber codon is 5'AUG3', which is most likely recognized by the anticodon 3'AUC5' of the amber suppressor tRNA. Thus, comparing this anticodon (3'AUC5') with the wild-type one (3'ACC5') shows that the change in the tRNA^{Trp} anticodon that created the amber suppressor was 3'ACC5' → 3'AUC5'. In other words, the change was a C → U transition.

CHAPTER 12

Problem

Here is the sequence of a portion of a bacterial gene:

5′CACTATGCTTGCGTGGACGCATTAAC3′
3′GTGATACGAACGCACCTGCGTAATTG5′

The top strand is the coding strand.

(*a*) Assuming that transcription starts with the first A and continues to the end, what would be the sequence of the mRNA contained in this gene fragment?

(*b*) What is the amino acid sequence encoded in this mRNA?

What effects would the following events have on this DNA sequence, and on its protein product, assuming that any mutations go unrepaired and finally affect *both* DNA strands?

(*c*) A tautomerization to the enol form of T in the first T of the coding strand during DNA replication.

(*d*) A tautomerization to the imino form of A in the second A of the anticoding strand during DNA replication.

(*e*) Deamination of the third C of the coding strand prior to DNA replication.

(*f*) O⁶-alkylation of the fourth G of the anticoding strand prior to DNA replication.

(*g*) Deletion of the third C in the coding strand.

Solution

(*a*) Since the top strand is the coding strand, it is the one that has the same polarity and sense as the mRNA. Therefore, this is the strand you should use as your model for the mRNA transcript of this gene. The only changes you need to bear in mind are that T's in the DNA coding strand will be replaced by U's in the mRNA, and that since transcription starts with the first A, you should delete the first C. Thus, the mRNA will read:

ACUAUGCUUGCGUGGACGCAUUAAC

(*b*) Using the genetic code in chapter 11, and the strategy in the first worked-out problem in that same chapter, we can find the first AUG, beginning with the fourth base in the mRNA. The reading frame and coding from there on is:

fMet Leu Ala Trp Thr His Stop
AUG   CUU GCG UGG ACG CAU UAA

Thus, the protein product will be fMet-Leu-Ala-Trp-Thr-His.

(*c*) A tautomerization to the enol form of the second T of the coding strand will make it pair like C, resulting in a T→C transition at that point. This will change the mRNA sequence from ACU, etc., to ACC, etc. However, this will have no effect on the coding properties of the mRNA, since the mutation occurred outside the coding region of the gene.

(*d*) A tautomerization to the imino form of A in the second A in the anticoding strand will make it pair like G, resulting in an A→G transition at that point. This will in turn cause a T→C transition in the coding strand. The change in the gene can be summarized as follows:

CACTATG--- → CACTA**C**G---
GTGATAC---   GTGAT**G**C---

where the boldface denotes the altered base pair. This will change the mRNA sequence from ACUAUG, etc., to ACUACG, etc. The change of the initiation codon from AUG to ACG will destroy its function, so the ribosome will search for another AUG (corresponding to ATG in DNA). Since there is none, no protein product will be made.

(*e*) Deamination of the third C of the coding strand will convert it to a U. If this change goes unrepaired, it will result in a C→T transition, and the new mRNA sequence and coding properties will be:

```
        fMet  Phe
ACU     AUG   UUU  etc.,
```

where the boldface indicates the altered base. Thus, the protein product will now be fMet-Phe-Ala-Trp-Thr-His, with a phenylalanine (Phe) in place of the original leucine (Leu).

(*f*) O⁶-alkylation of the fourth G of the anticoding strand will make it pair like A, resulting in a C→T transition in the coding strand. The change in the gene can be summarized as follows:

5′CACTATGCTTGCG--- →CACTATGCTTG**T**G---
3′GTGATACGAACGC---   GTGATACGAAC**A**C---

where the boldface denotes the altered base pair. The new mRNA sequence and coding properties will be:

```
        fMet  Leu  Val
ACU     AUG   CUU  GUG  etc.
```

Thus, the protein product will now be fMet-Leu-Val-Trp-Thr-His, with a valine (Val) in place of the original alanine (Ala).

(*g*) Deletion of the third C in the coding strand will result in the following DNA sequence:

CACTA**T**G**T**TGCGTGGACGCATTAAC
GTGATA**C**AACGCACCTGCGTAATTG

where the boldfaced bases flank the deleted base. The new mRNA sequence and coding properties will be:

```
        fMet  Leu  Arg  Gly  Arg  Ile  Asn
ACU     AUG   UUG  CGU  GGA  CGC  AUU  AAC
```

Thus, the protein product will now be fMet-Leu-Arg-Gly-Arg-Ile-Asn. Note that the frameshift has caused a drastic difference in the gene product.

CHAPTER 13

Problem

Consider an experiment in which we do not know anything about the order of three genes, X, Y, and Z, but we do know that they are close enough together to cotransduce. We use an $X^+ Y^+ Z^+$ donor and an $X^- Y^- Z^-$ recipient. We select for X^+ transductants and find the following frequencies of transduction:

103 are $X^+ Y^- Z^-$
315 are $X^+ Y^+ Z^-$
57 are $X^+ Y^+ Z^+$
0 are $X^+ Y^- Z^+$

We can ask the following major questions:

1. What is the order of these three markers?

2. What is the cotransduction frequency of X and Y? Of X and Z?

3. What are the DNA distances between X and Y? Between X and Z?

Solution

First, what is the marker order? We will see that all of the combinations but one can be achieved by two crossovers between donor and recipient DNA. The fourth can only be produced by four crossovers and will therefore be the rarest. The rarest combination is $X^+Y^-Z^+$, which appeared in none of the transductants. How can four crossovers between donor and recipient produce this combination? There is only one way, as shown in figure A.1, where Y is in the middle. Therefore, the order of markers is XYZ.

Here is another way of looking at this: If there are four crossovers, there must be one before and after each of the three markers. This will automatically give the transductant the donor alleles of the two end markers, and the recipient allele of the middle marker. Thus, a shortcut is to find the rarest combination ($X^+Y^-Z^+$ in this case) and determine which of the three loci has the recipient phenotype (Y^- in this case). Therefore, Y must be in the middle.

What are the cotransduction frequencies? For X and Y: $(315 + 57)/475 = 0.78$; for X and Z: $57/475 = 0.12$. Figure A.1 shows the crossovers that create each of the four marker combinations. Notice again that the first three involve only two crossovers, while the fourth involves four. Also note that a crossover before X is required in all cases, since only X^+ transductants were selected.

Finally, what are the DNA distances among the markers? Rearrangement of the expression gives:

$$d = L(1 - \sqrt[3]{x})$$

Given a length (L) of 91.5 kb, the distance between X and Y would be 91.5 kb $(1 - \sqrt[3]{.78})$, or 7.3 kb. Similarly, the distance between X and Z would be 91.5 kb $(1 - \sqrt[3]{.12})$, or 46.7 kb.

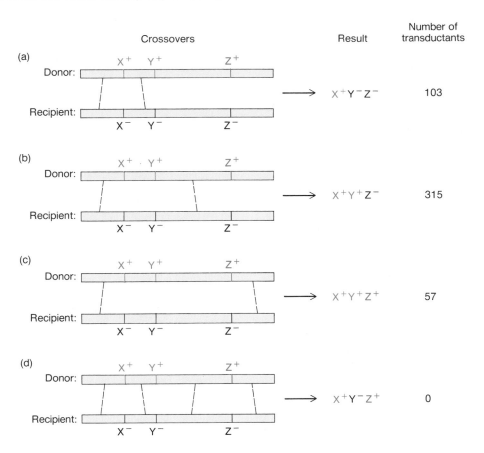

	Crossovers	Result	Number of transductants

(a) Donor: X^+ Y^+ Z^+ / Recipient: X^- Y^- Z^- → $X^+Y^-Z^-$ — 103

(b) Donor: $X^+ \cdot Y^+$ Z^+ / Recipient: X^- Y^- Z^- → $X^+Y^+Z^-$ — 315

(c) Donor: X^+ Y^+ Z^+ / Recipient: X^- Y^- Z^- → $X^+Y^+Z^+$ — 57

(d) Donor: X^+ Y^+ Z^+ / Recipient: X^- Y^- Z^- → $X^+Y^-Z^+$ — 0

FIGURE A.1

CHAPTER 15

Problem

Here is the amino acid sequence of part of a hypothetical protein whose gene you want to clone:

Arg-Leu-Met-Ile-Trp-Glu-Cys-Ser-Met-Leu

(a) What sequence of five amino acids would give a 17-mer probe (including two bases from the next codon) with the least degeneracy?

(b) How many different 17-mers would you have to synthesize to be sure your probe matches the corresponding sequence in your cloned gene perfectly?

(c) If you started your probe two codons to the right of the optimal one (the one you chose in part[a]), how many different 17-mers would you have to make?

Solution

(a) Begin by consulting the genetic code to determine the coding degeneracy of each amino acid in the sequence. This yields:

$$6 \quad 6 \quad 1 \quad 2 \quad 1 \quad 3 \quad 2 \quad 4 \quad 1 \quad 6$$
Arg-Leu-Met-Glu-Trp-Ile-Cys-Pro-Met-Leu

where the numbers above the amino acids represent the coding degeneracy for each. In other words, there are 6 codons for arginine, 6 for leucine, one for methionine, etc. Now the task is to find the contiguous set of five codons with the lowest degeneracy. A quick inspection shows that Met-Glu-Trp-Ile-Cys works best.

(b) To find how many different 17-mers we would have to prepare, we need to multiply the degeneracies at all positions within the region covered by our probe. For the five amino acids we have chosen, this is $1 \times 2 \times 1 \times 3 \times 2 = 12$. Note that we can use the first two bases (CC) in the proline (Pro) codons without encountering any degeneracy, since the 4-fold degeneracy in coding for proline is all due to the third base in the codon (CCU, CCA, CCC, CCG). Thus, our probe can be 17 bases long, instead of the 15 bases we get from the codons for the 5 amino acids we have selected.

(c) If we had started two amino acids farther to the right, starting with Trp, the degeneracy would have been $1 \times 3 \times 2 \times 4 \times 1 = 24$, so we would have had to prepare 24 different probes instead of just 12.

CHAPTER 17

Problem

A survey of an electrophoretic locus in a bass population observed 16 individuals of genotype C_1C_1, 57 of genotype C_1C_2, and 77 of genotype C_2C_2. What are the allelic frequencies of C_1 and C_2? What is the expected Hardy-Weinberg heterozygosity? Is this population statistically different from Hardy-Weinberg proportions?

Solution

There are $16 + 57 + 77 = 150$ individuals in the sample. Thus, the frequency of allele C_1 is:

$$p = [16 + (0.5)57]/150 = 0.297$$

and the frequency of C_2 is:

$$q = 1 - 0.297 = 0.703$$

The Hardy-Weinberg heterozygosity is:

$$2pq = 2(0.297)(0.703) = 0.418$$

To calculate the χ^2 value, we can set up a table as:

	Observed Numbers	Expected Numbers		$(0 - E)^2/E$
C_1C_1	16	$(0.297)^2 150 =$	13.23	0.58
C_1C_2	57	$2(0.297)(0.703)150 =$	62.64	0.51
C_2C_2	77	$(0.703)^2 150 =$	74.13	0.11
	150		150	$\chi^2 = 1.20$

There is one degree of freedom (3 categories minus 1 minus 1 for estimation of p), so this value is less than the cutoff value of 3.84 in table 2.5, making these observations consistent with Hardy-Weinberg proportions.

A·P·P·E·N·D·I·X

B

Answers to End-of-Chapter Problems and Questions

CHAPTER 2

1. 0.35, 0.61; with 1 degree of freedom, they are consistent.

2. 0.5; 0.33.

3. 0.0625; 0.00391; product rule.

4. (a)

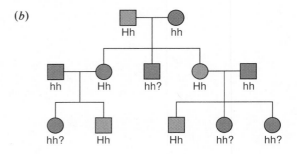

Aa (normal) — *Aa* (normal)

aa (albino)

   (b) 0.25; 0.5; 0.667.

5. 0.25; 0.240.

6. (a) 0.25 for each type. (b) 0.62; with 3 degrees of freedom, it is consistent.

7. 0.0625; 0.1875.

8. 0.0156; 0.1406.

9. Recessive; see table 2.7; *Aa, Aa, aa;* 0.5; 0.0.

10. Dominant; see table 2.7; *bb, Bb, bb;* 0.0.

11. Her husband is heterozygous; 0.5.

12. 0.052; with 1 degree of freedom, it is consistent.

13. 0.429; with 1 degree of freedom, it is consistent.

14. 0.125; 0.75.

15. (a)

	AB	*Ab*	*aB*	*ab*
AB	AABB	AABb	AaBB	AaBb
Ab	AABb	AAbb	AaBb	Aabb
aB	AaBB	AaBb	aaBB	aaBb
ab	AaBb	Aabb	aaBb	aabb

   (b) 0.5625, 0.1875, 0.1875, 0.0625.

16. 0.667; 0.25.

17. 0.333; 0.5.

18. Dominant. See table 2.7 for a complete list.

19. 12.5.

20. 0.00694; 0.0833.

CHAPTER 3

1. 0.26; with 2 degrees of freedom, they are consistent.

2. 0.28; with 1 degree of freedom, it is consistent.

3. A × A, A × B, A × AB, A × O, B × B, B × AB, B × O, AB × AB, AB × O, O × O.
$I^A I^A \times I^B I^B$ (all AB); $I^A I^A \times I^B I^O$ (½ AB, ½ A); $I^A I^O \times I^B I^B$ (½ AB, ½ B); $I^A I^O \times I^B$ (¼ A, ¼ AB, ¼ B, ¼ O).

4. 5.24—statistically significant at 0.05 level with 1 degree of freedom; 0.21—consistent with a 2:1 hypothesis.

5. To document the pleiotropic effects of a newly discovered genetic disease, assuming that the disease is due to a single gene, you would compare individuals with and without the disease for a number of different characteristics, especially those directly related to the genetic defect (as in PKU, where the basic defect is the inability to metabolize phenylalanine), and other characteristics that are possibly a consequence of the basic defect.

6. (a) The level of penetrance is the proportion of individuals of a given genotype who manifest a particular phenotype, while expressivity is the level of expression of a particular phenotype.

   (b)

Hh — hh

hh — Hh   hh?   Hh — hh

hh?   Hh   Hh   hh?   hh?

7. (a) Compare the phenotypes of groups of flies raised at different levels of temperature and diet, measuring closely the phenotypic and environmental factors. (b) Cross the mutation into different stocks and observe the offspring to determine if different crosses result in different phenotypes.

8. Type O is the universal donor; type AB is the universal recipient. Type O has no antigens and therefore is accepted by all types. Type AB has no antibodies and therefore can accept all types of blood.

9. $MM' \times mm.$

10. (a) ¼ $B_1 B_2$, ¼ $B_2 B_5$, ¼ $B_1 B_4$, ¼ $B_4 B_5$. (b) $B_1 B_4 \times B_2 B_2$.

11. (a) ½ ww^+ (red-eyed females), ½ wY (white-eyed males). (b) 0.36—with 3 degrees of freedom, they are consistent.

12.

$\frac{1}{4} A_1 A_1$ → $\frac{1}{4} B_1 B_1$, $\frac{1}{2} B_1 B_2$, $\frac{1}{4} B_2 B_2$

$\frac{1}{2} A_1 A_2$ → $\frac{1}{4} B_1 B_1$, $\frac{1}{2} B_1 B_2$, $\frac{1}{4} B_2 B_2$

$\frac{1}{4} A_2 A_2$ → $\frac{1}{4} B_1 B_1$, $\frac{1}{2} B_1 B_2$, $\frac{1}{4} B_2 B_2$

$\frac{1}{16}$ red ($\frac{1}{16} A_1 A_1 B_1 B_1$)

$\frac{1}{4}$ light red ($\frac{1}{8} A_1 A_1 B_1 B_2$, $\frac{1}{8} A_1 A_2 B_1 B_1$)

$\frac{3}{8}$ pink ($\frac{1}{16} A_1 A_1 B_2 B_2$, $\frac{1}{4} A_1 A_2 B_1 B_2$, $\frac{1}{16} A_2 A_2 B_1 B_1$)

$\frac{1}{4}$ light pink ($\frac{1}{8} A_1 A_2 B_2 B_2$, $\frac{1}{8} A_2 A_2 B_1 B_2$)

$\frac{1}{16}$ white ($\frac{1}{16} A_2 A_2 B_2 B_2$)

13. (a)

Frequency	Category	Genotypes
$\frac{1}{64}$	0	$(\frac{1}{64}A_1A_1B_1B_1C_1C_1)$
$\frac{3}{32}$	1	$(\frac{1}{32}A_1A_1B_1B_1C_1C_2, \frac{1}{32}A_1A_1B_1B_2C_1C_1, \frac{1}{32}A_1A_2B_1B_1C_1C_1)$
$\frac{15}{64}$	2	$(\frac{1}{64}A_1A_1B_1B_1C_2C_2, \frac{1}{64}A_1A_1B_2B_2C_1C_1, \frac{1}{64}A_2A_2B_1B_1C_1C_1,$ $\frac{1}{16}A_1A_1B_1B_2C_1C_2, \frac{1}{16}A_1A_2B_1B_1C_1C_2, \frac{1}{16}A_1A_2B_1B_2C_1C_1)$
$\frac{5}{16}$	3	$(\frac{1}{32}A_1A_1B_1B_2C_2C_2, \frac{1}{32}A_1A_2B_1B_1C_2C_2, \frac{1}{32}A_1A_1B_2B_2C_1C_2,$ $\frac{1}{32}A_1A_2B_2B_2C_1C_1, \frac{1}{32}A_2A_2B_1B_1C_1C_2, \frac{1}{32}A_2A_2B_1B_2C_1C_1,$ $\frac{1}{8}A_1A_2B_1B_2C_1C_2)$
$\frac{15}{64}$	4	$(\frac{1}{64}A_2A_2B_2B_2C_1C_1, \frac{1}{64}A_2A_2B_1B_1C_2C_2, \frac{1}{64}A_1A_1B_2B_2C_2C_2,$ $\frac{1}{16}A_2A_2B_1B_2C_1C_2, \frac{1}{16}A_1A_2B_2B_2C_1C_2, \frac{1}{16}A_1A_2B_1B_2C_2C_2)$
$\frac{3}{32}$	5	$(\frac{1}{32}A_1A_2B_2B_2C_2C_2, \frac{1}{32}A_2A_2B_1B_2C_2C_2, \frac{1}{32}A_2A_2B_2B_2C_1C_2)$
$\frac{1}{64}$	6	$(\frac{1}{16}A_2A_2B_2B_2C_2C_2)$

(b)

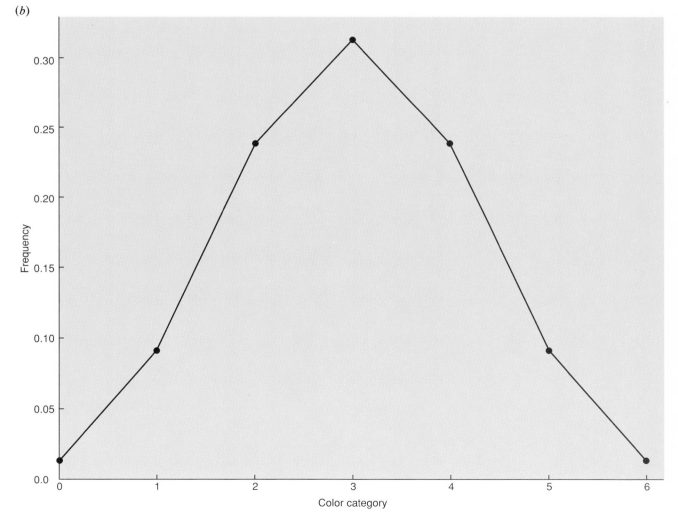

(c) Three genes have more classes and approach a bell-shaped curve.

14. Dominance concerns the interaction of two alleles at a given gene to determine phenotype. Epistasis involves the interaction of alleles at different genes. All of the traits examined by Mendel were determined at a single gene.

15. 1.56; with 1 degree of freedom, they are consistent.

16. 0.5.

17. Duplicate gene action could not result in white plants that differed only at one gene. In other words, a cross differing by two genes would produce all white progeny. Complementary gene action would give results as shown in figure 3.19.

18.

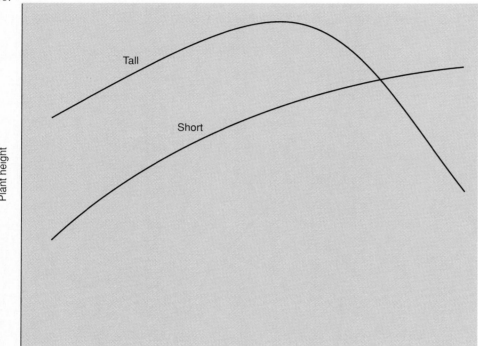

19. Polygenic means many genes affect the trait. Quantitative means that the trait is measured on a continuous scale. In some cases, polygenic traits may not have a continuous scale, while quantitative traits may not have a polygenic basis. These traits may be either single-gene or polygenic. Pedigree information would be needed in order to determine the genetic basis of such a trait.

20. 0.4375; 0.0.

CHAPTER 4

1. 0.04; with 1 degree of freedom, it is consistent.
   2.19; with 2 degrees of freedom, it is consistent.

2. 0.00118.

3. The synthesis of DNA and chromosomal events are not synchronous. See figure 4.10 for details.

4. In mitosis, chromosomes are aligned so that sister chromatids are on opposite sides of the metaphase plate. In meiosis I, homologues are paired and are on opposite sides of the plate. Meiosis II is like mitosis except that only one homologue is in each cell.

5. (*a*) 0.5 with *A*, 0.5 with *a;* 0.25 with *AB*, 0.25 with *Ab*, 0.25 with *aB*, and 0.25 with *ab*.

(*b*)

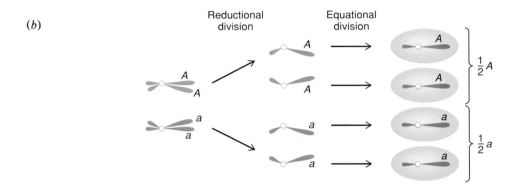

(*c*)

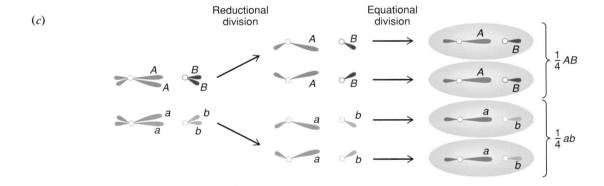

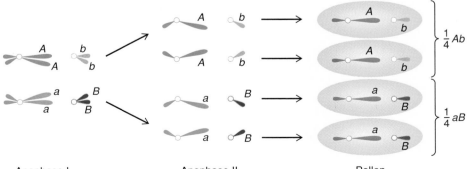

Anaphase I   Anaphase II   Pollen

6.

(a)

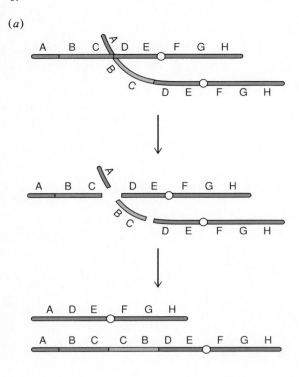

(b)

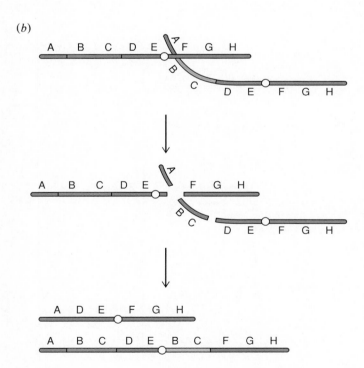

7. Genes *e* and *f*.

8. Pericentric inversions include the centromere, while paracentric inversions do not. The paracentric results in an acentric chromosome and a dicentric bridge, while the pericentric does not. Both have chromosomes with combinations of duplications and deletions, but those from the paracentric may be more severe.

9. (a)

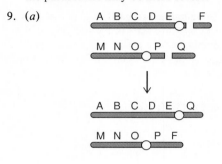

(b)

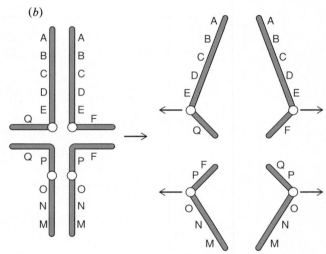

10. See figure 4.30. Many individuals with Down syndrome die before birth.

11. (a) See figure 4.34; XXY, XO. (b) XXX, XXY, XO (males), XXX, XXY, XO (females). (c) XXY, XY (males), XXX, XX, XO (females).

12. Chromosome 6.

13. 4 copies; *AAAA, AAAa, AAaa, Aaaa, aaaa;* 2 phenotypes.

14. 0.0625; 0.0625.

15. The smaller inversion is located just below the large one indicated by arrows (six bands long). The pericentric inversion starts below the third band from the top in humans and ends below the eighth band.

16. 0.25 $Z^A Z^A$ males, 0.25 $Z^A Z^a$ males, 0.25 $Z^A W$ females, 0.25 $Z^a W$ females.

17. Normal female; normal male. Equal numbers of X and autosomes result in females, while an X/A ratio of 0.5 results in males.

18. See figures 4.7 and 4.9.

19. The two parental genomes may cover up deleterious alleles in the other genome. Yes, normal meiosis.

20. 3 chromosomal fissions.

CHAPTER 5

1. 139.0; with 3 degrees of freedom, not consistent with hypothesis at 0.01 level.

2. *PL* (0.443), *Pl* (0.057), *pL* (0.057), *pl* (0.443).

3. $P-L-$ (0.696), $P-ll$ (0.054), $ppL-$ (0.054), $ppll$ (0.196); 1.42—with 3 degrees of freedom, they are consistent. (See chapter 4.)

4. 0.130. The estimate for the first testcross was 0.107, a lower estimate of recombination.

5.

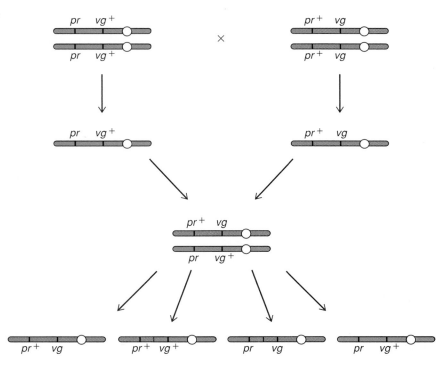

6. The gene order is *a–c–b*. 0.3 > 0.27 because some double crossovers have occurred.

7. The gene order would be *a–c–d–b*. *bcd*/+ + + × *bcd/bcd*.

8. *k*–0.141–*m*–0.083–*l*.

9. 0.0117. I = 1 − 0.00636/0.0117 = 0.46. Some interference because there are about half the proportion of doubles expected.

10.

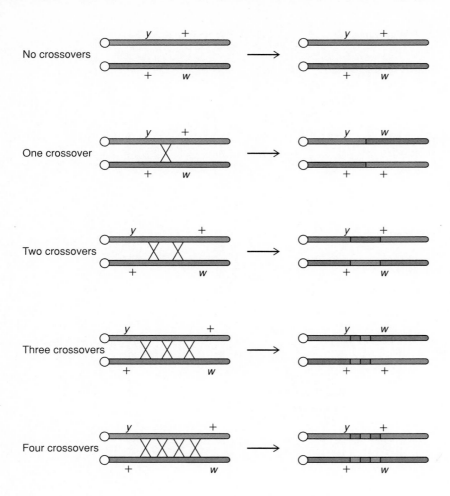

No crossovers

One crossover

Two crossovers

Three crossovers

Four crossovers

11. *2 h C/Y, 2 H c/Y, 1 H C/Y*; 0.2.

12. 0.214. Families having no recombinant male offspring are omitted here, causing a bias in the data.

13. 0.5 A_1A_2, 0.5 A_2—; all would be A_1A_2.

14. Chromosome 11.

15. X+Y, X*X+Y; wild-type male and female; white is a recessive allele.

16. *DDDDdddd; DDddDDdd, DDddddDD, ddDDDDdd,* or *ddDDddDD.*

17. Centromere–*F–E.*

18. 0.028; 0.0045.

19. Recombination may be localized into certain regions.

20. Mendel might not have formulated the principle of independent assortment correctly at all.

CHAPTER 6

1. The transforming principle induced a heritable change in *Pneumococci.* Since genes control heredity, it was very likely that the transforming principle was itself a gene or genes.

2. Avery et al. used cell-free extracts of virulent (S) cells as the transforming agent. Since no virulent cells were mixed with the avirulent ones, there was no possibility of resurrecting any virulent cells.

3. The Avery et al. approach offers the stronger argument for DNA as the genetic material. These workers carefully purified their DNA before using it to transform cells. In the Hershey-Chase approach, 10% of the phage protein was not accounted for and therefore could have entered the host cells along with the phage DNA. In principle, this protein could have been the genetic material.

4. No. Both protein and DNA contain C and H.

5. 5'-CGAATGCT-3'; 3'-TCGTAAGC-5'.

6. Thymine, 30%; cytosine, 20%.

7. (a) 12,000 helical turns; (b) 40.8 microns;
(c) 79.2 × 10⁶; (d) 240,000 phosphorous atoms
(120,000 in each strand).

8. 0.54 kb

CHAPTER 7

1. 100 proteins.

2. (a) Dispersive: 8 ‖‖

(b) Conservative: ‖ + 7‖

(c) Semiconservative: 2 ‖ + 6‖

3. (a) Dispersive:

14ₙ   15ₙ

(b) Conservative:

14ₙ   15ₙ

(c) Semiconservative:

14ₙ   15ₙ

4. Show that the mutant will grow on *c* but not on *b*. This shows that the *b* → *c* enzyme is defective. Therefore, its gene is defective.

5. About 53%. Because the T3 mRNAs are transcribed from most of the T3 DNA, and are therefore complementary to one of the T3 DNA strands, their G+C content should closely resemble the G+C content of the DNA template. Consider this example:

DNA: GAGCTACGGC
         CTCGATGCCG
               ↓ Transcription
mRNA: GAGCUACGGC

The G+C content of the DNA is 70% (14 out of 20 bases), and so is the G+C content of the mRNA (7 out of 10 bases).

6. No, you cannot predict the purine content of a transcript knowing only the purine content of the double-stranded DNA template. The purine content of the transcript will depend on which strand is transcribed. For example, if this strand (GGGCTA) is transcribed, its product will be CCCGAU (²⁄₆, or 33.3% purine). Transcription of the opposite strand yields GGGCUA, with a purine content of 66.7%. (Besides, the purine content of any double-stranded DNA is bound to be 50%).

7. (a)

5' ————————→3'

(b)

5' ——————— 3'
————————→

(c)

3' ——————— 5'
————————→

8. The proteins would be made with arginines where phenylalanines should have been. Most or all of these proteins would not function, because their altered primary structures would make them function abnormally.

9. (a) 5'GTCAATCGTAGCGGCCATAT3'
       3'CAGTTAGCATCGCCGGTATA5'

(b) 5'AUAUGGCCGCUACGAUUGAC3'

CHAPTER 8

1.

(a)

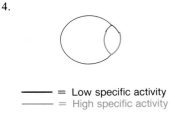

Progeny strands: ---- = Unlabeled
                   ——— = Labeled

(b)

2.

——— --- — ———

3. There are two classes of replicons; one begins replicating before the other.

4.

——— = Low specific activity
——— = High specific activity

5. (a) The gradient centrifugation of the pulse-labeled DNA would show only a peak of large DNA. No Okazaki fragments would appear, no matter how short the labeling period. (b) The DNA polymerase must be able to make DNA in both the 5' → 3' and 3' → 5' directions.

6. Helicases separate the two DNA strands in advance of the replicating fork. This process requires energy, which is provided by the ATPase's cleavage of ATP.

7. A helicase unwinds DNA, while a topoisomerase changes the topology of the DNA by increasing or decreasing its number of supercoils. Helicases break no DNA strands, whereas transient strand breakage is a necessary part of a topoisomerase's function.

8. SSB binds cooperatively to single-stranded DNA, but not to double-stranded DNA. This means that once one molecule of SSB binds to single-stranded DNA, it greatly facilitates the binding of other molecules, but this facilitation does not operate with double-stranded DNA.

9. Assuming that the linear DNA is not constrained somehow by proteins or other attached molecules, it should have no need for a swivel; since it has free ends, it can readily rotate to compensate for the separation of parental strands at the fork. Therefore, no strain will develop in the molecule.

10. Positive.

11. Negative. Because overwinding creates negative supercoils, unwinding will remove them.

12. Negatively supercoiled, because this will relax the DNA, and nature seeks the most relaxed state.

13. The primosome binds to the template for the trailing strand. This makes sense because the trailing strand needs a new primer for each new Okazaki fragment, whereas the leading strand replicates more or less continuously and rarely needs a new primer.

14. A replisome with two polymerase centers would be able to replicate both strands more or less at the same time and do so processively in the direction of the replicating fork, without having to waste time dissociating from and reassociating with the DNA. In order for this sort of double-headed replisome to work, the lagging strand template would have to loop 180 degrees around so that it entered the replisome in the same orientation as the leading strand template ($3' \rightarrow 5'$, bottom to top).

15. Primer synthesis cannot require a based-paired nucleotide to add to. If it did, the first nucleotide could never be incorporated. Therefore, the editing function cannot operate. If the wrong nucleotide is incorporated into a primer, it must remain there.

CHAPTER 9

1. RNA polymerase holoenzyme transcribes un-nicked T4 DNA actively in vitro. The transcription is specific (immediate early genes only) and asymmetric. The core enzyme transcribes such a template only weakly. What transcription occurs is nonspecific and symmetric.

2. (a) Up, because this change makes the sequence more like the consensus TATAAT; (b) down, because this change makes the sequence less like the consensus sequence.

3. The genes in the operon are all transcribed together, from a single promoter. Therefore, controlling transcription from that promoter controls all the genes simultaneously.

4. (a) The lac operon would always be turned on, because repressor cannot turn it off unless it can bind to the lac operator. (b) Same as (a). (c) The operon would be uninducible. Repressor would remain bound to operator even in the presence of inducer. (d) The operon would be transcribed only weakly. Even in the absence of glucose, CAP plus cAMP would be unable to facilitate polymerase binding to the lac promoter.

5. (a) The early genes only, because the gene 28 product is needed for the switch to middle transcription. (b) and (c) The early and middle genes only, because the products of both genes 33 and 34 are needed for the switch to late transcription.

6. Only the early (class I) genes would be transcribed, since gene 1 encodes the T7 RNA polymerase that transcribes the later genes.

7. The cells would not be able to sporulate, since σ^{29} and σ^{32} are needed to switch the transcriptional specificity of B. subtilis RNA polymerase from vegetative to sporulation genes.

8. The repressor gene (cI).

9. All but the repressor gene.

10. Transcripts made in the delayed early phase of infection all initiate at the immediate early promoters. Therefore, immediate early products, including pN, continue to be made throughout the delayed early phase.

11. It cannot result in cro expression because the cro gene is transcribed backwards, and therefore an anticoding mRNA is produced.

12. Positive. The expression of the delayed early genes is turned off unless pN acts positively to turn it on.

CHAPTER 10

1. 5S rRNA. It requires no processing.

2. Because a polymerase III promoter lies within the coding region of the gene.

3. First, nucleosomes are found in all eukaryotes that have been studied. Second, most of the histones are extremely well conserved. If they were not so vitally important, they would have evolved more rapidly.

4. The 11-nm fiber is essentially a string of nucleosomes. The nucleosomes contain all histones, but H1 is only bound to the outside of each nucleosome. Since H1 is not necessary for forming nucleosomes, its absence would not perturb the 11-nm fiber, beyond reducing its average diameter slightly. On the other hand, histone H1 plays an active role in coiling the nucleosome chain into the 25-nm fiber, or solenoid. Removing H1 would prevent formation of this chromatin structure.

5. The euchromatic chromosome.

6. Both stimulate transcription of specific genes. σ causes holoenzyme to bind tightly to promoters, but does not bind to them by itself. Eukaryotic transcription factors, on the other hand, bind independently to promoters.

7.

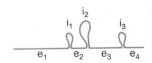

8.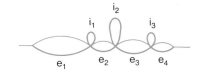

9.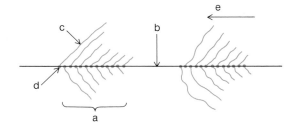

10.

11. GpppXpY . . .

12. Messenger RNA levels in the cytoplasm reflect much more than just transcription. They are also potentially influenced by capping, polyadenylation, splicing, transport, and RNA degradation.

13. Pulse-label cells with radioactive RNA precursor, then chase with unlabeled precursor. After various lengths of chase, hybridize the labeled total mRNA (polyadenylated RNA) to a cDNA that is complementary to the mRNA you are studying. The more labeled RNA that can still hybridize after a given time, the more stable the mRNA.

14. The first twenty-one amino acids in this protein constitute a signal peptide that is removed during maturation of the protein.

15. (a) Between −200 and −250. (b) Use a −200 to −250 fragment as competitor. It should compete best of all.

16. To show that poly(A) is present in the mRNAs, degrade them with RNase A and T1, then electrophorese the remaining RNA to see if a poly(A) has survived. To show that the poly(A) is at the 3′-end of the mRNA, digest the mRNA briefly with a 3′-specific RNase to see if the poly(A) is rapidly lost.

17. First, clone a gene containing the GAGA box, then introduce mutations into the GAGA sequence. Put the mutated gene into an in vitro or in vivo transcription system and measure the effect of the mutations on transcription. If the GAGA box behaves like the SV40 TATA box, mutations in this region should cause an increase in heterogeneity of transcription start sites.

18. 23 bases downstream from the second A of the TATA box (TATAAAT) lies the sequence CCCGTTA, where the boldfaced G is 26 bases downstream from the middle of the TATA box. This G is the most likely cap site, although the A preceding and the A and G following it are also candidates.

CHAPTER 11

1. The tRNAs bind together much more tightly than free trinucleotides because their anticodons are bent into a helical shape that facilitates base pairing.

2. The alanine would go where a methionine is supposed to go because the anticodon, not the amino acid, determines the response to a given codon.

3. The aminoacyl-tRNA synthetases are crucial to fidelity of translation; if they make mistakes, the translation machinery will incorporate incorrect amino acids.

4. First reaction: aminoacyl-AMP + pyrophosphate. Second reaction: aminoacyl-tRNA.

5. (a) Overlapping, 10 (AUG, UGG, GGC, etc.); nonoverlapping, 4 (AUG, GCA, GUG, CCA). (b) Overlapping, 3 (UGC, GCC, CCA); nonoverlapping, 1 (CCA).

6. (a) The reading frame would shift by two bases, and the gene would therefore not function. (b) The reading frame would not change, and the gene would therefore be functional, unless the extra three bases coded for an amino acid that disrupted the function of the gene product. (c) The reading frame would not change, and the gene would therefore be functional, unless the changes in the two adjacent codons were deleterious.

7. A repeating dipeptide. The two codons would be UUCGUU and CGUUCG, assuming that translation starts at the beginning of the message.

8. The proteins would be made with arginines where phenylalanines should have been. Most or all of these proteins would not function, because their altered primary structures would make them function abnormally.

9. It would eliminate initiation at that point. One of the AUGs in the middle of the message might then function as the start codon, but the protein product would be incomplete—truncated at the amino terminus.

10. (a) fMet-Phe-Lys-Met-Val-Thr-Trp. The next to last codon (UAA) causes termination, so the last codon (AUC) is not translated. (b) fMet-Val-Thr-Trp. Note that the base change destroyed the first start codon, so the second had to be used. (c) No change in amino acid sequence; UUU and UUC both code for Phe. (d) fMet-Leu-Lys-Met-Val-Thr-Trp. (e) fMet-Phe-Lys-Met-Val-Thr. The change converted the Trp codon to a stop signal.

11. (a) Leu, Ser, and Arg. (b) Trp and Met.

12. 256 (4^4).

13. CUU and CUC.

14. UAI.

15. Cloacin DF13 should profoundly inhibit translation, because it removes the part of the 16S rRNA that binds to Shine-Dalgarno sequences on mRNAs and therefore binds mRNA and ribosome together.

16. (*a*) Translocation; (*b*) aminoacyl-tRNA transfer to the ribosome; (*c*) peptide bond formation.

17. fMet-tRNA. This aminoacyl-tRNA must move to the P site to make room in the A site for the next aminoacyl-tRNA.

18. 402.

19. All QT proteins made in this suppressor strain would have an extra tyrosine at the carboxyl end, because the suppressor tRNA would recognize the first stop codon (UAG) in every gene as a tyrosine codon.

20. Wild-type Gln anticodon: 3′ GUU5′ (recognizes the Gln codon 5′CAA3′). Ochre suppressor anticodon: 3′AUU5′ (recognizes the ochre codon 5′UAA3′). The change was a G → A transition in the 3′-end base of the anticodon of the tRNA.

CHAPTER 12

1. A somatic mutation. If it were a germ-line mutation, the color would be uniform, or at least in a pattern of some sort. The unique spot of white was probably caused by a somatic mutation in this particular blossom that inactivated a gene responsible for the red color.

2. We know the yellow color is dominant, because the heterozygotes are yellow; the lethality of the yellow gene must be recessive, because only the homozygotes die.

3. No. Mutations in DNA polymerase III would surely be lethal, since they would block a function crucial to life: replication of DNA.

4. Conditional lethal mutations would be appropriate. Bacteria could be tested for the trait by attempting to grow them at normal temperature, but they could be maintained alive at lower temperature.

5. No, the product of a temperature-sensitive gene, not the gene itself, is temperature-sensitive.

6. After the patch of fur was shaved, the skin was relatively cool, so the temperature-sensitive, color-producing enzyme was active. This led to a patch of dark fur. The new fur warmed the skin, inactivating the color-producing enzyme, so the dark patch gradually turned white.

7. Transitions.

8. (*a*) Pyrimidine (especially thymine) dimers. (*b*) If this damage goes unrepaired, it can stall DNA replication or cause error-prone repair, which can induce missense mutations.

9. (*a*) DNA strand breaks. (*b*) Double-strand breaks are difficult to repair properly and therefore constitute lasting damage.

10. The mRNA becomes AUGGCCUAAAGAGG; the frameshift mutation has shifted the reading frame one base to the right. Now the mRNA codes for fMet-Ala. The next codon, UAA, codes for stop.

11. Add a base, preferably early in the mRNA, to compensate for the lost base. This will restore the reading frame.

12. Photoreactivating enzyme and O^6-methylguanine methyl transferase.

13. An exonuclease degrades a DNA molecule from the end, one base at a time; an endonuclease cuts within a DNA molecule.

14. A long gap in one of the strands. Long gaps do occur during repair in *E. coli,* suggesting that this bacterium uses the second, step-wise method.

15. Because the damage remains, even though DNA replication has occurred.

16. (*a*) Genes for transposase and terminal inverted repeats that the transposase recognizes. (*b*) Terminal inverted repeats that the other transposon's transposase recognizes.

17. Transposase is necessary for transposition.

18. The DNA does not always jump. In replicative transposition, it does not even leave its original location; instead, it replicates, and one of the copies goes to the new location.

19. (*a*) The inserted transposon will be surrounded by 5-bp direct repeats of host DNA. (*b*) See figure 12.29; imagine the cuts 5 bp apart, rather than 9 bp.

20. A "selfish" or "parasitic" DNA is a DNA element that replicates within a cell, but seems to provide nothing useful to the cell in return.

21. After allowing a certain time for transposition to occur, isolate the target plasmid and use it to transform cells with no inherent antibiotic resistance. Plate the transformed cells on medium containing both chloramphenicol and ampicillin. Only those cells that received a plasmid with both antibiotic resistance genes will survive; therefore, the number of survivors is related to the rate at which Tn3 transposed to the target plasmid.

22. Fewer. Homologous recombination requires some homology between the transposon and its target DNA, which automatically limits the number of target sites; illegitimate recombination requires little if any homology, so a transposon using this mechanism is free to go almost anywhere.

23. (*a*) Replicon fusion (cointegrate formation) and resolution. (*b*) A cointegrate.

24. (*a*) No product; the transposase is needed to form the cointegrate intermediate. (*b*) A cointegrate.

25. Several transposons conferring resistance to a variety of antibiotics can transpose to the same plasmid in a pathogenic bacterium. This bacterium will be resistant to all these antibiotics, and it can even share the plasmid with other dangerous bacteria, making them antibiotic-resistant. When these bacteria cause disease, they are very difficult to kill because they have built-in resistance to our drugs.

26. Two copies. At the boundaries between the two plasmids.

27. No. Tn3 transposes by replicating itself, so it would not leave the *C* gene during transposition, and reversion would not be nearly as frequent.

28. (*a*) Inhibitors of double-stranded DNA replication would block transposition of Tn3, but not Ty, because Tn3 transposition depends on this process, but the retrotransposon Ty does not. (*b*) and (*c*) Inhibitors of transcription and reverse transcription would block transposition of Ty, but not Tn3, because Ty transposition, just like retrovirus replication, requires these processes. Tn3 does not have to be transcribed, because all the gene products needed for its transposition are already present in the extract. (*d*) Inhibitors of translation will have no effect on either transposon; all proteins needed for transposition are supplied in the extract.

29. The intron would remain in the transposon.

30. The intron would be spliced out.

31. Xeroderma pigmentosum patients have defective DNA repair systems that cannot repair ultraviolet damage properly. As a result, these people respond to ultraviolet exposure by developing multiple skin cancers.

32. Chemicals, radiation, and viruses.

33. Transformation means the change in a cell's behavior from normal to tumor cell-like.

34. NIH 3T3 cells are not normal. They are already immortal and poised on the brink of transformation, needing just one final nudge to push them over. The mutant Ha-*ras* gene provides that nudge.

35. It is a somatic mutation. We know this because tumor cells have a mutant Ha-*ras* gene, while normal cells from the same patient have only the normal gene.

CHAPTER 13

1. Genotypes and phenotypes (respectively) are: (*a*) *lac⁻*, Lac⁻; (*b*) *leu⁺*, Leu⁺; (*c*) *thi⁻*, Thi⁻; (*d*) *strʳ*, Strʳ; (*e*) *purˢ*, Purˢ; (*f*) *bio⁺*, Bio⁺.

2. Prototrophs: (*a*), (*b*), (*d*), (*e*), and (*f*). Auxotroph: (*c*). This is the only one that cannot make a substance needed for growth on minimal medium containing glucose.

3. Yes, he was justified. If single mutants had been used, their reversion rate (one in a million) would have produced ten times as many prototrophs as their recombination rate (one in ten million). Therefore, the investigators would have observed almost as many prototrophs with the two mutant strains grown separately as when they were grown together; they would not have known whether they were observing recombination plus reversion, or just reversion.

4. It usually does not contain a full set of genes from the donor cell, so it is not a true diploid. We call it a merozygote or a merodiploid.

5. The Tonʳ phenotype is caused by a mutation. If it were due to an adaptation, all cultures would be about equal in resistance to the phage. (These were the actual results obtained by Luria and Delbrück in 1943.)

6. Integration of the F plasmid into the host genome. This mobilizes the host genome so that it behaves as the F plasmid normally would and transfers the chromosome (usually only in part) to a recipient cell.

7. Usually, the connection between conjugating cells breaks before the whole chromosome has a chance to transfer.

8.

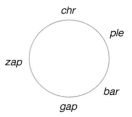

9. Conjugation is usually interrupted, just as it is in earthly bacteria. This lowers the likelihood of transfer of genes near the end of the chromosome; conjugation usually ceases before such transfer can take place.

10. 25 minutes.

11. *gap-bar-ple-chr-zap*

12. Place one mutant *zap* gene on an F-*zap* plasmid (by recombination) and conjugate this mutant with the other *zap⁻* mutant. This places both mutant genes in the same cell—one in the chromosome and the other on the F′ plasmid. If the mutations are in different genes, they will complement and give rise to a large number of Zap⁺ cells. If the mutations are in the same gene, very few Zap⁺ cells will arise.

13. They could be due to recombination between two DNAs with mutations in the same gene (either *zapA* or *zapB*). Alternatively, they could be due to intracistronic complementation, such as that observed in the *E. coli lacZ* gene.

14. λ phage heads can only accommodate about 10% extra DNA. Thus, the λ DNA will be able to pick up host genes from one side (*gal*) or the other (*bio*) of the prophage, but not from both.

15. Yes, they would cotransduce, because they are less than 2.4% of the *E. coli* genome apart, and this is the amount of DNA that can fit into a P1 phage head. Using the expression: frequency $= (1-d/L)^3$, we would predict a cotransduction frequency of 2%, not too far from the actual frequency of 1%.

16. First, infect a Zap⁺ Zip⁻ strain with Q2 and collect the progeny phage. Use these to infect a Zap⁻ Zip⁻ strain and screen for Zap⁺ transductants. Since these have received the *zap* gene, chances are good that they have also received the closely linked *zip* gene. Therefore, the Zap⁺ transductants can be screened for Zip⁺ transductants. The farther apart the mutations in the two damaged *zip* genes, the higher the frequency of these Zip⁺ transductants.

17. *zip zap bop*. If the order were *bop zip zap*, Bop⁺ Zip⁻ Zap⁻ transductants would be less common than Bop⁺ Zip⁺ Zap⁺.

18. The marker order is *awk nrd kat*. The cotransduction frequency between *awk* and *kat* is $(1 + 115)/1,063 = 0.11$. The cotransduction frequency between *awk* and *nrd* is $(309 + 115)/1,063 = 0.40$.

19. (250 plaques/0.05 ml) $\times$ 10^6 dilution factor $= 5 \times 10^9$ pfu/ml.

20. The curve will start to rise sooner and the rise will be more gradual. This is because infectious phages appear in infected cells before lysis occurs. These will be detected by breaking the cells open.

21. K M L (or L M K).

CHAPTER 14

1. The red blood cell itself can live without hemoglobin, but the whole organism would die without this product to carry oxygen to all its tissues.

2. (*a*) Determined only; (*b*) determined and differentiated.

3. After two cell divisions, there are only four cells, enough for identical quadruplets if the cells separate instead of staying together in one embryo, but not enough to produce quintuplets. One more cell division, or a total of three, would produce eight blastomeres. This is more than enough for quintuplets, but the cells must all be totipotent. If they have lost any of their genetic potential, they cannot give rise to a whole person.

4. You would observe neither DNase hypersensitivity nor activity. This is because the histones, which you added first, would bind throughout the gene and its 5′-flanking region, tying it up in nucleosomes. The factors in the fourteen-day extract would not be expected to remove these nucleosomes.

5. With a cloned gene, deletions of precisely defined length and position can be introduced, and the altered genes can then be tested for light sensitivity. Without a cloned gene, the deletions would have to be introduced by hit-and-miss genetic means, and their locations and extents would not be as well defined.

6. Expression of the gene with the mutated 5′-flanking region must be distinguishable from the unmutated gene. If the mutated pea rubisco gene were reintroduced into a pea plant, its expression would not be as easily distinguishable from expression of the resident pea rubisco gene.

7. The gene will be active because no histone H1 is present to cross-link the nucleosomes.

8. Early embryonic cells divide rapidly, thus they must produce a great deal of protein. 5S rRNA is one of the components of the ribosomes that must make this protein.

9. (*a*) The vitellogenin gene (ovalbumin is not produced in the liver); (*b*) neither.

10. Homeo boxes are found in genes that determine the body plan of *Drosophila*. Since the products of these genes probably influence the activity of several other genes, it is likely that the products of the homeo box genes in *Drosophila* bind to these other genes' control regions. This probably explains the importance of the DNA-binding domain.

11. A radioactive probe hybridizes to a given mRNA in a thin slice of an embryo. Wherever the gene for the mRNA is expressed, that mRNA will be available to hybridize to the probe. Hybridization is then detected by autoradiography.

12. When the fused *ftz-lacZ* DNA was injected into an embryo, β-galactosidase was expressed in a banded pattern, detected by cleavage of a colorless material to a blue dye by the enzyme. Since the only remaining element in the fused gene that was of *Drosophila* origin was the *ftz* promoter, it was very likely that the banded pattern of expression was due to a banded pattern of factors that could act on that promoter, namely transcription factors.

13. Mutants in these genes experience abnormal development.

CHAPTER 15

1. XgiI.

2. GATATC

3. Every $4^5 = 1,024$ bases.

4. BamHI and BglII. 5′-GATC-3′.

5. No; after ligating the two DNAs together, the site (GGATCT) is a hybrid—BamHI-like on one end and BglII-like on the other. Therefore, neither enzyme will recognize this site.

6. You would see a uniform lawn of bacteria instead of individual colonies. This is because *all* cells will grow, not just those containing plasmids with antibiotic resistance genes.

7. Tetracycline. Inserts in the BamHI site of pBR322 will inactivate the tetracycline gene.

8. No. The EcoRI site of pBR322 lies outside both the ampicillin and tetracycline resistance genes.

9. No. The pUC plasmids selectively accept small inserts. Therefore, it would take millions of pUC clones to build a complete genomic library from a complex genome.

10. Yes. The pUC plasmids are very convenient, and since the total amount of DNA we need to clone is not large, we do not have to worry about the fact that each clone will only have a limited amount of DNA.

11. Since the tobacco probe and the pea gene will almost certainly have different sequences, you should use a lower temperature (lower stringency) so that a stable hybrid can form between the two DNAs of similar, but not identical sequence.

12. (a) Tyr-Met-Cys-Trp-Ile. (Only the first two bases of the Ile codon would be used.) (b) $2 \times 1 \times 2 \times 1 \times 1 = 4$. (c) $6 \times 2 \times 1 \times 2 \times 1 = 24$.

13. When the vector and insert have tails long enough to form stable hybrids (approximately 12 bases). Then the ligation step can be completed within the host cell after transformation.

14. It prevents premature synthesis of the protein product of the cloned gene, which may be harmful to the host cell.

15. One in six; one-half will be in the right orientation and one-third of these will be in the right reading frame, since there are three possible reading frames; ($\frac{1}{2} \times \frac{1}{3} = \frac{1}{6}$).

16. T h r A s p S e r V a l G l y
A C C G A T T C A G T G G G C

Boldface type denotes the base change that alters the Ala codon to a Ser codon.

17. Cut the viral DNA with EcoRI, electrophorese the resulting five fragments, and Southern blot the fragments. Hybridize the Southern blot to a radioactive probe made from the cloned cDNA, and autoradiograph to see which of the five DNA fragments "light(s) up."

18. Collect mRNAs from all five tissues, electrophorese them, then Northern blot. Hybridize a radioactive probe (made from the cloned petunia gene) to the Northern blot and autoradiograph. The tissue that contained the highest concentration of the specific mRNA will hybridize best and will give the most intense band on the autoradiogram.

19. The size(s) of the mRNA(s) made from this gene.

20. 10.7 kb.

21. 8($2 \times 2 \times 2$).

22. Yes. Absence of this site would result in a longer fragment, but it would still be detected by the probe, because it would overlap the probe region by just as much.

23. No. The probe would not be able to hybridize to the right-hand fragment resulting from the presence of this site. Therefore, this fragment would not show up on the autoradiograph.

24. ATGAGTTGGCCAGCAGAGCGAGCATGGATGTAA

CHAPTER 16

1.

	p	q	p²	2pq	q²
a	0.4	0.6	0.16	0.48	0.36
b	0.68	0.32	0.462	0.435	0.102
c	0.467	0.533	0.218	0.498	0.284

	p²N	2pqN	q²N	χ²
a	16	48	36	2.78
b	23.1	21.8	5.1	20.1
c	62.1	141.9	80.9	5.63

With 1 degree of freedom, a is not significant, b is significant at the 0.01 level, and c is significant at the 0.05 level.

2. The expected numbers are 18, 84, and 98 for genotypes A_1A_1, A_1A_2, and A_2A_2. 0.45—with 1 degree of freedom, it is consistent.

3. A_1A_1, 0.49; A_2A_2, 0.09.

4. 0.01; 0.18.

5.

	P₁	P₂	P₃	χ²
a	0.3	0.3	0.35	1.87
b	0.2	0.2	0.6	400

With 3 degrees of freedom, a is not significant and b is significant at the 0.01 level.

6. 0.0064; 0.147.

7. 0.12, 0.16, 0.72; 0.01.

8. 0.408, 0.384, 0.208; f = 0.798.

9. (a)

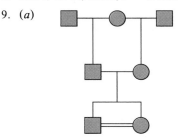

Inbreeding coefficient = 0.03125

(b)

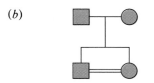

Inbreeding coefficient = 0.25

10. 10^{-6}, 7.25×10^{-7}, 4.5×10^{-7}, 1.75×10^{-7}, 10^{-7}. Δq is largest when $p = 1$ and $q = 0$ because all alleles are wild-type.

11. 0.3; 0.7. The probability of fixation is lower for the allele in lowest frequency because it is less likely to increase all the way to 1.0.

12. 0.0012, 0.035. The expected is much larger.

13. (a)

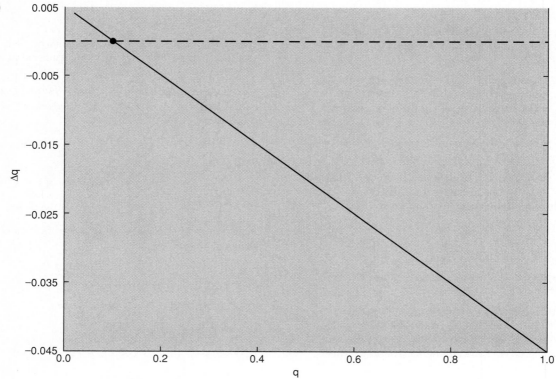

(b) 0.1.

14. 0.13. We assume that there has been no selection or genetic drift and that the allelic frequency estimates are correct.

15. (*a*) 0.0, −0.019, −0.056, −0.073, 0.0.

(*b*)

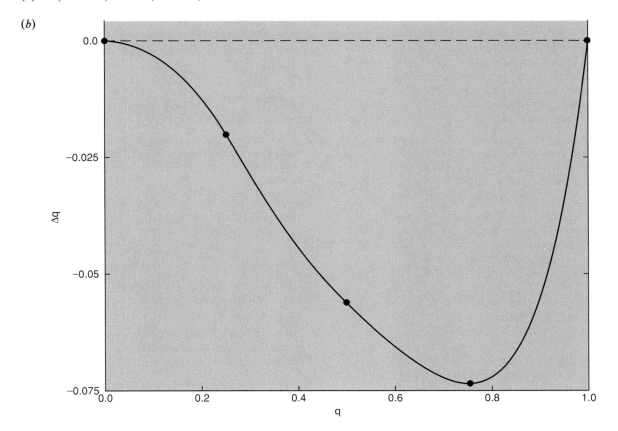

(*c*) 0.1968, 0.1937.

16. Melanic 1, typical 0.771; 0.4.

17. 0.2. 0.011, −0.2. There is a stable equilibrium at q = 0.2.

18. In general, selection and gene flow have the largest effects on Δq, and mutation has the smallest. Genetic drift can have a large effect in small populations. Genetic drift can either increase or decrease allelic frequency, while the other factors in particular situations will only change allelic frequencies in one direction.

19. Inbreeding increases the frequency of homozygotes and decreases the frequency of heterozygotes. Examine genotypic frequencies compared to Hardy-Weinberg expectations. Inbreeding in concert with selection may fix populations for favorable homozygotes.

20. Snails could be placed on different backgrounds to determine the rate of predation. Look at changes in allelic frequency from generation to generation to see if they vary more in small populations than in large.

CHAPTER 17

1. 0.0000167, 0.0041; 0.0000333, 0.0058.

2. 0.000001; 0.00002.

3. Malaria increased the frequency to equilibrium in Africa; gene flow and selection are reducing the frequency today. Mutation may be a limiting factor.

4. $u = sq^2 = 0.0004$. This value is much higher than known mutation rates, which are about 10^{-5} to 10^{-7}. Genetic drift, heterozygote advantage, or past selection have been offered as explanations.

5. Neutrality considers genetic variation a function of mutation and genetic drift, predicting alleles to be transient and fewer in small populations. Selection predicts constant alleles and more variation in small populations.

6. The molecular clock assumes a regular replacement of molecular variants over time. The rate of amino acid substitution is the change in amino acid sequence per unit time. Nucleotide substitution would be faster because there are fewer constraints and the third position changes fast.

7. 0.444, 0.25. There are more physical constraints on the second protein.

8. Species a and c. Species d.

9. If we assume that the molecular information gives us the true relationship, we can examine morphology and physiology, knowing these patterns and the time since divergence.

10. Selection is haploid, no dominance. Genetic drift is more likely because of only one copy versus four for diploid (male and female).

11. 0.00124; 0.03125. The second is larger because more recessive homozygotes are exposed to selection.

12. 0.478, 0.475; 0.456, 0.433. The reduction in heterozygosity is very low for the larger population and not too great for the smaller one.

13. Look at morphological, economic, and disease and pest resistance traits. Grow separately in their natural habitats, avoiding unnatural selection if possible.

14. 0.000312; 0.000064. Nearly five times as many progeny from half-first cousins are affected.

15. Avoid inbreeding, keep a large population, and avoid unnatural selection.

16. 0.337. Gene flow or mutation have not introduced the allele, and/or selection for resistance has not occurred. (See chapter 16.)

17. 0.81; 0.518. (See chapter 16.)

18. Mosquitoes have a high reproductive capacity, short generation length, and high dispersal. Furthermore, they adapt to DDT in several different ways.

19. All the new crops may be identical, resulting in loss of genetic variation and susceptibility to particular diseases.

CHAPTER 1

Creighton, H. B., and B. McClintock. 1931. A correlation of cytological and genetical crossing-over in *Zea mays. Proceedings of the National Academy of Sciences* 17:492–97.

Mirsky, A. E. 1968. The discovery of DNA. *Scientific American* 218(June):78–88.

Morgan, T. H. 1910. Sex-limited inheritance in *Drosophila. Science* 32:120–22.

Sturtevant, A. H. 1913. The linear arrangement of six sex-linked factors in *Drosophila,* as shown by their mode of association. *Journal of Experimental Zoology* 14:43–59.

CHAPTER 2

Olby, R. C. 1966. *Origins of Mendelism.* London: Constable.

Peters, J. A. 1962. *Classic papers in genetics.* Englewood Cliffs, NJ: Prentice-Hall.

Stern, C., and E. R. Sherwood. 1966. *The origin of genetics: A Mendel source book.* San Francisco: W. H. Freeman and Co.

Todd, N. B. 1977. Cats and commerce. *Scientific American* 237 (November):100–107.

CHAPTER 3

Bodmer, W. F., and L. L. Cavalli-Sforza. 1970. Intelligence and race. *Scientific American* 223(October):19–29.

Hedrick, P. W. 1984. *Population biology.* Boston: Jones and Bartlett Publishers.

Nakamura, Y., M. Leppert, P. O'Connell, R. Wolff, T. Holm, M. Culver, C. Martin, E. Fujimoto, M. Hoff, E. Kumlin, and R. White. 1987. Variable number of tandem repeat (VNTR) markers for human gene mapping. *Science* 235:1616–22.

Singer, S. 1978. *Human genetics.* San Francisco: W. H. Freeman and Co.

Swanson, C. P., T. Merz, and W. J. Young. 1981. *Cytogenetics: The chromosome in division, inheritance, and evolution.* Englewood Cliffs, N.J.: Prentice-Hall.

CHAPTER 4

Bridges, C. 1931. Nondisjunction as proof of the chromosome theory of heredity. *Genetics* 1:1–52; 107–63.

Hasold, T. J., and P. A. Jacobs. 1984. Trisomy in man. *Annual Review of Genetics* 18:69–98.

Selected
R·E·A·D·I·N·G·S

McLeish, J., and B. Snoad. 1958. *Looking at chromosomes.* New York: Macmillan.

Von Wettstein, D., et al. 1984. The synaptonemal complex in genetic segregation. *Annual Review of Genetics* 18:331–414.

CHAPTER 5

Finchman, J. R. S., P. R. Day, and A. Radford. 1979. *Fungal genetics.* London: Blackwell.

Ott, J. 1985. *Analysis of human genetic linkage.* Baltimore: Johns Hopkins University Press.

Ruddle, F. H. 1974. Hybrid cells and human genes. *Scientific American* 231(July):36–44.

CHAPTER 6

Adams, R. L. P., R. H. Burdon, A. M. Campbell, and R. M. S. Smellie, eds. 1976. *Davidson's The Biochemistry of the Nucleic Acids,* eighth edition. The structure of DNA, chapter 5. New York: Academic Press.

Avery, O. T., C. M. McLeod, and M. McCarty. 1944. Studies on the chemical nature of the substance-inducing transformation of pneumococcal types. *Journal of Experimental Medicine* 79:137–58.

Chargaff, E. 1950. Chemical specificity of nucleic acids and their enzymatic degradation. *Experientia* 6:201–209.

Dickerson, R. E. 1983. The DNA helix and how it is read. *Scientific American* 249(December):94–111.

Hershey, A. D., and M. Chase. 1952. Independent functions of viral protein and nucleic acid in growth of bacteriophage. *Journal of General Physiology* 36:39–56.

Watson, J. D., and F. H. C. Crick. 1953. Molecular structure of nucleic acids: A structure for deoxyribose nucleic acid. *Nature* 171:737–8.

————. 1953. Genetical implications of the structure of deoxyribonucleic acid. *Nature* 171:964–67.

CHAPTER 7

Beadle, G. W., and E. L. Tatum. 1941. Genetic control of biochemical reactions in *Neurospora. Proceedings of the National Academy of Sciences* 27:499–506.

Brenner, S., F. Jacob, and M. Meselson. 1961. An unstable intermediate carrying information from genes to ribosomes for protein synthesis. *Nature* 190:576–81.

Conceptual foundations of genetics: Selected readings. 1976. Edited by H. O. Corwin and J. B. Jenkins. Boston: Houghton Mifflin Co.

Crick, F. H. C. 1958. On protein synthesis. *Symposium of the Society for Experimental Biology* 12:138–63.

Meselson, M., and F. W. Stahl. 1958. The replication of DNA in *Escherichia coli. Proceedings of the National Academy of Sciences* 44:671–82.

Yanofsky, C., B. C. Carlton, J. R. Guest, D. R. Helinski, and U. Henning. 1964. On the colinearity of gene structure and protein structure. *Proceedings of the National Academy of Sciences* 51:266–72.

Yanofsky, C. 1967. Gene structure and protein structure. *Scientific American* 216(May):80–94.

CHAPTER 8

Cairns, J. 1963. The chromosome of *Escherichia coli. Cold Spring Harbor Symposia on Quantitative Biology* 28:43–46.

Callan, H. G. 1973. DNA replication in the chromosomes of eukaryotes. *Cold Spring Harbor Symposia on Quantitative Biology* 38:195–203.

Cox, M. M., and I. R. Lehman. 1987. Enzymes of general recombination. *Annual Review of Biochemistry* 56:229–62.

Dressler, D., and H. Potter. 1982. Molecular mechanism of genetic recombination. *Annual Review of Biochemistry* 51:727–62.

Kornberg, A. 1980. DNA replication. San Francisco: W. H. Freeman.

Kornberg, A. 1982. *Supplement to DNA replication.* San Francisco: W. H. Freeman.

McHenry, C. S. 1988. DNA polymerase III holoenzyme of *Escherichia coli. Annual Review of Biochemistry* 57:519–50.

Radman, M., and R. Wagner. 1988. The high fidelity of DNA replication. *Scientific American* 259(August):24–30.

Smith, G. R. 1987. Mechanism and control of homologous recombination in *Escherichia coli. Annual Review of Genetics* 21:179–201.

Wang, J. C. 1982. DNA topoisomerases. *Scientific American* 247(July):94–109.

CHAPTER 9

Beckwith, J. R., and D. Zipser, eds. 1970. The lactose operon. Cold Spring Harbor Laboratory.

Friedman, D. I., M. J. Imperiale, and S. L. Adhya. 1987. RNA 3′-end formation in the control of gene expression. *Annual Review of Genetics* 21:453–88.

Helmann, J. D., and M. J. Chamberlin. 1988. Structure and function of bacterial sigma factors. *Annual Review of Biochemistry* 57:839–72.

Hendrix, R. W., J. W. Roberts, F. W. Stahl, and R. A. Weisberg, eds. 1983. Lambda II. Cold Spring Harbor Laboratory.

Jacob, F., and J. Monod. 1961. Genetic regulatory mechanisms in the synthesis of proteins. *Journal of Molecular Biology* 3:318–56.

Jacob, F. 1966. Genetics of the bacterial cell (Nobel lecture). *Science* 152:1470–78.

Losick, R., and M. Chamberlin. 1976. RNA polymerase. Cold Spring Harbor Laboratory.

Miller, J. H., and W. S. Reznikoff, eds. 1978. The operon. Cold Spring Harbor Laboratory.

Monod, J. 1966. From enzymatic adaptation to allosteric transitions (Nobel lecture). *Science* 154:475–83.

Ptashne, M., and W. Gilbert. 1970. Genetic repressors. *Scientific American* 222(June):36–44.

Ptashne, M. 1989. How gene activators work. *Scientific American* 260(January):24–31.

Schleif, R. 1987. DNA binding proteins. *Science* 241:1182–87.

Zubay, G., D. Schwartz, and J. Beckwith. 1970. Mechanism of activation of catabolite-sensitive genes: A positive control system. *Proceedings of the National Academy of Sciences* 66:104–10.

CHAPTER 10

Burlingame, R. W., W. E. Love, B.-C. Wang, R. Hamlin, N.-H. Xuong, and E. N. Moudrianskis. 1985. Crystallographic structure of the octameric histone core of the nucleosome at a resolution of 3.3A. *Science* 228:546–53.

Chambon, P. 1981. Split genes. *Scientific American* 244(May):60–71.

Friedman, D. E., M. J. Imperiale, and S. L. Adhya. 1987. RNA 3'-end formation in the control of gene expression. *Annual Review of Genetics* 21:453–88.

Geiduschek, E. P., and G. P. Tocchini-Valentini. 1988. Transcription by RNA polymerase III. *Annual Review of Biochemistry* 57:873–914.

Kornberg, R. D., and A. Klug. 1981. The nucleosome. *Scientific American* 244(February):52–64.

McKnight, S., and R. Tjian. 1986. Transcriptional selectivity of viral genes in mammalian cells. *Cell* 46:795–805.

Miller, O. L. 1973. The visualization of genes in action. *Scientific American* 228(March):34–42.

Ptashne, M. 1989. How gene activators work. *Scientific American* 260(January):24–31.

Steitz, J. A. 1988. "Snurps." *Scientific American* 258(June):36–41.

CHAPTER 11

Bjork, G. R., et al. 1987. Transfer RNA modification. *Annual Review of Biochemistry* 56:263–87.

Crick, F. H. C., L. Barnett, S. Brenner, and R. J. Watts-Tobin. 1962. General nature of the genetic code for proteins. *Nature* 192:1227–32.

Crick, F. H. C. 1966. The genetic code. *Scientific American* 215(October):55–62.

Fox, T. D. 1987. Natural variation in the genetic code. *Annual Review of Genetics* 21:67–91.

Khorana. H. G. 1968. Synthesis in the study of nucleic acids. *Biochemical Journal* 109:709–25.

Lake, J. A. 1981. The ribosome. *Scientific American* 245(August):84–97.

Leder, P., and M. Nirenberg. 1964. RNA codewords and protein synthesis, II. Nucleotide sequence of a valine RNA codeword. *Proceedings of the National Academy of Sciences* 52:420–27.

Moore, P. B. 1988. The ribosome returns. *Nature* 331:223–27.

Nomura, M. 1984. The control of ribosome synthesis. *Scientific American* 250(January):102–14.

Quigley, G. J., and A. Rich. 1976. Structural domains of transfer RNA molecules. *Science* 194:796–806.

Rich, A., and S. H. Kim. 1978. The three-dimensional structure of transfer RNA. *Scientific American* 238(January):52–62.

CHAPTER 12

Ames, B. N., W. E. Durston, E. Yamasaki, and F. D. Lee. 1973. Carcinogens are mutagens: A simple test system combining liver homogenates for activation and bacteria for detection. *Proceedings of the National Academy of Sciences* 72:2423–27.

Baltimore, D. 1985. Retroviruses and retrotransposons: The role of reverse transcription in shaping the eukaryotic genome. *Cell* 40:481–82.

Barbacid, M. 1987. *ras* genes. *Annual Review of Biochemistry* 56:779–827.

Brenner, S., L. Barnett, F. H. C. Crick, and A. Orgel. 1961. The theory of mutagenesis. *Journal of Molecular Biology* 3:121–24.

Cohen, S. N., and J. A. Shapiro. 1980. Transposable genetic elements. *Scientific American* 242(February):40–49.

Doering, H.-P., and P. Starlinger. 1984. Barbara McClintock's controlling elements: Now at the DNA level. *Cell* 39:253–59.

Drake, J. W. 1970. *The molecular basis of mutation.* San Francisco: Holden-Day.

Engels, W. R. 1983. The P family of transposable elements in *Drosophila. Annual Review of Genetics* 17:315–44.

Fedoroff, N. V. 1984. Transposable genetic elements in maize. *Scientific American* 250(June):84–99.

Garfinkel, D. J., J. D. Boeke, and G. R. Fink. 1985. Ty element transposition: Reverse transcriptase and virus-like particles. *Cell* 42:507–17.

Little, J. W. 1982. The SOS regulatory system of *Escherichia coli. Cell* 29:11–12.

Loeb, L. A. 1985. Apurinic sites as mutagenic intermediates. *Cell* 40:483–84.

Miller, J. H. 1983. Mutational specificity in bacteria. *Annual Review of Genetics* 17:215–38.

Mizuuchi, K., and R. Craigie. 1986. Mechanism of bacteriophage Mu transposition. *Annual Review of Genetics* 20:385–429.

Radman, M., and R. Wagner. 1986. Mismatch repair in *Escherichia coli. Annual Review of Genetics* 20:523–38.

Sancar, A., and G. B. Sancar. 1988. DNA repair enzymes. *Annual Review of Biochemistry* 57:29–67.

Shapiro, J. A. 1983. *Mobile genetic elements.* New York: Academic Press.

Suzuki, D. T. 1970. Temperature-sensitive mutations in *Drosophila melanogaster. Science* 170:695–706.

Walker, G. C. 1985. Inducible DNA repair systems. *Annual Review of Biochemistry* 54:425–58.

Weinberg, R. A. 1988. Finding the anti-oncogene. *Scientific American* 259(September):34–41.

Wessler, S. R. 1988. Phenotypic diversity mediated by the maize transposable elements *Ac* and *Spm. Science* 242:399–405.

CHAPTER 13

Benzer, S. 1961. Genetic fine structure. In *Harvey Lectures,* Series 56. New York: Academic Press.

Hayes, W. 1968. *The genetics of bacteria and their viruses,* second edition. New York: Wiley.

Ippen-Ihler, K. A., and E. G. Minkley. 1986. The conjugation system of F, the fertility factor of *Escherichia coli. Annual Review of Genetics* 20:593–624.

Jacob, F., and E. L. Wollman. 1961. *Sexuality and the genetics of bacteria.* New York: Academic Press.

Morse, M. L., E. M. Lederberg, and J. Lederberg. 1956. Transduction in *Escherichia coli* K–12. *Genetics* 41:142–56.

Stent, G. S., and R. Calendar. 1978. *Molecular genetics: An introductory narrative,* chapters 9–13. San Francisco: W. H. Freeman and Co.

Wollman, E. L., F. Jacob, and W. Hayes. 1956. Conjugation and genetic recombination in *Escherichia coli* K–12. *Cold Spring Harbor Symposia on Quantitative Biology* 21:141–62.

CHAPTER 14

Borst, P., and D. R. Greaves. 1987. Programmed gene rearrangements altering gene expression. *Science* 235:658–67.

Brown, D. D. 1981. Gene expression in eukaryotes. *Science* 211:667–74.

Coulter, D., and E. Wieschaus. 1986. Segmentation genes and the distributions of transcripts. *Nature* 321:472–74.

DeRobertis, E. M., and J. B. Gurdon. 1979. Gene transplantation and the analysis of development. *Scientific American* 241(December):74–82.

Gehring, W. 1985. The homeo box: A key to the understanding of development? *Cell* 40:3–5.

———. 1985. The molecular basis of development. *Scientific American* 253(October):152B–162.

Gehring, W. J., and Y. Hiromi. 1986. Homeotic genes and the homeobox. *Annual Review of Genetics* 20:147–73.

Gurdon, J. B. 1968. Transplanted nuclei and cellular differentiation. *Scientific American* 219(December):24–35.

Holliday, R. 1987. The inheritance of epigenetic defects. *Science* 238:163–70.

———. 1989. A different kind of inheritance. *Scientific American* 260(June):40–48.

Levine, M., and T. Hoey. 1988. Homeobox proteins as sequence-specific transcription factors. *Cell* 55:537–40.

Moses, P. B., and N.-H. Chua. 1988. Light switches for plant genes. *Scientific American* 258(April):64–69.

Tonegawa, S. 1985. The molecules of the immune system. *Scientific American* 253(October):122–31.

Wolffe, A. P., and D. D. Brown. 1988. Developmental regulation of two 5S ribosomal RNA genes. *Science* 241:1626–32.

CHAPTER 15

Chilton, M.-D. 1983. A vector for introducing new genes into plants. *Scientific American* 248(June):50–59.

Cohen, S. 1975. The manipulation of genes. *Scientific American* 233(July):24–33.

Cohen, S., A. Chang, H. Boyer, and R. Helling. 1973. Construction of biologically functional bacterial plasmids *in vitro*. *Proceedings of the National Academy of Sciences* 70:3240–44.

Gilbert, W., and L. Villa-Komaroff. 1980. Useful proteins from recombinant bacteria. *Scientific American* 242(April):74–94.

Gusella, J. F., et al. 1983. A polymorphic DNA marker genetically linked to Huntington's disease. *Nature* 306:234–38.

Landegren, U., R. Kaiser, C. T. Caskey, and L. Hood. 1988. DNA diagnostics: Molecular techniques and automation. *Science* 242:229–37.

Nathans, D., and H. O. Smith. 1975. Restriction endonucleases in the analysis and restructuring of DNA molecules. *Annual Review of Biochemistry* 44:273–93.

Neufeld, P. J., and N. Colman. 1990. When science takes the witness stand. *Scientific American* 262(May):18–25.

Sanger, F., G. M. Air, B. G. Barrell, N. L. Brown, A. R. Coulson, J. C. Fiddes, C. A. Hutchison, P. M. Slocombe, and M. Smith. 1977. Nucleotide sequence of the ϕX174 DNA. *Nature* 265:687–95.

Watson, J. D., J. Tooze, and D. T. Kurtz. 1983. *Recombinant DNA: A Short Course.* New York: W. H. Freeman.

White, R., M. Leppert, D. T. Bishop, D. Barker, J. Berkowitz, C. Brown, P. Callahan, T. Holm, and L. Jerominski. 1985. Construction of linkage maps with DNA markers for human chromosomes. *Nature* 313:101–104.

White, R., and J.-M. Lalouel. 1988. Chromosome mapping with DNA markers. *Scientific American* 258(February):20–29.

CHAPTER 16

Falconer, D. S. 1981. *Introduction to quantitative genetics,* second edition. New York: Ronald Press.

Hedrick, P. W. 1983. *Genetics of populations.* Boston: Jones and Bartlett Publishers.

CHAPTER 17

Beadle, G. W. 1980. The ancestry of corn. *Scientific American* 242(January):112–19.

Clarke, B. 1975. The causes of biological diversity. *Scientific American* 233(August):50–60.

Hedrick, P. W. 1984. *Population biology.* Boston: Jones and Bartlett Publishers.

O'Brien, S. J., D. E. Widt, and Mitchell Bush. 1986. The cheetah in genetic peril. *Scientific American* 254(May):84–92.

Rick, C. M. 1978. The tomato. *Scientific American* 239(August):76–87.

A

Ac A transposable element found in corn, capable of transposing itself or directing the transposition of a defective *Ds* element.

acceptor stem The part of a tRNA involving the 5′- and 3′-ends base-paired together, which binds, or accepts, an amino acid.

acrocentric Chromosome with the centromere toward one end.

adenine (A) The purine base that pairs with thymine in DNA.

adenosine A nucleoside containing the base adenine.

adenovirus A DNA tumor virus in rodents, having a linear genome of moderate size.

admixture The amount of gene flow that has occurred from one parental population into a hybrid population.

Agrobacterium tumefaciens A bacterium that invades plants and causes crown gall tumors.

alcaptonuria A genetic disease characterized by an abnormal buildup in the urine of homogentisic acid, an intermediate in breakdown of the amino acid phenylalanine. When exposed to air, the urine turns black.

alkylation Addition of a carbon-containing alkyl group—to DNA, for example.

allele A particular form of a gene, presumably reflecting a certain DNA sequence, such as A_1 or A_2.

allelic frequency The proportion of alleles in the population that are of a given type.

allolactose The natural inducer of the *lac* operon.

allosteric protein A protein that changes shape and function when it binds to a small molecule.

Alu sequence A sequence about 300 bp long that is repeated about 300,000 times in the human genome and contains the recognition site (AGCT) for the restriction enzyme AluI.

amber codon UAG, coding for termination.

amber mutation *See* nonsense mutation.

amber suppressor A tRNA bearing an anticodon that can recognize the amber codon (UAG) and thereby suppress amber mutations.

Ames test A test for mutagens developed by Bruce Ames, in which the rate of reversion of auxotrophic bacterial strains *(Salmonella)* to prototrophy upon exposure to a chemical is taken as a measure of the chemical's mutagenicity.

amino acid The building block of proteins.

aminoacyl-tRNA synthetase The enzyme that links a tRNA to its cognate amino acid.

amino terminus The end of a polypeptide with a free amino group.

amplification Selective replication of a gene so that it is represented more than the normal once in a haploid genome.

anaphase A short stage of nuclear division in which the chromosomes move to the poles.

aneuploid Not having the normal diploid number of chromosomes.

annealing of DNA The process of bringing back together the two separate strands of denatured DNA.

Antennapedia (Antp) **mutants** Mutants of *Drosophila* in which a leg appears where an antenna ought to be.

Antennapedia **complex (ANT-C)** A large genetic locus in *Drosophila* that contains several homeotic genes, including *Antennapedia*.

anticoding strand The complement of the coding strand of a gene. Serves as template for an mRNA and therefore has polarity opposite that of the mRNA.

anticodon A three-base sequence in a tRNA that base-pairs with a specific codon.

antiparallel The relative polarities of the two strands in a DNA double helix; if one strand goes 5′–3′, top to bottom, the other goes 3′–5′. The same antiparallel relationship applies to any double-stranded polynucleotide or oligonucleotide, including the RNAs in a codon-anticodon pair.

antiterminator A protein, such as the λ *N* gene product, that overrides a terminator and allows transcription to continue.

apurinic site (AP site) A deoxyribose in a DNA strand that has lost its purine base.

assembly map A scheme showing the order of addition of ribosomal proteins during self-assembly of a ribosomal particle in vitro.

assortative mating Mating that is nonrandom with respect to the phenotypes in a population—that is, where the probability of individuals of two given phenotypes mating is not equal to the product of their frequencies in the population.

asymmetric transcription Transcription of only one strand of a given region of a double-stranded polynucleotide.

ATPase An enzyme that cleaves ATP, releasing energy for other cellular activities.

att **sites** Sites on phage and host DNA where recombination occurs, allowing integration of the phage DNA into the host genome as a prophage.

A-type particles Intracellular, noninfectious particles that resemble retroviruses. May be vehicles for transposition of retrotransposons.

autoradiography A technique in which a radioactive sample is allowed to expose a photographic emulsion, thus "taking a picture of itself."

autosome Any chromosome except the sex chromosomes.

auxotroph An organism that requires one or more substances in addition to minimal medium.

B

backcross A cross between an F₁ and an individual from one of the parental lines.

back-mutation *See* reversion.

bacteriophage *See* phage.

Barr body Structure formed when the inactivated X chromosome is heterochromatinized.

base A cyclic, nitrogen-containing compound linked to deoxyribose in DNA and to ribose in RNA.

base pair A pair of bases (A–T or G–C), one in each strand, that occur opposite one another in a double-stranded DNA or RNA.

B-DNA The standard Watson-Crick model of DNA favored at high relative humidity and in solution.

β-galactosidase An enzyme that breaks the bond between the two constituent sugars of lactose.

binomial formula The probability that in a group of a given size, a certain number of individuals will be one type and a certain number will be another type.

Bithorax **complex (BX-C)** A large genetic locus in *Drosophila* containing several homeotic genes that control segmentation in the fly's thorax and abdomen.

bivalent The structure formed when two homologous chromosomes are paired in meiosis I.

blastomere An early embryonic cell formed by cleavage of the egg.

blastula The hollow ball stage of animal embryogenesis.

branch migration Lateral motion of the branch of a chi structure during recombination.

bromodeoxyuridine (BrdU) An analogue of thymidine that can be incorporated into DNA in place of thymidine, and can then cause mutations by forming an alternate tautomer that base-pairs with guanine instead of adenine.

burst size The difference between the final and initial phage concentrations in a cell; the number of progeny phage produced by a single cell.

C

cap A methylated guanosine bound through a 5′–5′ triphosphate linkage to the 5′-end of an mRNA or an snRNA.

cap binding protein (CBP) A protein that associates with the cap on a eukaryotic mRNA and allows the mRNA to bind to a ribosome.

CAP (catabolite activator protein) A protein that, together with cAMP, activates operons that are subject to catabolite repression.

carboxyl terminus The end of a polypeptide with a free carboxyl group.

carcinogen A substance that causes cancer.

carcinogenesis The sequence of events that convert a normal cell to a cancer cell.

catabolite repression The repression of a gene or operon by glucose, or more likely, by a catabolite, or breakdown product of glucose.

cDNA A DNA copy of an RNA, made by reverse transcription.

centimorgan (cM) *See* map unit.

centromere Constricted region on the chromosome where spindle fibers are attached during cell division.

charging Coupling a tRNA with its cognate amino acid.

Charon phages A set of cloning vectors based on λ phage.

chiasma (chiasmata, pl.) The structure formed between nonsister chromatids during meiosis; physical evidence of recombination.

chi-square test A statistical test to determine if the observed numbers deviate from those expected under a particular hypothesis. Often expressed as χ^2 test.

chi structure The branched DNA structure formed by strand crossover during recombination.

chloramphenicol An antibiotic that kills bacteria by inhibiting the peptidyl transferase reaction catalyzed by 50S ribosomes.

chloroplast The photosynthetic organelle of green plants and other photosynthetic eukaryotes.

chromatids Copies of a chromosome produced in cell division.

chromatin The material of chromosomes, composed of DNA and chromosomal proteins.

***cis*-acting** A term that describes a genetic element, such as an enhancer, a promoter, or an operator, that must be on the same chromosome in order to influence a gene's activity.

***cis*-dominant** A mutation that is dominant, but only with respect to genes on the same piece of DNA (*in cis*).

***cis-trans* test** A test to see whether two mutations are true alleles (located in the same gene) or pseudoalleles (located in different genes). When the mutations are in *cis* position, wild-type behavior is always observed. When the mutations are in *trans* position, wild-type behavior is observed if the mutations are pseudoalleles, but mutant behavior is observed if they are alleles.

cistron A genetic unit defined by the *cis-trans* test. For all practical purposes, synonymous with the word "gene."

clastogen An agent that causes DNA strand breaks.

clones Individuals formed by an asexual process so that they are genetically identical to the original individual.

coding strand The strand of a gene that has the same polarity and coding properties as the mRNA produced by the gene.

codominant Two alleles whose phenotypic effects are both expressed in the heterozygote.

codon A three-base sequence in mRNA that causes the insertion of a specific amino acid into protein or causes termination of translation.

cointegrate An intermediate in transposition of a transposon such as Tn3 from one replicon to another. The transposon replicates, and the cointegrate contains the two replicons joined through the two transposon copies.

colinearity A relationship between a gene and its protein product in which the distance between any two mutations is directly proportional to the distance between the corresponding altered amino acids.

colony hybridization A procedure for selecting a bacterial clone containing a gene of interest. DNAs from a large number of clones are simultaneously hybridized to a labeled probe for the gene of interest.

common garden experiment Growing organisms of different types in the same environment to determine whether the differences are genetic or environmental.

complementation test *See cis-trans* test.

composite transposon A bacterial transposon composed of two types of parts: two arms containing IS or IS-like elements, and a central region comprised of the genes for transposition and one or more antibiotic resistance genes.

concatemers DNAs of multiple genome length.

conditional lethal A mutation that is lethal under certain circumstances, but not under others (e.g., a temperature-sensitive mutation).

conditional probability The probability of an event occurring, contingent upon given circumstances.

conjugation (of bacteria) The movement of genetic material from an F^+ to an F^- bacterium.

consanguineous mating A mating between related individuals.

conservation genetics An effort to maintain genetic variation in a species by managing natural or seminatural populations or by collecting and storing diverse genetic types of a species.

conservative replication DNA (or RNA) replication in which both parental strands remain together, producing a progeny duplex, both of whose strands are new.

conservative transposition Transposition in which both strands of the transposon DNA are conserved as they leave their original location and move to a new site.

copia The prototype of a group of *copia*-like transposable elements found in *Drosophila*, which transpose via a retrovirus-like mechanism.

cos The cohesive ends of the linear λ phage DNA.

cosmid A vector designed for cloning large DNA fragments. A cosmid contains the cos sites of λ phage, so it can be packaged into λ heads, and a plasmid origin of replication, so it can replicate as a plasmid.

cotransduction Transduction of two or more genetic markers in the same virus.

crossing over Physical exchange between homologous chromosomes that takes place in meiosis I, or during recombination in general.

crossover The joining of corresponding DNA strands on homologous chromosomes by strand breakage and rejoining.

CRP *See* CAP.

cyclic AMP (cAMP) An adenine nucleotide with a cyclic phosphodiester linkage between the 3' and 5' carbons. Implicated in a variety of control mechanisms in prokaryotes and eukaryotes.

cytidine A nucleoside containing the base cytosine.

cytosine (C) The pyrimidine base that pairs with guanosine in DNA.

defective virus A virus that is unable to replicate without a helper virus.

degenerate code A genetic code, such as the one employed by all life on Earth, in which more than one codon can stand for a single amino acid.

degrees of freedom An integer used to determine whether a chi-square value is statistically significant.

deleterious An allele that causes some reduction in viability.

deletion A mutation involving a loss of one or more base pairs; the case in which a chromosomal segment or gene is missing.

deletion mapping Genetic mapping that employs deletion mutants. For example, one of the mating partners may contain a well-defined deletion, and the other an unknown point mutation. If recombination occurs, we know the point mutation lies outside the deleted area.

denaturation (DNA) Separation of the two strands of DNA.

denaturation (protein) Disruption of the three-dimensional structure of a protein without breaking any covalent bonds.

deoxyribose The sugar in DNA.

determinant A substance in an embryonic cell that contributes to determination of that cell.

determination The irreversible commitment of a cell to differentiate along a given pathway.

dideoxyribonucleotide A nucleotide, deoxy at both $2'$ and $3'$ positions, used to stop DNA chain elongation in DNA sequencing.

differentiation Expression of the specialized characteristics of a given cell type.

dihybrid cross A cross between two individuals that are heterozygous at two genes; for example, $AaBb \times AaBb$.

diploid The chromosomal number of the zygote and other cells (except the gametes). Symbolized as 2N.

directional cloning Insertion of foreign DNA into two different restriction sites of a vector, such that the orientation of the insert can be predetermined.

dispersive replication A hypothetical mechanism in which the DNA becomes fragmented so that new and old DNA coexist in the same strand after replication.

D-loop A loop formed when a free DNA end "invades" a double helix, base-pairing with one of the strands and forcing the other to "loop out."

DNA (deoxyribonucleic acid) A polymer composed of deoxyribonucleotides linked together by phosphodiester bonds. The material of which most genes are made.

DNA fingerprints The use of highly variable genes (regions of DNA) to identify particular individuals.

DNA glycosylase An enzyme that breaks the glycosidic bond between a damaged base and its sugar.

DNA gyrase A topoisomerase that pumps negative superhelical turns into DNA. Relaxes the positive superhelical strain created by unwinding circular DNA during replication.

DNA ligase An enzyme that joins two double-stranded DNAs end-to-end.

DNA melting *See* denaturation (DNA).

DNA photolyase The enzyme that carries out light repair by breaking thymine dimers.

DNA polymerase I One of three different DNA-synthesizing enzymes in *E. coli*; used primarily in DNA repair.

DNA polymerase III holoenzyme The enzyme within the replisome, which actually makes DNA during replication.

DNase Deoxyribonuclease, an enzyme that degrades DNA.

dominant An allele or trait that expresses its phenotype when heterozygous with a recessive allele; for example, *A* is dominant over *a* because the phenotypes of *AA* and *Aa* are the same.

double helix The shape that two complementary DNA strands assume in a chromosome.

Drosophila melanogaster A species of fruit fly used widely by geneticists.

Ds A defective transposable element found in corn, which relies on an *Ac* element for transposition.

duplication The situation in which a chromosomal segment or gene is represented twice.

$$\boxed{E}$$

electrophile A molecule that seeks centers of negative charge in other molecules and attacks them there.

electrophoresis A procedure in which voltage is applied to charged molecules, inducing them to migrate. This technique can be used to separate DNA fragments. It can also be used to separate and identify proteins or enzymes resulting from the presence of different alleles.

elongation factor A protein that is necessary for either the aminoacyl-tRNA binding or the translocation step in the elongation phase of translation.

embryogenesis Embryonic development.

endonuclease An enzyme that makes cuts within a polynucleotide strand.

endoplasmic reticulum (ER) Literally, "cellular network"; a network of membranes in the cell upon which proteins destined for export from the cell are synthesized.

enhancer A DNA element that strongly stimulates transcription of a gene or genes. Enhancers are usually found upstream from the genes they influence, but they can also function if inverted or moved hundreds or even thousands of base pairs away.

enucleate egg An egg without a nucleus.

env The gene in a retrovirus that encodes the viral envelope (membrane) protein.

epistasis The situation in which the alleles at one gene cover up or alter the expression of alleles at another gene.

equilibrium A state at which there is no change in the allelic frequencies of the population.

error-prone repair A mechanism used by *E. coli* cells to replicate DNA that contains thymine dimers. DNA is replicated across from a dimer even though no base pairing is possible. This usually leads to mistakes.

Escherichia coli (E. coli) An intestinal bacterium; the favorite subject for bacterial genetics.

euchromatin Chromatin that is extended, accessible to RNA polymerase, and at least potentially active. These regions stain either lightly or normally and are thought to contain most of the genes.

excision repair A DNA repair mechanism that removes damaged nucleotides, then replaces them with normal ones.

exon A region of a gene that is ultimately represented in that gene's mature transcript. The word refers to both the DNA and its RNA product.

exon shuffling A hypothetical process in which exons from different genes can combine to form new genes over evolutionary time.

exonuclease An enzyme that degrades a polynucleotide from the end inward.

expression site A locus on a chromosome where a gene can be moved to be expressed efficiently. For example, the expression site in trypanosomes is at the end (telomere) of a chromosome.

expression vector A cloning vector that allows expression of the cloned gene.

expressivity The level of phenotypic expression for a particular genotype.

$$\boxed{F}$$

F_1 The progeny of a cross between two parental types that differ at one or more genes; the first filial generation.

F_2 The progeny of a cross between two F_1 individuals or the progeny of a self-fertilized F_1; the second filial generation.

F' An F plasmid that has picked up a piece of host DNA.

fate map A diagram that shows which parts of an embryo are destined to differentiate into each segment of the adult.

F-duction The transfer of bacterial genes from one cell to another on an F' plasmid.

F pilus The thin projection, carried on the surface of an F^+ bacterium, which allows bacterial cells to join during conjugation.

F plasmid The fertility plasmid that allows a donor bacterium to conjugate with a recipient.

fertility plasmid *See* F plasmid.

fine structure mapping Extensive genetic mapping within a gene, ideally down to the nucleotide level.

fingerprint (protein) The specific pattern of peptide spots formed when a protein is cut into pieces (peptides) with an enzyme (e.g., trypsin), and then the peptides are separated by chromatography.

5′-end The end of a polynucleotide with a free (or phosphorylated or capped) 5′-hydroxyl group.

fluctuation test A test to determine whether a given microbial trait is caused by a mutation. If it is, individual cultures should vary widely in the number of cells showing the trait, since mutation is a rare and random event and may occur early in the growth of some cultures and later in others. The test can also be used to measure mutation rates.

fMet *See* N-formyl methionine.

forced cloning *See* directional cloning.

forked-line approach A technique used to determine progeny types for multiple genes, based on independent assortment between the genes.

founder effect The chance changes in allelic frequency resulting from the initiation of a population by a small number of individuals.

frameshift mutation An insertion or deletion of one or two bases in the coding region of a gene, which changes the reading frame of the corresponding mRNA.

fushi tarazu (ftz) A homeo box-containing gene in *Drosophila* that helps define segmentation in the early embryo.

fusion protein A protein resulting from the expression of a recombinant DNA containing two open reading frames (ORFs) fused together. One or both of the ORFs can be incomplete.

gag The gene in a retrovirus that encodes the viral coat proteins.

gamete A haploid sex cell.

GC box A hexamer having the sequence GGGCGG on the coding strand, which occurs in a number of mammalian structural gene promoters. The binding site for the transcription factor Sp1.

gene The basic unit of heredity. Contains the information for making one RNA and, in most cases, one polypeptide.

gene cluster A group of related genes grouped together on a eukaryotic chromosome.

gene conversion The conversion of one gene's sequence to that of another.

gene expression The process by which gene products are made.

gene flow Movement of individuals from one population to another that results in the introduction of migrant alleles through mating and subsequent reproduction.

gene mutation A mutation confined to a single gene.

gene pool The total variety and number of alleles within a population or species.

generalized recombination Recombination that requires extensive sequence similarity between the recombining DNAs.

generalized transduction Use of P1 or other phage heads to carry bacterial genes from one cell to another.

genetic code The set of sixty-four codons and the amino acids (or terminations) they stand for.

genetic counseling Counseling based on the risks of producing genetically defective infants, usually given where some familial history of a genetic disease exists.

genetic drift The chance changes in allelic frequency that result from the sampling of gametes from generation to generation.

genetic marker A mutant gene or other peculiarity in a genome that can be used to "mark" a spot in a genome for mapping purposes.

genetic variance The variation of a phenotype in a population that results from genetic causes, symbolized by V_G. The extent of the additive genetic variance, symbolized by V_A, is important in determining the rate and amount of response to directional selection.

genome One complete set of genetic information from a genetic system; e.g., the single, circular chromosome of a bacterium is its genome.

genomic library A set of clones containing DNA fragments derived directly from a genome, rather than from mRNA.

genotype The allelic constitution of a given individual. The genotypes at locus *A* may be *AA, Aa,* or *aa,* assuming only two possible alleles at this locus.

genotype-environment interaction The effect on a phenotype value that results from a specific genotype and a specific environment and that is not predictable from either separately. Symbolized by GE_{ij}.

germinal mutation *See* germ-line mutation.

germ-line mutation A mutation that affects the sex cells, enabling it to be passed on to progeny.

Golgi apparatus A membranous organelle that packages newly synthesized proteins for export from the cell.

G protein A protein that binds GTP, which activates the protein to perform a function. When the G protein cleaves the GTP to GDP, it becomes inactivated until a new molecule of GTP binds.

guanine (G) The purine base that pairs with cytosine in DNA.

guanosine A nucleoside containing the base guanine.

$$\boxed{H}$$

hairpin A structure resembling a hairpin (bobby pin), formed by intramolecular base pairing in an inverted repeat of a single-stranded DNA or RNA.

half chiasma *See* chi structure.

half-life The time it takes for half of a population of molecules to disappear.

haploid The chromosomal number in the gamete. Symbolized as N.

haplotype The genetic constitution of a single chromosome, or region of a chromosome.

Hardy-Weinberg principle The principle stating that after one generation of random mating, single-locus genotypic frequencies can be represented as a function of the allelic frequencies.

helicase An enzyme that unwinds the DNA double helix.

helix-turn-helix A structural motif in certain DNA-binding proteins, especially those from prokaryotes, that fits into the DNA wide groove and gives the protein its binding capacity and specificity.

helper virus (or phage) A virus that supplies functions lacking in a defective virus, allowing the latter to replicate.

hemoglobin The red, oxygen-carrying protein in the red blood cells.

heritability The proportion of phenotypic variance that is the additive genetic component. Symbolized as h^2. Realized heritability is an estimate of h^2 from a selection experiment.

hermaphrodite An animal that has both male and female reproductive organs.

heterochromatin Chromatin that is condensed, inactive, and usually darkly staining.

heteroduplex A double-stranded polynucleotide whose two strands are not completely complementary.

heterogeneous nuclear RNA (hnRNA) A class of large, heterogeneous-sized RNAs found in the nucleus, including unspliced mRNA precursors.

heterokaryon A cell containing two or more nuclei of different origins.

heterokaryotype Individuals that are heterozygous for two chromosomal types.

heterosis Higher yield (or other phenotypic value) for F_1 individuals compared to the mean of the parents.

heterozygosity The proportion of heterozygotes in a population.

heterozygote A diploid genotype in which the two alleles for a given gene are different; for example, A_1A_2.

heterozygote advantage The situation in which the heterozygote has a higher fitness than either homozygote, leading to a stable equilibrium.

Hfr A strain of *E. coli* that experiences a high frequency of recombination because of integration of the F plasmid into the bacterial chromosome, which mobilizes the chromosome.

histones A class of five small, basic proteins intimately associated with DNA in most eukaryotic chromosomes.

hnRNA *See* heterogeneous nuclear RNA.

homeo box A sequence of about 180 base pairs found in homeotic genes and other development-controlling genes in *Drosophila* and other eukaryotes. Thought to encode a DNA-binding region of a protein.

homeotic gene A gene in which a mutation causes the transformation of one body part into another. Sometimes called a **homoeotic** gene.

homologous chromosomes Chromosomes that are identical in size, shape, and, except for allelic differences, in genetic composition.

homologous recombination *See* generalized recombination.

homozygote A diploid genotype in which both alleles for a given gene are identical; for example, A_1A_1 or *aa*.

host range The set of hosts that can be productively infected by a virus (or phage).

hot spots Highly mutable sites within a genome—sites that accumulate more than the average number of mutations.

housekeeping genes Genes that code for proteins used by all kinds of cells.

human immunodeficiency virus (HIV) The agent that is thought to cause acquired immune deficiency syndrome (AIDS).

hybrid cell A cell created by fusing two dissimilar cells. First, a heterokaryon forms; then, after nuclear fusion, a hybrid cell.

hybrid dysgenesis A phenomenon observed in *Drosophila* in which the hybrid offspring of two certain parental strains suffer so much chromosomal damage that they are sterile, or dysgenic.

hybridization (of polynucleotides) Forming a double-stranded structure from two polynucleotide strands (either DNA or RNA) from different sources.

hybrid polynucleotide The product of polynucleotide hybridization.

hybrid vigor *See* heterosis.

hyperchromic shift The increase in a DNA solution's absorbance of 260 nm light upon denaturation.

hyphae Strandlike chains of fungal cells.

illegitimate recombination
Recombination that requires little if any sequence similarity between the recombining DNAs and is therefore not site-specific.

imaginal disk A group of insect larval cells that give rise to a body part, such as a wing, leg, or antenna.

immunity region The control region of a λ or λ-like phage, containing the gene for the repressor as well as the operators recognized by this repressor.

immunoglobulin (antibody) A protein that binds very specifically to an invading substance and alerts the body's immune defenses to destroy the invader.

inbreeding Nonrandom mating in which the mating individuals are more closely related than individuals drawn by chance from the population.

inbreeding coefficient The probability that the two homologous alleles in an individual are identical by descent. Symbolized as f.

inbreeding depression The reduction of population fitness due to inbreeding.

in cis A condition in which two genes are located on the same chromosome.

incision Nicking a DNA strand with an endonuclease.

independent assortment A principle discovered by Mendel, which states that genes on different chromosomes behave independently.

independent culling levels Artificial selection in which animals must exceed a given value for each of several traits or be culled.

inducer A substance that releases negative control of an operon.

inducer (embryonic) A substance that one embryonic cell makes to influence the development of another cell.

initiation factor A protein that helps catalyze the initiation of translation.

inosine (I) A nucleoside containing the base hypoxanthine, which base-pairs with cytosine.

insertion sequence (IS) A simple type of transposon found in bacteria, containing only inverted terminal repeats and the genes needed for transposition.

interference The difference between the observed number of double recombinants and the number expected when recombination in one area does not influence that in another area.

interphase The stage of the cell cycle during which DNA is synthesized but the chromosomes are not visible.

intervening sequence (IVS) *See* intron.

intracistronic complementation Complementation of two mutations in the same gene. Can occur by cooperation among different defective monomers to form an active oligomeric protein.

in trans A condition in which two genes are located on separate chromosomes.

intron A region that interrupts the transcribed part of a gene. An intron is transcribed but is removed by splicing during maturation of the transcript. The word refers to the intervening sequence in both the DNA and its RNA product.

inversion An alteration in the sequence of genes in a chromosome. A pericentric inversion includes the centromere, while a paracentric inversion does not.

inverted repeat A symmetrical sequence of DNA, reading the same forward on one strand and backward on the opposite strand. For example:
GGATCC
CCTAGG

isoschizomers Two or more restriction endonucleases that recognize and cut the same restriction site.

J

joining region (J) The segment of an immunoglobulin gene encoding the last thirteen amino acids of the variable region. One of several joining regions is joined by a chromosomal rearrangement to the rest of the variable region, introducing extra variability into the gene.

karyotype A pictorial or photographic representation of all the chromosomes in a given individual.

Klenow fragment A fragment of DNA polymerase I, created by cleaving with a protease, that lacks the 5′→3′ exonuclease activity of the parent enzyme.

L

lac **operon** The operon encoding enzymes that permit a cell to metabolize the milk sugar lactose.

lariat The name given the lasso-shaped intermediate in certain splicing reactions.

leader A sequence of untranslated bases at the 5′-end of an mRNA.

leading strand The strand that is made continuously in semidiscontinuous DNA replication.

lethal An allele that causes mortality. A recessive lethal causes mortality only when homozygous.

light repair Direct repair of a thymine dimer by the enzyme DNA photolyase.

linkage The physical association of genes on the same chromosome.

locus (loci, pl.) The position of a gene on a chromosome, used synonymously with the term "gene" in many instances.

long terminal repeats (LTRs) Regions of several hundred base pairs of DNA found at both ends of the provirus of a retrovirus.

luxury genes Genes that code for specialized cell products.

Lyonized chromosome An X chromosome in a female mammal that is composed entirely of heterochromatin and is genetically inactive.

lysis Rupturing the membrane of a cell, as by a virulent phage.

lysogen A bacterium harboring a prophage.

map unit (centimorgan) The distance separating two genetic loci that recombine with a frequency of 1%.

maternal effect A phenotype governed by the genotype, not the phenotype, of the maternal parent, and unaffected by genes from the paternal parent.

maternal effect genes Genes active in an oocyte, which define a phenotype that is unaffected by genes from the male gamete.

maternal inheritance A nonMendelian pattern of inheritance that always depends only on the maternal parent. The paternal parent does not influence the F_1 or any succeeding generation.

maternal message An mRNA made in great quantity during oogenesis, then stored, complexed with protein in the egg cytoplasm.

mean The arithmetic average.

mean fitness The sum of the product of the relative fitness and the frequency of the genotypes in the population. Symbolized as w.

meiosis Cell division that produces gametes (or spores) having half the number of chromosomes of the parental cell.

merodiploid A bacterium that is only partially diploid, meaning diploid with respect to only some of its genes.

merozygote *See* merodiploid.

message *See* mRNA.

messenger RNA *See* mRNA.

metacentric Chromosome with the centromere in the middle.

metal finger *See* zinc finger.

metaphase Intermediate stage of nuclear division in which the chromosomes move to the equatorial plate.

microconidium The small, male gamete of the mold *Neurospora*.

minimal medium A growth medium for microbes that contains only the simple substances (salts and an energy source such as glucose) needed for growth of prototrophs.

mismatch repair The correction of a mismatched base incorporated by accident—in spite of the proofreading system—into a newly synthesized DNA.

missense mutation A change in a codon that results in an amino acid change in the corresponding protein.

mitosis Cell division that produces two daughter cells having nuclei identical to the parental cell.

modifier gene A gene that modifies the phenotype of another gene.

molecular clock A hypothesis suggesting that a regular turnover or replacement of molecular variants (amino acid or nucleotide substitution) takes place over time.

monohybrid cross A cross between two individuals that are heterozygous at one gene; for example, *Aa* × *Aa*.

monomorphic The situation in which all the individuals in a population are the same genetic type or have the same allele.

morphological mutation A mutation that affects the morphology, or appearance, of an organism.

mosaic embryo An embryo in which the early blastomeres are different from one another.

mRNA (messenger RNA) A transcript that bears the information for making one or more proteins.

multiple cloning site (MCS) A region in certain cloning vectors, such as the pUC plasmids and M13 phage DNAs, that contains several restriction sites in tandem. Any of these can be used for inserting foreign DNA.

multiplicity of infection (m.o.i.) The ratio of infectious virus (or phage) particles to infectable cells in an experiment.

mutagen A mutation-causing agent.

mutation The original source of genetic variation, caused by a change in a DNA base or a chromosome, for example. Spontaneous mutations are those that appear without explanation, while induced mutations are those attributed to a particular mutagenic agent.

mutation rate The rate of mutational change from one allelic form to another. Symbolized by u or v.

mutation-selection balance The equilibrium that results when mutation introduces detrimental alleles into a population and selection eliminates them.

mutator strain A strain of bacteria that makes more than the usual number of mistakes during DNA replication, yielding a higher than normal mutation rate.

mutually exclusive events Events such that if one occurs, the other cannot; for example, heads and tails on a coin.

myc A cellular oncogene, encoding a nuclear protein, that is rearranged in human Burkitt's lymphoma.

negative control A control system in which gene expression is turned off unless a controlling element (e.g., a repressor) is removed.

Neurospora crassa A common bread mold, developed by Beadle and Tatum as a subject for genetic investigation.

neutrality theory The theory used to explain the existence of molecular variation when there is no differential selection (alleles are neutral with respect to each other) and when new variants are introduced into the population by mutation and eliminated by genetic drift.

N-formyl methionine (fMet) The initiating amino acid in prokaryotic translation.

nick A single-stranded break in DNA.

nondisjunction Abnormal separation of chromosomes in meiosis (or mitosis).

nonsense mutation A mutation that creates a premature stop codon within a gene's coding region. Includes amber mutations (UAG), ochre mutations (UAA), and opal mutations (UGA).

norm of reaction The phenotypic pattern for a given genotype produced under different environmental conditions.

Northern blotting Transferring RNA fragments to a support medium. (*See* Southern blotting.)

nucleic acid A chainlike molecule (DNA or RNA) composed of nucleotide links.

nucleocapsid A structure containing a viral genome (DNA or RNA) with a coat of protein.

nucleoid A nonmembrane-bound area of a bacterial cell in which the chromosome resides.

nucleolus A cell organelle found in the nucleus that disappears during part of cell division. Contains the rRNA genes.

nucleoside A base bound to a sugar—either ribose or deoxyribose.

nucleosome A repeating structural element in eukaryotic chromosomes, composed of a core of eight histone molecules with about 200 base pairs of DNA wrapped around the outside.

nucleotide The subunit, or chainlink, in DNA and RNA, composed of a sugar, a base, and at least one phosphate group.

nutritional mutation A mutation that makes an organism dependent on a substance, such as an amino acid or vitamin, that it previously could make for itself.

ochre codon UAA, coding for termination.

ochre mutation *See* nonsense mutation.

ochre suppressor A tRNA bearing an anticodon that can recognize the ochre codon (UAA) and thereby suppress ochre mutations.

Okazaki fragment Small DNA fragments, 1,000–2,000 bases long, created by discontinuous synthesis of the lagging strand.

oligomeric protein A protein that contains more than one polypeptide subunit.

oligonucleotide A short piece of RNA or DNA.

oncogene A gene that participates in causing tumors.

oncogenesis The formation of tumors.

oncogenic Tumor-causing.

oocyte 5S rRNA genes The 5S rRNA genes (haploid number about 19,500 in *Xenopus laevis*) that are expressed only in oocytes.

oogenesis Development of an egg.

opal codon UGA, coding for termination.

opal mutation *See* nonsense mutation.

opal suppressor A tRNA bearing an anticodon that can recognize the opal codon (UGA) and thereby suppress opal mutations.

open reading frame (ORF) A reading frame that is uninterrupted by translation stop signals.

operator A DNA element found in prokaryotes that binds tightly to a specific repressor and thereby regulates the expression of adjoining genes.

operon A group of contiguous, coordinately controlled genes.

ORF *See* open reading frame.

origin of replication The unique spot in a replicon where replication begins.

pair-rule genes Genes that, when mutated, give rise to insects missing segment-sized body parts at two-segment intervals.

palindrome *See* inverted repeat.

parthenogenesis Reproduction in which organisms produce offspring identical to themselves without fertilization.

paternity exclusion The process of using genetic markers to exclude given men as fathers of a particular child.

pBR322 One of the original plasmid vectors for gene cloning.

pedigree A family tree illustrating the inheritance of particular genotypes or phenotypes.

penetrance The proportion of individuals with a given genotype that express it at the phenotypic level.

peptide bond The bond linking amino acids in a protein.

peptidyl transferase An enzyme that is an integral part of the large ribosomal subunit and catalyzes the formation of peptide bonds during protein synthesis.

phage A bacterial virus.

phenocopy Occurs when environmental factors induce a particular abnormal phenotype that resembles a genetically determined phenotype.

phenotype The morphological, biochemical, behavioral, or other properties of an organism. Often only a particular trait of interest, such as weight, is considered.

phosphodiester bond The sugar-phosphate bond that links the nucleotides in a nucleic acid.

photoreactivating enzyme *See* DNA photolyase.

photoreactivation *See* light repair.

phylogenetic tree A diagram that organizes the relationship between species or populations, putatively indicating ancestral relationships.

physical map A genetic map based on physical characteristics of the DNA, such as restriction sites, rather than on locations of genes.

plaque A hole that a virus (or phage) makes on a layer of host cells by infecting and killing the cells.

plaque assay An assay for virus (or phage) concentration in which the number of plaques produced by a given dilution of virus is determined.

plaque-forming unit (pfu) A virus (or phage) capable of forming a plaque in a plaque assay.

plaque hybridization A procedure for selecting a phage clone that contains a gene of interest. DNAs from a large number of phage plaques are simultaneously hybridized to a labeled probe for the gene of interest.

plasmid A circular DNA that replicates independently of the cell's chromosome.

pleiotropic mutation A mutation that affects the expression of several other genes.

pleiotropy The situation in which a single gene affects two or more characteristics.

point mutation An alteration of one, or a very small number, of contiguous bases.

pol The gene in a retrovirus that encodes the viral reverse transcriptase and RNase H.

poly(A) Polyadenylic acid. The string of about two hundred A's added to the end of a typical eukaryotic mRNA.

poly(A) polymerase The enzyme that adds poly(A) to an mRNA or to its precursor.

polycistronic message An mRNA bearing information from more than one gene.

polygenic trait A phenotypic trait determined by a number of genes.

polymorphism The existence of two or more genetically determined forms (alleles) in a population in substantial frequency. In practice, a polymorphic gene is one at which the frequency of the most common allele is less than 0.99.

polynucleotide A polymer composed of nucleotide subunits; DNA or RNA.

polypeptide A single protein chain.

polyploids Organisms with three or more sets of chromosomes. Autopolyploids receive all their chromosomes from the same species, while allopolyploids obtain their chromosomes from two (or more) species.

polytene chromosomes Large chromosomes in diptera that contain many parallel replicates tightly held together.

population A group of individuals that exist together in time and space, and that can interbreed.

positive control A control system in which gene expression depends on the presence of a positive effector such as CAP (and cAMP).

positive strand The strand of a simple viral genome with the same sense as the viral mRNAs.

postreplication repair *See* recombination repair.

posttranscriptional control Control of gene expression that occurs during the posttranscriptional phase when transcripts are processed by splicing, clipping, and modification.

posttranslational modification The set of changes that occur in a protein after it is synthesized.

pp60^{src} The product of the *src* gene.

Pribnow box (−10 box) An *E. coli* promoter element centered about ten bases upstream from the start of transcription.

primary structure The sequence of amino acids in a polypeptide, or of nucleotides in a DNA or RNA.

primase The enzyme within the primosome that actually makes the primer.

primer A small piece of RNA that provides the free end needed for DNA replication to begin.

primosome A complex of about twenty polypeptides, which makes primers for *E. coli* DNA replication.

probability The proportion of the time that a particular event occurs.

processed pseudogene A pseudogene that has apparently arisen by retrotransposonlike activity: transcription of a normal gene, processing of the transcript, reverse transcription, and reinsertion into the genome.

processing (of RNA) The group of cuts that occur in RNA precursors during maturation, including splicing, 5'- or 3'-end clipping, or cutting rRNAs out of a large precursor.

product rule A rule stating that the probability of both of two independent events occurring is the product of their individual probabilities.

promoter The part of a gene to which RNA polymerase binds prior to initiation of transcription—usually found just upstream from the coding region of a gene.

proofreading The process of checking each nucleotide for complementarity with its base-pairing partner as it is inserted during DNA replication.

prophage A phage genome integrated into the host's chromosome.

prophase Early stage of nuclear division in which chromosomes coil, condense, and become visible.

protein A polymer, or polypeptide, composed of amino acid subunits. Sometimes the term "protein" denotes a functional collection of more than one polypeptide (e.g., the hemoglobin protein consists of four polypeptide chains).

proteolytic processing Cleavage of a protein into pieces.

proto-oncogene A cellular counterpart of a viral oncogene. Proto-oncogenes do not ordinarily participate in oncogenesis unless they are overexpressed or mutated.

protoperithecium The large, female gamete of the mold *Neurospora*.

prototroph An organism that can grow on minimal medium containing only salts and an energy source.

provirus A double-stranded DNA copy of a retroviral RNA that inserts into the host genome.

pseudogene A nonallelic copy of a normal gene that is mutated so that it cannot function.

pulse-chase The process of giving a short period, or "pulse," of radioactive precursor so that a substance such as RNA becomes radioactive, then adding an excess of unlabeled precursor to "chase" the radioactivity out of the substance.

pure-breeding line Identical individuals that always produce progeny like themselves.

purine The parent base of guanine and adenine.

puromycin An antibiotic that resembles an aminoacyl-tRNA and kills bacteria by forming a peptide bond with a growing polypeptide and then releasing the incomplete polypeptide from the ribosome.

pyrimidine The parent base of cytosine, thymine, and uracil.

quantitative trait A trait that generally has a continuous distribution in a population and is usually affected by many genes and many environmental factors.

quaternary structure The way two or more polypeptides interact in a complex protein.

quenching Quickly chilling heat-denatured DNA to keep it denatured.

random mating population A group of individuals in which the probability of members mating with individuals of particular types is equal to their frequency in the population.

ras A family of oncogenes, found in several human tumors, that encode a protein called p21, which resembles a G protein.

rate of allelic substitution The rate of replacement of alleles in a population. If amino acid or nucleotide sequences are known, the rate of amino acid or nucleotide substitution can be calculated.

reading frame One of three possible ways the triplet codons in an mRNA can be translated. For example, the message CAGUGCUCGAC has three possible reading frames, depending on where translation begins: (1) CAG UGC UCG; (2) AGU GCU CGA; (3) GUG CUC GAC. A natural mRNA generally has only one correct reading frame.

recessive   An allele or trait that does not express its phenotype when heterozygous with a dominant allele; for example, *a* is recessive to *A* because the phenotype for *Aa* is like *AA* and not like *aa*.

reciprocal matings   Crosses that are identical genetically except that the male and female parents are switched.

recombinant DNA   The product of recombination between two (or more) fragments of DNA. Can occur naturally in a cell, or be fashioned by geneticists in vitro.

recombination   Reassortment of genes or alleles in new combinations. Can occur through physical exchange between homologous chromosomes.

recombination repair   A mechanism that *E. coli* cells use to replicate DNA containing thymine dimers. First, the two strands are replicated, leaving a gap across from the dimer. Next, recombination between the progeny duplexes places the gap across from normal DNA. Finally, the gap is filled in using the normal DNA as template.

recurrence risk   The probability that a given genetic disease will recur in a family, given that one individual (or more) in the family has already exhibited the disease.

regulative embryo   An embryo in which the early blastomeres are indistinguishable from one another.

relative fitness   The relative ability of different genotypes to pass on their alleles to future generations.

release factor   A protein that causes termination of translation at stop codons.

renaturation of DNA   *See* annealing of DNA.

repetitive DNA   DNA sequences that are repeated many times in a haploid genome.

rep helicase   The product of the *E. coli rep* gene; thought to be involved in unwinding the *E. coli* DNA during replication.

replicating fork   The point where the two parental DNA strands separate to allow replication.

replicative transposition   Transposition in which the transposon DNA replicates, so that one copy remains in the original location as another copy moves to the new site.

replicon   All the DNA replicated from one origin of replication.

replisome   The large complex of polypeptides, including the primosome, which replicates DNA in *E. coli*.

resolution   The second step in transposition through a cointegrate intermediate; it involves separation of the cointegrate into its two component replicons, each with its own copy of the transposon.

resolvase   The enzyme that catalyzes resolution of a cointegrate.

***res* sites**   Sites on the two copies of a transposon in a cointegrate, between which a crossover occurs to accomplish resolution.

restriction endonuclease   An enzyme that recognizes specific base sequences in DNA and cuts at or near those sites.

restriction fragment   A piece of DNA cut from a larger DNA by a restriction endonuclease.

restriction fragment length polymorphism (RFLP)   A variation from one individual to the next in the number of cut sites for a given restriction endonuclease in a given locus. This results in a variation in the lengths of restriction fragments generated by that enzyme cutting within that locus. These variable lengths, which are really alleles, can be treated like any other genetic marker and used in mapping studies.

restriction site   A sequence of bases recognized and cut by a restriction endonuclease.

retrotransposon   Transposable elements such as *copia* and Ty that transpose via a retroviruslike mechanism.

retrovirus   An RNA virus whose replication depends on formation of a provirus by reverse transcription.

reverse transcriptase   RNA-dependent DNA polymerase; the enzyme that catalyzes reverse transcription, commonly found in retroviruses.

reverse transcription   Synthesis of a DNA using an RNA template.

reversion   A mutation that cancels the effects of an earlier mutation in the same gene.

RF (replicative form)   The circular double-stranded form of the genome of a single-stranded DNA phage

such as φX174. The DNA assumes this form in preparation for rolling circle replication.

RFLP   *See* restriction fragment length polymorphism.

rho (ρ)   A protein that is needed for transcription termination at certain terminators in *E. coli*.

ribonuclease H (RNase H)   An enzyme that degrades the RNA part of an RNA-DNA hybrid.

ribose   The sugar in RNA.

ribosome   An RNA-protein particle that translates mRNAs to produce proteins.

RNA (ribonucleic acid)   A polymer composed of ribonucleotides linked together by phosphodiester bonds.

RNA-dependent DNA polymerase   *See* reverse transcriptase.

RNA polymerase   The enzyme that directs transcription, or synthesis of RNA.

RNA polymerase core   The collection of subunits of a prokaryotic RNA polymerase having basic RNA chain elongation capacity but no specificity of initiation; all the RNA polymerase subunits except the σ factor.

rolling circle replication   A mechanism of replication in which one strand of a double-stranded circular DNA remains intact and serves as the template for elongation of the other strand at a nick.

rRNA (ribosomal RNA)   The RNA molecules contained in ribosomes.

screen   A genetic sorting procedure that distinguishes desired organisms from unwanted ones, but does not automatically remove the latter.

secondary structure   The local folding of a polypeptide or RNA. In the case of RNA, the secondary structure is defined by intramolecular base pairing.

sedimentation coefficient   A measure of the rate at which a molecule or particle travels toward the bottom of a centrifuge tube under the influence of a centrifugal force.

segmentation genes   Genes that establish the segmentation pattern of an animal, especially an insect.

segregation A principle discovered by Mendel, which states that heterozygotes produce equal numbers of gametes having the two different alleles.

selection A genetic sorting procedure that eliminates unwanted individuals, usually by preventing their growth or by killing them.

selection limit The point beyond which artificial selection is unable to change the mean phenotypic value.

selective disadvantage The reduction in relative fitness of genotype A_2A_2 as compared with that of A_1A_1; also known as the selection coefficient, symbolized by s.

self-fertilization Fertilization of eggs with sperm or pollen from the same individual. Sometimes called **selfing.**

semiconservative replication DNA replication in which the two strands of the parental duplex separate completely and pair with new progeny strands. One parental strand is therefore conserved in each progeny duplex.

semidiscontinuous replication A mechanism of DNA replication in which one strand is made continuously and the other is made discontinuously.

sequencing Determining the amino acid sequence of a protein, or the base sequence of a DNA or RNA.

sex-duction *See* F-duction.

sex-influenced Traits that are expressed differently in the two sexes.

sex–limited Traits that are only expressed in one sex.

sex–linked Alleles that are on the sex chromosomes.

Shine-Dalgarno (SD) sequence A G-rich sequence (consensus = AGGAGGU) that is complementary to a sequence at the 3′-end of *E. coli* 16S rRNA. Base pairing between these two sequences helps the ribosome bind an mRNA.

shuttle vector A cloning vector that can replicate in two or more different hosts, allowing the recombinant DNA to shuttle back and forth between hosts.

sickle-cell disease A genetic disease in which abnormal β-globin is produced. Because of a single amino acid change, this blood protein tends to aggregate under low oxygen conditions, distorting red blood cells into a sickle shape.

sigma (σ) The prokaryotic RNA polymerase subunit that confers specificity of transcription—that is, the ability to recognize specific promoters.

signal peptide A stretch of about twenty amino acids, usually at the amino terminus of a polypeptide, that helps to anchor the nascent polypeptide and its ribosome in the endoplasmic reticulum. Polypeptides with a signal peptide are destined for packaging in the Golgi apparatus and are usually exported from the cell.

silent mutations Mutations that cause no detectable change in an organism, even in a haploid organism or in a homozygote.

single-copy DNA DNA sequences that are present once, or only a few times, in a haploid genome.

site-directed mutagenesis A method for introducing specific, predetermined alterations into a cloned gene.

site-specific recombination Recombination that always occurs by crossovers in the same place and depends on limited sequence similarity between the recombining DNAs.

small nuclear RNAs (snRNAs) A class of nuclear RNAs a few hundred nucleotides in length. These RNAs, together with tightly associated proteins, make up **snRNPs (small nuclear ribonucleoproteins)** that participate in splicing, polyadenylation, and 3′-end maturation of transcripts.

somatic 5S rRNA genes The 5S rRNA genes (haploid number about 400 in *Xenopus laevis)* that are expressed in both somatic cells and oocytes.

somatic cell genetics Genetic manipulations involving fusing somatic cells instead of mating organisms.

somatic cell hybridization Fusion of cells from two different species; used to determine the chromosomal location of genes.

somatic cells Nonsex cells.

somatic mutation A mutation that affects only somatic cells, so it cannot be passed on to progeny.

SOS response The activation of a group of genes, including *recA*, that helps *E. coli* cells respond to environmental insults such as chemical mutagens or radiation.

Southern blotting Transferring DNA fragments separated by gel electrophoresis to a suitable support medium such as nitrocellulose, in preparation for hybridization to a labeled probe.

spacer DNA DNA sequences found between, and sometimes within, repeated genes such as rRNA genes.

specialized transduction Use of λ or a similar phage to carry bacterial genes from one cell to another.

spliceosome The large RNA-protein body upon which splicing of nuclear mRNA precursors occurs.

splicing The process of linking together two exons while removing the intron that lies between them.

spore (1) A specialized haploid cell formed sexually by plants or fungi, or asexually by fungi. The latter can either serve as a gamete or germinate to produce a new haploid cell. (2) A specialized cell formed asexually by certain bacteria in response to adverse conditions. Such a spore is relatively inert and resistant to environmental stress.

sporulation Formation of spores.

src The oncogene of Rous sarcoma virus (RSV), which encodes a tyrosine protein kinase.

SSB Single-strand-binding protein, used during DNA replication. Binds to single-stranded DNA and keeps it from base-pairing with a complementary strand.

standard deviation The square root of the variance.

statistically significant A chi-square value (or other statistic) large enough to establish that the results are not consistent with the hypothesis used.

sticky ends Single-stranded ends of double-stranded DNAs that are complementary and can therefore base-pair and stick together.

stop codon One of three codons (UAG, UAA, and UGA) that code for termination of translation.

streptomycin An antibiotic that kills bacteria by causing their ribosomes to misread mRNAs.

stringency (of hybridization) The combination of factors (temperature, salt and organic solvent concentration) that influence the ability of two polynucleotide strands to hybridize. At high stringency, only perfectly complementary strands will hybridize. At reduced stringency, some mismatches can be tolerated.

structural genes Genes that code for proteins.

sum rule A rule stating that the combined probability of two (or more) mutually exclusive events occurring is the sum of their individual probabilities.

supercoil *See* superhelix.

superhelix A form of circular double-stranded DNA in which the double helix coils around itself like a twisted rubber band.

suppressor mutation A mutation that reverses the effects of a mutation in the same or another gene.

SV40 Simian virus 40; a DNA tumor virus with a small circular genome, capable of causing tumors in certain rodents.

T

TATA box An element with the consensus sequence TATAAAA, found about 25 base pairs upstream from the start of transcription in most eukaryotic promoters recognized by RNA polymerase II.

tautomeric shift The reversible change of one isomer of a DNA base to another by shifting the locations of hydrogen atoms and double bonds.

T-DNA The tumor-inducing part of the Ti plasmid.

telocentric Chromosome with the centromere very near one end.

telophase The last stage of nuclear division in which the nuclear membrane forms and encloses the chromosomes in the daughter cells.

temperate phage A phage that can enter a lysogenic phase in which a prophage is formed.

temperature-sensitive mutation A mutation that causes a defective product to be made at high temperature (the restrictive temperature) but yields a functional product at low temperature (the permissive temperature).

template A polynucleotide (RNA or DNA) that serves as the guide for making a complementary polynucleotide. For example, a DNA strand serves as the template for ordinary transcription.

terminal transferase An enzyme that adds deoxyribonucleotides, one at a time, to the 3′-end of a DNA.

terminator *See* transcription terminator.

tertiary structure The overall three-dimensional shape of a polypeptide or RNA.

testcross A cross between an F_1 individual (or an individual of unknown genotype) and the recessive parent of the F_1 (the tester).

tetrad analysis The use of the four meiotic products from a single meiosis (a tetrad) to determine the behavior of genes in meiosis.

TFIIIA A transcription factor that helps activate the vertebrate 5S rRNA genes.

thalassemia A genetic disease marked by failure to produce a functional mRNA for one of the two major adult hemoglobin proteins, α-globin or β-globin.

3′-end The end of a polynucleotide with a free (or phosphorylated) 3′-hydroxyl group.

three-point cross Eukaryotic: A testcross in which one parent is heterozygous for three genes. Prokaryotic: A cross in which the two parental genotypes differ at three genes.

thymidine A nucleoside containing the base thymine.

thymine (T) The pyrimidine base that pairs with adenine in DNA.

thymine dimer Two adjacent thymines in one DNA strand linked covalently, interrupting their base pairing with adenines in the opposite strand. This is the main genetic damage caused by ultraviolet light.

Ti plasmid The tumor-inducing plasmid from *Agrobacterium tumefaciens*.

topoisomerase An enzyme that changes a DNA's superhelical form, or topology.

totipotent cell A cell that retains full genetic capability, including the ability to give rise to a complete organism under appropriate circumstances.

trailing strand The strand that is made discontinuously in semidiscontinuous DNA replication.

***trans*-acting** A term that describes a genetic element, such as a repressor gene or transcription factor gene, that can be on a separate chromosome and still influence another gene. These *trans*-acting genes function by producing a diffusible substance that can act at a distance.

transcript An RNA copy of a gene.

transcription The process by which an RNA copy of a gene is made.

transcription terminator A specific DNA sequence that signals transcription to terminate.

transducing virus (or phage) A virus that can carry host genetic information from one cell to another.

transductants Cells that have received genetic information from a transducing virus.

transduction The use of a phage (or virus) to carry host genes from one cell to another cell of different genotype.

transfer RNA *See* tRNA.

transformation (genetic) An alteration in a cell's genetic makeup caused by introducing exogenous DNA.

transformation (malignant) The process by which a normal cell begins behaving like a tumor cell.

transformed cell An immortal cell that displays many of the characteristics of a cancer cell, but that may or may not be tumorigenic.

transgenic organism An organism into which a new gene or set of genes has been transferred.

transition A mutation in which a pyrimidine replaces a pyrimidine, or a purine replaces a purine.

translation The process by which ribosomes use the information in mRNAs to synthesize proteins.

translocation The translation elongation step following the peptidyl transferase reaction that involves moving an mRNA one codon's length through the ribosome and bringing a new codon into the ribosome's A site.

translocation (chromosomal) Transfer of a chromosome segment from one chromosome to another nonhomologous chromosome.

transposable element A DNA element that can move from one genomic location to another.

transposase The name for the collection of proteins, encoded by a transposon, that catalyze transposition.

transposition Transfer of a segment of DNA from one location to another, with or without retention of a copy of the segment in the original location.

transposon *See* transposable element.

transversion A mutation in which a pyrimidine replaces a purine, or vice versa.

trisomy Diploids having one extra chromosome ($2N + 1$).

tRNA (transfer RNA) Relatively small RNA molecules that bind amino acids at one end and "read" mRNA codons at the other, thus serving as the adapters that translate the mRNA code into a sequence of amino acids.

tumorigenicity The ability to cause tumors in whole animals.

Ty A yeast transposon that transposes via a retroviruslike mechanism.

ultimate carcinogen The most carcinogenic form to which a given substance can be metabolized.

undermethylated region A region of a gene or its flank that is relatively poor in, or devoid of, methyl groups.

unique DNA *See* single-copy DNA.

uracil (U) The pyrimidine base that replaces thymine in RNA.

uridine A nucleoside containing the base uracil.

variance A measure of the dispersion of the mean value in a population.

vector A piece of DNA (a plasmid or a phage DNA) that serves as a carrier in gene cloning experiments.

viroid An infectious agent (of plants) consisting simply of a small circular RNA.

virulent phage A phage that lyses its host.

wobble The ability of the third base of a codon to shift slightly to form a nonWatson-Crick base pair with the first base of an anticodon, thus allowing a tRNA to translate more than one codon.

wobble position The third base of a codon, where wobble base pairing is permitted.

Z-DNA A left-handed helical form of double-stranded DNA whose backbone has a zigzag appearance. This form is stabilized by stretches of alternating purines and pyrimidines.

zinc finger A finger-shaped protein structural motif found in DNA-binding proteins and characterized by two cysteines and two histidines (or four cysteines) bound to a zinc ion. By inserting into the wide groove of the DNA, these fingers help the proteins bind DNA.

zygote A fertilized egg.

A

ABO blood group, 40, 45–46, 114–15, 422
Abortion, spontaneous, 93, 96
Ac element, 300–302
Acetylcholinesterase gene, 418
Achondroplasia, 35, 40, 42, 282
Acid phosphatase gene, 418
Acridine dye, 256, 280, 283
Acrocentric chromosome, 73–74, 77–78, 88, 93
Adelberg, E. H., 323
Adenine, 9, 137–38, 140
    deamination of, 280–81
    tautomers of, 278
Adenosine deaminase gene, 418
Adenovirus, 223, 228–29, 232–33
Adenylate kinase gene, 418
Adjacent segregation, 91
Admixture, 430
African sleeping sickness, 364
Agriculture, cloned genes in, 397
Agrobacterium tumefaciens, 398–99
AIDS vaccine, 396
Albinism, 23, 34, 40, 117, 275, 427, 441
Albumin, 446–47
Alcaptonuria, 8, 156–58
Alcohol dehydrogenase genes, 418, 444–45
Alkaline phosphatase gene, 267, 418
Alkylation, of DNA, 283, 287–88
Alkyl group, 283
Allele, 21
    deleterious, 42, 455, 457
    detrimental (*see* Allele, deleterious)
    identical by descent, 424–26
    lethal (*see* Lethals)
    multiple, 45–48, 422
    nomenclature of, 30
Allelic frequency, 416–23
Allelic substitution, rate of, 445
Allolactose, 202
Allopolyploid, 85, 93–94
Allosteric protein, 202
Allotetraploid, 94
α-Globin gene, 115, 227, 364, 387, 448
α-Helix, 155–56
Alternate segregation, 91
AluI, 371
Alzheimer's disease
    mapping gene for, 406–7
    screening for, 407
Amber codon, 267–69, 276
Ames test, 293–94
Amino acid, 155
    activated, 254
    binding to transfer RNA, 254–55
    structure of, 154
Aminoacyl-AMP, 254
Aminoacyl-tRNA, 254–55, 257–58, 266–67
Aminoacyl-tRNA synthetase, 163, 254–55
Amino group, 133
Amish population, 443
Amniocentesis, 98, 406
Ampicillin resistance, 299
Anaphase
    meiosis I, 80
    meiosis II, 80
    mitotic, 77–78
Anemia
    Fanconi's, 292
    sickle-cell (*see* Sickle-cell disease)
Aneuploid, 85, 93, 95–98, 274
Animal
    aneuploidy in, 96
    polyploidy in, 93

transgenic, 397–98
wild vs. domestic, 60
Animal breeding, 61, 63, 451–57
Animal pole, 346
Antennapedia complex, 353, 359
Anthocyanin, 51–52
Antibiotic resistance, 459
    multiple, 298–99
Antibiotic resistance gene, 296–99, 373
Antibody, 46, 361–62. *See also* Immunoglobulin
Antibody probe, 378, 384
Anticodon, 164, 252, 254
Antigen
    blood group, 46
    red blood cell, 40
Antirepressor, lambda, 211
Antiterminator, lambda, 211–12
3′ AP endonuclease, 288–89
5′ AP endonuclease, 288–89
A peptide, 246
AP site, 283, 288–89
Apurinic site. *See* AP site
Apyrimidinic site. *See* AP site
ara operon, 205
Arber, Werner, 370
Artificial selection, 452–54
Ascomycete, 158
Ascospore, 118–21, 158
Ascus, 158
Asexual reproduction, 71
Aspergillus nidulans, chromosome number in, 72–73
Assortative mating
    negative, 423
    positive, 423
Ataxia telangiectasia, 292
ATP, in transfer RNA charging, 254
ATPase, 178
Atrazine resistance, 460
att B site, 326
Attenuation, 198
Attenuator, *trp,* 198
A-type particle, 303–4
Autoimmune disease, 47
Autopolyploid, 85, 93–95
Autoradiography, of DNA, 146–47, 172–75, 387–89
Autosome, 6
Autotetraploid, 93
AUUAAA motif, 240
Auxotroph, 276, 314
Avena fatua, inbreeding coefficient in, 425
Avery, Oswald, 8, 11, 134
Avise, John, 447–48
5–Azacytosine, 351–52

B

Bacillus subtilis
    replication in, 173–74
    sporulation in, 206–9
Backcross, 22
Back-mutation. *See* Reversion
Backward mutation rate, 426
Bacteria. *See also Escherichia coli*
    conjugation in, 313–22
    genetic notation for, 314
    growth on agar plate, 193
    mapping in, 322–30
    transduction in, 326–28
    transformation in, 132–37
Bacterial lawn, 332
Bacteriophage. *See* Phage
Baltimore, David, 302
BamHI, 371–72
Banana, ploidy of, 94
Barbacid, Mariano, 307

Barley, genetic variation in, 418
Barnett, Leslie, 256
Barr, Murray, 352
Barr body, 98, 352
Base. *See* Nitrogenous base
Base pairing, 140–41, 152, 164
    wobble, 258–59
72–Base-pair repeat, 227
Bateson, William, 23, 51, 106, 353
B chromosome. *See* Supernumerary
    chromosome
B-DNA, 143
Beadle, George, 9, 11, 158–59
Bee, chromosome number in, 72
Behavior, 55–56
Benzer, Seymour, 267, 326, 335–37
Benzo[a]pyrene, 306
Berg, Paul, 11
Bernstein, Harris, 268
β-Galactosidase, 200–203, 325,
    374–75, 381, 383–84
β-Galactosidase gene family, 378
β-Globin, 165–67
β-Globin gene, 115, 166–67, 223–24,
    227, 232–36, 347–48, 364,
    396, 415–16, 435, 442
β-Lactamase, 297
β-Lactoglobulin gene, 397–98
β-Pleated sheet, 155
bGH. *See* Bovine growth hormone
BglII, 371
bicoid gene, 355–57
Bidirectional replication, 172–75
Binomial coefficient, 27–28
Binomial formula, 27–28
Binomial probability, 27–29
Bird, sex determination in, 100
Biston betularia. See Peppered moth
Bisulfite, 283
Bithorax complex, 353
Bivalent, 80, 92
Black bread mold. *See Aspergillus
    nidulans*
Blakeslee, Alfred, 96
Blastoderm, 355, 357, 359
Blastomere, 343
Blastula, 344, 351
Blattner, Fred, 374–77
Blending inheritance, 16, 19–20
Blood transfusion, 46
Bloom's syndrome, 292
Bottleneck, 428
Bourgeois, Suzanne, 203
Boveri, Theodor, 11, 71
Bovine growth hormone (bGH), 397
–10 Box, 197, 209, 223
–35 Box, 197, 209
Boyer, Herbert, 11, 370–72
B peptide, 246
Branch migration, 188–89
BrdU. *See* 5–Bromodeoxyuridine
Brenner, Sydney, 11, 160, 268, 361
Bridges, Calvin, 11, 84, 100–101
Briggs, Robert, 344
Brinster, Ralph, 397
5–Bromodeoxyuridine (BrdU), 283
Brown, Donald, 226, 349
Burgess, Richard, 195
Burkitt's lymphoma, 364
Burst size, 332

C

cI gene, 210–13
cII gene, 212–13
cIII gene, 212
CAAT box, 223, 225. *See* CCAAT box
Cabbage, cross with radish, 94

Caenorhabditis elegans
    determination in, 360–61
    homeotic genes in, 361
Cairns, John, 172, 179, 321
Calf thymus DNA, 144
Callan, H. G., 175
cAMP. *See* Cyclic AMP
Cancer, 364
    DNA repair and, 292
    mutations and, 274–75, 305–10
    viruses and, 307–10
CAP. *See* Catabolite activator protein
Cap binding protein (CBP), 265
Capecchi, Mario, 269
Capping, of messenger RNA, 241,
    244, 265
Carbon, chemistry of, 132
Carbonyl group, 133
Carboxyl group, 132–33
Carcinogen, 284
    identification of, 293–94
    as mutagen, 305–7
    ultimate, 306
Carcinogenesis, 305
Carp, chromosome number in, 72–73
Carrier, 27, 35, 65
    identification of, 406–7
    proportion in population, 420
Casein, 244
Castle, William, 23, 419
Cat
    chromosome number in, 72–73
    coat color in, 28–30, 45, 276–77
    hair length in, 28–30
    karyotype of, 74
    sex chromosomes of, 74
    taillessness in, 42, 45
Catabolite activator protein (CAP),
    205–6, 214
Catabolite repression, 205
Cattle
    calf weight in, 457
    crossbred, 457
    heterosis in, 457
    karyotype of, 74
    milk production in, 452–53
    purebred, 457
    sex chromosomes of, 74
CBP. *See* Cap binding protein
CCAAT box, 223–24, 231. *See* CAAT
    box
CCAAT box transcription factor
    (CTF), 230–31
cDNA. *See* Complementary DNA
Cell cycle, 76
Cell differentiation, 342–49
Cell-free extract, 224
Cell morphology, 342–43
Centric fission, 93
Centric fusion, 93
Centromere, 73–75, 77–78, 231
Cepaea nemoralis. See Snail
Cesium chloride density gradient
    centrifugation, 152–53,
    294–95
Chain-counting technique, 426
Chain terminator, 393
Chamberlain, Michael, 196
Chambon, Pierre, 224–25, 229, 234,
    246–48
Chargaff, Erwin, 135, 139
Chargaff's rules, 139–40
Charon phage, 374–77, 383
Chase, Martha, 135–36
Cheetah
    genetic variation in, 418
    inbreeding in, 451
Chemical mutagenesis, 283–84, 386
Chiasma, 80, 108, 116
    half, 188–190
Chicken
    chromosome number in, 72–73
    development in, 347–48
    egg production by, 454–55

feather color in, 100
feather development in, 346
performance in, 454–55
sex determination in, 100
Chimney sweep, 305–6
Chimpanzee
    chromosome banding patterns in,
        99
    chromosome number in, 72–73
    phylogenetic relationship to other
        primates, 447
Chi-square test, 31–33, 421
Chi-square value, 32
Chi structure, 188–90
Chlamydomonas reinhardii, mutation
    rate in, 282
Chloramphenicol, 266
Chloramphenicol resistance, 298–99
Chlorophyll, 348
Chromatid, 77–78
Chromatin
    DNase-hypersensitivity sites in,
        347–48
    folding of
        radial loop model of, 221–22
        solenoid model of, 220–22
    histones in (*see* Histone)
    nonhistone proteins in, 222
    structure of, 218–22
        gene expression and, 231–32
Chromatosome, 219
Chromosomal fission, 93
Chromosomal fusion, 93
Chromosomal satellite, 73–74
Chromosome, 5–6
    acrocentric, 73–74, 77–78, 88, 93
    B (*see* Chromosome,
        supernumerary)
    banding patterns of, 74, 98–99
    breaks in, 85–86, 88, 108
    daughter, 78
    of eukaryotes, 218–22
    genes and, 82–85
    homeologous, 94
    homologous, 7, 72, 79–81
    metacentric, 73–74, 77–78, 88, 93,
        114
    morphology of, 73–76
    nonparental, 108
    parental, 108
    polytene, 75–76
    ring, 92
    secondary constriction in, 73–74
    sex (*see* Sex chromosome)
    structural changes in, 85–93
    supercoiled, 220–21
    supernumerary, 72–73
    telocentric, 73–74, 78
    variations in structure of, 414–15
Chromosome arm, 73–74
Chromosome number, 72–73, 81
    changes in, 85–98
Chromosome theory, of inheritance,
    5–6
Chua, Nam-Hai, 348
Cigarette tar, 306
Ciliate, genetic code in, 260–61
cis-acting element, 326
cis-dominant mutation, 203
cis-trans complementation test,
    324–26, 337
Cistron, 326–37
Clarkia, chromosomes of, 92
Classical genetics. *See* Mendelian
    genetics
Clastogen, 285
Clausen, J., 55
Cleft palate, 65
Clegg, Michael, 432
C locus, 300–302
Clone, 55, 344, 370

Cloned gene. *See also* Gene cloning
    in agriculture, 397
    expression of, 380–85
    identification with specific probe,
        378–79
    manipulation of, 385–95
    as probe, 387–92
    protein engineering with, 385–86
    sequencing of, 392–95
Codominance, 40–41, 47
Codon, 10, 164, 266. *See also* Genetic
    code
    nonsense (*see* Codon, stop)
    novel, 260–61
    stop, 258, 260–61, 267–69, 277
Codon-anticodon recognition, 164
Coefficient of coincidence, 114
Cohen, Stanley, 11, 370–72
Cohn, Melvin, 202
Coincidence, coefficient of, 114
Cointegrate, 297, 299
Colchicine, 94
Colicin E1, 175
Colinearity, of gene and protein,
    167–68
Collins, Francis, 406
Colon cancer, 308
Colony hybridization, 380
Color blindness, 48, 115
Common garden experiment, 59–60
Complementarity, in DNA, 9
Complementary DNA (cDNA), 243,
    387–88
    cloning of, 379–80
Complementation, 51, 337
    intracistronic, 326
Complementation test, *cis-trans,*
    324–27, 337
Complete dominance, 40
Complete penetrance, 43
Composite transposon, 298–99
Conalbumin gene, 229
Concatemer, 184
Conditional lethal, 44
Conditional mutation, 276–77
Conditional probability, 26–27
Conjugation, in bacteria, 313–22
    mapping by, 318–21
Conklin, E. G., 343
Conneally, Michael, 401, 403, 406
Consanguineous mating, 33, 426,
    442–43, 450
Consensus sequence, 197, 223, 234,
    240
Conservation genetics, 449–52
Conservative replication, 152–53
Conservative substitution, 219
Conservative transposition, 296
Constitutive heterochromatin, 74
Constitutive mutant, 203
copia, 302, 304
copia-like element, 302
Core particle, 219
Corn, 4
    chromosome number in, 72–73
    DNA of, 146
    ear length in, 63
    genetic variation in, 449
    height of, 63
    heterosis in, 456
    kernel color in, 25, 274, 300
    kernel shape in, 25
    mutation rate in, 282
    oil content of, 63
    protein content of, 63
    recombination in, 7
    root tip cells of, 217
    transposable elements of, 300–302
    yield of, 63
Correlation, 61
Correns, Carl, 11, 17, 23, 40
Cosmid, 377
cos site, 211, 377
Cotransduction, 328–30

Coumermycin, 180
Countable trait, 59
Coupling. *See* Linkage
Coupling double heterozygote, 109
Coupling gametes, 107
C peptide, 246
Creutzfeld-Jacob disease, 396
Crick, Francis, 9, 11, 135, 139–42, 159, 163, 256, 258
Cri du chat syndrome, 87
Criminal, identification of, 390–91
cro gene, 211–13
Crop plant
  genetic variation in, 449–50
  germ plasm collections, 449–50
  land races in, 449–50
Crossover, 7, 79–80, 187
  between two mutations, 329
  double, 111–12, 114, 121
  four-stranded, 117–22
  quadruple, 114
  triple, 114
  two-stranded, 117–18
  unequal, 122–25
Crown gall, 398–99
Cryptic mutant, 202–3
CTF. *See* CCAAT box transcription factor
Cuénot, Lucien, 23, 41–42
Culling level, 454
C-value paradox, 168–69
Cyclic AMP (cAMP), 205–6
Cystic fibrosis, 34, 401
  frequency of alleles causing, 440
  mapping gene for, 406–7
  screening for, 406
Cytochrome *c*, evolution of, 446
Cytosine, 9, 137, 140
  deamination of, 280–81
  methylation of, 281
  tautomers of, 278

Dalgarno, L., 264
Darbishire, 23
Darnell, James, 239
Darwin, Charles, 11, 414, 431
Datura, trisomy in, 96
Daughter chromosomes, 78
DDT resistance, 429, 457
Deaf-mutism, 53
Deamination, of nitrogenous base, 280–81
Deficiency. *See* Deletion
Degrees of freedom, 31–33, 421
Degu, inbreeding in, 451
Delbrück, Max, 317, 330, 332
Deleterious allele, 42, 455, 457
Deletion, 10, 85–87, 90–91, 122, 256, 299–300
  interstitial, 87
  single-base, 279–80
  terminal, 87
Deletion heterozygote, 87, 116
Deletion mapping, 116
Delta repeat, 302
Deoxyribonucleic acid. *See* DNA
Deoxyribose, 137–38
Deprotonation reaction, 132–33
Derepression, 201
DES. *See* Diethylstilbestrol
Determinant, 346
Determination, 346
  in *Caenorhabditis elegans*, 360–61
  in *Drosophila*, 352–59
  nature of, 349–61
Detrimental allele. *See* Deleterious allele

Development
  gene rearrangements during, 361–65
  mosaic, 343–44
  regulative, 344–45
Developmental genetics, 342–65
de Vries, Hugo, 11, 17
d'Herelle, F., 330
Diabetes, 59, 65, 395
Diakinesis, 80
Dideoxy method, of DNA sequencing, 392–94
Diethylstilbestrol (DES), 244
Differentiation, 342–49
Dihybrid cross, 23, 26, 29–31, 106
Dipeptidyl-tRNA, 266
Diploid, 5–6, 72–73
Diplotene, 80
Directional cloning, 374
Disaccharide, 200
Dispersive replication, 152–53
Displaced duplication, 86
Dizygotic twins, paternity of, 66
D-loop, 190
DNA
  alkylation of, 287–88
  annealing of, 145
  anticoding strand of, 164–65
  antiparallel strands in, 141–42
  autoradiography of, 146–47, 172–75, 387–89
  bacterial exchange of, 314–17
  base composition of, 135
  cesium chloride density gradient centrifugation of, 152–53, 294–95
  chemical nature of, 137–39
  circular, 146, 179, 184, 211, 316, 320–21
  coding strand of, 164–65
  complementary, 243, 379–80, 387–88
  complementary strands in, 142
  content per cell, 81, 168–69
  delta repeat in, 302
  denaturation of (*see* DNA, melting of)
  direct repeats in, 295–96, 305
  discovery of, 8
  DNase-hypersensitivity sites in, 347–48
  electron microscopy of, 146
  electron-rich centers in, 283
  electrophoresis of, 134, 146–48, 387–89, 393–94
  3′-end of, 139
  5′-end of, 139
  of eukaryotes, 146
  fingerprinting of, 66, 387–91
  GC content of, 144
  genetic capacity of, 168
  hairpin structure in, 198–99
  heavy isotope-labeled, 152–53
  heteroduplex, 188, 190
  homologous, 379
  hot spots in, 226–27, 281
  hybridization of (*see* Hybridization)
  interaction with proteins, 214
  inverted repeats in, 198–99, 295, 297, 301–2
  kilobases per map unit, 124–25
  long terminal repeats in, 302–3
  looping out of base in, 279–80
  melting of, 144, 161
  methylation of, 290, 351–52, 406
  minisatellite, 387–91
  of mitochondria, 146, 260–61, 447–48
  mutations in (*see* Mutation)
  nicked, 178–80, 185, 188, 190, 288–90

  noncoding, 168–69
  nucleotide sequence of, 444
  palindromes in, 371
  pericentromeric, 231
  of phage, 146, 331–32
  physical chemistry of, 143–48
  of plasmids, 369
  of prokaryotes, 146
  proof that it is genetic material, 8, 132–37
  pulse-labeled, 173–74
  radioactively labeled, 136–37
  recombinant, 294, 371–73, 378, 380
  renaturation of, 145
  repair of (*see* DNA repair)
  replication of (*see* Replication)
  selfish, 296
  sequencing of, 392–95
  size and shape of, 145–48, 168
  spacer, 237–38
  with sticky ends, 211, 371–73, 380
  structure of, 9, 139–48
  supercoiled, 146, 179–80, 221
  ultracentrifugation of, 134
  ultraviolet absorption spectrophotometry of, 134
  undermethylated, 351
  underwound, 179
  unwinding of, 178–80
  variable number tandem repeat in, 390
  X-ray diffraction pattern of, 139–40
DNA-binding protein, 214, 228–29, 353–55
dna genes, 181–82
DNA glycosylase, 288–89
DNA gyrase, 179–80, 185–86
DNA incising enzyme, 289–90
DNA ligase, 176–78, 183, 189, 290, 292, 371–73, 380
DNA photolyase, 287
DNA polymerase, 175, 177
  RNA-dependent (*see* Reverse transcriptase)
DNA polymerase I, 178, 183, 290
  Klenow fragment of, 379–80, 392–93
DNA polymerase III holoenzyme, 182, 186, 277–79, 291
DNA repair, 286–92
  defects in, 292
  error-prone, 291–92, 307
  excision, 288–90
  in human disease, 292, 306–7
  mismatch, 279, 290
  photoreactivation, 287
  postreplication (*see* DNA repair, recombination)
  recombination, 290–91
DNase, 134
DNase-hypersensitivity site, 347–48
Dobzhansky, Theodosius, 414–15
Dog
  chromosome number in, 72–73
  hip dysplasia in, 59
  karyotype of, 74
  sex chromosomes of, 74
Domain, in protein, 156–57
Dominance, 40–41
  complete, 40
  incomplete, 40–41, 43
Dominant trait, 5, 17, 19–21, 23–24
  in human beings, 35–36, 40
  pedigree for, 35–36
  X-linked, 49–50
Donkey
  chromosome number in, 72–73
  karyotype of, 74
  sex chromosomes of, 74
Dorcas gazelle, inbreeding in, 451
Double crossover, 111–12, 114, 121

Double helix, 9, 140–42
  right- vs. left-handed, 143
Down mutation, 197
Down syndrome, 92–93, 96–98, 116, 406
Drosophila melanogaster, 4
  alcohol dehydrogenase gene in, 444–45
  Bar eye mutant of, 118–19, 122–23
  bristle morphology in, 63, 122–23
  chromosome number in, 72–73
  chromosomes of, 98
  conditional lethals in, 44
  determination in, 352–59
  DNA of, 146
  egg production in, 63
  embryogenesis in, 354–57
  eye color in, 6–7, 83–85, 106–11, 275
  genetic map of, 109–13
  Glued mutant in, 432–33
  heat-shock genes in, 229, 231
  homeo box genes in, 355
  homeotic genes in, 353, 359
  hybrid dysgenesis in, 304–5
  maternal effect genes in, 355–57, 359
  miniature fork mutant in, 15
  modifier genes in, 45
  Morgan's experiments with, 6–7, 11, 83–84, 106–7, 109, 122
  pleiotropic mutants in, 43
  recombination in, 7, 116
  replication in, 174–75
  segmentation genes in, 353, 355, 357–59
  selection in, 432–33
  sex determination in, 100–101
  size of, 63
  transposons in, 302
  wing morphology in, 7, 30–31, 106–11, 122–23, 273, 275
  X chromosome of, 6–7, 48, 75–76
    attached, 118–19
  X-linked traits in, 83–85
Drosophila pseudoobscura
  chromosome variation in, 415
  genetic variation in, 417–18
  inversion heterokaryotypes in, 88
Drosophila simulans, chromosomes of, 98
Ds element, 300–302
Duffy blood group, 430
Dulbecco, Renato, 287
Duplicate gene action, 53
Duplication, 85–87, 90–91, 122–25
  displaced, 86
  reverse, 86
  tandem, 86
Dwarfism
  hypopituitary, 395–96
  six-fingered, 440, 443
Dystrophin, 407–8
Dystrophin gene, 232

East, Edward, 456
EcoRI, 371–72, 376–77, 383–84, 387
Ectoderm, 346
Edward syndrome, 97
EF. *See* Elongation factor
Egg
  enucleate, 344–45
  gradient of determinants in, 346
Electron microscopy, of DNA, 146
Electrophile, 283, 306

Electrophoresis
  of DNA, 134, 146–48, 387–89,
    393–94
  of proteins, 194, 415, 417
  pulsed-field, 148
  of ribosomal proteins, 261
  of RNA, 236, 392
Elephant, genetic variation in, 418
Elephant shrew, inbreeding in, 451
Ellis, E., 332
Ellis-van Crevald syndrome, 440, 443
Elongation factor-G (EF-G), 265–66
Elongation factor-Tu (EF-Tu), 265–66,
    308
Embryo
  mosaic, 343
  relative, 344
Embryogenesis, 344
  in *Drosophila*, 354–57
  homeo box genes in, 355
Empirical recurrence risk, 65
EMS. *See* Ethylmethane sulfonate
Endangered species, 428, 451–52
Endonuclease, uvrABC, 290
Endoplasmic reticulum, 245
Endospore, 208
engrailed gene, 353, 357
Enhancer, 225, 227–28
  immunoglobulin genes, 362–63
  light-responsive, 349
Enhancer-binding protein, 231
Enucleate egg, 344–45
Environment. *See* Genotype-
    environment interactions
Environmental variance, 60–61
Enzyme, 8, 156. *See also* Protein
  suicide, 288
Epigenesis, 342, 344
Epistasis, 51–53
EPSP synthase gene, 398–400
Equational division. *See* Meiosis II
Equatorial plate. *See* Metaphase plate
Equilibrium frequency, 433
  stable, 430
Error-prone repair, 291–92, 307
Erythrocyte, 347
Escherichia coli
  conjugation in, 314–22
  DNA of, 144, 146, 168
  genetic map of, 319–24
  mapping in, 322–30
  mutation rate in, 282
  nutritional mutants of, 276
  phage infection in, 135–36, 210–14
  promoters in, 197
  replication in, 172, 175, 181–84
  as research organism, 314
  ribosomes of, 163
  RNA polymerase of, 161
Esterase B1, 460
Estrogen, 246–48, 347
Ethical issues, in gene cloning, 402
Ethylmethane sulfonate (EMS), 284
Euchromatin, 74, 231
Eukaryote, 7
  chromosomes of, 218–22
  DNA of, 146
  expression vectors for, 384–85
  posttranscriptional control in, 242,
    244
  posttranslational control in, 242,
    245–46
  promoters in, 223–27
  replication in, 174–75
  ribosomes of, 260–61
  RNA polymerase of, 222–23
  transcriptional control in, 222,
    227–31, 242–44, 246–48
  translational control in, 242,
    244–45
  transposable elements of, 300–305
Euploidy, 85, 93

Evolution
  of cytochrome *c*, 446
  of fibrinopeptide, 446
  of hemoglobin, 446–47
  of histones, 219, 446
  of human beings, 447–48
  of mitochondrial DNA, 447–48
  molecular, 444–49
  of pesticide resistance, 459
  of sigma factors, 231
Excision repair, 288–90
Exon, 232
Exonuclease
  3′ → 5′, 186, 289
  5′ → 3′, 288–90
Expression vector, 381–85
  for eukaryotes, 384–85
  fusion protein production from,
    383–84
  inducible, 382–83
  promoter on, 381–82
Expressivity, 43–45
Extinction, 449–452

Factor IX gene, 397–98
Facultative heterochromatin, 74
Falconer, Douglas, 455
Fanconi's anemia, 292
Fate map, 359
F-duction, 323–24, 327
Fedoroff, Nina, 300
Felsenfeld, Gary, 347
Fertilization, 72–73, 82, 245
Fetal hemoglobin, 365
F₁ generation, 5–6, 17, 19, 23
F₂ generation, 5, 19–25, 106
F₃ generation, 20–21
Fibrinopeptide, evolution of, 446
Fine structure mapping, 335, 394–95
Fingerprinting
  of DNA, 66, 387–91
  of protein, 166
Fingerprint-ridge count, 61–64
Fink, Gerald, 303–4
First filial generation. *See* F₁
    generation
Fish, sex determination in, 99
Fitness
  mean, 431–32, 450–51
  relative, 431–32, 434–35, 451
F-*lac* plasmid, 323
Flax, rust resistance in, 460–61
Flour beetle, homeotic genes in, 123
Fluctuation test, 317–18
Forced cloning. *See* Directional cloning
Forensic science, 390–91
Forked-line approach, 29–31
Forward mutation, 286
Forward mutation rate, 426
Founder effect, 443, 451
Four-o'clock, flower color in, 40–41
Four-stranded crossover, 117–22
F pilus, 315, 317, 319
F plasmid, 316–319, 323
F′ plasmid, 323–324
Frameshift mutation, 255–56, 277,
    279–80, 283, 286–87
Frankel-Conrat, H., 255
Franklin, Rosalind, 9–11, 139–40
Free radical, 285
Frelinger, Jeffrey, 435
Frog
  chromosome number in, 72–73
  DNA of, 146
Fruit fly. *See Drosophila melanogaster*
Functional group, 132–33
Furuichi, Yasuhiro, 241
fushi tarazu gene, 353, 357–58
Fusion protein, 381, 383–84

Galactose-1-phosphate uridyl
    transferase gene, 418
Galactoside permease, 200, 203
Galactoside transacetylase, 200, 203
Galápagos Islands, 413, 450
gal operon, 205
Gamete, 3, 5, 21–23, 72–73, 79
  coupling, 107
  repulsion, 107
Gametogenesis, 72–73
Gamma radiation, as mutagen, 285
Garden pea
  anatomy of, 18
  chromosome number in, 72–73
  life cycle of, 18
  Mendel's experiments with, 5,
    16–19, 31–33, 43, 110–11
  seed shape in, 40, 42
Garen, Alan, 267
Garrod, Archibald, 8, 11, 156–57
Gastrulation, 347
G-band, 74
GC box, 224–25, 228, 230
GC content, of DNA, 144
Gehring, Walter, 355, 357–58
Gel electrophoresis, 415
Gene, 5, 16. *See also* DNA; *specific
    genes*
  chromosomes and, 82–85
  colinearity with protein, 167–68
  complementary (*see*
    Complementation)
  homeotic, 123, 353
  housekeeping, 342
  interrupted, 232–37
  jumping (*see* Transposon)
  light-induced, 348–49
  linked (*see* Linkage)
  luxury, 342
  master, 353
  maternal effect (*see* Maternal
    effect genes)
  modifier, 45
  monomorphic, 417
  multiple, 51–53
  nonallelic copies of, 365
  overlapping, 395
  pair-rule, 357
  polymorphic, 417–18
  rearrangement of, 361–65
  relationship to proteins, 8–9
  sequencing of, 392–95
  structural, 158, 200
  syntenic, 110
Gene cloning, 10, 370–85. *See also*
    Cloned gene
  cDNA cloning, 379–80
  directional, 374
  ethical issues in, 402
  expression of cloned genes, 380–85
  forced (*see* Gene cloning,
    directional)
  gal identification of specific clone,
    378–79
  manipulation of cloned genes,
    385–95
  practical applications of, 395–401
  safety of, 402
  using restriction endonucleases,
    370–72
  vectors for, 372–78
Gene cluster, 364
Gene expression, 158–65. *See also*
    Transcriptional control;
    Translational control
Gene family, 350, 387, 448
  sequential activation of members
    of, 364–65

Gene flow, 429–30, 452
  artificial, 453
Gene mutation, 274
Gene pool, 414
Gene probe. *See* Probe
Generalized recombination, 188–90
Generalized transduction, 326, 328–30
Gene therapy, 396–97
Genetic code, 10, 164, 255–61. *See
    also* Codon
  deciphering of, 257–58
  degeneracy in, 164–65, 257–58
  lack of gaps in, 255–56
  in mitochondria, 260–61
  nonoverlap of codons, 255
  triplet nature of, 256–57
  universality of, 259–61
Genetic counseling, 64–65, 93, 406–7,
    442
Genetic disease, 59, 450. *See also
    specific diseases*
  frequency of, 440–43
  intervening in, 396–97
  mapping of, 401–8
  mutation-selection balance for, 440
Genetic drift, 418, 427–29, 443–45,
    451–52
Genetic linkage. *See* Linkage
Genetic map, 109–11
  circular, 320–21
  of *Drosophila melanogaster*,
    109–13
  of *Escherichia coli*, 319–24
  of human being, 114–17, 124–25
  of phage λ, 211
  of phage Mu, 300
  of phage φX174, 395
Genetic mapping. *See* Mapping
Genetic recombination. *See*
    Recombination
Genetics
  classical (*see* Genetics, Mendelian)
  conservation, 449–52
  developmental, 342–65
  Mendelian, 4–8, 33
  molecular, 4, 8–10, 132
  population, 4, 414
  transmission (*see* Genetics,
    Mendelian)
Genetic screening, 406–7
Genetic variance, 60–61
  estimation of, 61–63
Genetic variation. *See* Variation
Genome, 93
Genomic library, 376–77
  of human genes, 376–77
Genotype, 6, 21
  norms of reaction for, 55–56
Genotype-environment interactions,
    54–56, 59–60
Genotypic frequency, 416–23
Genotypic value, 60
Genus, 4
Germinal mutation. *See* Germ-line
    mutation
Germ-line mutation, 274–75
Germ plasm collection, 449–50
Gilbert, Walter, 232, 392
Globin genes, 87, 123, 239, 243, 341,
    343
Globular protein, 155–56
Glucose, in catabolite repression, 205
Glued mutant, 432–33
Glutamate-oxaloacetate transaminase
    gene, 418
Glutamate-pyruvate transaminase
    gene, 418
Glycosidic bond, 288
Glyphosate resistance, 398–400
Goat
  karyotype of, 74
  sex chromosomes of, 74
Golden lion tamarin, inbreeding in,
    451

Golgi apparatus, 245
Goodness of fit, 31–33
Gorilla
    chromosome banding patterns in,
        99
    phylogenetic relationship to other
        primates, 447
gp28, 207, 209
gp33, 207, 209
gp34, 207, 209
G₁ phase, 76
G₂ phase, 76–77
G protein, 308–9
Grandpaternity exclusion, 67
Greater galago, inbreeding in, 451
Green revolution, 449
Green sea turtle, mitochondrial DNA
    of, 447–48
Griffith, Frederick, 8, 132–35
GTP
    in messenger RNA capping, 241
    in translation, 263–64, 266
GU—AG motif, 234
Guanine, 9, 137, 140
    alkylation of, 284
    tautomers of, 278
Gurdon, John, 344–45
Gusella, James, 403
Gyurasits, Elizabeth, 173

HaeIII, 371, 387–88
Hairy ear rims, 50–51
Haldane, J. B. S., 116
HAP. *See* Heat-shock activator protein
Haplodiploid, 72
Haploid, 5, 72–73
Haplotype, 66
Ha-*ras* oncogene, 307–10
Hardy, Godfrey, 11, 419
Hardy-Weinberg principle, 419–23
    for multiple alleles, 422
    for X-linked genes, 422–23
Harris, Harry, 417
Hayes, William, 316–17
Heat-shock activator protein (HAP),
    229, 231
Heat-shock genes, 209
HeLa cells, 228–29
Helicase, 178
Helix-turn-helix motif, 214, 353
Helper phage, 326
Hemoglobin, 156, 165, 342
    evolution of, 446–47
    fetal, 365
Hemoglobin S, 165–67
Hemophilia, 49–50, 115, 274, 397–98,
    422–23
Hemophilus influenzae, DNA of, 146
Herbicide resistance, 398–400
Heritability, 61, 452
    realized, 61
Hermaphrodite, 360
Hershey, A. D., 135–36, 334
Hessian fly, 460
Heterochromatin, 74–75, 231
    constitutive, 74
    facultative, 74
Heteroduplex DNA, 188, 190
Heterogametic sex, 100
Heterogenous nuclear RNA (hnRNA),
    233–34
Heterokaryotype, inversion, 88–90
Heterokaryotypic individual, 86–87
Heterosis, 456–57
Heterozygosity, average, 417–18
Heterozygote, 5–6, 21–22, 24–27
    coupling double, 109
    deletion, 87, 116
    inversion, 88

phenotype of, 40–41
repulsion double, 109
translocation, 90–93
Heterozygote advantage, 442, 459–60
Hexokinase gene, 418
Hexon gene, 232–33
Hfr strain, 317–21, 323
h gene, 334–35
hGH. *See* Human growth hormone
HindII, 370–71
HindIII, 371–72, 378, 382, 387, 402–5
Histone, 219–21, 350–51
    evolution of, 219, 446
    H1, 219–20, 350, 352
    H2A, 219
    H2B, 219
    H3, 219
    H4, 219
    IV, 446
Histone genes, 239, 244, 448
HIV. *See* Human immunodeficiency
    virus
HLA genes, 47–48, 66–67, 124, 391
hnRNA. *See* Heterogeneous nuclear
    RNA
Hogness, David, 359
Holliday, Robin, 189
Holmberg, Scott A., 298
Homeo box, 353–55, 359
    in vertebrates, 359–60
Homeo box genes, in embryogenesis,
    355
Homeologous chromosomes, 94
Homeotic gene, 123, 353
    in *Caenorhabditis elegans,* 361
    in *Drosophila,* 359
Homoeosis, 353
Homogametic sex, 100, 116
Homogentisic acid, 157–58
Homologous chromosomes, 7, 72,
    79–81
Homozygote, 5–6, 21–22
    selection against, 431–33, 440–41
Homunculus, 342
Hood, Leroy, 362
Hormone, 156, 347
    gene activation by, 246–48
Hormone receptor, 246–48
Horse
    chromosome number in, 72–73
    karyotype of, 74
    sex chromosomes of, 74
Horseshoe crab, genetic variation in,
    418
Horvitz, Robert, 361
Host-range mutant, of T-even phage,
    334
Hotchkiss, Rollin, 135, 351
Hot spot, 226–27, 281
Housefly, pesticide resistance in, 457
Housekeeping gene, 342
HpaI, 383, 442
HpaII, 371, 406
HTF island, 406
Huberman, J., 175
Human being
    admixture of races, 430
    aneuploidy in, 96–97
    average heterozygosity in, 418
    behavioral traits in, 55
    chromosome banding patterns in,
        99
    chromosome number in, 72–73
    chromosomes of, 105, 151, 218
    DNA of, 146
    dominant traits in, 35–36, 40
    evolution of, 447–48
    genetic diseases in (*see* Genetic
        disease)
    genetic map of, 114–17, 124–25
    genetic variation in, 418
    genomic library of, 376–77
    height of, 58–59, 61

karyotype of, 74–75
mapping in, 115–17, 401–8
Mendelian inheritance in, 33
mitochondrial DNA of, 448
mutation rate in, 281–82
phylogenetic relationship to other
    primates, 447
polymorphic genes in, 417–18
positive assortative mating in, 423
recessive trait in, 34–35
recombination in, 116
sex chromosomes of, 48, 97
transgenic, 402
X chromosome of, 115
Human growth hormone (hGH),
    395–97
Human immunodeficiency virus
    (HIV), 396
Human leukocyte antigen genes. *See*
    HLA genes
Huntington's disease, 35, 40–41,
    44–45, 276, 401
    mapping gene for, 403–5
    pedigree of family with, 405
    screening for, 406
Hybrid dysgenesis, 304–5
Hybridization
    colony, 380
    DNA-DNA, 145, 387–89, 402–5
    DNA-RNA, 145, 232–33, 242–43
    plaque, 377
    somatic cell, 116–17
    stringency of, 379
Hybrid population, 430
Hybrid vigor, 456
Hydrogen bond, in DNA, 140–41, 144
Hydroxyl group, 133
Hymenoptera, sex determination in,
    100
Hypoxanthine, 280–81

iab-2 gene, 359
ICR. *See* Internal control region
Identical twins. *See* Monozygotic twins
IF. *See* Initiation factor
Illegitimate recombination, 189, 297
Imaginal disk, 355
Immune electron microscopy, 261
Immunoglobulin, 361–62
    constant region of, 361–62
    domains of, 157
    heavy chain of, 157, 361–63
    light chain of, 157, 361–63
    variable region of, 361–62
Immunoglobulin genes, 228, 448
    joining region of, 362–63
    mutations in, 363
    rearrangement of, 361–63
Inborn errors of metabolism, 65
Inbred line, 454–56
Inbreeding, 423–24, 442–43, 452
Inbreeding coefficient, 424–26, 451
Inbreeding depression, 450–51
Incision, 288–89
Incomplete dominance, 40–41, 43
Incomplete penetrance, 43
Independent assortment, 83, 107, 109
    principle of, 23–25
Independent culling level, 454
Independent events, 25
Induced mutation, 426
Inducer
    embryonic, 346–47
    lac, 201–2, 204
Industrial melanism, 434–35
Ingram, Vernon, 165

Inheritance
    blending, 16, 19–20
    chromosome theory of, 5–6
    Mendel's laws of, 5
    particulate, 16
Initiation complex, 264
Initiation factor-1 (IF-1), 263–64
Initiation factor-2 (IF-2), 264
Initiation factor-3 (IF-3), 263–64
Inosine, 258–59
Inouye, Masayori, 381
Insect
    pesticide resistance in, 59, 460
    sex determination in, 100
Insertion, 10, 256
    single-base, 279–80
Insertion sequence, 294–96, 298–99
Insertion sequence-like element,
    298–99
Insulin, 342–43, 395
Insulin gene, 245–46, 343, 379
Integrated pest management, 460–61
Intelligence, race and, 64
Interference, 112–14
Interferon gene, 385
Intergenic region, 168, 286
Intergenic suppression, 286
Interkinesis, 80
Internal control region (ICR), 226–27
Interphase, 76, 79
Interrupted gene, 232–37
Intersex, 101
Interstitial deletion, 87
Interstitial translocation, 90
Intervening sequence. *See* Intron
Intracistronic complementation, 326
Intragenic suppression, 286
Intron, 168–69, 232–37, 286
Inversion, 85, 88–90, 299–300, 415
    paracentric, 88
    pericentric, 88
Inversion heterokaryotype, 88–90
Inversion heterozygote, 88
Inversion loop, 88–90
Ion exchange chromatography, of
    proteins, 222–23
IQ score, 61
Island population, 429–30
Islet of Langerhans, 342–43
Isoschizomer, 371

Jacob, François, 11, 160, 202–4,
    317–18, 320
Jacobs, W., 132
Japanese quail, inbreeding of, 450–51
Japanese serow, inbreeding in, 451
Jeffreys, Alec, 387, 390
Jones, Peter, 352
Jumping gene. *See* Transposon
Juvenile hormone resistance, 459

Kanamycin resistance, 299
Karpechenko, G., 94
Karyotype, 73–74
Kazazian, Haig, 305
Kel Kummer Taureg tribe, 442
Kelner, Albert, 287
Khorana, Har Gobind, 10–11, 256
Kidwell, Margaret, 305
Kilobase pair, 148
Kimura, Motoo, 444–45

Kinetochore, 73
King, Thomas, 344
Klenow fragment, 379–80, 392–93
Klinefelter syndrome, 97–98, 100
Klug, Aaron, 220
Kornberg, Arthur, 178, 182
KpnI, 371
Kunkel, L. M., 407

lacA gene, 200
lacI gene, 201, 204
lacY gene, 200, 203, 324–25
lacZ gene, 200, 324–25
lacZ' gene, 374–75
lac operon, 200–216
 on F-*lac* plasmid, 323
 negative control of, 201–4
 positive control of, 205–6
Lac^V phenotype, 322
Lactose, 200
Land race, 449–50
Leader region, 232
Leder, Philip, 234, 362
Lederberg, Joshua, 314
Leptotene, 79
Lesch-Nyhan syndrome, 397
Lethals, 41–42, 87, 276, 432–33, 455
 conditional, 44
Levan, Albert, 75
Levene, P., 132
Lewontin, Richard, 417
lexA gene, 291–92
Life cycle
 of garden pea, 18
 of *Neurospora,* 118–19, 158–59
Light-induced genes, 348–49
Light repair. *See* Photoreactivation
Lily, DNA of, 146
lin-12 gene, 361
Linkage, 25, 106
 in human beings, 114–17
 mechanism of, 106–9
Linkage disequilibrium, 442
Linkage map. *See* Genetic map
Linn, Stewart, 370
Livestock feed, antibiotics in, 299
Locus, 21
Long terminal repeat (LTR), 302–3
LTR. *See* Long terminal repeat
Luciferase gene, in tobacco, 400–401
Lung cancer, 307–8
Luria, Salvador, 317
Luxury gene, 342
Lwoff, André, 203
Lymphoma, Burkitt's, 364
Lyon, Mary, 352
Lysenko, Trofim, 458
Lysogen, 210
Lysogenic infection, 210, 212–14, 326
Lytic infection, 210–12, 326

McCarty, Maclyn, 11, 134
McClintock, Barbara, 7, 11, 300–302
MacLeod, Colin, 11, 134
Major histocompatibility complex
 (MHC) genes, 123
Malaria, 167, 415–16, 435, 442, 457
Mal de Meleda, 440, 442–43
Malic enzyme gene, 418
Manic-depression, 55
Manx cat, 42, 45

Map. *See* Genetic map
Map distance, 109, 168
 physical distance and, 124–25
Mapping, 7
 in bacteria, 322–30
 by conjugation, 318–21
 deletion, 116
 by DNA sequencing, 394–95
 by F-duction, 323
 fine structure, 335, 394–95
 in human beings, 115–17, 401–8
 in phage, 334–35
 with restriction fragment length
  polymorphisms, 124–25,
  401–5
 with three-point crosses, 111–12
 transduction, 328–30
Map unit, 109, 319, 335
Marker, 314
Master gene, 353
Maternal age, chromosome
  abnormalities and, 98
Maternal effect genes, in *Drosophila,*
  355–57, 359
Maternal message, 244–45, 346
Mating
 assortative, 423
 consanguineous, 33, 426, 442–43,
  450
 nonrandom, 423–24
 random, 419–20
 reciprocal, 17
Mating type genes, in yeast, 355
Matthaei, Johann Heinrich, 256
Maxam, Alan, 392
MboI, 371
MCS. *See* Multiple cloning site
M cytotype, 305
Mean, 58–59
Mean fitness, 431–32, 450–51
Megabase, 148
Megasporogenesis, 82
Meiosis, 22, 25, 72–73, 78–82
 in tetraploid, 95
 in triploid, 94
Meiosis I, 79–80
Meiosis II, 80–81
Melanin, 40, 275, 441
Melanism, industrial, 434–35
Melting temperature, of DNA, 144
Mendel, Gregor, 5, 11, 16, 23, 110–11
Mendelian genetics, 4–8
 in human beings, 33
Mercury resistance, 299
Meristic trait, 59
Merodiploid, 203–4, 316, 322–24,
  326–27
Merozygote. *See* Merodiploid
Meselson, Matthew, 11, 152, 160
Mesoderm, 346
Messenger, 160
Messenger RNA (mRNA), 10, 159,
  163
 bicoid, 356–57
 binding to ribosomes, 263–65
 capping of, 241, 244, 265
 casein, 244
 degradation of, 244
 discovery of, 159–61
 distribution in embryo, 356–58
 fushi tarazu, 357–58
 globin, 239, 243
 half-life of, 244
 histone, 239, 244
 immunoglobulin, 362–63
 lariat model of splicing of, 236–37
 maternal message, 244–45, 346
 ovalbumin, 243–44
 polyadenylation of, 239–40, 244
 polycistronic, 200
 quantitation of, 392
 RNA polymerase producing,
  222–23
 synthetic, 256–57

Messenger RNA (mRNA) genes, 232
Messing, Joachim, 378
Metabolic intermediate, 8
Metabolic pathway, 8, 157–58
Metabolism, inborn errors of, 52–53,
  65
Metacentric chromosome, 73–74,
  77–78, 88, 93, 114
Metafemale, 97, 100–101
Metal finger, 228–29
Metamale, 100–101
Metaphase
 meiosis I, 80
 meiosis II, 80
 mitotic, 71, 77–78
Metaphase plate, 78, 80
Methylation, of DNA, 290, 351–52,
  406
5–Methylcytosine, 281, 351–52
Methyl group, 133
Methylguanine, 288
Methylguanine methyl transferase, 288
Methyl transferase, 241
MHC genes. *See* Major
  histocompatibility complex
  genes
Microchromosome, 73
Micrococcal nuclease, 218
Microinjection, 344–45
2–Micron plasmid, 385
Microsporogenesis, 82
Microtubules, 78
Miescher, Friedrich, 11, 132
Migration, 429
Miller, Oscar, 237
Minichromosome, 218–20
Minimal medium, 275–76, 314
Minisatellite, 387–91
Minute (map unit), 319
Mismatch repair, 279, 290
Mispairing, 122
Missense mutation, 255, 277
Mitochondria
 DNA of, 146
 evolution of, 447–48
 genetic code in, 260–61
 mutations in, 447
 ribosomes of, 266
Mitochondrial Eve, 448
Mitosis, 71–73, 76–78, 81
MN blood group, 40–41, 66
Modifier gene, 45
Molecular clock, 445–46
Molecular evolution, 444–49
Molecular genetics, 4, 8–10, 132
Monoculture, 457
Monod, Jacques, 160, 202–4
Monohybrid cross, 17, 20–21
Monosomy, 85, 95
Monozygotic twins, 55, 61–62, 388–89
Morgan, Thomas Hunt, 6–7, 11,
  83–84, 106–7, 109, 122
Morphological mutation, 275
Mosaic development, 343–44
Mosquito, pesticide resistance in, 457
Mouse
 body weight in, 453
 chromosome number in, 72–73
 coat color in, 41–42, 44, 54,
  275–76
 DNA of, 146
 immunoglobulin genes of, 362
 inbred lines of, 455–56
 litter size in, 63, 455–56
 mitochondrial DNA of, 447
 mutation rate in, 282
 supermouse, 397
 tail length in, 63
 weight of, 63
M phase. *See* Mitosis
mRNA. *See* Messenger RNA
Muller, H. J., 11, 112
Multigene family, 125, 448

Multiple alleles, 45–48, 422
Multiple cloning site (MCS), 374–75,
  378, 381
Multiple drug resistance, 298–99
Multiple genes, 51–53
Multiplicity of infection, 334
Muscular dystrophy, 232, 401, 422
 identification of gene for, 407–8
Mutagen, 9, 277, 285
 as carcinogen, 305–7
 identification of, 293–94
 metabolism of, 294
Mutagenesis
 chemical, 283–84, 386
 site-directed, 223, 378, 385–86
 transposon, 298–300, 304–5
Mutant, 4
 constitutive, 203
 cryptic, 202–3
 fixation in population, 444
 loss from population, 444
Mutation, 9–10, 165–67, 186
 cancer and, 305–10
 chemically induced, 283–84
 cis-dominant, 203
 conditional, 276–77
 down, 197
 effect on DNA, 276–86
 forward, 286
 frameshift, 255–56, 277, 279–80,
  283, 286–87
 gene, 274
 genetic variation and, 426–27
 germ-line, 274–75
 in cis, 326
 induced, 426
 in trans, 326
 lethal (*see* Lethals)
 missense, 255, 277
 morphological, 275
 mutation-selection balance, 440
 nutritional, 275–76
 operator constitutive, 203–4
 point, 277
 in proto-oncogenes, 307–8
 radiation-induced, 283–84
 silent, 281, 285–86
 somatic, 274–75, 307, 363
 spontaneous, 277–82, 426
 temperature-sensitive, 276–77
 up, 197
 visible (*see* Mutation,
  morphological)
Mutation rate, 281–82, 442, 445–46
 backward, 426
 forward, 426
 in mitochondrial DNA, 447
 molecular evolution and, 444–45
Mutator, 277–79
mut genes, 277–79
Mutually exclusive events, 25
myc oncogene, 364
Mycoplasma
 DNA of, 144
 genetic code in, 260–61
Myeloma, 362
Myoglobin, 155–56, 251

Nail-patella syndrome, 115
Nalidixic acid, 180
Natal homing, 448
Natural selection, 414, 431–36
 against homozygote, 431–33,
  440–41
 heterozygote advantage, 433–36,
  442
 mutation-selection balance, 440
 purifying, 431

Nature-nurture dichotomy. *See* Genotype-environment interactions
Negative control, 201–2
Neurospora crassa
    Beadle and Tatum experiments with, 9, 158
    chromosome number in, 72–73
    DNA of, 146
    life cycle of, 118–19, 158–59
    mutation rate in, 282
    nutritional mutants of, 276
    tetrad analysis in, 118–21
Neutrality theory, 444–45
Neutron scattering, 261
Newt, replication in, 174–75
N-Formyl methionine, 263–64
N gene, 211–12
Nicholas II (Czar of Russia), 49–50
NIH 3T3 cells, 307–8
Nilsson-Ehle, Herman, 56
Nirenberg, Marshall, 10–11, 256–57
Nitrocellulose filter, 377, 384, 387–88
Nitrogenous base, 9, 137
    alkylation of, 283
    analogs of, 283
    deamination of, 280–81, 283
    modified, 252, 281
    tautomers of, 278–79
Nitrous acid, 283
Noll, Markus, 355
Nomura, Masayasu, 260
Nondisjunction, 84–85, 95, 97
Nonhistone protein, 222
Nonparental chromosomal type, 108
Nonrandom mating, 423–24
Nonrecombinant, 108
Nonsense codon. *See* Stop codon
Normal distribution, 57–58
Norm of reaction, 55–56
Northern blotting, 392
Northern elephant seal, genetic variation in, 418, 428
Novobiocin, 180
n' protein, 182
Nuclear membrane, 78
Nuclear polyhedrosis virus, 385
Nucleic acid, 132. *See also* DNA; RNA
Nuclein, 132
Nucleolus, 77–78, 238, 261
Nucleoside, 137–38
Nucleosome, 218–22, 347–48, 350–51
Nucleotide, 8, 138, 194
Nucleus, transplantation of, 344–45
Nullisomy, 95
Nutritional mutation, 275–76

O

Observation
    highly statistically significant, 32
    statistically significant, 32
Ochoa, Severo, 268
Ochre codon, 267–68
O gene, 212
Okazaki, Reiji, 175–76
Okazaki fragment, 176–78
Oligomeric protein, 325
Oligonucleotide primer, 178
O'Malley, Bert, 243
Oncogene, 307–10
    rearrangement of, 364
Oncogenesis, 307–8
One gene-one enzyme hypothesis, 9
One gene-one polypeptide hypothesis, 158
Oocyte, 98
    primary, 82
    secondary, 82
    5S ribosomal RNA genes in, 350–51

Oogenesis, 81–82, 244
Oogonium, 82
Opal codon, 267–68
Open reading frame (ORF), 394, 405
Operator, *lac*, 201, 203–4
Operator constitutive mutant, 203–4
Operon, 200–206. *See also specific operons*
Opine, 398
Orangutan, 99, 447
ORF. *See* Open reading frame
Organic molecules, 132
Organophosphate resistance, 460
Organ transplant, 47–48
Origin of replication, 173, 175
Orphan disease, 440
Outbreeding, 423
Ovalbumin gene, 246–48, 347
Ovary, 82
Overlapping genes, 395
Ozone layer, 284

P

p21, 307–9
Pachytene, 80
PAGE. *See* Polyacrylamide gel electrophoresis
Pair-rule gene, 357
Palindrome, 371
Palmiter, Richard, 397
Pancreatic amylase gene, 418
Pantothenic acid, biosynthesis of, 159
Paracentric inversion, 88
Pardee, Arthur, 203
Parentage, establishment of, 390
Parental chromosomal type, 108
Parthenogenesis, 71
Particulate inheritance, 16
Pascal's triangle, 27–28
Patau syndrome, 97
Paternity exclusion, 65–66
Pattern baldness, 50
Pauling, Linus, 139
PbsA protein, 460
PCR. *See* Polymerase chain reaction
Pedigree
    with consanguineous mating, 426, 443
    for dominant trait, 35–36
    for Huntington's disease family, 405
    for incompletely dominant allele, 43
    for recessive trait, 34–35
    for recessive trait caused by two genes, 53
    symbols used in, 33
    with two X-linked genes, 115
    for X-linked dominant trait, 50
    for X-linked recessive trait, 49
P element, 305
Penetrance, 43–45
    complete, 43
    incomplete, 43
Peppered moth
    color of, 47
    industrial melanism in, 434–35
Pepsinogen gene, 418
Peptidase A gene, 418
Peptidase C gene, 418
Peptidase D gene, 418
Peptide bond, 155
    formation of, 155, 266
Peptidyl-puromycin, 266–67
Peptidyl transferase, 265–67
Pere David's deer, genetic variation in, 428
Pericentric inversion, 88
Pericentromeric DNA, 231

Permissive temperature, 276–77
Pesticide resistance, 59, 429, 457–61
    control of, 460–61
    evolution of, 459
Petite yeast, 275
pfu. *See* Plaque-forming unit
P gene, 212
Phage. *See also* Transduction
    assay for, 332
    DNA of, 146, 331–32
    genetics of, 330–37
    helper, 326
    infection by
        burst size from, 332
        latent period in, 332–33
        rise period in, 332–33
    mapping in, 334–35
    promoters of, 197
    recombination in, 334–35
    temperate, 210
    T-even (*see* T-even phage)
    transducing, 326–28
    as vectors, 374–78
    virulent, 210
Phage fd, 331
Phage G4, 181
Phage λ, 189, 210–14, 294–95, 326, 331, 380, 384
    as carrier of host genes, 326–27
    DNA of, 146
    genetic map of, 211
    λgt11, 383–84
    lysogenic cycle of, 212–14
    lytic cycle of, 211–12
    replication of, 184
    as vector, 374–77
Phage M13, 177, 181, 378, 385–86
Phage MS2, 331
Phage Mu, 298–300
    genetic map of, 300
Phage P1, 326, 328–30
Phage φX174, 146–47, 181–84, 331, 394
    genetic map of, 395
Phage PM2, 146
Phage R17, 264, 269
Phage SPO1, 206–7
Phage T2, 131, 160–61, 168, 282, 331, 334
    Hershey-Chase experiment with, 135–36
Phage T4, 168, 195–96, 206, 267, 331–32, 335–37
    DNA of, 144
    rapid-lysis mutants of, 275
    replication in, 176–77
Phage T5, 331
Phage T6, 331
Phage T7, 208
Pharmaceuticals, genetically engineered, 395–96
Phenocopy, 45
Phenotype, 4, 17
Phenotypic distribution, of quantitative trait, 57
Phenotypic value, 59–61
Phenotypic variance, 60–61
Phenylalanine, breakdown of, 158
Phenylalanine-tRNA synthetase, 163
Phenylketonuria (PKU), 43, 55, 440–41
Phosphate group, 132–33, 138, 283
Phosphodiester bond, 139, 161
    formation of, 194–95
Phospho-ester bond, 138
Phosphoglucomutase-1 gene, 418
Phosphoglucomutase-3 gene, 418
6–Phosphogluconate dehydrogenase gene, 418
Photoreactivating enzyme, 287
Photoreactivation, 287
Phylogenetic tree, 446–47
Pigeon, transferrin gene in, 435–36

Pigweed, atrazine resistance in, 460
Pine, chromosome number in, 72–73
Pink bread mold. *See Neurospora crassa*
Pisum sativum. See Garden pea
PKU. *See* Phenylketonuria
Plant
    aneuploidy in, 96
    crop (*see* Crop plant)
    development in, 344
    genetic variation in, 457
    insect resistance in, 460
    light-induced genes in, 348–49
    pest resistance in, 457
    polyploidy in, 93–94
    seedless, 94
    self-fertilization in, 423–24
    sex determination in, 99–100
    transgenic, 348–49, 398–401
    vectors for, 398–99
Plant breeding, 61, 63, 94, 449–57
Plaque, 326, 332
Plaque assay, 332
Plaque-forming unit (pfu), 332–33
Plaque hybridization, 377
Plasmid. *See also* Conjugation
    with antibiotic resistance genes, 298–99
    DNA of, 369
    replication of, 175
    with transposon, 296–97
    as vector, 372–75
Plasmid pβ-gal13C, 383
Plasmid pBR322, 372–73, 377, 380, 385
Plasmid pIN, 381
Plasmid pKC30, 383
Plasmid pSC101, 371–72
Plasmid *ptrpLl,* 382
Plasmid pUC series, 374–75, 380–81
Plasmid RSF1010, 371–72
Plasmid Ti, 398–99
Platt, Terry, 198
Pleiotropy, 42–43, 325
P-M system, 304–5
Point mutation, 277
Polar body, 82
Pole cell, 355
pol genes, 303
Pollen, 82
Poly(AAG), 257
Polyacrylamide gel electrophoresis (PAGE), 194–95, 236
Polyadenylation, of messenger RNA, 239–40, 244
Poly(A) polymerase, 239–40
Polycistronic messenger RNA, 200
Polygenic trait, 56
Polyhedrin gene, 385
Polymerase chain reaction (PCR), 390–91
Polymorphism, 403, 417–18
Polynucleotide probe, 378–79
Polyoma virus, 185
Polypeptide. *See* Protein
Polyploidy, 93–95
Polytene chromosome, 75–76
Poly(U), 256
Poly(UAUC), 257
Poly(UC), 256–57
Poly(UUC), 256–57
Population, 416
    hybrid, 430
Population genetics, 4, 414
Positive control, 205–6
Postreplication repair. *See* Recombination repair
Posttranscriptional control
    of development, 347
    in eukaryotes, 242, 244
Posttranslational control, in eukaryotes, 242, 245–46

Potato, ploidy of, 93
Potentilla glandulosa, genotype-environment interactions in, 55–56, 60
Pott, Sir Percival, 305–6
Poultry. *See also* Chicken
  egg weight in, 63
  viability of, 63
  weight of, 63
Predation, 434–35
Preproinsulin, 245–46
Pribnow, David, 197
Pribnow box. *See* –10 Box
Primary transcript, 233
Primase, 181–83
Primer, 177–78, 181, 186
  removal of, 178, 183
  synthesis of, 181–82
  synthetic, 386
Primosome, 181–83
Primrose, flower color in, 45
Probability, 25
  binomial, 27–29
  conditional, 26–27
Proband, 34
Probe, 377–79, 402–5
  cloned gene as, 387–92
  counting genes with, 387–88
  fingerprinting DNA with, 387–91
  measurement of gene activity with, 392
  synthetic, 379
Product rule, 25–27
Proflavin, 283
Proinsulin, 246
Prokaryote, 7
  DNA of, 146
  promoters in, 197
  ribosomes of, 260
  transcriptional control in, 200–214
  transcription in, 194–99
  transposons in, 294–97
Prolactin, 244
Promoter, 161
  efficiency of, 197
  of *Escherichia coli,* 197
  of eukaryotes, 223–27
  on expression vector, 381–82
  fushi tarazu, 358
  immunoglobulin genes, 362–63
  lac, 204–5, 381–82
  lambda, 211–13, 382–83
  lpp, 381
  mannopine synthase, 398, 400
  mutations in, 197
  of phage, 197
  polyhedrin, 385
  of prokaryotes, 197
  RNA polymerase binding to, 196–97
  SV40, 224, 230
  thymidine kinase, 224–25, 230–31
  trp, 382
Promoter complex
  closed, 196–97
  open, 196–97
Prophage, 210
Prophase
  meiosis I, 79–80
  meiosis II, 80
  mitotic, 77–78
Protein
  allosteric, 202
  amino acid sequence of, 444–46
  colinearity with gene, 8–9, 167–68
  denaturation of, 194, 276
  in differentiated cells, 342–43
  domains in, 156–57
  electrophoresis of, 194, 415, 417
  export from cell, 245
  fingerprinting of, 166
  function of, 156–58

fusion, 381, 383–84
genetically-engineered, 385–86, 395–96
globular, 155–56
interaction with DNA, 214
ion exchange chromatography of, 222–23
oligomeric, 325
posttranslational modification of, 245–46
primary structure of, 155
production of, 10
quaternary structure of, 156–57
secondary structure of, 155
sequencing of, 165–66
synthesis of (*see* Translation)
tertiary structure of, 155–56
from transgenic organisms, 397–98
Protenor, sex determination in, 99–100
Proteolytic processing, of proteins, 245–46
Proto-oncogene, 307–10, 364
Prototroph, 276, 314
Provirus, 302–3
Pseudodominance, 87
Pseudogene, 365
Pseudouridine, 252
PstI, 371–73
Puberty, 82
Pulse-chase experiment, 244
Pulsed-field electrophoresis, 148
Punnett, R. C., 24, 51, 106
Punnett square, 24
Purifying selection, 431
Purine, 137
Puromycin, 266–67
PvuI, 371
Pygmy hippopotamus, inbreeding in, 451
Pyloric stenosis, 65
Pyrimidine, 137
Pyrimidine dimer, 291–92

Q

Q-band, 74
Q gene, 212
Quadrivalent, 95
Quadruple crossover, 114
Quantitative trait, 56–61
Queen Victoria (England), 49

R

Rabbit, coat color in, 45
Race
  admixture estimates, 430
  intelligence and, 64
Radiation
  cancer-causing, 306
  mutations induced by, 284–85
Radish, cross with cabbage, 94
Ralls, Kathy, 451
Random mating, 419–420
Rapid lysis mutant, of T-even phage, 275, 334
Rat
  maze-learning ability in, 55–56, 60
  pesticide resistance in, 457
  warfarin resistance in, 433, 439, 459–60
R-band, 74
Reading frame, 255–56, 279
  open, 394, 405
Realized heritability, 61
recA gene, 213–14, 291–92

RecA protein, 213–14, 291–92
Recessive trait, 5, 17, 20–21, 23–24
  in human beings, 34–35
  pedigree for, 34–35
  X-linked, 48–50
Reciprocal mating, 17
Reciprocal recombination, 187, 189
Reciprocal translocation, 90–93
Recognition helix, 214
Recombinant, 7, 108–9
Recombinant DNA, 294, 371–73, 378, 380
Recombination, 7, 80, 108
  in conjugation, 322
  in corn, 7
  in *Drosophila,* 7
  forms of, 187–89
  generalized, 188–90
  in human beings, 116
  illegitimate, 189, 297
  mechanism of, 187–90
  in phages, 334–35
  physical evidence for, 7
  rate of, 109
  reciprocal, 187, 189
  site-specific, 189
  within inversion loop, 89–90
Recombination repair, 290–91
Recon, 335
Recurrence risk, 64
  empirical, 65
Red blood cell
  antigens of, 40
  sickled, 165
Reductional division. *See* Meiosis I
Regulative development, 344–45
Relative fitness, 431–32, 434–35, 451
Relatives, phenotypic resemblance between, 63
Release factor-1 (RF-1), 269
Release factor-2 (RF-2), 269
Release factor-3 (RF-3), 269
Release factor eEF, 269
Replica plating, 373–74
Replicating bubble, 173–74
Replicating fork, 173–74
Replication, 9, 152–53, 171
  bidirectional, 172–75
  conservative, 152–53
  direction of, 175–76
  dispersive, 152–53
  DNA methylation after, 351–52
  DNA strand separation in, 178–80
  DNA structure and, 142
  elongation stage in, 182–85
  enzymology of, 178–86
  in eukaryotes, 174–75
  fidelity of, 186, 277–79
  initiation of, 181–82
  joining DNA fragments, 178
  lagging strand of DNA in, 175–76, 183
  leading strand of DNA in, 175–76
  mechanism of, 172–86
  origin of, 173, 175
  of plasmids, 175
  priming of, 177–78
  proofreading in, 186
  rate of, 175, 184
  rolling circle-style, 184, 319
  semiconservative, 152–53
  semidiscontinuous, 175–77, 186
  termination of, 185–86
  theta mode of, 172–73
  unidirectional, 173, 175
Replicative form (RF), 378, 386
Replicative transposition, 296–97
Replicon, 174–75
Replisome, 182–84
Repressor
  lac, 201–4, 382
  lambda, 203, 210–14, 383
  LexA, 291–92
  trp, 214

Reproduction
  asexual, 71
  sexual, 72–73
Reptile, sex determination in, 99–100
Repulsion, 106
Repulsion double heterozygote, 109
Repulsion gametes, 107
Resolution, 297
Resolvase, 297
Response, 61
res site, 297
Restriction endonuclease, 370–72
Restriction fragment length polymorphism (RFLP), 124–25
  mapping with, 401–5
Restriction site, 371–72
Restrictive temperature, 276–77
Retinoblastoma, 44
Retrotransposon, 304
Retrovirus, 302–4
Reverse duplication, 86
Reverse transcriptase, 243, 302–3, 379–80
Reversion, 267–68, 286–87
  second-site, 286
  true, 286
RF. *See* Release factor; Replicative form
RFLP. *See* Restriction fragment length polymorphism
r gene, 334–35
rho factor, 198–99
Ribonuclease H, 379–80
Ribonucleic acid. *See* RNA
Ribonucleoside triphosphate, 161
Ribose, 137
Ribosomal proteins, 162–63, 260–61
  electrophoresis of, 261
  functions of, 262
  of 30S particle, 261–62
Ribosomal RNA (rRNA), 159–62
  precursor of, 223
  RNA polymerase producing, 222–23
  5S, 162–63, 223, 226–28, 260
  16S, 162–63, 260–62
  23S, 162–63, 260
  trimming precursors of, 237–39
Ribosomal RNA (rRNA) genes, 232, 238, 448
  5S, 349–51
  5.8S, 261
  18S, 261
  28S, 261
Ribosome, 10, 159, 161–63
  A site on, 265–66
  of eukaryotes, 260–61
  messenger RNA binding to, 263–65
  of mitochondria, 266
  of prokaryotes, 260
  P site on, 265–67
  reconstitution of, 260, 262
  self-assembly of, 260–62
  30S particle of, 260–62
  50S particle of, 263–64
  subunits of, 162, 260
Ribulose bisphosphate carboxylase. *See* Rubisco
Rich, Alexander, 143, 252
Ricin gene, 379
Rifampicin, 177, 181
rII gene, 335–37
Ring chromosome, 92
Riordan, Jack, 406
R-looping, 232–35
RNA, 8
  chemical nature of, 137–39
  electrophoresis of, 236, 392
  heterogeneous nuclear, 233–34
  hybridization of (*see* Hybridization)
  messenger (*see* Messenger RNA)

physical chemistry of, 143–48
processing of, 237–41
pulsed-labeled, 243
ribosomal (*see* Ribosomal RNA)
small nuclear, 236
splicing of, 233–37
stable, 238
synthesis of (*see* Transcription)
transfer (*see* Transfer RNA)
RNA polymerase, 161–62, 177, 181, 195. *See also* Sigma factor
binding to promoter, 196–97
core enzyme, 195–96
of eukaryotes, 222–23
interaction with CAP-cAMP complex, 205–6
modification of, 206–7
mutations in, 276
of phage T7, 208
structure of, 194–96
subunits of, 195–96
RNA polymerase I, 222–23, 226
RNA polymerase II, 222–23, 240, 302–3
promoters recognized by, 223–25
transcription factors for, 228–31
RNA polymerase III, 222–23, 350
promoters recognized by, 226–27
transcription factor for, 228–29
RNA primer. *See* Primer
RNA transcript, hybridization assay for, 242–43
RNA virus, 142, 302–4
Roberts, Jeffrey, 198–99
Roberts, Richard, 453
Roeder, Robert, 222, 349
Rolling circle-style replication, 184, 319
Root tip cells, 217
Roundworm. *See* Caenorhabditis elegans
Roux, Wilhelm, 343
rRNA. *See* Ribosomal RNA
Rubisco, 348–49
Rust resistance, in flax, 460–61
Rutter, William, 222

S

Saccharomyces cerevisiae. See Yeast
S-Adenosyl methionine, 241
Safety, of gene cloning, 402
SalI, 371–72
Salmonella newport, 298–99
Salmonella typhimurium
in Ames test, 293–94
DNA of, 146
Salmonellosis, 298–99
Salmon sperm DNA, 144
Sanger, Frederick, 165, 392, 394
Satellite, chromosomal, 73–74
Scaffold, chromosomal, 221
Schizophrenia, 55
Screening, by replica plating, 373–74
Sea urchin egg, 245
Second filial generation. *See* F$_2$ generation
Second-site reversion, 286
Sedimentation coefficient, 162
Segmentation genes, in *Drosophila,* 353, 355, 357–59
Segregation, 82
adjacent, 91
alternate, 91
principle of, 16–20, 22
Selection, 328
artificial, 452–54
natural (*see* Natural selection)
Selection coefficient. *See* Selective disadvantage

Selection differential, 61, 452
Selection limit, 453
Selective disadvantage, 431–32, 435
Self-fertilization, 17–18, 20, 72, 423–25, 454–55
Selfish DNA, 296
Semiconservative replication, 152–53
Semidiscontinuous replication, 175–77, 186
Serratia, DNA of, 144
Sex chromosome, 6
of domestic animals, 74
of human beings, 48, 97
in sex determination, 99–102
Sex determination, 99–102
Sex-duction. *See* F-duction
Sex-influenced trait, 50
Sex-limited trait, 50
Sex-linked gene, 48–51
Sex pilus, 315
Sex-reversed individual, 100
Sexual reproduction, 72–73
Shadowing, of DNA, 146
Shapiro, James, 294
Shapiro, Lawrence, 352
Sharp, Phillip, 232
Shatkin, Aaron, 241
Sheep
polled, 50
transgenic, 397–98
Shepherd's purse, seed pod shape in, 53
Sheppard, Phillip, 434
Shine, J., 264
Shine-Dalgarno sequence, 264–65, 381–82
Shull, George, 456
Shuttle vector, 385
Siamese cat, 45, 276–77
Sickle-cell disease, 10, 41, 65, 165–67, 396–97, 415–17, 431, 433, 435
frequency of alleles causing, 440
heterozygote advantage in, 442
screening for, 407
Sigma factor, 195–97
evolution of, 231
heat-shock specific, 209
phage SPO1–specific, 206–7
sporulation-specific, 208–9
vegetative, 208–9
Signal peptide, 245
Silent mutation, 281, 285–86
Single-step growth experiment, 332–34
Single-strand-binding protein (SSB), 178–79, 182
Site-directed mutagenesis, 223, 378, 385–86
Site-specific recombination, 189
Skin cancer, 284, 292, 306
SmaI, 371
Small nuclear RNA (snRNA), 236
Smallpox vaccine, 396
Smith, Hamilton, 370
Snail, shell banding patterns in, 414, 431
snRNA. *See* Small nuclear RNA
Social maturity score, 61
Socioeconomic factors, 64
Solenoid, 220–22
Somatic cell hybridization, 116–17
Somatic mutation, 274–75, 307, 363
SOS response, 213–14, 291–92
Southern, E., 387
Southern blotting, 308, 387–89, 402–3, 405
Southern corn leaf blight, 449
Sp1, 228, 230
Spacer DNA, 237–38
Specialized transduction, 326–27
Species, 4
Sperm, 82

Spermatid, 82
Spermatocyte, 81
Spermatogenesis, 81–82
Spermatogonium, 81
S phase, 76–77
Spina bifida, 65
Spindle fiber, 77–78, 80
Splice junction, 234, 236
cryptic, 234–36
Spliceosome, 236
Splicing, of RNA, 233–37
Spontaneous mutation, 277–82, 426
Sporogenesis, 72–73
Sporulation
in *Bacillus subtilis,* 206–9
transcription during, 208–9
SSB. *See* Single-strand-binding protein
Stable equilibrium frequency, 430
Stable RNA, 238
Stahl, Franklin, 11, 152
Stalker, David, 398
Standard deviation, 58–59
Stasis, 418
Statistical significance, 32
Steitz, Joan, 264
Steroid hormone, 246–48
Steroid receptor, 246–48
Steward, F. C., 344
Stop codon, 258, 260–61, 267–69, 277
Strand invasion, 189–90
Strawberry, ploidy in, 93
Streptococcus pneumoniae
DNA of, 144
morphological mutations in, 275
transformation in, 132–35
Streptomycin, 262
Streptomycin resistance, 299
Stringency, of hybridization, 379
Structural gene, 158, 200
Sturtevant, A. H., 7, 11, 122
Styela, development in, 343–44, 346
Suicide enzyme, 288
Sulfonamide resistance, 299
Sum rule, 26–27
Supernumerary chromosome, 72–73
Suppression, 276
amber, 267
intergenic, 286
intragenic, 286
Suppressor, 267–70
Surf clam egg, 245
Sutton, Walter, 11, 71
SV40, 146, 218–20, 227, 230
Sweet pea
flower color in, 51–52, 106
pollen shape in, 106
Swine
body fat in, 63, 453–54
growth rate in, 453–54
karyotype of, 74
litter size in, 63
sex chromosomes of, 74
superpig, 397
weight of, 63
Synapsis, 79–80
Syntenic genes, 110

T

TAB. *See* TATA-binding protein
Tandem duplication, 86
T antigen, 227
TATA-binding protein (TAB), 231
TATA box, 223–25, 231, 348
Tatum, E. L., 9, 11, 158–59, 314
Tautomerization, 278
Taylor, Austin, 298
Tay-Sachs disease, 65, 440
T-DNA, 398–400

Telocentric chromosome, 73–74, 78
Telomere, 74
Telophase
meiosis I, 80
meiosis II, 81
mitotic, 77–78
Temin, Howard, 302
Temperate phage, 210
Temperature-sensitive mutation, 276–77
Template, DNA, 162
Terminal deletion, 87
Terminal transferase, 379–80
Terminator, 162, 198–99, 240, 381
rho-dependent, 199–200, 211–12
rho-independent, 199
Testcross, 22–23
Testes, 81
Testicular feminization, 100
Tetracycline resistance, 299
Tetrad, 121
Tetrad analysis, 118–21
Tetraploid, 93, 95
Tetrasomy, 95
T-even phage, 330–37
fine structure mapping in, 335
growth of, 331–34
host-range mutants of, 334
mutant, 334–35
rapid-lysis mutants of, 334
structure of, 331–34
TFIIIA, 228–29, 350–51
Thalassemia, 234–36, 448
Thalidomide, 45
Three-factor cross, 330
Three-point cross, 111–12
Thymidine kinase gene, 116–17
Thymine, 9, 137, 140
analogs of, 283
from methylcytosine, 281
tautomers of, 278
Thymine dimer, 284–85, 287, 289
Tjian, Robert, 228
Tjio, Joe-Hin, 75
tnp genes, 297
Tobacco
chromosome number in, 72–73
glyphosate-resistant, 398
luciferase gene in, 400–401
Tomato, genetic variation in, 450
Tomizawa, J., 322
Tonegawa, Susumo, 362
Topoisomerase, 179–80, 185
Tortoise, Galápagos, 413
Totipotent cell, 344
tra gene, 100
trans-acting element, 326
Transcription, 159, 161–62, 165
asymmetry of, 162
elongation step in, 161–62, 195–96
initiation of, 161, 195–97
measuring activity of specific gene, 392
mechanism of, 194–99
in prokaryotes, 194–99
reverse, 302
termination of, 161–62, 198–99
Transcriptional control
of development, 347–48
in eukaryotes, 222, 227–31, 242–44, 246–48
in prokaryotes, 200–214
temporal, 206–14
Transcription bubble, 162
Transcription factor, 198–99, 228–31, 346
in determination, 350–51
in development, 358
Transcription system
in vitro, 224
in vivo, 224

Transducing phage, 326–28
Transduction, 326–28. *See also* Phage
  generalized, 326, 328–30
  mapping by, 328–30
  specialized, 326–27
Transferrin gene, 435–36
Transfer RNA (tRNA), 159
  acceptor stem of, 252
  amino acid binding to, 254–55
  anticodon loop of, 252
  cloverleaf model of, 163–64,
    252–53
  dihydrouracil loop of, 252
  for initiation of translation, 264
  isoaccepting species of, 258
  modified bases in, 252, 258–59
  RNA polymerase producing,
    222–23
  structure of, 163–64, 252–54
  suppressor, 268–69, 276
  TψC loop of, 252
  in translation, 162–64
  variable loop of, 252
Transfer RNA (tRNA) genes, 232,
    238, 448
Transformation
  in bacteria, 8, 132–37
  cellular, 305, 307–10
Transgene, 397
Transgenic animal, 397–98
Transgenic plant, 348–49, 398–401
Transition, 277
Translation, 159, 162–65
  elongation step in, 265–67
  initiation of, 263–65
  termination of, 267–70
  translocation step in, 266
Translational control, in eukaryotes,
    242, 244–45
Translocation, 85, 90–93, 364
  interstitial, 90
  reciprocal, 90–93
Translocation heterozygote, 90–93
Transmission genetics. *See* Mendelian
    genetics
Transposable element, 274. *See also*
    Transposon
Transposase, 295–97
Transposase gene, 301–2
Transposition, 299
  conservative, 296
  mechanism of, 296–97
  public health and, 298–99
  replicative, 296–97
  Ty, 303–4
Transposon
  composite, 298–99
  in *Drosophila,* 302
  in eukaryotes, 300–305
  mutations and, 304–5
  in prokaryotes, 294–97
  in yeast, 302

Transposon Tn3, 296–97
Transposon Tn4, 299
Transposon Tn9, 298–99
Transposon Tn10, 298–99
Transposon Tn55, 299
Transposon Ty, 302–4
Transversion, 277
Travers, Andrew, 196
Tree frog, genetic variation in, 418
Triple crossover, 114
Triploid, 93–94
Trisomy, 85, 95–96, 116
Trisomy 21. *See* Down syndrome
Triturus, development in, 346–47
Trivalent, 94–95
tRNA. *See* Transfer RNA
trpA gene, 329
trp operon, 198
True-breeding line, 17
True reversion, 286
Trypanosome coat genes,
    rearrangement of, 364
Tryptophan synthetase, 167–68
Tsai, A., 175
Tschermak, Erich von, 11, 17, 23
Tsugita, A., 255
Tsui, Lap-Chee, 406
Tulipa, 39
Turner syndrome, 97–98, 100
Twins
  dizygotic, 66
  monozygotic, 55, 61–62, 388–89
Twort, F. W., 330
Two-stranded crossing over, 117–18
tyb gene, 303
Tyrosinase gene, 275

ubx gene, 359
Ultimate carcinogen, 306
Ultracentrifugation, of DNA, 134
Ultraviolet absorption
    spectrophotometry, of
    DNA, 134
Ultraviolet radiation, as mutagen,
    284–85
umuDC operon, 291–92, 294
Unequal crossover, 122–25
Unidirectional replication, 173, 175
Univalent, 94–95
Up mutation, 197
Upstream promoter element, 224–25
Uracil, 280–81
Uracil-DNA-glycosidase, 280–81
U1 RNA, 236
uvr genes, 290, 292

Vaccine, genetically-engineered, 396
Variable number tandem repeat
    (VNTR), 390
Variance, 58–59, 428–29
  environmental, 60–61
Variation
  in crop plants, 449–50
  effect of mutation of, 426–27
  effect of selection on, 431–36
  loss of, 451–52
  measurement of, 416–18
  types of, 414–16
Vector, 372–78
  expression, 381–85
  phage, 374–78
  plant, 398–99
  plasmid, 372–75
  shuttle, 385
  yeast, 385
Vegetal pole, 346
Vegetative state, 208
Vernalization, 458
Viroid, 142
Virulent phage, 210
Virus
  cancer-causing, 307–10
  RNA, 142, 302–4
Visible mutation. *See* Morphological
    mutation
VNTR. *See* Variable number tandem
    repeat

Wake, R. B., 173
Wang, D., 252
Warfarin resistance, 433, 439, 459–60
Wasp, chromosome number in, 72
Watson, James, 9, 11, 135, 139–42
W chromosome, 100
Weeds, pesticide resistance in, 457,
    460
Weigert, Martin, 267
Weinberg, Robert, 307
Weinberg, Wilhelm, 11, 419
Weintraub, Harold, 351
Weissmann, Charles, 234
Wexler, 403
Wheat
  Hessian fly resistance in, 460
  kernel color in, 56–57
  ploidy of, 94
Whittaker, J. R., 346
Wigler, Michael, 307
Wild-type, 30

Wilkins, Maurice, 9–11, 139–40
Williamson, Robert, 406
Wobble hypothesis, 258–59, 286
Wollman, Elie, 317–18
Worton, Ronald, 407
Wu, Carl, 229
Wu, T. T., 330

X chromosome, 48–51
  attached, 118–19
  of *Drosophila,* 6–7, 48, 75–76,
    118–19
  of human beings, 115
  inactivation of, 98, 352
  in sex determination, 99–102
Xenopus laevis
  nuclear transplantation in, 344–45
  ribosomal RNA genes of, 226–28,
    349–51
Xeroderma pigmentosum, 292, 306–7
X-gal, 374–75
X-linked trait, 115, 422–23
  dominant, 49–50
  in *Drosophila,* 83–85
  recessive, 48–50
XmaI, 371
X ray
  cancer and, 307
  as mutagen, 285
X-ray diffraction, of DNA, 139–40
XXX female, 97–98
XYY male, 97

Yanofsky, Charles, 167–68, 329
Y chromosome, 48–51
  in sex determination, 100
Yeast
  chromosome number in, 72–73
  chromosomes of, 148
  DNA of, 146
  expression vectors in, 385
  2–micron plasmid of, 385
  petite, 275
  transfer RNA of, 164
  transposons in, 302

Z chromosome, 100
Z-DNA, 143
Zea mays. See Corn
Zinc finger, 228–29
Zygote, 72–73, 344
Zygotene, 79–80